From Vision to Instrument: Creating a Next-Generation Event Horizon Telescope for a New Era of Black Hole Science

From Vision to Instrument: Creating a Next-Generation Event Horizon Telescope for a New Era of Black Hole Science

Editors

Michael D. Johnson
Shep Doeleman
Jose L. Gómez

Basel • Beijing • Wuhan • Barcelona • Belgrade • Novi Sad • Cluj • Manchester

Editors

Michael D. Johnson
Center for Astrophysics |
Harvard & Smithsonian
Cambridge, MA, USA

Shep Doeleman
Center for Astrophysics |
Harvard & Smithsonian
Cambridge, MA, USA

Jose L. Gómez
Instituto de Astrofísica de
Andalucía (IAA-CSIC)
Granada, Spain

Editorial Office
MDPI
St. Alban-Anlage 66
4052 Basel, Switzerland

This is a reprint of articles from the Special Issue published online in the open access journal *Galaxies* (ISSN 2075-4434) (available at: https://www.mdpi.com/journal/galaxies/special_issues/ngEHT_blackholes).

For citation purposes, cite each article independently as indicated on the article page online and as indicated below:

Lastname, A.A.; Lastname, B.B. Article Title. *Journal Name* **Year**, *Volume Number*, Page Range.

ISBN 978-3-0365-9354-8 (Hbk)
ISBN 978-3-0365-9355-5 (PDF)
doi.org/10.3390/books978-3-0365-9355-5

Cover image courtesy of Jose L. Gómez

Contents

Editorial

From Vision to Instrument: Creating a Next-Generation Event Horizon Telescope for a New Era of Black Hole Science

Michael D. Johnson [1,2,*], Sheperd S. Doeleman [1,2], José L. Gómez [3] and Avery E. Broderick [4,5,6]

1   Center for Astrophysics | Harvard & Smithsonian, 60 Garden Street, Cambridge, MA 02138, USA
2   Black Hole Initiative at Harvard University, 20 Garden Street, Cambridge, MA 02138, USA
3   Instituto de Astrofísica de Andalucía-CSIC, Glorieta de la Astronomía s/n, 18008 Granada, Spain
4   Perimeter Institute for Theoretical Physics, 31 Caroline Street North, Waterloo, ON N2L 2Y5, Canada
5   Department of Physics and Astronomy, University of Waterloo, 200 University Avenue West, Waterloo, ON N2L 3G1, Canada
6   Waterloo Centre for Astrophysics, University of Waterloo, Waterloo, ON N2L 3G1, Canada
*   Correspondence: mjohnson@cfa.harvard.edu

Citation: Johnson, M.D.; Doeleman, S.S.; Gómez, J.L.; Broderick, A.E. From Vision to Instrument: Creating a Next-Generation Event Horizon Telescope for a New Era of Black Hole Science. *Galaxies* **2023**, *11*, 92. https://doi.org/10.3390/galaxies11050092

Received: 9 August 2023
Accepted: 17 August 2023
Published: 22 August 2023

In April 2019, the Event Horizon Telescope (EHT) Collaboration successfully imaged a supermassive black hole (SMBH) for the first time, revealing the apparent "shadow" cast by the dark compact object M87* in the center of the elliptical galaxy Virgo A. More recent results include the first polarized images of M87* and the first images of the supermassive black hole in the center of the Milky Way, Sagittarius A*. Together, these results have defined the start of a new era in the detailed study of these exotic objects through images that directly reveal the deflection and capture of light in a strongly curved spacetime.

The next-generation EHT (ngEHT[1]) is a program to sharply increase the current EHT capabilities through longer observing campaigns, simultaneous multi-band observations, and the deployment of additional stations at optimal locations worldwide. These enhancements have the potential to again revolutionize our view of horizon-scale physics, enabling movies of black hole accretion and explosive transients, high-dynamic-range images that connect black holes directly to their galactic-scale relativistic jets, and powerful multi-wavelength and multi-messenger synergies with other next-generation facilities. A white paper describing some of these capabilities was submitted to the US Astro2020 Decadal Survey on Astronomy and Astrophysics.[2]

To develop scientific priorities and their associated requirements, the ngEHT has hosted a series of international meetings that have been entirely open to the scientific community. The first meeting (in February 2021) was virtual and featured 94 presentations, with over 500 participants from more than 30 countries. The second meeting (in November 2021) was also virtual with over 500 registrants, over 100 presentations, and a public outreach panel including prominent YouTube science popularizers. The most recent meeting (in June 2022) was hosted by the Instituto de Astrofísica de Andalucía (IAA) of the Consejo Superior de Investigaciones Científicas (CSIC). The meeting had over 140 participants who convened at the Parque de las Ciencias in Granada, Spain, from 22 to 25 June (see Figure 1). The meeting, titled "Assembling the ngEHT: Community-Driven Science to a Global Instrument", aspired to unite a large and growing community of black hole researchers and radio interferometry specialists and to find a common purpose that could define the next decade of discoveries.

Figure 1. The ngEHT meeting at the Parque de las Ciencias in Granada, Spain. This meeting included many leading experts in black hole and EHT science and was jointly coordinated with the EHT Collaboration.

This *Special Issue*[3] is an outgrowth of these meetings, with a collection of contributions that develop the scientific vision and architecture of the ngEHT in the following areas:

- Fundamental physics (studies of a black hole spacetime and tests of general relativity).
- Black holes and their cosmic context (SMBH formation and evolution, studies of SMBH binaries, multi-wavelength studies of black holes and jets, and large-scale jet collimation and kinematics).
- Accretion (probing accretion flow dynamics and structure, turbulence, and plasma studies near a SMBH).
- Jet launching (energy extraction from spinning black holes and jet kinematics and monitoring).
- Transients and impulsive phenomena.
- New horizons (terrestrial applications such as geodesy and synergies with other next-generation facilities).
- Algorithms and inference (imaging methods, model fitting to interferometric data, and synthetic data challenges).
- History, philosophy, and culture (implications of building new instruments in the current era).
- Advances in submillimeter VLBI instrumentation and software.
- VLBI array design and optimization.

In addition, this Special Issue contains two summary documents from the ngEHT. The first describes a reference array, instrumentation, and site selection. The second describes the key scientific goals and associated instrument requirements.

Together, the conference and this Special Issue paint an exciting picture of the future of black hole imaging, driven by an extraordinary community of scholars pursuing answers to some of nature's most intriguing secrets through technical breakthroughs and visionary science.

Acknowledgments: We are grateful to the hundreds of scientists and engineers who have contributed to the ngEHT project. In particular, we would like to acknowledge the organizers of the three international science meetings:

February 2021 Meeting (https://www.ngeht.org/ngeht-meeting-2021 (accessed on 21 August 2023))
Scientific Organizing Committee: Geoff Bower, Avery Broderick (co-chair), Alessandra Buonanno, Vitor Cardoso, Sheperd Doeleman, Charles Gammie, Daryl Haggard, David Hughes, Michael Johnson (co-chair), Chung-Pei Ma, Ramesh Narayan, Ue-Li Pen, Andy Strominger, and Anton Zensus
Local Organizing Committee: Lindy Blackburn, Nick Conroy (co-chair), Mina Himwich, Britt Jeter (co-chair), Tiffany Nichols, Daniel Palumbo, Alex Raymond, and Paul Tiede

November 2021 Meeting (https://www.ngeht.org/ngeht-meeting-november-2021 (accessed on 21 August 2023))
Scientific Organizing Committee: Lindy Blackburn, Avery Broderick (co-chair), Sheperd Doeleman, Garret Fitzpatrick, Jose L. Gómez, Kari Haworth, Paul Ho, Elizabeth Humphreys, Michael D. Johnson (co-chair), Ilya Mandel, Monika Moscibrodzka, Neil Nagar, Rob Selina, Zhiqiang Shen, Eva Silverstein, Ingrid Stairs, and Meg Urry
Local Organizing Committee: Richard Anantua, Lindy Blackburn (chair), Sandra Bustamante, Razieh Emami, Kotaro Moriyama, Rusen Lu, Elon Price, Venkatessh Ramakrishnan, Angelo Ricarte, Thalia Traianou, and Guang-Yao Zhao

June 2022 Meeting (https://www.ngeht.org/ngeht-meeting-june-2022 (accessed on 21 August 2023))
Scientific Organizing Committee: Avery Broderick, Ming-Tang Chen, Sheperd Doeleman, Heino Falcke, Garret Fitzpatrick, José L. Gómez (co-chair), Mareki Honma, Michael Johnson (co-chair), Svetlana Jorstad, Yuri Kovalev, Priyamvada Natarajan, and Marta Volonteri
Local Organizing Committee: Antxon Alberdi, Ilje Cho, Rohan Dahale, Marianna Foschi, Antonio Fuentes, Rocco Lico, Ioannis Myserlis, Alicia Pelegrina, Angelo Ricarte, Teresa Toscano, Thalia Traianou (chair), and Guang-Yao Zhao

Conflicts of Interest: The authors declare no conflict of interest.

Notes

1. https://www.ngeht.org/ (accessed on 21 August 2023).
2. https://baas.aas.org/pub/2020n7i256/release/1 (accessed on 21 August 2023).
3. https://www.mdpi.com/journal/galaxies/special_issues/ngEHT_blackholes (accessed on 21 August 2023).

Article

Reference Array and Design Consideration for the Next-Generation Event Horizon Telescope

Sheperd S. Doeleman [1,2,*], John Barrett [3], Lindy Blackburn [1,2], Katherine L. Bouman [4], Avery E. Broderick [5,6,7], Ryan Chaves [1], Vincent L. Fish [3], Garret Fitzpatrick [1], Mark Freeman [1], Antonio Fuentes [8], José L. Gómez [8], Kari Haworth [1], Janice Houston [1], Sara Issaoun [1,†], Michael D. Johnson [1,2], Mark Kettenis [9], Laurent Loinard [10], Neil Nagar [11], Gopal Narayanan [12], Aaron Oppenheimer [1], Daniel C. M. Palumbo [1,2], Nimesh Patel [1], Dominic W. Pesce [1,2], Alexander W. Raymond [1,13], Freek Roelofs [1,2], Ranjani Srinivasan [1], Paul Tiede [1,2], Jonathan Weintroub [1,2] and Maciek Wielgus [14]

1   Center for Astrophysics|Harvard & Smithsonian, 60 Garden Street, Cambridge, MA 02138, USA
2   Black Hole Initiative, Harvard University, 20 Garden Street, Cambridge, MA 02138, USA
3   Massachusetts Institute of Technology Haystack Observatory, 99 Millstone Road, Westford, MA 01886, USA
4   California Institute of Technology, 1200 East California Boulevard, Pasadena, CA 91125, USA
5   Perimeter Institute for Theoretical Physics, 31 Caroline Street North, Waterloo, ON N2L 2Y5, Canada
6   Department of Physics and Astronomy, University of Waterloo, 200 University Avenue West, Waterloo, ON N2L 3G1, Canada
7   Waterloo Centre for Astrophysics, University of Waterloo, Waterloo, ON N2L 3G1, Canada
8   Instituto de Astrofísica de Andalucía-CSIC, Glorieta de la Astronomía s/n, E-18008 Granada, Spain
9   Joint Institute for VLBI ERIC (JIVE), Oude Hoogeveensedijk 4, 7991 PD Dwingeloo, The Netherlands
10   Instituto de Radioastronomía y Astrofísica, Universidad Nacional Autónoma de México, Morelia 58089, Mexico
11   Astronomy Department, Universidad de Concepción, Casilla 160-C, Concepción 4030000, Chile
12   Department of Astronomy, University of Massachusetts, Amherst, MA 01003, USA; gopal@astro.umass.edu
13   Jet Propulsion Laboratory, California Institute of Technology 4800 Oak Grove Dr., Pasadena, CA 91109, USA
14   Max–Planck–Institut für Radioastronomie, Auf dem Hügel 69, D-53121 Bonn, Germany
*   Correspondence: sdoeleman@cfa.harvard.edu
†   NASA Hubble Fellowship Program, Einstein Fellow.

Citation: Doeleman, S.S.; Barrett, J.; Blackburn, L.; Bouman, K.L.; Broderick, A.E.; Chaves, R.; Fish, V.L.; Fitzpatrick, G.; Freeman, M.; Fuentes, A.; et al. Reference Array and Design Consideration for the Next-Generation Event Horizon Telescope. *Galaxies* **2023**, *11*, 107. https://doi.org/10.3390/galaxies11050107

Academic Editor: Wenwu Tian

Received: 17 August 2023
Revised: 15 September 2023
Accepted: 9 October 2023
Published: 18 October 2023

Abstract: We describe the process to design, architect, and implement a transformative enhancement of the Event Horizon Telescope (EHT). This program—the next-generation Event Horizon Telescope (ngEHT)—will form a networked global array of radio dishes capable of making high-fidelity real-time movies of supermassive black holes (SMBH) and their emanating jets. This builds upon the EHT principally by deploying additional modest-diameter dishes to optimized geographic locations to enhance the current global mm/submm wavelength Very Long Baseline Interferometric (VLBI) array, which has, to date, utilized mostly pre-existing radio telescopes. The ngEHT program further focuses on observing at three frequencies simultaneously for increased sensitivity and Fourier spatial frequency coverage. Here, the concept, science goals, design considerations, station siting, and instrument prototyping are discussed, and a preliminary reference array to be implemented in phases is described.

Keywords: black holes; supermassive black holes; general relativity; interferometry; accretion; relativistic jets; very-long-baseline interferometry; radio instrumentation; EHT; ngEHT

1. Introduction

On 10 April 2019, the Event Horizon Telescope project (EHT) released images of the supermassive black hole at the heart of galaxy M87 [1–6]. The observed ring of emission, formed by radio waves lensed in the gravitational field of a 6.5 billion solar mass black hole, has dimensions that match the predictions of General Relativity. Images of Sgr A*, the 4 million solar mass black hole at the center of the Milky Way, also exhibit a ring morphology with diameters anticipated by theory [7–12]. These results confirm that the EHT has observed the strong gravitational lensing signature of supermassive black holes [13–17], and these images have opened a new field of precision black hole studies on horizon scales.

This work is built upon decades of technical development and precursor observations. Pioneering first Very Long Baseline Interferometry (VLBI) experiments at wavelengths of 1.3 mm [18,19] demonstrated that observations with the required resolution were possible at frequencies where AGN are likely to be optically thin. The discovery of horizon-scale structure in both Sgr A* and M87 with purpose-built ultra-high bandwidth systems on early EHT arrays [20,21] confirmed that imaging these sources was feasible. Subsequent observations revealed time-variability and ordered magnetic fields on Schwarzschild radius dimensions [22,23]. The emergence of the EHT to a full imaging array grew from building community support through a decadal review processes [24], efforts to modify large-scale international facilities, such as ALMA, through global cooperation [25,26], and work to enable VLBI capability at the most remote observatories on the planet [27,28]. Over the course of two decades, all the technical, logistical, organizational, and analytical aspects of the full EHT were implemented by an expert team that grew from a few 10's to over 200 collaborators worldwide.

Building upon this legacy, the next-generation EHT (ngEHT) provides a roadmap to greatly accelerate the development of the EHT, envisaging a transformative new instrument capable of delivering real-time black hole movies. Where the EHT used existing mm/sub-mm facilities to form the first imaging array, the ngEHT will take the next step by designing and locating new dishes to optimize performance and scientific return. This vision offers excellent opportunities to engage the curious public on many levels. It is estimated that over a billion people have now seen the M87 image [29]. We anticipate that the long-term public and STEM education engagement as the ngEHT builds to its goal of black hole 'cinema' will be similar in scope.

For the purposes of this paper, the term "ngEHT" is used to describe a program to explore and define a long-term plan to enhance the EHT to realize a new set of transformative science goals. This paper describes that vision by outlining improvements in bandwidth, frequency range, new antenna deployment, and new operating modes that enable increases in angular resolution, Fourier spatial frequency coverage, sensitivity, and temporal resolution. For brevity, "ngEHT" will also be used as shorthand for the future arrays that will emerge through these plans, as well as for the constellation of improvements that constitute the ngEHT concept.

Technical advances in several areas make the design and implementation of the ngEHT within this decade a realistic goal. Over most of the past two decades, the bandwidth of VLBI systems has kept pace with Moore's Law—a doubling of capacity and speed approximately every 18 months (see, e.g., Event Horizon Telescope Collaboration et al. [2] and Figure 1 below). This is primarily due to the migration of VLBI instrumentation development to designs that adopt industry-standard components, including CPUs, Analog to Digital Converters, Field Programmable Gate Arrays (FPGA), and commercial data transmission protocols (e.g., Vertatschitsch et al. [30]). The increased bandwidth of these components and systems match the analog bandwidth improvements planned for international and national submm facilities, including ALMA [31] and the Submillimeter Array [32]. Meanwhile, the transport of larger data volumes captured by next-generation VLBI systems can be accommodated either by high-speed internet connections [33,34], or increased capacity of hard disk and solid-state disk, which can be shipped by commercial carriers. Once gathered at a central computing facility, the many 10s of Petabytes anticipated for next-generation EHT array observations can be correlated by purpose-built clusters, allocated time on national super-computing centers, or through virtual machine creation using cloud architectures (e.g., Gill et al. [35]). Once correlated, data analysis options include a growing number of video reconstruction algorithms that can render the dynamics of supermassive black hole activity on horizon scales [36–38]. These developments, combined with a positive mention of the ngEHT project [39] in the Radio/Millimeter/sub-Milimeter panel of the most recent US Astronomy Decadal Review [40], imply that implementing the ngEHT is both feasible and timely.

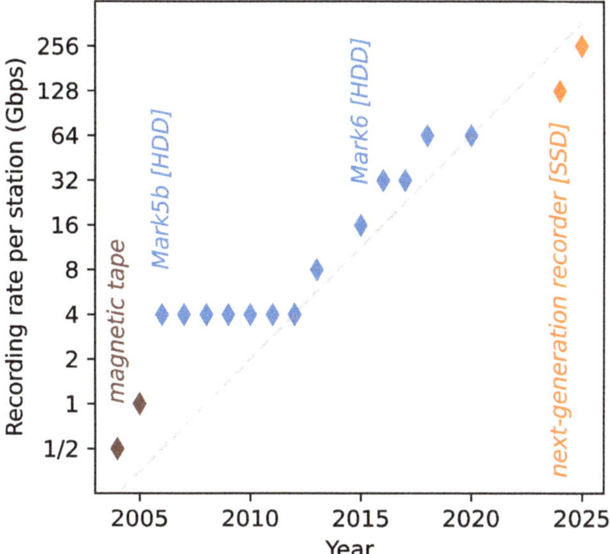

Figure 1. EHT/ngEHT data rate per station over two decades, roughly doubling every two years. The large bandwidths provide the EHT/ngEHT the necessary continuum sensitivity for ultra-high resolution VLBI imaging at (sub-)1 mm using a highly heterogeneous network of telescopes. Maintaining this trend has required the regular adoption of commercial technologies as they became available.

2. ngEHT Concept

The first images of M87 and Sgr A* revealed a clear ring morphology, but they achieved a dynamic range of only ∼10 [4,9]. Image fidelity from the 2017 data sets was primarily limited by sparse interferometric baseline coverage. The shortest baselines, between telescopes located at the same geographic location (ALMA-APEX in Chile and JCMT-SMA in Hawai'i), probe arc second scale structures. There is a large gap between these "intra-site" baselines and the first "inter-site" baseline that links LMT-SMT, which creates a baseline with an angular resolution corresponding to ∼150 µas. Furthermore, the 2017 observations included inter-site baselines between only five geographic locations for M87, and six locations for Sgr A*, fundamentally limiting the fidelity of image reconstruction on angular scales that resolve the black hole shadow.

The ngEHT concept focuses on overcoming these limits through several key developments. Foremost among these is the deployment of relatively modest-diameter radio dishes at optimized locations to increase baseline coverage. Figure 2 shows that even a 6 m diameter dish in marginal weather conditions can detect long baseline correlated fluxes from Sgr A* and M87 when paired with a large "anchor" aperture. This reflects the fact that the 2017 observations, though using fringe detection algorithms limited to 2 GHz bandwidth, achieved signal-to-noise ratios that were typically in excess of ∼10 and often reached ∼100. In other words, the current EHT is limited by baseline coverage and not sensitivity considerations. Through this increased baseline coverage, the ngEHT will reach image dynamic ranges that exceed 1000:1 for full Earth rotation aperture synthesis observations of M87 and other AGN. Time-lapse movies that capture the dynamics of M87's accretion flow and jet launch by combining bi-weekly observations will achieve similar dynamic ranges (pre-cursor multi-epoch observations are possible with the existing EHT at lower imaging dynamic range). For Sgr A*, which has an Innermost Stable Circular Orbit (ISCO) period of ∼$\frac{1}{2}$ h, the ngEHT snapshot baseline coverage in 5 min integrations will be sufficient for near real-time video reconstruction. Figure 3 shows the current EHT array and the location of potential new ngEHT sites.

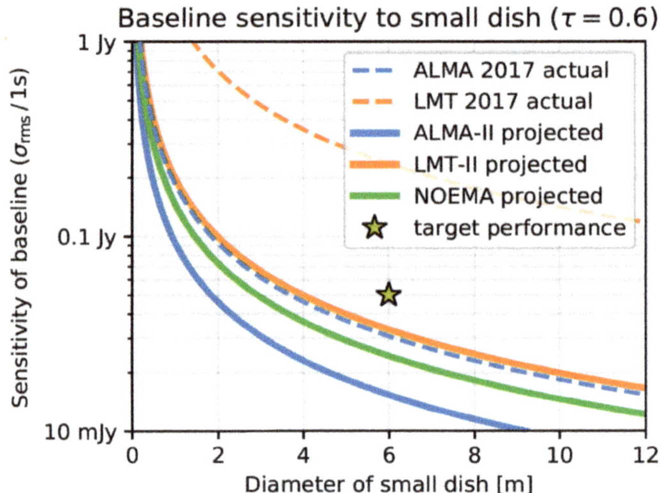

Figure 2. Interferometric baselines between key anchor stations and modest-diameter dishes have sufficient sensitivity to detect target flux densities on time scales of several seconds. A star marks the correlated flux expected for SgrA* and M87 over long ngEHT baselines. Performance for 2017 is taken over 2 GHz of bandwidth and the observed median sensitivity of ALMA and LMT during EHT April 2017 observations. ALMA-II assumes phase referencing using the entire 8 GHz (64 Gbps) of EHT bandwidth, while LMT-II assumes 16 GHz of bandwidth and aperture efficiency of $\eta_A = 0.37$. NOEMA is projected for a 12-element array under nominal weather conditions, and the small ngEHT remote site is evaluated at $\eta_A = 0.5$ and line-of-sight opacity $\tau = 0.6$. Atmospheric phase tracked on rapid timescales at 86 GHz or 230 GHz can be transferred to 345 GHz, allowing for longer coherent integration times and robust measurement at the highest ngEHT observing frequencies.

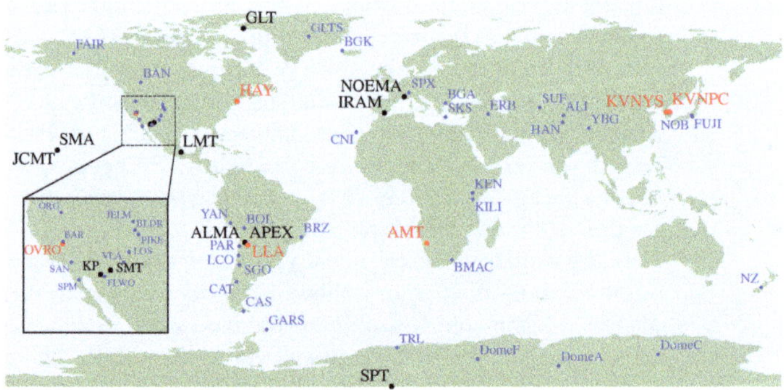

Figure 3. Current EHT sites (in black), other existing or near-future sites that may join global observations (in red), and potential new ngEHT sites (in blue).

Significant improvements in sensitivity will still be realized through the deployment of wider band receivers and backends, which can now typically digitize 8 GHz per sideband. For a given frequency band, the ngEHT targets dual-sideband and dual-polarization, for a potential Stokes I fringe detection that combines 32 GHz aggregate bandwidth. For any given baseline, this advance results in a net detection threshold that is four times lower than currently achievable.

In addition to this increase in overall received bandwidth, the ngEHT frequency coverage will include the 86 and 345 GHz bands. Routine multi-band operation has several important consequences for ngEHT capability. Each station pair probes distinct spatial frequencies when observing in different bands, and multi-frequency imaging algorithms can make use of the aggregate interferometric coverage to improve image fidelity (e.g., [41]). Observing in the 345 GHz band also improves the angular resolution of the global array by up to 50%. The EHT already offers 345 GHz observing capability on a subset of antennas [42], but not yet simultaneously with 230 GHz. Additional frequency bands also enable analyses and modeling that differentiate between gravitationally lensed achromatic features (e.g., the photon ring) and structures whose appearance have a spectral dependence (e.g., accretion flows and relativistic jets). Through the use of the frequency phase transfer technique (FPT; [43]), VLBI phase solutions determined at lower frequencies can be transferred to higher frequency observations, effectively removing atmospheric phase effects to extend coherent integration times for higher sensitivity. The full case for adding 86 GHz capability that leverages FPT through simultaneous multi-band systems is described in Issaoun et al. [44].

Combined, these enhancements lead to profound increases in array capability. The implementation roadmap for the ngEHT will proceed in two phases with the goal of ultimately adding ∼10 new dishes to the EHT. In Phase 1, a total of six new sites will be developed: four radio dishes will be deployed to new geographic locations (Section 4.4.1); and two existing facilities (the 37 m telescope at MIT Haystack Observatory and a 10 m telescope at the Owens Valley Radio Observatory) will be modified to participate in future observations (see, e.g., [45]). A Phase 2 will add four more telescopes, either by deploying additional new purpose-built telescopes, or by instrumenting planned single dish facilities due to come online by ∼2030 (Section 4.4.2). These phases, when complete, will double the number of dishes in the array recently fielded in the 2022 and 2023 annual EHT observing campaigns (see Section 4.4).

3. Next Generation Science Goals

The ngEHT design has been guided by a series of Key Science Goals (KSGs), developed through a community-driven process of exploration, evaluation, and prioritization. These goals and their associated instrument requirements are presented via a Science Traceability Matrix (STM) in a companion paper [46]; and a series of papers in a Special Issue of *Galaxies*[1] presents science topics in greater detail. Here, we briefly describe several of the ngEHT KSGs, which define the target baseline array architecture, and are summarized in Table 1.

Table 1. Select ngEHT Key Science Goals. For the full Science Traceability Matrix and additional details, see Johnson et al. [46].

Key Science Goal	Source	ngEHT Phase	References
Establish the existence of black hole horizons	M87*	Phase 1	Chael et al. [47]; Dokuchaev and Nazarova [48]
	Sgr A*	Phase 2	
Measure a SMBH's spin	M87*	Phase 2	Palumbo et al. [49]
	Sgr A*	Phase 2	Ricarte et al. [50]
Understanding Black Hole-Galaxy Formation, Growth and Coevolution	AGN Survey	Phase 1	Pesce et al. [51,52]; Ramakrishnan et al. [53]
Reveal how black holes accrete material	M87*	Phase 1	Balbus and Hawley [54]; Yuan and Narayan [55]
	Sgr A*	Phase 2	
Observe localized electron heating and acceleration	M87*	Phase 1	Rowan et al. [56]; Ball et al. [57]
	Sgr A*	Phase 2	
Determine if BH jets are powered by spin energy	M87*	Phase 2	Blandford and Znajek [58]; Tchekhovskoy et al. [59]
	Sgr A*	Phase 2	
Determine jet formation & launching mechanisms	M87*	Phase 1	Blandford et al. [60]
	Sgr A*	Phase 2	
Constraining Properties of the BH Photon Ring	M87*	Phase 2	Johnson et al. [61]; Tiede et al. [62]
	Sgr A*	Phase 2	

3.1. Existence and Properties of Black Hole Horizons

By characterizing the central brightness depression region in black hole images, the ngEHT can directly address the question of the existence of a black hole's horizon. For Magnetically Arrested Disk (MAD) accretion modes, which are favored for M87 [5], emission in the innermost part of the flow originates primarily in the equatorial plane, and the central depression (the "inner shadow") is defined by light paths that cross the event horizon without visiting the emitting region of the accretion system [47,48]. Measuring the shape of this "inner shadow" to be smaller than the photon orbit would correspond to observing the lensed event horizon, allowing estimates of the black hole's mass and spin [47]. For both M87 and Sgr A*, this measurement requires an imaging dynamic range of ~100:1. For Sgr A*, intrinsic variability presents an additional challenge that will require future algorithm development. Furthermore, enhancing the dynamic range of the images with the ngEHT will allow obtaining improved constraints on the brightness ratio between the black hole shadow interior and the observed emission ring. These constraints can be translated to an argument supporting the existence of the event horizon, by putting the most stringent limits on the albedo of the surface of an exotic compact object alternative to a black hole [12], ultimately limited only by the emission and absorption in the foreground by the gas located out of the equatorial plane (e.g., [63]).

3.2. Measurements of the Spin of a SMBH

General relativity predicts that astrophysical black holes are described solely by two properties: their mass and angular momentum (or "spin"). The ngEHT has the opportunity to produce direct secure measurements of a black hole spin through distinctive image features that reflect the imprint of the strongly curved spacetime near the horizon. In particular, images of GRMHD simulations show several robust indicators of spin (for a review, see [50]). The most promising of these is the spiraling polarization pattern around the emission ring [49]. By producing time-averaged polarimetric images of M87 and Sgr A* at both 230 and 345 GHz, the ngEHT will be able to securely measure this pattern and decouple the effects of the spacetime from those of the surrounding plasma (Faraday rotation and conversion), which are steeply chromatic.

3.3. Evolution of Supermassive Black Holes

Though the EHT has to date observed only two SMBHs (M87* and Sgr A*) with horizon-scale angular resolution, numerical simulations of black hole accretion flows (e.g., [5,11]) predict that the "shadow" structure seen toward these sources should be a generic image feature in sufficiently optically thin systems. Measurements of the size of the SMBH shadow can be used to constrain the black hole mass (e.g., [6,10]), and measurements of the near-horizon linear polarization structure may also be able to provide indirect constraints on the black hole spin [49,50,64]. Access to a population of shadow-resolved SMBHs would thus provide an opportunity to make uniquely self-consistent measurements of these spacetime properties, permitting corresponding studies of SMBH formation, growth, and co-evolution with their host galaxies.

The ngEHT is expected to be able to detect up to several dozen SMBHs with sufficient angular resolution and sensitivity to access their masses and spins through measurements of their horizon-scale structure [51,52]. A database of the most promising individual targets is being compiled within the ETHER sample [53].

3.4. Mechanisms of Black Hole Accretion

Despite decades of study, the mechanisms that drive accretion onto SMBHs are still poorly understood (for a review, see [55]). The ngEHT will make the first resolved movies of a black hole accretion flow, allowing a direct study of the dynamics of the turbulent plasma and the role of magnetic fields in providing an effective viscosity that drives infall [54,65].

3.5. Heating and Acceleration of Relativistic Electrons

In low density, low accretion rate systems, such as in M87* and Sgr A*, the Coulomb collision time for both electrons and protons is much larger than the dynamical (accretion) time scale. As a consequence, protons and electrons cannot redistribute their energy and a two-temperature plasma occurs. Assuming that the emission is mainly generated by electrons, their temperature is determined by the interplay between cooling and heating processes [66]. Given, that in low accretion rate systems, the cooling processes can be neglected (cooling time scale larger than the dynamical time scale), the impact of possible electron heating processes on the observed emission from M87* and Sgr A* can be probed.

Two of the main processes for the heating of electrons are turbulent heating (see, e.g., [67,68]) and magnetic reconnection heating (see, e.g., [56]). The results of two-temperature general relativistic magneto-hydrodynamic (GRMHD) simulations showed that magnetic reconnection heating leads to a disk-dominated emission structure while turbulent heating tends to a disk-jet structure [69–72]. The ngEHT with its improved u-v coverage and increased sensitivity will allow us to image and track at the same time the disk and faint jet structures on scales of $100r_g$ ($r_g = GM/c^2$). Together with the multi-frequency capabilities of the ngEHT movies, the total intensity and the spectral evolution can be produced. These movies, in close combination with detailed numerical simulations, can allow us to locate the heating sites and distinguish between the different electron heating processes in M87* and Sgr A⋆.

3.6. Energy Extraction from Black Holes

Energy from a spinning black hole can be extracted via the Blandford-Znajek (BZ) process [58], an electromagnetic analog of the classic Penrose process [73]. With the ngEHT we will probe this energy extraction mechanism via the generated jet power or more precisely via the so-called BZ-jet power. The BZ-jet power is proportional to the square of the black hole spin and to the square of the magnetic flux crossing the horizon [59]. In addition, the jet power can be measured from the observed spectral energy distribution or from the X-ray luminosity (see [74], for a detailed discussion on jet power estimates).

To compute a theoretical estimate for the BZ-jet power precise measurements of the black hole spin and the magnetic flux are necessary. As mentioned in Section 3.2 of this paper and in Broderick et al. [75], the combined ngEHT observations will provide the black hole spin and black hole mass with sufficient precision. The second quantity in the BZ-jet power, namely the magnetic flux across the horizon can be obtained either via polarimetric ngEHT observations [76] or via the frequency-dependent position of the core, i.e., the core-shift using multi-frequency observations [77]. In both cases the superior detection and imaging capabilities of the ngEHT will allow us to provide answers to this long-standing question of energy extraction from black holes. To perform this measurement for M87, Phase 2 of the ngEHT is required.

3.7. Jet Formation

Based on numerical GRMHD simulations we know that rotating black holes can launch jets via the BZ process (see, e.g., [59,78]). However, once launched the jets need to be accelerated and confined, whereas the associated physical processes behind this acceleration and collimation as well as the jet composition are still being debated (see [60], for a review). The jet composition electron-positron or electron-proton plasma can be probed via circular polarization and a detailed review can be found in the Special Issue by Emami et al. [79].

Details on the fluid structure and the formation process of the jet in M87 can be derived from the velocity field and the jet-to-counter-jet ratio [80,81]. The structure of the velocity field will allow us to probe the stratification of the jet into a fast inner spine and slow outer sheath (see, e.g., [82]). The ngEHT will enable such studies in objects other than M87 by resolving the transversal jet structure, e.g., in Centaurus A [83]. In addition to the poloidal velocity field (spine-sheath structure), the toroidal velocity field plays a crucial role in determining the formation process of the jet: Is the jet anchored in the accretion disk [84] or is the jet launched from the ergosphere of a rotating black hole [58]. Extracting the velocity field of the jet requires multi-frequency observations and a high cadence of observations. To avoid the "contamination" of the velocity field by secondary effects, i.e., by Kelvin-Helmholtz instabilities or re-collimation shocks (triggered by changes in the ambient medium) scales up to $100r_g$ are sufficient. In order to determine the velocity field in M87 the ngEHT in Phase 1 is required.

3.8. Constraining Properties of the Black Hole Photon Ring

One of the clearest predictions motivated by the first black hole images is that the observed ring of emission should exhibit a fine sub-structure: nested concentric rings, each formed by light rays that make successively more orbits around the photon shell region, located very close to the black hole's event horizon [61]. Each sub-ring is a lensed image of the surrounding accretion and jet emission with inner sub-rings becoming exponentially fainter and narrower. The structure of the primary ($n = 0$) ring, observed by the EHT, depends on a combination of the local spacetime and the detailed emission structure on Schwarzschild radius scales, while subsequent sub-rings ($n \geq 1$) asymptotically approach the true photon orbit, which is dependent exclusively on the spacetime metric [13]. Detection of the $n = 1$ ring, formed by photons that make a half-orbit around the black hole, would be important confirmation of this untested prediction of General Relativity and lead to new tests of GR in highly curved space–time [75,85]. Robust extraction of this feature with the ngEHT will require the longest Earth baselines at 345 GHz and geometric model fitting that uses multiple frequencies [62]. This science goal would be a target of the fully realized (Phase 2) ngEHT.

4. Optimizing the ngEHT Reference Array

The scientific performance of an array generically benefits from the addition of new stations, regardless of where those stations are located. However, when constrained by a fixed budget or a fixed number of new dishes to be added, determining the optimal placement of the new dishes is a challenge that requires finding a balance between many—often conflicting—objectives. For instance, science goals that require high angular resolution favor array configurations with many long baselines, while goals that involve high-fidelity imaging on large fields of view instead favor configurations containing dense short-baseline coverage. Similarly, while atmospheric opacity considerations favor the highest and driest locations, such sites are often remote and lack critical infrastructure, significantly driving up construction and operating costs. Any array configuration that one ultimately arrives at necessarily hinges on a non-unique choice about what exactly constitutes "optimality", and the result can depend sensitively on how one weighs the many relevant considerations when doing so.

In this section, we detail some of the considerations that are entering into the design process for the ngEHT array configuration. Section 4.1 describes how we have selected an initial pool of candidate sites to consider, Section 4.2 describes our procedure for simulating realistic ngEHT observations, and Section 4.3 details several metrics that we use to evaluate array quality. Section 4.4 describes our evaluation of the many different candidate arrays and discusses a strategy for translating array performance into site selection. Various details relevant to the site selection procedure are provided in Appendix A.

4.1. Candidate Sites

From most locations on the surface of the Earth, atmospheric opacity prevents observations at the primary ngEHT frequencies of 230 GHz and 345 GHz. We thus take our starting pool of candidate sites from [86], who identified sites with favorable atmospheric transmission properties for 230 GHz and 345 GHz observations during the March/April typical EHT observing season; the candidate sites are shown in Figure 3 and listed in Table A1.

Given the selection on atmospheric opacity performed in [86], the candidate sites are naturally situated in the highest and driest locations. Figure 4 shows the highest-elevation locations around the globe, and Figure 5 shows where the mean level of precipitable water vapor (PWV) is lowest throughout the year. We have computed the PWV using atmospheric data from the MERRA-2 database [87]. The PWV at a particular location is determined by integrating the water vapor through the column of atmosphere above that location (see, e.g., [88]),

$$\mathrm{PWV} = \frac{1}{\rho g} \int_0^{P_{\mathrm{surf}}} \frac{q(P)}{1-q(P)} dP. \tag{1}$$

Here, $q(P)$ is the specific humidity, P is the atmospheric pressure, P_{surf} is the atmospheric pressure at the surface, $\rho \approx 1\,\mathrm{g\,cm^{-3}}$ is the mass density of water, and $g \approx 9.81\,\mathrm{m\,s^{-2}}$ is the acceleration of gravity at the surface of the Earth. MERRA-2 provides both P and q in 42 different atmospheric layers as a function of geographic location and time.

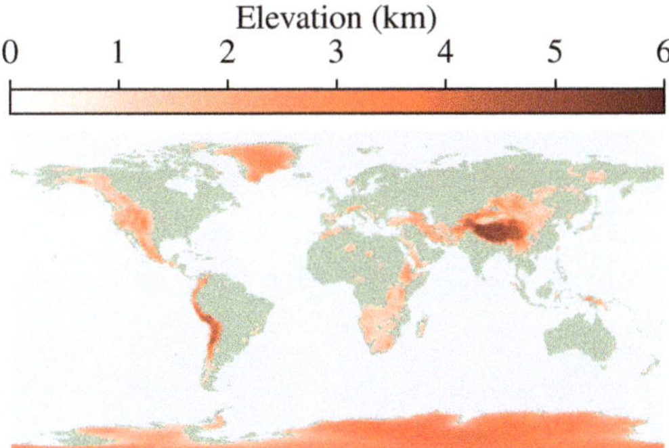

Figure 4. Global elevation map. Locations with elevations above 1000 m are shaded red, with darker colors indicating higher elevations.

4.2. Synthetic Data Generation

We evaluate candidate array performance using synthetic observations of the key science targets M87* and Sgr A*. For source models, we use the results of general relativistic magnetohydrodynamic (GRMHD) simulations that have been post-processed using ray-tracing and radiative transfer codes to produce images at the 230 GHz and 345 GHz observing frequencies appropriate for the ngEHT. Our M87* source model comes from the simulations carried out in [71], and our Sgr A* source model comes from the simulation library produced in [11].

We generate synthetic datasets using the ngehtsim[2] library. Given a candidate ngEHT array configuration and a source model, ngehtsim uses eht-imaging [89,90] to sample the Fourier transform of the source at the (u, v)-coverage corresponding to the array. Thermal noise σ_{ij} on a baseline between stations i and j is determined by the radiometer equation,

$$\sigma_{ij} = \frac{1}{\eta_q} \sqrt{\frac{\mathrm{SEFD}_i \mathrm{SEFD}_j}{2\Delta\nu\Delta t}}, \tag{2}$$

where $\Delta\nu$ is the observing bandwidth, Δt is the integration time, SEFD is the station system equivalent flux density, and $\eta_q = 0.88$ is an efficiency factor associated with 2-bit quantization during data collection [91]. We determine SEFDs for each station as a function of time using

$$\mathrm{SEFD} = \frac{2kT_{\mathrm{sys}}}{A_{\mathrm{eff}}} e^{\tau}, \tag{3}$$

where k is the Boltzmann constant, τ is the (time-dependent) line-of-sight atmospheric opacity, A_{eff} is the effective collecting area of the telescope,

$$T_{\text{sys}} = T_{\text{rx}} + T_{\text{atm}}\left(1 - e^{-\tau}\right) \tag{4}$$

is the system temperature, T_{rx} is the receiver temperature, and T_{atm} is the temperature of the atmosphere. We determine T_{atm} using historical atmospheric data from the MERRA-2 database [87], and τ is obtained by passing the atmospheric state information from MERRA-2 through the *am* radiative transfer code [92].

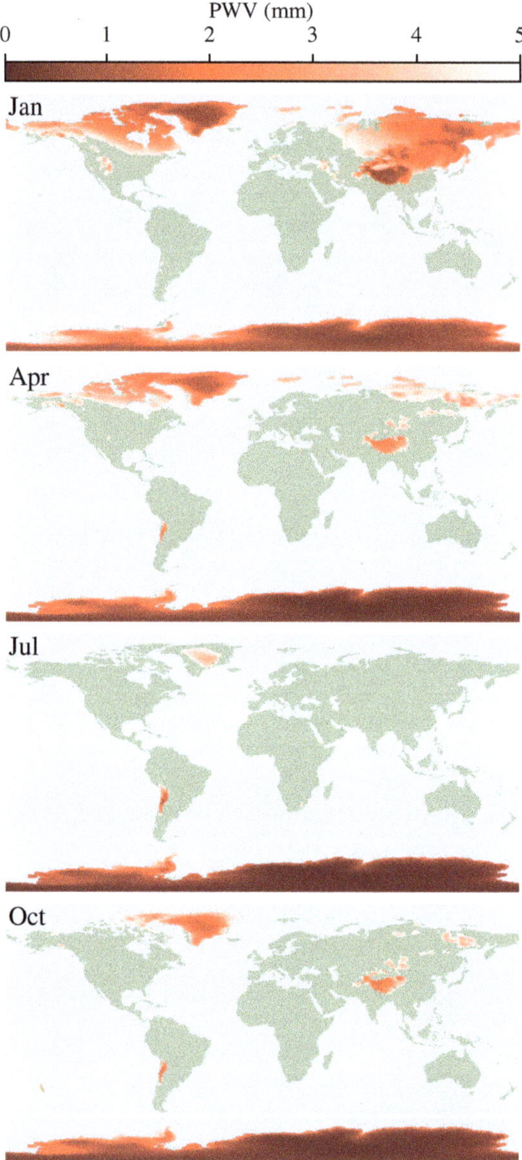

Figure 5. Locations around the globe with mean PWV less than 5 mm, in January (**top**), April (**second from top**), July (**second from bottom**), and October (**bottom**). A darker red coloring indicates a lower value of mean PWV, and only locations with elevations above 50 m are colored. We have determined the PWV via Equation (1) using atmospheric data from MERRA-2 [87], and the average is taken over all available data between 2012 and 2022.

For the synthetic datasets in this section, we use a Stokes I bandwidth of $\Delta \nu = 16\,\text{GHz}$ (for ngEHT) or $\Delta \nu = 2\,\text{GHz}$ (for EHT) at each of two frequency bands, one centered at $230\,\text{GHz}$ and the other centered at $345\,\text{GHz}$. We use an integration time of $\Delta t = 10\,\text{min}$, which is assumed to be enabled by suitable phase calibration, with a 50% duty cycle (i.e., 10 min on-source followed by 10 min off-source); the total duration of each observation is 24 h. We assume receiver temperatures T_{rx} of 50 K at $230\,\text{GHz}$ and 75 K at $345\,\text{GHz}$.

We emulate fringe-finding by applying a signal-to-noise ratio (SNR) thresholding scheme to the generated visibilities. The scheme employs a variant of the "fringe groups" strategy from Blackburn et al. [93] for assigning reliable measurements from a set of baseline visibilities: if visibility does not achieve an equivalent SNR of 5 (phase error of 11.5 degrees) on an integration time of 10 s (at $230\,\text{GHz}$) or 5 s (at $345\,\text{GHz}$), and if the stations comprising the baseline associated with that visibility do not participate in other baselines that achieve the requisite SNR, then that visibility is considered to be not measured and it is flagged from the dataset. Note that for the stations with dual-frequency capabilities, both of the frequency bands are checked simultaneously; if either one of the two frequency bands has an SNR that satisfies the threshold condition, then we assume that both bands can be detected. This multi-frequency fringe groups scheme emulates frequency phase transfer across the bands (see, e.g., [94]). We only emulate FPT when simulating ngEHT data; when simulating EHT data, we apply only the single-frequency fringe groups scheme.

4.3. Array Performance Metrics

The analysis methods utilized by the EHT for performing measurements of physical interest using VLBI data are in general highly computationally expensive to evaluate. Further, the added value of a particular set of new sites is non-linear in the number of sites; the number of new baselines is quadratic in the number of existing sites, and the value of an individual site is sensitive to its position with respect to existing dishes.

To evaluate the performance of candidate ngEHT array configurations without running computationally expensive analysis pipelines (such as imaging or model-fitting), we utilize metrics of array performance that are based on pre-analysis quantities. We primarily employ two metrics: one metric that quantifies the (u, v)-coverage and another metric that quantifies the aggregate baseline sensitivity. We compute the array performance metrics using synthetic observations at frequencies of $230\,\text{GHz}$ and $345\,\text{GHz}$, which drive the key science goals of the ngEHT. While $86\,\text{GHz}$ is an important addition that enables improved detection prospects at the higher frequencies (see Section 7.1), it serves primarily a calibration-related role and thus is not included in our (u, v)-coverage or baseline sensitivity metric computations. We note, though, that ngEHT sites could potentially be included in 3 mm wavelength VLBI networks (e.g., GMVA) or as part of ngVLA observations [44].

We use as our quantification of (u, v)-coverage quality the (u, v)-filling fraction metric (FF metric), defined in Palumbo et al. [95] as the fraction μ_{ff} of the area enclosed by a bounding circle in (u, v) of radius $1/\theta_{\text{res}}$ that is covered by the two-dimensional convolution of the coverage with a circular disk of radius $1/\theta_{\text{FOV}}$. Here, θ_{res} and θ_{FOV} are array performance specifications based on imaging expectations, and they are not predicted directly by the coverage; Figure 6 provides an illustration of how the FF metric is calculated. Palumbo et al. [95] found that as the filling fraction increases, imaging performance in compact imaging examples improves steadily until it flattens to a constant factor of the diffraction-limited image fidelity near $\mu_{\text{ff}} \gtrsim 0.5$. The FF metric naturally demands greater coverage for equivalent μ_{ff} as expectations of the imaging field of view θ_{FOV} increases; however, μ_{ff} does not capture the relative information density of the Fourier plane, and for many source morphologies, the importance of Fourier coverage decreases with radius from the (u, v)-coordinate origin. In this paper, we assume $\theta_{\text{res}} = 14\,\mu\text{as}$ (i.e., the angular resolution of an Earth-diameter baseline observing at $345\,\text{GHz}$) unless otherwise specified, but we use several different fields of view; when quoting FF metric values, we will thus specify the corresponding assumed field of view using the notation $\mu_{\text{ff}}(\theta_{\text{FOV}})$.

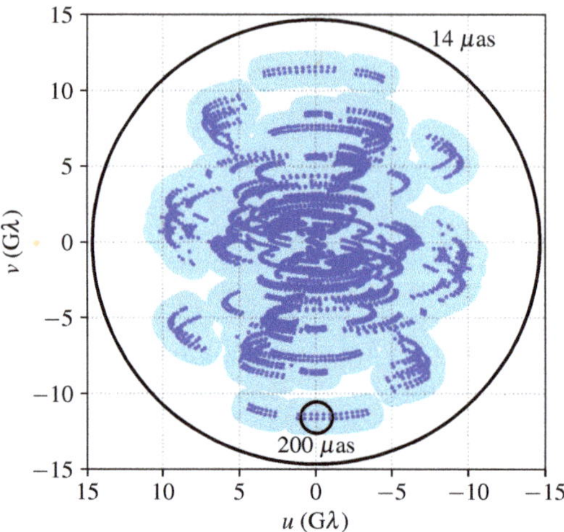

Figure 6. Illustration of the (u,v)-filling fraction metric described in Section 4.3 (see also Palumbo et al. [95]). Given some (u,v)-coverage—shown here by the blue points for a mock observation of M87 using the full ngEHT Phase 1 array in April—the FF metric is a measure of how much area within a circular region of radius $1/\theta_{\rm res}$ is sampled, after convolving the coverage with a disk of radius $1/\theta_{\rm FOV}$. In this case, $\theta_{\rm res} = 14\,\mu$as and $\theta_{\rm FOV} = 200\,\mu$as. The convolved coverage is shaded in light blue, and takes up a fraction $\mu_{\rm ff} = 0.5$ of the area of the outer circle.

For our quantification of aggregate array sensitivity, we use the point source sensitivity (PSS) metric,

$$\text{PSS} = \left(\sum_{i=1}^{N} \frac{1}{\sigma_i^2} \right)^{-1/2}, \tag{5}$$

where σ_i is the value of the thermal noise on visibility i (see Equation (2)), and the sum is taken over all visibilities in the dataset. The PSS metric, which has units of flux density, quantifies the sensitivity that the array could in principle achieve when measuring the flux density of a point source. It naturally folds in not only the observing bandwidth and diameter of each telescope in the array, but also the amount of mutual visibility that each site has with every other as well as the atmospheric transmission at each site.

4.4. Site Selection

The stringent atmospheric opacity requirements for observing at millimeter wavelengths means that only a small number of locations around the globe are suitable candidates (see Section 4.1). Given that the list of candidate sites presents a finite number of discrete locations on the globe where telescopes could be placed, we could in principle evaluate all possible new array configurations. The ability to confine the site search space to a finite number of options in this way is fairly unique to high-frequency VLBI, and it informs our optimization strategies below; the analogous site selection problem for connected-element arrays (e.g., VLA, ALMA, ngVLA) and low-frequency VLBI arrays (e.g., VLBA, SKA) presents a qualitatively different challenge.

In practice, though the number of possible new array configurations is finite, the space remains large and difficult to search comprehensively; the number of possible new array configurations that could be made using the 44 sites listed in Table A1 is approximately 1.8×10^{13}. Additionally, we would like to ensure that the selected sites enable the ngEHT array to perform well across all of the following situations:

- in observations of both M87* and Sgr A*;
- during observations that take place throughout the year;

- when observing alongside any subset of the EHT.

The performance of each candidate array must also be evaluated using several different quality metrics (see Section 4.3) that correspond to the various scientific goals. All of the above considerations result in multiplicative factors that further increase the expense of a comprehensive analysis.

Given the difficulty of comprehensively searching all possible combinations of new stations, we instead partition our site selection efforts into two stages corresponding to the two anticipated phases of ngEHT development. In the first stage—corresponding to ngEHT Phase 1—we consider the selection of three new sites from the pool of candidates. The availability of three 6.1-m BIMA dishes for refurbishment and relocation (see Section 7.5) provides a pathway to realizing a Phase 1 ngEHT array on a shorter ($\sim$few-year) timescale than it will take to field a larger array of newly constructed dishes. Optimizing for only three new sites at a time also reduces the number of site combinations to only $\binom{44}{3}$ = 13,244. In the second stage of the site selection analysis—corresponding to ngEHT Phase 2—we then consider the selection of five new sites from the remaining pool of candidates, corresponding to $\binom{41}{5}$ = 749,398 different site combinations. Dividing the optimization strategy into two stages in this way, and selecting a specific target number of new sites in each stage, substantially reduces the computational cost of optimizing the array configuration.

The sites and frequency configurations corresponding to the selected Phase 1 and Phase 2 ngEHT arrays are listed in Table 2.

Table 2. Site participation and frequency capabilities for the EHT and both phases of the ngEHT array. For the first column, EHT sites with existing 86 GHz capability are noted, but the EHT does not currently support 86 GHz operation; and some of these 86 GHz receivers cannot be used simultaneously with higher frequency receivers. In each of the three rightmost columns, sites that do not participate in the specified array are indicated with a "-" sign. Multi-frequency capabilities are indicated with a "+" sign; e.g., "230 + 345" indicates that the station can observe at both 230 GHz and 345 GHz simultaneously, whereas "230 345" indicates that it can only observe at each frequency separately. For completeness, we list in the rightmost column an alternative incarnation of the ngEHT Phase 2 array, in which we forgo the need to field new telescopes by relying instead on external facilities that are anticipated to come online in the next few years (see Section 4.4.3). For this alternate case, the JELM site would be added in Phase 1. [†] This site is being developed independently of the ngEHT.

Site	Status	EHT	ngEHT Phase 1	ngEHT Phase 2	ngEHT Phase 2 (alt.)
ALMA	existing	86 230 345	86 230 345	86 230 345	86 230 345
AMT	planned [†]	-	-	-	86 + 230 + 345
APEX	existing	230 345	86 230 345	86 230 345	86 230 345
BOL	planned	-	-	86 + 230 + 345	-
CNI	planned	-	86 + 230 + 345	86 + 230 + 345	86 + 230 + 345
GLT	existing	86 230 345	86 + 230 + 345	86 + 230 + 345	86 + 230 + 345
HAY	existing	-	86 + 230	86 + 230	86 + 230
IRAM	existing	86 + 230 345	86 + 230 345	86 + 230 345	86 + 230 345
JCMT	existing	86 230 345	86 + 230 + 345	86 + 230 + 345	86 + 230 + 345
JELM	planned	-	-	86 + 230 + 345	86 + 230 + 345
KILI	planned	-	-	86 + 230 + 345	-
KP	existing	86 230	86 + 230	86 + 230	86 + 230
KVNPC	planned [†]	-	-	-	86 + 230 + 345
KVNYS	existing	-	-	-	86 + 230 + 345
LCO	planned	-	86 + 230 + 345	86 + 230 + 345	86 + 230 + 345
LLA	planned [†]	-	-	-	86 + 230 + 345
LMT	existing	86 230	86 + 230 + 345	86 + 230 + 345	86 + 230 + 345
NOEMA	existing	86 230 345	86 + 230 345	86 + 230 345	86 + 230 345
OVRO	existing	-	86 + 230	86 + 230	86 + 230

Table 2. *Cont.*

Site	Status	EHT		ngEHT Phase 1	ngEHT Phase 2	ngEHT Phase 2 (alt.)
SGO	planned	-		-	86 + 230 + 345	-
SMA	existing	230	345	230 345	230 345	230 345
SMT	existing	230	345	86 + 230 + 345	86 + 230 + 345	86 + 230 + 345
SPM	planned	-		86 + 230 + 345	86 + 230 + 345	86 + 230 + 345
SPT	existing	230		86 + 230 + 345	86 + 230 + 345	86 + 230 + 345
SPX	planned	-		-	86 + 230 + 345	-

4.4.1. Phase 1

To determine the optimal locations for the three new 6.1 m Phase 1 dishes, we carry out a survey of all possible three-station combinations of the 44 sites listed in Table A1. For each candidate set of three sites, we explore the performance of the resulting array (1) for observations of both M87* and Sgr A*, (2) under weather conditions appropriate for January, April, July, and October, and (3) when observing alongside four different variants of the existing EHT array (specified in the top section of Table A2). These pre-existing array variants include various subsets of the EHT array, as well as the HAY and OVRO dishes that are expected to be outfitted with ngEHT equipment (see Section 2). We evaluate each candidate array using the metrics described in Section 4.3 for 100 instantiations of the weather conditions at each site, from which we then take median values to establish typical performance.

After evaluating all candidate arrays, we determine a "performance score" for each array according to its average ranking across the full suite of observing parameters, e.g., if a particular array is ranked first for one set of observing parameters, ranked third for a second set of observing parameters, and ranked fifth for a third set of observing parameters, then its performance score would be $(1 + 3 + 5)/3 = 3$. Arrays with smaller values of the performance score are those that have performed well across a range of observing parameters. Figure 7 shows the top 1% of all three-station candidate site combinations after ranking them by their performance scores, with each set of sites plotted as a connected three-baseline triangle. We identify six heavily populated clusters of high-performing site combinations:

- An "eastern cluster" containing two sites in South America and either CNI or, less frequently, one of the other mainland European sites (BGA, SKS, SPX).
- A "western cluster" containing two sites in South America and a site in North America, most typically either SPM, PIKE, or FAIR.
- A "northern cluster" containing two sites in South America and GLTS, or less commonly with BGK.
- A "southern cluster" containing two sites in South America and one of the Antarctic Dome sites.
- An "equatorial cluster" containing one site in South America, one site in North America, and CNI.
- A "polar cluster" containing one site in South America, GLTS, and one of the Antarctic Dome sites.

We see that the most favored sites tend to be those that are able to leverage simultaneous observability with existing sites. The overrepresentation of existing sites in the Western hemisphere means that sites in the Eastern hemisphere—particularly those in Asia and New Zealand—are correspondingly penalized.

To select from among the top-performing three-site combinations, we impose additional, more qualitative considerations. We disfavor the northern, southern, and polar clusters because they contain sites that are unable to observe either M87* (in the case of the Antarctic sites) or Sgr A* (in the case of GLTS and BGK). The Eastern and Western clusters suffer from a similar asymmetry, in that they include sites that have little mutual visibility with existing American and European stations, respectively. The equatorial cluster provides the most balance in terms of site geography, and it contains the three most favored regions for a new site: South America, North America, and CNI. Several of the South American sites are comparably well-represented among the top site combination candidates, as are a couple of the North American sites. After additionally accounting for initial site cost estimates and favoring lower-cost sites, we settle on the three-site combination of CNI, LCO, and SPM as our fiducial ngEHT Phase 1 additions.

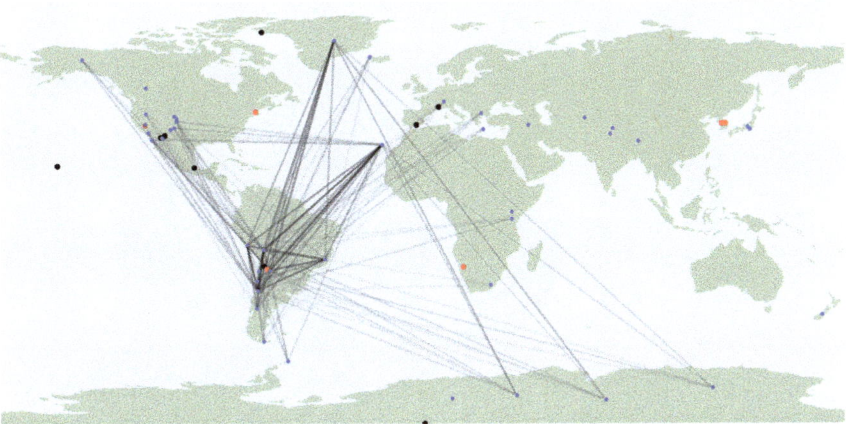

Figure 7. Top performing 1% of all three-station candidate site combinations from the Phase 1 exploration (see Section 4.4). Dots follow the same color convention as in Figure 3: black are current EHT sites, red are existing or near-future sites that may join global observations, and blue are potential new ngEHT sites. Three-station candidate site combinations are shown as connected black triangles.

4.4.2. Phase 2

In the second stage of our site selection analysis—corresponding to ngEHT Phase 2—we consider the addition of five new 9 m sites to the previous three 6.1 m sites determined from the Phase 1 selection. We explore the same observing targets and weather conditions as for the Phase 1 exploration, but we use updated pre-existing arrays that include the Phase 1 sites (see the bottom section of Table A2). We again evaluate each candidate array using the metrics described in Section 4.3 for 100 instantiations of the weather conditions at each site, and we use median metric values to establish typical performance.

The selection process for Phase 2 is ongoing, but preliminary results indicate that the combination of BOL, JELM, KILI, SGO, and SPX would provide a strong improvement to the array coverage. We thus take these sites to be our fiducial ngEHT Phase 2 additions for the purposes of this paper.

4.4.3. Alternate Staging of New Sites

Several new radio telescopes that could be used for ngEHT observations are planned to become operational in the coming years. Thus, an alternative staging approach would be to augment Phase 1 by adding JELM to the three new sites described in Section 4.4.1, and Phase 2 could then consist solely of the following planned telescopes: the LLAMA telescope in Argentina, the AMT in Namibia, the KVNYS telescope near Seoul, Korea, and the KVNPC telescope (currently under construction) near Pyeongchang, Korea. Together, the four Phase 1 sites (CNI, JELM, LCO, SPM) combined with OVRO, HAY, and the four planned telescopes (LLAMA, AMT, KVNYS, KVNPC) would constitute a near-doubling of the existing EHT array and would achieve comparable (u, v)-coverage to that provided by the array described in Section 4.4.2. This alternate pathway to a Phase 2 ngEHT would also provide capabilities sufficient to achieve all ngEHT Key Science Goals.

4.4.4. Baseline Coverage

Simulated EHT coverage for the array fielded during the 2023 observing campaign is shown in Figure 8. The enhanced baseline coverage that will be provided by the ngEHT Phases 1 and 2 is shown in Figures 9 and 10. In Figure 9, the JELM station has been added to reflect a Phase 1 array as described in Section 4.4.3, while Figure 10 shows the array as described in Section 4.4.2. Snapshot coverage for Sgr A* observations is shown in Figure 11, indicating that 1-min integrations produce spatial frequency sampling that can be used for increasingly detailed dynamical modeling.

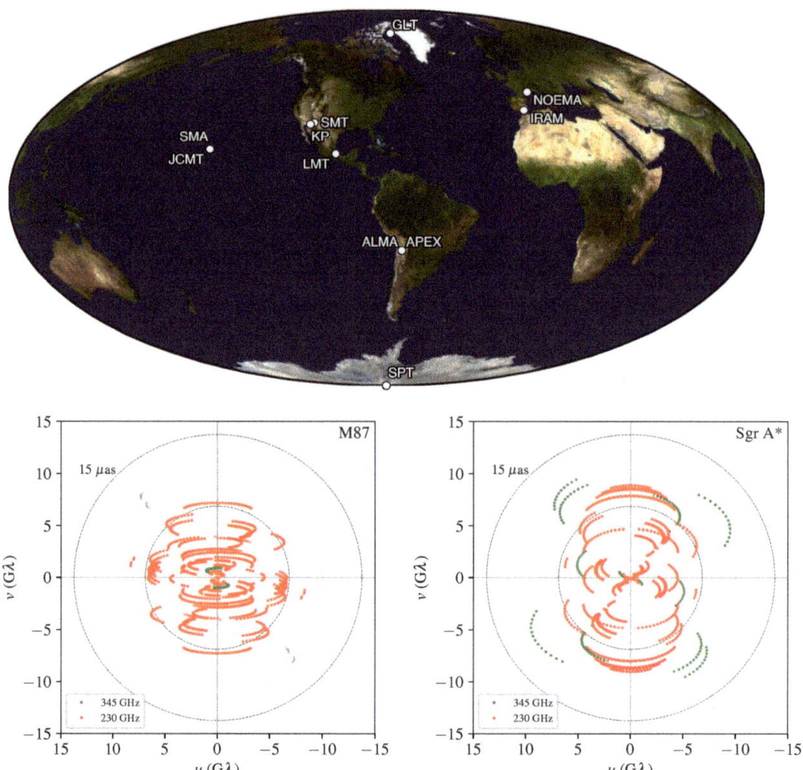

Figure 8. (**Top**) Current EHT array (2023). (**Bottom**) Interferometric coverage for M87* and Sgr A* at 230 & 345 GHz, assuming April observing conditions and a minimum observable elevation of 10 degrees. The coverage reflects estimated detections made through simulating M87* and Sgr A* models at both frequencies with the EHT array as fielded in 2023 (see Table 2, and Section 4.2). Note that for the EHT in 2023, 230 GHz and 345 GHz observations cannot be made simultaneously, so the coverage shown cannot be combined to form a full image (as is possible in the ngEHT Phase 1 and Phase 2 arrays). The opacity of each plotted data point is proportional to how frequently it is expected to be detected. The outer and inner dashed circles mark baseline lengths corresponding to angular scales of 15 µas and 30 µas, respectively.

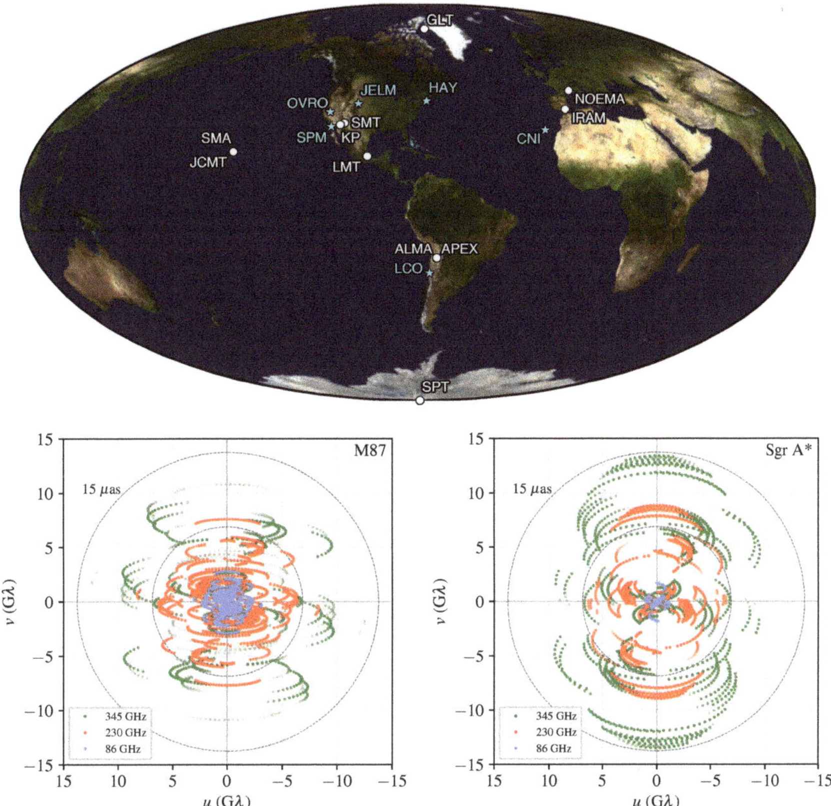

Figure 9. (**Top**) ngEHT Phase 1 array; white sites are current EHT dishes, blue sites are ngEHT sites. (**Bottom**) Interferometric coverage for M87* and Sgr A* at 86 GHz, 230 GHz, and 345 GHz, assuming April observing conditions and a minimum observable elevation of 10 degrees. The coverage reflects estimated detections made through simulating M87* and Sgr A* models at all three frequencies with the ngEHT Phase 1 array (see Table 2 and Section 4.2). Sites without multi-frequency capabilities are assumed to be observed only at their highest frequency. The opacity of each plotted data point is proportional to how frequently it is expected to be detected. The outer and inner dashed circles mark baseline lengths corresponding to angular scales of 15 μas and 30 μas, respectively.

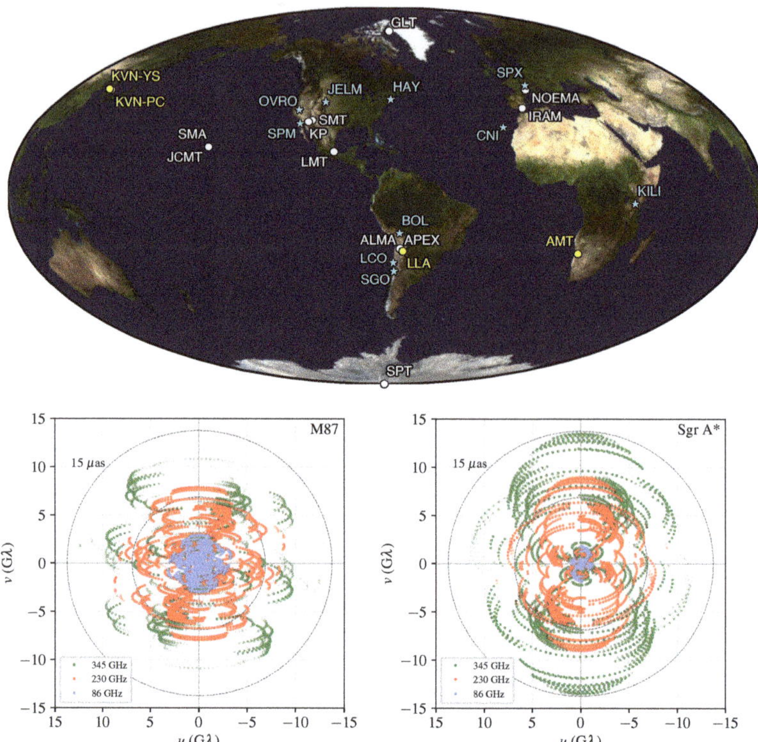

Figure 10. (**Top**) ngEHT Phase 2 array; white sites are current EHT dishes, blue sites are ngEHT sites, and yellow sites are planned or existing facilities that may join (ng)EHT observations and a minimum observable elevation of 10 degrees. (**Bottom**) Interferometric coverage for M87* and Sgr A* at 86 GHz, 230 GHz, and 345 GHz, assuming April observing conditions. The coverage reflects estimated detections made through simulating M87* and Sgr A* models at all three frequencies with the ngEHT Phase 2 array (see Table 2 and Section 4.2). Sites without multi-frequency capabilities are assumed to be observed only at their highest frequency. The opacity of each plotted data point is proportional to how frequently it is expected to be detected. The outer and inner dashed circles mark baseline lengths corresponding to angular scales of 15 μas and 30 μas, respectively.

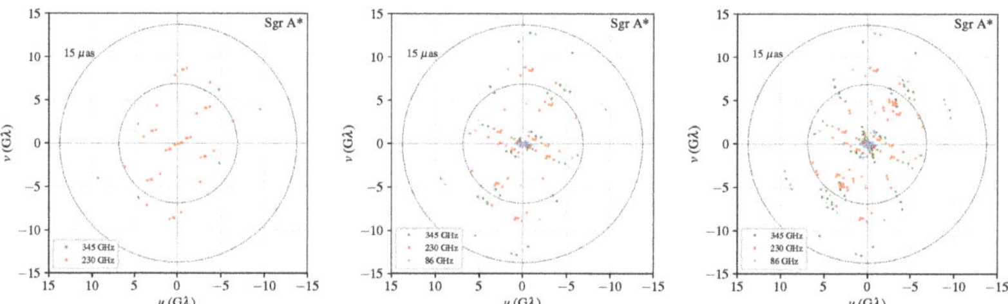

Figure 11. Representative snapshot coverage of SgrA*. For the EHT in 2023 (**left**), ngEHT Phase 1 (**middle**), and ngEHT Phase 2 (**right**), snapshot coverage for 1-min integrations on SgrA* is shown. The increase in spatial frequency sampling enables dynamic modeling of SgrA* with time resolution that is well matched to the dynamical time scales of the source. The outer and inner dashed circles mark baseline lengths corresponding to angular scales of 15 μas and 30 μas, respectively.

5. Operating Modes

Key Science Goals (KSGs) motivate five basic operation modes of the ngEHT, which enable specific science use cases. Details and constraints of each mode are defined by cost/benefit analyses and feasibility studies. Factors to be considered in this analysis include time allocation at various sites, weather, data throughput with implications for disk inventory and correlation, reliability and up-time, and maintenance strategy. The following five subsections provide a narrative summary for each of the five envisaged operating modes of the ngEHT, which are then summarized with salient characteristics in Table 3.

Table 3. The five ngEHT operating modes and selected salient characteristics of each.

OpsMode	Stations in Array	Cadence & Duration	Science Case
Campaign	14 to 21	once per year 7-day session	Sgr A*, M87 blazars, jets
Long term monitoring	5 to 20	once per 3 to 5 days, for 3 to 7 months	M87* & blazar kinematics, Sgr A* flares
Target of Opportunity	3 to 6	once per week during obs. season	flares, gravitational waves
CMF	14 to 21	during Campaign	AGNs, black hole binaries
Beyond ngEHT	1 to 10	dependent on science case	stellar birth, fast radio bursts

5.1. Campaign

This is a single epoch annual multi-day campaign, which is an extension of the standard annual campaign already executed by the Event Horizon Telescope (EHT). The 11 EHT sites defined by those used in the 2022 EHT array are assumed to participate with the ngEHT sites. In this mode, dedicated tracks are based on clearly defined, community-prioritized science cases, in some cases led by a principal investigator.

The campaign mode pursues M87 and Sgr A* science cases with enhanced capability relative to EHT due to improved sensitivity and great UV coverage from the larger 21 site array. This results in enhanced M87 imaging, snapshot sensitivity for Sgr A* movies, and studies of blazar jet collimation.

5.2. Long-Term Monitoring

The long-term monitoring mode uses extended duration and more frequent cadence observations with a smaller subset of the existing EHT sites participating. The ngEHT sites enable this mode through their purpose-designed flexibility and dedicated time allocation for VLBI.

Several multi-week observations over the course of the year once again have dedicated tracks based on clearly defined, community-prioritized science cases. These science cases are in the broad areas of M87* movies, blazar kinematic studies, and Sgr A* flaring activity monitoring. As an example, to continuously track changes in the M87* appearance (M87* movies), reconstructing images separated by the expected coherence timescale ($\sim 50\,GM/c^3$ ≈ 20 days) is needed. A single-year EHT campaign may only last about a week [1]—too short for a significant change in the source appearance, while combining results from separate years only provides uncorrelated source snapshots, without the ability to track continuous motion of the flow features [96]. Similarly, in the published EHT analyses of blazar observations [97–99] short duration of the EHT campaigns, and the lack of repeated observations on timescales of weeks or months, has been recognized as the main factor limiting the current EHT ability to study jet kinematics.

5.3. Target of Opportunity

Target of Opportunity (ToO) is an agile operational follow-up by ngEHT to an unpredictable event observed with another facility. It involves ad hoc subarrays of the 11 existing EHT sites—those that are available—while all of the ngEHT dedicated sites will be made available for suitably scientifically interesting ToO observations. Broad science areas are expected to be in the area of flares, gravitational waves, and fast radio burst counterparts.

5.4. Coordinated Multi-Facility

The Coordinated Multi-Facility (CMF) mode is characterized by coordinated, multi-facility, multi-messenger observations involving multiple ngEHT sites and at least one other ground or space instrument (e.g., Chandra, the GRAVITY instrument, and any of various optical/IR facilities). This CMF mode is a planned continuation of the successful EHT Multi-Wavelength campaigns (see [100]).

The broad science areas are expected to be multi-wavelength studies of Active Galactic Nuclei, binary and singular black holes.

5.5. Beyond ngEHT

This single-dish mode covers any observation that is performed outside the core ngEHT science mission, but will still be part of the ngEHT operating model due to local institutional requirements or synergies with other communities or facilities.

Science is expected to be in the broad area of star-forming regions, fast radio bursts, and astronomical maser studies of transitions in the ngEHT RF bands.

6. Data Processing

The next-generation (ngEHT) expands upon the existing 11-station EHT with around 10 additional small-dish antennas as well as simultaneous 230/345 GHz observations. In addition to the roughly ∼10-fold increase in aggregate data rate across the entire array, the ngEHT is expected to operate as a full-season agile observatory as opposed to the ∼few observing days per year of the current EHT. When all participating sites are observing, one night of ngEHT produces around ∼10 PB of raw data (around 0.5 PB per site), resulting in up to a ∼couple hundred PBs per year that must be processed. An efficient streamlined approach to data processing and management is required to facilitate media turn-over and to deliver quality assured science-ready data products in a timely manner.

The large data rates and volumes of the ngEHT motivate continued adoption and assimilation of new technologies, which has allowed a rough tracking of Moore's law over two decades of global mm-VLBI development (Figure 1). On the timescale of a ∼decade, we anticipate a transition from Hard Disk Drives (HDDs) to Solid State Disks (SSDs) for recording and eventually transport, which provides high-bandwidth, high-density, and power-efficient I/O. SSDs carry a gradually narrowing cost premium of 5–10 times that of HDDs (in \$/TB), but use of SSDs would allow ngEHT recording systems to keep up with the ngEHT data rates while staying within practical power, weight, and space footprints for efficient media handling, staging, and transport.

GPU's have become the platform of choice for massively parallel vector/tensor calculations due to their efficiency and ease of use, and they are being researched or already adopted for efficient VLBI correlation across several experiments. The "embarrassingly parallel" nature of VLBI correlation is suitable for high-throughput computing (HTC) workflows, and the irregular scheduling of VLBI observations means that on-demand scalable computational resources are desirable.

6.1. Data Transport

While observing, the ngEHT will produce an aggregate ∼5 Tbps of digital signal data that must ultimately be transported from remote sites to a central location for processing. Similar to the EHT, the only currently available means for moving such a large total volume of data from the (sometimes very-) remote locations in a reasonable amount of time is by physical transport of recorded media. Some VLBI arrays, such as the European VLBI Network[3] (EVN) are able to transport data electronically, due to considerably lower data rates and more accessible sites (typically at sea level) that are linked to a high-speed internet backbone. The ngVLA reference design [33] also includes real-time data transport (320 Gbps per antenna) and correlation via ground fiber (both dedicated and leased commercial), even for the longest baselines spanning the United States and territories. However because the ngEHT operates a (comparatively) small number of antennas at remote locations spanning the globe, shipment of physical media is expected to remain the fastest and most economical method of transferring 100 s PB of data for the foreseeable future. Consistent array-wide high-speed internet access, such as that provided by global commercial Satellite RF internet, will nevertheless be extremely useful for rapid transfer of small amounts ($\sim$1%) of data for interferometric validation and for obtaining near-realtime results where scientifically relevant.

The ngEHT is designed to operate full-season, and this motivates rapid processing and recycling of recording media to limit costs. Media are expected to be redeployed approximately once per two months (on average), versus once per 2–3 years as for the current EHT. As a result, there is less of a focus on media utility for economical long-term storage and more toward efficient recording and transport. Once data are brought to the correlation facility, they can be offloaded to local HDD-based storage if needed, for example, in the case of experiments including the South Pole Telescope which can incur several months of shipping delay. A rotating media library of 200 PB would be required to support a bimonthly turnaround of observations totaling 10 PB every three days while providing ample time for average shipping time and data offload.

6.2. Correlation

Correlation is the process of calculating pairwise correlation coefficients between the signals captured at each antenna. Because this is an operation on the PB of raw VLBI data, it is both I/O and computationally intensive and requires carefully matched computing platforms for effective processing. Correlation coefficients are typically calculated in the frequency domain using a so-called FX correlation architecture that enables efficient searching over unknown time delay via Fourier convolution. Frequency domain processing also allows for the convenient matching of signals from partially overlapping bandwidths as well as the application of linear and non-linear corrections to align the data. The consequences of an FX architecture is a large up-front cost to data channelization, scaling linearly with the number of antennas. For a 20-station network at ngEHT bandwidths, the $O(N)$ cost from data stream pre-processing and the $O(N^2)$ cost from calculating all pair-wise correlations are expected to be roughly comparable.

The current EHT records at 64 Gbps over 11 stations, for an aggregate rate of 0.7 Tbps. Data are correlated at dedicated computing clusters at MIT Haystack Observatory and the Max Planck Institute for Radioastronomy using the DiFX software correlator [101]. In aggregate, $\sim$2.5 k cores are able to process the full EHT bandwidth at about 10% real-time. Scaling linearly to the aggregate data rate of ngEHT requires $\sim$20 k cores to process ngEHT's $\sim$5 Tbps at 10% real-time (in comparison, 300 h of data per year, at 5 recorded hours per night on average, is a reasonable upper limit for ngEHT data throughput and reflects a duty cycle of 3.5% with respect to the total number of hours in a year). A quadratic scaling with the number of stations would imply double the requirement, but this can be balanced against $\sim$5–10% year-over-year improvements to single-core performance. CPU core density and efficiency are also increasing at a much faster rate, and GPU acceleration of both channelization and cross-multiply stages of correlation are expected to increase efficiency by another factor of $\sim$several. A detailed description and modeling of VLBI software correlation performance is presented in Vázquez et al. [102] alongside several benchmark results including those from the literature.

Approximately $\sim$60 M CPU core-hours would be required to correlate $\sim$680 PB (300 h) of raw data. VLBI data are taken non-continuously throughout the year and sometimes require multiple passes through correlation to iterate on a proper configuration. Thus, is it necessary to over-provision on-demand computational resources by a factor of $\sim$few in order to avoid backlogs and ensure regular turnaround of recording media. Around $\sim$100 k on-demand CPU cores would be appropriate to keep up with the largest projected ngEHT data volumes, which is the size of a large institutional research cluster or a few medium-sized clusters distributed geographically. Due to the over-provisioning, the resources are ideally time-shared with other computing requirements (calibration and imaging, other VLBI correlation, or other general uses).

6.3. Calibration and Reduction

Output from correlation is at a resolution of $\sim$1 MHz in bandwidth and $\sim$1 second in time, which is required to capture residual instrumental and environmental systematics that affect the measured correlation coefficients such as lines, frequency response, relative delays, and time-varying gains and atmospheric phase [2,3]. These products are smaller than the recorded VLBI signals by a factor of $>10^3$ due to the large amount of accumulation following cross-correlation. A calibration process then solves for a refined instrument model and folds in any additional priors on the instrument response.

A key element of the calibration process is "fringe-fitting" where a parameterized phase model (typically relative *delay* and *delay-rate* over a short time interval) is self-calibrated to the correlator output. The fitting process verifies that a correlated signal exists in the data, measures the correlation coefficient, and allows data to be further coherently averaged, reducing the overall data volume by another factor of $\sim$10^4. Dedicated fringe fitting and calibration pipelines [93,103] were developed for EHT data to address the heterogeneous nature of the array and unique data properties. Compared to correlation, the computing requirements to fit a basic phase calibration model are low. For example, the EHT 2017 campaign data set (5 nights, 8 stations) can be processed through a multi-stage calibration and reduction pipeline using $\sim$1.5 k CPU core-hours [93].

This initial stage of calibration and reduction is aimed at reducing the overall data volume and complexity for downstream data products while applying only well-determined calibration solutions. Since data are manipulated and averaged, it is important to avoid introducing calibration solution noise or detailed assumptions about the source. In cases where calibration solutions are under-determined or degenerate with source model parameters, they must be jointly modeled during analysis. The complexity and computational cost can increase dramatically due to the high dimensionality of an instrument model, particularly in the case of formal Bayesian inference [104,105].

7. Instrumentation Design

In this section, we describe the basic elements of the ngEHT system (see Figure 12). These are the result of several internal project reviews, including a Systems Requirements Review, held on 9–10 June 2022. At this stage of the project, the ngEHT team has developed initial instrumental requirements through a process of preliminary trade-off analysis. This process has enabled the development of several prototypes, which have been selected for deployment in Phase 1 of the project, and these specific elements of the ngEHT system are described below.

7.1. Receiver

In Figure 13, (left), we present a block diagram of a dual-frequency receiver being constructed for ngEHT and to be deployed at the LMT. A single cryostat will hold two different receivers and the two different frequency bands are sent to each receiver through a frequency diplexer. Each receiver is dual-polarized, and features sideband separation mixers (see Table 4). Both bands illuminate a single beam in the sky, and the overall dual-frequency receiver has eight IF outputs, each of which is 4–12 GHz wide.

In an effort to make the design highly modular and scalable to reproduce for additional new telescopes of the ngEHT array, considerable effort has been invested into making the mixer block compact and highly integrated. In Figure 13, (right), we show the components of this highly integrated block. Shown is a photo of the bottom block of a split-block mixer (bottom) and a schematic diagram of the components (top). A similar design will be employed for the 850 μm receiver as well. The 4 IF outputs from each of the mixer blocks are amplified cryogenically using commercially available low-noise amplifiers.

Each of the receiver bands is equipped with independent local-oscillator (LO) systems. YIG oscillators are lower frequencies (in the 18–30 GHz) range are multiplied up to the 3 mm wavelength band, and subsequently amplified using W-band power amplifiers. This is then fed through cryogenic triplers to produce the required LO signal. The drain currents of the last stage of the W-band power amplifiers can be adjusted to set the appropriate LO power for the mixers. The whole LO system is phase locked, and fully computer controlled with no mechanical moving parts.

Implementation of an additional 86 GHz capability to enable Frequency Phase Transfer (FPT) will proceed along multiple paths. For existing sites that already field 86 GHz receivers, these will be coupled where possible to higher frequency receivers using dichroics that enable simultaneous operation (e.g., GLT, JCMT). At existing sites that do not have 86 GHz receivers, or where existing 86 GHz systems cannot be used, new HEMT-based 86 GHz receivers, cooled to 20 K, will be added and coupled via dichroics. These new 86 GHz receivers will follow existing and proven designs. Finally, for the new ngEHT sites, a tri-band dewar that incorporates 86, 230, and 345 GHz receivers will be constructed, following existing designs and prototypes for the ongoing upgrade of the Submillimeter Array in Hawaii.

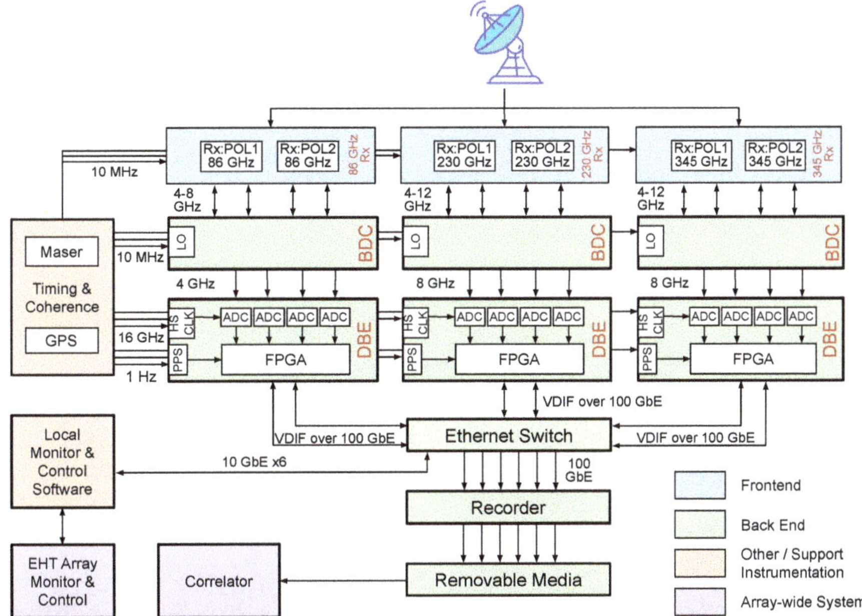

Figure 12. Functional block diagram of a next-generation EHT station. All elements shown in the figure are either commercially available (e.g., Hydrogen Maser), or in advanced prototyping stages, and suitable for deployment at ngEHT stations. The Timing & Coherence block consists of a Maser and GPS system, which provides ultra-stable clock signals for the DBE and references for the dual-polarization receivers and the BDC. A high-speed ethernet switch routes DBE packets to recorders with modular/removable media for shipment to the central correlator for interferometric processing. ngEHT Monitor and Control are handled by local and global systems.

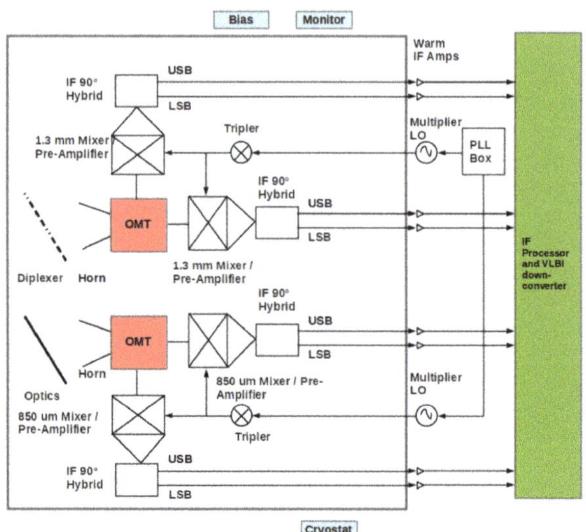

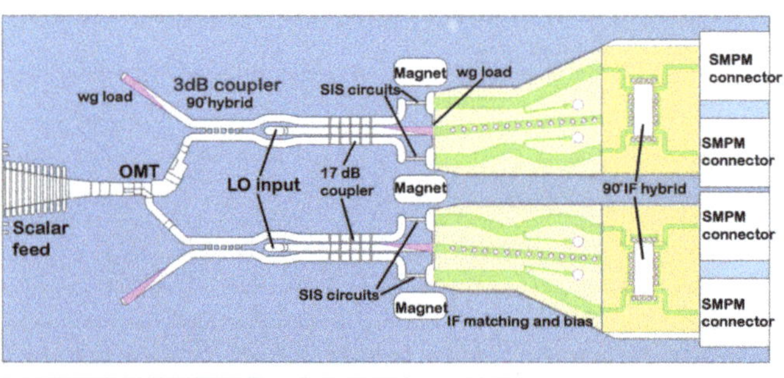

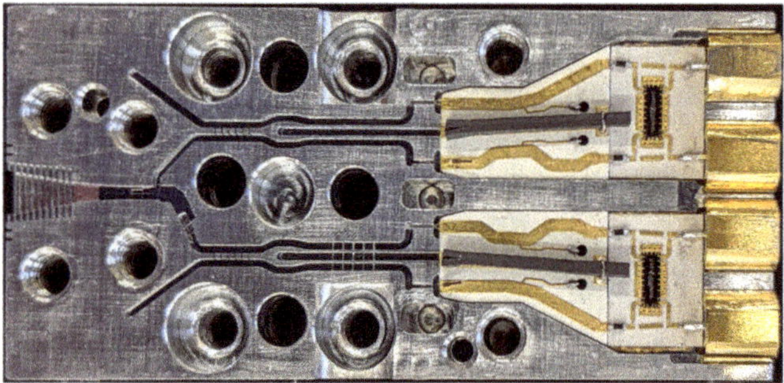

Figure 13. (**Top**) Block Diagram of the proposed dual-band SIS receivers. Both the 1.3 mm and 850 µm band receivers will be built inside a single cryostat. (**Middle**) Schematic outline of the 1.3 mm frontend receiver block. This block shows the cold section of the corrugated square feed-horn feeding an orthomode transducer (OMT) section that separates the input signal into two polarization channels, one in each of the top and bottom halves of the block. In each polarization, there is an RF 90° hybrid followed by LO couplers, ending in two SIS junctions. The IF outputs of the pair of SIS junctions pass through IF matching and bias tee to a superconducting IF 90° hybrid, which outputs the upper and lower sideband IF signals from that channel. In all 4 SIS junctions are used in each mixer block, with Cooper pair tunneling suppressed by permanent magnets. (**Bottom**) Photo of one half of an assembled 1.3 mm fronted receiver block.

Table 4. Specifications of the ngEHT multi-band Frequency Receiver.

Item	Description
3 mm RF Band	82–116 GHz
1 mm RF Band	210–280 GHz
850 µm RF Band	275–375 GHz
Polarizations	Dual pol in each band
Sidebands	2SB Receivers in each band
IF Frequency	4–12 GHz (1 mm, 0.85 µm)
	4–8 GHz (3 mm)
Receiver Noise	<50 K in 3 mm band
Temperature	60–70 K in 1 mm band
	70–80 K in 850 µm band

7.2. Backend

The ngEHT backend, consisting of the Block Down Converter (BDC) and the Digital BackEnd (DBE), will process twice the instantaneous bandwidth of the current EHT (reflecting the expansion of IF bandwidth from 4 GHz to 8 GHz).

The BDC performs a frequency translation and signal conditioning of the analog signal from the receivers. The Intermediate Frequency (IF) signal is converted to baseband, and output power levels are adjusted to optimally load the Analog to Digital Converter (ADC). The design of this BDC was initiated and functionality was implemented in a prototype, constructed by Xmicrowave LLC. The prototype was manufactured using drop-in PCB (Printed Circuit Board) modules instead of connected components, which is more representative of the final BDC PCB. A full characterization has been conducted and the results meet the required specifications. The final BDC will consist of integrated PCB units instead of discrete drop-ins.

The DBE prototype currently being used for testing and development is a two-board system. This prototype uses a custom circuit board holding four ADCs, which digitizes the analog signal from the BDC. This board sends the digital data stream to a commercial evaluation board, the VCU128 which houses the VU37P Field Programmable Gate Array (FPGA) from the Virtex Ultrascale+ family manufactured by Xilinx. Each 4-bit ADC is clocked at 16.384 GHz. The Nyquist bandwidth of this system is therefore 8.192 GHz, which is interoperable with the current EHT. The evaluation board is useful for current tests and development, and it will be replaced with a custom board design; the design of this new board is underway with an estimated one-year timeline to completion. Parts are being acquired to support a build of five units.

In addition to hardware (board) development, the initial firmware command set has been successfully completed, including an ADC interface module, a requantization block from 4 bits to 2 bits in the processing module, a packetization module, a 100 Gb transmission module, a Universal Asynchronous Receiver Transmitter(UART) monitor and control module, and a timing module. Further features that will be included in the firmware are channelization, 1 Gb monitor and control, and slope and ripple equalization.

With 2-bit quantization and Nyquist sampling, a single DBE can process the full IF bandwidth (8 GHz) from either the 1 mm or 0.85 mm band receiver, at a data rate of 128 Gb/s. For the 3 mm band, a narrower IF bandwidth (4 GHz) is sufficient to achieve Key Science Goals and Frequency Phase Transfer calibration. At 3 mm, the resulting data throughput is 64 Gb/s.

7.3. Recorder

The recorder is expected to be based around a set of commercial off-the-shelf (COTS) components hosted on a commodity multi-processor computer running a GNU/Linux operating system with a PCIe 4.0 interconnect. A single recording unit is matched to one or more streams from the digital back-end system (DBE), which is designed to output 64 Gbps data streams on 100 GbE interconnect using the VDIF transport protocol (VTP, [106]) over UDP. Specialized software on the recording unit provides efficient network capture at the required rates, simple packet inspection to ensure data continuity and integrity, distributed writing of VDIF file streams to disk, and an interface to the VLBI monitor and control system.

The host recording system will buffer the incoming data in system RAM, while simultaneously draining this data to persistent memory for storage. The persistent storage is expected to be a set of solid-state drives (SSDs) attached via PCIe/NVMe (integrated media). The total number, individual capacity and write performance of the component SSDs in the persistent memory pool will be selected such that they are sufficient to absorb the total aggregate data rate and meet the desired overall capacity and cost constraints. In order to facilitate playback of detachable data modules for subsequent correlation or transfer, the recorder will maintain a file system on the media so that may be mounted by separate machine. Comparison of specifications for the current Mark6 recorders used in the EHT and an ngRecorder is shown in Table 5.

Table 5. Specifications for a modular VLBI recorder, including those of the Mark6 [107] currently in use across the EHT. Reference specifications for a next-generation SSD-based recorder are based on common currently available COTS SSD storage servers.

	Mark6	**ngRecorder**
rack space	11U	2U
disks	32 HDD	24 SSD
capacity	512 TB	369 TB
interface	$4 \times 10/25$ GbE	2×100 GbE
rate	16/32 Gbps	128 Gbps
hours at rate	71.1/35.6	6.4
disk modules	yes	no

Although an SSD-based recorder has several advantages over an HDD-based system in terms of speed, power, density, weight, and latency, SSDs are anticipated to carry a significant cost premium to HDDs for the next decade. Moreover, a modular removable disk pack system analogous to the semi-custom Mark6 module [107] has yet to be designed, which limits the flexibility of current COTS SSD recorders. For this reason, large volume data storage and possibly transport may still rely on HDD-based solutions for some time, with SSD-to-HDD data offload capability at the site or at the correlator.

7.4. Array Monitoring and Control

The operations concept for the ngEHT extends beyond the single annual campaign of the current EHT:

- 60 nights of observing per year
- Up to 21 stations observing simultaneously
- Varied observation cadences and durations throughout the year
- Readiness for VLBI observing in 24 h or less to capture ToOs
- Multi-messenger campaigns
- Configurable subarrays
- As much remote operation as possible

This model and its increase in capability have a direct impact on the requirements and subsequent complexity of the M&C system for the ngEHT. As the M&C system serves as a main interface point for operations of the array, its design must be operator-centered and have due consideration for human factors concerns. As well, the operations concept is designed to address an explicit need, voiced at the ngEHT Operations Workshop (31 March 2022), to reduce the burden (relative to 2022 EHT operations) for on-site monitoring, control, and maintenance of VLBI equipment. The areas to address include differing methods of monitoring and control for each station and heavy reliance on local operations at each site, including the need for VLBI specialists on site.

As the first ngEHT sites are brought online, they will participate in the annual EHT observing campaign. To facilitate this participation, the M&C system will be compatible with the EHT operations plans and procedures by relaying data to the existing VLBI Monitor server, providing remote control of station subsystems, and providing status, logs, and metadata as required. Outside of the annual EHT campaign, the ngEHT operations concept calls for an annual monitoring campaign where the M&C system will be used to operate and monitor the entire array. It will provide a uniform and cohesive monitoring and control experience to the array operators while managing a heterogeneous array of ngEHT stations and stations that use the existing EHT VM&C system and backend equipment.

Collecting observation metadata from a heterogeneous array of telescopes that have non-standardized interfaces for M&C and data collection is a significant design and operational challenge. To take advantage of the opportunity presented by the ngEHT designing its own telescopes, it is expected that the M&C component of the telescopes for ngEHT sites will be designed in conjunction with the overall M&C system to make this interface as common as possible across the ngEHT sites.

As the number of stations and observations grows, providing on-site VLBI expertise will become increasingly challenging. The ngEHT design approach follows an operations model where station operators can remotely perform any required operations and maintenance, with specialist support being provided only when necessary. Remote operation is facilitated by the focus on human factors and operator-centered design, and leads to less reliance on manual operations and analysis. A cloud-based deployment of the array-level M&C system is envisioned as the way to provide "operations from anywhere" capability to the array operations staff. This is expected to include a server, database, and UI components that facilitate operation of the array. M&C capability at each station is still required to provide the control inputs to station subsystems and aggregate the local data for relay to the array-level system. Remote access to both the array- and station-level M&C systems are provided with appropriate security, authentication, and authorization methods.

To achieve all this, the M&C system architecture is expected to be built from off-the-shelf software components using open standards, including databases, message queueing and information exchange methods, and operator interface frameworks. This facilitates development and maintenance over the lifecycle of the array. A robustly defined software architecture allows isolation of site-specific dependencies to the smallest and fewest components necessary.

7.5. Antennas

The ngEHT concept adds ~10 new antennas to the existing EHT array. In Phase 1 the ngEHT program will deploy 3–4 modest-diameter antennas for the most rapid increase in next-generation science (see Section 4.4). To mitigate risk, the program has identified two possible paths toward this Phase 1 enhancement. The first would use three 6 m diameter antennas from the decommissioned Berkeley-Illinois-Maryland-Array (BIMA), which would be transported to the Las Campanas (LCO), San Pedro Martir (SPM), and Canary Island (CNI) sites.

The BIMA dishes have a surface accuracy specification of ~40 μm rms, sufficient for operation up to 345 GHz. Photogrammetry measurements will allow re-adjustment the surface to the required accuracy after re-assembly of the antenna. The panels of all three dishes are in good condition, as shown in Figure 14.

Figure 2 suggests that a 6 m diameter antenna with an aperture efficiency of 0.8 would allow us to reach the required sensitivity when paired with a large collecting area dish such as LMT or phased ALMA. But a larger diameter antenna will relax the requirement on long-distance baselines away from such anchor stations, and also have two additional advantages: easier calibration for pointing and focus measurements, and ability to carry out single-dish science projects while the antennas are not observing for ngEHT in VLBI mode.

Therefore, a second possible Phase 1 implementation path would be to use newly fabricated dishes of 9 m diameter. The specifications of the new antennas are summarized in Table 6. The ngEHT team is in discussions with several telescope vendors and it is clear that dishes with the required specifications can be procured within a reasonable cost envelope. In this case, Phase 1 would target four sites: the Mt. Jelm site in Wyoming (JELM), in addition to Canary Islands (CNI), San Pedro Martir (SPM), and Las Campanas (LCO).

Figure 14. Photograph showing the condition of the BIMA antenna dish surface (from March 2022).

Table 6. Specifications of the new ngEHT antennas.

Design Specifications	
Primary reflector diameter	9 m
Mount architecture	Alt-Az
Optics	Cassegrain
Sun avoidance zone	None
Operating Specifications	
Surface accuracy	30 μm rms
Frequency range	86–345 GHz
Aperture efficiency	0.8
Pointing accuracy	2″ rms (all sky, blind)
Tracking accuracy	0.2″
Aperture blockage	<5%
Gain variation with elevation	<5%
Range of motion in azimuth	$-180°$–$360°$,
Range of motion in elevation	$3°$–$90°$,
Slew speed	1°/s
Environmental Specifications	
Temperature	−15 to +35 °C operational
	−20 to +45 °C high
	−30 to +55 °C survival
Wind speed	10 m/s operational
	15 m/s high
	50 m/s survival

8. Summary and Conclusions

The ngEHT, described initially to the Astro2020 decadal survey review [39], is a program to plan extensions of the EHT array that will deliver high dynamic range imaging and movie-making capability for black hole studies on event horizon scales. It does so principally by deploying modest-diameter radio dishes at optimized geographical locations, which significantly increases interferometric baseline coverage (Figures 8–10), by implementing a simultaneous tri-band (86, 230, 345 GHz) receiver suite, and increasing the bandwidth of backend systems and data processing pipelines.

The process and initial results of optimizing site selection for ngEHT telescopes described here indicate two possible paths to achieve a next-generation EHT array.

In the first path, Phase 1 consists of adding dishes at two existing sites (OVRO and Haystack) to the current EHT, and available refurbished dishes from the BIMA array would be relocated to three sites (Las Campanas, Chile; San Pedro Martir, Mexico; Canary Islands, Spain). Then in Phase 2, additional sites would be developed; current analysis indicates that the combination of these locations: La Paz, Bolivia; Wyoming, US; Marangu, Tanzania; Santiago, Chile; and Bern, Switzerland, constitute an array that can deliver all of the threshold Key Science Goals. These Phase 2 sites should be considered possibilities at this stage; more work is required to assess them at all levels, including thorough consideration of cultural and environmental aspects.

In an alternate path, Phase 1 would again add both OVRO and Haystack to the EHT, and four new 9 m diameter dishes would be deployed to the Mt. Jelm site in Wyoming; Las Campanas, Chile; San Pedro Martir, Mexico; and Canary Islands, Spain. Then in Phase 2, planned new telescopes are added to the array as they become available, including the AMT, LLAMA and KVNYS, KVNPC facilities. Either of these approaches to realizing the ngEHT leads to the increases in global array capabilities that are required to achieve all ngEHT Key Science Goals.

Strategies for ngEHT data transport, correlation, calibration, and data reduction are all developed. Requirements for major instrumental sub-systems are specified, and details of prototypes to be used are described. In sum, this work brings the ngEHT project to the point of readiness for implementation.

Author Contributions: Conceptualization, S.S.D.; software, D.W.P.; writing—original draft preparation, S.S.D., D.W.P., R.C., V.L.F., A.E.B., J.B., L.B., G.F., K.H., J.H., G.N., D.C.M.P., N.P., A.W.R., F.R., R.S., P.T., J.W. and M.W.; writing—review and editing, all authors contributed to review and editing; visualization, S.S.D., D.W.P., L.B., D.C.M.P., G.F., G.N. and N.P.; project administration, A.O.; funding acquisition, S.S.D. All authors have read and agreed to the published version of the manuscript.

Funding: Support for this work was provided by the NSF through grants AST-1952099, AST-1935980, AST-1828513, and AST-1440254, and by the Gordon and Betty Moore Foundation through grant GBMF-10423. L.L. acknowledge DGAPA PAPIIT grant IN112820 and CONACYT-CF grant 263356. This work has also been supported in part by the Black Hole Initiative at Harvard University, which is funded by grants from the John Templeton Foundation and the Gordon and Betty Moore Foundation to Harvard University. N.N. acknowledges funding from ANID Chile via TITANs NCN19-058, Fondecyt 1221421, and BASAL FB210003.

Data Availability Statement: Not applicable.

Acknowledgments: We gratefully acknowledge the larger next-generation EHT community, which has, through several international meetings and workshops, helped to define ngEHT requirements and science goals. Special thanks to members of the ngEHT System Requirements Review panel: Crystal Brogan, Carolyn Crichton, Francois Kapp, Robert Laing, Aaron Parsons, Laura Vertatschitsch, and Andre Young. The review was conducted 9–10 June 2022, and the detailed feedback provided by the panel was invaluable in refining the ngEHT reference array design.

Appendix A. Additional Site Selection Details

Table A1 lists the sites considered for the ngEHT array optimization procedures described in Section 4. This pool of candidate sites has been taken from [86], and they have been selected for their favorable atmospheric transmission properties at 230 GHz and 345 GHz during the typical EHT observing season in March and April.

Table A2 specifies the pre-existing arrays assumed during the site selection procedure described in Section 4. Four different variants of pre-existing array are explored as parameters in the site selection procedure, and these variants are enumerated in the table.

Table A1. Existing or planned sites (top) and candidate ngEHT sites (bottom), updated from [86].

Site Code	Location	Latitude	Longitude	Elevation (m)
ALMA	Atacama, Chile	−23.032	−67.755	5040
AMT	Gamsberg, Namibia	−23.339	16.229	2340
APEX	Atacama, Chile	−23.005	−67.759	5060
GLT	Pituffik Space Base, Greenland	76.535	−68.686	70
HAY	Westford, Massachusetts, USA	42.624	−71.489	110
IRAM	Sierra Nevada, Spain	37.066	−3.393	2860
JCMT	Mauna Kea, Hawaii	19.823	−155.477	4070
KP	Arizona, USA	31.953	−111.615	1930
KVNPC	Pyeongchang, South Korea	37.534	128.450	500
KVNYS	Yonsei, South Korea	37.565	126.941	90
LLA	Salta, Argentina	−24.192	−66.475	4780
LMT	Sierra Negra, Mexico	18.986	−97.315	4620
NOEMA	Plateau de Bure, France	44.634	5.907	2550
OVRO	California, USA	37.231	−118.282	1210
SMA	Mauna Kea, Hawaii	19.824	−155.478	4070
SMT	Arizona, USA	32.702	−109.891	3170
SPT	South Pole, Antarctica	−90	0	2820
ALI	Hotan County, China	35.963	79.338	6080
BAN	Alberta, Canada	51.350	−116.206	3470
BAR	California, USA	37.634	−118.256	4340
BGA	Progled, Bulgaria	41.695	24.738	1730
BGK	Westfjords, Iceland	66.032	−23.052	830
BLDR	Colorado, USA	39.588	−105.643	4340
BMAC	Eastern Cape, South Africa	−31.096	27.889	2420
BOL	La Paz, Bolivia	−16.351	−68.131	5230
BRZ	Espírito Santo, Brazil	−20.439	−41.799	2850
CAS	Tierra del Fuego, Argentina	−54.790	−68.415	2850
CAT	Río Negro, Argentina	−41.170	−71.486	2100
CNI	La Palma, Canary Islands	28.299	−16.509	2360
DomeA	Upper ice sheet, Antarctica	−80.367	77.351	4090
DomeC	Upper ice sheet, Antarctica	−75.101	123.342	3230
DomeF	Upper ice sheet, Antarctica	−77.317	39.702	3700
ERB	Khalifan, Iraq	36.584	44.466	2110
FAIR	Alaska, USA	64.988	−147.599	620
FLWO	Arizona, USA	31.675	−110.951	1270
FUJI	Fujinomiya & Yamanashi, Japan	35.367	138.730	3750
GARS	Trinity Peninsula, Antarctica	−63.320	−57.895	20
GLTS	Ice sheet summit, Greenland	72.580	−38.449	3230
HAN	Ladakh, India	32.780	78.963	4500
JELM	Wyoming, USA	41.097	−105.977	2940
KEN	Meru, Kenya	−0.141	37.315	4260
KILI	Kilimanjaro, Tanzania	−3.088	37.406	4430
LCO	Coquimbo, Chile	−29.032	−70.685	2320
LOS	New Mexico, USA	35.880	−106.675	2000
NOB	Nagano, Japan	35.944	138.472	1370
NZ	Canterbury, New Zealand	−43.987	170.465	1010
ORG	Oregon, USA	42.635	−118.576	2970
PAR	Antofagasta, Chile	−24.628	−70.404	2640
PIKE	Colorado, USA	38.841	−105.041	4280
SAN	California, USA	34.099	−116.825	3500
SGO	Santiago, Chile	−33.3346	−70.270	3350

Table A1. *Cont.*

Site Code	Location	Latitude	Longitude	Elevation (m)
SKS	Crete, Greece	35.212	24.898	1740
SPM	Baja California, Mexico	31.045	−115.464	2800
SPX	Fieschertal, Switzerland	46.548	7.985	3510
SUF	Zaamin, Uzbekistan	39.623	68.468	2440
TRL	Jutulsessen, Antarctica	−72.010	2.540	1280
VLA	New Mexico, USA	34.079	−107.618	2120
YAN	Huanca Sancos, Peru	−13.938	−74.392	4230
YBG	Lhasa Tibet, China	30.006	91.027	5360

Table A2. The different pre-existing arrays considered as part of the site selection exploration (Section 4.4). Each of these combinations of stations is the starting set of sites for which the addition of three sites (for the Phase 1 analysis) or five sites (for the Phase 2 analysis) are explored. These starting arrays are chosen to generally represent the possible operating modes shown in Table 3. Set 1, for example, might be a minimal array useful for Target of Opportunity observations. Sets 2 and 3, with the addition of a large aperture, could provide flexible long-term monitoring capability. And set 4 includes all possible stations for a full campaign mode. The range of starting arrays also give some indication of optimal placement in the full campaign mode in the case where some sites are not available due to weather or technical issues.

Parameter Set	Pre-Existing Stations from EHT Array	Other Pre-Existing Stations Assumed
Phase 1 set 1	none	HAY, OVRO
Phase 1 set 2	LMT	HAY, OVRO
Phase 1 set 3	APEX, GLT, JCMT, LMT, SMT	HAY, OVRO
Phase 1 set 4	ALMA, APEX, GLT, IRAM, JCMT, KP, LMT, NOEMA, SMA, SMT, SPT	HAY, OVRO
Phase 2 set 1	none	CNI, HAY, LCO, OVRO, SPM
Phase 2 set 2	LMT	CNI, HAY, LCO, OVRO, SPM
Phase 2 set 3	APEX, GLT, JCMT, LMT, SMT	CNI, HAY, LCO, OVRO, SPM
Phase 2 set 4	ALMA, APEX, GLT, IRAM, JCMT, KP, LMT, NOEMA, SMA, SMT, SPT	CNI, HAY, LCO, OVRO, SPM

Notes

1   Available online: https://www.mdpi.com/journal/galaxies/special_issues/ngEHT_blackholes (accessed on 16 August 2023).

2   Available online: https://github.com/Smithsonian/ngehtsim (accessed on 16 August 2023).

3   Available online: https://www.evlbi.org/ (accessed on 16 August 2023).

References

1. Akiyama, K. et al. [Event Horizon Telescope Collaboration]. First M87 Event Horizon Telescope Results. I. The Shadow of the Supermassive Black Hole. *Astrophys. J. Lett.* **2019**, *875*, L1. [CrossRef]
2. Akiyama, K. et al. [Event Horizon Telescope Collaboration]. First M87 Event Horizon Telescope Results. II. Array and Instrumentation. *Astrophys. J. Lett.* **2019**, *875*, L2. [CrossRef]
3. Akiyama, K. et al. [Event Horizon Telescope Collaboration]. First M87 Event Horizon Telescope Results. III. Data Processing and Calibration. *Astrophys. J. Lett.* **2019**, *875*, L3. [CrossRef]
4. Akiyama, K. et al. [Event Horizon Telescope Collaboration]. First M87 Event Horizon Telescope Results. IV. Imaging the Central Supermassive Black Hole. *Astrophys. J. Lett.* **2019**, *875*, L4. [CrossRef]
5. Akiyama, K. et al. [Event Horizon Telescope Collaboration]. First M87 Event Horizon Telescope Results. V. Physical Origin of the Asymmetric Ring. *Astrophys. J. Lett.* **2019**, *875*, L5. [CrossRef]
6. Akiyama, K. et al. [Event Horizon Telescope Collaboration]. First M87 Event Horizon Telescope Results. VI. The Shadow and Mass of the Central Black Hole. *Astrophys. J. Lett.* **2019**, *875*, L6. [CrossRef]
7. Akiyama, K. et al. [Event Horizon Telescope Collaboration]. First Sagittarius A* Event Horizon Telescope Results. I. The Shadow of the Supermassive Black Hole in the Center of the Milky Way. *Astrophys. J. Lett.* **2022**, *930*, L12. [CrossRef]
8. Akiyama, K. et al. [Event Horizon Telescope Collaboration]. First Sagittarius A* Event Horizon Telescope Results. II. EHT and Multiwavelength Observations, Data Processing, and Calibration. *Astrophys. J. Lett.* **2022**, *930*, L13. [CrossRef]
9. Akiyama, K. et al. [Event Horizon Telescope Collaboration]. First Sagittarius A* Event Horizon Telescope Results. III. Imaging of the Galactic Center Supermassive Black Hole. *Astrophys. J. Lett.* **2022**, *930*, L14. [CrossRef]
10. Akiyama, K. et al. [Event Horizon Telescope Collaboration]. First Sagittarius A* Event Horizon Telescope Results. IV. Variability, Morphology, and Black Hole Mass. *Astrophys. J. Lett.* **2022**, *930*, L15. [CrossRef]

11. Akiyama, K. et al. [Event Horizon Telescope Collaboration]. First Sagittarius A* Event Horizon Telescope Results. V. Testing Astrophysical Models of the Galactic Center Black Hole. *Astrophys. J. Lett.* **2022**, *930*, L16. [CrossRef]
12. Akiyama, K. et al. [Event Horizon Telescope Collaboration]. First Sagittarius A* Event Horizon Telescope Results. VI. Testing the Black Hole Metric. *Astrophys. J. Lett.* **2022**, *930*, L17. [CrossRef]
13. Bardeen, J.M. Timelike and null geodesics in the Kerr metric. In *Black Holes (Les Astres Occlus)*; Gordon and Breach: New York, NY, USA, 1973; pp. 215–239.
14. Luminet, J.P. Image of a spherical black hole with thin accretion disk. *Astron. Astrophys.* **1979**, *75*, 228–235.
15. Falcke, H.; Melia, F.; Agol, E. Viewing the Shadow of the Black Hole at the Galactic Center. *Astrophys. J. Lett.* **2000**, *528*, L13–L16. [CrossRef]
16. Takahashi, R. Shapes and Positions of Black Hole Shadows in Accretion Disks and Spin Parameters of Black Holes. *Astrophys. J.* **2004**, *611*, 996–1004. [CrossRef]
17. Broderick, A.E.; Loeb, A. Imaging optically-thin hotspots near the black hole horizon of Sgr A* at radio and near-infrared wavelengths. *Mon. Not. R. Astron. Soc.* **2006**, *367*, 905–916. [CrossRef]
18. Padin, S.; Woody, D.P.; Hodges, M.W.; Rogers, A.E.E.; Emerson, D.T.; Jewell, P.R.; Lamb, J.; Perfetto, A.; Wright, M.C.H. 223 GHz VLBI Observations of 3C 273. *Astrophys. J. Lett.* **1990**, *360*, L11. [CrossRef]
19. Krichbaum, T.P.; Graham, D.A.; Witzel, A.; Greve, A.; Wink, J.E.; Grewing, M.; Colomer, F.; de Vicente, P.; Gomez-Gonzalez, J.; Baudry, A.; et al. VLBI observations of the galactic center source SGR A* at 86 GHz and 215 GHz. *Astron. Astrophys.* **1998**, *335*, L106–L110.
20. Doeleman, S.S.; Weintroub, J.; Rogers, A.E.E.; Plambeck, R.; Freund, R.; Tilanus, R.P.J.; Friberg, P.; Ziurys, L.M.; Moran, J.M.; Corey, B.; et al. Event-horizon-scale structure in the supermassive black hole candidate at the Galactic Centre. *Nature* **2008**, *455*, 78–80. [CrossRef]
21. Doeleman, S.S.; Fish, V.L.; Schenck, D.E.; Beaudoin, C.; Blundell, R.; Bower, G.C.; Broderick, A.E.; Chamberlin, R.; Freund, R.; Friberg, P.; et al. Jet-Launching Structure Resolved Near the Supermassive Black Hole in M87. *Science* **2012**, *338*, 355. [CrossRef]
22. Fish, V.L.; Doeleman, S.S.; Beaudoin, C.; Blundell, R.; Bolin, D.E.; Bower, G.C.; Chamberlin, R.; Freund, R.; Friberg, P.; Gurwell, M.A.; et al. 1.3 mm Wavelength VLBI of Sagittarius A*: Detection of Time-variable Emission on Event Horizon Scales. *Astrophys. J. Lett.* **2011**, *727*, L36. [CrossRef]
23. Johnson, M.D.; Fish, V.L.; Doeleman, S.S.; Marrone, D.P.; Plambeck, R.L.; Wardle, J.F.C.; Akiyama, K.; Asada, K.; Beaudoin, C.; Blackburn, L.; et al. Resolved magnetic-field structure and variability near the event horizon of Sagittarius A*. *Science* **2015**, *350*, 1242–1245. [CrossRef]
24. Doeleman, S.; Agol, E.; Backer, D.; Baganoff, F.; Bower, G.C.; Broderick, A.; Fabian, A.; Fish, V.; Gammie, C.; Ho, P.; et al. Imaging an Event Horizon: Submm-VLBI of a Super Massive Black Hole. *arXiv* **2009**, arXiv:0906.3899.
25. Doeleman, S. Building an event horizon telescope: (sub)mm VLBI in the ALMA era. In Proceedings of the 10th European VLBI Network Symposium and EVN Users Meeting: VLBI and the New Generation of Radio Arrays, Manchester, UK, 20–24 September 2010; p. 53.
26. Matthews, L.D.; Crew, G.B.; Doeleman, S.S.; Lacasse, R.; Saez, A.F.; Alef, W.; Akiyama, K.; Amestica, R.; Anderson, J.M.; Barkats, D.A.; et al. The ALMA Phasing System: A Beamforming Capability for Ultra-high-resolution Science at (Sub)Millimeter Wavelengths. *Publ. Astron. Soc. Pac.* **2018**, *130*, 015002. [CrossRef]
27. Inoue, M.; Algaba-Marcos, J.C.; Asada, K.; Blundell, R.; Brisken, W.; Burgos, R.; Chang, C.C.; Chen, M.T.; Doeleman, S.S.; Fish, V.; et al. Greenland telescope project: Direct confirmation of black hole with sub-millimeter VLBI. *Radio Sci.* **2014**, *49*, 564–571. [CrossRef]
28. Kim, J.; Marrone, D.P.; Beaudoin, C.; Carlstrom, J.E.; Doeleman, S.S.; Folkers, T.W.; Forbes, D.; Greer, C.H.; Lauria, E.F.; Massingill, K.D.; et al. A VLBI receiving system for the South Pole Telescope. In *Millimeter, Submillimeter, and Far-Infrared Detectors and Instrumentation for Astronomy IX*; Society of Photo-Optical Instrumentation Engineers (SPIE) Conference Series; Zmuidzinas, J., Gao, J.R., Eds.; SPIE: Bellingham, WA, USA, 2018; Volume 10708, p. 107082S.
29. Christensen, L.L.; Baloković, M.; Chou, M.Y.; Crowley, S.; Edmonds, P.; Foncea, V.; Hiramatsu, M.; Hunter, C.; Königstein, K.; Leach, S.; et al. An Unprecedented Global Communications Campaign for the Event Horizon Telescope First Black Hole Image. *Commun. Astron. Public J.* **2019**, *26*, 11.
30. Vertatschitsch, L.; Primiani, R.; Young, A.; Weintroub, J.; Crew, G.B.; McWhirter, S.R.; Beaudoin, C.; Doeleman, S.; Blackburn, L. R2DBE: A Wideband Digital Backend for the Event Horizon Telescope. *Publ. Astron. Soc. Pac.* **2015**, *127*, 1226. [CrossRef]
31. Carpenter, J.; Iono, D.; Kemper, F.; Wootten, A. The ALMA Development Program: Roadmap to 2030. *arXiv* **2020**, arXiv:2001.11076.
32. Grimes, P.; Blundell, R.; Leiker, P.; Paine, S.N.; Tong, E.C.Y.; Wilson, R.W.; Zeng, L. Receivers for the Wideband Submillimeter Array. In Proceedings of the 31st International Symposium on Space Terahertz Technology, Tempe, AZ, USA, 8–11 March 2020; pp. 60–66.
33. Selina, R.J.; Murphy, E.J.; McKinnon, M.; Beasley, A.; Butler, B.; Carilli, C.; Clark, B.; Durand, S.; Erickson, A.; Grammer, W.; et al. The ngVLA Reference Design. In *Science with a Next Generation Very Large Array*; Astronomical Society of the Pacific Conference Series; Murphy, E., Ed.; ASP: San Francisco, CA, USA, 2018; Volume 517, p. 15. [CrossRef]
34. Selina, R.; Murphy, E.; Beasley, A. The ngVLA: A technical development update. In *Ground-Based and Airborne Telescopes IX*; Marshall, H.K., Spyromilio, J., Usuda, T., Eds.; International Society for Optics and Photonics, SPIE: Bellingham, WA, USA, 2022; Volume 12182, p. 1218200. [CrossRef]

35. Gill, A.; Blackburn, L.; Roshanineshat, A.; Chan, C.K.; Doeleman, S.S.; Johnson, M.D.; Raymond, A.W.; Weintroub, J. Prospects for Wideband VLBI Correlation in the Cloud. *Publ. Astron. Soc. Pac.* **2019**, *131*, 124501. [CrossRef]

36. Johnson, M.D.; Bouman, K.L.; Blackburn, L.; Chael, A.A.; Rosen, J.; Shiokawa, H.; Roelofs, F.; Akiyama, K.; Fish, V.L.; Doeleman, S.S. Dynamical Imaging with Interferometry. *Astrophys. J.* **2017**, *850*, 172. [CrossRef]

37. Bouman, K.L.; Johnson, M.D.; Dalca, A.V.; Chael, A.A.; Roelofs, F.; Doeleman, S.S.; Freeman, W.T. Reconstructing Video of Time-Varying Sources From Radio Interferometric Measurements. *IEEE Trans. Comput. Imaging* **2018**, *4*, 512–527. [CrossRef]

38. Arras, P.; Frank, P.; Haim, P.; Knollmüller, J.; Leike, R.; Reinecke, M.; Enßlin, T. Variable structures in M87* from space, time and frequency resolved interferometry. *Nat. Astron.* **2022**, *6*, 259–269. [CrossRef]

39. Doeleman, S.; Blackburn, L.; Dexter, J.; Gomez, J.L.; Johnson, M.D.; Palumbo, D.C.; Weintroub, J.; Farah, J.R.; Fish, V.; Loinard, L.; et al. Studying Black Holes on Horizon Scales with VLBI Ground Arrays. *arXiv* **2019**, arXiv:1909.01411.

40. National Academies of Sciences, Engineering, and Medicine. *Pathways to Discovery in Astronomy and Astrophysics for the 2020s;* The National Academies Press: Washington, DC, USA, 2021. [CrossRef]

41. Chael, A.; Issaoun, S.; Pesce, D.w.; Johnson, M.D.; Ricarte, A.; Fromm, C.M.; Mizuno, Y. Multi-frequency Black Hole Imaging for the Next-Generation Event Horizon Telescope. *arXiv* **2022**, arXiv:2210.12226.

42. Crew, G.B.; Goddi, C.; Matthews, L.D.; Rottmann, H.; Saez, A.; Martí-Vidal, I. A Characterization of the ALMA Phasing System at 345 GHz. *Publ. Astron. Soc. Pac.* **2023**, *135*, 025002. [CrossRef]

43. Rioja, M.J.; Dodson, R.; Asaki, Y. The Transformational Power of Frequency Phase Transfer Methods for ngEHT. *Galaxies* **2023**, *11*, 16. [CrossRef]

44. Issaoun, S.; Pesce, D.W.; Roelofs, F.; Chael, A.; Dodson, R.; Rioja, M.J.; Akiyama, K.; Aran, R.; Blackburn, L.; Doeleman, S.S.; et al. Enabling Transformational ngEHT Science via the Inclusion of 86 GHz Capabilities. *Galaxies* **2023**, *11*, 28. [CrossRef]

45. Kauffmann, J.; Rajagopalan, G.; Akiyama, K.; Fish, V.; Lonsdale, C.; Matthews, L.D.; Pillai, T.G. The Haystack Telescope as an Astronomical Instrument. *Galaxies* **2023**, *11*, 9. [CrossRef]

46. Johnson, M.D.; Akiyama, K.; Blackburn, L.; Bouman, K.L.; Broderick, A.E.; Cardoso, V.; Fender, R.P.; Fromm, C.M.; Galison, P.; Gómez, J.L.; et al. Key Science Goals for the Next-Generation Event Horizon Telescope. *Galaxies* **2023**, *11*, 61. [CrossRef]

47. Chael, A.; Johnson, M.D.; Lupsasca, A. Observing the Inner Shadow of a Black Hole: A Direct View of the Event Horizon. *Astrophys. J.* **2021**, *918*, 6. [CrossRef]

48. Dokuchaev, V.I.; Nazarova, N.O. Event Horizon Image within Black Hole Shadow. *Sov. J. Exp. Theor. Phys.* **2019**, *128*, 578–585. [CrossRef]

49. Palumbo, D.C.M.; Wong, G.N.; Prather, B.S. Discriminating Accretion States via Rotational Symmetry in Simulated Polarimetric Images of M87. *Astrophys. J.* **2020**, *894*, 156. [CrossRef]

50. Ricarte, A.; Tiede, P.; Emami, R.; Tamar, A.; Natarajan, P. The ngEHT's Role in Measuring Supermassive Black Hole Spins. *Galaxies* **2023**, *11*, 6. [CrossRef]

51. Pesce, D.W.; Palumbo, D.C.M.; Narayan, R.; Blackburn, L.; Doeleman, S.S.; Johnson, M.D.; Ma, C.P.; Nagar, N.M.; Natarajan, P.; Ricarte, A. Toward Determining the Number of Observable Supermassive Black Hole Shadows. *Astrophys. J.* **2021**, *923*, 260. [CrossRef]

52. Pesce, D.W.; Palumbo, D.C.M.; Ricarte, A.; Broderick, A.E.; Johnson, M.D.; Nagar, N.M.; Natarajan, P.; Gómez, J.L. Expectations for Horizon-Scale Supermassive Black Hole Population Studies with the ngEHT. *Galaxies* **2022**, *10*, 109. [CrossRef]

53. Ramakrishnan, V.; Nagar, N.; Arratia, V.; Hernández-Yévenes, J.; Pesce, D.W.; Nair, D.G.; Bandyopadhyay, B.; Medina-Porcile, C.; Krichbaum, T.P.; Doeleman, S.; et al. Event Horizon and Environs (ETHER): A Curated Database for EHT and ngEHT Targets and Science. *Galaxies* **2023**, *11*, 15. [CrossRef]

54. Balbus, S.A.; Hawley, J.F. Instability, turbulence, and enhanced transport in accretion disks. *Rev. Mod. Phys.* **1998**, *70*, 1–53. [CrossRef]

55. Yuan, F.; Narayan, R. Hot Accretion Flows Around Black Holes. *Annu. Rev. Astron. Astrophys.* **2014**, *52*, 529–588. [CrossRef]

56. Rowan, M.E.; Sironi, L.; Narayan, R. Electron and Proton Heating in Transrelativistic Magnetic Reconnection. *Astrophys. J.* **2017**, *850*, 29. [CrossRef]

57. Ball, D.; Sironi, L.; Özel, F. Electron and Proton Acceleration in Trans-relativistic Magnetic Reconnection: Dependence on Plasma Beta and Magnetization. *Astrophys. J.* **2018**, *862*, 80. [CrossRef]

58. Blandford, R.D.; Znajek, R.L. Electromagnetic extraction of energy from Kerr black holes. *Mon. Not. R. Astron. Soc.* **1977**, *179*, 433–456. [CrossRef]

59. Tchekhovskoy, A.; Narayan, R.; McKinney, J.C. Efficient generation of jets from magnetically arrested accretion on a rapidly spinning black hole. *Mon. Not. R. Astron. Soc.* **2011**, *418*, L79–L83. [CrossRef]

60. Blandford, R.; Meier, D.; Readhead, A. Relativistic Jets from Active Galactic Nuclei. *Annu. Rev. Astron. Astrophys.* **2019**, *57*, 467–509. [CrossRef]

61. Johnson, M.D.; Lupsasca, A.; Strominger, A.; Wong, G.N.; Hadar, S.; Kapec, D.; Narayan, R.; Chael, A.; Gammie, C.F.; Galison, P.; et al. Universal interferometric signatures of a black hole's photon ring. *Sci. Adv.* **2020**, *6*, eaaz1310. [CrossRef]

62. Tiede, P.; Johnson, M.D.; Pesce, D.W.; Palumbo, D.C.M.; Chang, D.O.; Galison, P. Measuring Photon Rings with the ngEHT. *Galaxies* **2022**, *10*, 111. [CrossRef]

63. Vincent, F.H.; Gralla, S.E.; Lupsasca, A.; Wielgus, M. Images and photon ring signatures of thick disks around black holes. *Astron. Astrophys.* **2022**, *667*, A170. [CrossRef]

64. Ricarte, A.; Johnson, M.D.; Kovalev, Y.Y.; Palumbo, D.C.M.; Emami, R. How Spatially Resolved Polarimetry Informs Black Hole Accretion Flow Models. *Galaxies* **2023**, *11*, 5. [CrossRef]

65. Balbus, S.A.; Hawley, J.F. A powerful local shear instability in weakly magnetized disks. I—Linear analysis. II—Nonlinear evolution. *Astrophys. J.* **1991**, *376*, 214–233. [CrossRef]

66. Mahadevan, R. Scaling Laws for Advection-dominated Flows: Applications to Low-Luminosity Galactic Nuclei. *Astrophys. J.* **1997**, *477*, 585–601. [CrossRef]

67. Kawazura, Y.; Barnes, M.; Schekochihin, A.A. Thermal disequilibration of ions and electrons by collisionless plasma turbulence. *Proc. Natl. Acad. Sci. USA* **2019**, *116*, 771–776. [CrossRef]

68. Howes, G.G. A prescription for the turbulent heating of astrophysical plasmas. *Mon. Not. R. Astron. Soc.* **2010**, *409*, L104–L108. [CrossRef]

69. Ryan, B.R.; Ressler, S.M.; Dolence, J.C.; Gammie, C.; Quataert, E. Two-temperature GRRMHD Simulations of M87. *Astrophys. J.* **2018**, *864*, 126. [CrossRef]

70. Chael, A.; Rowan, M.; Narayan, R.; Johnson, M.; Sironi, L. The role of electron heating physics in images and variability of the Galactic Centre black hole Sagittarius A*. *Mon. Not. R. Astron. Soc.* **2018**, *478*, 5209–5229. [CrossRef]

71. Chael, A.; Narayan, R.; Johnson, M.D. Two-temperature, Magnetically Arrested Disc simulations of the jet from the supermassive black hole in M87. *Mon. Not. R. Astron. Soc.* **2019**, *486*, 2873–2895. [CrossRef]

72. Mizuno, Y.; Fromm, C.M.; Younsi, Z.; Porth, O.; Olivares, H.; Rezzolla, L. Comparison of the ion-to-electron temperature ratio prescription: GRMHD simulations with electron thermodynamics. *Mon. Not. R. Astron. Soc.* **2021**, *506*, 741–758. [CrossRef]

73. Penrose, R. Gravitational Collapse: The Role of General Relativity. *Nuovo C. Riv. Ser.* **1969**, *1*, 252.

74. Prieto, M.A.; Fernández-Ontiveros, J.A.; Markoff, S.; Espada, D.; González-Martín, O. The central parsecs of M87: Jet emission and an elusive accretion disc. *Mon. Not. R. Astron. Soc.* **2016**, *457*, 3801–3816. [CrossRef]

75. Broderick, A.E.; Tiede, P.; Pesce, D.W.; Gold, R. Measuring Spin from Relative Photon-ring Sizes. *Astrophys. J.* **2022**, *927*, 6. [CrossRef]

76. Akiyama, K. et al. [Event Horizon Telescope Collaboration]. First M87 Event Horizon Telescope Results. VIII. Magnetic Field Structure near The Event Horizon. *Astrophys. J. Lett.* **2021**, *910*, L13. [CrossRef]

77. Zamaninasab, M.; Clausen-Brown, E.; Savolainen, T.; Tchekhovskoy, A. Dynamically important magnetic fields near accreting supermassive black holes. *Nature* **2014**, *510*, 126–128. [CrossRef]

78. Gammie, C.F.; McKinney, J.C.; Tóth, G. HARM: A Numerical Scheme for General Relativistic Magnetohydrodynamics. *Astrophys. J.* **2003**, *589*, 444. [CrossRef]

79. Emami, R.; Anantua, R.; Ricarte, A.; Doeleman, S.S.; Broderick, A.; Wong, G.; Blackburn, L.; Wielgus, M.; Narayan, R.; Tremblay, G.; et al. Probing Plasma Composition with the Next Generation Event Horizon Telescope (ngEHT). *Galaxies* **2023**, *11*, 11. [CrossRef]

80. Mertens, F.; Lobanov, A.P.; Walker, R.C.; Hardee, P.E. Kinematics of the jet in M 87 on scales of 100–1000 Schwarzschild radii. *Astron. Astrophys.* **2016**, *595*, A54. [CrossRef]

81. Walker, R.C.; Hardee, P.E.; Davies, F.B.; Ly, C.; Junor, W. The Structure and Dynamics of the Subparsec Jet in M87 Based on 50 VLBA Observations over 17 Years at 43 GHz. *Astrophys. J.* **2018**, *855*, 128. [CrossRef]

82. Komissarov, S.S. Numerical simulations of relativistic magnetized jets. *Mon. Not. R. Astron. Soc.* **1999**, *308*, 1069–1076. [CrossRef]

83. Janssen, M.; Falcke, H.; Kadler, M.; Ros, E.; Wielgus, M.; Akiyama, K.; Baloković, M.; Blackburn, L.; Bouman, K.L.; Chael, A.; et al. Event Horizon Telescope observations of the jet launching and collimation in Centaurus A. *Nat. Astron.* **2021**, *5*, 1017–1028. [CrossRef]

84. Blandford, R.D.; Payne, D.G. Hydromagnetic flows from accretion disks and the production of radio jets. *Mon. Not. R. Astron. Soc.* **1982**, *199*, 883–903. [CrossRef]

85. Wielgus, M. Photon rings of spherically symmetric black holes and robust tests of non-Kerr metrics. *Phys. Rev. D* **2021**, *104*, 124058. [CrossRef]

86. Raymond, A.W.; Palumbo, D.; Paine, S.N.; Blackburn, L.; Córdova Rosado, R.; Doeleman, S.S.; Farah, J.R.; Johnson, M.D.; Roelofs, F.; Tilanus, R.P.J.; et al. Evaluation of New Submillimeter VLBI Sites for the Event Horizon Telescope. *Astrophys. J. Suppl. Ser.* **2021**, *253*, 5. [CrossRef]

87. Gelaro, R.; McCarty, W.; Suárez, M.J.; Todling, R.; Molod, A.; Takacs, L.; Randles, C.A.; Darmenov, A.; Bosilovich, M.G.; Reichle, R.; et al. The Modern-Era Retrospective Analysis for Research and Applications, Version 2 (MERRA-2). *J. Clim.* **2017**, *30*, 5419–5454. [CrossRef]

88. Salby, M.L. *Fundamentals of Atmospheric Physics*; Academic Press: Cambridge, MA, USA, 1996.

89. Chael, A.A.; Johnson, M.D.; Narayan, R.; Doeleman, S.S.; Wardle, J.F.C.; Bouman, K.L. High-resolution Linear Polarimetric Imaging for the Event Horizon Telescope. *Astrophys. J.* **2016**, *829*, 11. [CrossRef]

90. Chael, A.A.; Johnson, M.D.; Bouman, K.L.; Blackburn, L.L.; Akiyama, K.; Narayan, R. Interferometric Imaging Directly with Closure Phases and Closure Amplitudes. *Astrophys. J.* **2018**, *857*, 23. [CrossRef]

91. Thompson, A.R.; Moran, J.M.; Swenson, G.W., Jr. *Interferometry and Synthesis in Radio Astronomy*, 3rd ed.; Springer: Berlin/Heidelberg, Germany, 2017. [CrossRef]

92. Paine, S. *The Am Atmospheric Model*; Zenodo: Geneva, Switzerland, 2019. [CrossRef]

93. Blackburn, L.; Chan, C.k.; Crew, G.B.; Fish, V.L.; Issaoun, S.; Johnson, M.D.; Wielgus, M.; Akiyama, K.; Barrett, J.; Bouman, K.L.; et al. EHT-HOPS Pipeline for Millimeter VLBI Data Reduction. *Astrophys. J.* **2019**, *882*, 23. [CrossRef]

94. Rioja, M.J.; Dodson, R. Precise radio astrometry and new developments for the next-generation of instruments. *Astron. Astrophys. Rev.* **2020**, *28*, 6. [CrossRef]

95. Palumbo, D.C.M.; Doeleman, S.S.; Johnson, M.D.; Bouman, K.L.; Chael, A.A. Metrics and Motivations for Earth-Space VLBI: Time-resolving Sgr A* with the Event Horizon Telescope. *Astrophys. J.* **2019**, *881*, 62. [CrossRef]

96. Wielgus, M.; Akiyama, K.; Blackburn, L.; Chan, C.k.; Dexter, J.; Doeleman, S.S.; Fish, V.L.; Issaoun, S.; Johnson, M.D.; Krichbaum, T.P.; et al. Monitoring the Morphology of M87* in 2009–2017 with the Event Horizon Telescope. *Astrophys. J.* **2020**, *901*, 67. [CrossRef]

97. Kim, J.Y.; Krichbaum, T.P.; Broderick, A.E.; Wielgus, M.; Blackburn, L.; Gómez, J.L.; Johnson, M.D.; Bouman, K.L.; Chael, A.; Akiyama, K.; et al. Event Horizon Telescope imaging of the archetypal blazar 3C 279 at an extreme 20 microarcsecond resolution. *Astron. Astrophys.* **2020**, *640*, A69. [CrossRef]

98. Issaoun, S.; Wielgus, M.; Jorstad, S.; Krichbaum, T.P.; Blackburn, L.; Janssen, M.; Chan, C.k.; Pesce, D.W.; Gómez, J.L.; Akiyama, K.; et al. Resolving the Inner Parsec of the Blazar J1924-2914 with the Event Horizon Telescope. *Astrophys. J.* **2022**, *934*, 145. [CrossRef]

99. Jorstad, S.; Wielgus, M.; Lico, R.; Issaoun, S.; Broderick, A.E.; Pesce, D.W.; Liu, J.; Zhao, G.Y.; Krichbaum, T.P.; Blackburn, L.; et al. The Event Horizon Telescope Image of the Quasar NRAO 530. *Astrophys. J.* **2023**, *943*, 170. [CrossRef]

100. EHT MWL Science Working Group; Algaba, J.C.; Anczarski, J.; Asada, K.; Baloković, M.; Chandra, S.; Cui, Y.Z.; Falcone, A.D.; Giroletti, M.; Goddi, C.; et al. Broadband Multi-wavelength Properties of M87 during the 2017 Event Horizon Telescope Campaign. *Astrophys. J. Lett.* **2021**, *911*, L11. [CrossRef]

101. Deller, A.T.; Brisken, W.F.; Phillips, C.J.; Morgan, J.; Alef, W.; Cappallo, R.; Middelberg, E.; Romney, J.; Rottmann, H.; Tingay, S.J.; et al. DiFX-2: A More Flexible, Efficient, Robust, and Powerful Software Correlator. *Publ. Astron. Soc. Pac.* **2011**, *123*, 275. [CrossRef]

102. Vázquez, A.J.; Elosegui, P.; Lonsdale, C.J.; Crew, G.B.; Fish, V.L.; Ruszczyk, C.A. Model-based Performance Characterization of Software Correlators for Radio Interferometer Arrays. *Publ. Astron. Soc. Pac.* **2022**, *134*, 104501. [CrossRef]

103. Janssen, M.; Goddi, C.; van Bemmel, I.M.; Kettenis, M.; Small, D.; Liuzzo, E.; Rygl, K.; Martí-Vidal, I.; Blackburn, L.; Wielgus, M.; et al. rPICARD: A CASA-based calibration pipeline for VLBI data. Calibration and imaging of 7 mm VLBA observations of the AGN jet in M 87. *Astron. Astrophys.* **2019**, *626*, A75. [CrossRef]

104. Broderick, A.E.; Gold, R.; Karami, M.; Preciado-López, J.A.; Tiede, P.; Pu, H.Y.; Akiyama, K.; Alberdi, A.; Alef, W.; Asada, K.; et al. THEMIS: A Parameter Estimation Framework for the Event Horizon Telescope. *Astrophys. J.* **2020**, *897*, 139. [CrossRef]

105. Pesce, D.W. A D-term Modeling Code (DMC) for Simultaneous Calibration and Full-Stokes Imaging of Very Long Baseline Interferometric Data. *Astron. J.* **2021**, *161*, 178. [CrossRef]

106. Phillips, C.; Whitney, A.; Sekido, M.; Kettenis, M. *VTP: VDIF Transport Protocol*; Technical Report; Istituto di Radioastronomia: Bologna, Italy, 2012.

107. Whitney, A.R.; Beaudoin, C.J.; Cappallo, R.J.; Corey, B.E.; Crew, G.B.; Doeleman, S.S.; Lapsley, D.E.; Hinton, A.A.; McWhirter, S.R.; Niell, A.E.; et al. Demonstration of a 16 Gbps Station^{-1} Broadband-RF VLBI System. *Publ. Astron. Soc. Pac.* **2013**, *125*, 196. [CrossRef]

Article

Key Science Goals for the Next-Generation Event Horizon Telescope

Michael D. Johnson [1,2,*], Kazunori Akiyama [2,3,4], Lindy Blackburn [1,2], Katherine L. Bouman [5], Avery E. Broderick [6,7,8], Vitor Cardoso [9,10], Rob P. Fender [11,12], Christian M. Fromm [13,14,15], Peter Galison [2,16,17], José L. Gómez [18], Daryl Haggard [19,20], Matthew L. Lister [21], Andrei P. Lobanov [15], Sera Markoff [22,23], Ramesh Narayan [1,2], Priyamvada Natarajan [2,24,25], Tiffany Nichols [26], Dominic W. Pesce [1,2], Ziri Younsi [27], Andrew Chael [28], Koushik Chatterjee [1,2], Ryan Chaves [1], Juliusz Doboszewski [2,29], Richard Dodson [30], Sheperd S. Doeleman [1,2], Jamee Elder [2,29], Garret Fitzpatrick [1], Kari Haworth [1], Janice Houston [1], Sara Issaoun [1,†], Yuri Y. Kovalev [15,31,32], Aviad Levis [5], Rocco Lico [18,33], Alexandru Marcoci [34], Niels C. M. Martens [29,35,36], Neil M. Nagar [37], Aaron Oppenheimer [1], Daniel C. M. Palumbo [1,2], Angelo Ricarte [1,2], María J. Rioja [30,38,39], Freek Roelofs [1,2,40], Ann C. Thresher [41], Paul Tiede [1,2], Jonathan Weintroub [1,2] and Maciek Wielgus [15]

1   Center for Astrophysics | Harvard & Smithsonian, 60 Garden Street, Cambridge, MA 02138, USA
2   Black Hole Initiative, Harvard University, 20 Garden Street, Cambridge, MA 02138, USA
3   Massachusetts Institute of Technology Haystack Observatory, 99 Millstone Road, Westford, MA 01886, USA
4   National Astronomical Observatory of Japan, 2-21-1 Osawa, Mitaka, Tokyo 181-8588, Japan
5   California Institute of Technology, 1200 East California Boulevard, Pasadena, CA 91125, USA
6   Perimeter Institute for Theoretical Physics, 31 Caroline Street North, Waterloo, ON N2L 2Y5, Canada
7   Department of Physics and Astronomy, University of Waterloo, 200 University Avenue West, Waterloo, ON N2L 3G1, Canada
8   Waterloo Centre for Astrophysics, University of Waterloo, Waterloo, ON N2L 3G1, Canada
9   Niels Bohr International Academy, Niels Bohr Institute, Blegdamsvej 17, 2100 Copenhagen, Denmark
10 CENTRA, Departamento de Física, Faculdade de Ciências, Instituto Superior Técnico—IST, Universidade de Lisboa, Campo Grande, 1749-016 Lisboa, Portugal
11 Astrophysics, Department of Physics, University of Oxford, Keble Road, Oxford OX1 3RH, UK
12 Department of Astronomy, University of Cape Town, Private Bag X3, Rondebosch 7701, South Africa
13 Institut für Theoretische Physik und Astrophysik, Universität Würzburg, Emil-Fischer-Strasse 31, D-97074 Würzburg, Germany
14 Institut für Theoretische Physik, Goethe Universität, Max-von-Laue-Str. 1, D-60438 Frankfurt, Germany
15 Max-Planck-Institut für Radioastronomie, Auf dem Hügel 69, D-53121 Bonn, Germany
16 Department of History of Science, Harvard University, Cambridge, MA 02138, USA
17 Department of Physics, Harvard University, Cambridge, MA 02138, USA
18 Instituto de Astrofísica de Andalucía-CSIC, Glorieta de la Astronomía s/n, E-18008 Granada, Spain
19 Department of Physics, McGill University, 3600 rue University, Montreal, QC H3A 2T8, Canada
20 Trottier Space Institute at McGill, 3550 rue University, Montreal, QC H3A 2A7, Canada
21 Department of Physics and Astronomy, Purdue University, 525 Northwestern Avenue, West Lafayette, IN 47907, USA
22 Anton Pannekoek Institute for Astronomy, University of Amsterdam, Science Park 904, 1098 XH Amsterdam, The Netherlands
23 Gravitation and Astroparticle Physics Amsterdam (GRAPPA) Institute, University of Amsterdam, Science Park 904, 1098 XH Amsterdam, The Netherlands
24 Department of Astronomy, Yale University, 52 Hillhouse Avenue, New Haven, CT 06511, USA
25 Department of Physics, Yale University, P.O. Box 208121, New Haven, CT 06520, USA
26 Department of History, Princeton University, Dickinson Hall, Princeton, NJ 08544, USA
27 Mullard Space Science Laboratory, University College London, Holmbury St. Mary, Dorking, Surrey RH5 6NT, UK
28 Princeton Gravity Initiative, Princeton University, Jadwin Hall, Princeton, NJ 08544, USA
29 Lichtenberg Group for History and Philosophy of Physics, University of Bonn, 53113 Bonn, Germany
30 ICRAR, M468, The University of Western Australia, 35 Stirling Hwy, Crawley, WA 6009, Australia
31 Lebedev Physical Institute of the Russian Academy of Sciences, Leninsky prospekt 53, 119991 Moscow, Russia
32 Moscow Institute of Physics and Technology, Institutsky per. 9, 141700 Dolgoprudny, Russia
33 INAF-Istituto di Radioastronomia, Via P. Gobetti 101, I-40129 Bologna, Italy
34 Centre for the Study of Existential Risk, University of Cambridge, 16 Mill Lane, Cambridge CB2 1SB, UK
35 Freudenthal Institute, Utrecht University, 3584 CC Utrecht, The Netherlands
36 Descartes Centre for the History and Philosophy of the Sciences and the Humanities, Utrecht University, 3584 CC Utrecht, The Netherlands

Citation: Johnson, M.D.; Akiyama, K.; Blackburn, L.; Bouman, K.L.; Broderick, A.E.; Cardoso, V.; Fender, R.P.; Fromm, C.M.; Galison, P.; Gómez, J.L.; et al. Key Science Goals for the Next-Generation Event Horizon Telescope. *Galaxies* 2023, 11, 61. https://doi.org/10.3390/galaxies11030061

Academic Editor: Fulai Guo

Received: 25 March 2023
Revised: 9 April 2023
Accepted: 21 April 2023
Published: 24 April 2023

37   Astronomy Department, Universidad de Concepción, Casilla 160-C, Concepción 4030000, Chile
38   CSIRO Astronomy and Space Science, PO Box 1130, Bentley, WA 6102, Australia
39   Observatorio Astronómico Nacional (IGN), Alfonso XII, 3 y 5, 28014 Madrid, Spain
40   Department of Astrophysics, Institute for Mathematics, Astrophysics and Particle Physics (IMAPP), Radboud University, P.O. Box 9010, 6500 GL Nijmegen, The Netherlands
41   McCoy Family Center for Ethics, Stanford University, Stanford, CA 94305, USA
*   Correspondence: mjohnson@cfa.harvard.edu
†   NASA Hubble Fellowship Program, Einstein Fellow.

Abstract: The Event Horizon Telescope (EHT) has led to the first images of a supermassive black hole, revealing the central compact objects in the elliptical galaxy M87 and the Milky Way. Proposed upgrades to this array through the next-generation EHT (ngEHT) program would sharply improve the angular resolution, dynamic range, and temporal coverage of the existing EHT observations. These improvements will uniquely enable a wealth of transformative new discoveries related to black hole science, extending from event-horizon-scale studies of strong gravity to studies of explosive transients to the cosmological growth and influence of supermassive black holes. Here, we present the key science goals for the ngEHT and their associated instrument requirements, both of which have been formulated through a multi-year international effort involving hundreds of scientists worldwide.

Keywords: black holes; general relativity; interferometry; accretion; relativistic jets; very-long-baseline interferometry; EHT; ngEHT

1. Introduction

The Event Horizon Telescope (EHT) has produced the first images of the supermassive black holes (SMBHs) in the M87 galaxy ([1–8], hereafter M87* I–VIII) and at the center of the Milky Way ([9–14], hereafter Sgr A* I–VI). Interpretation of the EHT results for Sgr A* and M87* has relied heavily upon coordinated multi-wavelength campaigns spanning radio to gamma-rays ([10], ETH MWL Science Working Group et al. [15]) In addition, the EHT has made the highest resolution images to date of the inner jets of several nearby Active Galactic Nuclei (AGN), demonstrating the promise of millimeter VLBI in making major contributions to the studies of relativistic radio jets launched from SMBHs (Kim et al. [16], Janssen et al. [17], Issaoun et al. [18], Jorstad et al. [19]).

The EHT results were achieved using the technique of very-long-baseline interferometry (VLBI). In this approach, radio signals are digitized and recorded at a collection of telescopes; the correlation function between every pair of telescopes is later computed offline, with each correlation coefficient sampling one Fourier component of the sky image with angular frequency given by the dimensionless vector baseline (measured in wavelengths) projected orthogonally to the line of sight [20]. The EHT observations at 230 GHz are the culmination of pushing VLBI to successively higher frequencies across decades of development (e.g., [21–23]), giving a diffraction-limited angular resolution of $\sim$20 µas (for a review of mm-VLBI, see [24]). For comparison, the angular diameter of the lensed event horizon—the BH "shadow"—is $\theta_{\rm sh} \approx 10GM/(c^2D)$, where G is the gravitational constant, c is the speed of light, M is the BH mass, and D is the BH distance [25–28]. For M87*, $\theta_{\rm sh} \approx 40$ µas; for Sgr A*, $\theta_{\rm sh} \approx 50$ µas.

Despite the remarkable discoveries of the EHT, they represent only the first glimpse of the promise of horizon-scale imaging studies of BHs and of high-frequency VLBI more broadly. In particular, the accessible science in published EHT results is severely restricted in several respects:

- EHT images are effectively monochromatic. The currently published EHT measurements sample only 4 GHz of bandwidth, centered on 228 GHz. BH images are expected to have a complex structure in frequency, with changing synchrotron emissivity, optical depth, and Faraday effects, making multi-frequency studies a powerful source of physical insight (see, e.g., [29–32]). The EHT has successfully completed commissioning observations at 345 GHz [33], which is now a standard observing mode. However, 345 GHz observations will be strongly affected by atmospheric absorption, severely affecting sensitivity and likely restricting detections to intermediate baseline lengths among the most sensitive sites (e.g., [34]).

- EHT images have severely limited image dynamic range. Current EHT images are limited to a dynamic range of only ~10 [4,11], providing only modest information about image signatures that are related to the horizon and limiting the ability to connect the event-horizon-scale images to their relativistic jets seen until now only at larger scales, via lower wavelength observations.[1]

  For comparison, VLBI arrays operating at centimeter wavelengths routinely achieve a dynamic range of ~10^4 on targets such as M87* (e.g., [35]).

- EHT observations have only marginally resolved the rings in Sgr A* and M87*. The EHT only samples a few resolution elements across the images. For instance, the EHT has only determined an upper limit on the thickness of the M87* ring [6], and the azimuthal structure of the rings in both sources is poorly constrained.

- EHT images cannot yet study the dynamics of M87* or Sgr A*. The gravitational timescale is $t_g \equiv GM/c^3 \approx 9\,\mathrm{h}$ for M87* and is $t_g \approx 20\,\mathrm{s}$ for Sgr A*. In each source, the expected evolution timescale is ~$50\,t_g$ (e.g., [36])—approximately 20 days for M87* and 20 min for Sgr A*. Current EHT campaigns consist of sequential observing nights extending for only ~1 week, which is too short to study the dynamical evolution of M87*. Moreover, the current EHT baseline coverage is inadequate to meaningfully constrain the rapid dynamical evolution of Sgr A*, which renders standard Earth-rotation synthesis imaging inapplicable [11,12].

In short, published EHT images of M87* and Sgr A* currently sample only 5×5 spatial resolution elements, a single spectral resolution element, and a single temporal resolution element (snapshot for M87*; time-averaged for Sgr A*).

The next-generation EHT (ngEHT) is a project to design and build a significantly enhanced EHT array through the upgrade, integration, or deployment of additional stations (e.g., [37–44]), the use of simultaneous observations at three observing frequencies [45–47], and observations that extend over several months or more with a dense coverage in time [48]. The ngEHT currently envisions two primary development phases. In Phase 1, the ngEHT will add three or more dedicated telescopes to the current EHT, with primarily dual-band (230/345 GHz) observations over ~3 months per year.[2] In Phase 2, the ngEHT will add five or more additional dedicated telescopes, with simultaneous tri-band capabilities (86/230/345 GHz) at most sites and observations available year-round. The new ngEHT antennas are expected to have relatively modest diameters (6–10 m), relying on wide recorded bandwidths, strong baselines to large existing apertures, and long integrations enabled through simultaneous multi-band observations to achieve the needed baseline sensitivity. Figure 1 shows candidate ngEHT sites in each phase.

These developments will sharply improve upon the performance of the EHT. Relative to other premier and planned facilities that target high-resolution imaging (such as the SKA, ngVLA, ALMA, and ELTs), the defining advantage of the EHT is its unmatched angular resolution. However, relative to the imaging capabilities of the current EHT, the defining improvements of the ngEHT images will be in accessing *larger* angular scales through the addition of *shorter* interferometric baselines than those of the present array, and in expanding the simultaneous frequency coverage. In addition, the ngEHT will extend accessible timescales of the current EHT by ~5 orders of magnitude, enabling dynamic analysis with the creation of movies of Sgr A* (through improved "snapshot" imaging on ~minute timescales) and AGN including M87* (through densely sampled monitoring campaigns that extend from months to years). Figures 2 and 3 summarize these improvements.

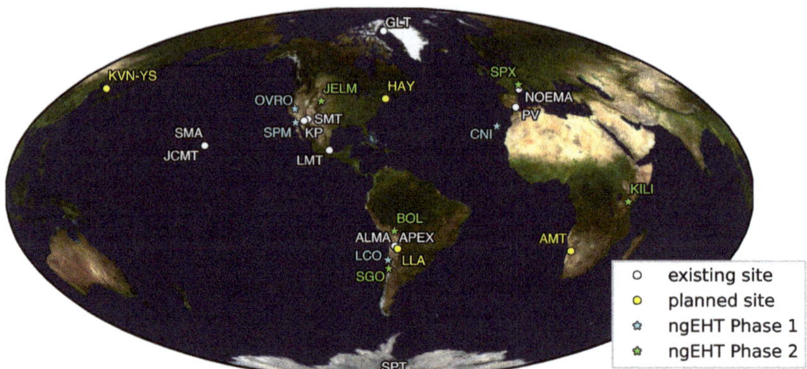

Figure 1. Distribution of EHT and ngEHT sites around the globe. Sites that have joined EHT campaigns are shown in white (see [2]), additional ngEHT Phase-1 sites are shown in cyan, and ngEHT Phase-2 sites are shown in green. Three of the EHT sites have joined since its initial observing campaign in 2017: the 12 m Greenland Telescope (GLT; [49]), the 12 m Kitt Peak Telescope (KP), the Northern Extended Millimeter Array (NOEMA) composed of twelve 15 m dishes. Several other existing or upcoming sites that plan to join EHT/ngEHT observations are shown in yellow: the 37 m Haystack Telescope (HAY; [41]), the 21 m Yonsei Radio Observatory of the Korea VLBI Network (KVN-YS; [42]), the 15 m Africa Millimetre Telescope (AMT; [43]), and the 12 m Large Latin American Millimeter Array (LLA; [44]). For additional details on the planned ngEHT specifications, see ngEHT Collaboration [50].

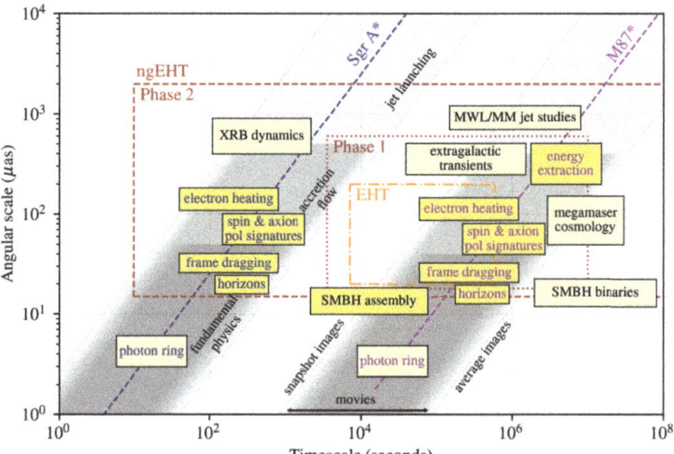

Figure 2. Comparison of image angular resolutions and timescales accessible to the EHT and ngEHT and the associated scientific opportunities. For M87* and Sgr A*, the ranges of angular resolution and timescale needed to study the three primary domains–fundamental physics, accretion, and jet launching–are indicated with the tilted shaded regions. These shaded regions are centered on the resolution-timescale for each source determined by the speed of light ($ct = D\theta$). Snapshot images require an array to form images on these timescales or shorter; average images require an array to form images over significantly longer timescales; movies require that an array can form images of the full range of timescales from snapshots to averages. The primary difference in M87* and Sgr A* is the factor of $\sim$1500 difference in the SMBH mass, which sets the system timescale. In contrast, the relevant angular scales in these systems are determined by the mass-to-distance ratio, which only differs by $\sim$20% for these two SMBHs. The approximate resolution-timescale pair to study each of the ngEHT Key Science Goals is indicated with the inset labeled boxes. Goals associated with Sgr A* or M87* are colored in blue or purple, respectively.

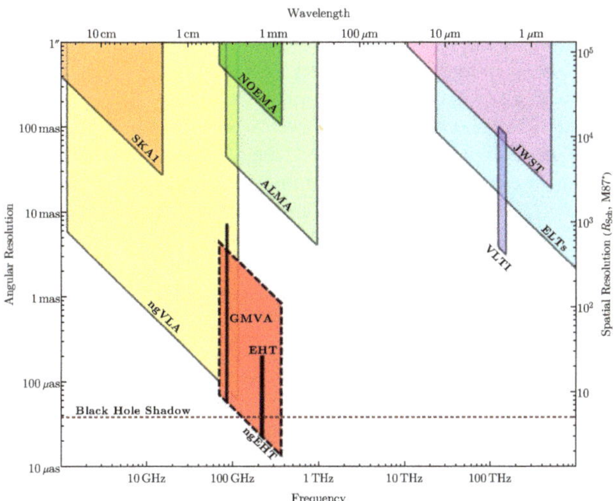

Figure 3. Range of observing frequency and angular resolution for selected current and upcoming facilities, from radio to the infrared. The ngEHT can achieve an imaging angular resolution that is significantly finer than any other planned facility or experiment. The ngEHT also envisions simultaneous multi-band observations, extending from 86 to 345 GHz, which will significantly expand the frequency coverage of currently published EHT data (black filled region). Figure adapted from Selina et al. [51].

To guide its design, the ngEHT has developed a set of Key Science Goals over the past two years, with contributions from hundreds of scientists. This process has included three international meetings[3,4,5], a Science Requirements Review (focused on identifying the most significant ngEHT science drivers), and a System Requirements Review (focused on identifying the associated instrument requirements). In addition, the ngEHT project has convened focused workshops on major topics, including assessing the motivation for adding 86 GHz capabilities to the ngEHT design to leverage phase transfer techniques[6], environmental and cultural issues related to ethical telescope siting, and the role of History, Philosophy, and Culture in the ngEHT (see Section 2.8 and [52] hereafter HPC White Paper). A series of papers present many aspects of these science cases in greater depth in a special issue of *Galaxies*[7].

In this paper, we summarize the Key Science Goals of the ngEHT and associated instrument requirements. We begin by discussing specific scientific objectives, organized by theme, in Section 2. We then summarize the prioritization and aggregated requirements of these science cases together with a condensed version of the ngEHT Science Traceability Matrix (STM) in Section 3. Details on the ngEHT concept, design, and architecture are presented in a companion paper [50].

2. Key Science Goals of the ngEHT

The ngEHT Key Science Goals were developed across eight Science Working Groups (SWGs). These goals span a broad range of targets, spatial scales, and angular resolutions (see Figure 2). We now summarize the primary recommendations from each of these working groups: Fundamental Physics (Section 2.1), Black Holes and their Cosmic Context (Section 2.2), Accretion (Section 2.3), Jet Launching (Section 2.4), Transients (Section 2.5), New Horizons (Section 2.6), Algorithms and Inference (Section 2.7), and History, Philosophy, and Culture (Section 2.8).

2.1. Fundamental Physics

BHs are an extraordinary prediction of general relativity: the most generic and simple macroscopic objects in the Universe. Among astronomical targets, BHs are unique in their ability to convert energy into electromagnetic and gravitational radiation with remarkable efficiency (e.g., [53–56]). Meanwhile, the study of BH stability and dynamics challenges our knowledge of partial differential equations, of numerical methods, and of the interplay between quantum field theory and the geometry of spacetime. The BH information paradox (e.g., [57]) and the existence of unresolved singularities in classical general relativity (e.g., [53,58]) point to deep inconsistencies in our current understanding of gravity and quantum mechanics. It is becoming clear that the main conceptual problems in BH physics hold the key to many current open foundational issues in theoretical physics.

Astrophysical BH systems are therefore an extraordinary test-bed for fundamental physics, although their extreme compactness renders them observationally elusive. Matter moving in the vicinity of an event horizon is subject to both extreme (thermo)dynamical conditions and intense gravitational fields, thereby providing a unique laboratory for the study of physical processes and phenomena mediated by the strongest gravitational fields in the Universe. Furthermore, by understanding the properties of matter and polarised electromagnetic radiation in this highly-nonlinear (and dynamical) regime, we can probe the underlying spacetime geometry of BHs and perform new tests of general relativity. The key to studying physics near the horizon is the capability to resolve, accurately extract, and precisely measure different features in BH images (see Figure 4). These image features can be periodic (e.g., oscillating fields), transient (e.g., reconnective processes and flares), persistent (the photon ring), or stochastic about a mean (e.g., polarization spiral patterns).

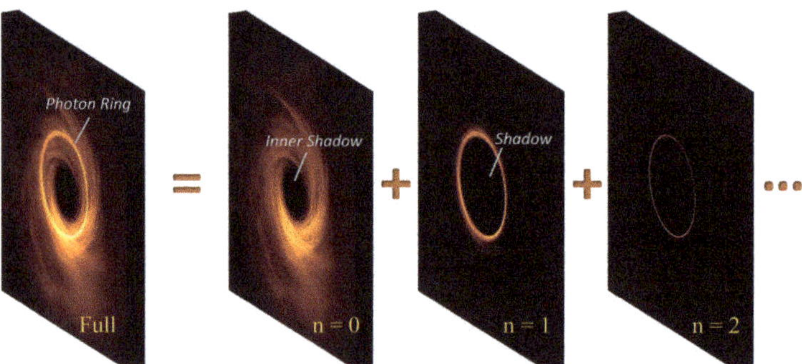

Figure 4. BH images display a series of distinctive relativistic features such as the BH apparent "shadow" (e.g., [27]), "inner shadow" (e.g., [59]), and "photon ring" (e.g., [60]).

The previous measurements of M87* and Sgr A* from the EHT provide compelling evidence for supermassive compact objects. The ngEHT has the capability to elevate existing EHT probes of the strong-field regime. We now describe four key science goals that target foundational topics in fundamental physics: studies of horizons (Section 2.1.1), measurements of SMBH spin (Section 2.1.2), studies of a BH photon ring (Section 2.1.3), and constraints on ultralight boson fields (Section 2.1.4). For a comprehensive discussion of these topics, see ngEHT Fundamental Physics SWG et al. [61].

2.1.1. Existence and Properties of Horizons

The formation of horizons (regions of spacetime that trap light) as gravitational collapse unfolds is one of the main and outstanding predictions of classical general relativity. Robust singularity theorems assert that BH interiors are also regions of breakdown of the classical Einstein equations, while quantum field theory is still associated with conundrums in the presence of horizons. Testing the existence and properties of horizons is therefore a key strong-field test of general relativity [62,63]. In astronomical terms, a horizon would be characterised by a complete absence of emission. It is clear that quantitative discussions of horizon physics will be strongly influenced both by the error in observational images and the modelling of matter and (spacetime) geometry at the core of simulated images.

For example, many models, especially those with spherical accretion onto BHs, tend to exhibit a pronounced apparent "shadow" (e.g., [27,64–66]). This feature shows a sharp brightness gradient at the boundary of the "critical curve" that corresponds to the boundary of the observer's line of sight into the BH. In contrast, models in which the emission is confined to a thin disk that extends to the horizon show a sharp brightness gradient in a smaller feature, the "inner shadow", which corresponds to the direct lensed image of the equatorial horizon [26,59,67]. The inner shadow gives the observer's line of sight into the BH that is unobscured by the equatorial emitting region.

Hence, BHs can give rise to a rich array of distinctive image features, but studies of horizons through imaging must account for potential degeneracies between the properties of the spacetime and those of the emitting material. Firm conclusions from imaging with the ngEHT will require significant improvements in both the image dynamic range and angular resolution of current EHT images, which have primarily demonstrated consistency with predictions of the Kerr metric (see Figure 5) and order-unity constraints on potential violations of general relativity (see, e.g., Sgr A* VI [14], Psaltis et al. [68], Kocherlakota et al. [69]). To leading order, the image dynamic range of the ngEHT will determine the luminosity of the features that can be studied, while the angular resolution will determine the size of the features that can be studied. Hence, quantitative statements about the existence of horizons will be primarily influenced by the dynamic range that can be achieved, while quantitative properties of the spacetime will be determined by the angular resolution [70,71]. Figure 6 shows an example of the improvement in both quantities that is possible using the ngEHT, enabling new studies of image signatures of the horizon. For additional discussion of potential ngEHT constraints on exotic horizonless spacetimes such as naked singularities and (non-hidden) wormholes, see ngEHT Fundamental Physics SWG et al. [61]. In addition to the necessity of image improvements, multi-frequency studies will be imperative to securely disentangle properties of the emission (which are chromatic) from features associated with the lensed horizon (which is achromatic). For all studies of horizons through imaging with the ngEHT, M87* and Sgr A* will be the primary targets because of their large angular sizes.

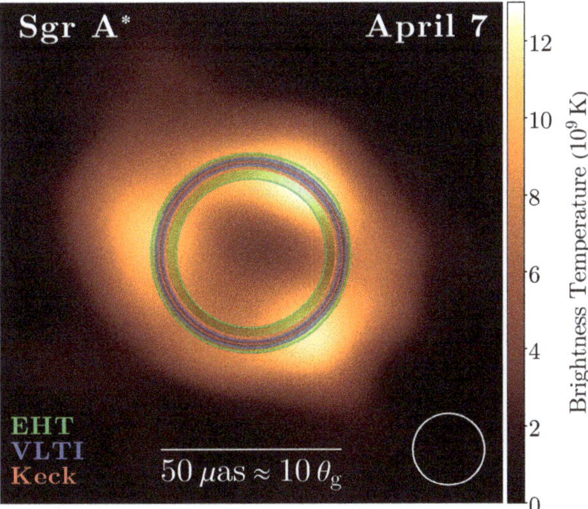

Figure 5. EHT representative average image of Sgr A* using data from 7 April 2017 [11]. The white circle in the lower-right shows a 20 µas beam that gives the approximate EHT resolution. The overlaid annuli show the predicted ranges of the Sgr A* critical curve using measurements of resolved stellar orbits using the VLTI (blue; [72]) and Keck (red; [73]); the ranges are dominated by the potential variation in size with spin, $d_{sh} = (9.6–10.4)\theta_g$ [25,74]. The green annulus shows the estimated range ($\pm 1\sigma$) of the critical curve using EHT measurements, which is consistent with these predictions [14]. However, because of the limited baseline coverage of the EHT, key image features such as the azimuthal brightness around the ring and the depth and shape of the central brightness depression are only weakly constrained with current observations.

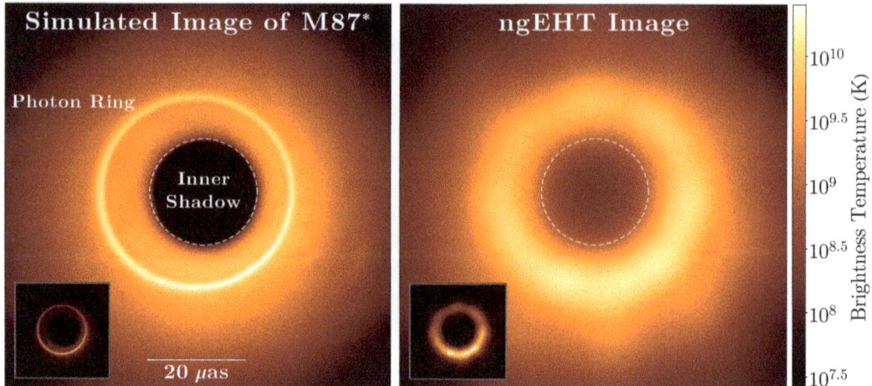

Figure 6. Accessing signatures of the event horizon with the ngEHT. Each panel shows an image on a logarithmic scale, with an inset shown with a linear scale. The left panel shows a time-averaged simulated image of M87*, which shows a prominent photon ring and inner shadow. The right panel shows a reconstructed ngEHT image using the Bayesian VLBI analysis package `Comrade.jl` [75] applied to simulated ngEHT phase-1 observations. The ngEHT provides both the angular resolution and dynamic range required to identify the deep brightness depression produced by the inner shadow in this simulated image.

2.1.2. Measuring the Spin of a SMBH

Astrophysical BHs are expected to be completely characterized by their mass and angular momentum [55,76,77]. Estimates of a SMBH spin through direct imaging would provide an invaluable complement to other techniques, such as the X-ray reflection method (see, e.g., [78]). However, the current EHT measurements provide only marginal, model-dependent constraints on the spins of M87* and Sgr A* [5,8,13].

The ngEHT has the opportunity to provide decisive measurements of spin through several approaches (for a summary of these methods, see [79]). The most compelling method would be to study the detailed structure of the lensing signatures such as the photon ring (see Section 2.1.3), or the (inner) shadow (see Section 2.1.1). However, while spin has a pronounced effect on these features, the effects of spin manifest on scales that are still much smaller than the nominal resolution of the ngEHT, so a conclusive detection may not be possible. Nevertheless, the effects of spin may be apparent in the emission structure on somewhat larger scales, particularly through the polarization structure in the emission ring (see Figure 7 and [80,81]). Finally, spin signatures are expected to be imprinted in the time-domain.

At least initially, ngEHT estimates of spin will likely rely on numerical simulations because unambiguous signatures of spin would require significantly finer angular resolution. These estimates will likely require confirmation through multiple lines of study—total intensity, polarization, and time-domain—and through a variety of modeling approaches including semi-analytic studies (e.g., [82]). Current studies indicate that the time-averaged polarized structure of M87* is the most reliable estimator of spin, with 345 GHz observations essential to improving angular resolution and also to quantify the potential effects of internal Faraday rotation on the polarized structure (see, e.g., [29,30,32]).

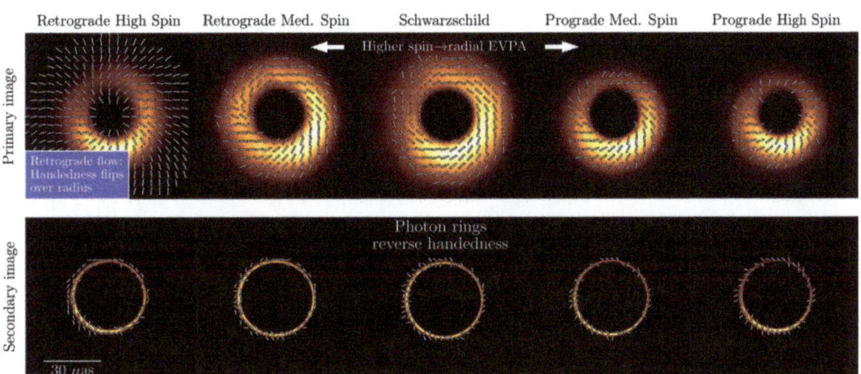

Figure 7. Summary of spin signatures in polarized images of time-averaged GRMHD simulations. In each panel, color indicates brightness and ticks show linear polarization direction. Rows show time-averaged primary (**top**) and secondary (**bottom**) images from MAD GRMHD simulations of M87*; columns show varying BH spin, ranging from a rapidly spinning BH with a retrograde accretion flow (**left**) to a non-spinning BH (**center**) to a rapidly spinning BH with a prograde accretion flow (**right**). The angular radius of the black hole, M/D, is identical in each panel. The polarization pattern becomes more radial at higher spin, as frame dragging enforces toroidal magnetic fields near the horizon. In retrograde flows, the spirals pattern reverses handedness over radius, indicating the transition from the prograde rotation within the ergosphere to the retrograde flow at larger radii. The handedness flips across sub-images, leading to depolarization in the photon ring of the full image (see [83,84]). By studying the polarized structure and its radial evolution, the ngEHT can estimate the spin of M87* and Sgr A* and quantify the effects of frame dragging. Adapted from Palumbo [85].

2.1.3. Constraining the Properties of a Black Hole's Photon Ring

The image of a BH is determined by two different factors: the complex astrophysical phenomena in its vicinity, which are the source of the emergent electromagnetic radiation, and the spacetime geometry, which introduces effects such as gravitational lensing and redshift. To isolate relativistic effects requires disentangling the complex, turbulent astrophysical environment from the comparatively simple spacetime dependence. Gravitational lensing is particularly useful in this context, as it gives rise to matter-independent ("universal") features, such as the "photon ring." The photon ring is a brightness enhancement along an approximately circular closed curve on the image, which arises from light rays undergoing multiple half-orbits around the BH before reaching the telescope [60]. These rays are small deviations from the unstable bound spherical orbits near a Kerr BH [25,86]. We index these half-orbits with the number n; the observer sees exponentially demagnified images of the accretion flow with each successive n (see Figure 4). At the resolution of Earth baselines at 230 GHz and 345 GHz, only $n = 0$ and $n = 1$ emission is likely to be detectable. Because the ngEHT cannot resolve the thickness of the primary ($n = 0$) ring, ngEHT studies of the photon ring necessarily require some degree of super-resolution, with associated model-dependent assumptions. In general, the principal challenge for ngEHT studies of the photon ring is to unambiguously disentangle the signals of the primary and secondary photon rings (see, e.g., [87]).

In the asymptotic ($n \to \infty$) limit, the photon ring has an intricate and universal structure which depends only on the spacetime geometry and acts as a lens for electromagnetic radiation (e.g., [74,88]). However, even at small-n, the photon ring carries information on the BH's mass and spin and provides a novel strong-field test of general relativity [89,90], especially if combined with a strong independent mass measurement (e.g., as is given by resolved stellar orbits of Sgr A*; see Figure 5). A clear goal for the ngEHT is to use the improved angular resolution and sensitivity to constrain the properties of the photon rings in M87* and Sgr A*.

Tests with both geometric model fitting of the sky intensity distribution and emissivity modelling in the BH spacetime suggest that the long baselines at 345 GHz are a strict requirement for detecting the $n = 1$ ring [82,87]. While intermediate baselines are required to support these model-fitting approaches, achieving the highest possible angular resolution is the driving requirement for studies of the photon ring. Photon ring detection using time-averaged images is likely most relevant to M87*, as Sgr A* observations are expected to be severely affected by scattering in the ionized interstellar medium [91–94]. Alternatively, signatures of the photon ring may be accessible in the time-domain, where "light echoes" can appear from either impulsive events such as flaring "hot spots" or from stochastic fluctuations in the accretion flow (see, e.g., [95–105]).

2.1.4. Constraining Ultralight Fields

The existence of ultralight boson fields with masses below the eV scale has been predicted by a plethora of beyond-Standard-Model theories (e.g., [106–110]). Such particles are compelling dark matter candidates and are, in general, extremely hard to detect or exclude with usual particle detectors. However, quite remarkably, rotating BHs can become unstable against the production of light bosonic particles through a process known as BH superradiance [111]. This process drives an exponential growth of the field in the BH exterior, while spinning the BH down. Superradiance is most effective for highly spinning BHs and when the boson's Compton wavelength is comparable to the BH's gravitational radius [111]. A BH of mass $\sim 10^{10} M_\odot$ such as M87* can be superradiantly unstable for ultralight bosons of masses 10^{-21} eV (this particular value leads to "fuzzy" dark matter, predicting a flat distribution that is favored by some observations; [112]).

For very weakly interacting particles, the process depends primarily on the mass and spin of the BH, and on the mass and spin of the fundamental boson. By requiring the predicted instability timescale to be smaller than the typical accretion timescale (that tends to spin up the BH instead), one can then draw regions in the parameter space where highly spinning BHs should not reside, if bosons within the appropriate mass range exists in nature. Thus, BH spin measurement can be used to constrain the existence of ultralight bosons. In particular, obtaining a lower limit on the BH spin is enough to place some constraints on boson masses (with the specific boson mass range constraint dependent on the BH spin). This approach is practically the only means to constrain weakly interacting fundamental fields in this mass range. Davoudiasl and Denton [113] used this line of argument to constrain masses of ultralight boson dark matter candidates with the initially reported EHT measurements that the SMBH must be spinning to produce sufficient jet power [5].

Among all the families of suggested ultralight particles, axions are one of the best studied and most highly motivated from a particle physics perspective. For axions with strong self-interactions, the super-radiance process will end up with a weakly saturating phase where the axion field saturates the highest possible density in the Universe. Due to the axion-photon coupling, the coherently oscillating axion field that forms around the BH due to superradiance can give rise to periodic rotation of the electric vector position angle (EVPA) of the linearly polarized emission. The amplitude of the EVPA oscillation is proportional to the axion-photon coupling constant and is independent of the photon frequency. The variations of the EVPA behave as a propagating wave along the photon ring for a nearly face-on BH. For instance, using the 4 days of polarimetric measurements of M87* published by the EHT collaboration in 2021 [7], one can already constrain the axion-photon coupling to previously unexplored regions [114,115]. The upper bound on the axion mass window is determined by the spin of the BH via the condition for superradiance to occur.

For improved constraints on these fundamental fields and their electromagnetic couplings, the ngEHT must observe polarimetric images of M87* in a series of at least 3 days over a 20-day window (the expected oscillation period). As for other cases that rely on polarimetry, observations at both 230 and 345 GHz are imperative to isolate the potential effects of Faraday rotation, and repeated observations will be needed to distinguish periodic oscillations from stochastic variability [116].

2.2. Black Holes and Their Cosmic Context

The growth of SMBHs is driven primarily by gas accretion and BH-BH mergers. Mergers are expected to dominate low-redshift SMBH growth in dense environments, especially in the high mass range to which the ngEHT will be most sensitive [117–120]. Gas accretion onto SMBHs is a critical piece of the current galaxy formation paradigm, in which feedback from accreting SMBHs is required to regulate gas cooling and star formation in massive galaxies (e.g., [121–123]). At present, however, the details of the feedback processes are poorly understood and are currently the largest source of uncertainty in understanding the combined mass assembly history and evolution of galaxies and their central SMBHs.

The ngEHT will provide unique observational access to both modes of SMBH growth through studies that extend over a vast range of scales (see Figure 8). By beginning to resolve the accretion flows of dozens of AGNs, the ngEHT will enable the extraction of information on their masses, spins, and accretion rates, providing crucial insights into their mass assembly history and growth (Section 2.2.1). In addition, the ngEHT will have sufficient angular resolution to identify sub-parsec binary SMBHs at any redshift, providing a powerful complement to gravitational wave observations of galaxy mergers (Section 2.2.2). In addition, the ngEHT will provide new insights into how SMBHs influence their galactic environments via feedback through multi-wavelength and multi-messenger studies of their relativistic outflows. (Section 2.2.3). We now discuss the goals and requirements associated with each of these objectives.

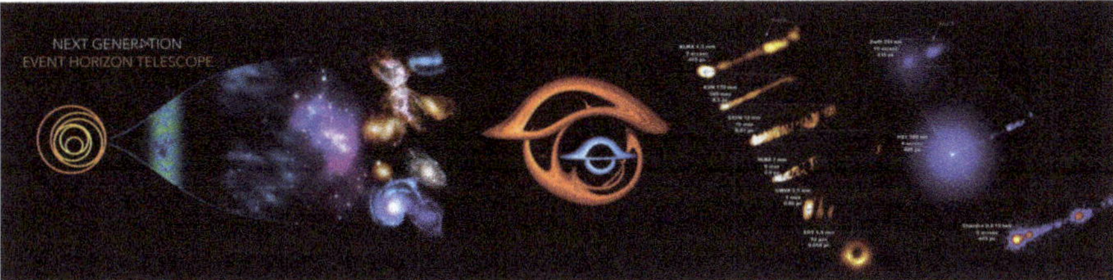

Figure 8. Conceptual illustration of the science cases explored within the "Black holes and their cosmic context" SWG: BH growth, binary BHs and gravitational waves, and MWL studies of BHs and jets. Credits from left to right: Perimeter Institute, NASA's Goddard Space Flight Center/Jeremy Schnittman and Brian P. Powell, J. C. Algaba for the EHT Collaboration [15]. Composition: Thalia Traianou, IAA-CSIC.

2.2.1. Understanding Black Hole-Galaxy Formation, Growth and Coevolution

The masses and spins of SMBHs encode their assembly history. SMBH masses trace this assembly history in a statistical fashion, with the distribution of SMBH masses—i.e., the BH mass function (BHMF)—capturing the population-level growth and evolution over cosmic time [124]. Measurements of SMBH spin can trace the growth histories of individual objects. For instance, BHs accreting from a thin disk with a steady rotation axis can be spun up to a maximum value of $a = 0.998$ [125], while discrete accretion episodes from disks with random rotation axes will tend to spin a BH down [126]. In addition, BHs accreting at low-Eddington rates for Gyrs can also be spun down due from the energy extraction that is required to power their jets via the Blandford-Znajek process [55,127].

The ngEHT will provide access to SMBH masses by observing the sizes of their horizon-scale emitting regions at (sub)millimeter wavelengths. EHT observations of M87* have demonstrated that measurements of the diameter of the ring-like emission structure can be used to constrain the SMBH mass [6]. The ~11% mass measurement precision achieved using the initial EHT observations of M87*—and even the comparatively modest ~25% precision achieved for the more challenging observations of Sgr A* [12]—establish the "shadow size technique" as among the most precise means of measuring SMBH masses (see, e.g., [128]). With the additional angular resolution and sensitivity provided by the ngEHT, Pesce et al. [129] estimate that ∼50 SMBH masses will be measurable for nearby AGNs distributed throughout the sky (see Figure 9). These measurements will substantially increase the number of SMBHs with precisely-measured masses, improving our understanding of the BHMF in the local Universe.

Relative to mass measurements, observational spin measurements for SMBHs are currently scarce; only roughly three dozen spin measurements are available for nearby SMBHs, with the majority obtained from X-ray diagnostics of the iron K-alpha line [78]. These iron-line measurements are uncertain because the method is highly sensitive to the orbital radius at which the accretion disk's inner edge truncates, which is typically assumed to occur at the innermost stable circular orbit [130]. In addition to their large uncertainties, current X-ray measurements are also biased towards high Eddington ratio objects.

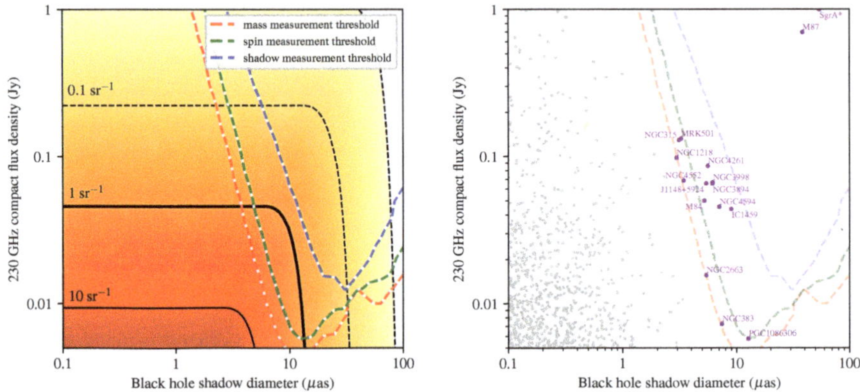

Figure 9. SMBH population studies with the ngEHT. (**left**) Black contours show the estimated cumulative number density of SMBHs as a function of shadow diameter and 230 GHz flux density. Colored contours indicate threshold values at which the ngEHT Phase-1 could plausibly measure the SMBH mass (red), spin (green), and shadow (blue) in a superresolution regime. Reproduced from Pesce et al. [129]. (**right**) Estimated 230 GHz compact flux density and BH shadow diameter for a subset of bright VLBI-detected SMBHs in the ETHER database. Colored lines again indicate the approximate measurement thresholds for the ngEHT Phase-1 array to measure the BH mass, spin, and shadow as shown on the left. Adapted from Ramakrishnan et al. [131].

The ngEHT will provide access to SMBH spins by observing the polarized radio emission emitted by the horizon-scale accretion flows around nearby AGNs. Current EHT observations have provided only modest constraints on the spin of M87* [5,8], but recent and ongoing advances in our theoretical understanding of near-horizon accretion flows will soon enable more precise spin quantifications from similar observations. As detailed in Ricarte et al. [79], linear polarimetric observations made by the ngEHT will provide estimates of SMBH spins by tracing the near-horizon magnetic field structures. The curl of the linear polarization pattern in the emission ring near a SMBH has been shown to correlate with SMBH spin in GRMHD simulations [80,81]. Ongoing studies indicate that this correlation originates from changes in the magnetic field geometry that are associated with frame dragging, which becomes stronger as spin increases [132]. Pesce et al. [129] estimate that the ngEHT will be able to constrain ∼30 SMBH spins through measurements of their horizon-scale polarized radio emission. Moreover, the spin measurements enabled by the ngEHT will offer fundamentally new insights by constraining the spins of low Eddington ratio SMBHs—rather than the high Eddington ratio SMBHs preferentially measured using X-ray techniques—which is a regime that is more representative of the overall SMBH population in the Universe.

The estimates from Pesce et al. [133] and Pesce et al. [129] for the number of SMBHs for which the ngEHT can make mass and/or spin measurements are based on statistical considerations, using our current understanding of the local BHMF and the distribution of SMBH accretion rates to predict how many objects should fall within the observable window. However, identifying the specific objects to target with the ngEHT for these measurements requires dedicated observational surveys of AGN to determine which sources are sufficiently bright, massive, and nearby. To this end, the Event Horizon and Environs ([ETHER; [131]]) database aims to provide a standardized catalog of ngEHT targets. Currently, the ETHER sample includes $\sim 10^3$ SMBHs that have been previously observed to have mas-scale structure at cm wavelengths and which have predicted 230 GHz flux densities greater than a few mJy. Of these sources, $\sim$ten have bright 8–86 GHz VLBI detections (from jet emission) and are predicted to be bright enough to image their jet bases at $\lesssim 100\,\mathrm{R_g}$ with the ngEHT (see Figure 9). The identification of ngEHT targets with bright accretion inflows but without detected cm-wave jets is ongoing; the currently known ~ 200 BHs with estimated ring sizes $\geq 5\,\mu\mathrm{as}$ primarily have (observed arcsec-scale and/or predicted mas-scale) 230 GHz flux densities less than 1 mJy, with the brightest falling in the 1 to 10 mJy range. The upcoming release of the e-ROSITA all-sky hard X-ray survey (with SDSS V and 4MOST spectroscopic followups for BH mass estimates) is expected to significantly expand the list of potential targets in this accretion-inflow-only sample, which will permit definitive specifications on the sensitivity requirement for the ngEHT to measure a large population of horizon-resolved sources.

2.2.2. Understanding How SMBHs Merge through Resolved Observations of Sub-Parsec Binaries

Binary SMBHs are generic products of galaxy mergers, that are thought to drive structure formation in our dark energy-driven cold dark matter Universe. During SMBH mergers, dynamical friction and stellar mass segregation act to draw the two resident massive objects to the center of the merger remnant [134]. The environmental interactions that drive the binary to separations of ~ 0.1–10 pc are understood, but the mechanism(s) that drive continued inspiral beyond this point—and in particular, to the sub-parsec regime in which gravitational wave emission is expected to efficiently complete the merger process— remain unclear (e.g., [135]). A number of solutions to this long-standing and so-called "final parsec problem" [136,137] have been proposed; for instance, interactions with gas in a circumbinary disk, and three-body interactions with stars could all contribute and have significant influence on the shape and evolutionary timescale of the binary. Uncovering the details of the physics in this last parsec informs the science cases of future gravitational-wave detectors such as Pulsar Timing Arrays (PTAs) and space-based gravitational-wave interferometry (e.g., LISA).

The ngEHT will have a nominal angular resolution of 15 μas, which implies a linear resolution of ≤ 0.13 pc across all redshifts. The effective resolving power may be further improved by a factor of several through the use of "super-resolution" techniques (e.g., Chael et al. [138], Akiyama et al. [139], Broderick et al. [140]). The ngEHT can therefore *spatially resolve* SMBH binaries that have entered their steady-state gravitational wave emission phase. The orbital period at this stage is typically short (months to years), which makes it accessible to multi-epoch observations with the ngEHT. Furthermore, D'Orazio and Loeb [141] estimate that between ~ 1 and 30 sub-parsec SMBH binaries should have millimeter flux densities in the $\gtrsim 1$ mJy regime that will be accessible with the ngEHT.

2.2.3. Multi-Wavelength and Multi-Messenger Studies of SMBHs and Their Relativistic Outflows

The EHT has already demonstrated the immense value of extensive multi-wavelength campaigns to augment horizon-scale imaging (e.g., [10,15]), and the ngEHT will similarly benefit from coordinated observations (for a review, see [142]). In particular, the relativistic jets launched by SMBHs extend the gravitational influence of BHs to galactic scales, convert-ing and transporting immense amounts of energy across the full electromagnetic spectrum.

These jets act as powerful particle accelerators that are thought to produce ultra-high energy cosmic rays and have also been implicated in the production of high-energy neutrinos (e.g., [143–148]).

The ngEHT can directly image flaring regions, creating an opportunity to shed light on the physical mechanisms that drive acceleration of protons to PeV energies and generation of high-energy neutrinos. Moreover, crucial insights into the jet composition can be obtained by combining information about the jet dynamics with information about the accretion power (e.g., from X-ray observations). Ideally, this will involve both triggered and monitoring ngEHT observations. Triggered observations would happen when a neutrino arrives within an error region from a strong blazar. Limiting the trigger on both the neutrino energy (above $\sim$100 GeV) and VLBI flux density (above $\sim$0.5 Jy) would increase the probability of association and ensure sufficiently high dynamic range of the ngEHT images, respectively. The initial trigger would be followed by $\sim$monthly monitoring for a year. In addition to this mode, observing a large sample of the strongest blazars with $\sim$monthly monitoring, supplemented by additional single-dish flux monitoring, will provide an opportunity to study the evolution of sources before neutrino production and to characterize the features that are associated with neutrino production.

In addition, by observing a population of blazars, the ngEHT will be able to measure the jet profile from the immediate vicinity of a SMBH through the acceleration and collimation zones and past the Bondi radius (e.g., [149]). By observing with coordinated multi-wavelength campaigns, the ngEHT will provide decisive insights into the nature of the bright, compact "core" feature that is seen in many blazars (e.g., [150]). Current EHT images of blazars show complex, multi-component emission [16,18,19], so ngEHT observations extending over multiple months to study the evolution of these components will be imperative.

Multi-wavelength and multi-messenger studies of flaring activity in blazar jets will require ngEHT monitoring campaigns with triggering capabilities followed by a cadence of the order of weeks. Full Stokes polarization capabilities with high accuracy (systematic errors on polarization $\lesssim$0.1%) and high imaging dynamic range ($\gtrsim$1000:1 to detect faint jet emission) will be required for mapping the magnetic field in the jet regions through Faraday rotation analyses. Close coordination with other next-generation instruments, such as the Cherenkov Telescope Array (CTA), LISA, SKA, ngVLA, and Athena will significantly enrich the potential for multi-wavelength and multi-messenger studies with the ngEHT.

2.3. Accretion

Electromagnetic radiation from SMBHs such as M87* and Sgr A* originates in hot gas, which is brought close to the BH by an accretion disk (for a review of hot accretion flows, see [56]). Some of the same gas is also expelled in relativistic jets or winds. Spatially resolved images of the disk and its associated dynamics provide a remarkable new opportunity to study accretion physics.

BH accretion disks are believed to operate with the help of the magnetorotational instability,[8] which amplifies the magnetic field in the plasma and uses the associated shear stress to transport angular momentum outward [154,155]. Signatures of the magnetic field are revealed via linear and circular polarization of the emitted radiation. Yet, while spatially-resolved and time-resolved spectropolarimetric observations are thus exceptional tools for studying the inner workings of BH accretion, we do not at present have even a single spatially-resolved image of any BH accretion disk.

The closest current results are through EHT observations of M87* and Sgr A*. The ring-shaped 230 GHz emission surrounding a central brightness depression confirms strong light deflection and capture near these BHs. However, the angular resolution and dynamic range achieved so far by the EHT are modest, and it is unclear what part of the observed radiation is from the accretion disk and what is from the jet (see, e.g., [5,13]). The ngEHT will have the sensitivity to image out to larger radii from the BH and to make time-resolved movies in all Stokes parameters. These advances will enable progress on three broad fronts in accretion physics: revealing the physical mechanism that drives accretion onto SMBHs (Section 2.3.1), observing localized electron heating and dissipation (Section 2.3.2), and measuring signatures of frame dragging near a rotating black hole (Section 2.3.3).

2.3.1. Revealing the Driver of Black Hole Accretion

Our current understanding of accretion close to a BH is largely guided by ideal general relativistic magnetohydrodynamical (GRMHD) numerical simulations (see, e.g., [156,157]). These simulations suggest that the strength and topology of the magnetic field play an important role. When the field is weak and scrambled, the accreting gas becomes turbulent, with eddies over a wide range of length scales (e.g., [158]). When the field is strong, and especially when it also has a dipolar configuration (this is called a "magnetically arrested disk" or MAD; [159]), accretion occurs via large discrete inflowing streams punctuated by episodic outward eruptions of magnetic flux. The ngEHT will be able to identify these and other dynamical patterns in the accretion flow by making real-time movies. Flux-tube eruptions [102,151,152,160], orbiting spiral patterns (e.g., [161]), and bubbling turbulence, could all be accessible to observations. Crucially, spatially-resolved measurements of the linear polarization fraction, degree of circular polarization, and Faraday rotation, will provide rich detail on the magnetic field topology and its strength (e.g., [32]). Different target sources will presumably have different dynamics and field configurations, opening up a fruitful area of research. In the specific case of a MAD system, it is unknown exactly how the strong field originates. One proposal posits that the field is generated in situ by a radiation-driven battery mechanism (e.g., [162]). It predicts a specific relative orientation of the dipolar magnetic field with respect to the accretion disk angular velocity vector. If any of ngEHT's targets is MAD (EHT observations suggest M87* and Sgr A* may both be such systems), testing the predictions of the radiation battery model would be an important secondary goal.

Accretion-related ngEHT science will be primarily enabled through observations of M87* and Sgr A*, with two major associated challenges. First, the most interesting effects occur in regions of the disk within a few event horizon radii. However, this is precisely where the observed image is highly distorted by the gravitational lensing action of the BH, the same effect which produces the ring image of M87*. Disentangling lensing to reveal the true underlying structure of the accretion disk will require new image processing techniques. Second, the observed image will often be a superposition of radiation from the accretion disk and the jet. The two components will need to be separated. One promising method is to utilize dynamics and variability, which can be quite different in the disk and in the jet. Observations with a cadence of t_g would be ideal to study the most rapid variability, and interesting variations are expected on all timescales up to 10^3–$10^4\,t_g$. Full-night observations with sufficient baseline coverage for snapshot imaging on sub-minute timescales will be needed for Sgr A*, while a monitoring campaign with a sub-week cadence and extending for at least 3 months (and, ideally, over multiple years) will be ideal for M87*.

2.3.2. Localized Heating and Acceleration of Relativistic Electrons

The radiation emitted from an accretion disk is produced by hot electrons, which receive their heat energy via poorly-understood plasma processes in the magnetized gas. The most promising idea for heating is magnetic reconnection, which can occur in regions with large-scale topological reversals of the magnetic field, or in regions with large shear, or where small-scale turbulent eddies dissipate their energy. All of these processes are at their most extreme in the relativistic environment found in BH accretion disks.

Our current understanding of relativistic magnetic reconnection is based on particle-in-cell (PIC) simulations (e.g., [163–166]). These numerical studies show clear evidence for unequal heating of electrons and ions, as well as acceleration of both into a non-thermal distribution with a power-law tail at high energies. Electron heating in large flares in BH disks would be especially interesting for ngEHT observations. A flare may initially appear as a bright localized region in the image. It will subsequently move around the image, will also likely spread to become more diffuse, and will show effects from graviational lensing (see, e.g., [95,97,102,167]). Both the ordered motion of the heated region and its spreading will provide fundamental information on the microscopic plasma physics processes. The heated electrons will also cool as they radiate, causing the electron distribution function (eDF) to evolve. Multi-wavelength imaging will provide a handle on both the dynamics and the eDF evolution.

Less dramatic steady heating should also be present, and it will likely show strong variations as a function of radius in both amplitude and eDF. With the enhanced dynamic range of the ngEHT, these spatial variations should be accessible over a factor of 10 range of radius. Particle acceleration and heating is relevant for a wide range of astrophysical phenomena. While we have some information on low energy processes from laboratory experiments and measurements in the solar wind, there is currently no observational technique for direct study of heating in relativistic settings. Imaging BH accretion disks with the ngEHT can reveal localized heating and acceleration on astrophysical scales and will track the evolution of the energized electrons. Lessons from such observations would have a widespread impact in many other areas of astrophysics.

2.3.3. Dynamical Signatures of Frame Dragging Near a Rotating Black Hole

Direct observations of the inner region of the accretion disk provide an opportunity to study the object at the center, namely, the BH itself. While the most significant effect on large scales is the immense gravitational pull of a BH, another gravitational property of these objects is arguably even more interesting. Namely, a spinning BH has the remarkable property that it drags space around it in the direction of its spin. This so-called frame-dragging effect is felt by all objects outside the BH, including the accretion disk. The effect is strongest in regions within a few event horizon radii of the BH.

Spatially-resolved and time-resolved imaging has the potential to confirm the frame-dragging effect and to study its details (see, e.g., [168]). Since the accretion disk is fed by gas at a large distance from the BH, the outer disk's angular momentum vector is likely to be randomly oriented with respect to the BH spin axis. Only when gas comes close to the BH does it feel the spin direction of the BH via frame-dragging. The manner in which the disk adjusts its orientation can provide direct confirmation of the frame-dragging phenomenon. If the disk is tilted with respect to the BH spin vector, it is expected to precess and align with the BH inside a certain radius (see, e.g., [5]). Both the precession and alignment can be observed and studied by the ngEHT. In the special case of a retrograde accretion flow (i.e., when the disk's orbital motion is in the opposite direction to the BH spin), the angular velocity of the disk gas will reverse direction close to the horizon. There will be a related effect also in the orientation of the projected magnetic field, which may be visible in polarimetric ngEHT images. Observing these effects directly with the ngEHT would be a breakthrough achievement and would provide a new tool to study a central prediction of the Kerr spacetime (see also Section 2.1.2).

2.4. Jet Launching

Relativistic jets are among the most energetic phenomena in our universe, emitting radiation throughout the entire electromagnetic spectrum from radio wavelengths to the gamma-ray regime, and even accelerating particles to highest measured energies (for a review, see [169]). The most powerful jets are those that are anchored by nuclear SMBHs in AGN, as emphatically demonstrated through the images of M87* with the EHT. Yet, despite this impressive breakthrough, the actual jet launching mechanism and power source is still uncertain. The ngEHT has the potential to make pivotal discoveries related to the power source of relativistic jets (Section 2.4.1), and to the physical conditions that launch, collimate, and accelerate these jets (Section 2.4.2).

2.4.1. Jet Power and Black Hole Energy Extraction

According to our current theories, jets can either be powered by the liberation of gravitational potential energy in the accreting material (e.g., [170]) or by directly extracting the rotational energy of a spinning BH [55]. In both processes, magnetic fields must play a crucial role. Therefore, measuring the velocity field of the innermost jet regions and comparing de-projected rotation of the magnetic fields with the rotation of the BH ergosphere will probe whether jets are launched by rotating BHs.

The ideal target to address this question is M87* because of its large BH mass ($M \approx 6.5 \times 10^9 \, M_\odot$), proximity ($D \approx 16.8 \, \mathrm{Mpc}$), and prominent jet ($P_{\mathrm{jet}} \gg 10^{42} \, \mathrm{erg/s}$). Sgr A* also provides an important target to study—despite decades of VLBI observations, there is no firm evidence for a jet in Sgr A* at any wavelength. Nevertheless, there are compelling reasons to continue the search for a jet in Sgr A* with the ngEHT, including the potential for interstellar scattering to obscure the jet at longer wavelengths (e.g., [171]), evidence for an outflow in frequency-dependent time lags during flares (e.g., [172,173]), and the fact that favored GRMHD models for Sgr A* based on constraints from EHT observations predict the presence of an efficient jet outflow [13]. Comparing the jets in M87* and Sgr A*, together with knowledge of their respective BH properties, will provide fundamental insights into the role of the BH and its environment in producing a jet.

Current EHT observations are limited both in terms of the baseline coverage and image dynamic range, which prohibits estimates of physical parameters in the critical region just downstream of the BH. The ngEHT will provide superior baseline coverage and increased dynamic range, allowing reconstructed movies that simultaneously resolve horizon scale structure and the jet base in M87* and Sgr A*. To identify the source of the jet's power with the ngEHT will require estimates of the magnetic flux threading the SMBH, the spin of the SMBH (see Section 2.1.2), and the total jet power. These estimates will require high-fidelity polarized and multi-frequency images with an angular resolution of $\sim$15 µas (a spatial resolution of $\approx 4GM/c^2$) and with sufficient dynamic range to simultaneously study both the near-horizon magnetosphere and the jet over many dynamical timescales.

2.4.2. Physical Conditions and Launching Mechanisms for Relativistic Jets

The ngEHT has the potential to substantially improve our understanding of the mechanisms that launch, collimate, and accelerate relativistic jets by measuring the physical conditions at the jet base. For instance, multi-frequency VLBI observations at cm-wavelength mainly probe the extended jet regions and have revealed that the energy distributions of relativistic electrons responsible for the emission follow power-laws. This is in marked contrast to the recent EHT observations of the horizon scale structure around M87* and Sgr A*, which has been successfully modeled using thermal distributions of electrons [5,13]. Important questions therefore arise regarding which physical mechanisms are able to accelerate the thermal particles, and where this particle energization occurs. Using multi-frequency observations at 86 GHz and 230 GHz while making use of VLBI synergies with the next-generation Very Large Array (ngVLA), the spectral index distribution of the radio emission can be mapped at high resolution, allowing estimates of the underlying eDF and indicating possible particle acceleration sites. In addition, linear polarization studies will reveal the magnetic field structure and strength in the jet, and circular polarization will reveal the plasma composition (leptonic/hadronic), opening a window to more detailed understanding of jet microphysics (e.g., [174]).

According to recent GRMHD models, a dynamic range of $\sim 10^4$ will enable us to probe the jet in M87* at a wavelength of 1.3 mm on scales of hundreds of microarcseconds and to reliably measure the velocity profile. In addition to the aforementioned array requirements, monitoring of the jet with cadences of days to weeks is required (for M87*, 1 day corresponds to roughly $3t_g$). Figure 10 shows simulated ngEHT reconstructions of the M87* jet, illustrating the ability of the ngEHT to conclusively identify and track kinematic structure throughout the jet. Finally, in addition to M87*, there are several other potential AGN targets (e.g., Cen A, 3C120, 3C84) of comparable BH mass and distance, which can also serve as laboratories to study jet launching activity. More distant AGN ($z > 0.1$) would require imaging on a ~monthly basis.

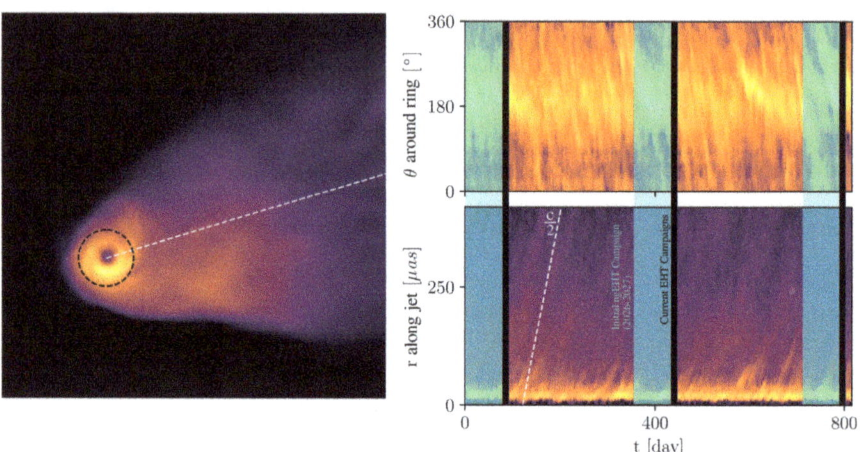

Figure 10. Studying accretion and jet dynamics with the ngEHT. (**left**) A frame from a simulated movie of M87* [175]. (**right**) Azimuthal (**top**) and radial (**bottom**) brightness variations in a reconstructed movie of M87* using ngEHT Phase-1 coverage. The top panel shows how azimuthal variations around the black dashed circle track orbital dynamics near the BH, evident here as diagonal striations with sub-Keplerian angular velocity. The bottom panel shows how radial variations along the white dashed line will reveal the SMBH-jet connection and measure acceleration within the jet-launching region. Initial ngEHT monitoring campaigns (light blue vertical bands) will span 3 months per year with a dense (sub-week) observing cadence; for comparison, current EHT campaigns (dark vertical bands) only span ~2 weeks per year, which is insufficient to measure the dynamics of the accretion disk or jet.

2.5. Transients

Astrophysical transients are the sites of some of the most extreme physics in the present-day universe, including accreting sources such as BH X-ray binaries and Tidal Disruption Events, explosive events such as supernova as well as the LIGO/VIRGO gravitational wave bursts associated with neutron star-neutron star mergers such as GW170817 [176,177].

In essentially all cases, the radio emission from these transients corresponds to synchrotron emission from relativistic electrons spiralling in magnetic fields either in a jet or in structures which have been energised by a jet associated with the transient (e.g., [178–182]). As with supermassive BHs in AGN, probing the formation, propagation and ultimate energetics of these jets is central to understanding the physics of BHs and how they convert gravitational potential energy of infalling matter into powerful collimated outflows.

Because the field of astrophysical transients is so diverse, we have chosen to focus the ngEHT key science goals and associated requirement related to transients on two sets of objects: BH X-ray binaries (Section 2.5.1) and extragalactic transients (Section 2.5.2). Together, these categories span most of the range both in the astrophysics under study and in the technical requirements for the ngEHT.

2.5.1. Dynamics of Black Hole X-ray Binaries

Black hole X-ray binaries (BHXRBs) represent the bright end of the population of massive stellar remnants in our galaxy. They are expected to number in the few thousands among a likely population of ~10^8 stellar mass BHs in our galaxy, with a typical mass around $7M_\odot$. They accrete, usually intermittently, from a close binary companion and often reach accretion rates close to the Eddington limit. In other words, they are around five (eight) orders of magnitude less massive than Sgr A* (M87*) and are accreting at $> 10^7$ times higher Eddington-ratioed rates. There are good reasons, and indeed much circumstantial evidence, to suggest that the coupling between accretion 'states' and jet formation at high Eddington ratios are similar between supermassive and stellar-mass BHs, so their study genuinely, and dramatically, extends the parameter space of study of BHs (e.g., [183,184]).

The event horizons of these stellar-mass BHs will likely never be resolvable by conventional telescopes, but remarkably it has been established that high-time resolution X-ray variability studies of BHXRBs probe the same range of scales in gravitational radii, $r_g \equiv GM/c^2$, as the direct EHT imaging of M87* and Sgr A*. Furthermore, decades of work has established good, but not yet precise enough, connections between characteristic patterns of variability, arising from within $100r_g$, and the formation and launch of the most powerful jets.

With the ngEHT, we will be able to probe BHXRB jets on scales around $10^6 r_g$, at which scales bright (sub-)mm flares often have flux densities in excess of 1 Jy and evolve considerably on timescales of *minutes* (e.g., [185–187]). VLBI studies of jets at ten times larger angular scales have provided the most precise determination of jet launch time (and the corresponding activity in the accretion flow), evidence for strong directional variation and precession of the jet, and circumstantial evidence for interactions and—presumably—internal shocks between components moving at different speeds. This is also the region in r_g in which the jets of M87* and other AGN have been seen to switch from an initially parabolic to a later conical cross section (e.g., [149,188–191]). With the ngEHT, we can directly test if this same collimation is occurring in BHXRB jets. Finally, we now know from the ThunderKAT project on MeerKAT [192] that large-scale jets from BHXRBs which decelerate and terminate in the ISM on timescales of $\sim$1 year are common (rate of 2–4/year): therefore only in this class of object can we track events from their creation and launch in the accretion flow through to their termination, providing an opportunity for precise calorimetry of their kinetic power.

2.5.2. Extragalactic Transients

The broad term of extragalactic transients encompasses sources including Gamma Ray Bursts (GRBs), Tidal Disruption Events (TDEs), neutron star mergers, supernovae, fast radio bursts (FRBs), fast blue optical transients (FBOTs) and other related phenomena. The origin of the radio emission from these objects is often within relativistic jets, but it may also be more (quasi-)spherical.

Some of these phenomena remain optically thick and bright at (sub-)mm wavelengths for a considerable period of time (months; e.g., [193,194]) which places far less stringent requirements for response and scheduling of ngEHT. Nevertheless, there is a wide range of important physics which could conceivably be tackled, such as whether or not jets are being produced commonly (very topical for TDE jets, which may even be associated with neutrino production) and how much kinetic power was released in the event. Thus, the ngEHT could make significant discoveries by measuring the kinetic power, physical structure, and velocity in extragalactic transients such as GRBs, GW events, TDEs (e.g., [195]), FRBs, and FBOTs.

2.6. New Horizons

The "New Horizons" SWG was formed to explore and assess non-traditional avenues for ngEHT scientific breakthroughs. This group has examined topics including terrestrial applications such as planetary radar science, geodesy, and improved celestial reference frames [196,197]; studies of coherent sources including magnetars, masers, and fast radio bursts; and precise astrometry of AGN [198]. We now describe the two key science goals that have been identified by this SWG, both with cosmological applications: measurements of proper motion and parallax for a sample of AGN at distances up to $\sim$80 Mpc (Section 2.6.1), and studies of SMBHs and their accretion disks using water vapor megamasers, which can provide accurate measurements of the Hubble constant up to distances of $\sim$50 Mpc (Section 2.6.2).

2.6.1. Proper Motions and Secular (CMB) Parallaxes of AGN

The multi-band capabilities of the ngEHT will enable the use of the source-frequency phase referencing ([SFPR; [199]) technique, potentially achieving $\sim\mu$as-level astrometry for targets that are sufficiently bright and close to known reference sources [46,47]. In addition to many other scientific applications such as measurements of (chromatic) AGN jet core shifts (e.g., [200–202]) and the (achromatic) orbital motions of binary SMBH systems (e.g., [141]), one of the opportunities afforded by this astrometric precision is a measurement of the so-called "cosmological proper motion" [203] or "secular extragalactic parallax" [204]. Because the Solar System is moving with respect to the cosmic microwave background (CMB) with a speed of $\sim$370 km s^{-1} [205], extragalactic objects in the local Universe should exhibit a contribution, $\mu_{\rm sec}$, to their proper motion from the Solar Systems's peculiar motion:

$$\mu_{\rm sec} \approx \left(0.018\,\mu{\rm as\,year}^{-1}\right)\left(\frac{H_0}{70\,{\rm km\,s\,Mpc}^{-1}}\right)\frac{|\sin(\beta)|}{z}, \tag{1}$$

where z is the object's cosmological redshift, H_0 is the Hubble constant, and β is the angle between the location of the source and the direction of the Solar System's motion with respect to the CMB [203]. An object located at a distance of 10 Mpc ($z \approx 0.0023$) is thus expected to have $\mu_{\rm sec} \sim 8\,\mu{\rm as\,year}^{-1}$, while an object located at a distance of 100 Mpc ($z \approx 0.023$) is expected to have a proper motion of $\mu_{\rm sec} \sim 0.8\,\mu{\rm as\,year}^{-1}$. By measuring the proper motion of many objects and using multi-frequency observations to mitigate chromatic effects in time-variable core shift effects (see, e.g., [206]), the ngEHT could thus isolate the contribution of $\mu_{\rm sec}$ and provide coarse estimates of H_0 that are independent of standard methods (e.g., [207–210]).

2.6.2. Studies of Black Hole Masses and Distances with Megamasers

Water vapor megamasers residing in the molecular disks around nearby AGNs on scales of $\sim$0.1 pc ($\sim$10^5 r_g) have proven to be powerful tools for making precise measurements of SMBH masses (e.g., [211,212]), geometric distances to their host galaxies (e.g., [213,214]), and the Hubble constant (e.g., [209,215]). While the majority of the research carried out to date has utilized the 22 GHz rotational transition of the water molecule, other transitions are expected to exhibit maser activity under similar physical conditions as those that support 22 GHz masers [216,217]. In particular, both the 183 GHz [218,219] and the 321 GHz [220–223] transitions have been observed as masers towards AGN. The latter transition falls in the ngEHT observing band, as does another tranisition at 325 GHz that is also expected to exhibit maser activity [224].

Observations of water megamaser systems with the ngEHT will necessarily target transitions such as those at 321 GHz and 325 GHz, rather than the transition at 22 GHz. If the submillimeter systems are as bright as those at 22 GHz, then the >order-of-magnitude improvement in angular resolution brought about by the ngEHT will impart a corresponding improvement in the precision of maser position measurements in these systems. However, the typical brightness of submillimeter megamaser systems is currently unknown, and the two sources that have to date been observed at 321 GHz both exhibit fainter emission at 321 GHz than at 22 GHz [220,221]; it is thus possible that systematically fainter submillimeter transitions (relative to 22 GHz) will offset the improvement in position measurement precision through reduced signal-to-noise ratios. Nevertheless, even comparable measurement precisions for submillimeter transitions will provide a statistical improvement in the mass and distance constraints for systems observed in multiple transitions. Furthermore, because the optimal physical conditions (e.g., gas temperature and density) for pumping maser activity differ between the different transitions, simultaneous measurements of multiple transitions in a single source may be used to provide constraints on those physical conditions [216,217]. It is also possible that future surveys will uncover populations of AGN that exhibit submillimeter maser activity but no 22 GHz emission, thereby increasing the sample of sources for which the megamaser-based measurement techniques can be applied.

2.7. Algorithms and Inference

The results produced by the EHT collaboration have been enabled by a suite of new calibration, imaging, and analysis softwares, many of which were custom-built to tackle the unique challenges associated with the sparsity and instrumental corruptions present in EHT data as well as with the rapid source evolution and scattering in Sgr A* (e.g., [138–140,225–239]). Many of the difficulties that motivated imaging developments for the EHT are expected to be compounded in ngEHT observations, with a large increase in data volume (increased bandwidth, more stations, and faster observing cadence), dimensionality (multi-frequency and multi-epoch), and requisite imaging fidelity (larger reconstructible field of view and higher imaging dynamic range). The next generation of algorithmic development is already underway, with new data processing [240,241], imaging [31,75,242], machine learning [81], and full spacetime [81,82,97,234,243] methods being designed to address the challenges and opportunities associated with ngEHT data.

To assess the scientific potential of the ngEHT, inform array design, and prompt the development of new algorithms, the ngEHT has launched a series of Analysis Challenges [34]. For each challenge, synthetic (ng)EHT datasets are generated from theoretical source models. These datasets are made available through the ngEHT Analysis Challenge website[9] and are accessible to anyone upon request. Participants then analyze the data by, e.g., reconstructing an image or fitting a model, and submit their results through the website. All submissions are evaluated with metrics quantifying, e.g., data fit quality or similarity of image reconstructions to the ground truth source model.

Challenge 1 focused on static source models of Sgr A* and M87* at 230 and 345 GHz, and was set up mainly to test the challenge process and infrastructure. Challenge 2 was more science oriented, and focused on movie reconstructions from realistic synthetic observations of Sgr A* and M87* at 86, 230, and 345 GHz. Both challenges received submissions from a broad array of reconstruction methods. Figure 11 shows two submitted movie reconstructions from Challenge 2. The M87* reconstruction shows the ngEHT's ability to reconstruct both the BH shadow and extended jet dynamics at high dynamic range, allowing detailed studies of jet launching. The Sgr A* shearing hotspot reconstruction, based on [97] and motivated by the observational results of GRAVITY Collaboration et al. [244], shows the ngEHT's ability to reconstruct rapid (intra-hour) accretion dynamics, even in moderate weather conditions at 230 GHz. In general, Roelofs et al. [34] found that standalone 345 GHz imaging of the M87* jet or Sgr A* dynamics is challenging due to severe atmospheric turbulence and optical depth effects. However, multi-frequency reconstructions showed that by utilizing information from 86 and 230 GHz, the M87* jet may be reconstructed at 345 GHz (see also [31]). Additionally, while the Sgr A* shearing hotspot orbit could be reconstructed well, variability in GRMHD simulations was found to be more challenging to reconstruct due to the more turbulent nature of the plasma.

Two additional challenges are being run. Challenge 3 focuses on polarimetric movie reconstructions, and Challenge 4 will focus on science extraction, particularly to attempt measurements of the BH photon ring and the spacetime parameters. The merit of frequency phase transfer techniques for multi-frequency imaging will also be investigated (see also [45]).

2.8. History, Philosophy, and Culture

The History, Philosophy, and Culture (HPC) SWG includes scholars from the humanities, social sciences, and sciences. HPC Key Science Goals were developed across four focus groups: Responsible Siting (Section 2.8.1), Algorithms, Inference, and Visualization (Section 2.8.2), Foundations (Section 2.8.3), and Collaborations (Section 2.8.4). We will now briefly summarize a selection of these goals that have been prioritized; for a more complete description, see HPC White Paper [52].

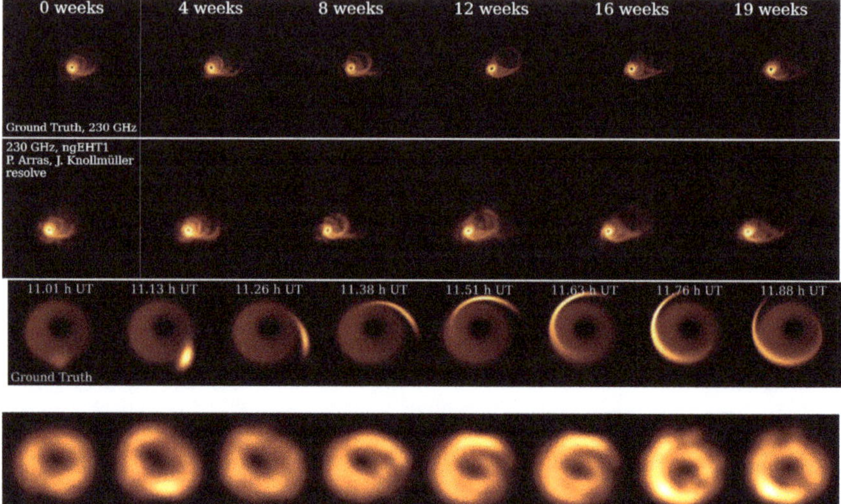

Figure 11. Example ngEHT reconstructions for Sgr A* (top two rows) and M87* (bottom two rows), using submissions for the second ngEHT Analysis Challenge [34]. For each source, upper panels show ground truth movie frames, and lower panels show example reconstructions. The M87* ground truth movie is a GRMHD simulation generated with H-AMR [245] and ray-traced with `ipole` [246]; the reconstructed movie was produced using `resolve` [225]. The Sgr A* simulation is a semi-analytic accretion flow with a shearing hot spot [97,247]; the reconstructed movie was produced using `StarWarps` [248]. Panels are reproduced from Roelofs et al. [34].

2.8.1. Responsible Siting

Telescope siting has, historically, relied almost entirely upon ensuring that sites meet technical specifications required for observation including weather, atmospheric clarity, accessibility, and cost. As the issues at Mauna kea in Hawai'i show,[10] telescopes exist within a broader context and, as they choose sites, scientific collaborations incur the obligation to address ethical, social, and environmental specifications alongside technical ones.

The ngEHT has already hosted a workshop dedicated to advancing responsible siting practices, which drew together experts in a wide range of fields including philosophy, history, sociology, advocacy, science, and engineering.[11] This workshop was run by a dedicated siting focus group within the ngEHT HPC SWG, aimed at addressing the broader impacts of constructing and operating the chosen sites, with the goal of guiding short- and long-term siting decisions. Of particular interest to the group is consultation with areas outside of astronomy which also face questions of responsible siting including biotechnology[12], archaeology and paleontology[13], physics[14], and nuclear technologies[15]. Ultimately, the goal is to model the decision-making process by joining technical, environmental, and community concerns, and to arrive at explicit guidelines that could assist with future siting challenges.

For the ngEHT to achieve its goals on responsible siting, a number of concrete steps are necessary. First, the collaboration must integrate social and environmental impacts into its siting decisions, initially, via the inclusion of ethicists, social scientists, environmental experts, and local community advocates in siting meetings who will contribute to the decision-making process as well as the inclusion of explicit cultural, social and environmental factors in siting decision metrics; later, via the creation and performance of explicit community impact studies, in addition to reviewing the environmental impact studies already performed as part of the standard siting. These studies will embrace surveys of local social factors for sites to aid in the decision process and will involve on-site community consultation as well as work with local government and academic structures.

Second, the collaboration must ensure that when telescopes are built, the building process is collaborative and non-extractive as well as sensitive to the history and culture of local communities and the lands in consideration. This goal will require establishing an ongoing dialogue with local community groups as early as possible in the siting process, and setting up explicit agreements that are mutually beneficial to all parties. As such, funding for community consultation and projects is a central part of the funding structure for the ngEHT; the aim is to ensure that local educational, scientific, and economic opportunities are built into the project from the out-set. This will involve examining local relationships with existing sites to be supplemented with new technology, as well as forging new relationships where un-developed sites are under consideration. The ngEHT project will be carefully considering who is at the table, and ensuring all local groups that may be impacted have a voice in the process. The ngEHT will also aim to work to integrate local and traditional knowledge into this process, recognizing that these are not in tension with scientific knowledge, but are continuous with it. Moreover, each site will be unique, with different needs and histories that will inform the kinds of relationships that will develop. As such, part of the community impact study will need to detail what sort of benefits local communities may want from, as well as offer to, the ngEHT collaboration. Possibilities include improved infrastructure, education funding, outreach, and knowledge exchange under terms and conditions that meet the needs of the communities in question.

The ngEHT must also accept the fact that community, environmental, and cultural aspects may prevent a site from being developed, and that a 'no' from locals is a legitimate outcome. A clear goal, then, is to work with community siting experts from both inside and outside astronomy to establish what a 'no' looks like, as well as a 'yes', and to develop norms and practices which can help survey local groups to ensure their voices are being heard.

Third, the ngEHT aims to minimize its environmental impact, including careful consideration of how construction and development of sites may impact native ecosystems as well as actively planning for what the eventual decommissioning and subsequent environmental repair of a site will look like. The ngEHT is committed, wherever possible, to using environmentally friendly techniques, technology, and materials, including in energy-efficient data-storage and computing.

Finally, a major goal of this focus group will be the production of one or more papers detailing current best-practices for responsible telescope siting. Here, the initial three to five sites (i.e., those in the ngEHT Phase-1) will be treated as proof-of-concept sites where norms can be designed and established, and experts from both inside and outside astronomy will be brought in to help guide the paper writing process.

2.8.2. Algorithms, Inference, and Visualization

The ngEHT is a long-term project which will heavily rely on software supported modes of reasoning, including imaging algorithms for image reconstruction, and GRMHD simulations and relativistic ray tracing codes for parameter extraction. Philosophers of other sciences relying on computer simulations (including climate sciences and theoretical cosmology) diagnosed that problematic features might arise in such situations. These include [266,267] (i) kludging: temporary and ad hoc choices (concerning, e.g., values of parameters, or a manner of merging together two pre-existing fragments of code) made for convenience and without principled justification; (ii) generative entrenchment: contingent choices made during code development in order to deal with problems arising in particular contexts are baked in and transferred to future versions; over time, awareness of the origin of various fragments might be lost; (iii) confirmation holism: assigning success or failure of a numerical model as a whole to a particular fragment of code becomes very hard. Some of these problematic features may have positive elements—for example, feature (i) makes code development faster than if it were properly documented; feature (ii) might represent consensus of the collaboration. Awareness of these features and development of active means of preventing their negative effects will make inference methods of the ngEHT more reliable.

Further, new inferential methods based on various forms of machine learning and artificial intelligence are becoming increasingly widespread, including in astronomy. Such methods come with many benefits, including much faster data processing times, but also with drawbacks, including a lack of epistemic transparency (the inner workings of a machine learning model are not easily available or even understood by its users, in contrast with the steps taken by a more traditional imaging or parameter extraction algorithm), and risk of building in bias through training on data sets containing untested assumption about the target system. Frameworks for mitigating these risks, so-called explainable artificial intelligence, have been developed (e.g., [268–270]). We will systematically evaluate these methods and motivations behind them, isolating those which can and which should be applied to future ngEHT data analysis pipelines.

Reception of astronomical images takes place in a broader context of visual culture, and we will consider the importance of aesthetic choices made during production, such as assignment of color to underlying physical parameters or landscape associations invoked by the resulting image (e.g., Kessler [271]; HPC White Paper [52]). As for the EHT, images produced by the ngEHT will shape public perception of black holes and astronomy. Analysis of such cultural factors will help with being intentional about the impact and perception of images—inside and outside the technical community. Accordingly, procedures for systematically including such choices and for testing whether an image succeeds in conveying the intended connotations will be developed, and applied, for example, to future polarization data and multi-frequency images.

The requirement to achieve long-term reliability of ngEHT inferences will necessitate the identification of inference methods deemed undesirable, and development of software evaluation tests to ameliorate those features. Improving image presentation will focus attention on cultural factors that shape audience reactions to visualizations—in turn, we will need to develop comprehension tests probing audience responses.

2.8.3. Foundations

The Foundations focus group complements the Fundamental Physics working group, providing a different, critical lens for thinking about what the ngEHT observations can tell us about fundamental physics. The ngEHT results will both be informed by, and inform, philosophical and historical perspectives on issues such as scientific representation and modeling, idealization, underdetermination, theory testing (confirmation), and more (Section 3, [52]). This focus group facilitates ongoing interdisciplinary discussions of foundational issues, in parallel with discussions in fundamental physics.

2.8.4. Collaborations

A fully integrated working group of scholars from the social sciences and humanities within a STEM collaboration provides an unprecedented opportunity to optimize the collaboration structure from the very beginning. Our main goal is a structure that enables, encourages, and emphasizes transparent decision-making, diversity, fair credit assignment and accountability (Section 4, [52]).[16] This translates directly into various requirements for the ngEHT collaboration, as detailed further below.

In addition, a long-term Forecasting Tournament [287,288] will clarify the ngEHT decision-making process. Participants' judgments about the outcome of ngEHT experiments and observations will reveal the novelty of eventual results and will elucidate the process of hypothesis generation and testing. By systematically collecting predictions, we will be able to track the return on testing different hypotheses, identify unresolved ambiguities within the design or implementation of an experiment (which may lead to new areas of investigation) and develop a more (cost) effective research management strategy [289]. There are also direct epistemic advantages to surveying predictions and expectations. The EHT already went to great lengths to counteract a quite natural tendency to halt image reconstruction when the images coincided with anticipated results (for example, blind trials with known, simulated data; autonomous imaging groups who did not share intermediate results; [4]). In the ngEHT, imaging programs will become more elaborate, use of AI more extensive, and data volumes will expand rapidly. Therefore it will be increasingly important for the collaboration to be aware of forecasted results—precisely to avoid premature confirmation.

These goals require frequent monitoring and evaluation of the internal communication structure and climate. This will be achieved via the complementary methods of surveys, interviews, and network analysis tools from the digital humanities (HPC White Paper [52]).[17] This will require that at least some collaboration members be available for interviews and surveys. Only against the backdrop of this ongoing feedback loop will it be possible to ensure the long-term effectiveness of the following further requirements: (a) a governance structure that includes a central, representative, elected body, as well as a standing ethics committee responsible for the creation, adherence to, and updating of the collaboration's Community Principles and Code of Conduct—see HPC White Paper [52] for a tentative proposal; (b) an authorship and membership model tailored to the needs of a modern collaboration involving members from a diverse group of (scientific and non-scientific) cultures, i.e., a model that accounts for fair distribution of credit and accountability and allows for and realizes the value of dissenting opinions.[18] A dedicated task force has begun developing such a model.

3. Summary

The ngEHT project has undergone a multi-year design process to define community-driven science priorities for the array. This process has identified breakthrough science related to studies of BH spacetimes, as well as a wealth of new opportunities beyond what has been explored with past EHT experiments. These science opportunities arise from the potential to substantially expand upon the currently explored parameter space:

- Improved angular resolution and image fidelity through increased sensitivity and baseline coverage. These enhancements are the most significant requirements for studies of fundamental physics with the ngEHT.
- Expanding from independent multi-band observations to simultaneous multi-band observations at 86, 230, and 345 GHz. This upgrade will substantially improve the EHT's sensitivity to observe faint sources, dim extended emission, and compact structure on the longest baselines at 345 GHz, especially through the use of multi-frequency phase transfer.
- Adding more sites to enable "snapshot" imaging of variable sources including Sgr A*, and extending observing campaigns over multiple years. Together, these upgrades will improve the temporal sensitivity of current EHT observations by $\sim$5 orders of magnitude, enabling a wealth of new variability studies (see Figure 2).

We have classified each of the key science goals discussed in Section 2 as either *Threshold* or *Objective*. Threshold science goals define the minimum target that the array concept is designed to meet. Objective science goals are additional major science opportunities or stretch target for the array concept to meet. This classification does not indicate the relative merit of the science objective; some goals are assigned as objective because they are considered to be too speculative or high-risk (e.g., studies of the photon ring and frame dragging), insufficiently unique to the ngEHT (e.g., studies of axions and SMBH binaries), or too poorly understood to define a precise associated instrument requirement that will guarantee success (e.g., studies of extragalactic transients). Table 1 provides the categorization of each goal. In addition, we have developed a set of homogeneous array requirements for the science goals in the framework of a Science Traceability Matrix (STM). A representative subset of the STM is given in Figure 12.

Table 1. Key Science Goals of the ngEHT.

Threshold Science Goals

- Establish the existence and properties of black hole horizons
- Measure the spin of a SMBH
- Reveal black hole-galaxy formation, growth and coevolution
- Reveal how BHs accrete material using resolved movies on event horizon scales
- Observe localized heating and acceleration of relativistic electrons on astrophysical scales
- Determine whether jets are powered by energy extraction from rotating BHs
- Determine the physical conditions and launching mechanisms for relativistic jets

Objective Science Goals

- Constrain the properties of a BH's photon ring
- Constrain ultralight boson fields
- Determine how SMBHs merge through observations of sub-parsec binaries
- Connect SMBHs to high-energy and neutrino events within their jets
- Detect frame dragging within the ergosphere of a rotating BH
- Measure the inner jet structure and dynamics in BH X-ray binaries
- Detect the kinetic power, physical structure, and velocity in extragalactic transients
- Detect proper motions and secular (CMB) parallaxes of AGN up to ~80 Mpc distances
- Leverage AGN accretion disk megamasers to measure their AGN host properties

Key Science Goal († = threshold science goal)	M87*	Sgr A*	Other SMBH	Transient	Other	Physical Parameter	Observable	Mode	Cadence	Frequency (GHz)	Freq. Phase Transfer	ngEHT Phase
Fundamental Physics												
†Establish the existence and properties of black hole horizons	X	X				Lensed image of the horizon	Measure brightness and shape of the dimmest region of the apparent shadow	Single Observation	Single campaign with full array	230+345	Yes	M87*: 1 Sgr A*: 2
†Measure the spin of a SMBH	X	X				SMBH dimensionless spin	Average polarization spiral (β phase) over 10 epochs at 230 and 345 GHz	Multiple Observations	M87*: 10 observations separated by >1 month; Sgr A*: 10 full nights with full array	230+345	Yes	2
Constrain the properties of a black hole's photon ring	X	X				$n=1$ photon ring	Statistically significant detection of persistent thin ring feature	Multiple Observations	M87*: 3 observations separated by >1 month; Sgr A*: 3 full nights with full array	230+345	Yes	M87*: 1 Sgr A*: 2
Constrain ultralight boson fields	X	X				Superradiance from clouds of sub-eV ultralight bosons	Polarization angle oscillation along the photon ring and spin measurement	Multiple Observations	M87*: 3 observations within 20 days; Sgr A*: 3 full nights with full array	230+345	No	1
Black Holes & their Cosmic Context												
†Reveal Black Hole-Galaxy Formation, Growth and Coevolution	X	X	X			SMBH masses and indirect estimates of their spins	SMBH emission ring and its polarized structure in a sample of >10 sources	Multiple Observations	One observation (~one night) per target, repeated twice	230	No	1
Determine how SMBHs merge through observations of sub-parsec binaries					X	SMBH binary orbit, masses, and (indirect) spins	SMBH spatial separation & evolution of that spatial separation	Periodic Monitoring	Several measurements taken over at least half of the orbital period (months to years)	230	No	1
Connect SMBHs to high-energy and neutrino events within their jets	X		X	X	X	Neutrinos produced in regions with PeV protons	Mapping of the jet (imaging), neutrino emission location	Multiple Observations	~Monthly observations of >20 bright blazars and those with neutrino triggers	86+230+345	Yes	1
Black Hole Accretion												
†Reveal how black holes accrete material using resolved movies on event horizon scales	X	X				Accreting plasma properties	Surface brightness and spectral index of the direct image near the photon ring	Periodic Monitoring	M87*: Every 3 days for 3 months (250GM/c³); Sgr A*: One full night at least 3 times	86+230+345; 230+345	Yes	M87*: 1 Sgr A*: 2
†Observe localized heating and acceleration of relativistic electrons on astrophysical scales	X	X				Time-dependent temp., $|B|$, and density in flaring regions	Spatially and time-resolved compact flaring structures in sub-mm movies	Periodic Monitoring	M87*: Every 3 days for 3 months (250GM/c³); Sgr A*: One full night at least 3 times	86+230+345; 230+345	Yes	M87*: 1 Sgr A*: 2
Detect frame dragging within the ergosphere of a rotating black hole	X	X				Direction of accretion flow rotation on scales of 2-10M	Radial evolution of resolved polarization structure and dynamics on scales of 2-10M	Periodic Monitoring	M87*: Every 3 days for 3 months (250GM/c³); Sgr A*: One full night at least 3 times	86+230+345; 230+345	Yes	2
Jet Launching												
†Determine whether jets are powered by energy extraction from rotating black holes	X					Magnetic flux threading BH, BH spin, and total jet power	Polarized, multi-frequency images on horizon scales and SMBH spin estimate	Multiple Observations	Every 3 days for 3 months (250GM/c³)	86+230+345	Yes	2
†Determine the physical conditions and launching mechanisms for relativistic jets	X					Jet/counter-jet composition, B-field structure, and velocity field on scales of 5-100M	Full polarization, multi-frequency movies with spectral index and rotation measure	Periodic Monitoring	Every 3 days for 3 months (250GM/c³)	86+230	No	1
		X						Multiple Observations	One full night at least 3 times	230+345	Yes	1
Transients												
Measure the inner jet structure and dynamics in black hole X-ray binaries				X	X	Jet collimation profile and velocity at 10^5-10^8 M	Motion, brightness, and size of ejected components during flares	Target of Opportunity	Triggered ~10-hr observation with 1-2 follow ups on ~days timescale. 2-4 targets per year	86+230	Yes	1
Detect the kinetic power, physical structure, and velocity in extragalactic transients				X	X	Kinetic power, structure, and velocity of transient outflows	Temporally and spatially resolved morphology of transient outflows	Target of Opportunity	Monthly observations following initial detection for 1-2 years. 2-3 targets per year	86+230	Yes	1
New Horizons												
Detect proper motions and secular (CMB) parallaxes of AGN up to ~80 Mpc distances					X	Proper motions and secular (CMB) parallaxes	Multi-year tracking of many sources across the sky with 1ps (~5 µas) delay fidelity	Multiple Observations	Multiple observations spread over >3 years per source for >10 sources	86+230	Yes	2
Leverage AGN accretion disk megamasers to measure their AGN host properties					X	SMBH masses and distances; Hubble constant	Spectral lines of megamasers	Multiple Observations	Monthly observations of ~10 sources	300-325	No	1

Figure 12. Representative subset of the ngEHT Science Traceability Matrix (STM). Daggers (†) indicate threshold science goals. The STM is used to guide the array design and to inform decisions about the multi-phase deployment.

In conclusion, the ngEHT scientific community has identified a series of science objectives, with associated observational advances that are feasible over the coming decade. Taken together, they offer a remarkable opportunity to push the frontiers of VLBI and to enable a series of new discoveries that will elucidate the extraordinary role of BHs across all astrophysical scales.

Author Contributions: Conceptualization, M.D.J.; software, A.E.B., M.D.J. and D.W.P.; writing—original draft preparation, M.D.J., L.B., V.C., R.P.F., C.M.F., P.G., J.L.G., D.H., M.L.L., A.P.L., S.M., R.N., P.N., D.W.P., Z.Y., K.C., J.D., R.D., J.E., Y.Y.K., R.L., A.M., N.C.M.M., N.M.N., D.C.M.P., A.R., M.J.R., F.R. and A.C.T.; writing—review and editing, M.D.J., L.B., V.C., R.P.F., C.M.F., P.G., J.L.G., D.H., M.L.L., A.P.L., S.M., R.N., P.N., T.N., D.W.P., Z.Y., K.C., J.D., R.D., S.S.D., J.E., G.F., S.I., Y.Y.K., R.L., A.M., N.C.M.M., N.M.N., D.C.M.P., A.R., M.J.R., F.R., A.C.T., P.T., K.A., K.L.B., A.C., R.C., K.H.,

J.H., A.O., J.W. and M.W.; visualization, M.D.J., D.W.P., D.W.P., A.L. and F.R. All authors have read and agreed to the published version of the manuscript.

Funding: The ngEHT design studies are funded by National Science Foundation grants AST-1935980 and AST-2034306 and the Gordon and Betty Moore Foundation (GBMF-10423). This work was supported by the Black Hole Initiative at Harvard University, which is funded by grants from the John Templeton Foundation and the Gordon and Betty Moore Foundation to Harvard University. This work was supported by Volkswagen Foundation, VILLUM Foundation (grant no. VIL37766) and the DNRF Chair program (grant no. DNRF162) by the Danish National Research Foundation. We acknowledge financial support provided under the European Union's H2020 ERC Advanced Grant "Black holes: gravitational engines of discovery" grant agreement no. Gravitas–101052587. NCMM acknowledges support from the European Union's Horizon Europe research and innovation programme for the funding received under the Marie Skłodowska-Curie grant agreement No. 101065772 (PhilDarkEnergy) and the ERC Starting Grant agreement No. 101076402 (COSMO-MASTER). Views and opinions expressed are however those of the author only and do not necessarily reflect those of the European Union or the European Research Council. Neither the European Union nor the granting authority can be held responsible for them. NN acknowledges funding from TITANs NCN19-058 and Fondecyt 1221421.

Data Availability Statement: Not applicable.

Acknowledgments: We are pleased to acknowledge the hundreds of scientists and engineers worldwide who have contributed to the ngEHT science case and have helped define the associated instrument requirements. In particular, we are grateful to the panelists of the ngEHT Science Requirements Review: Jim Moran, Mariafelicia De Laurentis, Eric Murphy, Geoff Bower, Katherine Blundell, and Randy Iliff. The detailed feedback from this review substantially sharpened the science goals and informed the specification of threshold and objective requirements. We also thank the EHTC internal referee, Laurent Loinard, the MPIfR internal referee, Eduardo Ros, and the anonymous referees for their detailed feedback and suggestions, which significantly improved the manuscript.

Conflicts of Interest: The authors declare no conflict of interest.

Notes

[1] Since its first observing campaign, three sites have joined the EHT (see Figure 1). These additions are expected to substantially improve upon the dynamic range of published EHT images.

[2] In contrast, most telescopes of the present EHT are astronomical facilities that only commit a small fraction of their total observing time to VLBI.

[3] https://www.ngeht.org/ngeht-meeting-2021 (accessed on 20 April 2023).

[4] https://www.ngeht.org/ngeht-meeting-november-2021 (accessed on 20 April 2023).

[5] https://www.ngeht.org/ngeht-meeting-june-2022 (accessed on 20 April 2023).

[6] https://www.ngeht.org/broadening-horizons-2022 (accessed on 20 April 2023).

[7] https://www.mdpi.com/journal/galaxies/special_issues/ngEHT_blackholes (accessed on 20 April 2023).

[8] Angular momentum transport may also occur in magnetic flux eruptions (see, e.g., [151]), which would also have distinctive signatures in ngEHT images and movies (see, e.g., [102,152,153]).

[9] https://challenge.ngeht.org/ (accessed on 20 April 2023).

[10] Two excellent doctoral dissertations offer fine-grained analysis of the mountaintop dispute, and are a good entry point into this issue. Ref. [249] focuses on the triply conflicting astronomical, environmental and indigenous narratives that collided at Mt. Graham, Mauna Kea, and Kitt Peak; Ref. [250] addresses the Kanaka rights claim, specifically about the Thirty Meter Telescope (TMT), in opposition to a framing of the dispute as one of "stakeholders" or a "multicultural" ideal. Ref. [251] focuses on Mauna Kea in a subsequent article, also on the TMT. An important current Hawaiian-led impact assessment of the TMT, including further links, is [252]; other Native Hawaiian scientists, including [253] have spoken for a much-changed process and against the notion that opposition to the TMT is against science.

[11] The workshop was held on the 4th of November 2022. Workshop Speakers included C. Prescod-Weinstein, K. Kamelamela, H. Nielson, M. Johnson, J. Havstad, T. Nichols, R. Chiaravalloti, S. Doeleman, G. Fitzpatrick, J. Houston, A. Oppenheimer, P. Galison, A. Thresher and P. Natarajan. Much of the work being performed by the responsible siting group owes its genesis in the excellent contributions of the speakers and attendees of the workshop and we are grateful for their past and ongoing contributions.

[12] For a detailed discussion of siting and community guidelines for gene-drive technology, for example, see Singh [254].

[13] There is much discussion within these fields of how we ought to think about community-led and non-extractive science. Good starting places for the literature include Watkins [255], Supernant and Warrick [256].

[14] An outstanding example of joint concern crossing environmental, cultural, epistemic, and technical concerns, in the case of LIGO, can be found in Nichols [257]. Another instanced of community participation by (here in relation to NASA for their Asteroid Redirect Mission): Tomblin et al. [258]. On the siting of the Superconducting Supercollider, Riordan et al. [259]; an historical-anthropological study of the placement of the French/European launch center, Redfield [260].

[15] Consent, and environmental justice, have been at the center of siting nuclear facilities, including power generation, weapons testing, accident sites, and waste disposal. The literature is vast, but a starting point with many further references can be found in sources including: Gerrard [261] addresses community concerns about siting from the perspective on an environmental lawyer; Kuletz [262] focuses on Western US nuclear sites of waste; Masco [263] attends to the quadruple intersection of weapons scientists, Pueblo Indian nations, nuevomexicano communities, and activists as they live amidst and confront the legacy of Los Alamos. On consent-based siting rather than top-down imposition, see Hamilton et al. [264]; and for a recent development and analysis of consent-based siting, Richter et al. [265].

[16] For lessons learnt regarding knowledge formation, governance, organisational structure, decision-making, diversity, accountability, creativity, credit assignment and the role of consensus, from a range of perspectives across the humanities and social sciences, see e.g., (a) in general: Galison and Hevly [272], Knorr Cetina [273], Sullivan [274], Shrum et al. [275], Boyer-Kassem et al. [276] and references therein; (b) for specific collaborations and institutions: Collins [277], Nichols [278] on LIGO; Boisot et al. [279], Ritson [280], Sorgner [281], Merz and Sorgner [282] on ATLAS and/or CERN; Jebeile [283] on the IPCC; Smith et al. [284], Vertesi [285] on NASA; and Traweek [286] on SLAC and KEK.

[17] Regarding network analysis, communication structures and epistemic communities, see for instance the following texts and references therein: Kitcher [290,291], Zollman [292,293,294], Longino [295], Lalli et al. [296,297], Light and Moody [298], Wüthrich [299], Šešelja [300].

[18] Regarding authorship challenges and possible solutions relevant to the ngEHT context, see e.g., Resnik [301], Boyer-Kassem et al. [276], Rennie et al. [302], Cronin [303], Galison [304], Wray [305], McNutt et al. [306], Bright et al. [307], Heesen [308], Dang [309], Nogrady [310], Habgood-Coote [311] and www.icmje.org/icmje-recommendations.pdf (accessed on 20 April 2023).

References

1. Akiyama, K. et al. [Event Horizon Telescope Collaboration]. First M87 Event Horizon Telescope Results. I. The Shadow of the Supermassive Black Hole. *Astrophys. J. Lett.* **2019**, *875*, L1. [CrossRef]
2. Akiyama, K. et al. [Event Horizon Telescope Collaboration]. First M87 Event Horizon Telescope Results. II. Array and Instrumentation. *Astrophys. J. Lett.* **2019**, *875*, L2. [CrossRef]
3. Akiyama, K. et al. [Event Horizon Telescope Collaboration]. First M87 Event Horizon Telescope Results. III. Data Processing and Calibration. *Astrophys. J. Lett.* **2019**, *875*, L3. [CrossRef]
4. Akiyama, K. et al. [Event Horizon Telescope Collaboration]. First M87 Event Horizon Telescope Results. IV. Imaging the Central Supermassive Black Hole. *Astrophys. J. Lett.* **2019**, *875*, L4. [CrossRef]
5. Akiyama, K. et al. [Event Horizon Telescope Collaboration]. First M87 Event Horizon Telescope Results. V. Physical Origin of the Asymmetric Ring. *Astrophys. J. Lett.* **2019**, *875*, L5. [CrossRef]
6. Akiyama, K. et al. [Event Horizon Telescope Collaboration]. First M87 Event Horizon Telescope Results. VI. The Shadow and Mass of the Central Black Hole. *Astrophys. J. Lett.* **2019**, *875*, L6. [CrossRef]
7. Akiyama, K. et al. [Event Horizon Telescope Collaboration]. First M87 Event Horizon Telescope Results. VII. Polarization of the Ring. *Astrophys. J. Lett.* **2021**, *910*, L12. [CrossRef]
8. Akiyama, K. et al. [Event Horizon Telescope Collaboration]. First M87 Event Horizon Telescope Results. VIII. Magnetic Field Structure near The Event Horizon. *Astrophys. J. Lett.* **2021**, *910*, L13. [CrossRef]
9. Akiyama, K. et al. [Event Horizon Telescope Collaboration]. First Sagittarius A* Event Horizon Telescope Results. I. The Shadow of the Supermassive Black Hole in the Center of the Milky Way. *Astrophys. J. Lett.* **2022**, *930*, L12. [CrossRef]
10. Akiyama, K. et al. [Event Horizon Telescope Collaboration]. First Sagittarius A* Event Horizon Telescope Results. II. EHT and Multiwavelength Observations, Data Processing, and Calibration. *Astrophys. J. Lett.* **2022**, *930*, L13. [CrossRef]
11. Akiyama, K. et al. [Event Horizon Telescope Collaboration]. First Sagittarius A* Event Horizon Telescope Results. III. Imaging of the Galactic Center Supermassive Black Hole. *Astrophys. J. Lett.* **2022**, *930*, L14. [CrossRef]
12. Akiyama, K. et al. [Event Horizon Telescope Collaboration]. First Sagittarius A* Event Horizon Telescope Results. IV. Variability, Morphology, and Black Hole Mass. *Astrophys. J. Lett.* **2022**, *930*, L15. . [CrossRef]
13. Akiyama, K. et al. [Event Horizon Telescope Collaboration]. First Sagittarius A* Event Horizon Telescope Results. V. Testing Astrophysical Models of the Galactic Center Black Hole. *Astrophys. J. Lett.* **2022**, *930*, L16. [CrossRef]
14. Akiyama, K. et al. [Event Horizon Telescope Collaboration]. First Sagittarius A* Event Horizon Telescope Results. VI. Testing the Black Hole Metric. *Astrophys. J. Lett.* **2022**, *930*, L17. [CrossRef]
15. Algaba, J.C. et al. [EHT MWL Science Working Group]. Broadband Multi-wavelength Properties of M87 during the 2017 Event Horizon Telescope Campaign. *Astrophys. J. Lett.* **2021**, *911*, L11. [CrossRef]

16. Kim, J.Y.; Krichbaum, T.P.; Broderick, A.E.; Wielgus, M.; Blackburn, L.; Gómez, J.L.; Johnson, M.D.; Bouman, K.L.; Chael, A.; Akiyama, K.; et al. Event Horizon Telescope imaging of the archetypal blazar 3C 279 at an extreme 20 microarcsecond resolution. *Astron. Astrophys.* **2020**, *640*, A69. [CrossRef]
17. Janssen, M.; Falcke, H.; Kadler, M.; Ros, E.; Wielgus, M.; Akiyama, K.; Baloković, M.; Blackburn, L.; Bouman, K.L.; Chael, A.; et al. Event Horizon Telescope observations of the jet launching and collimation in Centaurus A. *Nat. Astron.* **2021**, *5*, 1017–1028. [CrossRef]
18. Issaoun, S.; Wielgus, M.; Jorstad, S.; Krichbaum, T.P.; Blackburn, L.; Janssen, M.; Chan, C.k.; Pesce, D.W.; Gómez, J.L.; Akiyama, K.; et al. Resolving the Inner Parsec of the Blazar J1924-2914 with the Event Horizon Telescope. *Astrophys. J. Suppl.* **2022**, *934*, 145. [CrossRef]
19. Jorstad, S.; Wielgus, M.; Lico, R.; Issaoun, S.; Broderick, A.E.; Pesce, D.W.; Liu, J.; Zhao, G.Y.; Krichbaum, T.P.; Blackburn, L.; et al. The Event Horizon Telescope Image of the Quasar NRAO 530. *Astrophys. J. Suppl.* **2023**, *943*, 170. [CrossRef]
20. Thompson, A.R.; Moran, J.M.; Swenson, G.W. *Interferometry and Synthesis in Radio Astronomy*, 3rd ed.; Springer: Berlin/Heidelberg, Germany, 2017. [CrossRef]
21. Padin, S.; Woody, D.P.; Hodges, M.W.; Rogers, A.E.E.; Emerson, D.T.; Jewell, P.R.; Lamb, J.; Perfetto, A.; Wright, M.C.H. 223 GHz VLBI Observations of 3C 273. *Astrophys. J. Lett.* **1990**, *360*, L11. [CrossRef]
22. Krichbaum, T.P.; Graham, D.A.; Witzel, A.; Greve, A.; Wink, J.E.; Grewing, M.; Colomer, F.; de Vicente, P.; Gomez-Gonzalez, J.; Baudry, A.; et al. VLBI observations of the galactic center source SGR A* at 86 GHz and 215 GHz. *Astron. Astrophys.* **1998**, *335*, L106–L110.
23. Doeleman, S.S.; Weintroub, J.; Rogers, A.E.E.; Plambeck, R.; Freund, R.; Tilanus, R.P.J.; Friberg, P.; Ziurys, L.M.; Moran, J.M.; Corey, B.; et al. Event-horizon-scale structure in the supermassive black hole candidate at the Galactic Centre. *Nature* **2008**, *455*, 78–80. [CrossRef]
24. Boccardi, B.; Krichbaum, T.P.; Ros, E.; Zensus, J.A. Radio observations of active galactic nuclei with mm-VLBI. *Astron. Astrophys. Rev.* **2017**, *25*, 4. [CrossRef]
25. Bardeen, J.M. Timelike and null geodesics in the Kerr metric. In *Black Holes (Les Astres Occlus)*; Gordon and Breach: New York, NY, USA, 1973; pp. 215–239.
26. Luminet, J.P. Image of a spherical black hole with thin accretion disk. *Astron. Astrophys.* **1979**, *75*, 228–235.
27. Falcke, H.; Melia, F.; Agol, E. Viewing the Shadow of the Black Hole at the Galactic Center. *Astrophys. J. Lett.* **2000**, *528*, L13–L16. [CrossRef] [PubMed]
28. de Vries, A. The apparent shape of a rotating charged black hole, closed photon orbits and the bifurcation set A_4. *Class. Quant. Gravity* **2000**, *17*, 123–144. [CrossRef]
29. Mościbrodzka, M.; Dexter, J.; Davelaar, J.; Falcke, H. Faraday rotation in GRMHD simulations of the jet launching zone of M87. *Mon. Not. RAS* **2017**, *468*, 2214–2221. [CrossRef]
30. Ricarte, A.; Prather, B.S.; Wong, G.N.; Narayan, R.; Gammie, C.; Johnson, M.D. Decomposing the internal faraday rotation of black hole accretion flows. *Mon. Not. RAS* **2020**, *498*, 5468–5488. [CrossRef]
31. Chael, A.; Issaoun, S.; Pesce, D.W.; Johnson, M.D.; Ricarte, A.; Fromm, C.M.; Mizuno, Y. Multi-frequency Black Hole Imaging for the Next-Generation Event Horizon Telescope. *arXiv* **2022**, arXiv:2210.12226.
32. Ricarte, A.; Johnson, M.D.; Kovalev, Y.Y.; Palumbo, D.C.M.; Emami, R. How Spatially Resolved Polarimetry Informs Black Hole Accretion Flow Models. *Galaxies* **2023**, *11*, 5. [CrossRef]
33. Crew, G.B.; Goddi, C.; Matthews, L.D.; Rottmann, H.; Saez, A.; Martí-Vidal, I. A Characterization of the ALMA Phasing System at 345 GHz. *Publ. ASP* **2023**, *135*, 025002. [CrossRef]
34. Roelofs, F.; Blackburn, L.; Lindahl, G.; Doeleman, S.S.; Johnson, M.D.; Arras, P.; Chatterjee, K.; Emami, R.; Fromm, C.; Fuentes, A.; et al. The ngEHT Analysis Challenges. *Galaxies* **2023**, *11*, 12. [CrossRef]
35. Walker, R.C.; Hardee, P.E.; Davies, F.B.; Ly, C.; Junor, W. The Structure and Dynamics of the Subparsec Jet in M87 Based on 50 VLBA Observations over 17 Years at 43 GHz. *Astrophys. J.* **2018**, *855*, 128. [CrossRef]
36. Wielgus, M.; Akiyama, K.; Blackburn, L.; Chan, C.K.; Dexter, J.; Doeleman, S.S.; Fish, V.L.; Issaoun, S.; Johnson, M.D.; Krichbaum, T.P.; et al. Monitoring the Morphology of M87* in 2009–2017 with the Event Horizon Telescope. *Astrophys. J.* **2020**, *901*, 67. [CrossRef]
37. Raymond, A.W.; Palumbo, D.; Paine, S.N.; Blackburn, L.; Córdova Rosado, R.; Doeleman, S.S.; Farah, J.R.; Johnson, M.D.; Roelofs, F.; Tilanus, R.P.J.; et al. Evaluation of New Submillimeter VLBI Sites for the Event Horizon Telescope. *Astrophys. J. Suppl.* **2021**, *253*, 5. [CrossRef]
38. Bustamante, S.; Blackburn, L.; Narayanan, G.; Schloerb, F.P.; Hughes, D. The Role of the Large Millimeter Telescope in Black Hole Science with the Next-Generation Event Horizon Telescope. *Galaxies* **2023**, *11*, 2. [CrossRef]
39. Yu, W.; Lu, R.S.; Shen, Z.Q.; Weintroub, J. Evaluation of a Candidate Site in the Tibetan Plateau towards the Next Generation Event Horizon Telescope. *Galaxies* **2023**, *11*, 7. [CrossRef]
40. Akiyama, K.; Kauffmann, J.; Matthews, L.D.; Moriyama, K.; Koyama, S.; Hada, K. Millimeter/Submillimeter VLBI with a Next Generation Large Radio Telescope in the Atacama Desert. *Galaxies* **2023**, *11*, 1. [CrossRef]
41. Kauffmann, J.; Rajagopalan, G.; Akiyama, K.; Fish, V.; Lonsdale, C.; Matthews, L.D.; Pillai, T.G. The Haystack Telescope as an Astronomical Instrument. *Galaxies* **2023**, *11*, 9. [CrossRef]

42. Asada, K.; Kino, M.; Honma, M.; Hirota, T.; Lu, R.S.; Inoue, M.; Sohn, B.W.; Shen, Z.Q.; Ho, P.T.P.; Akiyama, K.; et al. White Paper on East Asian Vision for mm/submm VLBI: Toward Black Hole Astrophysics down to Angular Resolution of 1_S. *arXiv* **2017**, arXiv:1705.04776.
43. Backes, M.; Müller, C.; Conway, J.E.; Deane, R.; Evans, R.; Falcke, H.; Fraga-Encinas, R.; Goddi, C.; Klein Wolt, M.; Krichbaum, T.P.; et al. The Africa Millimetre Telescope. In Proceedings of the 4th Annual Conference on High Energy Astrophysics in Southern Africa (HEASA 2016), Cape Town, South Africa, 25–27 August 2016; p. 29. [CrossRef]
44. Romero, G.E. Large Latin American Millimeter Array. *arXiv* **2020**, arXiv:2010.00738.
45. Issaoun, S.; Pesce, D.W.; Roelofs, F.; Chael, A.; Dodson, R.; Rioja, M.J.; Akiyama, K.; Aran, R.; Blackburn, L.; Doeleman, S.S.; et al. Enabling Transformational ngEHT Science via the Inclusion of 86 GHz Capabilities. *Galaxies* **2023**, *11*, 28. [CrossRef]
46. Rioja, M.J.; Dodson, R.; Asaki, Y. The Transformational Power of Frequency Phase Transfer Methods for ngEHT. *Galaxies* **2023**, *11*, 16. [CrossRef]
47. Jiang, W.; Zhao, G.Y.; Shen, Z.Q.; Rioja, M.J.; Dodson, R.; Cho, I.; Zhao, S.S.; Eubanks, M.; Lu, R.S. Applications of the Source-Frequency Phase-Referencing Technique for ngEHT Observations. *Galaxies* **2023**, *11*, 3. [CrossRef]
48. Doeleman, S.; Blackburn, L.; Dexter, J.; Gomez, J.L.; Johnson, M.D.; Palumbo, D.C.; Weintroub, J.; Farah, J.R.; Fish, V.; Loinard, L.; et al. Studying Black Holes on Horizon Scales with VLBI Ground Arrays. *Bull. Am. Astron. Soc.* **2019**, *51*, 256.
49. Inoue, M.; Algaba-Marcos, J.C.; Asada, K.; Blundell, R.; Brisken, W.; Burgos, R.; Chang, C.C.; Chen, M.T.; Doeleman, S.S.; Fish, V.; et al. Greenland telescope project: Direct confirmation of black hole with sub-millimeter VLBI. *Radio Sci.* **2014**, *49*, 564–571. [CrossRef]
50. Doeleman, S. et al. [ngEHT Collaboration]. Reference Array and Design Consideration for the next-generation Event Horizon Telescope. *Galaxies* **2023**, *in prepration*.
51. Selina, R.J.; Murphy, E.J.; McKinnon, M.; Beasley, A.; Butler, B.; Carilli, C.; Clark, B.; Durand, S.; Erickson, A.; Grammer, W.; et al. The ngVLA Reference Design. In *Science with a Next Generation Very Large Array*; Murphy, E., Ed.; Astronomical Society of the Pacific Conference Series; NASA/ADS: Cambridge, MA, USA, 2018; Volume 517, p. 15,
52. Galison, P.; Doboszewski, J.; Elder, J.; Martens, N.C.M.; Ashtekar, A.; Enander, J.; Gueguen, M.; Kessler, E.A.; Lalli, R.; Lesourd, M.; et al. The Next Generation Event Horizon Telescope Collaboration: History, Philosophy, and Culture. *Galaxies* **2023**, *11*, 32. [CrossRef]
53. Penrose, R. Gravitational Collapse: the Role of General Relativity. *Nuovo Cim. Riv. Ser.* **1969**, *1*, 252.
54. Shakura, N.I.; Sunyaev, R.A. Black holes in binary systems. Observational appearance. *Astron. Astrophys.* **1973**, *24*, 337–355.
55. Blandford, R.D.; Znajek, R.L. Electromagnetic extraction of energy from Kerr black holes. *Mon. Not. RAS* **1977**, *179*, 433–456. [CrossRef]
56. Yuan, F.; Narayan, R. Hot Accretion Flows Around Black Holes. *Ann. Rev. Astron. Astrophys.* **2014**, *52*, 529–588. [CrossRef]
57. Harlow, D. Jerusalem lectures on black holes and quantum information. *Revi. Modern Phys.* **2016**, *88*, 015002. [CrossRef]
58. Senovilla, J.M.M.; Garfinkle, D. The 1965 Penrose singularity theorem. *Class. Quant. Gravity* **2015**, *32*, 124008. [CrossRef]
59. Chael, A.; Johnson, M.D.; Lupsasca, A. Observing the Inner Shadow of a Black Hole: A Direct View of the Event Horizon. *Astrophys. J.* **2021**, *918*, 6. [CrossRef]
60. Johnson, M.D.; Lupsasca, A.; Strominger, A.; Wong, G.N.; Hadar, S.; Kapec, D.; Narayan, R.; Chael, A.; Gammie, C.F.; Galison, P.; et al. Universal interferometric signatures of a black hole's photon ring. *Sci. Adv.* **2020**, *6*, eaaz1310. [CrossRef]
61. Ayzenberg, D.; Brito, R.; Britzen, S.; Broderick, A.E.; Carballo-Rubio, R.; Cardoso, V.; Chael, A.; Chen, Y.; Cunha, P.V.P.; Eichhorn, A.; et al. Fundamental Physics Opportunities with the Next-Generation Event Horizon Telescope. 2023. Available online: https://www.ngeht.org/hpc (accessed on 20 April 2023).
62. Carballo-Rubio, R.; Di Filippo, F.; Liberati, S.; Visser, M. Phenomenological aspects of black holes beyond general relativity. *Phys. Rev. D* **2018**, *98*, 124009. [CrossRef]
63. Cardoso, V.; Pani, P. Testing the nature of dark compact objects: A status report. *Liv. Rev. Relat.* **2019**, *22*, 4. [CrossRef]
64. Jaroszynski, M.; Kurpiewski, A. Optics near Kerr black holes: spectra of advection dominated accretion flows. *Astron. Astrophys.* **1997**, *326*, 419–426. [CrossRef]
65. Narayan, R.; Johnson, M.D.; Gammie, C.F. The Shadow of a Spherically Accreting Black Hole. *Astrophys. J. Lett.* **2019**, *885*, L33. [CrossRef]
66. Younsi, Z.; Psaltis, D.; Özel, F. Black Hole Images as Tests of General Relativity: Effects of Spacetime Geometry. *Astrophys. J.* **2023**, *942*, 47. [CrossRef]
67. Dokuchaev, V.I.; Nazarova, N.O. Event Horizon Image within Black Hole Shadow. *Sov. J. Exp. Theor. Phys.* **2019**, *128*, 578–585. [CrossRef]
68. Psaltis, D.; Medeiros, L.; Christian, P.; Özel, F.; Akiyama, K.; Alberdi, A.; Alef, W.; Asada, K.; Azulay, R.; Ball, D.; et al. Gravitational Test beyond the First Post-Newtonian Order with the Shadow of the M87 Black Hole. *Phys. Rev. Lett.* **2020**, *125*, 141104. [CrossRef]
69. Kocherlakota, P.; Rezzolla, L.; Falcke, H.; Fromm, C.M.; Kramer, M.; Mizuno, Y.; Nathanail, A.; Olivares, H.; Younsi, Z.; Akiyama, K.; et al. Constraints on black-hole charges with the 2017 EHT observations of M87*. *Phys. Rev. D* **2021**, *103*, 104047. [CrossRef]
70. Vincent, F.H.; Wielgus, M.; Abramowicz, M.A.; Gourgoulhon, E.; Lasota, J.P.; Paumard, T.; Perrin, G. Geometric modeling of M87* as a Kerr black hole or a non-Kerr compact object. *Astron. Astrophys.* **2021**, *646*, A37. [CrossRef]
71. Carballo-Rubio, R.; Cardoso, V.; Younsi, Z. Toward very large baseline interferometry observations of black hole structure. *Phys. Rev. D* **2022**, *106*, 084038. [CrossRef]

72. Abuter, R. et al. [GRAVITY Collaboration]. Mass distribution in the Galactic Center based on interferometric astrometry of multiple stellar orbits. *Astron. Astrophys.* **2022**, *657*, L12. [CrossRef]
73. Do, T.; Hees, A.; Ghez, A.; Martinez, G.D.; Chu, D.S.; Jia, S.; Sakai, S.; Lu, J.R.; Gautam, A.K.; O'Neil, K.K.; et al. Relativistic redshift of the star S0-2 orbiting the Galactic Center supermassive black hole. *Science* **2019**, *365*, 664–668. [CrossRef] [PubMed]
74. Takahashi, R. Shapes and Positions of Black Hole Shadows in Accretion Disks and Spin Parameters of Black Holes. *Astrophys. J.* **2004**, *611*, 996–1004. [CrossRef]
75. Tiede, P. Comrade: Composable Modeling of Radio Emission. *J. Open Source Softw.* **2022**, *7*, 4457. [CrossRef]
76. Robinson, D.C. Uniqueness of the Kerr Black Hole. *Phys. Rev. Lett.* **1975**, *34*, 905–906. . [CrossRef]
77. Gibbons, G.W. Vacuum polarization and the spontaneous loss of charge by black holes. *Commun. Math. Phys.* **1975**, *44*, 245–264. [CrossRef]
78. Reynolds, C.S. Observational Constraints on Black Hole Spin. *Ann. Rev. Astron. Astrophys.* **2021**, *59*, 117–154. [CrossRef]
79. Ricarte, A.; Tiede, P.; Emami, R.; Tamar, A.; Natarajan, P. The ngEHT's Role in Measuring Supermassive Black Hole Spins. *Galaxies* **2022**, *11*, 6. [CrossRef]
80. Palumbo, D.C.M.; Wong, G.N.; Prather, B.S. Discriminating Accretion States via Rotational Symmetry in Simulated Polarimetric Images of M87. *Astrophys. J.* **2020**, *894*, 156. [CrossRef]
81. Qiu, R.; Ricarte, A.; Narayan, R.; Wong, G.N.; Chael, A.; Palumbo, D. Using Machine Learning to Link Black Hole Accretion Flows with Spatially Resolved Polarimetric Observables. *arXiv* **2022**, arXiv:2212.04852.
82. Palumbo, D.C.M.; Gelles, Z.; Tiede, P.; Chang, D.O.; Pesce, D.W.; Chael, A.; Johnson, M.D. Bayesian Accretion Modeling: Axisymmetric Equatorial Emission in the Kerr Spacetime. *Astrophys. J.* **2022**, *939*, 107. [CrossRef]
83. Jiménez-Rosales, A.; Dexter, J.; Ressler, S.M.; Tchekhovskoy, A.; Bauböck, M.; Dallilar, Y.; de Zeeuw, P.T.; Drescher, A.; Eisenhauer, F.; von Fellenberg, S.; et al. Relative depolarization of the black hole photon ring in GRMHD models of Sgr A* and M87*. *Mon. Not. RAS* **2021**, *503*, 4563–4575. [CrossRef]
84. Palumbo, D.C.M.; Wong, G.N. Photon Ring Symmetries in Simulated Linear Polarization Images of Messier 87*. *Astrophys. J.* **2022**, *929*, 49. [CrossRef]
85. Palumbo, D.P. Spin Signatures of Rotating Black Holes. 2023.
86. Teo, E. Spherical Photon Orbits Around a Kerr Black Hole. *Gener. Relat. Gravit.* **2003**, *35*, 1909–1926. .:1026286607562. [CrossRef]
87. Tiede, P.; Johnson, M.D.; Pesce, D.W.; Palumbo, D.C.M.; Chang, D.O.; Galison, P. Measuring Photon Rings with the ngEHT. *Galaxies* **2022**, *10*, 111. [CrossRef]
88. Johannsen, T.; Psaltis, D. Testing the No-hair Theorem with Observations in the Electromagnetic Spectrum. II. Black Hole Images. *Astrophys. J.* **2010**, *718*, 446–454. [CrossRef]
89. Wielgus, M. Photon rings of spherically symmetric black holes and robust tests of non-Kerr metrics. *Phys. Rev. D* **2021**, *104*, 124058. [CrossRef]
90. Broderick, A.E.; Tiede, P.; Pesce, D.W.; Gold, R. Measuring Spin from Relative Photon-ring Sizes. *Astrophys. J.* **2022**, *927*, 6. [CrossRef]
91. Psaltis, D.; Johnson, M.; Narayan, R.; Medeiros, L.; Blackburn, L.; Bower, G. A Model for Anisotropic Interstellar Scattering and its Application to Sgr A*. *arXiv* **2018**, arXiv:1805.01242.
92. Johnson, M.D.; Narayan, R.; Psaltis, D.; Blackburn, L.; Kovalev, Y.Y.; Gwinn, C.R.; Zhao, G.Y.; Bower, G.C.; Moran, J.M.; Kino, M.; et al. The Scattering and Intrinsic Structure of Sagittarius A* at Radio Wavelengths. *Astrophys. J.* **2018**, *865*, 104. [CrossRef]
93. Issaoun, S.; Johnson, M.D.; Blackburn, L.; Brinkerink, C.D.; Mościbrodzka, M.; Chael, A.; Goddi, C.; Martí-Vidal, I.; Wagner, J.; Doeleman, S.S.; et al. The Size, Shape, and Scattering of Sagittarius A* at 86 GHz: First VLBI with ALMA. *Astrophys. J.* **2019**, *871*, 30. [CrossRef]
94. Zhu, Z.; Johnson, M.D.; Narayan, R. Testing General Relativity with the Black Hole Shadow Size and Asymmetry of Sagittarius A*: Limitations from Interstellar Scattering. *Astrophys. J.* **2019**, *870*, 6. [CrossRef]
95. Broderick, A.E.; Loeb, A. Imaging optically-thin hotspots near the black hole horizon of Sgr A* at radio and near-infrared wavelengths. *Mon. Not. RAS* **2006**, *367*, 905–916. [CrossRef]
96. Moriyama, K.; Mineshige, S.; Honma, M.; Akiyama, K. Black Hole Spin Measurement Based on Time-domain VLBI Observations of Infalling Gas Clouds. *Astrophys. J.* **2019**, *887*, 227. [CrossRef]
97. Tiede, P.; Pu, H.Y.; Broderick, A.E.; Gold, R.; Karami, M.; Preciado-López, J.A. Spacetime Tomography Using the Event Horizon Telescope. *Astrophys. J.* **2020**, *892*, 132. [CrossRef]
98. Chesler, P.M.; Blackburn, L.; Doeleman, S.S.; Johnson, M.D.; Moran, J.M.; Narayan, R.; Wielgus, M. Light echos and coherent autocorrelations in a black hole spacetime. *Class. Quant. Gravity* **2021**, *38*, 125006. [CrossRef]
99. Hadar, S.; Johnson, M.D.; Lupsasca, A.; Wong, G.N. Photon ring autocorrelations. *Phys. Rev. D* **2021**, *103*, 104038. [CrossRef]
100. Wong, G.N. Black Hole Glimmer Signatures of Mass, Spin, and Inclination. *Astrophys. J.* **2021**, *909*, 217. [CrossRef]
101. Gelles, Z.; Himwich, E.; Johnson, M.D.; Palumbo, D.C.M. Polarized image of equatorial emission in the Kerr geometry. *Phys. Rev. D* **2021**, *104*, 044060. [CrossRef]
102. Gelles, Z.; Chatterjee, K.; Johnson, M.; Ripperda, B.; Liska, M. Relativistic Signatures of Flux Eruption Events near Black Holes. *Galaxies* **2022**, *10*, 107. [CrossRef]
103. Wielgus, M.; Moscibrodzka, M.; Vos, J.; Gelles, Z.; Martí-Vidal, I.; Farah, J.; Marchili, N.; Goddi, C.; Messias, H. Orbital motion near Sagittarius A* . Constraints from polarimetric ALMA observations. *Astron. Astrophys.* **2022**, *665*, L6. [CrossRef]

104. Vos, J.; Mościbrodzka, M.A.; Wielgus, M. Polarimetric signatures of hot spots in black hole accretion flows. *Astron. Astrophys.* **2022**, *668*, A185. [CrossRef]

105. Emami, R.; Tiede, P.; Doeleman, S.S.; Roelofs, F.; Wielgus, M.; Blackburn, L.; Liska, M.; Chatterjee, K.; Ripperda, B.; Fuentes, A.; et al. Tracing Hot Spot Motion in Sagittarius A* Using the Next-Generation Event Horizon Telescope (ngEHT). *Galaxies* **2023**, *11*, 23. [CrossRef]

106. Peccei, R.D.; Quinn, H.R. Constraints imposed by CP conservation in the presence of pseudoparticles. *Phys. Rev. D* **1977**, *16*, 1791–1797. [CrossRef]

107. Preskill, J.; Wise, M.B.; Wilczek, F. Cosmology of the invisible axion. *Phys. Lett. B* **1983**, *120*, 127–132. [CrossRef]

108. Abbott, L.F.; Sikivie, P. A cosmological bound on the invisible axion. *Phys. Lett. B* **1983**, *120*, 133–136. [CrossRef]

109. Dine, M.; Fischler, W. The not-so-harmless axion. *Phys. Lett. B* **1983**, *120*, 137–141. [CrossRef]

110. Arvanitaki, A.; Dimopoulos, S.; Dubovsky, S.; Kaloper, N.; March-Russell, J. String Axiverse. *Phys. Rev. D* **2010**, *81*, 123530. [CrossRef]

111. Brito, R.; Cardoso, V.; Pani, P. Superradiance: New Frontiers in Black Hole Physics. *Lect. Notes Phys.* **2015**, *906*, 1–237. [CrossRef]

112. Hu, W.; Barkana, R.; Gruzinov, A. Cold and fuzzy dark matter. *Phys. Rev. Lett.* **2000**, *85*, 1158–1161. [CrossRef] [PubMed]

113. Davoudiasl, H.; Denton, P.B. Ultralight Boson Dark Matter and Event Horizon Telescope Observations of M 87*. *Phys. Rev. Lett.* **2019**, *123*, 021102. [CrossRef] [PubMed]

114. Chen, Y.; Shu, J.; Xue, X.; Yuan, Q.; Zhao, Y. Probing Axions with Event Horizon Telescope Polarimetric Measurements. *Phys. Rev. Lett.* **2020**, *124*, 061102. [CrossRef] [PubMed]

115. Chen, Y.; Liu, Y.; Lu, R.S.; Mizuno, Y.; Shu, J.; Xue, X.; Yuan, Q.; Zhao, Y. Stringent axion constraints with Event Horizon Telescope polarimetric measurements of M87*. *Nat. Astron.* **2022**, *6*, 592–598. [CrossRef]

116. Chen, Y.; Li, C.; Mizuno, Y.; Shu, J.; Xue, X.; Yuan, Q.; Zhao, Y.; Zhou, Z. Birefringence tomography for axion cloud. *J. Cosmol. Aatrop. Phys.* **2022**, *2022*, 073. [CrossRef]

117. Kulier, A.; Ostriker, J.P.; Natarajan, P.; Lackner, C.N.; Cen, R. Understanding Black Hole Mass Assembly via Accretion and Mergers at Late Times in Cosmological Simulations. *Astrophys. J.* **2015**, *799*, 178. [CrossRef]

118. Weinberger, R.; Springel, V.; Pakmor, R.; Nelson, D.; Genel, S.; Pillepich, A.; Vogelsberger, M.; Marinacci, F.; Naiman, J.; Torrey, P.; et al. Supermassive black holes and their feedback effects in the IllustrisTNG simulation. *Mon. Not. RAS* **2018**, *479*, 4056–4072. [CrossRef]

119. Ricarte, A.; Natarajan, P. Exploring SMBH assembly with semi-analytic modelling. *Mon. Not. RAS* **2018**, *474*, 1995–2011. [CrossRef]

120. Pacucci, F.; Loeb, A. Separating Accretion and Mergers in the Cosmic Growth of Black Holes with X-ray and Gravitational-wave Observations. *Astrophys. J.* **2020**, *895*, 95. [CrossRef]

121. Haehnelt, M.G.; Natarajan, P.; Rees, M.J. High-redshift galaxies, their active nuclei and central black holes. *Mon. Not. RAS* **1998**, *300*, 817–827. [CrossRef]

122. Di Matteo, T.; Springel, V.; Hernquist, L. Energy input from quasars regulates the growth and activity of black holes and their host galaxies. *Nature* **2005**, *433*, 604–607. [CrossRef]

123. Croton, D.J.; Springel, V.; White, S.D.M.; De Lucia, G.; Frenk, C.S.; Gao, L.; Jenkins, A.; Kauffmann, G.; Navarro, J.F.; Yoshida, N. The many lives of active galactic nuclei: Cooling flows, black holes and the luminosities and colours of galaxies. *Mon. Not. RAS* **2006**, *365*, 11–28. [CrossRef]

124. Kelly, B.C.; Merloni, A. Mass Functions of Supermassive Black Holes across Cosmic Time. *Adv. Astron.* **2012**, *2012*, 970858. [CrossRef]

125. Thorne, K.S. Disk-Accretion onto a Black Hole. II. Evolution of the Hole. *Astrophys. J.* **1974**, *191*, 507–520. [CrossRef]

126. King, A.R.; Pringle, J.E.; Hofmann, J.A. The evolution of black hole mass and spin in active galactic nuclei. *Mon. Not. RAS* **2008**, *385*, 1621–1627. [CrossRef]

127. Narayan, R.; Chael, A.; Chatterjee, K.; Ricarte, A.; Curd, B. Jets in Magnetically Arrested Hot Accretion Flows: Geometry, Power and Black Hole Spindown. *arXiv* **2021**, arXiv:2108.12380.

128. Kormendy, J.; Ho, L.C. Coevolution (Or Not) of Supermassive Black Holes and Host Galaxies. *Ann. Rev. Astron. Astrophys.* **2013**, *51*, 511–653. [CrossRef]

129. Pesce, D.W.; Palumbo, D.C.M.; Ricarte, A.; Broderick, A.E.; Johnson, M.D.; Nagar, N.M.; Natarajan, P.; Gómez, J.L. Expectations for Horizon-Scale Supermassive Black Hole Population Studies with the ngEHT. *Galaxies* **2022**, *10*, 109. [CrossRef]

130. Brenneman, L. Measuring Supermassive Black Hole Spins in AGN. *Acta Polytech.* **2013**, *53*, 652. [CrossRef]

131. Ramakrishnan, V.; Nagar, N.; Arratia, V.; Hernández-Yévenes, J.; Pesce, D.W.; Nair, D.G.; Bandyopadhyay, B.; Medina-Porcile, C.; Krichbaum, T.P.; Doeleman, S.; et al. Event Horizon and Environs (ETHER): A Curated Database for EHT and ngEHT Targets and Science. *Galaxies* **2023**, *11*, 15. [CrossRef]

132. Emami, R.; Ricarte, A.; Wong, G.N.; Palumbo, D.; Chang, D.; Doeleman, S.S.; Broaderick, A.; Narayan, R.; Weintroub, J.; Wielgus, M.; et al. Unraveling Twisty Linear Polarization Morphologies in Black Hole Images. *arXiv* **2022**, arXiv:2210.01218.

133. Pesce, D.W.; Palumbo, D.C.M.; Narayan, R.; Blackburn, L.; Doeleman, S.S.; Johnson, M.D.; Ma, C.P.; Nagar, N.M.; Natarajan, P.; Ricarte, A. Toward Determining the Number of Observable Supermassive Black Hole Shadows. *Astrophys. J.* **2021**, *923*, 260. [CrossRef]

134. Merritt, D.; Milosavljević, M. Massive Black Hole Binary Evolution. *Liv. Rev. Relat.* **2005**, *8*, 8. [CrossRef]

135. Begelman, M.C.; Blandford, R.D.; Rees, M.J. Massive black hole binaries in active galactic nuclei. *Nature* **1980**, *287*, 307–309. [CrossRef]

136. Armitage, P.J.; Natarajan, P. Accretion during the Merger of Supermassive Black Holes. *Astrophys. J.* **2002**, *567*, L9. [CrossRef]

137. Milosavljević, M.; Merritt, D. The Final Parsec Problem. In Proceedings of the The Astrophysics of Gravitational Wave Sources, College Park, MA, USA, 24–26 April 2003; Centrella, J.M., Ed.; American Institute of Physics Conference Series; Volume 686, pp. 201–210. [CrossRef]

138. Chael, A.A.; Johnson, M.D.; Narayan, R.; Doeleman, S.S.; Wardle, J.F.C.; Bouman, K.L. High-resolution Linear Polarimetric Imaging for the Event Horizon Telescope. *Astrophys. J.* **2016**, *829*, 11. [CrossRef]

139. Akiyama, K.; Kuramochi, K.; Ikeda, S.; Fish, V.L.; Tazaki, F.; Honma, M.; Doeleman, S.S.; Broderick, A.E.; Dexter, J.; Mościbrodzka, M.; et al. Imaging the Schwarzschild-radius-scale Structure of M87 with the Event Horizon Telescope Using Sparse Modeling. *Astrophys. J.* **2017**, *838*, 1. [CrossRef]

140. Broderick, A.E.; Pesce, D.W.; Tiede, P.; Pu, H.Y.; Gold, R. Hybrid Very Long Baseline Interferometry Imaging and Modeling with THEMIS. *Astrophys. J.* **2020**, *898*, 9. [CrossRef]

141. D'Orazio, D.J.; Loeb, A. Repeated Imaging of Massive Black Hole Binary Orbits with Millimeter Interferometry: Measuring Black Hole Masses and the Hubble Constant. *Astrophys. J.* **2018**, *863*, 185. [CrossRef]

142. Lico, R.; Jorstad, S.G.; Marscher, A.P.; Gómez, J.L.; Liodakis, I.; Dahale, R.; Alberdi, A.; Gold, R.; Traianou, T.; Toscano, T.; et al. Multi-Wavelength and Multi-Messenger Studies Using the Next-Generation Event Horizon Telescope. *Galaxies* **2023**, *11*, 17. [CrossRef]

143. Aartsen, M.G. et al. [IceCube Collaboration]. Neutrino emission from the direction of the blazar TXS 0506+056 prior to the IceCube-170922A alert. *Science* **2018**, *361*, 147–151. [CrossRef]

144. Aartsen, M.G. et al. [IceCube Collaboration]. Multimessenger observations of a flaring blazar coincident with high-energy neutrino IceCube-170922A. *Science* **2018**, *361*, eaat1378. [CrossRef]

145. Plavin, A.; Kovalev, Y.Y.; Kovalev, Y.A.; Troitsky, S. Observational Evidence for the Origin of High-energy Neutrinos in Parsec-scale Nuclei of Radio-bright Active Galaxies. *Astrophys. J.* **2020**, *894*, 101. [CrossRef]

146. Plavin, A.V.; Kovalev, Y.Y.; Kovalev, Y.A.; Troitsky, S.V. Directional Association of TeV to PeV Astrophysical Neutrinos with Radio Blazars. *Astrophys. J.* **2021**, *908*, 157. [CrossRef]

147. Giommi, P.; Padovani, P. Astrophysical Neutrinos and Blazars. *Universe* **2021**, *7*, 492. [CrossRef]

148. Plavin, A.V.; Kovalev, Y.Y.; Kovalev, Y.A.; Troitsky, S.V. Growing evidence for high-energy neutrinos originating in radio blazars. *arXiv* **2023**, arXiv:2211.09631.

149. Kovalev, Y.Y.; Pushkarev, A.B.; Nokhrina, E.E.; Plavin, A.V.; Beskin, V.S.; Chernoglazov, A.V.; Lister, M.L.; Savolainen, T. A transition from parabolic to conical shape as a common effect in nearby AGN jets. *Mon. Not. RAS* **2020**, *495*, 3576–3591. [CrossRef]

150. Marscher, A.P.; Jorstad, S.G.; D'Arcangelo, F.D.; Smith, P.S.; Williams, G.G.; Larionov, V.M.; Oh, H.; Olmstead, A.R.; Aller, M.F.; Aller, H.D.; et al. The inner jet of an active galactic nucleus as revealed by a radio-to-γ-ray outburst. *Nature* **2008**, *452*, 966–969. [CrossRef] [PubMed]

151. Chatterjee, K.; Narayan, R. Flux Eruption Events Drive Angular Momentum Transport in Magnetically Arrested Accretion Flows. *Astrophys. J.* **2022**, *941*, 30. [CrossRef]

152. Ripperda, B.; Liska, M.; Chatterjee, K.; Musoke, G.; Philippov, A.A.; Markoff, S.B.; Tchekhovskoy, A.; Younsi, Z. Black Hole Flares: Ejection of Accreted Magnetic Flux through 3D Plasmoid-mediated Reconnection. *Astrophys. J. Lett.* **2022**, *924*, L32. [CrossRef]

153. Jia, H.; Ripperda, B.; Quataert, E.; White, C.J.; Chatterjee, K.; Philippov, A.; Liska, M. Millimeter Observational Signatures of Flares in Magnetically Arrested Black Hole Accretion Models. *arXiv* **2023**, arXiv:2301.09014.

154. Balbus, S.A.; Hawley, J.F. A Powerful Local Shear Instability in Weakly Magnetized Disks. I. Linear Analysis. *Astrophys. J.* **1991**, *376*, 214. [CrossRef]

155. Balbus, S.A.; Hawley, J.F. Instability, turbulence, and enhanced transport in accretion disks. *Rev. Modern Phys.* **1998**, *70*, 1–53. [CrossRef]

156. Gammie, C.F.; McKinney, J.C.; Tóth, G. HARM: A Numerical Scheme for General Relativistic Magnetohydrodynamics. *Astrophys. J.* **2003**, *589*, 444–457. [CrossRef]

157. Porth, O.; Chatterjee, K.; Narayan, R.; Gammie, C.F.; Mizuno, Y.; Anninos, P.; Baker, J.G.; Bugli, M.; Chan, C.K.; Davelaar, J.; et al. The Event Horizon General Relativistic Magnetohydrodynamic Code Comparison Project. *Astrophys. J. Suppl.* **2019**, *243*, 26. [CrossRef]

158. Narayan, R.; Sądowski, A.; Penna, R.F.; Kulkarni, A.K. GRMHD simulations of magnetized advection-dominated accretion on a non-spinning black hole: role of outflows. *Mon. Not. RAS* **2012**, *426*, 3241–3259. [CrossRef]

159. Narayan, R.; Igumenshchev, I.V.; Abramowicz, M.A. Magnetically Arrested Disk: An Energetically Efficient Accretion Flow. *Publ. ASJ* **2003**, *55*, L69–L72. [CrossRef]

160. Chan, C.K.; Psaltis, D.; Özel, F.; Medeiros, L.; Marrone, D.; Sadowski, A.; Narayan, R. Fast Variability and Millimeter/IR Flares in GRMHD Models of Sgr A* from Strong-field Gravitational Lensing. *Astrophys. J.* **2015**, *812*, 103. [CrossRef]

161. Guan, X.; Gammie, C.F.; Simon, J.B.; Johnson, B.M. Locality of MHD Turbulence in Isothermal Disks. *Astrophys. J.* **2009**, *694*, 1010–1018. [CrossRef]

162. Contopoulos, I.; Nathanail, A.; Sądowski, A.; Kazanas, D.; Narayan, R. Numerical simulations of the Cosmic Battery in accretion flows around astrophysical black holes. *Mon. Not. RAS* **2018**, *473*, 721–727. [CrossRef]

163. Sironi, L.; Spitkovsky, A. Relativistic Reconnection: An Efficient Source of Non-thermal Particles. *Astrophys. J. Lett.* **2014**, *783*, L21. [CrossRef]

164. Rowan, M.E.; Sironi, L.; Narayan, R. Electron and Proton Heating in Transrelativistic Magnetic Reconnection. *Astrophys. J.* **2017**, *850*, 29. [CrossRef]

165. Werner, G.R.; Uzdensky, D.A.; Begelman, M.C.; Cerutti, B.; Nalewajko, K. Non-thermal particle acceleration in collisionless relativistic electron-proton reconnection. *Mon. Not. RAS* **2018**, *473*, 4840–4861. [CrossRef]

166. Ball, D.; Sironi, L.; Özel, F. Electron and Proton Acceleration in Trans-relativistic Magnetic Reconnection: Dependence on Plasma Beta and Magnetization. *Astrophys. J.* **2018**, *862*, 80. [CrossRef]

167. Doeleman, S.S.; Fish, V.L.; Broderick, A.E.; Loeb, A.; Rogers, A.E.E. Detecting Flaring Structures in Sagittarius A* with High-Frequency VLBI. *Astrophys. J.* **2009**, *695*, 59–74. [CrossRef]

168. Ricarte, A.; Palumbo, D.C.M.; Narayan, R.; Roelofs, F.; Emami, R. Observational Signatures of Frame Dragging in Strong Gravity. *Astrophys. J. Lett.* **2022**, *941*, L12. [CrossRef]

169. Blandford, R.; Meier, D.; Readhead, A. Relativistic Jets from Active Galactic Nuclei. *Ann. Rev. Astron. Astrophys.* **2019**, *57*, 467–509. [CrossRef]

170. Salpeter, E.E. Accretion of Interstellar Matter by Massive Objects. *Astrophys. J.* **1964**, *140*, 796–800. [CrossRef]

171. Markoff, S.; Bower, G.C.; Falcke, H. How to hide large-scale outflows: Size constraints on the jets of Sgr A*. *Mon. Not. RAS* **2007**, *379*, 1519–1532. [CrossRef]

172. Yusef-Zadeh, F.; Roberts, D.; Wardle, M.; Heinke, C.O.; Bower, G.C. Flaring Activity of Sagittarius A* at 43 and 22 GHz: Evidence for Expanding Hot Plasma. *Astrophys. J.* **2006**, *650*, 189–194. [CrossRef]

173. Brinkerink, C.; Falcke, H.; Brunthaler, A.; Law, C. Persistent time lags in light curves of Sagittarius A*: evidence of outflow. *arXiv* **2021**, arXiv:2107.13402.

174. Emami, R.; Anantua, R.; Ricarte, A.; Doeleman, S.S.; Broderick, A.; Wong, G.; Blackburn, L.; Wielgus, M.; Narayan, R.; Tremblay, G.; et al. Probing Plasma Composition with the Next Generation Event Horizon Telescope (ngEHT). *Galaxies* **2023**, *11*, 11. [CrossRef]

175. Chael, A.; Narayan, R.; Johnson, M.D. Two-temperature, Magnetically Arrested Disc simulations of the jet from the supermassive black hole in M87. *Mon. Not. RAS* **2019**, *486*, 2873–2895. [CrossRef]

176. Abbott, B.P.; Abbott, R.; Abbott, T.D.; Acernese, F.; Ackley, K.; Adams, C.; Adams, T.; Addesso, P.; Adhikari, R.X.; Adya, V.B.; et al. GW170817: Observation of Gravitational Waves from a Binary Neutron Star Inspiral. *Phys. Rev. Lett.* **2017**, *119*, 161101. [CrossRef]

177. Abbott, B.P.; Abbott, R.; Abbott, T.D.; Acernese, F.; Ackley, K.; Adams, C.; Adams, T.; Addesso, P.; Adhikari, R.X.; Adya, V.B.; et al. Gravitational Waves and Gamma-Rays from a Binary Neutron Star Merger: GW170817 and GRB 170817A. *Astrophys. J. Lett.* **2017**, *848*, L13. [CrossRef]

178. Sari, R.; Piran, T.; Halpern, J.P. Jets in Gamma-Ray Bursts. *Astrophys. J. Lett.* **1999**, *519*, L17–L20. [CrossRef]

179. Mirabel, I.F.; Rodríguez, L.F. Sources of Relativistic Jets in the Galaxy. *Ann. Rev. Astron. Astrophys.* **1999**, *37*, 409–443. [CrossRef]

180. Fender, R. Jets from X-ray binaries. In *Compact Stellar X-ray Sources*; Cambridge University Press: Cambridge, UK, 2006; Volume 39, pp. 381–419.

181. Mooley, K.P.; Deller, A.T.; Gottlieb, O.; Nakar, E.; Hallinan, G.; Bourke, S.; Frail, D.A.; Horesh, A.; Corsi, A.; Hotokezaka, K. Superluminal motion of a relativistic jet in the neutron-star merger GW170817. *Nature* **2018**, *561*, 355–359. [CrossRef] [PubMed]

182. Alexander, K.D.; van Velzen, S.; Horesh, A.; Zauderer, B.A. Radio Properties of Tidal Disruption Events. *Space Sci. Rev.* **2020**, *216*, 81. [CrossRef]

183. Falcke, H.; Körding, E.; Markoff, S. A scheme to unify low-power accreting black holes. Jet-dominated accretion flows and the radio/X-ray correlation. *Astron. Astrophys.* **2004**, *414*, 895–903. .:20031683. [CrossRef]

184. Körding, E.G.; Jester, S.; Fender, R. Accretion states and radio loudness in active galactic nuclei: analogies with X-ray binaries. *Mon. Not. RAS* **2006**, *372*, 1366–1378. [CrossRef]

185. Miller-Jones, J.C.A.; Blundell, K.M.; Rupen, M.P.; Mioduszewski, A.J.; Duffy, P.; Beasley, A.J. Time-sequenced Multi-Radio Frequency Observations of Cygnus X-3 in Flare. *Astrophys. J.* **2004**, *600*, 368–389. [CrossRef]

186. Tetarenko, A.J.; Sivakoff, G.R.; Miller-Jones, J.C.A.; Rosolowsky, E.W.; Petitpas, G.; Gurwell, M.; Wouterloot, J.; Fender, R.; Heinz, S.; Maitra, D.; et al. Extreme jet ejections from the black hole X-ray binary V404 Cygni. *Mon. Not. RAS* **2017**, *469*, 3141–3162. [CrossRef]

187. Tetarenko, A.J.; Sivakoff, G.R.; Miller-Jones, J.C.A.; Bremer, M.; Mooley, K.P.; Fender, R.P.; Rumsey, C.; Bahramian, A.; Altamirano, D.; Heinz, S.; et al. Tracking the variable jets of V404 Cygni during its 2015 outburst. *Mon. Not. RAS* **2019**, *482*, 2950–2972. [CrossRef]

188. Asada, K.; Nakamura, M. The Structure of the M87 Jet: A Transition from Parabolic to Conical Streamlines. *Astrophys. J. Lett.* **2012**, *745*, L28. [CrossRef]

189. Hada, K.; Kino, M.; Doi, A.; Nagai, H.; Honma, M.; Hagiwara, Y.; Giroletti, M.; Giovannini, G.; Kawaguchi, N. The Innermost Collimation Structure of the M87 Jet Down to ∼10 Schwarzschild Radii. *Astrophys. J.* **2013**, *775*, 70. [CrossRef]

190. Tseng, C.Y.; Asada, K.; Nakamura, M.; Pu, H.Y.; Algaba, J.C.; Lo, W.P. Structural Transition in the NGC 6251 Jet: an Interplay with the Supermassive Black Hole and Its Host Galaxy. *Astrophys. J.* **2016**, *833*, 288. [CrossRef]

191. Okino, H.; Akiyama, K.; Asada, K.; Gómez, J.L.; Hada, K.; Honma, M.; Krichbaum, T.P.; Kino, M.; Nagai, H.; Bach, U.; et al. Collimation of the Relativistic Jet in the Quasar 3C 273. *Astrophys. J.* **2022**, *940*, 65. [CrossRef]

192. Fender, R.; Woudt, P.A.; Corbel, S.; Coriat, M.; Daigne, F.; Falcke, H.; Girard, J.; Heywood, I.; Horesh, A.; Horrell, J.; et al. ThunderKAT: The MeerKAT Large Survey Project for Image-Plane Radio Transients. In Proceedings of the MeerKAT Science: On the Pathway to the SKA, Stellenbosch, South Africa, 25–27 May 2016; p. 13. [CrossRef]

193. Ho, A.Y.Q.; Phinney, E.S.; Ravi, V.; Kulkarni, S.R.; Petitpas, G.; Emonts, B.; Bhalerao, V.; Blundell, R.; Cenko, S.B.; Dobie, D.; et al. AT2018cow: A Luminous Millimeter Transient. *Astrophys. J.* **2019**, *871*, 73. [CrossRef]
194. Margutti, R.; Metzger, B.D.; Chornock, R.; Vurm, I.; Roth, N.; Grefenstette, B.W.; Savchenko, V.; Cartier, R.; Steiner, J.F.; Terreran, G.; et al. An Embedded X-ray Source Shines through the Aspherical AT 2018cow: Revealing the Inner Workings of the Most Luminous Fast-evolving Optical Transients. *Astrophys. J.* **2019**, *872*, 18. [CrossRef]
195. Curd, B.; Emami, R.; Roelofs, F.; Anantua, R. Modeling Reconstructed Images of Jets Launched by SANE Super-Eddington Accretion Flows around SMBHs with the ngEHT. *Galaxies* **2022**, *10*, 117. [CrossRef]
196. Eubanks, T.M. Anchored in Shadows: Tying the Celestial Reference Frame Directly to Black Hole Event Horizons. *arXiv* **2020**, arXiv:2005.09122.
197. Charlot, P.; Jacobs, C.S.; Gordon, D.; Lambert, S.; de Witt, A.; Böhm, J.; Fey, A.L.; Heinkelmann, R.; Skurikhina, E.; Titov, O.; et al. The third realization of the International Celestial Reference Frame by very long baseline interferometry. *Astron. Astrophys.* **2020**, *644*, A159. [CrossRef]
198. Reid, M.J.; Honma, M. Microarcsecond Radio Astrometry. *Ann. Rev. Astron. Astrophys.* **2014**, *52*, 339–372. [CrossRef]
199. Rioja, M.; Dodson, R. High-precision Astrometric Millimeter Very Long Baseline Interferometry Using a New Method for Atmospheric Calibration. *Astron. J.* **2011**, *141*, 114. [CrossRef]
200. Sokolovsky, K.V.; Kovalev, Y.Y.; Pushkarev, A.B.; Lobanov, A.P. A VLBA survey of the core shift effect in AGN jets. I. Evidence of dominating synchrotron opacity. *Astron. Astrophys.* **2011**, *532*, A38. [CrossRef]
201. Pushkarev, A.B.; Hovatta, T.; Kovalev, Y.Y.; Lister, M.L.; Lobanov, A.P.; Savolainen, T.; Zensus, J.A. MOJAVE: Monitoring of Jets in Active galactic nuclei with VLBA Experiments. IX. Nuclear opacity. *Astron. Astrophys.* **2012**, *545*, A113. [CrossRef]
202. Jiang, W.; Shen, Z.; Martí-Vidal, I.; Wang, X.; Jiang, D.; Kawaguchi, N. Millimeter-VLBI Observations of Low-luminosity Active Galactic Nuclei with Source-frequency Phase Referencing. *Astrophys. J. Lett.* **2021**, *922*, L16. [CrossRef]
203. Kardashev, N.S. Cosmological proper motion. *Astron. Zhurnal* **1986**, *63*, 845–849.
204. Paine, J.; Darling, J.; Graziani, R.; Courtois, H.M. Secular Extragalactic Parallax: Measurement Methods and Predictions for Gaia. *Astrophys. J.* **2020**, *890*, 146. [CrossRef]
205. Hinshaw, G.; Weiland, J.L.; Hill, R.S.; Odegard, N.; Larson, D.; Bennett, C.L.; Dunkley, J.; Gold, B.; Greason, M.R.; Jarosik, N.; et al. Five-Year Wilkinson Microwave Anisotropy Probe Observations: Data Processing, Sky Maps, and Basic Results. *Astrophys. J. Suppl.* **2009**, *180*, 225–245. [CrossRef]
206. Plavin, A.V.; Kovalev, Y.Y.; Pushkarev, A.B.; Lobanov, A.P. Significant core shift variability in parsec-scale jets of active galactic nuclei. *Mon. Not. RAS* **2019**, *485*, 1822–1842. [CrossRef]
207. Riess, A.G.; Casertano, S.; Yuan, W.; Macri, L.M.; Scolnic, D. Large Magellanic Cloud Cepheid Standards Provide a 1% Foundation for the Determination of the Hubble Constant and Stronger Evidence for Physics beyond ΛCDM. *Astrophys. J.* **2019**, *876*, 85. [CrossRef]
208. Aghanim, N. et al. [Planck Collaboration]. Planck 2018 results. VI. Cosmological parameters. *Astron. Astrophys.* **2020**, *641*, A6. [CrossRef]
209. Pesce, D.W.; Braatz, J.A.; Reid, M.J.; Riess, A.G.; Scolnic, D.; Condon, J.J.; Gao, F.; Henkel, C.; Impellizzeri, C.M.V.; Kuo, C.Y.; et al. The Megamaser Cosmology Project. XIII. Combined Hubble Constant Constraints. *Astrophys. J. Lett.* **2020**, *891*, L1. [CrossRef]
210. Wong, K.C.; Suyu, S.H.; Chen, G.C.F.; Rusu, C.E.; Millon, M.; Sluse, D.; Bonvin, V.; Fassnacht, C.D.; Taubenberger, S.; Auger, M.W.; et al. H0LiCOW-XIII. A 2.4 per cent measurement of H_0 from lensed quasars: 5.3σ tension between early- and late-Universe probes. *Mon. Not. RAS* **2020**, *498*, 1420–1439. [CrossRef]
211. Miyoshi, M.; Moran, J.; Herrnstein, J.; Greenhill, L.; Nakai, N.; Diamond, P.; Inoue, M. Evidence for a black hole from high rotation velocities in a sub-parsec region of NGC4258. *Nature* **1995**, *373*, 127–129. [CrossRef]
212. Kuo, C.Y.; Braatz, J.A.; Condon, J.J.; Impellizzeri, C.M.V.; Lo, K.Y.; Zaw, I.; Schenker, M.; Henkel, C.; Reid, M.J.; Greene, J.E. The Megamaser Cosmology Project. III. Accurate Masses of Seven Supermassive Black Holes in Active Galaxies with Circumnuclear Megamaser Disks. *Astrophys. J.* **2011**, *727*, 20. [CrossRef]
213. Herrnstein, J.R. Observations of the Sub-Parsec Maser Disk in NGC 4258. Ph.D. Thesis, Harvard University, Cambridge, MA, USA, 1997.
214. Braatz, J.A.; Reid, M.J.; Humphreys, E.M.L.; Henkel, C.; Condon, J.J.; Lo, K.Y. The Megamaser Cosmology Project. II. The Angular-diameter Distance to UGC 3789. *Astrophys. J.* **2010**, *718*, 657–665. [CrossRef]
215. Reid, M.J.; Braatz, J.A.; Condon, J.J.; Lo, K.Y.; Kuo, C.Y.; Impellizzeri, C.M.V.; Henkel, C. The Megamaser Cosmology Project. IV. A Direct Measurement of the Hubble Constant from UGC 3789. *Astrophys. J.* **2013**, *767*, 154. [CrossRef]
216. Yates, J.A.; Field, D.; Gray, M.D. Non-local radiative transfer for molecules: modelling population inversions in water masers. *Mon. Not. RAS* **1997**, *285*, 303–316. [CrossRef]
217. Gray, M.D.; Baudry, A.; Richards, A.M.S.; Humphreys, E.M.L.; Sobolev, A.M.; Yates, J.A. The physics of water masers observable with ALMA and SOFIA: model predictions for evolved stars. *Mon. Not. RAS* **2016**, *456*, 374–404. [CrossRef]
218. Humphreys, E.M.L.; Greenhill, L.J.; Reid, M.J.; Beuther, H.; Moran, J.M.; Gurwell, M.; Wilner, D.J.; Kondratko, P.T. First Detection of Millimeter/Submillimeter Extragalactic H_2O Maser Emission. *Astrophys. J. Lett.* **2005**, *634*, L133–L136. [CrossRef]
219. Humphreys, E.M.L.; Vlemmings, W.H.T.; Impellizzeri, C.M.V.; Galametz, M.; Olberg, M.; Conway, J.E.; Belitsky, V.; De Breuck, C. Detection of 183 GHz H_2O megamaser emission towards NGC 4945. *Astron. Astrophys.* **2016**, *592*, L13. [CrossRef]

220. Hagiwara, Y.; Miyoshi, M.; Doi, A.; Horiuchi, S. Submillimeter H_2O Maser in Circinus Galaxy—A New Probe for the Circumnuclear Region of Active Galactic Nuclei. *Astrophys. J. Lett.* **2013**, *768*, L38. [CrossRef]

221. Pesce, D.W.; Braatz, J.A.; Impellizzeri, C.M.V. Submillimeter H_2O Megamasers in NGC 4945 and the Circinus Galaxy. *Astrophys. J.* **2016**, *827*, 68. [CrossRef]

222. Hagiwara, Y.; Horiuchi, S.; Doi, A.; Miyoshi, M.; Edwards, P.G. A Search for Submillimeter H_2O Masers in Active Galaxies: The Detection of 321 GHZ H_2O Maser Emission in NGC 4945. *Astrophys. J.* **2016**, *827*, 69. [CrossRef]

223. Hagiwara, Y.; Horiuchi, S.; Imanishi, M.; Edwards, P.G. Second-epoch ALMA Observations of 321 GHz Water Maser Emission in NGC 4945 and the Circinus Galaxy. *Astrophys. J.* **2021**, *923*, 251. [CrossRef]

224. Kim, D.J.; Fish, V. Spectral Line VLBI Studies Using the ngEHT. *Galaxies* **2023**, *11*, 10. [CrossRef]

225. Arras, P.; Frank, P.; Haim, P.; Knollmüller, J.; Leike, R.; Reinecke, M.; Enßlin, T. Variable structures in M87* from space, time and frequency d interferometry. *Nat. Astron.* **2022**, *6*, 259–269. [CrossRef]

226. Fish, V.L.; Johnson, M.D.; Lu, R.S.; Doeleman, S.S.; Bouman, K.L.; Zoran, D.; Freeman, W.T.; Psaltis, D.; Narayan, R.; Pankratius, V.; et al. Imaging an Event Horizon: Mitigation of Scattering toward Sagittarius A*. *Astrophys. J.* **2014**, *795*, 134. [CrossRef]

227. Lu, R.S.; Roelofs, F.; Fish, V.L.; Shiokawa, H.; Doeleman, S.S.; Gammie, C.F.; Falcke, H.; Krichbaum, T.P.; Zensus, J.A. Imaging an Event Horizon: Mitigation of Source Variability of Sagittarius A*. *Astrophys. J.* **2016**, *817*, 173. [CrossRef]

228. Johnson, M.D. Stochastic Optics: A Scattering Mitigation Framework for Radio Interferometric Imaging. *Astrophys. J.* **2016**, *833*, 74. [CrossRef]

229. Akiyama, K.; Ikeda, S.; Pleau, M.; Fish, V.L.; Tazaki, F.; Kuramochi, K.; Broderick, A.E.; Dexter, J.; Mościbrodzka, M.; Gowanlock, M.; et al. Superresolution Full-polarimetric Imaging for Radio Interferometry with Sparse Modeling. *Astron. J.* **2017**, *153*, 159. [CrossRef]

230. Johnson, M.D.; Bouman, K.L.; Blackburn, L.; Chael, A.A.; Rosen, J.; Shiokawa, H.; Roelofs, F.; Akiyama, K.; Fish, V.L.; Doeleman, S.S. Dynamical Imaging with Interferometry. *Astrophys. J.* **2017**, *850*, 172. [CrossRef]

231. Chael, A.A.; Johnson, M.D.; Bouman, K.L.; Blackburn, L.L.; Akiyama, K.; Narayan, R. Interferometric Imaging Directly with Closure Phases and Closure Amplitudes. *Astrophys. J.* **2018**, *857*, 23. [CrossRef]

232. Blackburn, L.; Chan, C.K.; Crew, G.B.; Fish, V.L.; Issaoun, S.; Johnson, M.D.; Wielgus, M.; Akiyama, K.; Barrett, J.; Bouman, K.L.; et al. EHT-HOPS Pipeline for Millimeter VLBI Data Reduction. *Astrophys. J.* **2019**, *882*, 23. [CrossRef]

233. Janssen, M.; Goddi, C.; van Bemmel, I.M.; Kettenis, M.; Small, D.; Liuzzo, E.; Rygl, K.; Martí-Vidal, I.; Blackburn, L.; Wielgus, M.; et al. rPICARD: A CASA-based calibration pipeline for VLBI data. Calibration and imaging of 7 mm VLBA observations of the AGN jet in M 87. *Astron. Astrophys.* **2019**, *626*, A75. [CrossRef]

234. Broderick, A.E.; Gold, R.; Karami, M.; Preciado-López, J.A.; Tiede, P.; Pu, H.Y.; Akiyama, K.; Alberdi, A.; Alef, W.; Asada, K.; et al. THEMIS: A Parameter Estimation Framework for the Event Horizon Telescope. *Astrophys. J.* **2020**, *897*, 139. [CrossRef]

235. Sun, H.; Bouman, K.L. Deep Probabilistic Imaging: Uncertainty Quantification and Multi-modal Solution Characterization for Computational Imaging. *arXiv* **2020**, arXiv:2010.14462.

236. Park, J.; Byun, D.Y.; Asada, K.; Yun, Y. GPCAL: A Generalized Calibration Pipeline for Instrumental Polarization in VLBI Data. *Astrophys. J.* **2021**, *906*, 85. [CrossRef]

237. Pesce, D.W. A D-term Modeling Code (DMC) for Simultaneous Calibration and Full-Stokes Imaging of Very Long Baseline Interferometric Data. *Astron. J.* **2021**, *161*, 178. [CrossRef]

238. Sun, H.; Bouman, K.L.; Tiede, P.; Wang, J.J.; Blunt, S.; Mawet, D. α-deep Probabilistic Inference (α-DPI): Efficient Uncertainty Quantification from Exoplanet Astrometry to Black Hole Feature Extraction. *Astrophys. J.* **2022**, *932*, 99. [CrossRef]

239. Janssen, M.; Radcliffe, J.F.; Wagner, J. Software and Techniques for VLBI Data Processing and Analysis. *Universe* **2022**, *8*, 527. [CrossRef]

240. Hoak, D.; Barrett, J.; Crew, G.; Pfeiffer, V. Progress on the Haystack Observatory Postprocessing System. *Galaxies* **2022**, *10*, 119. [CrossRef]

241. Yu, W.; Romein, J.W.; Dursi, L.J.; Lu, R.S.; Pope, A.; Callanan, G.; Pesce, D.W.; Blackburn, L.; Merry, B.; Srinivasan, R.; et al. Prospects of GPU Tensor Core Correlation for the SMA and the ngEHT. *Galaxies* **2023**, *11*, 13. [CrossRef]

242. Müller, H.; Lobanov, A.P. DoG-HiT: A novel VLBI multiscale imaging approach. *Astron. Astrophys.* **2022**, *666*, A137, [CrossRef]

243. Levis, A.; Srinivasan, P.P.; Chael, A.A.; Ng, R.; Bouman, K.L. Gravitationally Lensed Black Hole Emission Tomography. In Proceedings of the 2022 IEEE/CVF Conference on Computer Vision and Pattern Recognition (CVPR), New Orleans, LA, USA, 21–24 June 2022; pp. 19809–19818. [CrossRef]

244. Abuter, R. et al. [GRAVITY Collaboration]. Detection of orbital motions near the last stable circular orbit of the massive black hole SgrA*. *Astron. Astrophys.* **2018**, *618*, L10. [CrossRef]

245. Liska, M.T.P.; Chatterjee, K.; Issa, D.; Yoon, D.; Kaaz, N.; Tchekhovskoy, A.; van Eijnatten, D.; Musoke, G.; Hesp, C.; Rohoza, V.; et al. H-AMR: A New GPU-accelerated GRMHD Code for Exascale Computing with 3D Adaptive Mesh Refinement and Local Adaptive Time Stepping. *Astrophys. J. Suppl.* **2022**, *263*, 26. [CrossRef]

246. Mościbrodzka, M.; Gammie, C.F. IPOLE-semi-analytic scheme for relativistic polarized radiative transport. *Mon. Not. RAS* **2018**, *475*, 43–54. [CrossRef]

247. Broderick, A.E.; Fish, V.L.; Johnson, M.D.; Rosenfeld, K.; Wang, C.; Doeleman, S.S.; Akiyama, K.; Johannsen, T.; Roy, A.L. Modeling Seven Years of Event Horizon Telescope Observations with Radiatively Inefficient Accretion Flow Models. *Astrophys. J.* **2016**, *820*, 137. [CrossRef]

248. Bouman, K.L.; Johnson, M.D.; Dalca, A.V.; Chael, A.A.; Roelofs, F.; Doeleman, S.S.; Freeman, W.T. Reconstructing Video of Time-Varying Sources from Radio Interferometric Measurements. *IEEE Trans. Comput. Imaging* **2018**, *4*, 512–527. [CrossRef]
249. Swanner, L.A. *Mountains of Controversy: Narrative and the Making of Contested Landscapes in Postwar American Astronomy*; Harvard University: Cambridge, MA, USA, 2013.
250. Salazar, J.A. Multicultural settler colonialism and indigenous struggle in Hawai'i: The politics of astronomy on Mauna a Wākea. Ph.D. Thesis, University of Hawai'i at Manoa, Honolulu, HI, USA, 2014.
251. Swanner, L. Instruments of science or conquest? Neocolonialism and modern American astronomy. *Hist. Stud. Natl. Sci.* **2017**, *47*, 293–319. [CrossRef]
252. Kahanamoku, S.; Alegado, R.; Kagawa-Viviani, A.; Kamelamela, K.L.; Kamai, B.; Walkowicz, L.M.; Prescod-Weinstein, C.; Reyes, M.A.D.L.; Neilson, H. A Native Hawaiian-led summary of the current impact of constructing the Thirty Meter Telescope on Maunakea. *arXiv* **2020**, arXiv:2001.00970.
253. Alegado, R. Telescope opponents fight the process, not science. *Nature* **2019**, *572*, 7. [CrossRef]
254. Singh, J.A. Informed consent and community engagement in open field research: Lessons for gene drive science. *BMC Med. Ethics* **2019**, *20*, 54. [CrossRef] [PubMed]
255. Watkins, J. Through wary eyes: Indigenous perspectives on archaeology. *Annu. Rev. Anthropol.* **2005**, *34*, 429–449. [CrossRef]
256. Supernant, K.; Warrick, G. Challenges to critical community-based archaeological practice in Canada. *Can. J. Archaeol. J. Can. d'Archéol.* **2014**, *38*, 563–591.
257. Nichols, T. Hidden in Plain Sight Merging the Physics Laboratory with the Surrounding Environment. 2023. *Unpublished manuscript*. Available online: https://www.scientificamerican.com/article/hidden-in-plain-sight/ (accessed on 20 April 2023).
258. Tomblin, D.; Pirtle, Z.; Farooque, M.; Sittenfeld, D.; Mahoney, E.; Worthington, R.; Gano, G.; Gates, M.; Bennett, I.; Kessler, J.; et al. Integrating public deliberation into engineering systems: Participatory technology assessment of NASA's Asteroid Redirect Mission. *Astropolitics* **2017**, *15*, 141–166. [CrossRef]
259. Riordan, M.; Hoddeson, L.; Kolb, A.W. *Tunnel Visions: The Rise and Fall of the Superconducting Super Collider*; University of Chicago Press: Chicago, IL, USA, 2015.
260. Redfield, P. *Space in the Tropics: From Convicts to Rockets in French Guiana*; University of California Press: Auckland City, CA, USA, 2000.
261. Gerrard, M.B. *Whose Backyard, Whose Risk: Fear and Fairness in Toxic and Nuclear Waste Siting*; MIT Press: Cambridge, MA, USA, 1996.
262. Kuletz, V.L. *The Tainted Desert*; Routledge: New York, NY, USA, 1998.
263. Masco, J. The nuclear borderlands. In *The Nuclear Borderlands*; Princeton University Press: Princeton, NJ, USA, 2013.
264. Hamilton, L.; Scowcroft, B.; Ayers, M.; Bailey, V.; Carnesale, A.; Domenici, P.; Eisenhower, S.; Hagel, C.; Lash, J.; Macfarlane, A.; et al. *Blue Ribbon Commission on America's Nuclear Future: Report to the Secretary of Energy*; Blue Ribbon Commission on America's Nuclear Future (BRC): Washington, DC, USA, 2012.
265. Richter, J.; Bernstein, M.J.; Farooque, M. The process to find a process for governance: Nuclear waste management and consent-based siting in the United States. *Energy Res. Soc. Sci.* **2022**, *87*, 102473. [CrossRef]
266. Frigg, R.; Thompson, E.; Werndl, C. Philosophy of climate science part II: Modelling climate change. *Philos. Compass* **2015**, *10*, 965–977. [CrossRef]
267. Winsberg, E. *Philosophy and Climate Science*; Cambridge University Press: Cambridge, UK, 2018.
268. Samek, W.; Montavon, G.; Vedaldi, A.; Hansen, L.K.; Müller, K.R. *Explainable AI: Interpreting, Explaining and Visualizing Deep Learning*; Springer Nature: Berlin/Heidelberg, Germany, 2019; Volume 11700.
269. Zednik, C. Solving the black box problem: A normative framework for explainable artificial intelligence. *Philos. Technol.* **2021**, *34*, 265–288. [CrossRef]
270. Beisbart, C.; Räz, T. Philosophy of science at sea: Clarifying the interpretability of machine learning. *Philos. Compass* **2022**, *17*, e12830. [CrossRef]
271. Kessler, E.A. *Picturing the Cosmos: Hubble Space Telescope Images and the Astronomical Sublime*; University of Minnesota Press: Minneapolis, MN, USA, 2012.
272. Galison, P.; Hevly, B. (Eds.) *Big Science: The Growth of Large-Scale Research*; Stanford University Press: Stanford, CA, USA, 1992.
273. Knorr Cetina, K. *Epistemic Cultures: How the Sciences Make Knowledge*; Harvard University Press: Cambridge, MA, USA, 1999.
274. Sullivan, W.T. *Cosmic Noise: A History of Early Radio Astronomy/Woodruff T. Sullivan, III*; Cambridge University Press: Cambridge, UK; New York, NY, USA, 2009; p. xxxii.
275. Shrum, W.; Chompalov, I.; Genuth, J. Trust, Conflict and Performance in Scientific Collaborations. *Soc. Stud. Sci.* **2001**, *31*, 681–730. [CrossRef]
276. Boyer-Kassem, T.; Mayo-Wilson, C.; Weisberg, M. *Scientific Collaboration and Collective Knowledge*; Oxford University Press: New York, NY, USA, 2017.
277. Collins, H. *Gravity's Kiss: The Detection of Gravitational Waves*; MIT Press: Cambridge, MA, USA, 2017.
278. Nichols, T. Constructing Stillness: Theorization, Discovery, Interrogation, and Negotiation of the Expanded Laboratory of the Laser Interferometer Gravitational-Wave Observatory. Ph.D. Thesis, Harvard University, Harvard, MA, USA, 2022.
279. Boisot, M.; Nordberg, M.; Yami, S.; Nicquevert, B. *Collisions and Collaboration: The Organization of Learning in the ATLAS Experiment at the LHC*; Oxford University Press: Oxford, UK, 2011. [CrossRef]
280. Ritson, S. Creativity and modelling the measurement process of the Higgs self-coupling at the LHC and HL-LHC. *Synthese* **2021**, *199*, 11887–11911. [CrossRef]

281. Sorgner, H. Constructing 'Do-Able'Dissertations in Collaborative Research: Alignment Work and Distinction in Experimental High-Energy Physics Settings. *Sci. Technol. Stud.* **2022**, *35*, 38–57.
282. Merz, M.; Sorgner, H. Organizational Complexity in Big Science: Strategies and Practices. *Synthese* **2022**, *200*, 211. [CrossRef] [PubMed]
283. Jebeile, J. Values and Objectivity in the Intergovernmental Panel on Climate Change. *Soc. Epistemol.* **2020**, *34*, 453–468. . [CrossRef]
284. Smith, R.W.; Hanle, P.A.; Kargon, R.H.; Tatarewicz, J.N. *The Space Telescope. A study of NASA, Science, Technology, and Politics*; Cambridge University Press: Cambridge, MA, USA, 1993.
285. Vertesi, J. *Shaping Science: Organizations, Decisions, and Culture on NASA's Teams*; University of Chicago Press: Chicago, IL, USA, 2020. [CrossRef]
286. Traweek, S. *Beamtimes and Lifetimes: The World of High Energy Physicists*; Harvard University Press: Harvard, MA, USA, 1988.
287. Mellers, B.; Stone, E.; Murray, T.; Minster, A.; Rohrbaugh, N.; Bishop, M.; Chen, E.; Baker, J.; Hou, Y.; Horowitz, M.; et al. Identifying and cultivating superforecasters as a method of improving probabilistic predictions. *Perspect. Psychol. Sci.* **2015**, *10*, 267–281. [CrossRef]
288. Camerer, C.F.; Dreber, A.; Holzmeister, F.; Ho, T.H.; Huber, J.; Johannesson, M.; Kirchler, M.; Nave, G.; Nosek, B.A.; Pfeiffer, T.; et al. Evaluating the replicability of social science experiments in Nature and Science between 2010 and 2015. *Nat. Hum. Behav.* **2018**, *2*, 637–644. [CrossRef] [PubMed]
289. DellaVigna, S.; Pope, D.; Vivalt, E. Predict science to improve science. *Science* **2019**, *366*, 428–429. [CrossRef] [PubMed]
290. Kitcher, P. The Division of Cognitive Labor. *J. Philos.* **1990**, *87*, 5–22. [CrossRef]
291. Kitcher, P. *The Advancement of Science: Science Without Legend, Objectivity Without Illusions*; Oxford University Press: Oxford, UK, 1993.
292. Zollman, K.J.S. The Communication Structure of Epistemic Communities. *Philos. Sci.* **2007**, *74*, 574–587. [CrossRef]
293. Zollman, K.J.S. The Epistemic Benefit of Transient Diversity. *Erkenntnis* **2010**, *72*, 17–35. [CrossRef]
294. Zollman, K.J. Network Epistemology: Communication in Epistemic Communities. *Philos. Compass* **2013**, *8*, 15–27. [CrossRef]
295. Longino, H. The Social Dimensions of Scientific Knowledge. In *The Stanford Encyclopedia of Philosophy*, Summer 2019 ed.; Zalta, E.N., Ed.; Metaphysics Research Lab, Stanford University: Stanford, MA, USA, 2019.
296. Lalli, R.; Howey, R.; Wintergrün, D. The dynamics of collaboration networks and the history of general relativity, 1925–1970. *Scientometrics* **2020**, *122*, 1129–1170. [CrossRef]
297. Lalli, R.; Howey, R.; Wintergrün, D. The Socio-Epistemic Networks of General Relativity, 1925–1970. In *The Renaissance of General Relativity in Context*; Blum, A.S., Lalli, R., Renn, J., Eds.; Einstein Studies; Springer International Publishing: Cham, Switzerland, 2020; pp. 15–84. [CrossRef]
298. Light, R.; Moody, J. *The Oxford Handbook of Social Networks*; Oxford University Press: Oxford, UK, 2021. [CrossRef]
299. Wüthrich, A. Characterizing a Collaboration by Its Communication Structure. *Synthese, unpublish work.*
300. Šešelja, D. Agent-based models of scientific interaction. *Philos. Compass* **2022**, *17*, e12855. [CrossRef]
301. Resnik, D.B. A Proposal for a New System of Credit Allocation in Science. *Sci. Eng. Ethics* **1997**, *3*, 237–243. [CrossRef]
302. Rennie, D.; Yank, V.; Emanuel, L. When Authorship Fails: A Proposal to Make Contributors Accountable. *JAMA* **1997**, *278*, 579–585. [CrossRef] [PubMed]
303. Cronin, B. Hyperauthorship: A postmodern perversion or evidence of a structural shift in scholarly communication practices? *J. Am. Soc. Inf. Sci. Technol.* **2001**, *52*, 558–569. [CrossRef]
304. Galison, P., The Collective Author. *Scientific Authorship: Credit and Intellectual Property in Science*; Galison, P., Biagioli, M., Eds.; Routledge: New York, NY, USA; Oxford, UK, 2003; pp. 325–353.
305. Wray, K.B. Scientific Authorship in the Age of Collaborative Research. *Stud. Hist. Philos. Sci. Part A* **2006**, *37*, 505–514. [CrossRef]
306. McNutt, M.K.; Bradford, M.; Drazen, J.M.; Hanson, B.; Howard, B.; Jamieson, K.H.; Kiermer, V.; Marcus, E.; Pope, B.K.; Schekman, R.; et al. Transparency in authors' contributions and responsibilities to promote integrity in scientific publication. *Proc. Natl. Acad. Sci. USA* **2018**, *115*, 2557–2560. [CrossRef]
307. Bright, L.K.; Dang, H.; Heesen, R. A Role for Judgment Aggregation in Coauthoring Scientific Papers. *Erkenntnis* **2018**, *83*, 231–252. [CrossRef]
308. Heesen, R. Why the Reward Structure of Science Makes Reproducibility Problems Inevitable. *J. Philos.* **2018**, *115*, 661–674. [CrossRef]
309. Dang, H. Epistemology of Scientific Collaborations. Ph.D. Thesis, University of Pittsburgh, Pittsburgh, PA, USA, 2019.
310. Nogrady, B. Hyperauthorship: The publishing challenges for 'big team' science. *Nature* **2023**, *615*, 175–177. [CrossRef]
311. Habgood-Coote, J. What's the Point of Authors? *Br. J. Philos. Sci.* 2021, *forthcoming.* [CrossRef]

Article

The Next Generation Event Horizon Telescope Collaboration: History, Philosophy, and Culture

Peter Galison [1,2,3,*], Juliusz Doboszewski [1,4,*], Jamee Elder [1,4,*], Niels C. M. Martens [5,4,6,7,*], Abhay Ashtekar [8], Jonas Enander [9], Marie Gueguen [10], Elizabeth A. Kessler [11], Roberto Lalli [12,13], Martin Lesourd [1], Alexandru Marcoci [14], Sebastián Murgueitio Ramírez [15], Priyamvada Natarajan [1,16,17], James Nguyen [18], Luis Reyes-Galindo [19], Sophie Ritson [20], Mike D. Schneider [21], Emilie Skulberg [22,23], Helene Sorgner [7,24], Matthew Stanley [25], Ann C. Thresher [26], Jeroen Van Dongen [22,23], James Owen Weatherall [27], Jingyi Wu [27] and Adrian Wüthrich [7,28]

1 Black Hole Initiative, Harvard University, Cambridge, MA 02138, USA
2 Department of History of Science, Harvard University, Cambridge, MA 02138, USA
3 Department of Physics, Harvard University, Cambridge, MA 02138, USA
4 Lichtenberg Group for History and Philosophy of Physics, University of Bonn, 53113 Bonn, Germany
5 Freudenthal Institute, Utrecht University, 3584 CC Utrecht, The Netherlands
6 Institute for Theoretical Particle Physics and Cosmology, RWTH Aachen University, 52074 Aachen, Germany
7 Epistemology of the LHC Research Unit, University of Wuppertal, IZWT, Gaußstraße 20, 42119 Wuppertal, Germany
8 Physics Department & Institute for Gravitation and the Cosmos, Pennsylvania State University, State College, PA 16802, USA
9 Independent scholar, Stockholm, Sweden
10 Institute of Physics, University of Rennes 1, 35042 Rennes, France
11 American Studies Program, Stanford University, Stanford, CA 94305, USA
12 Max Planck Institute for the History of Science, 14195 Berlin, Germany
13 Dipartimento di Ingegneria Meccanica e Aerospaziale, Politecnico di Torino, 10129 Turin, Italy
14 Centre for the Study of Existential Risk, University of Cambridge, Cambridge CB2 1SB, UK
15 Department of Philosophy, Purdue University, West Lafayette, IN 47907, USA
16 Department of Astronomy, Yale University, New Haven, CT 06511, USA
17 Department of Physics, Yale University, New Haven, CT 06520, USA
18 Department of Philosophy, Stockholm University, 10691 Stockholm, Sweden
19 Knowledge Technology and Innovation/Philosophy Groups, Wageningen University, 6708 WG Wageningen, The Netherlands
20 Research Innovation Commercialisation, The University of Melbourne, Parkville 3010, Australia
21 Department of Philosophy, University of Missouri, Columbia, MO 65201, USA
22 Institute of Physics, University of Amsterdam, 1090 GL Amsterdam, The Netherlands
23 Vossius Center for History of Humanities and Sciences, University of Amsterdam, 1090 GL Amsterdam, The Netherlands
24 Department of Science Communication and Higher Education Research & Department of Science, Technology and Society Studies, University of Klagenfurt, 9020 Klagenfurt, Austria
25 Gallatin School of Individualized Study, New York University, New York, NY 10003, USA
26 McCoy Family Center for Ethics in Society and the Stanford Doerr School for Sustainability, Stanford University, Stanford, CA 94305, USA
27 Department of Logic and Philosophy of Science, University of California, Irvine, CA 92697, USA
28 Institute of History and Philosophy of Science, Technology, and Literature, Technische Universität Berlin, 10623 Berlin, Germany
* Correspondence: galison@fas.harvard.edu (P.G.); jdobosze@uni-bonn.de (J.D.); jelder@fas.harvard.edu (J.E.); n.c.m.martens@uu.nl (N.C.M.M.)

Citation: Galison, P.; Doboszewski, J.; Elder, J.; Martens, N.C.M.; Ashtekar, A.; Enander, J.; Gueguen, M.; Kessler, E.A.; Lalli, R.; Lesourd, M.; et al. The Next Generation Event Horizon Telescope Collaboration: History, Philosophy, and Culture. *Galaxies* **2023**, *11*, 32. https://doi.org/10.3390/galaxies11010032

Academic Editor: Christian Corda

Received: 15 November 2022
Revised: 8 December 2022
Accepted: 19 January 2023
Published: 15 February 2023

Abstract: This white paper outlines the plans of the History Philosophy Culture Working Group of the Next Generation Event Horizon Telescope Collaboration.

Keywords: black holes; ngEHT; robustness; no hair theorems; scientific collaborations; philosophy; history; social sciences; governance; visualization

---

1. Introduction

Coordinating author: Galison, P.; Contributing authors: Elder, J. and Thresher, A.C.

Deep in the development of physics lie crucial intersections of science and philosophy. When Isaac Newton released his *Principia Mathematica* to the world, he included a "Scholium" on space and time. It contains no diagrams, mathematical expressions, experimental reports, theorems, or specific laws of motion or gravity. Instead, the Scholium sets out the starting terms of the inquiry itself, delving into the nature of space, time, and place. "I must observe", Newton insisted, "that the common people conceive those quantities under no other notions but from the relation they bear to sensible objects. And thence arise certain prejudices, for the removing of which it will be convenient to distinguish them into absolute and relative, true and apparent, mathematical and common." Rulers and clocks, calendars and sunrises, all the motions we use to tell time: these were merely the observable, "sensible" aspects of our basic concepts. How, asked Newton, are we "to obtain the true motions from their causes, effects, and apparent differences, and the converse." These deeply philosophical questions motivated the writing of the *Principia* ([1], pp. 6–12).

For Einstein, too, philosophical analysis was essential to subverting conformist tendencies in approaching central questions of physics. Nowhere was this more important than in his relativity theories: first, in his revision of space, time, and simultaneity in special relativity, leading to the unified spacetime introduced by Hermann Minkowski; and second, in Einstein's far deeper 1915 reconfiguration of spacetime in general relativity.[1] Einstein drew on a range of philosophical influences: from his youth forward, Einstein maintained a persisting interest in the work of Immanuel Kant and the neo-Kantians; he and his "Olympia Academy" dug line-by-line into Henri Poincaré's work on conventionalism; he sustained an abiding, if critical, interest in the work of the Vienna Circle; he also borrowed from Ernst Mach, who was deeply suspicious of an absolute, sense-independent notion of space and time. Throughout his life, Einstein believed that epistemology—the study of the formation, nature and justification of knowledge—and science "are dependent upon each other. Epistemology without contact with science becomes an empty scheme. Science without epistemology is—insofar as it is thinkable at all—primitive and muddled." ([7], pp. 683–684).[2]

We are now in a golden age of astronomy replete with extraordinary astrophysical objects. Of these, none has elicited as much fascination as black holes. The one-way membrane of the event horizon, the inner region where spacetime trajectories can cross themselves, and the singular breakdown of spacetime structure are just some of the provocations that black holes have presented to history and philosophy of science. Black holes thus present an opportunity to continue the tradition of intertwining groundbreaking physics with historical, philosophical, and cultural analysis.

From its start in 2015/16, the Black Hole Initiative (BHI) has set the history and philosophy of black holes alongside mathematics, physics, and astronomy as a crucial disciplinary ingredient.[3] Though based at Harvard, the BHI has drawn on collaborators far beyond its halls. Many of the scientists within the BHI have also been involved with the Event Horizon Telescope (EHT), a long-running, planetary-scale virtual telescope composed of widely-dispersed observatories.

The EHT observatories, eight on six sites as of 2017, and expanded to eleven observatories since, register millimeter electromagnetic waves from the same source by putting the data on hard drives with precise time stamps given by a hydrogen maser. The drives are then transported to central computing facilities where supercomputer "correlators" align the recorded signals. These aligned data can then be used to create images. In April 2019, the EHT Collaboration released the first ever picture of a black hole, M87*, the 6.5 billion solar mass compact object at the center of the elliptical galaxy M87 in the constellation Virgo [19]. Three years later, the EHT issued an image of the supermassive black hole, Sgr A*, at the center of the Milky Way [20]. Extending this work, the next generation EHT (ngEHT) aims to supplement the EHT network of telescopes with an additional ten or more sites that would fill out the virtual telescope and bring in new hardware and software, that together would make possible higher-resolution pictures and even movies.[4]

In the imaging campaign leading to the first pictures of M87* and Sgr A*, cross-fertilisation of science studies with the work of the EHT imaging group placed black hole images within a broader historical-epistemic context of pictorial argumentation. This allowed the objectivity of the black hole images to be framed in terms of longer-term and analytic approaches to the objectivity of images [21]. The goal now is to expand this imbrication in the next generation Event Horizon Telescope, setting the History, Philosophy, and Culture (HPC) Working Group as one of the eight science working groups of the collaboration as of 2022. These working groups will bring to bear on the study of black holes the resources of the history and philosophy of science along with the panoply of disciplines that compose Science and Technology Studies (STS). More specifically, the goal is to put this interdisciplinary working group into productive conversation with the other science and technical working groups—in the process of research and not as a post hoc account. Parallel to the other working groups, HPC will divide into four focus groups:

1. Algorithms, Inference, and Visualization,
2. Foundations,
3. Collaborations,
4. Siting, Education, In- and Outreach.

The Algorithms, Inference and Visualization (AIV) focus group aims to understand the epistemic and aesthetic choices that will guide ngEHT image production. To do so, the group will work closely with the Algorithms and Inference Working Group of the ngEHT. The AIV focus group provides a philosophical, historical, and social scientific complement to this working group, providing a space for a comparative discussion of inference methods and the broader social context of image dissemination. In this article (Section 2) we will report on the power and limits of "robustness" as an analytic virtue, and on the visual conventions of the EHT and ngEHT to come.

The Foundations focus group builds on the existing BHI Foundations Seminar, which draws historians, philosophers, and scientists to its meetings on topics ranging from the thermodynamics of black holes to the nature of singularities. In this article (Section 3) we discuss the relationship between theory and observation, through selected topics of foundational interest (e.g., no hair theorems) that illustrate the often-complex nature of this relationship.

Alongside these bridges between history, philosophy and scientific work are questions about the constitution of the ngEHT. What structure should its governance have? How should the collaboration ensure transparency, choose scientific goals, and assure representation in decision-making? What rules of the road should guide comportment in the collaboration, ranging from authorship and credit to collegiality, diversity, equity, and inclusion? Such questions will be addressed by the Collaborations focus group. Here we include a preliminary discussion of these issues (Section 4), drawing not only on the History and Philosophy of Science (HPS) but on the broader mix of Science and Technology Studies (STS) (including sociological and ethnographic work). To these questions, we offer initial reflections on the broad range of topics within the purview of the AIV, Foundations, and Collaborations focus groups—*initial, not final*, as befits these early, formative days of the ngEHT.

One important note: we acknowledge the cultural, historical, epistemic, political, environmental, and economic issues that surround the siting of telescopes. These problems have recently been at the fore of both academic and public interest due to ongoing conflicts at places like the Thirty Meter Telescope in Hawai'i, and the Square Kilometre Array in South Africa and Australia, where local communities have protested the projects for reasons including a lack of inclusion, concern for religious, cultural, and environmental sites, and the ongoing role of science within the longer history of colonialism and self-determination.[5] These sites, and others, highlight the need for careful discussions of our ethical obligations towards local communities, individuals, and the environment when building instruments.[6]

Given the importance of such topics, we have decided more serious work is required before we comment on the normative aspects of siting. As such, we will not be discussing siting in this paper, but are instead determined to build and maintain a broadly-diverse, appropriately interdisciplinary focus group dedicated to the topic, drawing on community members, scientists, philosophers, humanists, and social scientists to frame these issues. We anticipate producing publications dedicated solely to this topic in the near future.

2. Algorithms, Inference, and Visualization

Coordinating author: Doboszewski, J.; Contributing authors: Elder, J.; Enander, J.; Galison, P.; Gueguen, M.; Kessler, E.A.; Nguyen, J.; Skulberg, E.; Stanley, M. and Van Dongen, J.

2.1. Introduction

The Algorithms, Inference, and Visualization (AIV) focus group is a space for a general and comparative discussion of inference methods. The overarching goal is to analyse (and also contribute to) the epistemic and aesthetic choices that will guide ngEHT image production and interpretation. Many lessons can be learned from other computationally heavy areas of science (such as climate science or cosmological simulations) and other large experiments in physics. Here we discuss two example clusters of questions of interest to the AIV: robustness and reliability of imaging methods, and aesthetic choices in black hole imaging. A broader look at such issues will allow us to keep track of the range of factors contributing to decision-making, leading to better-informed choices in the long run.

2.2. Robustness and Reliability of Imaging

"Robustness" is often used in discussions of EHT data and results, including the analyses of both M87* and Sgr A*. Here we offer a short guide to its different uses in the scientific and philosophical literature, before we turn to discussing the use of robustness in justifying EHT and ngEHT results.

The robustness of a result can be characterized as the claim that if a variety of derivations, tests, or lines of evidence converge on a result, then that result is more secure than if it were obtained with only a single line of evidence. For that boost in confidence to hold, lines of evidence should be, in some sense, independent: convergence should not be attributable to some mistaken or irrelevant assumption shared by all lines of evidence (although see [28] for a discussion of the difficulty explicating what this amounts to).[7]

Experimental results are robust in the above sense when aspects of the experimental setup are varied, but results nonetheless converge—for example, when multiple independent measurements of Avogadro's number produce consistent results, these results are considered to be robust. In typical experimental situations, many factors can be varied, including the sample population or control group, initial or boundary conditions, and the measurement apparatus. Many such variations are impossible in the (ng)EHT, which will deal with a small number of sources, initially sparse sampling, lack of control over sources, and a lack of alternative instruments capable of performing the same measurements. However, multiple redundancies are built into the EHT measurements. For example, the use of varied calibration pipelines builds confidence that the result is not due to idiosyncratic factors in a particular pipeline. For some purposes like mass measurements other means of accessing the system (e.g., observations of S stars orbiting Sgr A*) also contribute to the robustness of the EHT results.

The results of modeling and data analysis methods can also be called robust when they are consistent across variations in modeling assumptions, analysis methods, or parameter choices. The robust occurrence of some features (e.g., the temperature increase for a range of climate simulations, or ring size for a range of EHT imaging methods) increases confidence in that aspect of the modeling outcomes, while other, less stable features (e.g., regional precipitation for climate simulations, the positions of bright 'knot' structures in EHT images of Sgr A*) should be treated with caution.

Among the main lines of criticism formulated against robustness arguments, two seem especially relevant in the context of the ngEHT. Such criticisms envisage the ensemble of models containing (1) a shared core of assumptions, which make the models comparable, and (2) an unshared part, deemed problematic (e.g., modeling assumptions, idealizations, parametrizations or measurement apparatus), whose possible impact on the models' output must be understood and eliminated.

The first criticism argues that in numerical models the shared core common to all models tends to include problematic assumptions. Common idealizations, such as iterative and discretization errors, are unavoidable to numerically solve the problem but are also important sources of numerical artifacts. Hence, their impact cannot be determined through robustness reasoning. The second criticism points out that the mere convergence of results cannot by itself indicate a reliability or partial truth: something else is needed. Gueguen [29] examines a number of cases where convergent results across N-body simulations may be attributable to numerical artifacts. For example, Baushev et al. [30] point out that N-body cosmological simulations predict a "cuspy" profile for dark matter halo density for galaxy center regions (in conflict with observations). They argue that the convergence of simulations on such predictions is produced by numerical artifacts rather than by a physically realistic process captured by the simulations. This case shows how the apparent robustness of simulation results may not indicate that the results are reliable. As emphasized by [31] in their response to the seminal paper by [32] on robustness, from a purely logical point of view, robustness can guarantee reliability only in those cases where we already know that one of the models in the set is correct.[8] This condition is rarely satisfied when robustness is the most needed, i.e, when it is used to supplement the absence of analytic solutions or experimental measures that could determine whether one of the models is indeed correct. Hence there is a clear need to analyze when robustness is an efficient tracer of reliability within the ngEHT program, and when it needs to be supplemented or substituted.

In the suite of papers that the EHT issued on M87* [19,33–39] and Sgr A* [20,40–44], the collaboration's overwhelming concern was to establish, with confidence, the existence of a ring surrounding the black hole shadow. That is, the EHT Collaboration did not want to issue a false positive. For that reason, in the M87* image work *robustness* was key; the collaboration: varied the priors to make sure those choices were not forcing the image to be a ring; isolated four image-making groups to avoid cross-contaminating expectations based on others' results; and varied image reconstruction methods to ensure that the observed ring was not an artifact of any one imaging method. These measures constituted a determined drive to be sure that in the image of M87* the ring and bright crescent in the south were as unshakeable as possible. The commitment to robustness came with an unavoidable cost: other, valid effects—observations outside the ring, for example—might have been omitted. However, especially for this first, momentous publication, the collective desire was for an appropriately robust, and therefore conservative, claim.

Yet, robustness is not the only possible epistemic desideratum. Over the course of the next generation of work, we may well want to pursue other, complementary ambitions. With highly specific models, physicists and observers could explore other predicted phenomena that might otherwise be lost in the noise. More unsteady, delicate phenomena in the accretion disk and jet formation, for example, could be detected using models and templates of various kinds. In particle physics, such targeted searches are common—this is what triggers do when they pluck a particular signal, interaction, particle, or phenomenon out of the vast sea of other results. Indeed, in many domains of physics, initial statements of groundbreaking results are more statistically fragile.[9] Robustness is thus a core epistemic virtue, but not the only one: too strong an emphasis on it could lead to false negatives by blinding us to hard-to-see phenomena just above the noise. Selectivity, pushed too hard, can produce false positives, giving us back what we hope and expect to see. We need *both* robustness *and* model-based selectivity. However, there are epistemic trade-offs to be made between these different epistemic virtues. Future work with the ngEHT will involve decisions about which virtues to prioritize in which contexts.

2.3. Science and Aesthetics in Black Hole Imaging

All images have style expressed through "shared visual features" ([46], p. 4). Graphs, for instance, tend to avoid detail. Certain color schemes are more used than others. Including or removing artifacts is another choice. Astronomical images, whether based on empirical data or simulations, reflect an array of choices and decisions, and they also participate in their larger historical and cultural contexts. Their creation and interpretation rely on pre-existing visual traditions that establish the norms, expectations, and methods by which a scientific image is given meaning. The AIV focus group will draw on the extensive scholarship on imaging in astronomy and physics to reflect on such image-making choices and decisions by the ngEHT, as well as how the results are received and understood both within the scientific community and beyond [21,45,47–63].

Over the last several decades, images have furthered scientific understandings of black holes. However, until the EHT images, these representations were based on astronomers' calculations and simulations rather than observations. In the early 1970s, researchers visualized the basic outlines of black holes but their images were still in a schematic style [64–66]. Later in the same decade, more detailed and naturalistic visualizations of black holes emerged: a film clip by Leigh Palmer, Maurice Pryce, and William Unruh (unpublished, but shared in multiple lectures) and Jean-Pierre Luminet's black and white drawing of a black hole accretion disk [67]. Yet later, color visualizations, such as those by Heino Falcke, Fulvio Melia, and Eric Agol of Sgr A*, theorized how the black hole shadow might look if observed using VLBI [68] (see also [69] for the first visualizations in color).

Simulations remained a critical part of the EHT imaging process, resulting in observations that integrated theory in interesting ways. New data imaging pipelines were developed and used together with a library of synthetic images produced by general relativistic magnetohydrodynamic simulations and general relativistic ray tracing [36,39,43]. Comparing the observations with theoretical simulations was key for establishing that the observed ring was created by synchrotron emission from a hot plasma orbiting near the black hole. Although these specific techniques were novel, astronomers have long been aware of the dependence of their observations on theory. The need to reduce collected data to a more concise and tractable form in order to account for phenomena such as stellar aberration, atmospheric refraction, or the so-called "personal equation" (variations due to a specific observer's idiosyncrasies) means that astronomy as a discipline has reflected on the role of theory in making raw data into useful depictions of celestial bodies for generations [70,71]. There is a long intellectual ancestry of ever-more complex reliance on theory to allow for increasingly powerful forms of observation and imaging. These increases in scope and depth, however, also required more delicate conceptual and social scaffolding, increasing the possible influence of bias and blind-spots [72,73]. The EHT Imaging Group was keenly aware of concerns about bias and systematic error; from the beginning, the imaging process was shaped by these concerns, in order to ensure the validity of the image [19].

Another concern for the EHT was the legibility of their images for a wide audience—particularly for the first image of M87*, given its novelty. The color palette—a ring in orange-red hues against a black background—was chosen with this in mind; orange was believed more likely to signify heat than blue (even though blue has shorter wavelengths and is therefore "hotter" than orange). Because the EHT Collaboration wanted to share one image with audiences of varying degrees of specialization (see [74], on the basis of [75]), a single averaged image was created from multiple images based on different imaging methods. Notably, the averaging of the Sgr A* image was different than that of M87*, with the former averaging process being more complex than the latter (see [19] for M87* and [20] for Sgr A*). These averaging techniques also connect to historical practices going back to the very beginning of technology-assisted scientific images with Galileo, Hooke, and Hevelius.

Such figures used compositing techniques to make their early telescopic and microscopic images legible to wide, non-specialist audiences (particularly those who did not have access to the relevant instruments). Even through the nineteenth century, and well into the twentieth, it was accepted that astronomers would need to synthesize many individual observations in order to produce a reliable drawn or photographic image [58,76,77].

Given that more than a billion people saw the M87* image within days of its release [78], EHT imaging choices will continue to influence how black holes are perceived and understood. The next generation of images produced by the ngEHT will build on these perceptions while introducing new considerations; increasing the bandwidth, including other frequencies, and adding telescope sites, will allow for greater resolution, and the production of moving images (movies). This means that further choices will need to be made about how to convey this information in an image.

The history of astronomical images (and their reception) offers models to consider. Many existing astronomical images use color to distinguish between different wavelengths, and the hues often signify both physical properties and evoke aesthetic responses. For example, color in many Hubble Space Telescope images indicates relative temperatures while also creating a resemblance to the sublime nineteenth-century paintings of the western regions of the USA [79]. Such seemingly naturalistic color choices elicit questions from viewers, who assume color corresponds to human perception. In other instances—remote sensing of the Earth and some planetary images—more obviously engineered color choices enhance morphology yet emphasize the reliance on technology to extend human vision [63,79]. Looking forward, ngEHT might also find it valuable to seek models beyond the history of scientific images when making decisions on how to represent data. This could include representation of movement in film or video games, or examining the work of artists who use scientific data as the basis of their aesthetic explorations [80]. EHT images of M87* and Sgr A* have elicited a range of responses (from awe to disappointment) and have already shaped the iconography of black holes [74]. ngEHT imaging represents an opportunity to consider once again how imaging decisions, whether motivated by scientific or aesthetic concerns, shape the scientific and public perception of black holes.

3. Foundations

Coordinating author: Elder, J.; Contributing authors: Ashtekar, A.; Doboszewski, J.; Enander, J.; Lesourd, M.; Murgueitio Ramírez, S.; Schneider, M.D.; Thresher, A.C. and Weatherall, J.O.

3.1. Introduction

The Foundations focus group is an extension of the existing Foundations Seminar at the Black Hole Initiative (BHI). This seminar provides a venue for discussion of foundational issues relating to black holes. Previous themes of the seminar include: singularities, black hole thermodynamics, the analytic extension of the exterior Kerr metric, and theory vs. observation in astrophysics (among others). As we take on a new role as a focus group of the HPC working group, we will aim to facilitate further discussion of these themes in the context of the ngEHT.

In what follows, we illustrate issues that arise from such discussions. To do so, we narrow the focus to the final theme in the above list: bridging the gap between theory and observation. In Section 3.2, we provide some examples of where challenges arise for the applicability of theoretical results to real-world black holes. This includes a discussion of the no-hair theorems in Section 3.2.1 and a discussion of the relationship between concepts like mass, charge, and angular momentum in cosmological settings with and without a (positive) cosmological constant, in Section 3.2.2. Then, in Section 3.3, we sketch some philosophical responses to these apparent challenges.

The key questions that we seek to address in this section are these: how do we (or should we) apply formal mathematical results to a messy world where many of the assumptions behind those results are not, strictly speaking, realized? Furthermore, how can empirical results be brought to bear on theory in such cases? Our goal is to address such questions in the context of (supermassive) black holes such as those observed by the EHT and ngEHT. While the discussion of these questions here is only a beginning, answering such questions in the future will have important consequences for our understanding of applications (and tests) of theoretical results using the ngEHT array.

Overall, this section serves as an example of the kinds of discussion that will continue to take place within the Foundations seminar as it takes on a second, complementary role as a focus group within the HPC working group. In addition to the theme discussed here, singular spacetimes, black hole thermodynamics, and other foundational topics concerning black holes will be the subject of ongoing philosophical discussion.

3.2. Challenges for the Applicability of Theory to Astrophysical Black Holes: Two Examples

Astrophysicists and astronomers often refer to exact solutions of the Einstein field equations—especially the Schwarzschild and Kerr metrics—when describing and interpreting their observations but there are potential problems with this. The Schwarzschild and Kerr metrics are highly idealized, involving assumptions that might not be physically realistic (see [81] for a discussion of this point in the context of the notion of an event horizon). For example, astrophysical black holes exist in the presence of matter fields, in a universe whose expansion is characterized by a positive cosmological constant, whereas these two metrics are solutions of the vacuum Einstein field equations and are asymptotically flat. It is therefore imperative to investigate the domain of applicability of these descriptions for astrophysical black holes. This means carefully explicating the ways that these solutions are used and examining the conditions under which the idealizations inherent in these solutions may or may not be problematic.

For illustrative purposes, we briefly consider two examples: the physical relevance of the no-hair theorems and the applicability of quantities such as mass, charge, and angular momentum for $\Lambda > 0$, where Λ is the cosmological constant.

3.2.1. No-Hair Theorems

It is widely assumed that the geometry around astrophysical black holes is well described by the Kerr (or Kerr–Newman) family of metrics.[10] The justification for this is based on the application of so-called 'no hair' theorems, according to which stationary black hole spacetimes solving the Einstein field equations in vacuum, or the Maxwell–Einstein field equations with an electromagnetic stress-energy tensor, are exhausted by the Kerr and Kerr–Newman families of metrics, respectively.

However, this line of reasoning depends on a range of assumptions that may be called into question for physically realistic black holes. First, the no-hair theorems apply to stationary black holes (see Section 1 of [83]), so their application relies on the assumption that astrophysical black holes eventually settle down to a stationary state. Second, existing no-hair theorems rely on various mathematical assumptions that are highly unrealistic. In the standard formulation, analyticity of the spacetime metric is required in order to show the existence of the appropriate Killing vector fields; but astrophysical modeling of gas and plasma strongly suggests the presence of shocks in the vicinity of a black hole, making analyticity an implausible assumption. Third, the no-hair theorems are known to fail in the presence of matter fields (other than electric fields); see Section 5 of [83] for a variety of examples arising if the source side of the Einstein's field equations is a (classical) Yang–Mills term.

This illustrates some important concerns about the applicability of no-hair theorems for astrophysical black holes. Given that several of the assumptions behind the theorem do not, strictly speaking, hold in reality, to what extent should we expect real black holes to be well-described by the Kerr(–Newman) metrics? Furthermore, are there ngEHT observations that might provide evidence of deviations from Kerr(–Newman)? No such deviations have been observed by the EHT to date. However, ref. [44] provides constraints on potential deviations from the Kerr metric based on the 2017 observations of Sgr A*.

3.2.2. Mass, Charge, and Angular Momentum in $\Lambda > 0$

If we study the Einstein field equations with $\Lambda = 0$, adopting certain assumptions about global spacetime structure (e.g., that the underlying manifold is simply connected at infinity and spacetime geometry is asymptotic to Minkowski spacetime), the theory of general relativity seems to single out a small number of global quantities—ADM mass, charge, and angular momentum[11]—which play a central role in understanding and quantifying basic astrophysical phenomena. However, cosmological observations support the conclusion that the accelerated expansion of the universe is well described by a positive cosmological constant, i.e., $\Lambda > 0$ [84]. Some have taken this empirical finding to signify the need for a better understanding of the character of global quantities in an asymptotically de Sitter universe, to replace the ones currently in use (see [85] for discussion on this inference, including caveats). Recent progress in defining and understanding counterparts of the ADM quantities in the $\Lambda > 0$ case has been made by Abhay Ashtekar and collaborators [86–89]. Unlike the familiar ADM quantities noted above, these new ones take for granted different assumptions about global spacetime structure.

It would be prudent to clarify the relationships between the global quantities in the $\Lambda = 0$ and $\Lambda > 0$ cases, including the role of the flat case in characterizing black holes in the $\Lambda > 0$ context. In particular, doing so seems necessary to interpret what astronomers are observing when they measure the mass, spin, etc. of real astrophysical black holes under an idealizing assumption that the cosmological constant vanishes. The central issue here is a general question about how to interpret global properties and asymptotic spacetime assumptions as relevant to astrophysical modeling. However, a further issue arises: how to interpret specific asymptotic assumptions within idealized models in a situation where physical expectations about the expected asymptotic spacetime structure ($\Lambda > 0$) are very different from that of an idealized model? One might hope that a systematic understanding of isolated systems includes an interpretation of asymptotic spacetime assumptions as becoming approximately true 'far away' [90]. However, in a $\Lambda > 0$ context, there would seem to be such a thing as moving 'too far away' from the isolated system under study (due to the presence of cosmological horizons that separate distant observers from the system). Therefore, considering the particular case of $\Lambda > 0$ complicates any such story about asymptotically flat structure becoming approximately true.

3.3. Theory and Observation: Bridging the Gap

Purported problems like the examples above elicit a range of responses from strict to pragmatic. One guiding question for the Foundations group moving forward is the following: how can we apply theory to observations (and vice versa) when strictly speaking there is a mismatch (e.g., the conditions of theorem are not met in the real world)? Furthermore, how can this be justified? Answering these questions will generally be sensitive to the details of the case—including the precision of the description needed and the stability of the theoretical results across changes in assumptions. For now, we defer detailed consideration of the above examples to future work. Here, we instead outline some different perspectives on the general theme along with some guiding philosophical morals.

A strict approach, prioritizing mathematical rigor, is to adopt a cautious stance and not apply concepts or models outside the domain in which the assumptions behind them are true. If the assumptions behind a theorem are not true then it is not considered to be physically relevant. This approach embodies a conservatism toward epistemic risk, prioritizing the avoidance of errors over pursuing potentially fruitful (but risky) avenues of reasoning. On such a view, the issues described above are indeed considered to be problematic, amounting to a pressing need for further study and understanding. The no-hair theorem case, for example, suggests a need for a better understanding of black holes beyond the Kerr–Newman family. The manifest failure of no-hair theorems in the presence of matter fields means that we should absolutely expect to see deviations from the Kerr metric in the near-horizon regime. A better understanding of what these deviations could be and how we might test for them should be part of the scientific landscape.

A more pragmatic approach to these issues is based on a different conception of the roles of models in scientific inferences. Indeed, a recent 'pragmatic turn' [91] in the philosophy of modeling and measurement has led to a greater emphasis on epistemic goods such as reliability (e.g., [92]) and adequacy for purpose (e.g., [93]) over truth.

For Cartwright et al. [92] the mismatch between models and the real world is resolved by noting that science gets to truth via reliability. Indeed, the vast majority of science is not the kind of thing that takes truth values. Models, along with things like measures, experiments, codes, narratives and techniques are essential parts of science; and yet, what would it mean to say a code or technique is true? Instead, we can ask the more important question—are they reliable? If so, for what? In what context? On this view, reliability, far more than truth, captures the actual goals and structures of science and helps explain why models are useful for black hole physics—because our goal is to create reliable systems for capturing black hole dynamics and properties: systems that, in turn, give us reliable results for the particular job at hand. We have only to look at processes of model-building and model-selection to see this in action—particular idealizations are chosen, and values set, that get us closest to useful results. This, in turn, is taken to be a proxy for truth that works provided we remain within the context the model is built or adapted to be useful for.[12]

From this perspective, asking whether the assumptions underlying various foundational results are true is misguided, and better questions would be whether the assumptions are reasonably clear and the results are useful for various purposes. This seems to be the attitude adopted by many working physicists. However, even if one grants that a pragmatic, or even instrumentalist, attitude to foundational issues is justified for many practical purposes, one might think simply dismissing the foundational worries raised above is too fast. One reason is that a way in which our models can be useful and even reliable is by identifying points of tension in our understanding of a given physical system—in this case, black holes. Those points of tension, where models with apparently overlapping domains disagree, or where it is unclear whether the assumptions of this or that theorem truly apply to a given case, have historically been catalysts for developing new physics that can explain why different, apparently inconsistent, models nonetheless work in different contexts. A too-radical form of instrumentalism about scientific modeling would presumably reject the demand to make our models consistent, or to at least resolve the tensions that may arise between them [94,95].

Future ngEHT observations will play a mediating role, bridging the gap between real astrophysical black holes and idealized theoretical descriptions of them. Doing so will mean scrutinizing the reliability of our best models of black holes and the domains of applicability of theoretical results pertaining to them.

4. Collaborations

Coordinating author: Martens, N.C.M.; Contributing authors: Doboszewski, J.; Elder, J.; Galison, P.; Lalli, R.; Marcoci, A.; Nguyen, J.; Ritson, S.; Schneider, M.D.; Skulberg, E.; Sorgner, H.; Van Dongen, J.; Wu, J. and Wüthrich, A.

4.1. Introduction

The Collaborations focus group lies at the intersection of various approaches within the humanities and social sciences, including history, philosophy, sociology, science and technology studies, integrated history and philosophy of science, and law. As a result, we combine a mix of different methodologies, including literature analysis, comparative case studies (e.g., ATLAS, LIGO-Virgo, IPCC, Hubble, JWST; see below), tools from the digital humanities, interviews and surveys. This will enhance our ability to engage with the rest of the ngEHT collaboration in a way that includes a diversity of opinions, all with the aim of supporting a constant dialogue to provide real-time recommendations to the ngEHT collaboration, qua collaboration, at each of its various stages of development and operation.

The focus group concerns itself with the relationship between individuals and the ngEHT collaboration as a whole. To address ngEHT's social epistemology (i.e., how knowledge is produced in social groups such as scientific collaborations), we delve into how knowledge is conditioned by the collaborative production of data, images, and text, and what the process of negotiation entails for its claims about the world.[13] It is clear from previous large-scale collaborations—and the ngEHT will be no exception—that *the establishment of fact,* and *what constitutes a fact* are to some extent the result of the social negotiation of consensus.[14] Thus, group structure and the distribution of authority play a direct role in what counts as knowledge. For instance, the particle physics community has converged on near-universal conventions regarding the determination of facts—five sigmas are required for a discovery—whereas only two sigmas are required to exclude new physics hypotheses.[15] In contrast, there is currently no such shared standard in astrophysics.

The importance of human judgments was evident throughout the EHT imaging process. For example, multiple imaging pipelines produced their own images of M87* based on a range of choices (imaging methods, specific algorithms, priors and other inputs, etc.). These results then had to be aggregated in order to present a result that represented the collective judgment of the collaboration. The averaged image of M87* released in 2019 reflects a particular choice about how to do this aggregation (cf., Section 2).[16]

4.2. Knowledge Formation and Governance: Top-Down vs. Bottom-Up

Large-scale scientific collaboration can take place within a variety of governance/organizational structures, ranging from top-down hierarchical structures to more loosely organized bottom-up collaboration in the absence of a formal governing structure. We, the ngEHT collaboration, see ourselves as (ideally) being located somewhere in the middle of this spectrum—in particular somewhat closer to the bottom-up extreme than the EHT collaboration. In this section, we briefly illustrate this claim by contrasting the ngEHT with instances of scientific collaboration found at either extreme—specifically, the particle physics collaborations ATLAS and CMS associated with the Large Hadron Collider (LHC), the gravitational-wave-detecting LIGO–Virgo collaboration (LVC), and the Intergovernmental Panel on Climate Change (IPCC).

At the top-down extreme of the spectrum, partially exemplified by ATLAS and CMS, as well as the LVC, we find hierarchical structures with a centralized, physical headquarters and funding stream, with one or a few main instruments or purposes, and a large number of committees that decide which collaboration papers are published and how, which members get to present at which conference, etc. The collaboration is prioritized over the individual member; consensus is prioritized over dissent and diversity of opinions, with dissent being procedurally dealt with internally before (consensus) results are published. This structure facilitates a strong group identity, obtaining a large amount of funding for a dedicated, coordinated purpose, and achieving that purpose in the most efficient way possible. However, there is a risk that individual credit and creativity are lost to some extent. In contrast to these top-down examples, the ngEHT is a loosely organized, informally scripted, yet formally documented collaboration. Although workshops and conferences bring together researchers for short periods of time, observations will take place from different continents, researchers usually work from different geographical locations, and no building has been constructed for the purpose of housing ngEHT research. Instead, the asynchronous electronic infrastructure ([105], p. 159) of Overleaf, Slack, Google documents, slides, and telecons will be used to coordinate matters.

At the bottom-up end of the spectrum, we do not find formal collaborations per se but instead entire scientific communities with a common subject and a more or less uniform research culture. In such cases, authors coalesce in and out of projects, with members of the community communicating via conferences and peer-reviewed publications rather than in a physical headquarters. In this bottom-up model, individual groups can pursue any research direction that they themselves consider fruitful—as long as they manage to get funding—and publish dissenting results. A coherent, negotiated narrative connecting all these results and delineating the *facts* is more likely to be established later (if at all), through review papers and review presentations in textbooks. Particularly striking examples are meta-analyses in medical communities or the recent report by the Intergovernmental Panel on Climate Change (endnote 5) which synthesizes 14,000 papers from the climate science community. In contrast to such extreme bottom-up examples, some sustained collaboration is required to achieve the ngEHT's main goals: financing and building additional telescopes and coordinating the whole network of telescopes so that it has access to; the joint reduction of data; and, finally, reporting its findings in publications. Moreover, it is important to stress that maximizing the benefits of bottom-up approaches does not come for free; it is not a mere matter of the absence of a top-down governance structure, but also the implementation of positive measures that bring out the advantages of bottom-up approaches, such as room for diversity and individual creativity.

One important challenge for the ngEHT then, regarding the spectrum of bottom-up versus top-down approaches to social epistemology and governance, is to be the best rather than the worst of both worlds. In the remainder of this section, we outline some preliminary thoughts on how this can be achieved. In particular, we discuss the need to facilitate dissent (Section 4.3) and to adopt a governance charter (Section 4.4).

4.3. Knowledge Formation: Differences of Opinion

Should large scientific collaborations aim for consensus? The extent to which consensus is ideal for a scientific collaboration depends on how consensus is construed. First, we can consider the *unit* of consensus: should the group agree on individual propositions or collections of (logically connected) propositions?[17] Second, we can consider the *bearer* of consensus. In the first instance, whether a group should aim at consensus may depend on the nature of the collaboration: what ties the individuals together?[18] In the second instance, when we attribute consensus to a group, are we "summarising" the attitudes of the individuals, or does the collaborative aspect add something to this—possibly in the sense of a "plural subject" or a "group agent" [108–110]? Third, we can distinguish between at least two *attitudes* relevant to the consensus: if a group is in consensus does it (or each member of it) hold a consensual belief, or a consensual acceptance, where different epistemic norms are associated with each attitude (e.g., belief requires a commitment to truth while acceptance may not) [111–114].[19] Fourth, we can ask about the *extent* of consensus: at one extreme consensus might be identified with unanimity, but some level of dissent may be consistent with consensus, and indeed, as we discuss below, even encouraged [116]. Clarifying each of these dimensions allows us to ask more fine-grained questions about the nature and desirability of consensus (e.g., we can attribute a consensus belief to the group without necessarily requiring that all, or even any, of the individuals, believe all, or even any, of the propositions the group believes, although they may accept them in virtue of being in the collaboration).

The above requires us to take a step back and ask what *being in the ngEHT collaboration* actually means. Issues such as who may be a member of the ngEHT collaboration and an author of the collaboration's papers need to be made explicit. How is membership established, and what does it imply to be a member? Which rights, responsibilities and credits follow from membership? Who may become a member? Is a vetting procedure required, and which members get to decide who else may become a member? Should there be different types of membership? Are all members also on the author list of collaboration papers? Is it possible to be a member without being an author? Might different types of authorship (e.g., data compiler, data analyst, text writer) be desirable? How are papers written and what epistemic goals might be favored by such a process? What happens if the collaboration is succeeded by another, or splits up: who owns the collaborative knowledge? Answers to these questions make clear who is a party to making knowledge; and thus also, what constitutes knowledge.[20] In the near future we aim to survey how different modalities of membership and authorship have been crafted in comparable yet different collaborations (ATLAS at the LHC, LVC, and IPCC), and make an inventory of current practices in the EHT and ngEHT collaborations, including an analysis of their advantages and drawbacks.

Returning to consensus, some construal of consensus is *prima facie* valuable and to be expected in scientific collaborations. First, because epistemic peers presented with the same evidence are, on the first approach, expected to reach the same conclusions [120]. Second, the higher the number of independent and competent scientists who believe a particular claim, the more likely it is to be true (the relevant result is a generalized Condorcet's Jury Theorem [121,122]). Third, the stronger the consensus for a claim, the more likely it is for the general public to accept it [123]. Finally, a lack of consensus is often what politicians and lobbyists use to undermine the findings of scientific collaborations [124].

On the other hand, there are reasons to be wary of some construals of consensus. Consensus between individuals may be impossible to achieve in contexts where the collaboration involves individuals with different values and/or disparate areas of expertise. Furthermore, the fact that epistemic peers *may* reasonably disagree on substantive issues motivates the applicability of judgment aggregation theory to scientific collaborations [125,126]. Finally, when consensus is enforced through a collaboration's policies in a top-down fashion (cf. Section 4.2), this may disincentivize deliberation and the exploration of competing hypotheses [127]; it may also produce the appearance of agreement when there is none [120,128–130].

It is thus important to find a good balance between top-down and bottom-up approaches to structuring an organization (cf. Section 4.2) that promotes consensus-building without prematurely suppressing dissent. Having a diversity of beliefs and practices among team members can be epistemically beneficial to science. For example, individuals in collaboration may draw on different (and possibly even competing) sources of evidence and theories in order to justify their conclusions [131]. Moreover, if all team members test the same hypothesis (and especially, by means of the same methods), they may prematurely settle for false beliefs. Several authors (notably [132]) have advocated for a period of *transient diversity* during scientific research when different epistemic options are sufficiently tested before the community settles on a consensus.

Mechanisms that allow for or encourage transient diversity thus present strategies to promote a desired kind of creativity at the group level within bottom-up research contexts [133,134]. While the influence of (diverging) non-cognitive values in science is unavoidable, it is not necessarily pernicious [135], and transient diversity could provide one such mechanism. Indeed, a more inclusive representation of values and perspectives is expected to produce epistemically more robust results [136]. Increasing transient epistemic diversity may also be helped by incorporating perspectives from marginalized groups into the scientific inquiry [137–140]. Furthermore, facilitating minority views and carefully publicizing (partial) dissent increases transparency and enhances rather than erodes the credibility of the collaborations' conclusions [116,120]. One motivation for this bottom-up line of reasoning stems from the social turn in the philosophy of science [141]: emphasizing the political, social, and psychological aspects of scientific collaborations encourages the idea that trustworthy decisions in science, as in other social institutions, requires deliberation, transparency and openness. Enforcing consensus goes against these norms.

In light of the above, what techniques and policies should guide collaboration within the ngEHT? Firstly, there are several mechanisms that can generate (transient) diversity. Of particular interest are modeling results [142] showing that the less connected the epistemic community is, the more likely it is to converge to the true belief—but the slower it is at doing so [132,143–145]. For high stakes frontier research where it is important to be correct, it may be warranted to temporarily limit communication between team members. For instance, the limited communications between the imaging teams at the EHT may have epistemically benefited the final results [142,146].[21]

Moreover, there already is evidence regarding the benefits of including groups traditionally excluded from knowledge production; some local and Indigenous communities on EHT's sites would have relevant scientific knowledge that other team members do not (see [147] for collaboration with Indigenous communities). However, empirical and simulation results show that marginalized and minoritized people often receive less credit in scientific collaborations [148–150]. Given this, collaborative teams should consider explicit, ongoing discussions about credit assignment procedures, being particularly vigilant about assigning fair credit to marginalized knowers' contributions—this will be one of the roles of the ethics committee proposed in the next subsection. A related concern is that creative research is stifled and individuals are prevented from developing diverse and novel ideas. Large research collaborations may tend towards conservatism, in part stemming from multiple requirements for collective approval [151] and a preference for well-tested over novel approaches [152]. When considering how we might ideally organize a research collaboration, it is thus important to consider creativity from both an individual and a collective perspective [153], including the opportunities for researchers to publish individual contributions to collaborative research, such as PhD theses [154].

The Collaborations focus group will also explore how ngEHT members interact with one another. Methodologically, we can use concepts and tools from network theory to quantitatively investigate the structure of the collaboration and its change over time. By using a multi-layered network perspective of socio-epistemic networks we can investigate how the social structure is related to the production of new knowledge [155,156]. Network approaches also allow us to understand the flow of information within the collaboration. An illustrative example in this regard is recent work analyzing more than 20,000 emails sent via internal mailing lists of a major particle physics collaboration [157]. This analysis revealed a pronounced sub-structure of the communication network featuring smaller "communities" within the collaboration. The communication network is also relatively dense and, in a network-theoretical sense, less hierarchical than most such networks, which is surprising given the top-down governance structures in place. Such analyses of communications networks may provide insight into how large-scale collaborations collectively produce knowledge.

Similar network analyses could also be done for the ngEHT. This descriptive project could also inform the normative guidance that we provide to the collaboration; the analyses could be used to test hypotheses about what communication structures might be particularly conducive to epistemic success, and which mechanisms and governance structures would foster such communication. This work could then be connected to the rich body of literature spanning decision theory, social psychology, and mathematics that explores the advantages and drawbacks of different ways of structuring deliberation between, and eliciting judgments from, experts [158,159], as well as formal frameworks for conceptualizing the relationship between the attitudes of individuals and the attitudes of the group [106]. For the ngEHT, the exact balance between seeking collectivist consensus from the outset or operating via integration and trade-offs between autonomous viewpoints will depend on how data and responsibilities are shared among members, whether there are distinct organizational sub-units within the collaboration, and what the final arbiter is in cases of conflict (e.g., whether an appeal to a higher authority is possible, and how that authority is legitimized). The authorship of publications (whether they are mainly collectively authored or authored by distinct groups within the collaboration) will likely reflect these organizational norms [160].

In sum, it is clear that it would be beneficial for the ngEHT not to enforce consensus in the top-down fashion known from, among others, the various LHC collaborations. The Collaborations focus group aims to enrich the somewhat abstract existing literature by investigating concrete mechanisms and organizational structures that can maximize the benefits of epistemic diversity, applicable to the ngEHT context via a detailed analysis of the practice of the ngEHT collaboration with tools from the digital humanities and with internal surveys. It is crucial that these organizational structures are geared towards representation, diversity, sufficient freedom for individual creativity, the appropriate balance between transparency and epistemic distance at various stages of the collaboration, and appropriate assignment of credit, as elaborated upon in the following subsection.

4.4. Governance

Well-structured governance is key to the future of collaboration. A main task for the Collaborations focus group will be to systematically analyze the organizational structure of various similar collective entities, including LIGO-Virgo, EHT, ATLAS, CMS, CERN, IPCC, the UN, Hubble and JWST, to identify their main benefits and drawbacks. Surveys conducted among ngEHT members, based on a similar survey conducted within the EHT collaboration, will also provide valuable data moving forward. These lessons will be synthesized into the optimal governance model for the ngEHT, keeping in mind the desiderata and worries described in the previous subsections.

To give the reader a tentative first impression of what such a governance model might look like, we sketch here an initial suggestion. We view this as the beginning of an ongoing conversation about the optimal governance model for the ngEHT collaboration. This model will then be iteratively tested and improved, especially with regards to how it facilitates knowledge formation and adapted as circumstances change. Given that the nascent ngEHT collaboration has already begun to take shape, it is crucial that this group make what recommendations we can—however preliminary—at this early stage. We are now in a position to influence organizational structures that may become increasingly entrenched as the ngEHT project gains momentum.

The core of the collaboration is its eleven working groups—eight science working groups (including HPC) and three technical working groups. In other collaborations, working groups have worked particularly well to generate a sense of community and strong science. The major Principal Investigators that lead the working groups alongside (and overlapping with) the Management Team—including the ngEHT director, chief scientist, and chief engineer—take on the dual responsibilities of fiscal probity (fulfilling the contracts) and keeping a steady hand on the tiller to keep the collaboration in line with its founding goals. They would be guided and supported by a small number of governance structures (Figure 1): a central Scientific Council, a Project Advisory Committee, a Facilities Advisory Board, an Ethics Committee and a Publication Committee. These structures are not intended to provide top-down constraints by appointees, such as forcing consensus, but are instead (partially) elected, representative bodies that streamline the collaboration in a way that celebrates diversity and raises ethical scientific comportment to a primary aim.

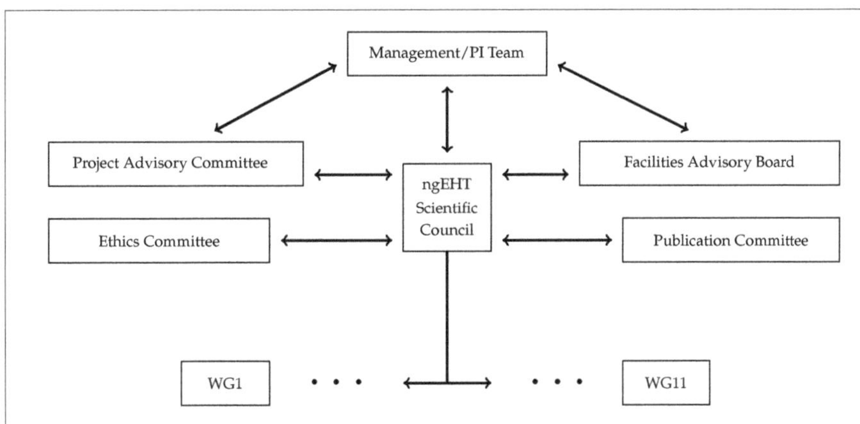

Figure 1. Tentative governance structure of the ngEHT collaboration.

Scientific Council & Project Advisory Committee. The ngEHT, like LIGO, includes multiple sites and dozens of scientific groups. To run its program, LIGO established a scientific council (LSC) that determines the scientific priorities and the overall mission—responsible too for science, instrumentation, communication, and operation. Composing the LSC are representatives of the various groups, in proportion to their membership size.[22]

The ngEHT might follow a modified version of that model which offers a way for the membership to shape scientific and technological policy and to facilitate decisions about priorities (such as targets, observation cadence, instrument standards, and aims). The ngEHT Scientific Council would be composed of representatives chosen by the constituent groups—no such representative body exists within the EHT collaboration. Where participating institutions or other stakeholders, including local communities and junior members, are too small to field separate representatives, they could be grouped together to form a larger body. The elected council would receive advice from the already existing Project Advisory Committee/Science Advisory Board), consisting of appointed, experienced and mostly external scholars, including Nobel laureates.

Ethics Committee & Transparent Ethical Charter. In founding the ngEHT, a charter specifying structure is desired, but should equally include transparent record keeping, voting procedures, and appointments as well as principles of membership, publication, authorship, credit, and conflict resolution. Along with these procedures, the charter would lay down a guiding, forceful commitment to diversity, equity and inclusion, as well as to ethical comportment regarding fairness, respectful interactions, and accountability. Putting this in the founding charter would give it the weight it deserves, showing these values are foundational, not *pro forma*. As groups join the ngEHT, it would be essential, in addition, to have a Memorandum of Understanding underscoring commitment to the charter and to the particular roles and responsibilities of the group. A high-level ethics committee—ideally its members would include several members of the History Philosophy Culture Working Group—would be tasked with drafting this charter to be sent to the rest of the collaboration for feedback, with overseeing the adherence to this charter once in place, and with updating the charter based on continuous feedback. It should maintain and publicize policies to promote an equitable, inclusive, and welcoming workplace. This committee could also include or run elections to identify ombudspersons and mediators as part of a broader mandate to do all in its power to stop intolerable actions visited upon collaborators such as harassment, bullying or marginalization on the basis of race, gender, nationality, or identity.

Facilities Advisory Board. The ngEHT will use some established facilities and so, in part, resembles an experiment at a particular facility telescope—the ngEHT will apply for time. Essential to realizing its mission, the ngEHT aims to build approximately ten additional sites beyond the existing telescope facilities made use of by the EHT: five in a first phase with an additional five to follow. The Facilities Advisory Board would consist of representatives of some of the telescopes or groups of telescopes, and, if needed, scientifically-relevant facilities (e.g., large-scale computation/correlators) even if they are not direct stakeholders. Note that the Facilities Advisory Board and Project Advisory Committee are separate entities, in contrast to the structure of the EHT collaboration.

Publication Committee. The aim of the publication committee would not be to provide negative constraints beyond standard checks regarding the use of proprietary data. It is not to be a gatekeeper that approves the official opinions and results of the members of the collaboration. Instead, its aim is positive: to streamline the process of publications through the collaboration and work of smaller subsets of members that relates to the ngEHT, by coordinating internal review in cases where this may be helpful, by ensuring that credit is given where credit is due, and by coordinating the ngEHT science book and other strategies that enhance the overall visibility of ngEHT related outputs, all in line with policies set out in the ngEHT's charter.

The ngEHT, like the IPCC, is an overarching framework for dozens of institutes across the world, each funded in different ways. Like the IPCC, the ngEHT has working groups. In contrast to the ngEHT, the IPCC was formed by an international compact, offering not novel research but a mechanism for collective, reliable assessment of existing research—including evaluators of different career stages, genders, and geographical regions. The ngEHT could learn from the way the IPCC has honed methods of assembling expert judges to assess both scientific/technical questions and to assist in effective final write-ups of the work.

Similarly, the LHC detectors ATLAS and CMS have elaborated effective (but different) means of evaluating their own work before publication, which could serve as inspiration.

In sum, a governance model like the model proposed above would serve to support the working groups and help them excel, not by providing constraints that prioritise the collaboration over the individual working group members, but in a way that streamlines their work by ensuring diverse representation of the various stakeholders.

5. Conclusions

This white paper has presented some—but by no means all—of the plans of the History Philosophy and Culture Working Group of the ngEHT collaboration. It is unprecedented for scholars from the humanities and social sciences to be integrated into a physics collaboration of this size, from the very beginning and with the same standing within the collaboration structure as its STEM members. We would like to cordially invite other scholars from the humanities and social sciences to join us in this exciting endeavor of making the ngEHT a prime model for interdisciplinary collaboration and recording high-quality videos of a black hole together.

Author Contributions: Writing—original draft preparation, all authors; writing—review and editing, all authors; supervision, P.G.; project administration, P.G., J.D., J.E. and N.C.M.M. All authors have read and agreed to the published version of the manuscript.

Funding: J. Doboszewski, J. Elder and N.C.M. Martens would like to thank the Volkswagen Foundation for its support in providing the funds to create the Lichtenberg Group for History and Philosophy of Physics at the University of Bonn. M. Gueguen and N.C.M. Martens would like to thank the European Union's Horizon 2020 research and innovation programme for the funding received under the Marie Skłodowska-Curie grant agreements No. 101026214 and No. 101065772, respectively. P. Galison, J. Doboszewski, J. Elder, M. Lesourd, and P. Natarajan also acknowledge the support of the Black Hole Initiative, which is funded by grants from the John Templeton Foundation and the Gordon and Betty Moore Foundation (although the opinions expressed in this work are those of the authors and do not necessarily reflect the views of these Foundations). J. Elder also acknowledges the support of the "Inductive Metaphysics" project funded by the Deutsche Forschungsgemeinschaft (DFG), Research Unit FOR 2495 (specifically subproject B6: "The Role of Inference to the Best Explanation in the Discovery of Gravitational Waves"). N.C.M. Martens, H. Sorgner, and A. Wüthrich's contribution was made possible by funding from the DFG (FOR 2063)/FWF (I 4410-G) Research Unit "Epistemology of the LHC", and A. Wüthrich's contribution furthermore by funding from the European Union (ERC, Project NEPI, No. 101044932). Views and opinions expressed are however those of the authors only and do not necessarily reflect those of the European Union or the European Research Council. Neither the European Union nor the granting authority can be held responsible for them.

Data Availability Statement: Not applicable.

Acknowledgments: We want to recognize the early and important contributions of T. Nichols, N. Conway (on outreach), A. Raymond, G. Fitzpatrick, M. Johnson (on technical siting) to the formation of the HPC working group, which in turn built on the already long-running BHI Foundations seminar (with many thanks e.g., to F. Azhar, M. Lesourd, E. Curiel and the participants of that seminar). In framing the scope of the still-developing Siting focus group that will report in subsequent publications, the Siting Workshop conveners and framers, including A. Thresher and P. Natarajan (later joined by D. Palumbo), thank the presenters at the first Siting Workshop which has helped guide subsequent developments: C. Prescod-Weinstein, K. Kamelamela, H. Nielson, M. Johnson, K. Fox, J. Havstad, T. Nichols, R. Chiaravalloti, S. Doeleman, G. Fitzpatrick, J. Houston, A. Oppenheimer. We would like to thank Jonas Enander, Luis Reyes-Galindo, Mike Schneider & Jeroen van Dongen for their internal review of this white paper. We are grateful for valuable discussions with the attendees of the HPC Kick-Off Workshop (Black Hole Initiative, Harvard, Feb–March 2021), with the attendees of the ngEHT meeting (Granada, June 2022), and with the other members of the HPC working group.

Conflicts of Interest: The authors declare no conflict of interest.

Notes

[1] A very helpful framing of the history of general relativity can be found in [2]. On Einstein's special theory of relativity, focusing on his redefinition of simultaneity, see [3]. On the eclipse expedition of 1919 and its surround—as a historical example of observational history, see [4,5]. On Einstein's own trajectory to general relativity, see [6].

[2] On the philosophically-inflected work of Einstein, see, as an entrée into the literature [8–12]; and for a launch into the philosophy in Einstein's physics see [9,13]. Of books on the philosophy of spacetime, Earman's have been a grounding point of many discussions [14,15], as has the (physics-based) lapidary take on general relativity by Wald [16]. For a fine example of a more recent conceptual analysis, see [17].

[3] On the long-term history of relativity as it opened up into the science of black holes in particular, see [18].

[4] See ([20], Sections 4.4 and 9) for discussion of "dynamic imaging", which results in a movie of the source (i.e., a series of images or frames) instead of a single image.

[5] Two excellent doctoral dissertations offer fine-grained analysis of the mountaintop dispute, and include a wide range of further references. Swanner [22] focuses on the triply conflicting astronomical, environmental and indigenous narratives that collided at Mt. Graham, Mauna Kea, and Kitt Peak; Salazar [23] addresses the Kanaka rights claim, specifically addressing the Thirty Meter Telescope (TMT), in opposition to a framing of the dispute as one of "stakeholders" or a "multicultural" ideal. Swanner focuses on Mauna Kea in a subsequent article, also on the TMT [24]. For an important current Hawaiian-led impact assessment of the TMT including additional references, see Kahanamoku et al. [25]. Many further references across a wide cross-disciplinary range including archaeology, biology, among others, will be given in a subsequent paper directed toward siting.

[6] Highlighting the environmental, social, experimental, and ethical implications of locating scientific facilities through a robust history of locating LIGO's sites, see Nichols, T. [26,27]

[7] If "secure" is understood in terms of degrees of belief (expressed by some function satisfying the Kolmogorov axioms of probability), then "boost in confidence" can be understood as (something like) the statement that the conjunction $E_1 \& \ldots \& E_n$ confirms R to a greater extent than E_i alone, for any i; where R is the result, and E_i are lines of evidence.

[8] Here we retain Orzack and Sober's terminology, describing models as "true" or "correct". Note, however, that this terminology is controversial (see Section 3.3) with some recent philosophical treatments of models suggesting that models themselves are neither true nor false.

[9] On the contrast between inclusive and selective instrumental demonstration in particle physics, see Galison [45].

[10] Or, in the context of a positive cosmological constant (see Section 3.2.2), perhaps instead one assumes a Kerr-de Sitter (or Kerr–Newman–de Sitter) metric. A good recent discussion of black holes with positive cosmological constant is in ([82], ch.5). One way to give these metrics is by writing them in Boyer-Lindquist coordinates, including some functions δ and σ, which are functions of radius, spin, mass, and Λ. The mass read off from such a solution is the same as the mass of the Kerr metric.

[11] ADM stands for Richard Arnowitt, Stanley Deser and Charles W. Misner, authors of the Hamiltonian formulation of general relativity known as the ADM formalism, within which the ADM quantities are defined.

[12] This perspective also has implications for how we think about the use of robustness reasoning discussed in Section 2.2.

[13] For instance, it is well known that the more authors a scientific paper has, the more conservative the claims in the paper may be, and the longer (on average) the paper, as well as its title, tend to be [96]. Single-authored blogs tend to be more readable than blogs authored by two authors, as measured by the Flesch readability score, despite no difference in average sentence length [97]. If this can be extrapolated to journal papers with large numbers of authors, the ngEHT may want to experiment with breaking up papers into separate papers, each of which is written by a smaller set of authors, and/or for the writing to be done by the smallest possible number of people with other members of the project providing input in other ways/at other stages (e.g., everyone is involved in outlining the structure of the paper and the eventual editing, but not in the writing process in between). The latest report by the Intergovernmental Panel on Climate Change (IPCC) provides a model of such a practice. A first draft by one of their working groups (WG1) was written by just the working group, comprising 240 scientists (Assessment Report [AR] 1 WG1 IPCC, 2021). After this, a much larger number of scientists from around the world provided comments that were incorporated into subsequent drafts. The ngEHT could consider writing papers following this model, scaled down according to the smaller number of scientists involved.

[14] On the historical contingency of our notion of fact, see [98–101].

[15] On the role of 'sigma's' in modern physics, see [102].

[16] On the practice of averaging over black hole images as epistemic practice, see [103,104].

[17] Work in judgment aggregation theory highlights the impact these relations can have on the consistency of the group attitude, see [106].

[18] See [107]'s distinction between the "commitment" and "distributed" models of group knowledge.

[19] The distinction between belief and acceptance can also help us conceptualize the role of idealization in science, as discussed in Section 3.3, see, for instance, [115].

[20] Compare, e.g., with discussions on including string theorists as physicists [117–119].

[21] Interesting in this regard is the current ngEHT analysis challenge, where part of the collaboration creates a training set from simulated signals with noise added to them (and potentially also some fake signals), with another part of the collaboration honing their analysis tools on this training data without knowing how it was created.

[22] On the LIGO Scientific Collaboration Charter [161].

References

1. Newton, I. Newton's Scholium on Time, Space, Place and Motion. In *Philosohiae Naturalis Principia Mathematica*; Motte, A., Translator; University of California Press: Berkely, CA, USA, 1934. Available online: https://plato.stanford.edu/entries/newton-stm/scholium.html (accessed on 23 August 2022).
2. Eisenstaedt, J. *The Curious History of Relativity: How Einstein's Theory of Gravity Was Lost and Found Again*; Princeton University Press: Princeton, NJ, USA, 2006.
3. Galison, P. *Einstein's Clocks, Poincaré's Maps. Empires of Time*; W.W. Norton: New York, NY, USA, 2003.
4. Kennefick, D. *No Shadow of a Doubt. The 1919 Eclipse That Confirmed Einstein's Theory of Relativity*; Princeton University Press: Princeton, NJ, USA, 2019.
5. Stanley, M. *Einstein's War. How Relativity Triumphed Amid the Vicious Nationalism of World War I*; Dutton: New York, NY, USA, 2019.
6. Renn, J. *Albert Einstein: Chief Engineer of the Universe. Documents of a Life's Pathway*; Wiley-VCH: Hoboken, NJ, USA, 2005.
7. Einstein, A. Remarks to the Essays Appearing in this Collective Volume. In *Albert Einstein Philosopher-Scientist*; Schilpp, P.A., Ed.; MJF Books: New York, NY, USA, 1970; pp. 663–688.
8. Holton, G. Mach, Einstein, and the search for reality. In *Thematic Origins of Scientific Thought*; Harvard University Press: Camrbidge, MA, USA, 1988; pp. 237–277.
9. Ryckman, T. *The Reign of Relativity. Philosophy in Physics 1915–1925*; Oxford University Press: Oxford, UK, 2005.
10. van Dongen, J. *Einstein's Unification*; Cambridge University Press: Cambridge, UK, 2010.
11. Janssen, M.; Lehner, C. *The Cambridge Companion to Einstein*; Cambridge University Press: Cambridge, UK, 2014.
12. Howard, D.A.; Giovanelli, M. Stanford Encyclopedia of Philosophy. In *The Stanford Encyclopedia of Philosophy*; Chapter Einstein's Philosophy of Science; Metaphysics Research Lab, Stanford University: Stanford, CA, USA, 2019.
13. Norton, J.D. Philosophy in Einstein's science. In *Idealist Alternatives to Materialist Philosophies of Science*; MacEwen, P., Ed.; Brill: Leiden, The Netherlands, 2019; pp. 95–127.
14. Earman, J. *World Enough and Space-Time: Absolute versus Relational Theories of Space and Time*; MIT Press: Cambridge, MA, USA, 1992.
15. Earman, J. *Bangs, Crunches, Whimpers, and Shrieks: Singularities and Acausalities in Relativistic Spacetimes*; Oxford University Press: New York, NY, USA, 1995.
16. Wald, R.M. *Space, Time, and Gravity: The theory of the Big Bang and Black Holes*; University of Chicago Press: Chicago, IL, USA, 1992.
17. Curiel, E. The many definitions of a black hole. *Nat. Astron.* **2019**, *3*, 27–34. [CrossRef]
18. Thorne, K. *Black Holes and Time Warps. Einstein's Outrageous Legacy*; W.W. Norton: New York, NY, USA, 1995.
19. Akiyama, K.; et al. [The Event Horizon Telescope Collaboration]. First M87 event horizon telescope results. IV. Imaging the central supermassive black hole. *Astrophys. J. Lett.* **2019**, *875*, L4. [CrossRef]
20. Akiyama, K.; et al. [The Event Horizon Telescope Collaboration]. First Sagittarius A* Event Horizon Telescope Results. III. Imaging of the Galactic Center Supermassive Black Hole. *Astrophys. J. Lett.* **2022**, *930*, L14. [CrossRef]
21. Daston, L.; Galison, P. *Objectivity*; Zone Books: Brooklyn, NY, USA, 2007.
22. Swanner, L.A. Mountains of Controversy: Narrative and the Making of Contested Landscapes in Postwar American Astronomy. Ph.D. Thesis, Harvard University, Cambridge, MA, USA, 2013.
23. Salazar, J.A. Multicultural Settler Colonialism and Indigenous Struggle in Hawai'i: The Politics of Astronomy on Mauna a Wākea. Ph.D. Thesis, University of Hawai'i at Manoa, Honolulu, HI, USA, 2014.
24. Swanner, L.A. Instruments of Science or Conquest? Neocolonialism and Modern American Astronomy. *Hist. Stud. Nat. Sci.* **2017**, *47*, 293–319. [CrossRef]
25. Kahanamoku, S.; Alegado, R.A.; Kagawa-Viviani, A.; Kamelamela, K.L.; Kamai, B.; Walkowicz, L.M.; Prescod-Weinstein, C.; Reyes, M.A.D.L.; Neilson, H. A Native Hawaiian-led summary of the current impact of constructing the Thirty Meter Telescope on Maunakea. *arXiv* **2020**, arXiv:2001.00970.
26. Nichols, T. Constructing Stillness: Theorization, Discovery, Interrogation, and Negotiation of the Expanded Laboratory of the Laser Interferometer Gravitational-Wave Observatory. Ph.D. Dissertation, Harvard University, Cambridge, MA, USA, 2022.
27. Nichols, T. Hidden in Plain Sight: Merging the Physics Laboratory with the Surrounding Environment. Unpublished manuscript, *submitted*.
28. Schupbach, J.N. Robustness analysis as explanatory reasoning. *Br. J. Philos. Sci.* **2018**, *69*, 275–300. [CrossRef]
29. Gueguen, M. On Robustness in Cosmological Simulations. *Philos. Sci.* **2020**, *87*, 1197–1208.
30. Baushev, A.; del Valle, L.; Campusano, L.; Escala, A.; Muñoz, R.; Palma, G. Cusps in the center of galaxies: A real conflict with observations or a numerical artefact of cosmological simulations? *J. Cosmol. Astropart. Phys.* **2017**, *2017*, 042. [CrossRef]
31. Orzack, S.H.; Sober, E. A Critical Assessment of Levins's The Strategy of Model Building in Population Biology (1966). *Q. Rev. Biol.* **1993**, *68*, 533–546. [CrossRef]
32. Levins, R. The Strategy of Model Building in Population Biology. *Am. Sci.* **1966**, *54*, 421–431.

33. Akiyama, K.; et al. [The Event Horizon Telescope Collaboration]. First M87 event horizon telescope results. I. The shadow of the supermassive black hole. *Astrophys. J. Lett.* **2019**, *875*, L1. [CrossRef]
34. Akiyama, K.; et al. [The Event Horizon Telescope Collaboration]. First M87 event horizon telescope results. II. Array and instrumentation. *Astrophys. J. Lett.* **2019**, *875*, L2. [CrossRef]
35. Akiyama, K.; et al. [The Event Horizon Telescope Collaboration]. First M87 event horizon telescope results. III. Data processing and calibration. *Astrophys. J. Lett.* **2019**, *875*, L3. [CrossRef]
36. Akiyama, K.; et al. [The Event Horizon Telescope Collaboration]. First M87 event horizon telescope results. V. Physical origin of the asymmetric ring. *Astrophys. J. Lett.* **2019**, *875*, L5. [CrossRef]
37. Akiyama, K.; et al. [The Event Horizon Telescope Collaboration]. First M87 Event Horizon Telescope Results. VI. The Shadow and Mass of the Central Black Hole. *Astrophys. J. Lett.* **2019**, *875*, L6. [CrossRef]
38. Akiyama, K.; et al. [The Event Horizon Telescope Collaboration]. First M87 Event Horizon Telescope Results. VII. Polarization of the Ring. *Astrophys. J. Lett.* **2021**, *910*, L12. [CrossRef]
39. Akiyama, K.; et al. [The Event Horizon Telescope Collaboration]. First M87 Event Horizon Telescope Results. VIII. Magnetic Field Structure near The Event Horizon. *Astrophys. J. Lett.* **2021**, *910*, L13. [CrossRef]
40. Akiyama, K.; et al. [The Event Horizon Telescope Collaboration]. First Sagittarius A* Event Horizon Telescope Results. I. The Shadow of the Supermassive Black Hole in the Center of the Milky Way. *Astrophys. J. Lett.* **2022**, *930*, L12. [CrossRef]
41. Akiyama, K.; et al. [The Event Horizon Telescope Collaboration]. First Sagittarius A* Event Horizon Telescope Results. II. EHT and Multiwavelength Observations, Data Processing, and Calibration. *Astrophys. J. Lett.* **2022**, *930*, L13. [CrossRef]
42. Akiyama, K.; et al. [The Event Horizon Telescope Collaboration]. First Sagittarius A* Event Horizon Telescope Results. IV. Variability, Morphology, and Black Hole Mass. *Astrophys. J. Lett.* **2022**, *930*, L15. [CrossRef]
43. Akiyama, K.; et al. [The Event Horizon Telescope Collaboration]. First Sagittarius A* Event Horizon Telescope Results. V. Testing Astrophysical Models of the Galactic Center Black Hole. *Astrophys. J. Lett.* **2022**, *930*, L16. [CrossRef]
44. Akiyama, K.; et al. [The Event Horizon Telescope Collaboration]. First Sagittarius A* Event Horizon Telescope Results. VI. Testing the Black Hole Metric. *Astrophys. J. Lett.* **2022**, *930*, L17. [CrossRef]
45. Galison, P. *Image and Logic: A Material Culture of Microphysics*; University of Chicago Press: Chicago, IL, USA, 1997.
46. Kemp, M. *Visualizations: The Nature Book of Art and Science*; University of California Press: San Diego, CA, USA, 2000.
47. Bigg, C. Travelling Scientist, Circulating Images and the Making of the Modern Scientific Journal: Norman Lockyer's Visual Communication of Astrophysics in Nature. *Nuncius* **2015**, *30*, 675–698. [CrossRef] [PubMed]
48. Bigg, C. The view from here, there and nowhere? Situating the observer in the planetarium and in the solar system. *Early Pop. Vis. Cult.* **2017**, *15*, 204–226. [CrossRef]
49. Elkins, J. *Six Stories from the End of Representation: Images in Painting, Photography, Astronomy, Microscopy, Particle Physics, and Quantum Mechanics, 1980–2000*; Stanford University Press: Stanford, CA, USA, 2008.
50. Fineman, M.; Saunders, B. *Apollo's Muse: The Moon in the Age of Photography*; Yale University Press: New Haven, CT, USA, 2019.
51. Hentschel, K.; Whittmann, A.D. *The Role of Visual Representations in Astronomy: History and Research Practice: Contributions to a Colloquium Held at Göttingen in 1999*; Deutsch: Thun Frankfurt am Main, Germany, 2000.
52. Hentschel, K. *Mapping the Spectrum: Techniques of Visual Representation in Research and Teaching*; Oxford University Press: New York, NY, USA, 2002.
53. Kaiser, D. *Drawing Theories Apart: The Dispersion of Feynman Diagrams in Postwar Physics*; University of Chicago Press: Chicago, IL, USA, 2005.
54. Lane, K.M.D. *Geographies of Mars*; University of Chicago Press: Chicago, IL, USA, 2010.
55. Messeri, L. *Placing Outer Space: An Earthly Ethnography of Other Worlds*; Duke University Press: Durham, UK; London, UK, 2016.
56. Nall, J. *News From Mars: Mass Media and the Forging of a New Astronomy, 1860–1910*; University of Pittsburgh Press: Pittsburgh, PA, USA, 2019.
57. Nasim, O. The 'Landmark' and 'Groundwork' of stars: John Herschel, photography and the drawing of nebulae. *Stud. Hist. Philos. Sci. Part A* **2011**, *42*, 67–84.
58. Nasim, O. *Observing by Hand: Sketching the Nebulae in the Nineteenth Century*; University of Chicago Press: Chicago, IL, USA, 2013.
59. Schaffer, S. On Astronomical Drawing. In *Picturing Science, Producing Art*; Galison, P.L., Jones, C.A., Eds.; Routledge: New York, NY, USA, 1998.
60. Pang, A.S.K. 'Stars should henceforth register themselves': Astrophotography at the early Lick Observatory. *Br. J. Hist. Sci.* **1997**, *30*, 177–202. [CrossRef]
61. Stanley, M. Merging the Sun and the Stars: The hybrid images of the 1919 eclipse. In Proceedings of the Presentation, American Physical Society Meeting, New York, NY, USA, 10 April 2022.
62. Tai, C.; van der Steen, B.; van Dongen, J. *Anton Pannekoek: Ways of Viewing Science and Society*; Amsterdam University Press: Amsterdam, The Netherlands, 2019.
63. Vertesi, J. *Seeing Like a Rover: How Robots, Teams, and Images Craft Knowledge of Mars*; University of Chicago Press: Chicago, IL, USA, 2015.
64. Godfrey, B.B. Mach's Principle, the Kerr Metric, and Black-Hole Physics. *Phys. Rev. D* **1970**, *1*, 2721.
65. Bardeen, J.M. Timelike and Null Geodesics in the Kerr Metric. In *Black Holes (Les Astres Occlus)*; DeWitt, C., DeWitt, B.S., Eds.; Gordon & Breach: New York, NY, USA, 1973; Volume 23, pp. 215–239.

66. Cunningham, C.T.; Bardeen, J.M. The optical appearance of a star orbiting an extreme Kerr black hole. *Astrophys. J.* **1973**, *183*, 237–264.
67. Luminet, J.P. Image of a spherical black Hole with thin accretion disk. *Astron. Astrophys.* **1979**, *75*, 228–235.
68. Falcke, H.; Melia, F.; Agol, E. Viewing the Shadow of the Black Hole at the Galactic Center. *Astrophys. J. Lett.* **2000**, *528*, L13–L16.
69. Fukue, J.; Yokoyama, T. Color photographs of an accretion disk around a black hole. *Publ. Astron. Soc. Jpn.* **1988**, *40*, 15–24.
70. Schaffer, S. Astronomers mark time: Discipline and the personal equation. *Sci. Context* **1988**, *2*, 115–145. [CrossRef]
71. Stanley, M. Where Is That Moon, Anyway? The Problem of Interpreting Historical Solar Eclipse Observations. In *Raw Data are an Oxymoron*; Gitelman, L., Ed.; MIT Press: Cambridge, MA, USA, 2013; pp. 77–88.
72. Shapere, D. The Concept of Observation in Science and Philosophy. *Philos. Sci.* **1982**, *49*, 485–525.
73. Pinch, T. Towards an Analysis of Scientific Observation: The Externality and Evidential Significance of Observational Reports in Physics. *Soc. Stud. Sci.* **1985**, *15*, 3–36. [CrossRef]
74. Skulberg, E. The Event Horizon as a Vanishing Point: A History of the First Image of a Black Hole Shadow from Observation. Ph.D. Thesis, University of Cambridge, Cambridge, UK, 2021.
75. Issaoun, S. (Radboud University, Nijmegen, The Netherlands). Interview by Emilie Skulberg, 12 September 2019.
76. Pang, A.S.K. The Social Event of the Season: Solar Eclipse Expeditions and Victorian Culture. *Isis* **1993**, *84*, 252–277.
77. Tai, C. Left Radicalism and the Milky Way: Connecting the Scientific and Socialist Virtues of Anton Pannekoek. *Hist. Stud. Nat. Sci.* **2017**, *47*, 200–254.
78. Kessler, E.; Galison, P. To See the Unseeable. *Aperture* **2019**, *237*, 75–78.
79. Kessler, E.A. *Picturing the Cosmos: Hubble Space Telescope Images and the Astronomical Sublime*; University of Minnesota Press: Minneapolis, MN, USA, 2012.
80. Clarke, V. *Universe: Exploring the Astronomical World*; Phaidon: London, UK, 2017.
81. Booth, I. Black-hole boundaries. *Can. J. Phys.* **2005**, *83*, 1073–1099. [CrossRef]
82. Chruściel, P.T. *Geometry of Black Holes*; Oxford University Press: Oxford, UK, 2020.
83. Chruściel, P.T.; Costa, J.L.; Heusler, M. Stationary black holes: Uniqueness and beyond. *Living Rev. Relativ.* **2012**, *15*, 1–73.
84. Aghanim, N.; Akrami, Y.; Ashdown, M.; Aumont, J.; Baccigalupi, C.; Ballardini, M.; Banday, A.J.; Barreiro, R.B.; Bartolo, N.; Basak, S.; et al. Planck 2018 results. *Astron. Astrophys.* **2020**, *641*, A6. [CrossRef]
85. Schneider, M.D. Empty Space and the (Positive) Cosmological Constant. *PhilSci Archive*. 2022. Available online: http://philsci-archive.pitt.edu/21076/ (accessed on 1 November 2022).
86. Ashtekar, A.; Bahrami, S. Asymptotics with a positive cosmological constant. IV. The no-incoming radiation condition. *Phys. Rev. D* **2019**, *100*, 024042. [CrossRef]
87. Ashtekar, A.; Bonga, B.; Kesavan, A. Asymptotics with a positive cosmological constant: I. Basic framework. *Class. Quantum Gravity* **2014**, *32*, 025004. [CrossRef]
88. Ashtekar, A.; Bonga, B.; Kesavan, A. Gravitational Waves from Isolated Systems: Surprising Consequences of a Positive Cosmological Constant. *Phys. Rev. Lett.* **2016**, *116*, 051101. [CrossRef] [PubMed]
89. Ashtekar, A. Implications of a positive cosmological constant for general relativity. *Rep. Prog. Phys.* **2017**, *80*, 102901. [CrossRef] [PubMed]
90. Wallace, D. Isolated systems and their symmetries, part II: Local and global symmetries of field theories. *Stud. Hist. Philos. Sci.* **2022**, *92*, 249–259. [CrossRef] [PubMed]
91. Bokulich, A.; Parker, W. Data models, representation and adequacy-for-purpose. *Eur. J. Philos. Sci.* **2021**, *11*, 1–26.
92. Cartwright, N.; Hardie, J.; Montuschi, E.; Soleiman, M.; Thresher, A. *The Tangle of Science: Reliability Beyond Method, Rigour, and Objectivity*; Oxford University Press: Oxford, UK, 2022.
93. Parker, W.S. Model Evaluation: An Adequacy-for-Purpose View. *Philos. Sci.* **2020**, *87*, 457–477. [CrossRef]
94. Stein, H. Yes, but... Some Skeptical Remarks on Realism and Anti-Realism. *Dialectica* **1989**, *43*, 47–65. [CrossRef]
95. Mitsch, C. An Examination of Some Aspects of Howard Stein's Work. *Stud. Hist. Philos. Mod. Phys.* **2019**, *66*, 1–13. [CrossRef]
96. Lewison, G.; Hartley, J. What is in a title? Numbers of words and presence of colons. *Scientometrics* **2005**, *63*, 341–356. [CrossRef]
97. Hartley, J.; Cabanac, G. Are two authors better than one? Can writing in pairs affect the readability of academic blogs? *Scientometrics* **2016**, *109*, 2119–2122. [CrossRef]
98. Daston, L.; Müller Wille, S.; Sibum, H.O. *A History of Facts*; Max Planck Institute for the History of Science: Berlin, Germany, 2001.
99. Poovey, M. *A History of the Modern Fact*; University of Chicago Press: Chicago, IL, USA, 1998.
100. ten Hagen, S.L. How "Facts" Shaped Modern Disciplines: The Fluid Concept of Fact and the Common Origins of German Physics and Historiography. *Hist. Stud. Nat. Sci.* **2019**, *49*, 300–337.
101. de Waal, E.; ten Hagen, S.L. The Concept of Fact in German Physics around 1900: A Comparison between Mach and Einstein. *Phys. Perspect.* **2020**, *22*, 55–80. [CrossRef]
102. Franklin, A. *Shifting Standards. Experiments in Particle Physics in the Twentieth Century*; University of Pittsburgh Press: Pittsburgh, PA, USA, 2013.
103. Galison, P. Philosophy of the Shadow. 2019. Available online: https://www.youtube.com/watch?v=BofWFoiKARQ (accessed on 23 January 2023).
104. Galison, P. (Director) The Edge of All We Know. *Netflix*, 2020.

105. Galison, P.; Jones, C.A. Factory, laboratory, studio: Dispersing sites of production. In *The Architecture of Science*; MIT Press: Cambridge, MA, USA, 1999; pp. 497–540.
106. List, C. The Theory of Judgment Aggregation: An Introductory Review. *Synthese* **2012**, *187*, 179–207. [CrossRef]
107. Bird, A. When Is There a Group that Knows?: Distributed Cognition, Scientific Knowledge, and the Social Epistemic Subject. In *Essays in Collective Epistemology*; Oxford University Press: Oxford, UK, 2014; pp. 42–63.
108. Quinton, A. Social Objects. *Proc. Aristot. Soc.* **1976**, *76*, 1–27. [CrossRef]
109. Gilbert, M. *On Social Facts*; Routledge: New York, NY, USA, 1989.
110. List, C.; Pettit, P. *Group Agency: The Possibility, Design, and Status of Corporate Agents*; Oxford University Press: Oxford, UK, 2011.
111. Wray, K.B. Collective Belief Furthermore, Acceptance. *Synthese* **2001**, *129*, 319–333. [CrossRef]
112. Gilbert, M.; Pilchman, D. Belief, Acceptance, and What Happens in Groups: Some Methodological Considerations. In *Essays in Collective Epistemology*; Lackey, J., Ed.; Oxford University Press: Oxford, UK, 2014.
113. Dang, H.; Bright, L.K. Scientific Conclusions Need Not Be Accurate, Justified, or Believed by Their Authors. *Synthese* **2021**, *199*, 8187–8203. [CrossRef]
114. Dethier, C. Science, Assertion, and the Common Ground. *Synthese* **2022**, *200*, 1–19. [CrossRef]
115. Elgin, C. *True Enough*; MIT Press: Cambridge, MA, USA, 2017.
116. Dellsén, F. Consensus versus Unanimity: Which Carries More Weight? *Br. J. Philos. Sci.* **2021**.
117. Galison, P. Mirror symmetry: Persons, values, and objects. In *Growing Explanations: Historical Explanations on Recent Science*; Wise, M.N., Ed.; Duke University Press: Durham, NC, USA, 2004; pp. 23–63.
118. van Dongen, J. String theory, Einstein, and the identity of physics: Theory assessment in absence of the empirical. *Stud. Hist. Philos. Sci. Part A* **2021**, *89*, 164–176. [CrossRef]
119. Greenspan, L. Holography, application, and string theory's changing nature. *Stud. Hist. Philos. Sci. A* **2022**, *94*, 72–86. [CrossRef]
120. Hulme, M. Lessons from the IPCC: Do scientific assessments need to be consensual to be authoritative. In *Future Directions for Scientific Advice in Whitehall*; Doubleday, R., Wilsdon, J., Eds.; Centre for Science and Policy Cambridge: Cambridge, UK, 2013; pp. 142–147.
121. List, C.; Goodin, R.E. Epistemic Democracy: Generalizing the Condorcet Jury Theorem. *J. Political Philos.* **2001**, *9*, 277–306. [CrossRef]
122. Dietrich, F.; Spiekermann, K. Jury Theorems. In *The Stanford Encyclopedia of Philosophy*, Summer 2022 ed.; Zalta, E.N., Ed.; Metaphysics Research Lab, Stanford University: Stanford, CA, USA, 2022.
123. Lewandowsky, S.; Gignac, G.E.; Vaughan, S. The pivotal role of perceived scientific consensus in acceptance of science. *Nat. Clim. Chang.* **2013**, *3*, 399–404. [CrossRef]
124. Oreskes, N.; Conway, E.M. *Merchants of Doubt: How a Handful of Scientists Obscured the Truth on Issues from Tobacco Smoke to Global Warming*; Bloomsbury Publishing: London, UK, 2010.
125. Bright, L.K.; Dang, H.; Heesen, R. A role for judgment aggregation in coauthoring scientific papers. *Erkenntnis* **2017**, *83*, 231–252. [CrossRef]
126. Marcoci, A.; Nguyen, J. Judgement aggregation in scientific collaborations: The case for waiving expertise. *Stud. Hist. Philos. Sci. Part A* **2020**, *84*, 66–74. [CrossRef] [PubMed]
127. Beatty, J.; Moore, A. Should We Aim for Consensus? *Episteme* **2010**, *7*, 198–214. [CrossRef]
128. Fuller, S. The Elusiveness of Consensus in Science. *PSA Proc. Bienn. Meet. Philos. Sci. Assoc.* **1986**, *1986*, 106–119. [CrossRef]
129. Weatherall, J.O.; O'Connor, C. Conformity in scientific networks. *Synthese* **2021**, *198*, 7257–7278.
130. Fazelpour, S.; Steel, D. Diversity, Trust, and Conformity: A Simulation Study. *Philos. Sci.* **2022**, *89*, 209–231. [CrossRef]
131. Dang, H. Do Collaborators in Science Need to Agree? *Philos. Sci.* **2019**, *86*, 1029–1040. [CrossRef]
132. Zollman, K.J. The epistemic benefit of transient diversity. *Erkenntnis* **2010**, *72*, 17. [CrossRef]
133. Currie, A. Existential risk, creativity & well-adapted science. *Stud. Hist. Philos. Sci. Part A* **2019**, *76*, 39–48.
134. Schneider, M.D. Creativity in the social epistemology of science. *Philos. Sci.* **2021**, *88*, 882–893. [CrossRef]
135. Parker, W. Values and uncertainties in climate prediction, revisited. *Stud. Hist. Philos. Sci. Part A* **2014**, *46*, 24–30. [CrossRef]
136. Longino, H.E. *Science as Social Knowledge: Values and Objectivity in Scientific Inquiry*; Princeton University Press: Princeton, NJ, USA, 1990.
137. Wylie, A. Why Standpoint Matters. In *Science and Other Cultures: Issues in Philosophies of Science and Technology*; Figueroa, R., Harding, S.G., Eds.; Routledge: New York, NY, USA, 2003; pp. 26–48.
138. Mills, C. White Ignorance. In *Race and Epistemologies of Ignorance*; Sullivan, S., Tuana, N., Eds.; State University of New York Press: Albany, NY, USA, 2007; pp. 11–38.
139. Du Bois, W.E.B. *The Souls of Black Folk*; Oxford University Press: Oxford, UK, 2008.
140. Wu, J. Epistemic Advantage on the Margin. *Philos. Phenomenol. Res.* **2022**, 1–23. [CrossRef]
141. Longino, H.E. *The Fate of Knowledge*; Princeton University Press: Princeton, NJ, USA, 2018.
142. Wu, J.; O'Connor, C. How Should We Promote Transient Diversity in Science? *Synthese* **2023**, *201*, 37. [CrossRef]
143. Zollman, K.J. The communication structure of epistemic communities. *Philos. Sci.* **2007**, *74*, 574–587.
144. Lazer, D.; Friedman, A. The network structure of exploration and exploitation. *Adm. Sci. Q.* **2007**, *52*, 667–694. [CrossRef]
145. Fang, C.; Lee, J.; Schilling, M.A. Balancing exploration and exploitation through structural design: The isolation of subgroups and organizational learning. *Organ. Sci.* **2010**, *21*, 625–642. [CrossRef]

146. Galison, P.; Newman, W.E. Interview with Peter Galison: On Method. *Technol. | Archit.+ Des.* **2021**, *5*, 5–9. [CrossRef]
147. Wylie, A.; Gonzalez, S.L.; Ngandali, Y.; Lagos, S.; Miller, H.K.; Fitzhugh, B.; Haakanson, S.; Lape, P. Collaborations in Indigenous and Community-Based Archaeology: Preserving the Past Together. *Assoc. Wash. Archaeol.* **2020**, *19*, 15–33.
148. Ross, M.B.; Glennon, B.M.; Murciano-Goroff, R.; Berkes, E.G.; Weinberg, B.A.; Lane, J.I. Women are Credited Less in Science than are Men. *Nature* **2022**, *608*, 135–145. [CrossRef]
149. Sarsons, H.; Gërxhani, K.; Reuben, E.; Schram, A. Gender differences in recognition for group work. *J. Political Econ.* **2021**, *129*, 101–147. [CrossRef]
150. Rubin, H.; O'Connor, C. Discrimination and collaboration in science. *Philos. Sci.* **2018**, *85*, 380–402. [CrossRef]
151. Ritson, S. Something from Nothing: 'Non-discovery' and Transformations in High Energy Experimental Physics at the Large Hadron Collider. In *Aesthetics of Experiments*; Ivanova, M., Murphy, A., Eds.; Routledge: Abingdon, UK, 2023.
152. Merz, M.; Sorgner, H. Organizational complexity in big science: Strategies and practices. *Synthese* **2022**, *200*, 1–21. [CrossRef]
153. Ritson, S. Creativity and modeling the measurement process of the Higgs self-coupling at the LHC and HL-LHC. *Synthese* **2021**, *199*, 11887–11911. [CrossRef]
154. Sorgner, H. Constructing 'Do-Able' Dissertations in Collaborative Research: Alignment Work and Distinction in Experimental High-Energy Physics Settings. *Sci. Technol. Stud.* **2022**, *35*. [CrossRef]
155. Lalli, R.; Howey, R.; Wintergrün, D. The dynamics of collaboration networks and the history of general relativity, 1925–1970. *Scientometrics* **2020**, *122*, 1129–1170. [CrossRef]
156. Lalli, R.; Howey, R.; Wintergrün, D. The Socio-Epistemic Networks of General Relativity, 1925–1970. In *The Renaissance of General Relativity in Context*; Blum, A.S., Lalli, R., Renn, J., Eds.; Einstein Studies; Springer International Publishing: Cham, Switzerland, 2020; pp. 15–84. [CrossRef]
157. Wüthrich, A. Characterizing a collaboration by its communication structure. *Synthese, accepted.*
158. Morgan, M.G. Use (and abuse) of expert elicitation in support of decision making for public policy. *Proc. Natl. Acad. Sci. USA* **2014**, *111*, 7176–7184. [CrossRef]
159. Burgman, M.A. *Trusting Judgements: How to Get the Best Out of Experts*; Cambridge University Press: Cambridge, UK, 2016.
160. Vertesi, J. *Shaping Science: Organizations, Decisions, and Culture on NASA's Teams*; University of Chicago Press: Chicago, IL, USA, 2020.
161. LIGO Scientific Collaboration Charter, 13 April 2020, LIGO–M2000029-v3. Available online: https://dcc.ligo.org/public/0166/M2000029/003/M2000029-v3.pdf (accessed on 14 November 2022).

Article

The ngEHT Analysis Challenges

Freek Roelofs [1,2,*], **Lindy Blackburn** [1,2], **Greg Lindahl** [1], **Sheperd S. Doeleman** [1,2], **Michael D. Johnson** [1,2], **Philipp Arras** [3,4], **Koushik Chatterjee** [1,2], **Razieh Emami** [1], **Christian Fromm** [5,6,7], **Antonio Fuentes** [8], **Jakob Knollmüller** [4,9], **Nikita Kosogorov** [10,11], **Hendrik Müller** [7], **Nimesh Patel** [1], **Alexander Raymond** [1,2], **Paul Tiede** [1,2], **Efthalia Traianou** [8] and **Justin Vega** [1,12]

[1] Center for Astrophysics | Harvard & Smithsonian, 60 Garden Street, Cambridge, MA 02138, USA
[2] Black Hole Initiative, Harvard University, 20 Garden Street, Cambridge, MA 02138, USA
[3] Technical University Munich (TUM), Boltzmannstr. 3, 85748 Garching, Germany
[4] Max-Planck Institute for Astrophysics, Karl-Schwarzschild-Str. 1, 85748 Garching, Germany
[5] Institut für Theoretische Physik und Astrophysik, Universität Würzburg, Emil-Fischer-Str. 31, 97074 Würzburg, Germany
[6] Institut für Theoretische Physik, Goethe-Universität Frankfurt, Max-von-Laue-Straße 1, 60438 Frankfurt am Main, Germany
[7] Max-Planck-Institut für Radioastronomie, Auf dem Hügel 69, 53121 Bonn, Germany
[8] Instituto de Astrofísica de Andalucía-CSIC, Glorieta de la Astronomía s/n, E-18008 Granada, Spain
[9] Excellence Cluster ORIGINS, Boltzmannstr. 2, 85748 Garching, Germany
[10] Moscow Institute of Physics and Technology, Institutsky per. 9, 141700 Dolgoprudny, Russia
[11] Lebedev Physical Institute of the Russian Academy of Sciences, Leninsky Prospekt 53, 119991 Moscow, Russia
[12] Department of Physics, Northeastern University, 360 Huntington Ave, Boston, MA 02115, USA
[*] Correspondence: freek.roelofs@cfa.harvard.edu

Abstract: The next-generation Event Horizon Telescope (ngEHT) will be a significant enhancement of the Event Horizon Telescope (EHT) array, with $\sim$10 new antennas and instrumental upgrades of existing antennas. The increased uv-coverage, sensitivity, and frequency coverage allow a wide range of new science opportunities to be explored. The ngEHT Analysis Challenges have been launched to inform the development of the ngEHT array design, science objectives, and analysis pathways. For each challenge, synthetic EHT and ngEHT datasets are generated from theoretical source models and released to the challenge participants, who analyze the datasets using image reconstruction and other methods. The submitted analysis results are evaluated with quantitative metrics. In this work, we report on the first two ngEHT Analysis Challenges. These have focused on static and dynamical models of M87* and Sgr A* and shown that high-quality movies of the extended jet structure of M87* and near-horizon hourly timescale variability of Sgr A* can be reconstructed by the reference ngEHT array in realistic observing conditions using current analysis algorithms. We identify areas where there is still room for improvement of these algorithms and analysis strategies. Other science cases and arrays will be explored in future challenges.

Keywords: very long baseline interferometry; black holes; active galactic nuclei; radio astronomy; imaging; instrument design; telescopes; algorithms; data analysis

Citation: Roelofs, F.; Blackburn, L.; Lindahl, G.; Doeleman, S.S.; Johnson, M.D.; Arras, P.; Chatterjee, K.; Emami, R.; Fromm, C.; Fuentes, A.; et al. The ngEHT Analysis Challenges. *Galaxies* **2023**, *11*, 12. https://doi.org/10.3390/galaxies11010012

Academic Editor: Margo Aller

Received: 10 November 2022
Revised: 16 December 2022
Accepted: 21 December 2022
Published: 10 January 2023

1. Introduction

1.1. The ngEHT

The Next-Generation Event Horizon Telescope (ngEHT) [1,2] will build on the success of the Event Horizon Telescope (EHT), the mm VLBI array, which has imaged the black hole shadows of M87* and Sgr A* [3–16]. The array will be transformatively enhanced with the current design envisioning $\sim$10 additional stations, a quadrupled bandwidth, and frequency coverage, including 86 [17], 230, and 345 GHz. Multiple operating modes will make it suitable for a wide array of science cases. The primary science goals will involve making movies of M87* and Sgr A* resolving the plasma dynamics on event horizon scales, providing black hole photon ring measurements sufficiently accurate to put constraints on black hole spin, and increasing the sample of black hole shadows imaged [18,19].

1.2. Challenge Motivation

End-to-end science simulations, which cover the full source physics, observation, calibration, and analysis processes, are of great value for the design and optimization of new instrumentation in astrophysics. These simulations realistically predict what the capabilities of the new instrument will be and which science questions it will be able to answer and can help guide the instrument design and analysis algorithm development. The ngEHT Analysis Challenges aim to provide such end-to-end simulations, bringing together expertise in all relevant areas to be applied to a well-defined set of problems. The challenge concept was inspired by the EHT Imaging Challenges [20]. In these challenges, EHT imaging experts imaged synthetic EHT datasets of different source models, which led to the rapid development of imaging algorithms and strategies tailored to the specifics of EHT datasets. While the EHT Imaging Challenges were aimed at maximizing the image quality that can be obtained from a known instrument, the ngEHT Analysis Challenges aim to help guide the development of a new instrument. Additionally, the ngEHT concept allows for the expansion of the imaging into two new dimensions, which are frequency (the ngEHT will operate at 2–3 distinct frequency bands simultaneously) and time (movie making). While not the focus of the challenges reported in this work, we aim to extend the ngEHT Analysis Challenges to model fitting and parameter estimation as well.

1.3. Challenge Procedure

For each challenge, we generate synthetic datasets from a set of source models. The source models (see also [21]) are representative of a specific ngEHT science case, may be static or time-variable, and may be generated for different (potential) ngEHT frequencies (86, 230, and 345 GHz in this work). The synthetic datasets are generated for different arrays (here, the 2022 EHT array and an ngEHT reference array) and contain different levels of data complexities (e.g., systematic weather or instrument noise). For each challenge, the synthetic datasets and other information and instructions are released to the challenge participants through the ngEHT Analysis Challenge website[1]. Participants then upload their analysis results through the same website before a pre-set deadline. Image and movie reconstructions are then uniformly plotted for visual comparison and evaluated using quantitative metrics (see Section 3). Participation is open to anyone, with access to the downloads provided upon request to the organizers.

1.4. Outline

In this work, we report on the first two ngEHT Analysis Challenges. Section 2 details the reconstruction algorithms used by the challenge participants, and Section 3 describes the submission evaluation metrics. Challenge 1 and 2 source models, synthetic data generation, and results are presented in Sections 4 and 5, respectively, and the conclusions and outlook are discussed in Section 6.

2. Reconstruction Methods

In radio interferometry, image reconstruction is an underconstrained problem, as the finite number of telescopes and baselines cause only a limited number of Fourier components of the image (visibilities) to be measured. Hence, an infinite number of images could fit the data, and additional assumptions need to be made in order to arrive at a unique image solution. Different image reconstruction algorithms tackle this problem in different ways. The algorithms can be divided into inverse modeling, regularized maximum likelihood (RML), and Bayesian methods. The section below describes the algorithms used for the image reconstructions in this work, separated into methods reconstructing static images and methods reconstructing movies.

An alternative method to reconstruct the sky brightness distribution is fitting (geometrical) models to the interferometric data. Since such reconstructions have not been submitted for the challenges described in this paper, we do not discuss them here. We aim to explore these methods in future challenges, aimed at measuring specific black holes and accretion parameters.

2.1. Static Imaging

2.1.1. CLEAN

The CLEAN algorithm is a well-known inverse modeling imaging technique. The basic algorithm was developed by Högbom [22], with other variants developed later. CLEAN deconvolves a sampling function (known as the dirty beam) from the measured brightness (or dirty map) of a radio source. The imaging procedure via CLEAN involves a number of iterations, where in each iteration, the algorithm creates a point-source component, the CLEAN component, at the position of the brightness peak in the dirty image. Then, it convolves the CLEAN component with the dirty beam, subtracting it from the dirty image and transferring it to the clean map [22]. The cleaning iterations continue until a specific cleaning halting requirement is met. In the case of noisy data, the user can steer the process by limiting the searching area with CLEAN windows. Finally, the generated set of CLEAN components is convolved with a Gaussian restoring beam. The image quality can be further enhanced via self-calibration and references thereafter [23], which corrects the amplitude and phase information using the current image estimate. The residual dirty image, representing the image noise level, may be added to the clean map as the final step.

During the last decades, this technique has been widely used for imaging astronomical targets, as well as for a broad range of other applications [24]. Together with eht-imaging and SMILI (Section 2.1.2), it was one of the methods used for reconstructing the first EHT images of M87* [3–8] and Sgr A* [11–16].

The strategy followed for the ngEHT Analysis Challenges used a semi-scripted approach, in a similar fashion to the one in [13], employing the CLEAN algorithm via the software DIFMAP and references thereafter [25].

2.1.2. RML Methods: EHT-Imaging and SMILI

RML methods calculate each pixel of the source image $\mathbf{I}$ by fitting directly to the data $\mathbf{D}$, with the fidelity of the final image to be adjusted by specific regularization terms, e.g., [26–28]. The data $\mathbf{D}$ consists of separate data products d. These are typically visibility amplitudes, closure phases, or (log) closure amplitudes, see, e.g., [29]. These regularizers R could entail the entropy, sparsity, smoothness, or other properties of the image. For more details on regularizer definitions, see Appendix A of Event Horizon Telescope Collaboration et al. [6]. RML methods find an image that minimizes a specified objective function,

$$J(\mathbf{I}) = \sum_d \alpha_d \chi_d^2(\mathbf{I}) - \sum_R \beta_R S_R(\mathbf{I}), \tag{1}$$

consisting of goodness-of-fit (χ_d^2) and regularization (S_R) terms, weighted by hyperparameters (α_d and β_R).

Both the eht-imaging [29,30] and SMILI [31,32] frameworks are suitable for directly using the closure phases and (closure) amplitudes, making them ideal for high-frequency interferometric imaging [33]. As for CLEAN, multiple rounds of self-calibration are often performed.

Various submitters used different regularizers and weights for producing reconstructions of Challenge 1 and 2 data. For the M87 datasets, an informed prior was often used consisting of a small Gaussian with most of the flux, corresponding to the core, and a large disk with little flux to capture the extended emission from the jet. For the Sgr A* images, a disk or Gaussian prior was often used, deblurring the data and, in some cases, applying a constant noise floor to mitigate the intergalactic scattering before imaging. eht-imaging was also used to produce multi-frequency images, regularizing the spectral index map [34].

2.2. Dynamical Imaging

2.2.1. EHT-Imaging

The dynamical imaging module of `eht-imaging` (abbreviated to `ehtim-di` in this work) generalizes static imaging using a regularized maximum likelihood approach to reconstruct movies of time-variable sources [35]. Specifically, the reconstruction consists of a series of N_t images (movie), $\mathbf{M} = \{\mathbf{I}_1, \mathbf{I}_2, \ldots, \mathbf{I}_{N_t}\}$. These images are determined by minimizing an objective function,

$$J \equiv \sum_d \alpha_d \chi_d^2(\mathbf{M}) - \sum_R \beta_{\mathrm{R}} \left[\frac{1}{N_t} \sum_{j=1}^{N_t} S_R(\mathbf{I}_j) \right] + \sum_x \gamma_x \mathcal{R}_x(\mathbf{M}). \tag{2}$$

The objective function consists of three components:

- Like for static RML imaging, a data term which defines the log-likelihood of the reconstruction with respect to whatever data products are fit.
- A spatial regularization term, where for each regularizer, we compute a weighted sum over individual image regularization terms, $S_R(\mathbf{I}_j)$.
- A dynamical regularization term with temporal regularizers $\mathcal{R}_x(\mathbf{M})$ with associated hyperparameters γ_x. This term computes a penalty function that can be used to favor reconstructions that evolve smoothly in time ($\mathcal{R}_{\Delta t}$), that have small variations relative to the mean ($\mathcal{R}_{\Delta I}$), or that evolve according to fluid motion with a steady flow ($\mathcal{R}_{\mathrm{flow}}$).

For the dynamical imaging reconstructions in the analysis challenges, we first fit a simple geometrical model to the full dataset, i.e., a thick "m-ring"; see [14,36]. We then used this model as both a prior (for relative entropy of individual images) and initialization of reconstructed movies, with a typical frame separation of 1 min for Sgr A*. We fit amplitudes and closure phases, with iterative self-calibration of the visibility amplitudes. The imaging was performed using gradient descent with the limited-memory Broyden–Fletcher–Goldfarb–Shanno (BFGS) algorithm [37], as implemented in `Scipy` [38].

2.2.2. StarWarps

The `StarWarps` algorithm [39] reconstructs time-variable sources by simultaneously reconstructing both the image and its time evolution. `StarWarps` reconstructs N_t images $\mathbf{M} = \{\mathbf{I}_1, \mathbf{I}_2, \ldots, \mathbf{I}_{N_t}\}$, using the observational data snapshots $\mathbf{D} = \{\mathbf{D}_1, \mathbf{D}_2, \ldots, \mathbf{D}_{N_t}\}$ at the corresponding timestamps. It employs a dynamical imaging model φ at each timestamp j:

$$\varphi_{\mathbf{D}_j|\mathbf{I}_j} = \mathcal{N}_{\mathbf{D}_j}(f_j(\mathbf{I}_j), R_j), \tag{3}$$

$$\varphi_{\mathbf{I}_j} = \mathcal{N}_{\mathbf{I}_1}(\boldsymbol{\mu}_j, \boldsymbol{\Lambda}_j), \tag{4}$$

$$\varphi_{\mathbf{I}_j|\mathbf{I}_{j-1}} = \mathcal{N}_{\mathbf{I}_j}(\mathbf{A}\mathbf{I}_{j-1}, \mathbf{Q}), \tag{5}$$

where $\mathcal{N}_{\mathbf{D}_j}(f_j(\mathbf{I}_j), R_j)$ refers to the multivariate normal distribution of $\mathbf{D}_j$ with mean $f_j(\mathbf{I}_j)$ and the covariance R_j. $\boldsymbol{\Lambda}_i = \mathrm{diag}[\boldsymbol{\mu}_i]^T \boldsymbol{\Lambda}' \mathrm{diag}[\boldsymbol{\mu}_i]$, where $\boldsymbol{\Lambda}'$ is defined in terms of the priors; see Equation (13) of [39] for more details. $\boldsymbol{\mu}_i$ is the mean of a multivariate Gaussian distribution and $\boldsymbol{\Lambda}$ describes the covariance, setting the spatial regularization. $f_j(\mathbf{I}_j)$ describes the functional relationship between the source image $\mathbf{I}_j$ and the contemporaneous observed data $\mathbf{D}_j$. The global time evolution of the source between timestamps $j-1$ and j is described by the evolution matrix $\mathbf{A}$, so that $\mathbf{I}_j \approx \mathbf{A}\mathbf{I}_{j-1}$, and any additional perturbations of the source are constrained by the covariance matrix $\mathbf{Q}$. The process hence reduces to static imaging for ($\mathbf{A} = \mathbf{1}, \mathbf{Q} = \mathbf{0}$). The joint probability distribution is then given by:

$$p(\mathbf{M}, \mathbf{D}; \mathbf{A}) \propto \prod_{j=1}^{N_t} \varphi_{\mathbf{D}_j|\mathbf{I}_j} \prod_{j=1}^{N_t} \varphi_{\mathbf{I}_j} \prod_{j=2}^{N_t} \varphi_{\mathbf{I}_j|\mathbf{I}_{j-1}}. \tag{6}$$

In StarWarps, we jointly solve for the image reconstructions as well as $\mathbf{A}$. First, we learn $\mathbf{A}$ using the Expectation-Maximization (EM) algorithm and then reconstruct the images with that $\mathbf{A}$.

For the analysis challenge image reconstruction with `StarWarps`, we used the visibility amplitudes, log closure amplitudes, and the bispectrum (triple amplitudes and closure phases) as our data products, with 2% added systematic noise. We used the EHT 2017 image of Sgr A* [11] blurred with a 25 µas Gaussian kernel as a prior; see also [40].

2.2.3. Resolve

The algorithm `resolve`[2] approaches the imaging task for the (ng)EHT from a probabilistic, Bayesian perspective. It is based on Bayes' theorem:

$$\mathcal{P}(\mathbf{M}|\mathbf{D}) = \frac{\mathcal{P}(\mathbf{D}|\mathbf{M})\,\mathcal{P}(\mathbf{M})}{\mathcal{P}(\mathbf{D})}, \tag{7}$$

where $\mathbf{D}$ refers to the measured data, and $\mathbf{M}$ denotes the time-varying sky brightness distribution. The quantity $\mathcal{P}(\mathbf{M}|\mathbf{D})$ is called posterior probability densityand contains all information on $\mathbf{M}$ after taking the information from the data $\mathbf{D}$ into account. In the case of the dynamic Sgr A* model, we consider $\mathbf{M}$ to be a discretized quantity with spatial dimensions 200×200 and a temporal axis of length 720—in total, $2.88 \cdot 10^7$ degrees of freedom. Therefore, the posterior can be considered to be a function $\mathbb{R}^{28,800,000} \to \mathbb{R}^{>0}$.

The prior probability density $\mathcal{P}(\mathbf{M})$ represents our knowledge of the source before the data are considered. Since the sky brightness distribution represents a flux density, we can safely assume that its values are non-negative. Additionally, we know a priori that the emission is correlated in both the spatial and temporal directions. Thus, we assume generic homogeneous and isotropic spatial and temporal correlation structures whose specific form is learned from the data alongside $\mathbf{M}$. For more details on the prior, refer to Arras et al. [41,42].

The likelihood $\mathcal{P}(\mathbf{D}|\mathbf{M})$ encodes our knowledge of the measurement process. In general, the calibration pipeline of the (ng)EHT provides visibilities whose phases suffer from temporally uncorrelated station-based effects. The amplitudes of the visibilities are approximately correct and only subject to small time-correlated station-based effects. Therefore, we use closure phases and self-calibrated (non-closure) amplitudes in the likelihood.

After combining prior and likelihood, our best guess for the time-variable behavior of the source is given by the expectation value of the posterior: $\int \mathbf{M}\mathcal{P}(\mathbf{M}|\mathbf{D})\,d\mathbf{M}$. Since evaluating such high-dimensional integrals directly is virtually impossible, we use Metric Gaussian Variational Inference [43] that provides approximate solutions to Bayes' theorem efficiently. As a result, we obtain a collection of approximate posterior samples that can be averaged to obtain an approximate posterior mean. Additionally, the variability of the approximate posterior samples represents the uncertainty of the computed solution. This uncertainty could be propagated to downstream analyses of the time-variable reconstruction of the source.

For the ngEHT analysis challenges, we adopted the implementation of `resolve` of Arras et al. [42] where `resolve` has been applied to the 2017 EHT observation of M87*. For evaluating the interferometry measurement equation (i.e., the non-equidistant Fourier transform), we use the implementation presented in Arras et al. [44]. Variations of `resolve` have been verified and tested against standard methods in various contexts [41,42]. To enable comparisons to results of algorithms that do not quantify uncertainties, we will depict only the posterior mean of the sky brightness distribution s in the following.

2.2.4. DoG-HiT

`DoG-HiT` is a multi-scale RML imaging algorithm [45]. `DoG-HiT` models the image by a set of wavelets (a dictionary Γ) constructed by the difference of the Gaussian method: $\mathbf{I} = \Gamma\mathscr{I}$ [45,46]. With `DoG-HiT`, we aim to recover the array of wavelet coefficients $\mathscr{I}$ that represents the true sky brightness distribution best. The wavelets define filters in the

Fourier domain that are ring-like. Hence, every wavelet compresses the spatial information from a specific band of baselines in the uv-coverage. For DoG-HiT, the scales are fitted to the uv-coverage, thus giving rise to wavelets most sensitive to gaps in the uv-coverage and wavelets most sensitive to Fourier coefficients sampled by baselines. In the spirit of compressed sensing, DoG-HiT utilizes a sparsity promoting penalization by a l_0 penalty term on the wavelet coefficients. In detail, we solve the following optimization problem consisting of data fidelity terms for the closure quantities (χ^2_{cp} and χ^2_{camp}), the l_0 penalty term and a total flux constraint by an updated forward-backward splitting approach [45]:

$$\mathscr{I} \in \text{argmin}_{\mathscr{I}} \left[\chi^2_{cp}(\Gamma\mathscr{I}) + \chi^2_{camp}(\Gamma\mathscr{I}) + \alpha \cdot \|\mathscr{I}\|_{l_0} + R_{flux}(\mathscr{I}, f) \right], \tag{8}$$

where R_{flux} is a total flux indicator function with flux f, and Γ denotes the wavelet dictionary. DoG-HiT is a data-driven, automatic imaging pipeline that depends on only one hyper-parameter (the relative weighting of the penalization term α). It has been demonstrated to produce high-quality, super-resolved reconstructions for static sources in a relatively short time with minimal manual interaction and without the need for extensive parameter surveys. e.g., compare the Challenge 1 reconstructions with DoG-HiT in Section 6 of [45].

The dynamic reconstructions are based rather straightforwardly upon the success of this static imaging. We utilize the automatic static imaging pipeline to construct a mean image from the full length of the observation without taking the dynamics of the source into account. DoG-HiT computes a set of statistically significant wavelet coefficients from the mean image as a byproduct (the multiresolution support). Then the mean image (with a relatively bad fit to the data due to not respecting the source dynamics) is subtracted from the self-calibrated visibilities, and the observation is cut into frames of six minutes. The residuals are minimized frame by frame with StarWarps with implicit dynamic variability imposed by StarWarps. A small StarWarps internal regularization parameter is used, but, in contrast to StarWarps, the reconstruction is performed in a multiscalar constrained minimization framework (multiresolution support constraint), i.e., only the wavelet coefficients classified as significant during the static image reconstruction are allowed to vary. This introduces a correlation between frames and consistency to the mean static image.

DoG-HiT is still under development and is currently extended to dynamic, polarimetric reconstructions [47] with promising first results on synthetic data (see upcoming Challenge 3 reconstructions). A finer set of directional-dependent wavelet functions [46] allows for dynamic reconstructions in a constrained minimization reconstruction on frames independently, thus replacing StarWarps during the current DoG-HiT dynamic imaging pipeline and relying on a completely unsupervised, automatic wavelet approach only. On the one hand, such an unsupervised, automatic imaging procedure is desired as it reduces the human bias in the reconstruction; on the other hand, driving by an astronomer could be crucial to address data issues, in particular for challenging data sets such as what will be produced by the ngEHT.

3. Submission Evaluation Metrics

Submitted reconstructions were evaluated with several quantitative quality metrics. These metrics, which all probe different aspects of what makes a high-quality reconstruction, are summarized below.

3.1. Data Fit Quality

The goodness-of-fit of the submitted reconstructions to the provided synthetic data was quantified by computing the reconstruction visibilities using the synthetic data uv-coverage and then calculating the χ^2-metric on closure quantities, χ^2_{cphase} and χ^2_{lcamp}. These quantities are the closure phases, which are the sum of visibility phases measured simultaneously on a closed triangle of baselines, and the (log) closure amplitudes, respectively, which are ratios of visibility amplitudes on a baseline quadrangle; see, e.g., [48]. Closure quantities are robust against station-based calibration errors.

3.2. Ground Truth Image Similarity

Apart from the goodness-of-fit to the synthetic data, another important quality metric is the similarity of the reconstruction to the ground truth source model. We quantify this similarity using the normalized cross-correlation. The normalized cross-correlation between two images X and Y is

$$\rho_{\text{NX}} = \frac{1}{N} \sum_{i=1}^{N} \frac{(X_i - \langle X \rangle)(Y_i - \langle Y \rangle)}{\sigma_X \sigma_Y}. \tag{9}$$

Here, N is the number of pixels in the images, X_i and Y_i are the pixel values of images X and Y, respectively, $\langle \ldots \rangle$ denotes an average, and σ_X and σ_Y are the standard deviations of the pixel values of images X and Y, respectively. The value of ρ_{NX} will be equal to 1 for identical images (maximal correlation), 0 for completely uncorrelated images, and -1 for perfectly anticorrelated images. Using the implementation in `eht-imaging`, the two images are regridded to contain the same number of pixels with equal pixel size and aligned to maximize ρ_{NX}. For M87 reconstructions, we are often most interested in the arrays' ability to reconstruct the large-scale and low-surface brightness jet emission. Therefore, we also compute ρ_{NX} on the log pixel values ($\rho_{\text{NX,log}}$). In order to suppress the influence of low-surface brightness image noise, which may appear at a certain flux level depending on the particularities of the reconstruction algorithm, we limit the dynamic range of the ground truth and reconstructed images to 10^4 in this case.

3.3. Effective Resolution

ρ_{NX} also provides a way to compute the effective angular resolution obtained by the reconstructed image. Following Event Horizon Telescope Collaboration et al. [6], we blur the ground truth model images with a circular Gaussian with varying FWHM and calculate $\rho_{\text{NX,FWHM}}$ with respect to the ground truth model images using Equation (9). For a submitted reconstruction with a particular $\rho_{\text{NX,rec}}$ with respect to the ground truth, the effective resolution θ_{eff} is then the (interpolated) FWHM for which $\rho_{\text{NX,rec}}$ is equal to $\rho_{\text{NX,FWHM}}$.

3.4. Dynamic Range

Dynamic range is usually defined as the ratio between the brightest and dimmest pixel value in an image and has been frequently used in radio astronomy to assess the ability of an array to reconstruct low-surface brightness features. For images reconstructed with `CLEAN` algorithms, the dynamic range can be naturally calculated as the ratio between the brightest `CLEAN` component and the noise floor (Section 2.1.1). However, for images reconstructed with other algorithms (e.g., RML-based approaches), formally defining a dynamic range metric that works universally and reflects our intuitive sense of dynamic range is non-trivial. This difficulty has two main causes. First, not all imaging methods naturally produce a noise floor, such as `CLEAN` and have many (near)-zero pixel values. Second, many imaging algorithms produce spurious structures due to, e.g., sparse uv-coverage, so that the lowest reconstructed pixel brightness cannot be used to robustly define a dynamic range metric.

To evaluate the challenge reconstructions, we use a dynamic range proxy following Bustamante et al. [49]. This metric considers the ratios between the brightest pixel of the ground truth image $\mathbf{I}_{\text{groundtruth}}$ and the absolute pixel residuals of the reconstructed image $\mathbf{I}_{\text{reconstructed}}$ with respect to the ground truth,

$$\mathcal{D} = \frac{\max(\mathbf{I}_{\text{groundtruth}} * \mathcal{G}^{2\text{D}}_{\theta_{\text{eff}}})}{\left| \mathbf{I}_{\text{reconstructed}} - \mathbf{I}_{\text{groundtruth}} * \mathcal{G}^{2\text{D}}_{\theta_{\text{eff}}} \right|}. \tag{10}$$

Here, $*$ denotes convolution, $|\ldots|$ indicates that we take the absolute values, and $\mathcal{G}^{2D}_{\theta_{\text{eff}}}$ represents a two-dimensional circular Gaussian with an FWHM equal to the effective resolution θ_{eff} of the reconstructed image. From $\mathcal{D}$, which has the form of an image, we can calculate a dynamic range proxy by selecting the qth quantile of the pixel values:

$$\mathcal{D}_q = \text{quantile}(\mathcal{D}, q). \tag{11}$$

By using the residuals, this metric penalizes spurious structures in the reconstructed image and does not rely on a noise floor being calculated as part of the imaging process. A disadvantage is that it requires the ground truth image and hence cannot be applied to real data. Setting q too low will make the metric dominated by outliers in the residual image, while setting it too high will not penalize high residuals strongly enough. We set $q = 0.1$ for our dynamic range proxy $\mathcal{D}_{0.1}$. Because of its sensitivity to q and the chosen blurring kernel, we emphasize that this metric should not be regarded as giving *the* dynamic range of the image but rather as a dynamic range proxy that can be used to compare different reconstructions of the same source model.

4. Challenge 1

4.1. Rationale and Charge

The primary objectives of the first challenge were to set up a framework for the generation of synthetic ngEHT data based on theoretical source models, to conduct the organized submission and cross-comparison of reconstruction results from multiple people, and to obtain a first idea of the benefits and challenges of ngEHT datasets as compared to the current EHT. The model and data properties were therefore kept relatively simple. Participants were asked to submit image reconstructions for each provided synthetic dataset. The challenge was not blind, i.e., the participants had access to the input source models and synthetic data generation script. The challenge was launched on 18 June 2021, and the submission deadline was 16 July 2021. It was advertised to the ngEHT simulations group. All information is available on the challenge website: https://challenge.ngeht.org/challenge1/ (accessed on 19 December 2022).

4.2. Source Models

For Challenge 1, we used two static, unpolarized models of M87 and Sgr A*, respectively. Both models are displayed in Figure 1 and described below. More detailed descriptions and comparisons of the source models used for Challenge 1 and 2 can be found in Chatterjee et al. [21].

4.2.1. M87

The Challenge 1 M87 model is a magnetically arrested disk (MAD) general relativistic magnetohydrodynamics (GRMHD) frame from a rapid spinning black hole $a_* = 0.94$ with electron thermodynamics from reconnection heating; see [50] for details. The GRMHD simulation was performed with the BHAC code [51] using three levels of adaptive mesh refinement (AMR) in logarithm Kerr-Schild coordinates. The numerical grid covers $384 \times 4192 \times 192$ cells in radial, azimuthal and theta directions and extends up to 2500 gravitational radii (GM/c^2, where G is Newton's gravitational constant, M is the black hole mass, and c is the speed of light) in the radial direction. The mass accretion rate and MAD parameter (see [52]) were monitored, and after obtaining a steady state, we performed the general relativistic radiative transfer (GRRT) calculations with the radiative transfer code BHOSS [53,54].

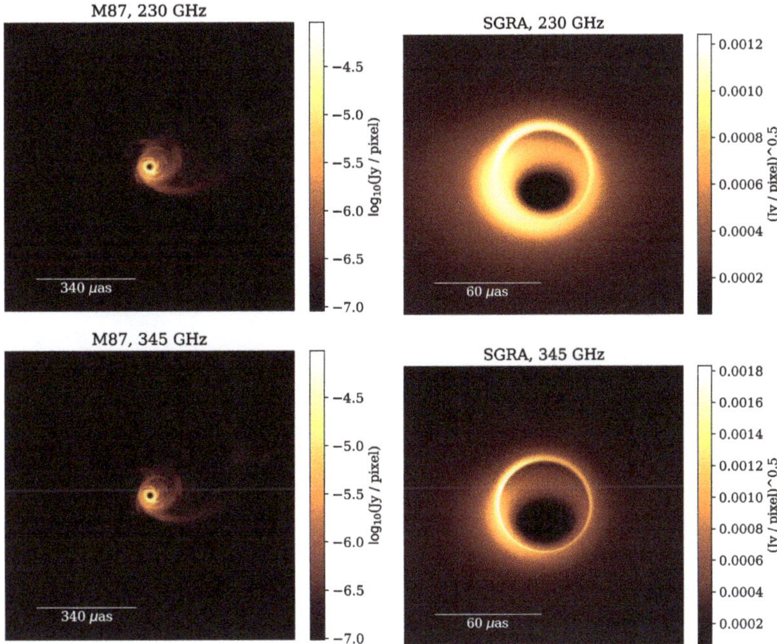

Figure 1. Source models used for Challenge 1.

During the radiative transport, we included non-thermal particles via the kappa electron distribution (see [55]) in the jet sheath while excluding the highly magnetized spine by using a cut in the magnetization at a value of 1 (typically referred to as a sigma cut). The power-law slope of the kappa distribution was set by a particle in cell (PIC)-motivated sub-grid model depending on the local magnetization and plasma-beta following Ball et al. [56]. In addition, we included a fraction of the magnetic energy density to accelerate the non-thermal particles (see [57,58]). In the jet wind and disk region, we used a thermal electron distribution, where the electron temperature is directly obtained from the GRMHD simulation. In order to guarantee capturing a small scale structure on the horizon scale and at the same time the large-scale jet structure, we used a field of view (FOV) of 1 mas using a resolution of 4096×4096 pixels. Since the GRMHD simulations are scale-free, we normalized our GRRT simulations by setting the mass ($6.5 \times 10^9 M_\odot$) and distance (16.9 Mpc) of the black hole in M87 and iterated the mass accretion rate until a compact flux density of 0.8 Jy at 230 GHz was obtained.

4.2.2. Sgr A*

The Challenge 1 Sgr A* model is a stationary semi-analytic radiatively inefficient accretion flow (RIAF) model, e.g., [59]. This model can be used to test the capabilities of next-generation arrays in precision modeling of black hole parameters. High resolution is needed to capture the unique signature from a subring structure. This model does not capture any variability due to turbulence in the system. The basic model has $a_* = 0$ (Schwarzschild) at an inclination of i = 130 deg and includes non-thermal particles. The model includes disk height following [60], sub-Keplerian flow properties ($\kappa = 0.5$, $\alpha = 0.5$), following the notation of Tiede et al. [61]—e.g., Equations (10) and (11), and fitted to the observed data of [62–65]. Images were ray-traced at 230 and 345 GHz with 4096×4096 pixels and a FOV of 128 GM/c^2, using a distance of 8.178 kpc and mass of $4.14 M_\odot$ [66]. Finally, the 230 and 345 GHz images were scattered with the same realization of the Johnson et al. [35] interstellar scattering model before generating the synthetic data.

4.3. Synthetic Data

4.3.1. Station Locations

Two arrays were used to generate the Challenge 1 synthetic data. EHT2022 consists of the 11 stations that participated in the 2022 EHT observations. In the ngEHT reference array 1 (ngEHT1), 10 stations are added to this array. The station locations were chosen based on a *uv*-coverage analysis (A. Raymond, priv. comm.), investigating which combination of sites from Raymond et al. [67] provided optimal *uv*-coverage by performing a Monte Carlo simulation involving telescope dropouts due to weather conditions. The LMT, SPT, and KP were not included in the 345 GHz observations with EHT2022. The station locations are shown in Figure 2.

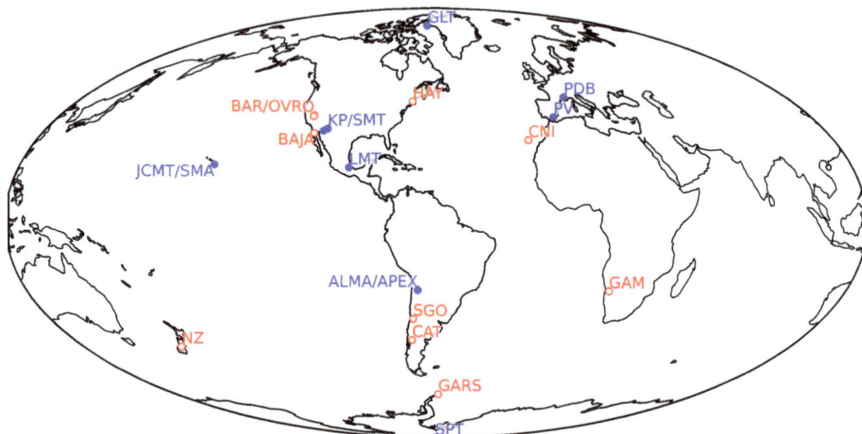

Figure 2. Station locations for Challenge 1 and 2. Stations in blue form the EHT2022 array, and stations in red are added to EHT2022 to form a reference array ngEHT1. The new station locations are near the National Astronomical Observatory in Baja California, Mexico (BAJA); Barcroft Field Station in California, USA (BAR); Cerro Catedral in R'io Negro in Argentina (CAT), La Palma, part of the Canary Islands, Spain (CNI); the Gamsberg in Namibia (GAM), the German Antarctic Receiving Station O'Higgins in Antarctica (GARS); Haystack Observatory in Westford, MA, USA; Canterbury, New Zealand (NZ); Owens Valley Radio Observatory in California, USA (OVRO); and Santiago, Chile (SGO). See Raymond et al. [67] for details, e.g., weather statistics for each site.

4.3.2. Data Properties

A 24-h observing track was simulated for each array, source, and frequency, resulting in eight separate datasets. Each track consists of 10-min scans interleaved with 10-min gaps and is identical for each dataset. A single frequency channel with a time resolution of 10 s was provided. Thermal noise expected from the receiver and atmospheric opacity were added to the complex visibilities calculated using `eht-imaging` [29,30]. The following assumptions were made for all sites:

- Receiver temperature: 60 K for 230 GHz; 100 K for 345 GHz
- Aperture efficiency: 0.68 for 230 GHz; 0.42 for 345 GHz
- Bandwidth: 8 GHz
- Quantization efficiency: 0.88
- Dish diameter: 6 m for new sites, actual diameter for existing sites
- Opacity: median values in April as extracted from the MERRA-2 data by Raymond et al. [67], at 30-degree elevation. The opacities were set constant throughout and across the different datasets but are frequency-dependent.

Visibility phases were scrambled but stabilized across scans. No further systematic errors were added. After data generation, data points with a signal-to-noise ratio less than 1 were flagged. Figure 3 shows the resulting *uv*-coverage for all datasets.

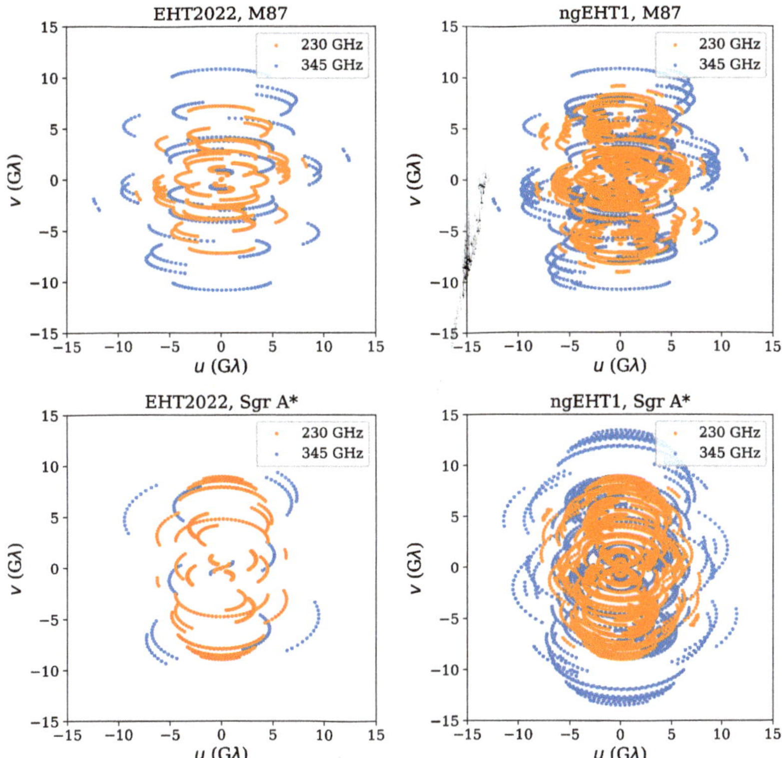

Figure 3. *uv*-coverage for the Challenge 1 datasets.

4.4. Results

Challenge 1 image reconstructions were submitted by three individual submitters and one team, using CLEAN, SMILI, and eht-imaging. For M87, one submitter performed a multi-frequency image reconstruction with eht-imaging, using information from the 230 GHz data to reconstruct the image at 345 GHz, and vice versa. All submitted reconstructions are displayed in Figures 4 and 5 for M87 and Sgr A*, respectively. Reconstruction quality metrics (Section 3) are shown in Table 1. While it should be kept in mind that the synthetic data was idealized in certain aspects (no systematic amplitude errors, and a static Sgr A* source model), these results show some interesting trends.

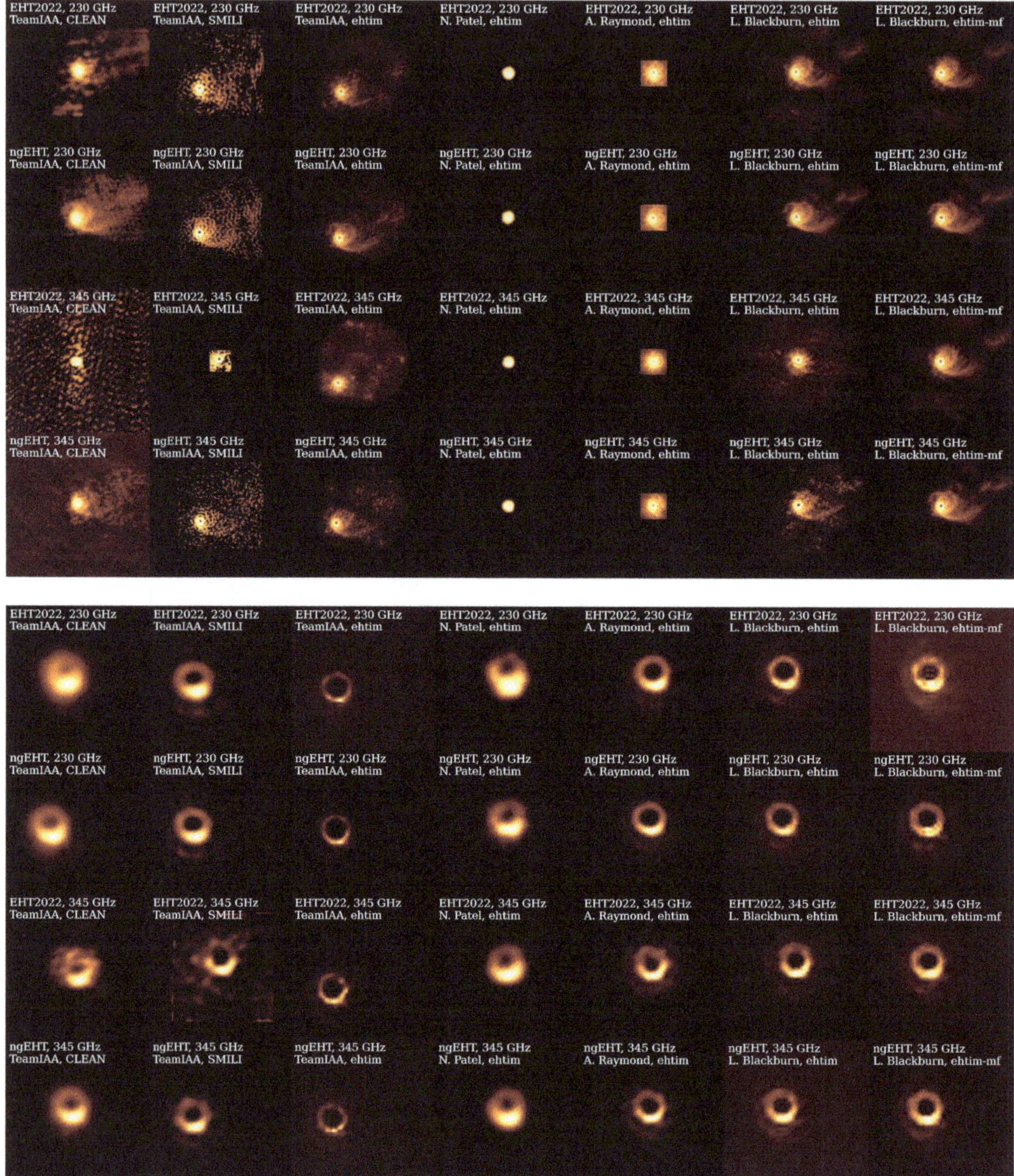

Figure 4. M87 reconstructions submitted for Challenge 1. Images are shown on a log scale with a 1 mas field of view in the top set of panels. The same images are shown on a linear scale with a 200 µas field of view in the bottom set of panels.

All M87 reconstructions recover the black hole shadow, whereas the jet features are only recovered by some. The low surface brightness structure in the M87 jet is already visible in some EHT2022 reconstructions. The jet reconstructions improve significantly with ngEHT1 coverage, as attested by both visual inspection of the images and the quality metrics. The `eht-imaging` submissions perform best, although the ρ_{NX} ranking of individual submissions changes depending on whether the linear or log-scale images are used for the comparison. $\rho_{NX,log}$ and $\mathcal{D}_{0.1}$ are more sensitive to the reconstruction of the extended jet structure and generally show a clearer improvement of ngEHT1 versus EHT2022 reconstructions. The `CLEAN` and `SMILI` reconstructions show poorer jet structure recovery than `eht-imaging`, although they may potentially be improved by adapting the specific scripts used for these reconstructions. The reconstruction quality is generally better for 230 GHz than for 345 GHz due to the higher flux and better *uv*-coverage at 230 GHz. Reconstructions with relatively high χ^2 values often have relatively low ρ_{NX} values. The multi-frequency analysis (`ehtim-mf`) is an exception with relatively good reconstruction quality (especially as shown by $\rho_{NX,log}$ and $\mathcal{D}_{0.1}$) for relatively high χ^2 values. The multi-frequency analysis is especially useful for reconstructing the jet features at 345 GHz, as these are reconstructed significantly more poorly with other methods.

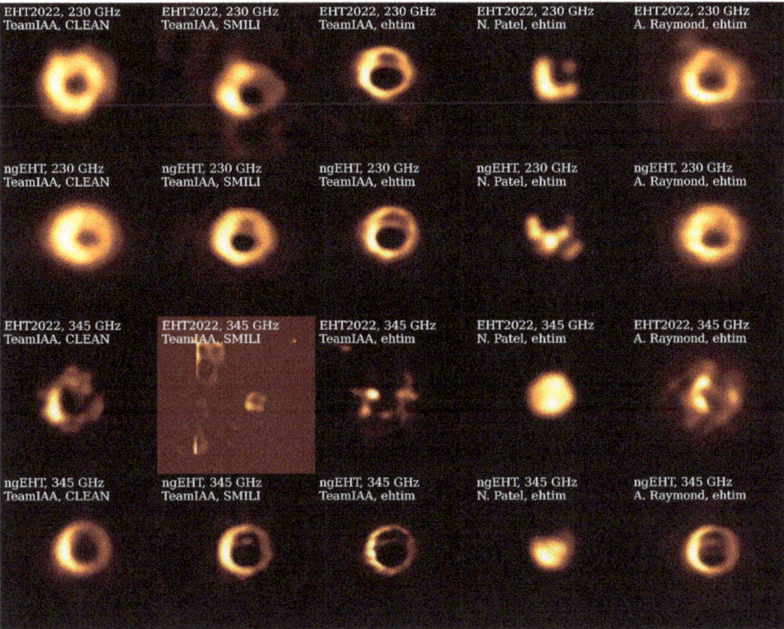

Figure 5. Sgr A* reconstructions submitted for Challenge 1. Images are shown on a linear scale with a 200 μas field of view.

For the Sgr A* model, the black hole shadow is recovered by all arrays except the EHT2022 array at 345 GHz: the *uv*-coverage is too sparse for a high-fidelity image reconstruction in this case (Kitt Peak, the LMT, and the SPT cannot observe at 345 GHz yet). ngEHT reconstructions at 345 GHz are significantly sharper than EHT2022 and ngEHT reconstructions at 230 GHz. ngEHT reconstructions at 230 GHz are generally less noisy than EHT2022 reconstructions at the same frequency, but for Sgr A*, the real value of ngEHT coverage will be in dynamical reconstructions (Section 5). The χ^2_{lcamp} are generally high for Sgr A* reconstructions, which is likely due to the comparison with the provided synthetic data, which includes interstellar scattering, while submitters may have deblurred the visibility amplitudes in the reconstruction process.

Table 1. Reconstruction quality metrics for Challenge 1.

Source	Array	ν (GHz)	Submitter	Method	χ^2_{cp}	χ^2_{lcamp}	ρ_{NX}	$\rho_{NX,log}$	θ_{eff}	$\mathcal{D}_{0.1}$
M87	EHT2022	230	L. Blackburn	ehtim	1.1	1.01	0.93	0.87	5.4	856
M87	EHT2022	230	L. Blackburn	ehtim-mf	5.17	4.36	0.88	0.9	9.8	797
M87	EHT2022	230	N. Patel	ehtim	3.66	1159.56	0.77	0.52	21.2	418
M87	EHT2022	230	TeamIAA	SMILI	0.99	1.06	0.83	0.79	14.6	409
M87	EHT2022	230	TeamIAA	CLEAN	2.94	879.77	0.8	0.8	17.7	529
M87	EHT2022	230	TeamIAA	ehtim	1.79	1.03	0.89	0.91	8.9	564
M87	EHT2022	230	A. Raymond	ehtim	2.28	1.77	0.9	0.72	8.0	291
M87	ngEHT	230	L. Blackburn	ehtim-mf	2.62	1.43	0.89	0.96	8.9	1681
M87	ngEHT	230	L. Blackburn	ehtim	1.07	1.01	0.93	0.95	5.4	1604
M87	ngEHT	230	N. Patel	ehtim	3.5	89.74	0.83	0.52	14.6	640
M87	ngEHT	230	TeamIAA	SMILI	1.01	1.03	0.87	0.85	10.8	708
M87	ngEHT	230	TeamIAA	CLEAN	1.32	138.45	0.84	0.91	13.6	1828
M87	ngEHT	230	TeamIAA	ehtim	1.08	1.01	0.91	0.97	7.1	1727
M87	ngEHT	230	A. Raymond	ehtim	1.65	2.14	0.92	0.73	6.2	532
M87	EHT2022	345	L. Blackburn	ehtim-mf	2.36	1.06	0.91	0.87	5.7	1403
M87	EHT2022	345	L. Blackburn	ehtim	1.19	0.62	0.91	0.72	5.7	984
M87	EHT2022	345	N. Patel	ehtim	1.2	7.29	0.79	0.53	16.7	734
M87	EHT2022	345	TeamIAA	SMILI	1.19	0.62	0.79	0.66	16.7	645
M87	EHT2022	345	TeamIAA	ehtim	1.22	0.62	0.88	0.81	8.2	700
M87	EHT2022	345	TeamIAA	CLEAN	3.34	2.77	0.82	0.38	13.7	320
M87	EHT2022	345	A. Raymond	ehtim	1.19	0.62	0.88	0.74	8.2	546
M87	ngEHT	345	L. Blackburn	ehtim	1.15	0.97	0.92	0.89	4.9	1570
M87	ngEHT	345	L. Blackburn	ehtim-mf	1.25	1.13	0.91	0.94	5.7	2244
M87	ngEHT	345	N. Patel	ehtim	1.2	9.99	0.79	0.54	16.7	853
M87	ngEHT	345	TeamIAA	CLEAN	1.31	4.39	0.84	0.75	11.8	651
M87	ngEHT	345	TeamIAA	SMILI	1.16	1.0	0.85	0.71	10.9	766
M87	ngEHT	345	TeamIAA	CLEAN	1.31	4.39	0.84	0.75	11.8	651
M87	ngEHT	345	TeamIAA	ehtim	1.16	0.98	0.9	0.92	6.5	1638
M87	ngEHT	345	A. Raymond	ehtim	1.17	1.0	0.91	0.75	5.7	782
Sgr A*	EHT2022	230	N. Patel	ehtim	6.08	347.88	0.8	-	45.5	-
Sgr A*	EHT2022	230	TeamIAA	ehtim	1.11	33.13	0.95	-	14.3	-
Sgr A*	EHT2022	230	TeamIAA	CLEAN	140.97	130.2	0.9	-	23.4	-
Sgr A*	EHT2022	230	TeamIAA	SMILI	1.47	23.19	0.85	-	32.6	-
Sgr A*	EHT2022	230	A. Raymond	ehtim	3.02	8.27	0.89	-	25.2	-
Sgr A*	ngEHT	230	N. Patel	ehtim	20.23	122.65	0.65	-	100.0	-
Sgr A*	ngEHT	230	TeamIAA	SMILI	1.4	8.81	0.95	-	14.3	-
Sgr A*	ngEHT	230	TeamIAA	CLEAN	2.3	23.3	0.9	-	23.4	-
Sgr A*	ngEHT	230	TeamIAA	ehtim	1.06	10.61	0.97	-	10.1	-
Sgr A*	ngEHT	230	A. Raymond	ehtim	1.14	1.87	0.93	-	18.1	-
Sgr A*	EHT2022	345	N. Patel	ehtim	1.03	20.32	0.64	-	61.9	-
Sgr A*	EHT2022	345	TeamIAA	CLEAN	71.44	66.33	0.79	-	24.5	-
Sgr A*	EHT2022	345	TeamIAA	ehtim	1.03	1.95	0.65	-	57.8	-
Sgr A*	EHT2022	345	TeamIAA	SMILI	1.63	1.7	0.34	-	100.0	-
Sgr A*	EHT2022	345	A. Raymond	ehtim	1.03	0.85	0.78	-	26.0	-
Sgr A*	ngEHT	345	N. Patel	ehtim	2.18	15.58	0.64	-	61.9	-
Sgr A*	ngEHT	345	TeamIAA	ehtim	1.14	1.19	0.93	-	7.5	-
Sgr A*	ngEHT	345	TeamIAA	CLEAN	2.24	4.48	0.87	-	14.0	-
Sgr A*	ngEHT	345	TeamIAA	SMILI	1.17	1.23	0.89	-	11.7	-
Sgr A*	ngEHT	345	A. Raymond	ehtim	1.14	1.15	0.9	-	10.6	-

5. Challenge 2

5.1. Rationale and Charge

With the challenge infrastructure and initial participant imaging efforts set up in the first challenge, the second challenge was more realistic and science-oriented and different from the first challenge in two aspects.

First, the ground truth source models were dynamic instead of static. For Sgr A*, with variability on timescales of $\sim$ minutes, the charge to the participants was to reconstruct a movie of the source evolving across a single day of observations. We used two source models to test reconstruction capabilities for different variability properties: a GRMHD model with turbulent variability, and a shearing hotspot in a RIAF disk, exhibiting a more coherent variable structure. Hints of such coherent variability of Sgr A* at 230 GHz consistent with an orbiting hotspot have been observed by Wielgus et al. [68]. For M87*, with variability on timescales of $\sim$ days, we used a bright jet GRMHD model like in the first challenge but evolved it over a period of five months, simulating a full-day observation every week. The charge was to reconstruct a movie of the large-scale and low-surface brightness jet emission, connecting it to the dynamics near the black hole shadow.

The second aspect in which this challenge differed from the previous one is that the synthetic observations included significantly more realistic effects. Contrary to the idealized data generated for Challenge 1, which only includes thermal noise, the Challenge 2 data sets have been generated under the assumption of realistic observing conditions and include data systematics originating from weather, instrumental, and calibration effects (see Section 5.3). The results of this challenge thus reflect what would actually be seen by an array built with the described specifics, using current reconstruction algorithms.

Challenge 2 was launched on 25 October 2021. It was advertised more broadly than the first challenge to the full ngEHT community. The first submission comparisons were completed in January 2022, and due to the complexity of the datasets and ongoing development of reconstruction algorithms, reconstructions were submitted until August 2022.

5.2. Source Models

This section describes the dynamical source models used for Challenge 2. See Chatterjee et al. [21] for more detailed model descriptions and comparisons.

5.2.1. M87

The Challenge 2 M87 model is a GRMHD movie with 20 frames that are spaced $20GM/c^3$ ($\sim$1 week) apart. The pixel resolution is 2048×2048, with a field of view of 1 mas. The images were ray-traced from a H-AMR [69] simulation (MAD, spin 0.94) using `ipole` [70]. R_{high} was set to 160 and accelerated electron heating was included, setting $\kappa = 3.5$ [57]. We only use the Stokes I information from the model. The model is shown in Figure 6.

5.2.2. Sgr A*

For Challenge 2, two dynamic source models were used for Sgr A*: a GRMHD model exhibiting turbulent variability and a RIAF + shearing hotspot model with more ordered variability properties. Sample frames of the source models used for Challenge 2 are shown in Figure 7.

The GRMHD model is a MAD model with a spin of 0.5. The images were ray-traced in Stokes I with BHOSS [71], assuming thermal electrons. The 500 frames are spaced $10\,GM/c^3$ (221 s) apart. The pixel resolution is 2048×2048, with a field of view of 400 µas.

The second Sgr A* model is a RIAF [72] plus shearing hotspot [61] semi-analytical model. The hotspot parameters are inspired by [73], and the black hole spin was set to 0.1. The pixel resolution is 313×313, with a field of view of 315 µas. The frames are spaced 30 s apart and form a 4-h movie of a hotspot shearing and falling in, which is repeated a few times over the course of the 24-h observation.

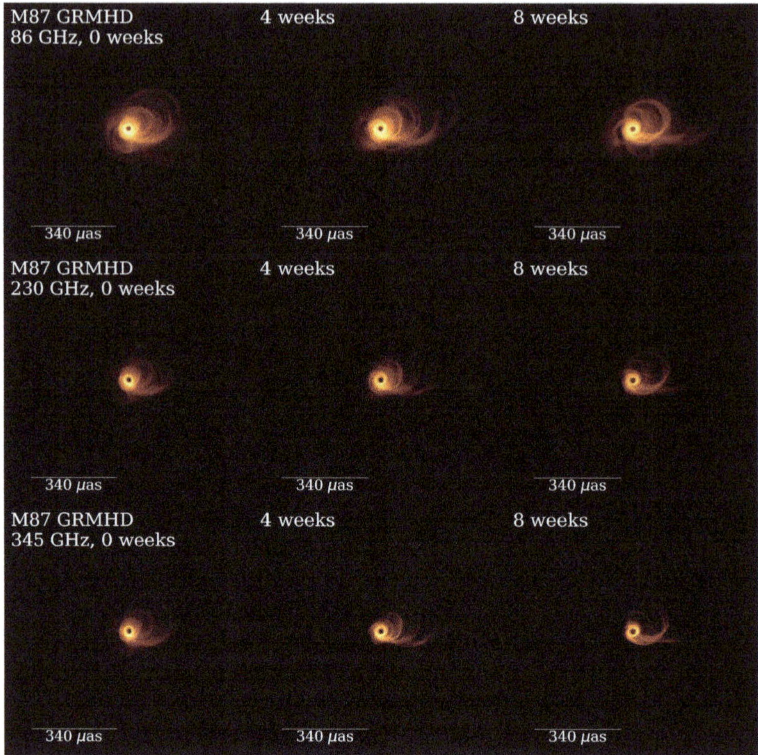

Figure 6. Ground truth M87 source models used for Challenge 2. For each frequency, three movie frames are shown. Images are shown on a log scale, which is normalized to the brightest pixel value across each set of three movie frames, with a dynamic range of $10^{3.5}$.

5.3. Synthetic Data

The synthetic data for Challenge 2 includes significantly more systematic effects than Challenge 1 (see also Section 5.1). In the SYMBA pipeline [74], atmospheric absorption, emission, delays and turbulence are simulated, and antenna pointing offsets are added to the simulated datasets with MeqSilhouette [75,76]. These are then calibrated by performing a fringe fit, a priori amplitude calibration with rPICARD [77], and network calibration with eht-imaging.

The station locations, dish sizes, aperture efficiencies, and receiver temperatures are identical to those used in Challenge 1 (Figure 2, Section 4.3). For each site, the input precipitable water vapor (PWV), ground temperature, and ground pressure were calculated from the Modern-Era Retrospective Analysis for Research and Applications, version 2 (MERRA-2) from the NASA Goddard Earth Sciences Data and Information Services Center GES DISC, [78], processed with the am atmospheric model software [79] (see [74] for details. All weather quantities were based on median conditions on 1 April (2000–2020) as registered in the MERRA-2 climatological data. The atmospheric coherence time at 230 GHz was assumed to be 20 s for a PWV of 1 mm, 3 s for a PWV of 15 mm, interpolated linearly between these values for the different sites and scaled linearly with frequency. Pointing offsets were assumed to be stable across each 10-min scan and drawn randomly from a Gaussian distribution with an RMS of 2 arc-seconds.

5.4. Results

Seven submitters or teams provided dynamical reconstructions of the Challenge 2 datasets. Reconstruction quality metrics for all submissions are shown in Table 2.

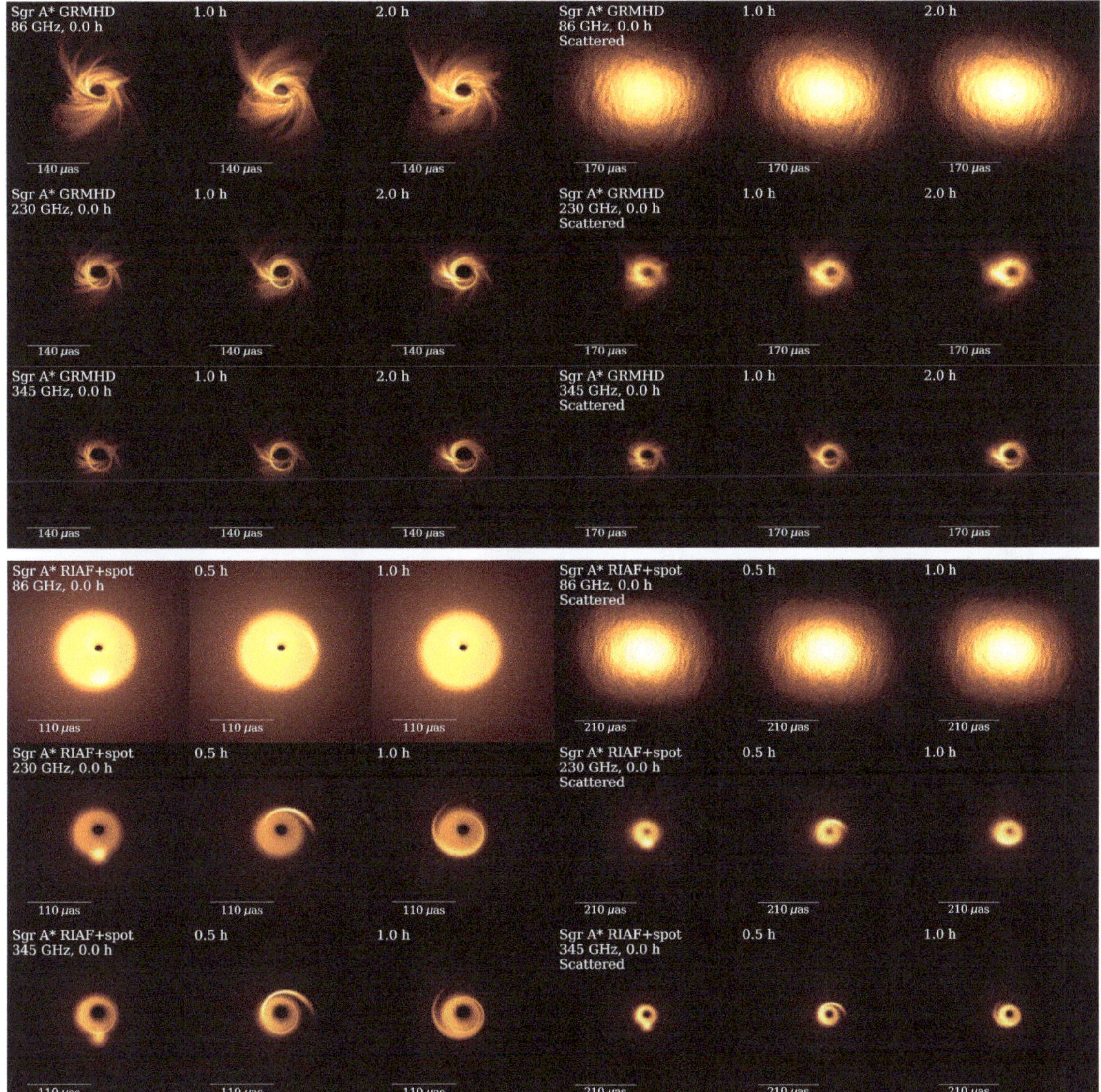

Figure 7. Ground truth Sgr A* source models used for Challenge 2. For each model and frequency (rows), three movie frames (columns) are shown, with interstellar scattering applied to the first frame in the rightmost column. Images are shown on a square root scale, which is normalized to the brightest pixel value across each set of three movie frames. The scattered movies were used as inputs for the Challenge 2 synthetic data generation.

5.4.1. M87 GRMHD

Six frames of the 86 and 230 GHz M87 reconstructions are shown in Figures 8 and 9, respectively. At 86 GHz, EHT2022 coverage allows reconstruction of the central component and overall shape of the extended jet emission, but all reconstructions contain spurious artifacts. With ngEHT1 coverage, these artifacts become far less severe or disappear completely, and the jet dynamics can be imaged as the jet features move outwards over the course of several weeks. The 230 GHz reconstructions provide significantly more detail, both in the extended jet and in the visibility of the black hole shadow. The reconstructions again show a strong improvement of ngEHT1 compared to EHT2022, although the EHT2022 reconstructions from the `resolve` algorithm already show some jet dynamics with EHT2022 coverage. Figures 10 and 11 show spectral index and individual frequency image reconstructions, respectively, from `resolve` when solving for all frequencies simultaneously and imposing a prior on the spectral index map (`resolve-mf`) for the first movie frame only. This method leads to remarkably high-quality images for all frequencies and arrays, even showing the black hole's central brightness depression at 86 GHz. These results demonstrate that utilizing information from simultaneous multi-frequency observations can significantly boost the reconstruction quality; see also [17].

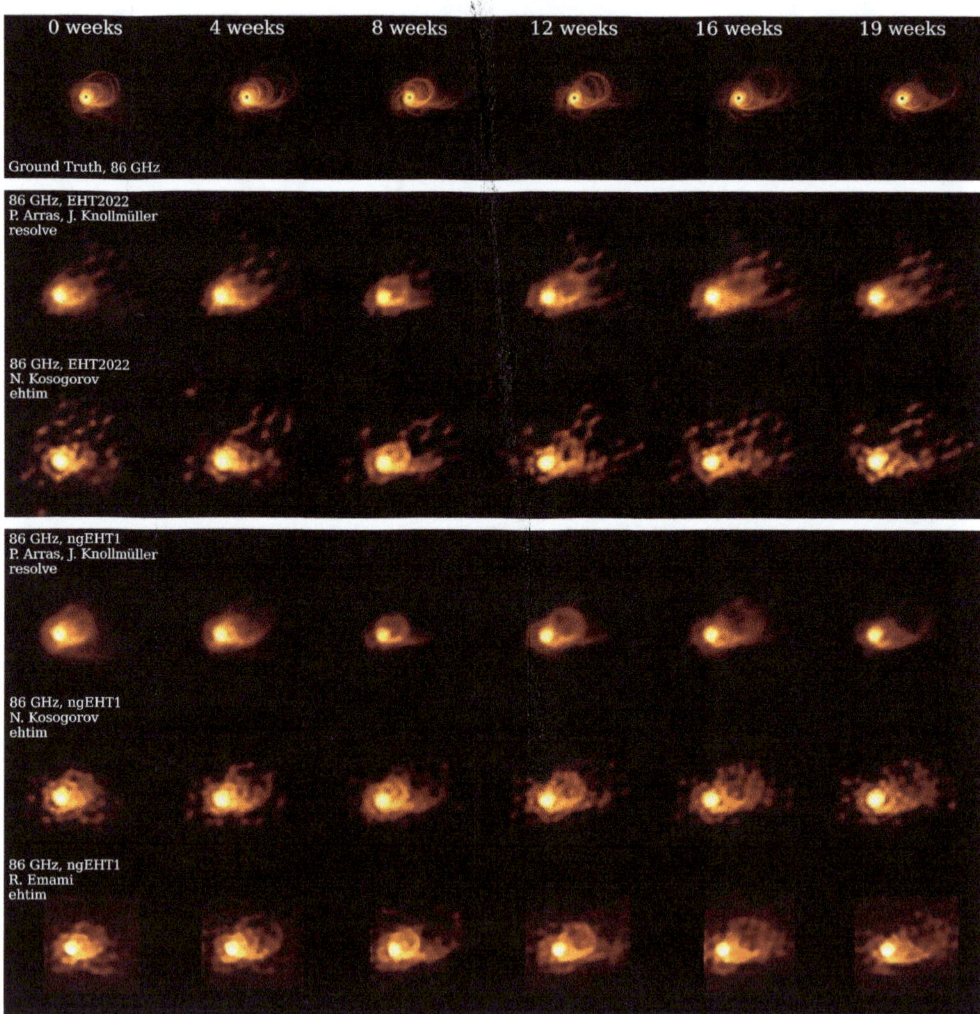

Figure 8. Selection of Challenge 2 M87 86 GHz submissions. Images are shown on a log scale, which is normalized to the brightest pixel value across each set of three movie frames, with a dynamic range of $10^{3.5}$ and field of view of 1 mas.

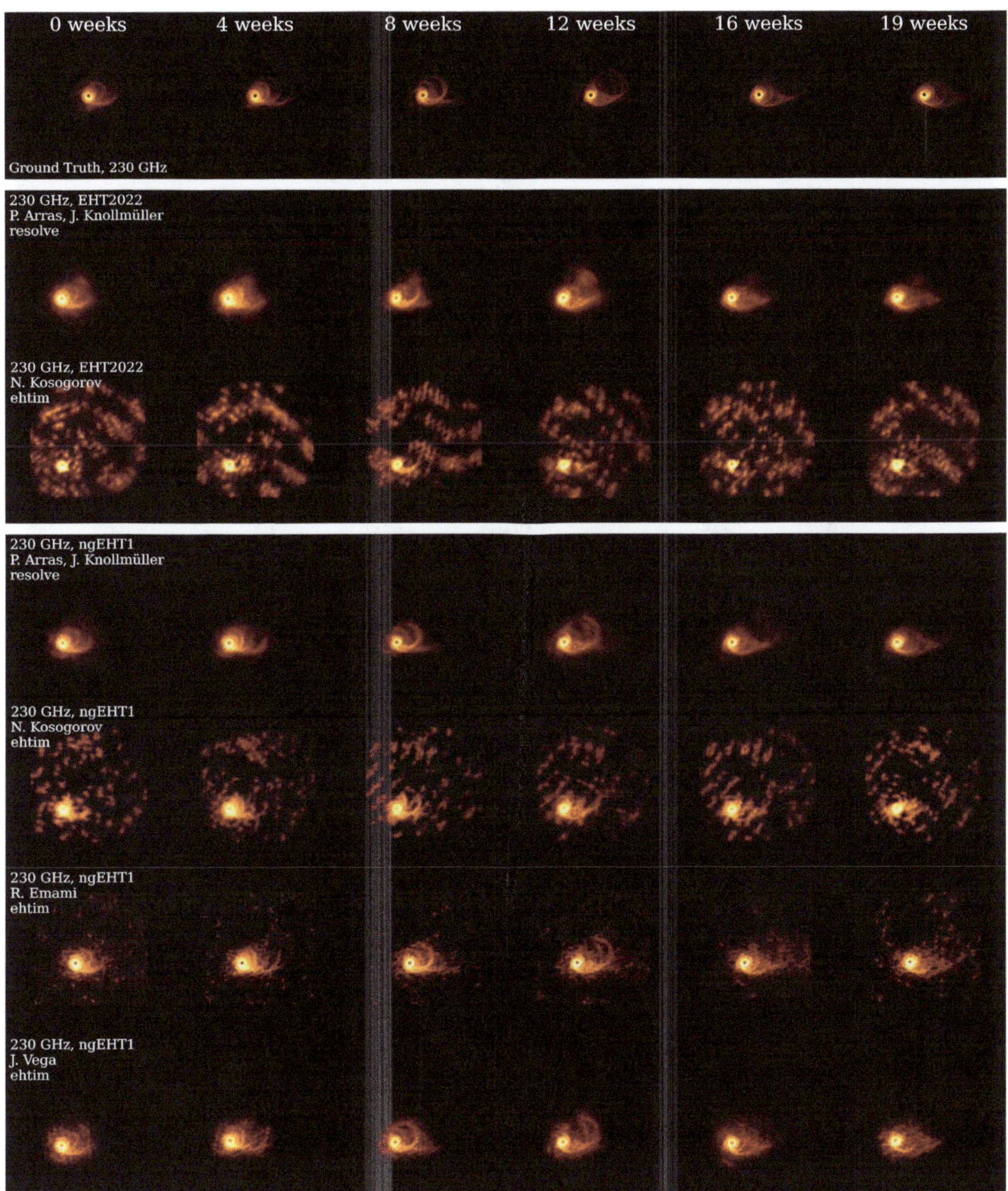

Figure 9. Selection of Challenge 2 M87 230 GHz submissions. Images are shown on a log scale, which is normalized to the brightest pixel value across each submitted set of movie frames, with a dynamic range of $10^{3.5}$, on a field of view of 1 mas.

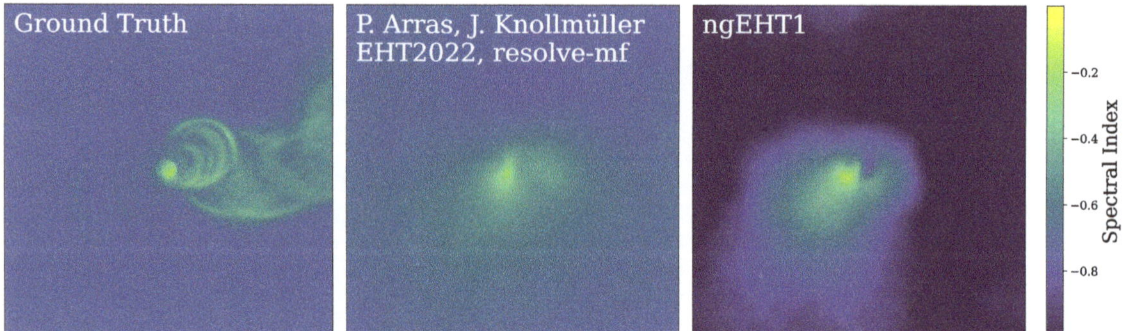

Figure 10. Spectral index maps of the Challenge 2 M87 ground truth model (first frame) and `resolve` reconstructions of the spectral index map with the EHT2022 and ngEHT1 arrays. The ground truth spectral index map was blurred with a Gaussian with a FWHM of 9.4 µas.

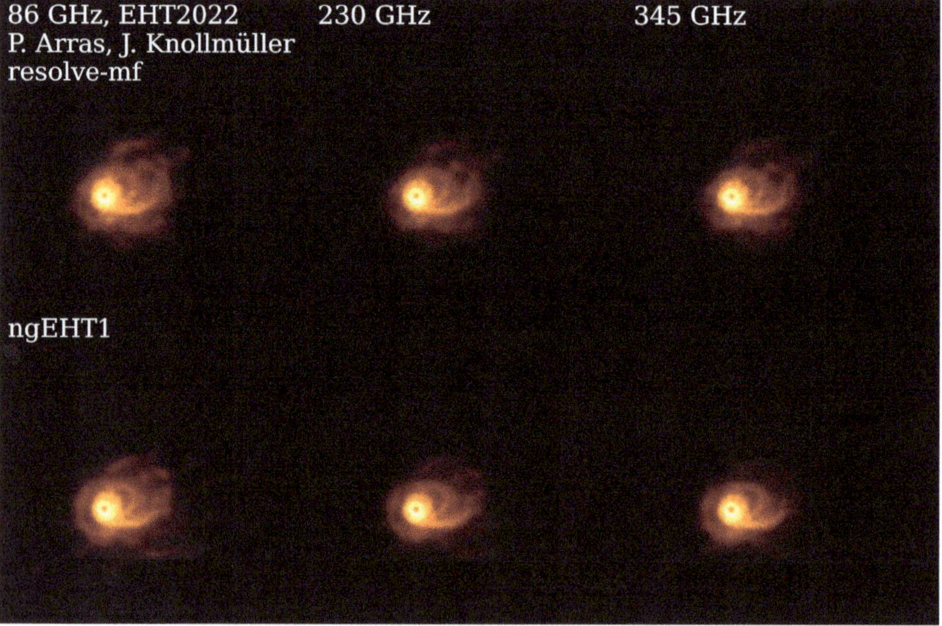

Figure 11. Multi-frequency `resolve` reconstructions of the Challenge 2 M87 model at 86, 230, and 345 GHz (first frame) with the EHT2022 and ngEHT1 arrays.

The reconstruction quality metrics show that images with low ρ_{NX} or $\rho_{NX,\log}$ often have relative high χ^2 and low θ_{eff} and $\mathcal{D}_{0.1}$, with the `resolve` and especially the `resolve-mf` reconstructions performing best overall, with the caveat that the `resolve-mf` reconstructions were only performed for the first movie frame. For single-frequency reconstructions, θ_{eff} reaches 21.2 µas at 86 GHz and 6.6 µas at 230 GHz (median values across the 20 reconstructed frames); the super-resolution with respect to the nominal array resolution (60 and 23 µas for 86 and 230 GHz, respectively) is significant (up to a factor 3.5) for most reconstructions. For multi-frequency reconstructions, the supper-resolution factor increases even further, up to 8.6 at 86 GHz. χ^2-values generally increase as a function of frequency, which is likely due to increased data complexity with more severe systematics. The 345 GHz reconstructions often showed difficulty in reconstructing the extended jet structure, which could be due to the sparser coverage and more severe corruptions and noise.

5.4.2. Sgr A* RIAF+hotspot

Figure 12 shows eight frames of all Sgr A* RIAF+hotspot submissions at 230 GHz. These frames span the 11–12 h UT window. This time window was chosen for the visual and metric submission comparisons as it corresponds to the first full rotation of the hotspot after it forms, and it is the hotspot rotation we aimed to reconstruct for this model. The 11–12 UT window also has a strong overlap with the ngEHT1 "best times", e.g., [80] window for Sgr A*, with 14 stations observing Sgr A* simultaneously from 11.3 until 13.5 h UT. Furthermore, all submissions contained frames in this window, whereas the total UT ranges reconstructed varied strongly between submissions.

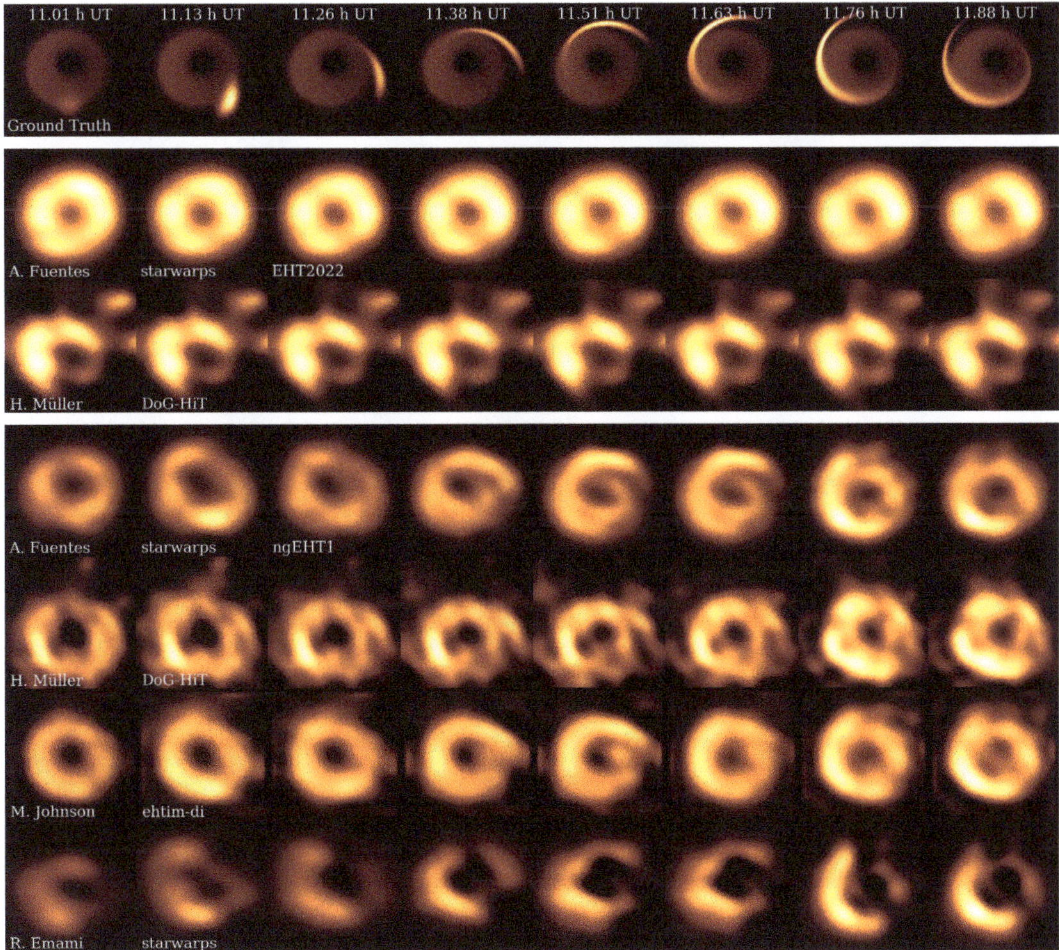

Figure 12. Challenge 2 Sgr A* RIAF+hotspot 230 GHz submissions. Images are shown on a linear scale, which is normalized to the brightest pixel value across each submitted set of movie frames on a field of view of 126 μas.

None of the EHT2022 reconstructions show a significantly variable source structure, although a ring-like structure is recovered. The ngEHT1 reconstructions vary in quality, with especially the StarWarps and `eht-imaging` dynamical imaging algorithms recovering the shearing hotspot. This particular hotspot model was particularly challenging because of its quick shearing. Furthermore, the data sampling with 10-min scans interleaved with 10-min gaps was relatively sparse compared to the hotspot period, giving only ∼3 scans per full hotspot rotation. These features make the reconstruction quality obtained by some methods remarkable.

The χ^2 metrics (Table 2) are remarkably high for the Sgr A* reconstructions, which has several causes. First, in order to provide a uniform comparison, the metrics were calculated on the 11–12 h UT window only. The χ^2 are lower when considering the full UT ranges submitted ($\sim$3–4 for the highest-quality reconstructions). Considering that the source is particularly variable and hence more difficult to reconstruct during the 11–12 UT window, it is not surprising that the χ^2 are higher here. Second, the χ^2 were calculated with respect to the original synthetic data, which have a 10-s resolution within 10-min scans. Many submitters added systematic noise and time-averaged the data down to $\sim$minutes before imaging, which included averaging of rapidly variable structures in the visibility domain. For example, closure phases may swing by well over 120 degrees within a single 10-min scan. Combined with the high signal-to-noise ratios of the ngEHT visibilities, the fit quality to the original input data is then significantly poorer than seen during the imaging process. Finally, the submissions are compared to the data, which includes interstellar scattering, while many submitters deblurred the data before imaging. This process does not affect the closure phases, but the closure amplitudes are affected. The effective resolution θ_{eff} for the best reconstructions is comparable to the nominal array resolution, reflecting the increased difficulty of recovering intraday time-variable structures compared to static reconstructions, which often reached significant super-resolution.

Figure 13 shows the average image position angle as calculated by the Ring Extractor algorithm REx [14,81], which characterizes the properties of ring-like images for the ground truth and a few reconstructions in the 11–12 UT window. The ngEHT1 StarWarps reconstruction, in particular, shows excellent agreement with the ground truth, and the ngEHT1 ehtim-di reconstruction performs well after about 11.4 UT. The EHT2022 StarWarps reconstruction shows a stable position angle (note the 2π ambiguity) that is generally offset from the ground truth.

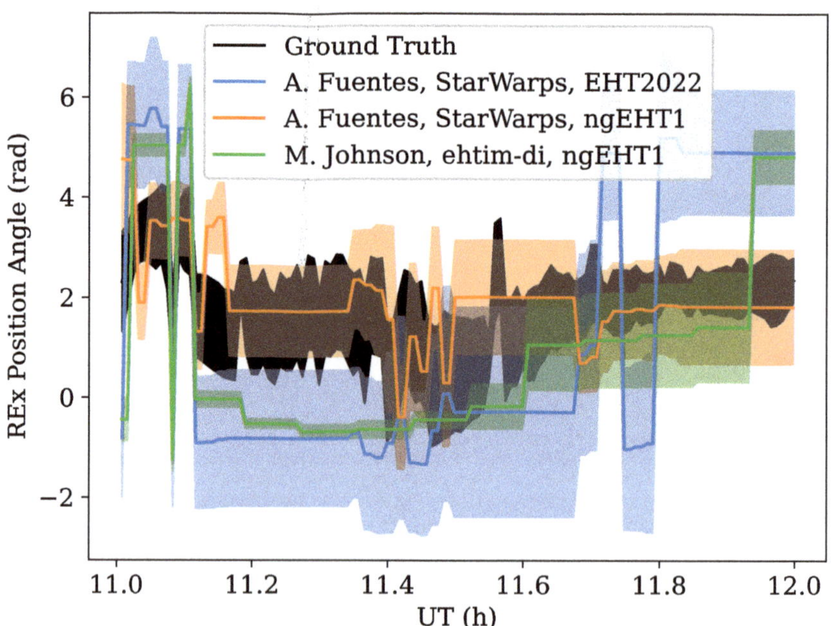

Figure 13. REx position angle fits with 1σ uncertainties for three Challenge 2 Sgr A* RIAF+hotspot 230 GHz submissions, compared to the ground truth.

Table 2. Reconstruction quality metrics for Challenge 2. Tabulated values are medians across the reconstructed frames except for the `resolve-mf` reconstructions, which were only done for the first movie frame. For the Sgr A* models, the metrics were evaluated on a common UT range for all reconstructions (see text for details).

Model	Array	ν (GHz)	Submitter	Method	$\chi^2_{\rm cp}$	$\chi^2_{\rm lcamp}$	$\rho_{\rm NX}$	$\rho_{\rm NX,log}$	$\theta_{\rm eff}$	$\mathcal{D}_{0.1}$
M87 GRMHD	EHT2022	86	P. Arras, J. Knollmüller	resolve	1.94	2.01	0.83	0.92	24.5	1156
M87 GRMHD	EHT2022	86	P. Arras, J. Knollmüller	resolve-mf	1.71	4.82	0.96	0.97	7.0	3970
M87 GRMHD	EHT2022	86	N. Kosogorov	ehtim	7.16	2.69	0.8	0.82	32.0	585
M87 GRMHD	ngEHT1	86	P. Arras, J. Knollmüller	resolve	1.45	1.4	0.85	0.96	21.2	3054
M87 GRMHD	ngEHT1	86	P. Arras, J. Knollmüller	resolve-mf	1.43	1.73	0.95	0.99	8.2	7248
M87 GRMHD	ngEHT1	86	R. Emami	ehtim	1.84	1.69	0.8	0.89	30.4	1315
M87 GRMHD	ngEHT1	86	N. Kosogorov	ehtim	2.06	1.58	0.8	0.93	30.4	919
M87 GRMHD	ngEHT1	86	N. Kosogorov	CLEAN	193.55	10,266.39	0.75	0.74	46.8	749
M87 GRMHD	EHT2022	230	P. Arras, J. Knollmüller	resolve	2.03	3.25	0.92	0.96	7.3	3881
M87 GRMHD	EHT2022	230	P. Arras, J. Knollmüller	resolve-mf	2.03	6.67	0.93	0.97	7.1	6424
M87 GRMHD	EHT2022	230	N. Kosogorov	ehtim	4.16	3.15	0.88	0.54	12.6	429
M87 GRMHD	ngEHT1	230	P. Arras, J. Knollmüller	resolve	2.53	2.35	0.92	0.98	6.6	8742
M87 GRMHD	ngEHT1	230	P. Arras, J. Knollmüller	resolve-mf	2.57	3.12	0.93	0.99	7.1	12,154
M87 GRMHD	ngEHT1	230	J. Vega	ehtim	2.55	2.3	0.91	0.97	8.1	4807
M87 GRMHD	ngEHT1	230	R. Emami	ehtim	2.55	2.47	0.89	0.83	10.9	2061
M87 GRMHD	ngEHT1	230	N. Kosogorov	ehtim	2.85	2.81	0.89	0.71	11.4	1060
M87 GRMHD	ngEHT1	230	N. Kosogorov	CLEAN	325.47	385.28	0.79	0.6	22.6	226
M87 GRMHD	EHT2022	345	P. Arras, J. Knollmüller	resolve-mf	5.26	5.79	0.93	0.97	7.2	6994
M87 GRMHD	ngEHT1	345	P. Arras, J. Knollmüller	resolve-mf	6.39	6.89	0.92	0.98	7.3	9732
M87 GRMHD	ngEHT1	345	R. Emami	ehtim	6.14	5.38	0.59	0.42	61.8	61
M87 GRMHD	ngEHT1	345	N. Kosogorov	ehtim	5.99	4.94	0.81	0.47	16.6	563
M87 GRMHD	ngEHT1	345	N. Kosogorov	CLEAN	12.41	16.28	0.83	0.67	14.4	1157
Sgr A* RIAFSPOT	EHT2022	230	A. Fuentes	StarWarps	1.85	1.78	0.83	-	37.3	-
Sgr A* RIAFSPOT	EHT2022	230	H. Müller	DoG-HiT	5.61	5.12	0.77	-	56.8	-
Sgr A* RIAFSPOT	ngEHT1	230	M. Johnson	ehtim-di	7.39	11.78	0.87	-	24.4	-
Sgr A* RIAFSPOT	ngEHT1	230	A. Fuentes	StarWarps	4.24	3.05	0.89	-	23.0	-
Sgr A* RIAFSPOT	ngEHT1	230	R. Emami	StarWarps	6.87	11.98	0.83	-	43.3	-
Sgr A* RIAFSPOT	ngEHT1	230	H. Müller	DoG-HiT	33.31	38.91	0.84	-	33.0	-
Sgr A* RIAFSPOT	ngEHT1	345	A. Fuentes	StarWarps	5.37	3.63	0.85	-	28.6	-
Sgr A* RIAFSPOT	ngEHT1	345	R. Emami	StarWarps	5.7	3.86	0.74	-	56.8	-
Sgr A* GRMHD	EHT2022	230	A. Fuentes	StarWarps	9.49	3.61	0.68	-	56.0	-
Sgr A* GRMHD	EHT2022	230	H. Müller	DoG-HiT	153.81	32.15	0.68	-	57.4	-
Sgr A* GRMHD	ngEHT1	230	M. Johnson	ehtim-di	3.99	7.14	0.87	-	18.4	-
Sgr A* GRMHD	ngEHT1	230	A. Fuentes	StarWarps	3.97	7.47	0.85	-	21.1	-
Sgr A* GRMHD	ngEHT1	230	R. Emami	StarWarps	4.0	6.91	0.87	-	17.5	-
Sgr A* GRMHD	ngEHT1	230	H. Müller	DoG-HiT	13.88	8.18	0.8	-	29.0	-
Sgr A* GRMHD	ngEHT1	230	P. Arras, J. Knollmüller	resolve	5.57	4.52	0.84	-	21.9	-
Sgr A* GRMHD	ngEHT1	345	R. Emami	StarWarps	4.94	4.19	0.61	-	56.9	-

5.4.3. Sgr A* GRMHD

Finally, Figure 14 shows eleven frames of all Sgr A* GRMHD submissions at 230 GHz, spanning the best times window (11.3–13.5 h UT). Like for the Sgr A* RIAF+hotspot model, the EHT2022 reconstructions are static. In the StarWarps reconstruction, the ring morphology is recovered, but the detailed emission along the ring is not. The ngEHT1 reconstructions are generally much sharper, and the azimuthal brightness variations are reconstructed accurately, with the `ehtim-di` and StarWarps submissions showing the best quality metric values (Table 2). Due to the relatively stable and turbulent nature of the variability in this model, the reconstruction of temporal variations is more difficult to assess than for the other source models.

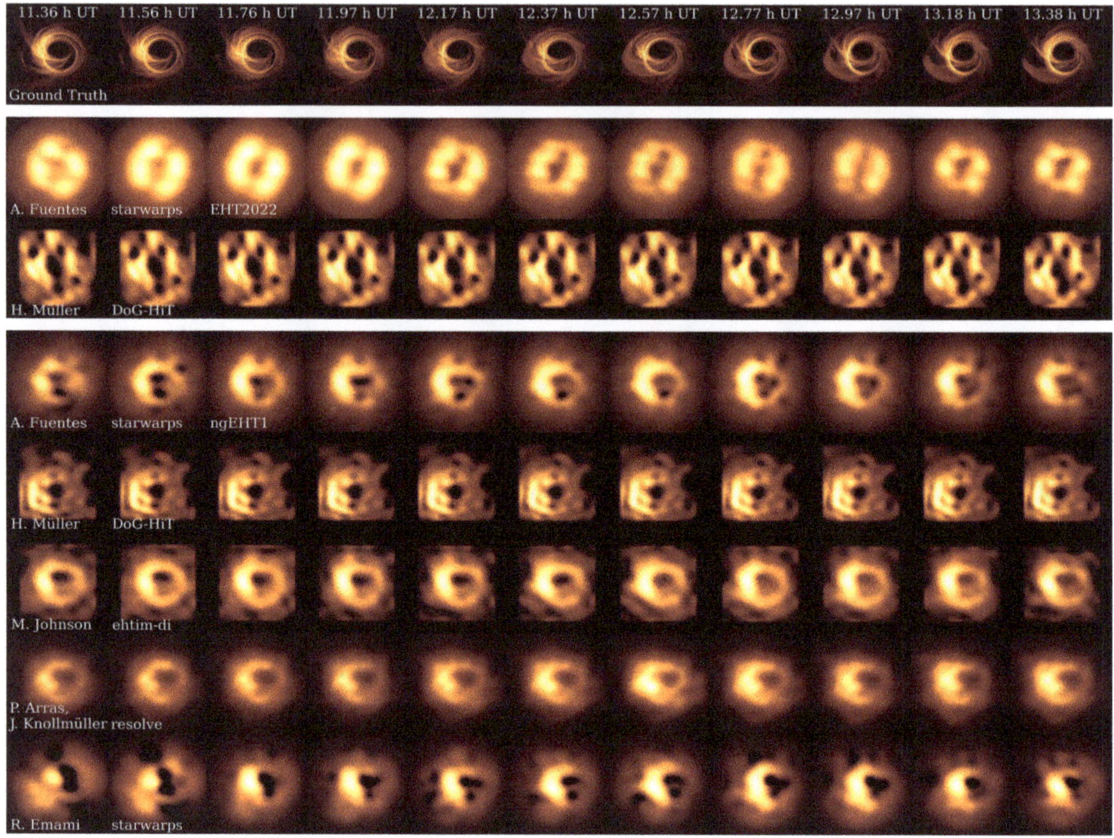

Figure 14. Challenge 2 Sgr A* GRMHD 230 GHz submissions. Images are shown on a square root scale, which is normalized to the brightest pixel value across each submitted set of movie frames on a field of view of 126 µas.

6. Conclusions and Outlook

The first two ngEHT Analysis Challenges have provided a number of useful insights. First, current imaging algorithms are capable of reconstructing high-resolution and high-fidelity movies of M87* and Sgr A*, revealing the jet dynamics of M87* on mas scales down to the event horizon and intra-hour dynamics of the Sgr A* accretion flow on event-horizon scales. This work provides high-quality reconstructions from synthetic data that include fully realistic observation and calibration effects under mediocre weather conditions (median conditions for April), showing excellent prospects for ngEHT performance.

The sources have been reconstructed with a breadth of traditional (EHT) imaging algorithms and newer algorithms that are under active development, such as `resolve` and `DoG-HiT`. Different algorithms showed different performances for different datasets. For example, the `StarWarps` and `eht-imaging` algorithms showed strong performance on dynamical reconstructions of Sgr A* (see [40,82] for other examples), while the `resolve` algorithm did remarkably well on recovering the extended jet structure of M87. Multi-frequency reconstructions gave the best M87 jet reconstruction results for both challenges, providing an opportunity to produce high-resolution 86 GHz images showing the central brightness depression related to the black hole (inner) shadow [83,84] and faithful reconstructions of the extended jet at 345 GHz. Both these features are difficult to reconstruct using the individual frequencies alone. `DoG-HiT` reconstructions of the Sgr A* RIAF+hotspot model resolved the hotspot orbit and shearing, albeit with a lower quality than with other methods. Since `DoG-HiT` is the most recently developed algorithm used in the challenges and is completely automatic without special manual adaption to the data sets, these results are promising and can inform further development. `SMILI` and `CLEAN` have been applied to Challenge 1 data only, where they did not perform as well as `eht-imaging` in reconstructing the extended M87 jet.

The reconstructions from any algorithm do not necessarily show its maximum potential performance. Between algorithms, there are differences in the freedom that the user has to steer the reconstruction process. `CLEAN` traditionally requires significant user input and steering (e.g., defining `CLEAN` windows) but has been adapted to a more automated approach for EHT analysis [14]. RML methods such as `eht-imaging` and `SMILI` require setting regularizers and weights but also allow some input on the reconstruction procedure by setting, e.g., convergence criteria and blurring steps between image rounds. On the other hand, `DoG-HiT` depends on just one free parameter. The outcome of Bayesian methods generally depends on the set priors. From the results of these challenges, each method's dependence on user input is difficult to assess and would require dedicated parameter surveys. Based on, e.g., the `eht-imaging` submissions, the results can certainly depend strongly on the user. However, for submissions reconstructed with the same method but with strongly different resulting images, the χ^2 are a good indicator of the reconstruction quality. For the lower-quality image reconstructions, either the used parameters or the script setup often did not allow a good fit to the data.

Regarding data generation, one lesson learned is that the used schedule of 10-min scans and 10-min gaps makes reconstructing the rapid variability of Sgr A* challenging, and in fact, it is remarkable that dynamical imaging algorithms were able to reconstruct the one-hour period and rapidly shearing Sgr A* hotspot orbit with just three scans and a duty cycle of 50%. A denser schedule with shorter gaps could potentially improve these reconstructions significantly and also help in reconstructing the rapid variability from GRMHD simulations. Furthermore, the Challenge 2 345 GHz data have proven difficult to image, which is likely attributable to the severe atmospheric effects considering the weather parameters were medians for April at all sites. Since, in reality, 345 GHz observations would likely only be scheduled on days with excellent weather at suitable sites, a next challenge should be performed with more optimistic 345 GHz weather conditions. The experience from these challenges has shown that both `eht-imaging` and `SYMBA` are viable and well-performing pathways for generating synthetic ngEHT data. User-friendly tools to generate synthetic ngEHT data from a centralized repository of instrument and weather parameters using both pathways are under development [2].

ngEHT Analysis Challenge 3[3] is an extension of Challenge 2 to full Stokes and will show how well various algorithms can reconstruct dynamics in polarization. Challenge 4 will focus on more specific ngEHT science goals, such as measuring the photon ring size and black hole spin, involving modeling methods as well, e.g., [85]. The merit of simultaneous multi-frequency observations allowing for frequency phase transfer, e.g., [86,87], will be explored in this challenge as well (see also [17]). Future challenges could also involve varying the number and locations of stations. While the impact of a single station's presence or location diminishes as the array grows and becomes more robust against station losses, the effect of using partial instead of the full array or using different sets of new sites could be tested in end-to-end simulations (see also [2]).

The ngEHT Analysis Challenges have brought together expertise in theoretical modeling, synthetic data generation, and image reconstruction, spurring development in all these areas. The continued challenges will involve polarization, model fitting, and science interpretation to form a complete and end-to-end process of ngEHT simulations, which helps in maximizing the science potential of the array.

Author Contributions: Conceptualization, L.B., S.S.D., M.D.J. and F.R.; methodology, P.A., L.B., K.C., R.E., C.F., A.F., M.D.J., J.K., N.K., H.M., N.P., A.R., F.R., P.T., E.T. and J.V.; software, G.L., P.A., K.C., C.F., J.K., H.M., M.D.J. and F.R.; validation, L.B. and F.R.; formal analysis, P.A., L.B., K.C., R.E., C.F., A.F., M.D.J., J.K., N.K., H.M., N.P., A.R., F.R., P.T., E.T. and J.V.; data curation, L.B., G.L. and F.R.; writing—original draft preparation, P.A., R.E., M.D.J., H.M., F.R. and E.T.; writing—review and editing, P.A., L.B., K.C., S.S.D., R.E., C.F., A.F., M.D.J., J.K., N.K., G.L., H.M., N.P., A.R., F.R., P.T., E.T. and J.V.; visualization, F.R.; supervision, S.S.D.; funding acquisition, S.S.D. All authors have read and agreed to the published version of the manuscript.

Funding: This research was supported by NSF grants AST-1935980 and AST-2034306. This work was supported by the Black Hole Initiative at Harvard University, made possible through the support of grants from the Gordon and Betty Moore Foundation and the John Templeton Foundation. The opinions expressed in this publication are those of the author(s) and do not necessarily reflect the views of the Moore or Templeton Foundations. Hendrik Müller received financial support for this research from the International Max Planck Research School (IMPRS) for Astronomy and Astrophysics at the Universities of Bonn and Cologne. This research is supported by the DFG research grant "Jet physics on horizon scales and beyond" (Grant No. FR 4069/2-1), the ERC synergy grant "BlackHoleCam: Imaging the Event Horizon of Black Holes" (Grant No. 610058) and ERC advanced grant "JETSET: Launching, propagation and emission of relativistic jets from binary mergers and across mass scales" (Grant No. 884631). Jakob Knollmüller acknowledges funding by the Deutsche Forschungsgemeinschaft (DFG, German Research Foundation) under Germany´s Excellence Strategy—EXC 2094—390783311. Razieh Emami acknowledges the support by the Institute for Theory and Computation at the Center for Astrophysics as well as grant numbers 21-atp21-0077, NSF AST-1816420 and HST-GO-16173.001-A for very generous support.

Data Availability Statement: The synthetic data generated for this work can be downloaded from https://challenge.ngeht.org/ (accessed on 19 December 2022), with access provided upon request to the corresponding author.

Conflicts of Interest: The authors declare no conflict of interest. The funders had no role in the design of the study; in the collection, analyses, or interpretation of data; in the writing of the manuscript, or in the decision to publish the results.

Notes

[1] https://challenge.ngeht.org/ (accessed on 19 December 2022)
[2] https://gitlab.mpcdf.mpg.de/ift/resolve (accessed on 19 December 2022)
[3] https://challenge.ngeht.org/challenge3/ (accessed on 19 December 2022).

References

1. Doeleman, S.; Blackburn, L.; Dexter, J.; Gomez, J.L.; Johnson, M.D.; Palumbo, D.C.; Weintroub, J.; Farah, J.R.; Fish, V.; Loinard, L.; et al. Studying Black Holes on Horizon Scales with VLBI Ground Arrays. *Proc. Bull. Am. Astron. Soc.* **2019**, *51*, 256.
2. Doeleman, S. Reference Array and Design Consideration for the next-generation Event Horizon Telescope. *Galaxies* 2023, *in prep.*
3. Event Horizon Telescope Collaboration; Akiyama, K.; Alberdi, A.; Alef, W.; Asada, K.; Azulay, R.; Baczko, A.K.; Ball, D.; Baloković, M.; Barrett, J.; et al. First M87 Event Horizon Telescope Results. I. The Shadow of the Supermassive Black Hole. *Astrophys. J. Lett.* **2019**, *875*, L1. [CrossRef]
4. Event Horizon Telescope Collaboration; Akiyama, K.; Alberdi, A.; Alef, W.; Asada, K.; Azulay, R.; Baczko, A.K.; Ball, D.; Baloković, M.; Barrett, J.; et al. First M87 Event Horizon Telescope Results. II. Array and Instrumentation. *Astrophys. J. Lett.* **2019**, *875*, L2. [CrossRef]
5. Event Horizon Telescope Collaboration; Akiyama, K.; Alberdi, A.; Alef, W.; Asada, K.; Azulay, R.; Baczko, A.K.; Ball, D.; Baloković, M.; Barrett, J.; et al. First M87 Event Horizon Telescope Results. III. Data Processing and Calibration. *Astrophys. J. Lett.* **2019**, *875*, L3. [CrossRef]
6. Event Horizon Telescope Collaboration; Akiyama, K.; Alberdi, A.; Alef, W.; Asada, K.; Azulay, R.; Baczko, A.K.; Ball, D.; Baloković, M.; Barrett, J.; et al. First M87 Event Horizon Telescope Results. IV. Imaging the Central Supermassive Black Hole. *Astrophys. J. Lett.* **2019**, *875*, L4. [CrossRef]
7. Event Horizon Telescope Collaboration; Akiyama, K.; Alberdi, A.; Alef, W.; Asada, K.; Azulay, R.; Baczko, A.K.; Ball, D.; Baloković, M.; Barrett, J.; et al. First M87 Event Horizon Telescope Results. V. Physical Origin of the Asymmetric Ring. *Astrophys. J. Lett.* **2019**, *875*, L5. [CrossRef]

8. Event Horizon Telescope Collaboration; Akiyama, K.; Alberdi, A.; Alef, W.; Asada, K.; Azulay, R.; Baczko, A.K.; Ball, D.; Baloković, M.; Barrett, J.; et al. First M87 Event Horizon Telescope Results. VI. The Shadow and Mass of the Central Black Hole. *Astrophys. J. Lett.* **2019**, *875*, L6. [CrossRef]

9. Event Horizon Telescope Collaboration; Akiyama, K.; Algaba, J.C.; Alberdi, A.; Alef, W.; Anantua, R.; Asada, K.; Azulay, R.; Baczko, A.K.; Ball, D.; et al. First M87 Event Horizon Telescope Results. VII. Polarization of the Ring. *Astrophys. J. Lett.* **2021**, *910*, L12. [CrossRef]

10. Event Horizon Telescope Collaboration; Akiyama, K.; Algaba, J.C.; Alberdi, A.; Alef, W.; Anantua, R.; Asada, K.; Azulay, R.; Baczko, A.K.; Ball, D.; et al. First M87 Event Horizon Telescope Results. VIII. Magnetic Field Structure near The Event Horizon. *Astrophys. J. Lett.* **2021**, *910*, L13. [CrossRef]

11. Event Horizon Telescope Collaboration; Akiyama, K.; Alberdi, A.; Alef, W.; Algaba, J.C.; Anantua, R.; Asada, K.; Azulay, R.; Bach, U.; Baczko, A.K.; et al. First Sagittarius A* Event Horizon Telescope Results. I. The Shadow of the Supermassive Black Hole in the Center of the Milky Way. *Astrophys. J. Lett.* **2022**, *930*, L12. [CrossRef]

12. Event Horizon Telescope Collaboration; Akiyama, K.; Alberdi, A.; Alef, W.; Algaba, J.C.; Anantua, R.; Asada, K.; Azulay, R.; Bach, U.; Baczko, A.K.; et al. First Sagittarius A* Event Horizon Telescope Results. II. EHT and Multiwavelength Observations, Data Processing, and Calibration. *Astrophys. J. Lett.* **2022**, *930*, L13. [CrossRef]

13. Event Horizon Telescope Collaboration; Akiyama, K.; Alberdi, A.; Alef, W.; Algaba, J.C.; Anantua, R.; Asada, K.; Azulay, R.; Bach, U.; Baczko, A.K.; et al. First Sagittarius A* Event Horizon Telescope Results. III. Imaging of the Galactic Center Supermassive Black Hole. *Astrophys. J. Lett.* **2022**, *930*, L14. [CrossRef]

14. Event Horizon Telescope Collaboration ; Akiyama, K.; Alberdi, A.; Alef, W.; Algaba, J.C.; Anantua, R.; Asada, K.; Azulay, R.; Bach, U.; Baczko, A.K.; et al. First Sagittarius A* Event Horizon Telescope Results. IV. Variability, Morphology, and Black Hole Mass. *Astrophys. J. Lett.* **2022**, *930*, L15. [CrossRef]

15. Event Horizon Telescope Collaboration; Akiyama, K.; Alberdi, A.; Alef, W.; Algaba, J.C.; Anantua, R.; Asada, K.; Azulay, R.; Bach, U.; Baczko, A.K.; et al. First Sagittarius A* Event Horizon Telescope Results. V. Testing Astrophysical Models of the Galactic Center Black Hole. *Astrophys. J. Lett.* **2022**, *930*, L16. [CrossRef]

16. Event Horizon Telescope Collaboration; Akiyama, K.; Alberdi, A.; Alef, W.; Algaba, J.C.; Anantua, R.; Asada, K.; Azulay, R.; Bach, U.; Baczko, A.K.; et al. First Sagittarius A* Event Horizon Telescope Results. VI. Testing the Black Hole Metric. *Astrophys. J. Lett.* **2022**, *930*, L17. [CrossRef]

17. Issaoun, S.; Pesce, D.W.; Roelofs, F.; Chael, A.; Dodson, R.; Rioja, M.J.; Akiyama, K.; Aran, R.; Blackburn, L.; Doeleman, S.S.; et al. Enabling transformational ngEHT science via the inclusion of 86 GHz capabilities. *Galaxies* 2023, *submitted*.

18. Pesce, D.W.; Palumbo, D.C.M.; Narayan, R.; Blackburn, L.; Doeleman, S.S.; Johnson, M.D.; Ma, C.P.; Nagar, N.M.; Natarajan, P.; Ricarte, A. Toward Determining the Number of Observable Supermassive Black Hole Shadows. *Astrophys. J. Lett.* **2021**, *923*, 260. [CrossRef]

19. Pesce, D.W.; Palumbo, D.C.M.; Ricarte, A.; Broderick, A.E.; Johnson, M.D.; Nagar, N.M.; Natarajan, P.; Gómez, J.L. Expectations for Horizon-Scale Supermassive Black Hole Population Studies with the ngEHT. *Galaxies* **2022**, *10*, 109. [CrossRef]

20. Bouman, K.L. Extreme Imaging via Physical Model Inversion: Seeing around Corners and Imaging Black Holes. Ph.D. Thesis, Massachusetts Institute of Technology, Cambridge, MA, USA, 2017.

21. Chatterjee, K.; Chael, A.; Tiede, P.; Mizuno, Y.; Emami, R.; Fromm, C.; Ricarte, A.; Blackburn, L.; Roelofs, F.; Johnson, M.D.; et al. Comparing accretion flow morphology in numerical simulations of black holes from the ngEHT Model Library: The impact of radiation physics. *arXiv* **2022**, arXiv:2212.01804.

22. Högbom, J.A. Aperture Synthesis with a Non-Regular Distribution of Interferometer Baselines. *Astron. Astrophys. Suppl. Ser.* **1974**, *15*, 417.

23. Wilkinson, P.N.; Readhead, A.C.S.; Purcell, G.H.; Anderson, B. Radio structure of 3C 147 determined by multi-element very long baseline interferometry. *Nature* **1977**, *269*, 764–768. [CrossRef]

24. Cornwell, T.J. Hogbom's CLEAN algorithm. Impact on astronomy and beyond. Commentary on: Högbom J. A., 1974, A&AS, 15, 417. *Astron. Astrophys.* **2009**, *500*, 65–66. [CrossRef]

25. Shepherd, M.C.; Pearson, T.J.; Taylor, G.B. DIFMAP: An interactive program for synthesis imaging. *Proc. Bull. Am. Astron. Soc.* **1995**, *27*, 903.

26. Frieden, B.R. Restoring with Maximum Likelihood and Maximum Entropy. *J. Opt. Soc. Am.* **1972**, *62*, 511–518. [CrossRef] [PubMed]

27. Coughlan, C.P.; Gabuzda, D.C. High resolution VLBI polarization imaging of AGN with the maximum entropy method. *Mon. Not. R. Astron. Soc.* **2016**, *463*, 1980–2001. [CrossRef]

28. Narayan, R.; Nityananda, R. Maximum Entropy Image Restoration in Astronomy. *Annu. Rev. Astron. Astrophys.* **1986**, *24*, 127–170. [CrossRef]

29. Chael, A.A.; Johnson, M.D.; Bouman, K.L.; Blackburn, L.L.; Akiyama, K.; Narayan, R. Interferometric Imaging Directly with Closure Phases and Closure Amplitudes. *Astrophys. J.* **2018**, *857*, 23. [CrossRef]

30. Chael, A.A.; Johnson, M.D.; Narayan, R.; Doeleman, S.S.; Wardle, J.F.C.; Bouman, K.L. High-resolution Linear Polarimetric Imaging for the Event Horizon Telescope. *Astrophys. J.* **2016**, *829*, 11. [CrossRef]

31. Akiyama, K.; Kuramochi, K.; Ikeda, S.; Fish, V.L.; Tazaki, F.; Honma, M.; Doeleman, S.S.; Broderick, A.E.; Dexter, J.; Mościbrodzka, M.; et al. Imaging the Schwarzschild-radius-scale Structure of M87 with the Event Horizon Telescope Using Sparse Modeling. *Astrophys. J.* **2017**, *838*, 1. [CrossRef]

32. Akiyama, K.; Ikeda, S.; Pleau, M.; Fish, V.L.; Tazaki, F.; Kuramochi, K.; Broderick, A.E.; Dexter, J.; Mościbrodzka, M.; Gowanlock, M.; et al. Superresolution Full-polarimetric Imaging for Radio Interferometry with Sparse Modeling. *Astrophys. J.* **2017**, *153*, 159. [CrossRef]

33. Chael, A.A.; Bouman, K.L.; Johnson, M.D.; Narayan, R.; Doeleman, S.S.; Wardle, J.F.C.; Blackburn, L.L.; Akiyama, K.; Wielgus, M.; Chan, C.k.; et al. ehtim: Imaging, analysis, and simulation software for radio interferometry. *Astrophys. Source Code Libr.* **2019**, ascl-1904.

34. Chael, A.; Issaoun, S.; Pesce, D.W.; Johnson, M.D.; Ricarte, A.; Fromm, C.M.; Mizuno, Y. Multi-frequency Black Hole Imaging for the Next-Generation Event Horizon Telescope. *arXiv* **2022**, arXiv:2210.12226.

35. Johnson, M.D.; Narayan, R.; Psaltis, D.; Blackburn, L.; Kovalev, Y.Y.; Gwinn, C.R.; Zhao, G.Y.; Bower, G.C.; Moran, J.M.; Kino, M.; et al. The Scattering and Intrinsic Structure of Sagittarius A* at Radio Wavelengths. *Astrophys. J.* **2018**, *865*, 104. [CrossRef]

36. Johnson, M.D.; Lupsasca, A.; Strominger, A.; Wong, G.N.; Hadar, S.; Kapec, D.; Narayan, R.; Chael, A.; Gammie, C.F.; Galison, P.; et al. Universal interferometric signatures of a black hole's photon ring. *Sci. Adv.* **2020**, *6*, eaaz1310. [CrossRef] [PubMed]

37. Byrd, R.H.; Lu, P.; Nocedal, J. A Limited Memory Algorithm for Bound Constrained Optimization. *SIAM J. Sci. Stat. Comput.* **1995**, *16*, 1190–1208. [CrossRef]

38. Jones, E.; Oliphant, T.; Peterson, P. SciPy: Open Source Scientific Tools for Python. 2001. Available online: https://github.com/takluyver/scipy.org-new/blob/master/www/scipylib/citing.rst (accessed on 25 August 2015).

39. Bouman, K.L.; Johnson, M.D.; Dalca, A.V.; Chael, A.A.; Roelofs, F.; Doeleman, S.S.; Freeman, W.T. Reconstructing Video from Interferometric Measurements of Time-Varying Sources. *arXiv* **2017**, arXiv:1711.01357.

40. Emami, R.; Tiede, P.; Doeleman, S.S.; Roelofs, F.; Wielgus, M.; Blackburn, L.; Liska, M.; Chatterjee, K.; Ripperda, B.; Fuentes, A.; et al. Tracing the hot spot motion using the next generation Event Horizon Telescope (ngEHT). *arXiv* **2022**, arXiv:2211.06773.

41. Arras, P.; Bester, H.L.; Perley, R.A.; Leike, R.; Smirnov, O.; Westermann, R.; Enßlin, T.A. Comparison of classical and Bayesian imaging in radio interferometry. *Astron. Astrophys.* **2021**, *646*, A84. [CrossRef]

42. Arras, P.; Frank, P.; Haim, P.; Knollmüller, J.; Leike, R.; Reinecke, M.; Enßlin, T. Variable structures in M87* from space, time and frequency d interferometry. *Nat. Astron.* **2022**, *6*, 259–269. [CrossRef]

43. Knollmüller, J.; Enßlin, T.A. Metric Gaussian Variational Inference. *arXiv* **2019**, arXiv:1901.11033. [CrossRef]

44. Arras, P.; Reinecke, M.; Westermann, R.; Enßin, T.A. Efficient wide-field radio interferometry response. *Astron. Astrophys.* **2021**, *646*, A58. [CrossRef]

45. Müller, H.; Lobanov, A.P. DoG-HiT: A novel VLBI multiscale imaging approach. *Astron. Astrophys.* **2022**, *666*, A137. [CrossRef]

46. Müller, H.; Lobanov, A. Multi-scale and Multi-directional VLBI Imaging with CLEAN. *Astron. Astrophys.* 2022, *submitted*.

47. Müller, H.; Lobanov, A. Dynamic polarimetry with the multiresolution support. *Astron. Astrophys.* 2022, *submitted*.

48. Thompson, A.R.; Moran, J.M.; Swenson, G.W., Jr. *Interferometry and Synthesis in Radio Astronomy*, 3rd ed.; Springer Nature: Berlin/Heidelberg, Germany, 2017. [CrossRef]

49. Bustamante, S.; Blackburn, L.; Narayanan, G.; Schloerb, F.P.; Hughes, D. The Role of the Large Millimeter Telescope in Black Hole Science with the Next-Generation Event Horizon Telescope. *Galaxies* **2023**, *11*, 2. [CrossRef]

50. Mizuno, Y.; Fromm, C.M.; Younsi, Z.; Porth, O.; Olivares, H.; Rezzolla, L. Comparison of the ion-to-electron temperature ratio prescription: GRMHD simulations with electron thermodynamics. *Mon. Not. R. Astron. Soc.* **2021**, *506*, 741–758. [CrossRef]

51. Porth, O.; Olivares, H.; Mizuno, Y.; Younsi, Z.; Rezzolla, L.; Moscibrodzka, M.; Falcke, H.; Kramer, M. The black hole accretion code. *Comput. Astrophys. Cosmol.* **2017**, *4*, 1. [CrossRef]

52. Tchekhovskoy, A.; McKinney, J.C. Prograde and retrograde black holes: Whose jet is more powerful? *Mon. Not. R. Astron. Soc. Lett.* **2012**, *423*, L55–L59. [CrossRef]

53. Younsi, Z.; Wu, K.; Fuerst, S.V. General relativistic radiative transfer: Formulation and emission from structured tori around black holes. *Astron. Astrophys.* **2012**, *545*, A13. [CrossRef]

54. Younsi, Z.; Porth, O.; Mizuno, Y.; Fromm, C.M.; Olivares, H. Modelling the polarised emission from black holes on event horizon-scales. In *Proceedings of the Perseus in Sicily: From Black Hole to Cluster Outskirts*; Asada, K., de Gouveia Dal Pino, E., Giroletti, M., Nagai, H., Nemmen, R., Eds.; Cambridge University Press: Cambridge, UK, 2020; Volume 342, pp. 9–12. [CrossRef]

55. Pandya, A.; Zhang, Z.; Chandra, M.; Gammie, C.F. Polarized Synchrotron Emissivities and Absorptivities for Relativistic Thermal, Power-law, and Kappa Distribution Functions. *Astrophys. J.* **2016**, *822*, 34. [CrossRef]

56. Ball, D.; Sironi, L.; Özel, F. Electron and Proton Acceleration in Trans-relativistic Magnetic Reconnection: Dependence on Plasma Beta and Magnetization. *Astrophys. J.* **2018**, *862*, 80. [CrossRef]

57. Davelaar, J.; Olivares, H.; Porth, O.; Bronzwaer, T.; Janssen, M.; Roelofs, F.; Mizuno, Y.; Fromm, C.M.; Falcke, H.; Rezzolla, L. Modeling non-thermal emission from the jet-launching region of M 87 with adaptive mesh refinement. *Astron. Astrophys.* **2019**, *632*, A2. [CrossRef]

58. Fromm, C.M.; Cruz-Osorio, A.; Mizuno, Y.; Nathanail, A.; Younsi, Z.; Porth, O.; Olivares, H.; Davelaar, J.; Falcke, H.; Kramer, M.; et al. Impact of non-thermal particles on the spectral and structural properties of M87. *Astron. Astrophys.* **2022**, *660*, A107. [CrossRef]

59. Broderick, A.E.; Loeb, A. Imaging optically-thin hotspots near the black hole horizon of Sgr A* at radio and near-infrared wavelengths. *Mon. Not. R. Astron. Soc.* **2006**, *367*, 905–916. [CrossRef]
60. Pu, H.Y.; Broderick, A.E. Probing the Innermost Accretion Flow Geometry of Sgr A* with Event Horizon Telescope. *Astrophys. J.* **2018**, *863*, 148. [CrossRef]
61. Tiede, P.; Pu, H.Y.; Broderick, A.E.; Gold, R.; Karami, M.; Preciado-López, J.A. Spacetime Tomography Using the Event Horizon Telescope. *Astrophys. J.* **2020**, *892*, 132. [CrossRef]
62. Zhao, J.H.; Young, K.H.; Herrnstein, R.M.; Ho, P.T.P.; Tsutsumi, T.; Lo, K.Y.; Goss, W.M.; Bower, G.C. Variability of Sagittarius A*: Flares at 1 Millimeter. *Astrophys. J.* **2003**, *586*, L29–L32. [CrossRef]
63. Bower, G.C.; Dexter, J.; Markoff, S.; Gurwell, M.A.; Rao, R.; McHardy, I. A Black Hole Mass-Variability Timescale Correlation at Submillimeter Wavelengths. *Astrophys. J. Lett.* **2015**, *811*, L6. [CrossRef]
64. Bower, G.C.; Dexter, J.; Asada, K.; Brinkerink, C.D.; Falcke, H.; Ho, P.; Inoue, M.; Markoff, S.; Marrone, D.P.; Matsushita, S.; et al. ALMA Observations of the Terahertz Spectrum of Sagittarius A*. *Astrophys. J. Lett.* **2019**, *881*, L2. [CrossRef]
65. Liu, H.B.; Wright, M.C.H.; Zhao, J.H.; Brinkerink, C.D.; Ho, P.T.P.; Mills, E.A.C.; Martín, S.; Falcke, H.; Matsushita, S.; Martí-Vidal, I. Linearly polarized millimeter and submillimeter continuum emission of Sgr A* constrained by ALMA. *Astron. Astrophys.* **2016**, *593*, A107. [CrossRef]
66. Gravity Collaboration; Abuter, R.; Amorim, A.; Bauböck, M.; Berger, J.P.; Bonnet, H.; Brandner, W.; Clénet, Y.; Coudé Du Foresto, V.; de Zeeuw, P.T.; et al. A geometric distance measurement to the Galactic center black hole with 0.3% uncertainty. *Astron. Astrophys.* **2019**, *625*, L10. [CrossRef]
67. Raymond, A.W.; Palumbo, D.; Paine, S.N.; Blackburn, L.; Córdova Rosado, R.; Doeleman, S.S.; Farah, J.R.; Johnson, M.D.; Roelofs, F.; Tilanus, R.P.J.; et al. Evaluation of New Submillimeter VLBI Sites for the Event Horizon Telescope. *Astrophys. J. Suppl. Ser.* **2021**, *253*, 5. [CrossRef]
68. Wielgus, M.; Moscibrodzka, M.; Vos, J.; Gelles, Z.; Martí-Vidal, I.; Farah, J.; Marchili, N.; Goddi, C.; Messias, H. Orbital motion near Sagittarius A* . Constraints from polarimetric ALMA observations. *Astron. Astrophys.* **2022**, *665*, L6. [CrossRef]
69. Liska, M.; Chatterjee, K.; Tchekhovskoy, A.; Yoon, D.; van Eijnatten, D.; Hesp, C.; Markoff, S.; Ingram, A.; van der Klis, M. H-AMR: A New GPU-accelerated GRMHD Code for Exascale Computing With 3D Adaptive Mesh Refinement and Local Adaptive Time-stepping. *arXiv* **2019**, arXiv:1912.10192.
70. Mościbrodzka, M.; Gammie, C.F. IPOLE—Semi-analytic scheme for relativistic polarized radiative transport. *Mon. Not. R. Astron. Soc.* **2018**, *475*, 43–54. [CrossRef]
71. Younsi, Z.; Psaltis, D.; Özel, F. Black Hole Images as Tests of General Relativity: Effects of Spacetime Geometry. *arXiv* **2021**, arXiv:2111.01752.
72. Broderick, A.E.; Fish, V.L.; Johnson, M.D.; Rosenfeld, K.; Wang, C.; Doeleman, S.S.; Akiyama, K.; Johannsen, T.; Roy, A.L. Modeling Seven Years of Event Horizon Telescope Observations with Radiatively Inefficient Accretion Flow Models. *Astrophys. J.* **2016**, *820*, 137. [CrossRef]
73. Gravity Collaboration; Abuter, R.; Amorim, A.; Anugu, N.; Bauböck, M.; Benisty, M.; Berger, J.P.; Blind, N.; Bonnet, H.; Brandner, W.; et al. Detection of the gravitational redshift in the orbit of the star S2 near the Galactic centre massive black hole. *Astron. Astrophys.* **2018**, *615*, L15. [CrossRef]
74. Roelofs, F.; Janssen, M.; Natarajan, I.; Deane, R.; Davelaar, J.; Olivares, H.; Porth, O.; Paine, S.N.; Bouman, K.L.; Tilanus, R.P.J.; et al. SYMBA: An end-to-end VLBI synthetic data generation pipeline. Simulating Event Horizon Telescope observations of M 87. *Astron. Astrophys.* **2020**, *636*, A5. [CrossRef]
75. Blecher, T.; Deane, R.; Bernardi, G.; Smirnov, O. MEQSILHOUETTE: A mm-VLBI observation and signal corruption simulator. *Mon. Not. R. Astron. Soc.* **2017**, *464*, 143–151. [CrossRef]
76. Natarajan, I.; Deane, R.; Martí-Vidal, I.; Roelofs, F.; Janssen, M.; Wielgus, M.; Blackburn, L.; Blecher, T.; Perkins, S.; Smirnov, O.; et al. MeqSilhouette v2: Spectrally resolved polarimetric synthetic data generation for the event horizon telescope. *Mon. Not. R. Astron. Soc.* **2022**, *512*, 490–504. [CrossRef]
77. Janssen, M.; Goddi, C.; van Bemmel, I.M.; Kettenis, M.; Small, D.; Liuzzo, E.; Rygl, K.; Martí-Vidal, I.; Blackburn, L.; Wielgus, M.; et al. rPICARD: A CASA-based calibration pipeline for VLBI data. Calibration and imaging of 7 mm VLBA observations of the AGN jet in M 87. *Astron. Astrophys.* **2019**, *626*, A75. [CrossRef]
78. Gelaro, R.; McCarty, W.; Suárez, M.J.; Todling, R.; Molod, A.; Takacs, L.; Randles, C.A.; Darmenov, A.; Bosilovich, M.G.; Reichle, R.; et al. The Modern-Era Retrospective Analysis for Research and Applications, Version 2 (MERRA-2). *J. Clim.* **2017**, *30*, 5419–5454. [CrossRef]
79. Paine, S. The am Atmospheric Model. 2019. https://zenodo.org/record/3406496#.Y7017nZByUk (accessed on 19 December 2022).
80. Farah, J.; Galison, P.; Akiyama, K.; Bouman, K.L.; Bower, G.C.; Chael, A.; Fuentes, A.; Gómez, J.L.; Honma, M.; Johnson, M.D.; et al. Selective Dynamical Imaging of Interferometric Data. *Astrophys. J. Lett.* **2022**, *930*, L18. [CrossRef]
81. Chael, A.A. Simulating and Imaging Supermassive Black Hole Accretion Flows. Ph.D. Thesis, Harvard University, Cambridge, MA, USA, 2019.
82. Bella, N.L.; Issaoun, S.; Roelofs, F.; Fromm, C.; Falcke, H. Expanding Sgr A* dynamical imaging capabilities with an African extension to the Event Horizon Telescope. *Astron. Astrophys.* 2023, *submitted*.
83. Chael, A.; Johnson, M.D.; Lupsasca, A. Observing the Inner Shadow of a Black Hole: A Direct View of the Event Horizon. *Astrophys. J.* **2021**, *918*, 6. [CrossRef]

84. Bronzwaer, T.; Falcke, H. The Nature of Black Hole Shadows. *Astrophys. J.* **2021**, *920*, 155. [CrossRef]
85. Tiede, P.; Johnson, M.D.; Pesce, D.W.; Palumbo, D.C.M.; Chang, D.O.; Galison, P. Measuring Photon Rings with the ngEHT. *arXiv* **2022**, arXiv:2210.13498.
86. Dodson, R.; Rioja, M.J.; Jung, T.; Goméz, J.L.; Bujarrabal, V.; Moscadelli, L.; Miller-Jones, J.C.A.; Tetarenko, A.J.; Sivakoff, G.R. The science case for simultaneous mm-wavelength receivers in radio astronomy. *New Astron. Rev.* **2017**, *79*, 85–102. [CrossRef]
87. Rioja, M.J.; Dodson, R. Precise radio astrometry and new developments for the next-generation of instruments. *Astron. Astrophys. Rev.* **2020**, *28*, 6. [CrossRef]

Article

Enabling Transformational ngEHT Science via the Inclusion of 86 GHz Capabilities

Sara Issaoun [1,2,*], Dominic W. Pesce [1,3], Freek Roelofs [1,3], Andrew Chael [2,4], Richard Dodson [5], María J. Rioja [5,6,7], Kazunori Akiyama [3,8,9], Romy Aran [1,10], Lindy Blackburn [1,3], Sheperd S. Doeleman [1,3], Vincent L. Fish [8], Garret Fitzpatrick [1], Michael D. Johnson [1,3], Gopal Narayanan [11], Alexander W. Raymond [1,3] and Remo P. J. Tilanus [12,13,14,15]

[1] Center for Astrophysics Harvard & Smithsonian, 60 Garden Street, Cambridge, MA 02138, USA
[2] NASA Hubble Fellowship Program, Einstein Fellow
[3] Black Hole Initiative, Harvard University, 20 Garden Street, Cambridge, MA 02138, USA
[4] Princeton Center for Theoretical Science, Jadwin Hall, Princeton University, Princeton, NJ 08544, USA
[5] International Centre for Radio Astronomy Research, M468, The University of Western Australia, 35 Stirling Hwy, Crawley, WA 6009, Australia
[6] CSIRO Astronomy and Space Science, P.O. Box 1130, Bentley, WA 6102, Australia
[7] Observatorio Astronómico Nacional (IGN), Alfonso XII, 3 y 5, 28014 Madrid, Spain
[8] Haystack Observatory, Massachusetts Institute of Technology, 99 Millstone Rd., Westford, MA 01886, USA
[9] National Astronomical Observatory of Japan, 2-21-1 Osawa, Mitaka, Tokyo 181-8588, Japan
[10] Harvard College, Harvard University, Cambridge, MA 02138, USA
[11] Department of Astronomy, University of Massachusetts, Amherst, MA 01003, USA
[12] Steward Observatory and Department of Astronomy, University of Arizona, 933 N. Cherry Ave., Tucson, AZ 85721, USA
[13] Department of Astrophysics, Institute for Mathematics, Astrophysics and Particle Physics (IMAPP), Radboud University, P.O. Box 9010, 6500 GL Nijmegen, The Netherlands
[14] Leiden Observatory, Leiden University, Postbus 2300, 9513 RA Leiden, The Netherlands
[15] Netherlands Organisation for Scientific Research (NWO), Postbus 93138, 2509 AC Den Haag, The Netherlands
* Correspondence: sara.issaoun@cfa.harvard.edu

Citation: Issaoun, S.; Pesce, D.W.; Roelofs, F.; Chael, A.; Dodson, R.; Rioja, M.J.; Akiyama, K.; Aran, R.; Blackburn, L.; Doeleman, S.S.; et al. Enabling Transformational ngEHT Science via the Inclusion of 86 GHz Capabilities. *Galaxies* **2023**, *11*, 28. https://doi.org/10.3390/galaxies11010028

Academic Editor: Alok Chandra Gupta

Received: 9 November 2022
Revised: 23 January 2023
Accepted: 28 January 2023
Published: 10 February 2023

Abstract: We present a case for significantly enhancing the utility and efficiency of the ngEHT by incorporating an additional 86 GHz observing band. In contrast to 230 or 345 GHz, weather conditions at the ngEHT sites are reliably good enough for 86 GHz to enable year-round observations. Multi-frequency imaging that incorporates 86 GHz observations would sufficiently augment the (u, v) coverage at 230 and 345 GHz to permit detection of the M87 jet structure without requiring EHT stations to join the array. The general calibration and sensitivity of the ngEHT would also be enhanced by leveraging frequency phase transfer techniques, whereby simultaneous observations at 86 GHz and higher-frequency bands have the potential to increase the effective coherence times from a few seconds to tens of minutes. When observation at the higher frequencies is not possible, there are opportunities for standalone 86 GHz science, such as studies of black hole jets and spectral lines. Finally, the addition of 86 GHz capabilities to the ngEHT would enable it to integrate into a community of other VLBI facilities—such as the GMVA and ngVLA—that are expected to operate at 86 GHz but not at the higher ngEHT observing frequencies.

Keywords: very long baseline interferometry; black holes; active galactic nuclei; radio astronomy; instrument design

1. Introduction

Building upon the success of the Event Horizon Telescope (EHT; [1–14]), the next-generation EHT (ngEHT) is a proposed global very long baseline interferometry (VLBI) telescope network that aims to carry out horizon-scale observations of M87 and Sgr A* at (sub)millimeter wavelengths [15]. By adding ∼10 new VLBI stations to the EHT array and increasing the overall array sensitivity, the ngEHT will be able to achieve high-fidelity imaging and even movie-making capabilities. The primary scientific goals of the ngEHT require an angular resolution of $\lesssim 20\,\mu$as and thus motivates observing at the highest VLBI frequencies, currently 230 and 345 GHz (see Bustamante et al. [16] for dual-band receiver details). However, the design specifications for the ngEHT have yet to be finalized, and the addition of an 86 GHz band is under consideration. Adding 86 GHz capabilities to the ngEHT would provide numerous prospects for improving the performance of the array, expanding its science applications and permitting it to operate jointly with other existing or near-future facilities. In this work, we explore a number of these motivating factors and present a case for including 86 GHz capabilities as part of the ngEHT array's design.

2. Science Drivers

2.1. Black Hole Shadow and Jet Physics

Observations of horizon-scale targets at 86 GHz supplement a primary science driver of the ngEHT: connecting dynamics and properties of black hole shadows with the creation and launching of astrophysical jets. The horizon-scale jet emission is typically brighter at 86 GHz than at 230 or 345 GHz, owing to the negative spectral index and increased optical depth at 86 GHz. The inference of jet structure from 86 GHz observations to 230 and 345 GHz images, with the use of multi-frequency imaging techniques [17], will enable the recovery of faint large-scale jet emissions in horizon-scale images with higher fidelity compared to high-frequency imaging alone. In Figure 1, we show example reconstructions of simulated emission from M87 at 230 GHz with and without multi-frequency information from 86 GHz (and 345 GHz) observations. Imaging with the full EHT+ngEHT array was done using visibility amplitudes and closure phases, and ngEHT-only reconstructions also made use of closure amplitudes to improve convergence. The ground-truth models are shown in Figure 2. These reconstructions assume that the telescopes are well pointed, well focused, and amplitude-calibrated, another avenue in which 86 GHz capabilities can improve the overall array performance. Relative registration of images across frequencies was performed via the algorithm, although additional observing techniques can provide that information more robustly (see Sections 2.2 and 3.4). While the best imaging results are obtained using the full core EHT and new ngEHT stations with multi-frequency imaging, the addition of 86 GHz information dramatically improves 230 GHz images with the new ngEHT stations alone (or in conjunction with a single sensitive core-EHT site, such as the LMT or ALMA). The inclusion of 86 GHz observing guarantees good coverage and station operation the entire year; see Section 3. This would, for example, enable images and movies of the shadow and jet in M87 with high cadence over long periods of time, see Figure 1.

The 128 Gbps recording rate of the ngEHT brings about a factor of 32 increase compared to the current 86 GHz VLBI recording rate of 4 Gbps (currently limited by the recording capability of the Very Long Baseline Array). The increase in sensitivity brought by the ngEHT at 86 GHz would greatly improve polarimetric imaging of jet structure and inner accretion flows, particularly for low-polarization sources such as M87 and Sgr A*. Polarimetric imaging is essential for the discrimination between magnetic field configurations [7,8], thereby tightening our constraints on black hole astrophysical models.

For Sgr A*, the ngEHT will be sufficiently sensitive for detecting refractive scattering in the interstellar medium toward the Galactic Center, for which the effects are most dominant on long VLBI baselines. Detections of refractive scattering in the millimeter regime were detected on north–south baselines, where the diffractive scattering is weakest [18,19], and detections in the centimeter regime were detected on east–west baselines [20]. While these detections allowed us to eliminate the more extreme models of magnetic field wander (i.e., variations transverse to the line of sight across a range of viewing angles), further constraints of the model parameter space require a wider coverage of position angles at many refractive timescales. A higher baseline sensitivity and increased observing cadence at 86 GHz would provide detections of refractive scattering along baselines in all directions, which could definitively discriminate between different models for the underlying magnetic field wander in the interstellar medium that predicts varying levels of refractive noise along different baseline directions [21]. Joint modeling of the screen across the three simultaneous observing bands will also improve the scattering mitigation of Sgr A* observations.

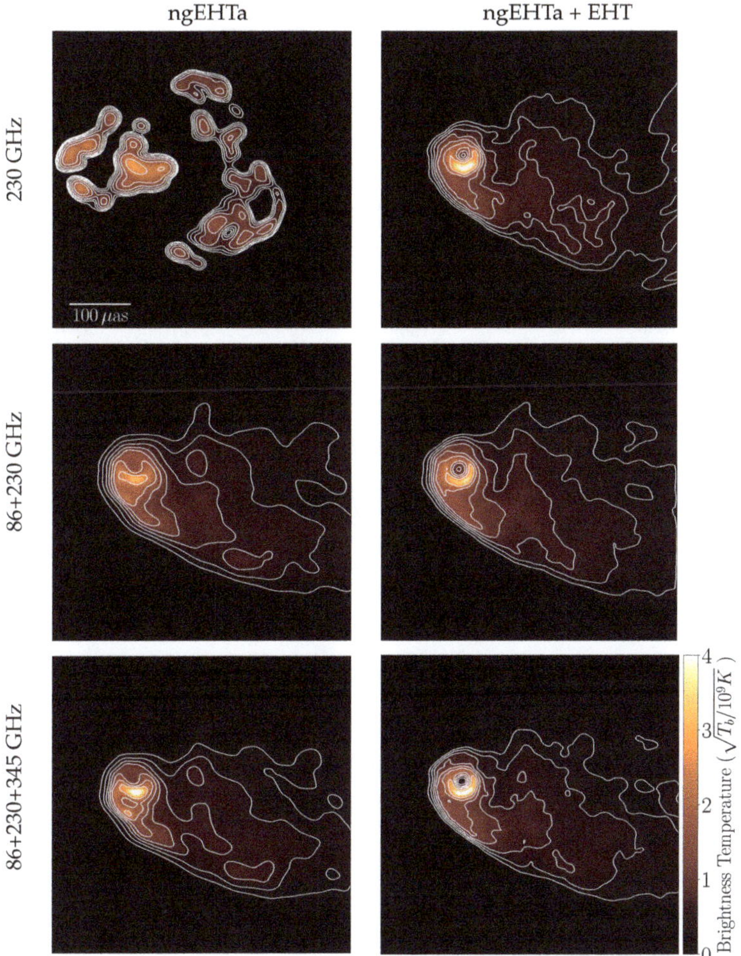

Figure 1. Demonstration of multi-frequency capabilities with simulated observations of the M87 jet at 230 GHz reconstructed with and without 86 GHz complementary observations. The left column shows reconstructions using the new ngEHTa configuration stations alone (see Section 3.1 for array specifications), and the right column shows reconstructions with the full ngEHT array, including the core EHT stations. Top: Simulated reconstructions of 230 GHz only observations. The contours are spaced logarithmically, starting at 1% of the peak value and increasing by factors of 2. Middle: Simulated reconstructions at 230 GHz using multi-frequency 86 and 230 GHz observations. Bottom: Simulated reconstructions at 230 GHz using multi-frequency 86, 230 and 345 GHz observations. This demonstration offers a compelling view of the imaging advantages of adding 86 GHz receivers to the ngEHT, enabling flexible high-fidelity imaging of the black hole shadow and jet during times when the core EHT sites would not be readily available.

Furthermore, the addition of simultaneous 86 GHz observing in the 230 and 345 GHz configuration will provide wide frequency coverage for rotation measure and Faraday rotation studies, core shift studies, and time-lag measurements in the event of flares. Due to the variability of Sgr A*, simultaneity of 86 GHz observations is essential for relative astrometry between the frequency bands, and the registration of the three bands for accurate spectral index and rotation measure mapping.

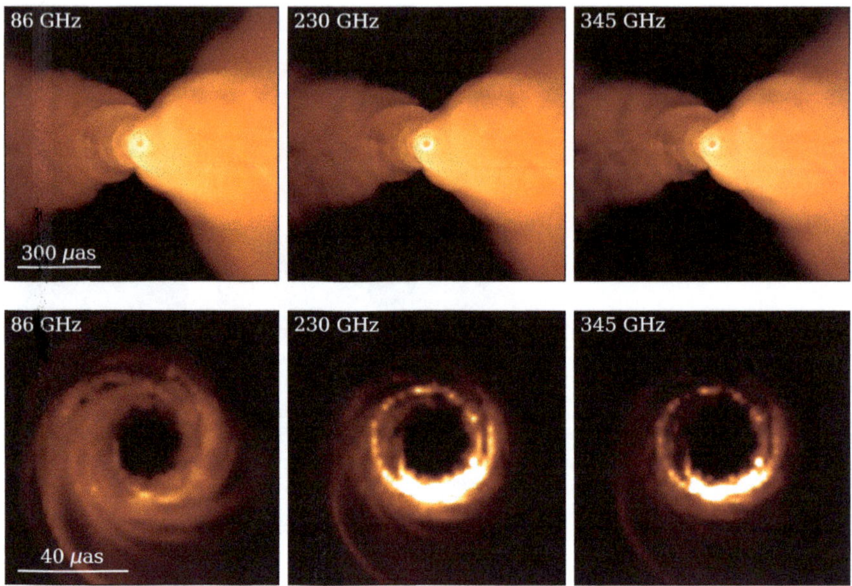

Figure 2. Input images used for the M87 synthetic data in this work, produced from the GRMHD simulations in [22]. The upper panels show a 1 mas field of view with a logarithmic colorscale to highlight the extended jet emission, and the lower panels show the central 100 μas around the black hole using a linear color scale to more easily see the photon ring region.

2.2. Telescope Calibration and Astrometry

At high frequencies, single-frequency phase referencing using nearby calibrators is challenging due to the varying atmospheric conditions between the calibrator and the main target. Even at 86 GHz, atmospheric coherence limits the switching time between the target and the calibrator reference to 15 s or less. While a single demonstration at 86 GHz exists [23], single-frequency phase referencing at high frequencies is, in general, only conceivable using simultaneous observations in multiple directions, either with multiple beams or more likely paired antennas; ngVLA would have this capability by having multiple antennas per site location. Single-frequency astrometry at 230 and 345 GHz is thus difficult without a proper phase reference position on the sky.

Frequency phase transfer (FPT) is a powerful tool for 230 GHz calibration in marginal weather conditions. The FPT approach was initially developed for (sub)mm compact arrays (e.g., [24]) and first used in VLBI by Middelberg et al. [25] to extend the coherence time at 86 GHz, using the VLBA. The work of Rioja and Dodson enabled bona-fide astrometry (source frequency phase referencing; SFPR) in addition (Dodson and Rioja [26] and Rioja and Dodson [27] using the VLBA up to 86 GHz) and has been demonstrated up to 130 GHz using the Korean VLBI Network (KVN [28,29]). With FPT techniques applied to simultaneous KVN observations at 22, 43, 87, and 130 GHz, coherence times at 130 GHz were increased from tens of seconds to ∼20 min. Adding SFPR using a separate reference source, the coherence time was further increased to many hours (20% loss after eight hours of integration). FPT has been successfully applied to other KVN observations as well. Application to MOnitoring of GAmma-ray Bright AGN (iMOGABA) observations led to the imaging of several sources at 86 and 129 GHz that were not detected without FPT [30]. Zhao et al. [31] applied the technique to simultaneous 22 and 43 GHz observations with the KVN and VERA in Japan (combined as KaVA), increasing the coherence time at 43 GHz from ~1 min to tens of minutes. As demonstrated by Zhao et al. [32], a second round of FPT applied to the residuals from the first round (FPT-square), was successful at taking out ionospheric effects without the need to observe a reference source as in the SFPR technique, increasing coherence times from 20 min to more than eight hours at 86 GHz.

As shown in Section 3.1, 86 GHz observing is possible throughout the entire year at all sites. On observing days with marginal or poor 230 GHz weather, the phase stability at the lower frequency of 86 GHz would allow us to solve for phase offsets on a particular station at the lower frequency and transfer them to a higher frequency (230 and/or 345 GHz). This technique requires at minimum simultaneous 86 and 230 GHz observing to be effective, due to the short coherent time at the higher frequencies. Another requirement is that sources at 86 GHz be bright and compact enough to be detectable on all baselines, which is not a limiting factor for most sources apart from Sgr A*, which is scatter-broadened at this frequency [33–36]. Finally, an integer frequency ratio is preferred between the different frequency bands in order to optimally use these techniques and should allow bonafide astrometry between the bands (i.e., between 86 and 230 and/or 345 GHz). Astrometry at these frequencies is expected to yield residual systematic errors of ~3 µas [27,37–39], though we note that such analyses will additionally need to account for frequency-dependent structural changes such as the well-known core-shift effect (e.g., [40]). These specifications are discussed in more detail in Section 3.4.

With simultaneous receivers at 86 GHz, the ngEHT will have a frequency overlap with a large network of well-located VLBI stations that enable astrometric observations. Using SFPR, it would be possible to astrometrically connect these to the highest frequencies and should be able to provide relative astrometry at 345 GHz. With certain array configurations, this should be possible with just ngEHT observations. The frequency phase transfer technique allows for correction of the fast tropospheric variations and increases the coherence at the higher frequencies. This enables the switching time between the target and calibrator to be longer (of order ~ minutes) and allows for the calibrator to be at a larger distance (up to 10°–20°, as demonstrated with the KVN). This technique, called "source frequency phase referencing", allows for relative astrometry to directly register observations of the target at the three observed frequencies with respect to the calibrator. Explanations of the various flavors of frequency phase transfer and their applications to the ngEHT are provided in a complementary publication by Rioja et al. [38].

In addition to scientific input, 86 GHz capabilities enhance technical specifications of the ngEHT array. Individual station calibration will be improved with the ability to point and focus at a lower frequency, where calibrator sources have higher flux density and the atmosphere is more stable. The calibrator sky is limited at 230 and 345 GHz, and the catalog of sources available at 86 GHz, both for continuum and spectral-line pointing, is significantly larger. Calibration operations at 86 GHz would enable station participation in marginal 230 GHz weather conditions.

2.3. Stand-Alone 86 GHz Science

Over the past few decades, VLBI observations at 86 GHz have provided high-quality images of AGN sources, spatially and temporally resolving their innermost structure in total intensity and polarization, thereby providing new insights into the origins, collimation, and dynamics of relativistic jets (e.g., [41]).

For example, recent observations with the Global Millimeter VLBI Array (GMVA) have imaged the core region of 3C84 and OJ287 [42–45], measured jet collimation profiles of various AGNs (e.g., [46,47]), provided a survey of AGN core brightness temperature measurements [48], and imaged jet dynamics associated with gamma ray flares and source variability across the electromagnetic spectrum (e.g., [49–53]). Multi-wavelength (mm-) VLBI is also useful for studying the dynamics of X-ray binary jets, where the jet formation occurs on much smaller timescales (e.g., [54]).

Spectral line VLBI observations at 86 GHz can also be used to study SiO maser emission near stars. In stars on the asymptotic giant branch (AGB), SiO maser emission probes the highly dynamical circumstellar gas regions where dust grains are formed and accelerated outwards with the gas, driving the formation of planetary nebulae (e.g., [55], and references therein). In high-mass star formation regions, imaging SiO maser emission spots can probe the dynamics of the protostellar accretion disk and outflow, shedding more light on the star formation process (e.g., [56,57]).

The planned next-generation Very Large Array (ngVLA) will be able to observe at 86 GHz with unprecedented sensitivity ([58,59]; see also Section 4). However, as the ngVLA sites are limited to the American continent, its maximum baseline length and hence angular resolution are significantly smaller than those of the current GMVA. The ngEHT operating at 86 GHz could provide up to $\sim$3 times longer baselines than the ngVLA while providing significantly more sensitivity than the GMVA due to its increased bandwidth. The ngEHT and ngVLA operating together would form an extremely high-sensitivity and high-resolution array at 86 GHz, providing unprecedented images of AGN jet sources (Section 4).

3. Array Specifications and Performance

The ngEHT will be comprised of up to 10 new (sub)millimeter radio telescopes distributed around the globe and operating as a VLBI network [15]. There are two primary operating modes that the ngEHT is expected to employ. The first operating mode is a campaign mode, during which the ngEHT dishes will observe alongside the current EHT dishes as part of a large and sensitive array. Owing to the need to coordinate such observations among many telescopes, a number of which are themselves facility instruments, the campaign mode will likely only operate during a small number of observing windows within any given calendar year. The second operating mode is a standalone ngEHT mode that will be more versatile and which is expected to operate throughout the year.

3.1. Array

In this article, we consider two different reference array configurations for the ngEHT, which we refer to as the ngEHTa and ngEHTb arrays (Doeleman et al., in prep; Roelofs et al. [60]). The individual sites contained in each of these arrays are listed in Table 1, and world maps illustrating their global distributions are shown in Figure 3. The ngEHTa configuration contains 10 stations, and the ngEHTb configuration contains 8 stations with a slightly larger average dish diameter.

Figure 3. Maps of the two ngEHT array configurations, with the current EHT sites shown in cyan and new ngEHT sites in yellow. The ngEHTa configuration with 6 m dishes is shown on the left, and the ngEHTb configuration with 10 m dishes is shown on the right.

For the ngEHTa reference configuration, 7 of the 10 new sites were assumed to be equipped with 6-meter dishes, and the remaining three sites were the Haystack Observatory (HAY; 37-meter dish), the Gamsberg mountain (GAM; the primary candidate site for the Africa Millimetre Telescope, which will refurbish the 15-meter SEST dish currently in Chile), and Owens Valley Radio Observatory (OVRO; 10-meter dish). For the ngEHTb configuration, 7 out of 8 new sites were assumed to be equipped with 10-meter dishes, and the remaining site was GAM with a 15-meter dish. The bottom row of panels in Figure 4 shows the (u, v) coverage for both of these arrays as seen from M87, and the top row of panels shows the expected signal-to-noise ratio as a function of baseline length.

Table 1. Station participation in current and future VLBI arrays.

Station	EHT	ngEHTa	ngEHTb	GMVA
ALMA	X	-	-	X
APEX	X	-	-	-
SMA	X	-	-	-
JCMT	X	-	-	-
LMT	X	-	-	X
SMT	X	-	-	-
KPTO	X	-	-	-
NOEMA	X	-	-	X
PV	X	-	-	X
SPT	X	-	-	-
GLT	X	-	-	X
BAJA	-	X	X	-
BAR	-	X	-	-
CAS	-	-	X	-
CAT	-	X	-	-
CNI	-	X	X	-
GAM	-	X	X	-
GARS	-	X	-	-
HAY	-	X	-	-
LLA	-	-	X	-
NZ	-	X	X	-
OVRO	-	X	-	-
PIKE	-	-	X	-
SGO	-	X	X	-
GBT	-	-	-	X
BR	-	-	-	X
FD	-	-	-	X
KP	-	-	-	X
LA	-	-	-	X
MK	-	-	-	X
NL	-	-	-	X
OV	-	-	-	X
PT	-	-	-	X
EF	-	-	-	X
YS	-	-	-	X
ONS	-	-	-	X
MET	-	-	-	X
KVN	-	-	-	X

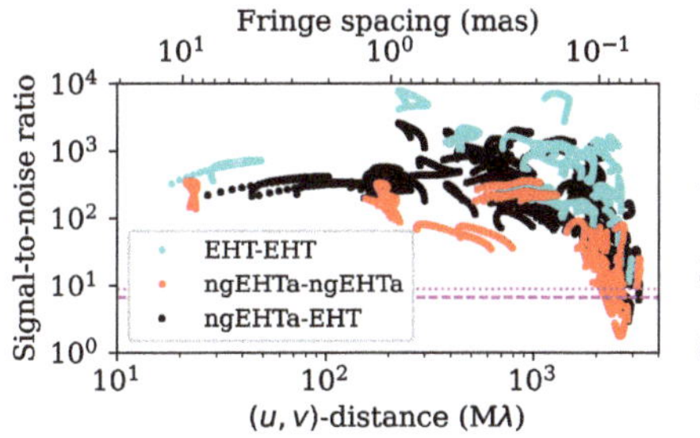

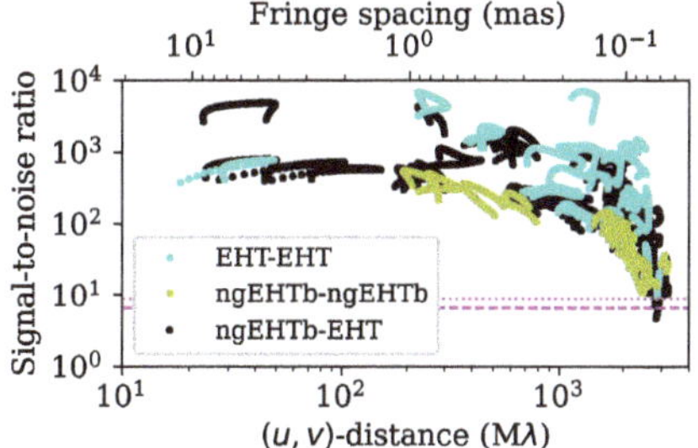

Figure 4. *Cont.*

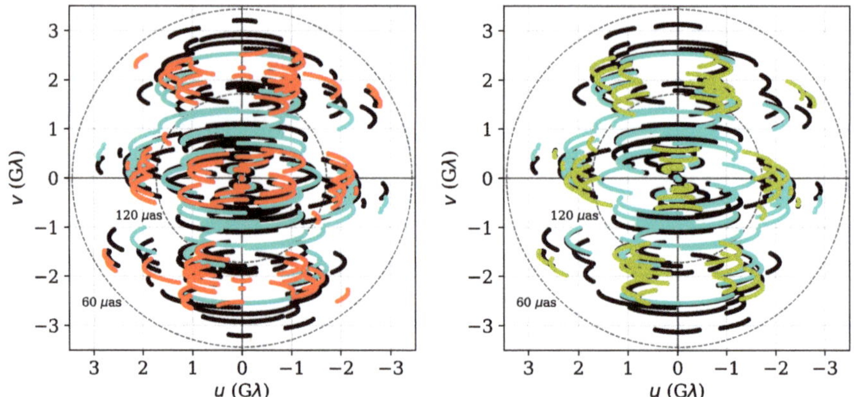

Figure 4. Coverage at 86 GHz (u, v) (**bottom row**) and signal-to-noise ratio versus (u, v) distance (**top row**) for the two ngEHT array configurations, as viewed from M87. The left panels show baselines between EHT stations in cyan, baselines between ngEHTa stations in red, and baselines between ngEHTa and EHT stations in black. The right panels show baselines between EHT stations in cyan, baselines between ngEHTb stations in yellow, and baselines between ngEHTb and EHT stations in black. The horizontal dashed and dotted lines in the top panels show the 86 GHz S/N levels necessary to achieve 90% phase coherence at 230 and 345 GHz, respectively.

The various new ngEHT sites were selected primarily because of their suitability for observations at 230 and 345 GHz [61], and as a result, they tend to have excellent prospects for 86 GHz observations. Figure 5 shows the median 86 GHz zenith opacities at each of the EHT and ngEHT sites as a function of month during the year. We can see that the majority of the sites exhibit median 86 GHz opacities less than 0.1 throughout the entire year, corresponding to $\gtrsim$90% atmospheric transmission.

3.2. Synthetic Data

For the various explorations carried out in this article, we have generated synthetic interferometric datasets using the `eht-imaging` library [62,63]. As the input source structure for M87, we used images generated from the GRMHD simulations described in [22] and ray-traced at observer frequencies of 86, 230, and 345 GHz. Figure 2 shows the images of the source at these three frequencies.

During synthetic data generation, SEFDs for each of the stations have been determined following the procedure and atmospheric parameters from [61]. We carried out a Monte Carlo weather sampling procedure, whereby 100 versions of each synthetic dataset were generated using independent instantiations of the atmospheric temperature and zenith opacity at every site. All results presented in this article were then computed using the statistics of these 100 samples for each synthetic observation.

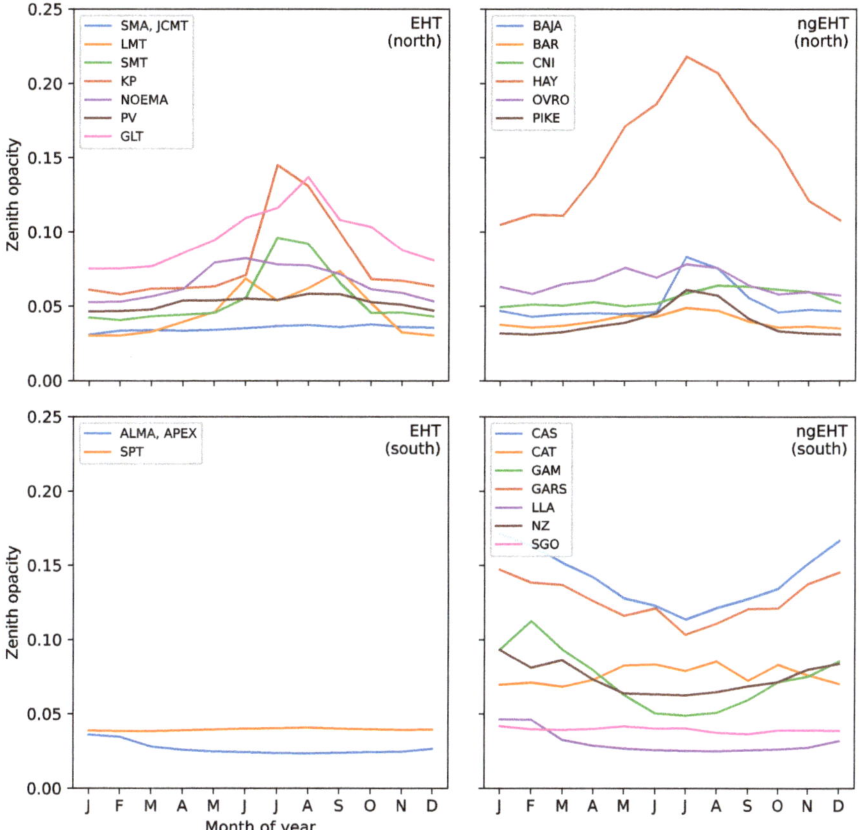

Figure 5. Median-zenith atmospheric opacity at 86 GHz as a function of time at each of the sites in the EHT (**left panels**) and ngEHT (**right panels**) arrays; the sites are split by hemisphere. The top row shows northern hemisphere sites, and the bottom row shows southern hemisphere sites. We note that for frequencies below ∼130 GHz, the atmosphere over Chajnantor is typically more transparent even than that over the South Pole (e.g., [64]).

3.3. Performance Metrics

We employed two different metrics to assess the performance of a particular array. Our selected metrics can be computed directly from visibility measurements, so as to be independent of the various specific algorithmic and procedural choices that go into image reconstruction.

Our first metric is the "point source sensitivity", or PSS, which is a measure of overall array sensitivity. For a set of N complex visibilities with thermal noises σ_i, the PSS is given by

$$\mathrm{PSS} = \left(\sum_{i=1}^{N} \frac{1}{\sigma_i^2} \right)^{-1/2}. \tag{1}$$

The PSS has units of flux density, and for a perfectly-calibrated array, it would be equal to the measurement uncertainty in the flux density of an observed point source. A smaller value for the PSS thus indicates a more sensitive array.

Our second metric is the "(u, v)-filling fraction", or FF, which was developed by [65] and is a measure of how completely-filled the Fourier coverage is. Computation of the FF depends not only on the (u, v) coverage, but also on the specifications of two additional values: an angular resolution and a field of view. Given a circle in the (u, v) plane with radius determined by the specified angular resolution, the FF metric value is taken to be the fraction of this circle's area that is occupied by the (u, v) coverage after convolution with a circular tophat function with a radius determined by the specified field of view. For all FF calculations in this article, we specified a 56.4 µas angular resolution (equal to that of an Earth-diameter baseline observed at 86 GHz), and we specified a field of view of 1 mas for M87 observations. The FF metric value is normalized to fall between zero and one: a value of zero indicates no coverage, and a value of one indicates a fully covered Fourier plane.

Figure 6 shows the PSS and FF metric behavior for the EHT, ngEHT, and composite arrays throughout the year, relative to their median values. We see that for both the PSS and FF metrics, the 86 GHz behavior is substantially more stable in time than the corresponding metrics at 230 GHz or 345 GHz. The arrays may suffer from substantial performance degradation when observing at 230 GHz or 345 GHz in the northern summer relative to the northern winter, but observations at 86 GHz should not be significantly impacted.

3.4. Multi-Frequency Calibration Techniques

One promising calibration technique made possible by the addition of 86 GHz capabilities is FPT, in which atmospheric phase fluctuations are tracked at 86 GHz and then transferred to the higher-frequency bands. FPT results in increased coherence time at the higher frequencies. Bona fide astrometry can be added using SFPR, by interleaving observations of a second source to remove the remaining FPT dispersive residual terms. A comprehensive error analysis formulation of FPT and SFPR was initially presented in Rioja and Dodson [27]. Overviews and discussions can be found in Dodson et al. [55] and Rioja and Dodson [37].

FPT requires observing a source simultaneously in (at least) two different frequencies. The key assumption underlying FPT is that phase tracking and calibration at one of the frequencies—almost always taken to be the lower frequency—is easier than at the other frequency. There are at least three reasons for why phase calibration is easier at lower frequencies:

1. At (sub)millimeter observing wavelengths, the most rapidly-fluctuating contribution to the visibility phase comes from the troposphere, whose timescale typically decreases with increasing observing frequency and whose magnitude is proportional to ν.
2. Atmospheric absorption and receiver noise temperatures are lower at 86 GHz than they are at higher frequencies, essentially making each telescope more sensitive and permitting higher S/N to be achieved within any given integration time.
3. Dimensionless baseline lengths are proportional to ν, meaning that the spatial scales probed by any given baseline are larger when observing at lower frequencies. As many VLBI sources (e.g., AGN) are resolved at (sub)millimeter observing wavelengths, shorter baselines typically have higher correlated flux densities at lower frequencies in this regime, again permitting higher S/N to be achieved within any given integration time.

To successfully carry out FPT, the source must be detectable at the lower frequency within a timescale that is approximately equal to the phase coherence timescale at the higher frequency, such that the phase variations can be tracked over time. As the tropospheric term dominates these phase variations, and because the magnitude of the tropospheric variations is proportional to ν, we can apply a frequency-scaled version of the lower frequency phase solution to the higher-frequency data. For periods of time over which the intrinsic source phases are only slowly varying, the removal of the dominant phase corruption permits substantially increased integration times.

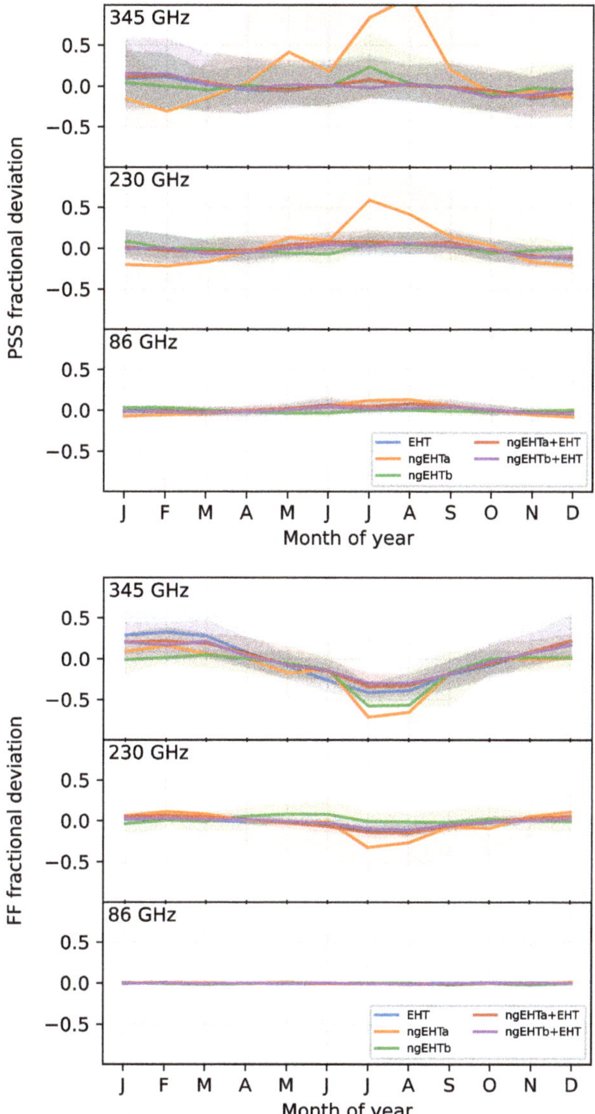

Figure 6. Deviation of the PSS (**top panel**) and FF (**bottom panel**) metrics from their yearly median values, each plotted as a fraction of that median value versus time for five different array configurations. The EHT, ngEHTa, ngEHTb, ngEHTa+EHT, and ngEHTb+EHT arrays are each plotted in a different color, as indicated in the legend. The top, middle, and bottom rows in each panel correspond to observing frequencies of 345, 230, and 86 GHz, respectively. The light shaded region around each line encompasses the inter-quartile range determined by the weather Monte Carlo procedure. A target field-of-view of 1000 μas has been assumed for all (u, v)-filling fraction computations.

FPT imposes more demanding S/N requirements for the low-frequency detection than would typically be necessary for single-frequency phase calibration. The S/N of a detection is related to the RMS phase fluctuations σ_ϕ by

$$S/N \approx \frac{1}{\sigma_\phi}. \tag{2}$$

The RMS phase fluctuations in turn determine the coherence, η, which for Gaussian variations is given by

$$\eta = e^{-\sigma_\phi^2/2} = e^{-1/2(S/N)^2}. \tag{3}$$

The RMS phase fluctuations at the lower frequency are also scaled by the frequency ratio when transferred to the higher frequency, meaning that the effective S/N at the higher frequency is smaller by the same factor. Thus, achieving a coherence of $\eta \geq 0.9$ at a frequency of 345 GHz requires S/N $\gtrsim$ 2.2. However, if the phase at 345 GHz is being determined by FPT from 86 GHz, then the equivalent S/N at 86 GHz must be S/N $\gtrsim$ 8.8 to achieve the same 345 GHz coherence. The horizontal dashed and dotted lines in the top panels of Figure 4 show the 86 GHz S/N levels necessary to achieve $\eta \geq 0.9$ at 230 GHz and 345 GHz.

The scaling of the phase variations at the lower frequency before applying them to the higher frequency can also result in phase-wrapping ambiguities. Such ambiguities are avoided only if the frequency ratio between the lower and higher frequencies is an integer [66]. For the more general non-integer case, these ambiguities can introduce seemingly random phase jumps whenever the lower frequency phase wraps. Attempting to "unwrap" the phases prior to transferring can improve the performance, but this is an imperfect solution that will perform increasingly poorly as SNR decreases. It is thus preferable to maintain an integer value R when employing the FPT technique, which motivates particular choices of frequency bands. Figure 7 illustrates how such constraints manifest for the proposed ngEHT frequency configuration containing three bands. One "optimal" arrangement is highlighted in blue and corresponds to a ~4 GHz bandwidth in the lowest frequency band (spanning ~82.5–86.5 GHz), a ~12 GHz bandwidth in the middle frequency band (spanning ~248–260 GHz), and a ~16 GHz bandwidth in the highest frequency band (spanning ~330–346 GHz).

Applications at lower frequencies have demonstrated that FPT-aided coherence times can extend to tens of minutes [67], and adding in a third frequency to remove residual ionospheric phase fluctuations can potentially extend the coherence times to multiple hours [32]. Integration times of minutes have also been achieved by the EHT using phase stabilization at 230 GHz alone for sources with $\gtrsim$Jy-level flux densities [3]. An FPT from 86 GHz would enable similarly increased coherence times for substantially weaker sources than would otherwise be observable with the ngEHT.

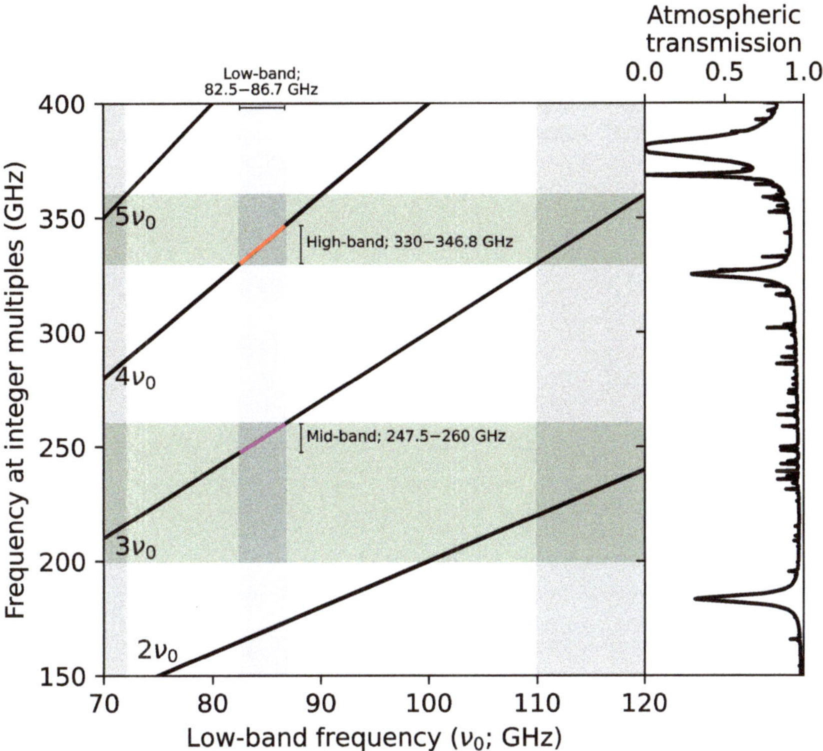

Figure 7. Frequency coverage constraints imposed by the desire to transfer phase information from a low-frequency band ("low-band", around ∼86 GHz) to two different higher-frequency bands ("mid-band" around ∼230 GHz and "high-band" around ∼345 GHz). The black curves in the left panel show integer multiples of the low-band frequency, and the green shaded regions indicate approximate available spectral windows for the mid- and high-frequency receivers. The three vertical shaded regions indicate low-band frequency ranges where an integer multiple of that frequency passes through both the mid- and high-band spectral windows. The middle vertical shaded region (highlighted in blue) corresponds to our proposed spectral arrangement; the available frequency ranges for each of the three bands are labeled, and the corresponding segments of the black curves are highlighted. For reference, the right panel shows the atmospheric transmission as a function of frequency.

4. Technical Interoperability

In the previous section we have shown that the ngEHT stations are able to observe at 86 GHz for the entire year. This offers great flexibility to enhance observing time at the higher frequencies, but also to provide stand-alone 86 GHz observing time with the array when observation conditions at the higher frequencies are poor.

The addition of 86 GHz capabilities to the ngEHT telescopes creates an opportunity for interoperability with major current and upcoming facilities. The new ngEHT dishes will most likely have small (≤10 m) diameters, meaning that the sensitivities of baselines between ngEHT dishes will be comparatively modest relative to, e.g., many EHT baselines. Substantial increases in both sensitivity and coverage at 86 GHz could be achieved by jointly observing with the ngEHT and one or more other arrays. In this section, we explore the capabilities of the ngEHT stations on their own and in combination with two external facilities (see Tables 1 and 2): the Global Millimeter VLBI Array (GMVA), the current leading 86 GHz VLBI array, and the next generation Very Large Array (ngVLA), a future major facility that is expected to become the most sensitive array observing at 86 GHz.

Table 2. Station overview and receiver capabilities, showing which sites are capable (or expected to be capable) of observing in which frequency bands.

Station	86 GHz	230 GHz	345 GHz
ALMA	X	X	X
APEX	-	X	X
SMA	-	X	X
JCMT	X	X	X
LMT	X	X	planned
SMT	-	X	X
KPTO	-	X	-
NOEMA	X	X	X
PV	X	X	X
SPT	-	X	X
GLT	X	X	X
BAJA	X	X	X
BAR	X	X	X
CAS	X	X	X
CAT	X	X	X
CNI	X	X	X
GAM	X	X	-
GARS	X	X	X
HAY	X	X	X
LLA	X	X	X
NZ	X	X	X
OVRO	X	X	X
PIKE	X	X	X
SGO	X	X	X
ngVLA	X	-	-
GBT	X	-	-
BR	X	-	-
FD	X	-	-
KP	X	-	-
LA	X	-	-
MK	X	-	-
NL	X	-	-
OV	X	-	-
PT	X	-	-
EF	X	-	-
YS	X	-	-
ONS	X	-	-
MET	X	-	-
KVN	X	X	-

The baseline sensitivity of the GMVA is currently limited by the recording bandwidth of the VLBA array (4 Gbps), and the ngEHT is planning to operate with a bandwidth of 256 Gbps. Currently, EHT sites with 86 GHz receivers are able to observe, as part of the GMVA, by only correlating a fraction of the observed frequency band. They have potential applications also for sub-arraying stations that are able to record at a higher rate. An alternative to the GMVA would be to make use of the high bandwidth of the ngEHT+EHT 86 GHz sites. This alternative offers a significant increase in baseline sensitivity due to the higher recording rate (especially valuable for weak polarization signals) and comparable coverage and point-source sensitivity to the GMVA. However, to more accurately reflect the capabilities of the current GMVA, all synthetic datasets labeled "GMVA" in this paper are limited to 512 MHz. The synthetic datasets for the other arrays use the bandwidths specified in Figure 7.

Figure 8 shows histograms of the baseline signal-to-noise ratio (S/N) for a number of potential array combinations. When observing as a standalone array, the ngEHT achieves a typical baseline S/N in tens or hundreds. By jointly observing with the EHT, GMVA, and/or ngVLA, a typical baseline S/N of hundreds or thousands is achieved; some baselines have an S/N in excess of 10^4. A measure of the total array sensitivity is captured by the PSS metric plotted in Figure 9, which improves by more than an order of magnitude when observing with the EHT, GMVA, and/or ngVLA alongside the ngEHT. Figure 9 also shows the improvement in the FF metric that is achieved by joint observations; we can see that joint observations substantially improve the FF to make it superior to that of the standalone ngEHT and the standalone ngVLA.

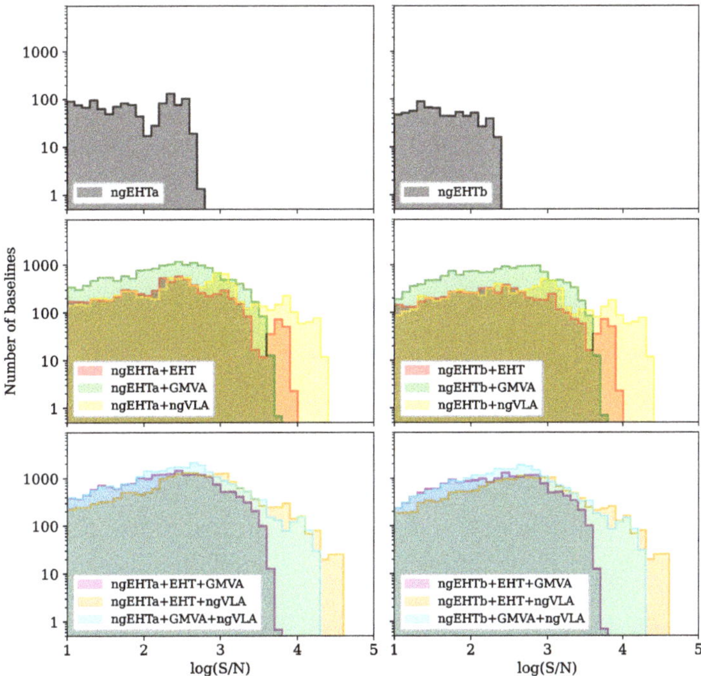

Figure 8. Histograms of baseline signal-to-noise ratios for the different arrays observing M87 at 86 GHz. Each histogram is the result of averaging over 100 Monte Carlo realizations of weather at every site, so the resulting histogram bins do not necessarily contain integer numbers of baselines.

We also prepared a demonstration of the imaging capabilities with the ngVLA, which is shown in Figure 10. We used the long-baseline ngVLA array configuration (LBA), which consists of 30 18-meter dishes on 10 sites across the United States. Co-located stations were modeled as a single site with properly scaled sensitivity, and in addition, we modeled the ngVLA core in New Mexico as a single highly sensitive site (with an SEFD of 10 Jy). We used the underlying model for M87 shown in Section 2, and obtained reconstructions at 86 GHz with the ngVLA alone, with the ngEHT+EHT combined array (here acting as a high-sensitivity alternative to the GMVA), and with the ngEHT+EHT+ngVLA combination. We notice that the ngVLA greatly improves the dynamic range for higher-fidelity reconstructions of the jet, whereas the ngEHT+EHT increases the resolution of the reconstruction and enables the imaging of the central brightness depression related to the black-hole shadow. Owing to the increased optical depth at 86 GHz compared to, e.g., 230 and 345 GHz (see also Figure 2), the appearance of this central brightness depression probes the "inner shadow" rather than the photon ring, which provides opportunities for measuring accretion flow and black hole properties [68]. While the optical depth of M87 at 86 GHz is uncertain, our simulation results inform the potential for long-term monitoring with the ngEHT+EHT+ngVLA at 86 GHz to study dynamics of both the (inner) shadow and the jet.

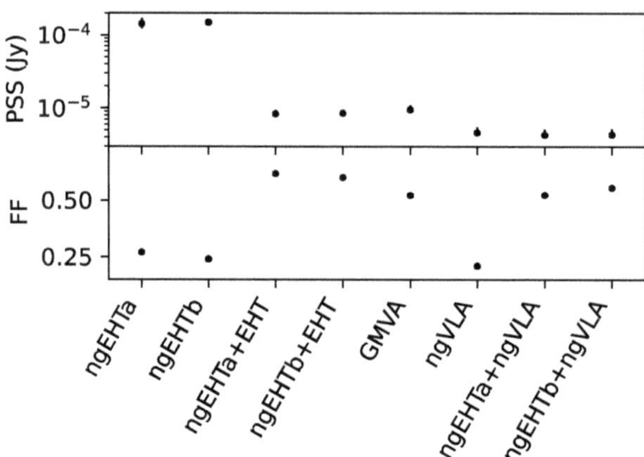

Figure 9. Comparison of the point source sensitivity (PSS, top panel, in units of Jy) and (u, v)-filling fraction (FF, bottom panel, unitless) metrics for several different array configurations. Each point shows the median metric value determined for 86 GHz M87 observations across 100 Monte Carlo realizations of weather at every site and across each month of the year, and the errorbars indicate the 16th to 84th percentile ranges.

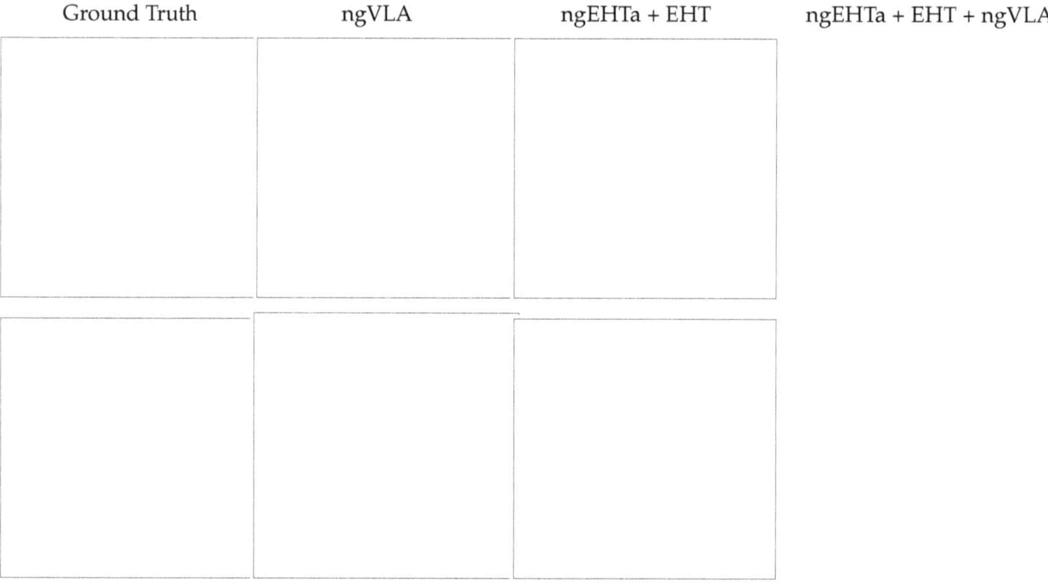

Figure 10. Demonstration of interoperability with ngVLA using simulated observations of the M87 jet at 86 GHz. The columns from left to right are: the ground truth image; the reconstruction with the ngVLA only; the reconstruction with the full ngEHT array; and the reconstruction with the full ngEHT array combined with the ngVLA. Top: Images plotted in square-root scale to emphasize the jet reconstruction; the contours are spaced logarithmically, starting at 1% of the peak value and increasing by factors of 2. Bottom: Images on a linear scale to emphasize the shadow reconstruction. The sensitivity of the ngVLA improves the dynamic range of the ngEHT, and the ngEHT boosts the resolution of the ngVLA, enabling imaging of both the shadow and the jet at 86 GHz with high fidelity.

5. Summary and Conclusions

We argue that supplementing the ngEHT with 86 GHz observing capabilities—particularly simultaneous multi-frequency capabilities at 86, 230, and 345 GHz—would improve the overall performance and flexibility of the array, helping it to achieve its primary scientific goals of high-dynamic-range images and movies of the M87 jet base region and the evolving accretion flow around Sgr A*.

One of the main benefits afforded by the addition of 86 GHz capabilities to the ngEHT would be the ability to carry out agile (i.e., rapid response or target of opportunity) and/or year-round observations. The primary observing frequencies of the ngEHT (230 and 345 GHz) require very stable atmospheric conditions, which renders consistent and agile observing difficult. We have shown that observing at 86 GHz should be possible year-round, as there are reliable weather conditions at all sites, allowing for more flexibility to observe transient sources and other targets.

Through multi-frequency imaging, 86 GHz observations can be combined with 230/345 GHz observations to permit imaging of the M87 jet structure with high fidelity. A standalone ngEHT array (i.e., without the addition of EHT or other sites) can only reliably reconstruct the M87 shadow and jet if 86 GHz information is present in multi-frequency imaging. Observations at 86 GHz would thus open up a significant fraction of time in which some or all of the EHT sites may not be available, but core ngEHT science remains achievable.

Simultaneous observations at 86 GHz, together with 230/345 GHz, are a practical requirement for absolute phase calibration and astrometry. Observing at 86 GHz would permit the ngEHT to connect with a well-established astrometric network of facilities around the world. Furthermore, simultaneous multi-frequency observing capabilities would enable full-array enhancements in calibration and sensitivity via multi-frequency phase reference techniques. By transferring the simultaneously measured phases at 86 GHz to the higher frequencies, the effective coherence times at 230 and 345 GHz could be increased from a few seconds to tens of minutes (and perhaps even longer).

In addition to simultaneous observations with the higher frequency bands, standalone 86 GHz capabilities of the ngEHT could also be used communally and to connect with other facilities, such as the currently operating GMVA and the upcoming ngVLA. Joint observations with the ngEHT and ngVLA at 86 GHz would leverage their substantial complementarity; the ngVLA provides sensitivity and the ngEHT provides angular resolution. We demonstrate that if the accretion flow in M87 is sufficiently optically thin at 86 GHz, then joint observations with the ngEHT and ngVLA could image both the horizon-scale shadow structure and the extended jet emission.

Author Contributions: Conceptualization, S.S.D., S.I., D.W.P. and F.R.; methodology, R.A., L.B., A.C., R.D., S.I., M.D.J., D.W.P., M.J.R. and F.R.; software, R.A., A.C., S.I., D.W.P., A.W.R. and F.R.; writing—original draft preparation, A.C., S.I., D.W.P. and F.R.; writing—review and editing, K.A., R.A., L.B., A.C., R.D. and S.S.D., V.L.F., G.F., S.I., M.D.J., G.N., D.W.P., A.W.R., M.J.R., F.R. and R.P.J.T.; visualization, A.C., S.I., D.W.P. and F.R.; supervision, S.S.D.; project administration, S.S.D. and G.F. All authors have read and agreed to the published version of the manuscript.

Funding: Support for this work was provided by the NSF through grants AST-1440254, AST-1935980, and AST-2034306; and by the Gordon and Betty Moore Foundation through grant GBMF-10423. This work has been supported in part by the Black Hole Initiative at Harvard University, which is funded by grants from the John Templeton Foundation and the Gordon and Betty Moore Foundation to Harvard University. S.I. is supported by the NASA Hubble Fellowship grant HST-HF2-51482.001-A awarded by the Space Telescope Science Institute, which is operated by the Association of Universities for Research in Astronomy, Inc., for NASA, under contract NAS5-26555. A.C. was supported by the NASA Hubble Fellowship grant HST-HF2-51431.001-A awarded by the Space Telescope Science Institute, which is operated by the Association of Universities for Research in Astronomy, Inc., for NASA, under contract NAS5-26555. K.A. has been also financially supported by following NSF grants: OMA-2029670, AST-1614868, AST-2107681, AST-2132700.

Data Availability Statement: No observational data were used in this publication.

Conflicts of Interest: The authors declare no conflict of interest.

References

1. Event Horizon Telescope Collaboration, et al. First M87 Event Horizon Telescope Results. I. The Shadow of the Supermassive Black Hole. *Astron. J. Lett.* **2019**, *875*, L1. [CrossRef]
2. Event Horizon Telescope Collaboration, et al. First M87 Event Horizon Telescope Results. II. Array and Instrumentation. *Astron. J. Lett.* **2019**, *875*, L2. [CrossRef]
3. Event Horizon Telescope Collaboration, et al. First M87 Event Horizon Telescope Results. III. Data Processing and Calibration. *Astron. J. Lett.* **2019**, *875*, L3. [CrossRef]
4. Event Horizon Telescope Collaboration, et al. First M87 Event Horizon Telescope Results. IV. Imaging the Central Supermassive Black Hole. *Astron. J. Lett.* **2019**, *875*, L4. [CrossRef]
5. Event Horizon Telescope Collaboration, et al. First M87 Event Horizon Telescope Results. V. Physical Origin of the Asymmetric Ring. *Astron. J. Lett.* **2019**, *875*, L5. [CrossRef]
6. Event Horizon Telescope Collaboration, et al. First M87 Event Horizon Telescope Results. VI. The Shadow and Mass of the Central Black Hole. *Astron. J. Lett.* **2019**, *875*, L6. [CrossRef]
7. Event Horizon Telescope Collaboration, et al. First M87 Event Horizon Telescope Results. VII. Polarization of the Ring. *Astron. J. Lett.* **2021**, *910*, L12. [CrossRef]
8. Event Horizon Telescope Collaboration, et al. First M87 Event Horizon Telescope Results. VIII. Magnetic Field Structure near The Event Horizon. *Astron. J. Lett.* **2021**, *910*, L13. [CrossRef]
9. Event Horizon Telescope Collaboration, et al. First Sagittarius A* Event Horizon Telescope Results. I. The Shadow of the Supermassive Black Hole in the Center of the Milky Way. *Astron. J. Lett.* **2022**, *930*, L12. [CrossRef]
10. Event Horizon Telescope Collaboration, et al. First Sagittarius A* Event Horizon Telescope Results. II. EHT and Multiwavelength Observations, Data Processing, and Calibration. *Astron. J. Lett.* **2022**, *930*, L13. [CrossRef]
11. Event Horizon Telescope Collaboration, et al. First Sagittarius A* Event Horizon Telescope Results. III. Imaging of the Galactic Center Supermassive Black Hole. *Astron. J. Lett.* **2022**, *930*, L14. [CrossRef]
12. Event Horizon Telescope Collaboration, et al. First Sagittarius A* Event Horizon Telescope Results. IV. Variability, Morphology, and Black Hole Mass. *Astron. J. Lett.* **2022**, *930*, L15. [CrossRef]
13. Event Horizon Telescope Collaboration, et al. First Sagittarius A* Event Horizon Telescope Results. V. Testing Astrophysical Models of the Galactic Center Black Hole. *Astron. J. Lett.* **2022**, *930*, L16. [CrossRef]
14. Event Horizon Telescope Collaboration, et al. First Sagittarius A* Event Horizon Telescope Results. VI. Testing the Black Hole Metric. *Astron. J. Lett.* **2022**, *930*, L17. [CrossRef]
15. Doeleman, S.; Blackburn, L.; Dexter, J.; Gomez, J.L.; Johnson, M.D.; Palumbo, D.C.; Weintroub, J.; Farah, J.R.; Fish, V.; Loinard, L.; et al. Studying Black Holes on Horizon Scales with VLBI Ground Arrays. *Bull. Am. Astron. Soc.* **2019**, *51*, 256.
16. Bustamante, S.; Blackburn, L.; Narayanan, G.; Schloerb, F.P.; Hughes, D. The Role of the Large Millimeter Telescope in Black Hole Science with the Next-Generation Event Horizon Telescope. *Galaxies* **2023**, *11*, 2. [CrossRef]
17. Chael, A.; Issaoun, S.; Pesce, D.W.; Johnson, M.D.; Ricarte, A.; Fromm, C.M.; Mizuno, Y. Multi-frequency Black Hole Imaging for the Next-Generation Event Horizon Telescope. *arXiv* **2022**, arXiv:2210.12226.
18. Issaoun, S.; Johnson, M.D.; Blackburn, L.; Brinkerink, C.D.; Mościbrodzka, M.; Chael, A.; Goddi, C.; Martí-Vidal, I.; Wagner, J.; Doeleman, S.S.; et al. The Size, Shape, and Scattering of Sagittarius A* at 86 GHz: First VLBI with ALMA. *Astron. J.* **2019**, *871*, 30. [CrossRef]
19. Issaoun, S.; Johnson, M.D.; Blackburn, L.; Broderick, A.; Tiede, P.; Wielgus, M.; Doeleman, S.S.; Falcke, H.; Akiyama, K.; Bower, G.C.; et al. Persistent Non–Gaussian Structure in the Image of Sagittarius A* at 86 GHz. *Astron. J.* **2021**, *915*, 99. [CrossRef]
20. Gwinn, C.R.; Kovalev, Y.Y.; Johnson, M.D.; Soglasnov, V.A. Discovery of Substructure in the Scatter-broadened Image of Sgr A*. *Astron. J. Lett.* **2014**, *794*, L14. [CrossRef]
21. Psaltis, D.; Johnson, M.; Narayan, R.; Medeiros, L.; Blackburn, L.; Bower, G. A Model for Anisotropic Interstellar Scattering and its Application to Sgr A*. *arXiv* **2018**, arXiv:1805.01242.
22. Chael, A.; Narayan, R.; Johnson, M.D. Two-temperature, Magnetically Arrested Disc simulations of the jet from the supermassive black hole in M87. *Mon. Not. R. Astron. Soc.* **2019**, *486*, 2873–2895. [CrossRef]
23. Porcas, R.W.; Rioja, M.J. VLBI phase-reference investigations at 86 GHz. In Proceedings of the 6th EVN Symposium, Bonn, Germany, 25–28 June 2002; p. 65.
24. Asaki, Y.; Saito, M.; Kawabe, R.; Morita, K.I.; Sasao, T. Phase compensation experiments with the paired antennas method. *Radio Sci.* **1996**, *31*, 1615–1626. [CrossRef]
25. Middelberg, E.; Roy, A.L.; Walker, R.C.; Falcke, H. VLBI observations of weak sources using fast frequency switching. *A&A* **2005**, *433*, 897–909. [CrossRef]
26. Dodson, R.; Rioja, M.J. VLBA Scientific Memorandum n. 31: Astrometric calibration of mm-VLBI using "Source/Frequency Phase Referenced" observations. *arXiv* **2009**, arXiv:0910.1159.

27. Rioja, M.; Dodson, R. High-precision Astrometric Millimeter Very Long Baseline Interferometry Using a New Method for Atmospheric Calibration. *Astron. J.* **2011**, *141*, 114. [CrossRef]
28. Han, S.T.; Lee, J.W.; Kang, J.; Oh, C.S.; Byun, D.Y.; Je, D.H.; Chung, M.H.; Wi, S.O.; Song, M.; Kang, Y.W.; et al. Korean VLBI Network Receiver Optics for Simultaneous Multifrequency Observation: Evaluation. *Publ. Astron. Soc. Pac.* **2013**, *125*, 539. [CrossRef]
29. Rioja, M.J.; Dodson, R.; Jung, T.; Sohn, B.W. The Power of Simultaneous Multifrequency Observations for mm-VLBI: Astrometry up to 130 GHz with the KVN. *Astron. J.* **2015**, *150*, 202. [CrossRef]
30. Algaba, J.C.; Zhao, G.Y.; Lee, S.S.; Byun, D.Y.; Kang, S.C.; Kim, D.W.; Kim, J.Y.; Kim, J.S.; Kim, S.W.; Kino, M.; et al. Interferometric Monitoring of GAMMA–RAY Bright Active Galactic Nuclei II: Frequency Phase Transfer. *J. Korean Astron. Soc.* **2015**, *48*, 237–255. [CrossRef]
31. Zhao, G.Y.; Jung, T.; Sohn, B.W.; Kino, M.; Honma, M.; Dodson, R.; Rioja, M.; Han, S.T.; Shibata, K.; Byun, D.Y.; et al. Source-Frequency Phase-Referencing Observation of AGNS with KAVA Using Simultaneous Dual-Frequency Receiving. *J. Korean Astron. Soc.* **2019**, *52*, 23–30. [CrossRef]
32. Zhao, G.Y.; Algaba, J.C.; Lee, S.S.; Jung, T.; Dodson, R.; Rioja, M.; Byun, D.Y.; Hodgson, J.; Kang, S.; Kim, D.W.; et al. The Power of Simultaneous Multi-frequency Observations for mm-VLBI: Beyond Frequency Phase Transfer. *Astron. J.* **2018**, *155*, 26. [CrossRef]
33. Lee, S.S.; Lobanov, A.P.; Krichbaum, T.P.; Witzel, A.; Zensus, A.; Bremer, M.; Greve, A.; Grewing, M. A Global 86 GHz VLBI Survey of Compact Radio Sources. *Astron. J.* **2008**, *136*, 159–180. [CrossRef]
34. Hada, K.; Doi, A.; Kino, M.; Nagai, H.; Hagiwara, Y.; Kawaguchi, N. An origin of the radio jet in M87 at the location of the central black hole. *Nature* **2011**, *477*, 185–187. [CrossRef]
35. Hada, K.; Kino, M.; Doi, A.; Nagai, H.; Honma, M.; Akiyama, K.; Tazaki, F.; Lico, R.; Giroletti, M.; Giovannini, G.; et al. High-sensitivity 86 GHz (3.5 mm) VLBI Observations of M87: Deep Imaging of the Jet Base at a Resolution of 10 Schwarzschild Radii. *Astron. J.* **2016**, *817*, 131. [CrossRef]
36. Kim, J.Y.; Krichbaum, T.P.; Lu, R.S.; Ros, E.; Bach, U.; Bremer, M.; de Vicente, P.; Lindqvist, M.; Zensus, J.A. The limb-brightened jet of M87 down to the 7 Schwarzschild radii scale. *A&A* **2018**, *616*, A188. [CrossRef]
37. Rioja, M.J.; Dodson, R. Precise radio astrometry and new developments for the next-generation of instruments. *A&A* **2020**, *28*, 6. [CrossRef]
38. Rioja, M.J.; Dodson, R.; Asaki, Y. The Transformational Power of Frequency Phase Transfer Methods for ngEHT. *Galaxies* **2023**, *11*, 16. [CrossRef]
39. Jiang, W.; Zhao, G.Y.; Shen, Z.Q.; Rioja, M.J.; Dodson, R.; Cho, I.; Zhao, S.S.; Eubanks, M.; Lu, R.S. Applications of the Source-Frequency Phase-Referencing Technique for ngEHT Observations. *Galaxies* **2023**, *11*, 3. [CrossRef]
40. Marcaide, J.M.; Shapiro, I.I. VLBI study of 1038+528A and B: Discovery of wavelength dependence of peak brightness location. *Astron. J.* **1984**, *276*, 56–59. [CrossRef]
41. Boccardi, B.; Krichbaum, T.P.; Ros, E.; Zensus, J.A. Radio observations of active galactic nuclei with mm-VLBI. *A&A* **2017**, *25*, 4. [CrossRef]
42. Kim, J.Y.; Krichbaum, T.P.; Marscher, A.P.; Jorstad, S.G.; Agudo, I.; Thum, C.; Hodgson, J.A.; MacDonald, N.R.; Ros, E.; Lu, R.S.; et al. Spatially resolved origin of millimeter-wave linear polarization in the nuclear region of 3C 84. *A&A* **2019**, *622*, A196. [CrossRef]
43. Paraschos, G.F.; Kim, J.Y.; Krichbaum, T.P.; Zensus, J.A. Pinpointing the jet apex of 3C 84. *A&A* **2021**, *650*, L18. [CrossRef]
44. Oh, J.; Hodgson, J.A.; Trippe, S.; Krichbaum, T.P.; Kam, M.; Paraschos, G.F.; Kim, J.Y.; Rani, B.; Sohn, B.W.; Lee, S.S.; et al. A persistent double nuclear structure in 3C 84. *Mon. Not. R. Astron. Soc.* **2022**, *509*, 1024–1035. [CrossRef]
45. Gómez, J.L.; Traianou, E.; Krichbaum, T.P.; Lobanov, A.; Fuentes, A.; Lico, R.; Zhao, G.Y.; Bruni, G.; Kovalev, Y.Y.; Lahteenmaki, A.; et al. Probing the innermost regions of AGN jets and their magnetic fields with RadioAstron. V. Space and ground millimeter-VLBI imaging of OJ 287. *arXiv* **2021**, arXiv:2111.11200.
46. Boccardi, B.; Perucho, M.; Casadio, C.; Grandi, P.; Macconi, D.; Torresi, E.; Pellegrini, S.; Krichbaum, T.P.; Kadler, M.; Giovannini, G.; et al. Jet collimation in NGC 315 and other nearby AGN. *A&A* **2021**, *647*, A67. [CrossRef]
47. Casadio, C.; MacDonald, N.R.; Boccardi, B.; Jorstad, S.G.; Marscher, A.P.; Krichbaum, T.P.; Hodgson, J.A.; Kim, J.Y.; Traianou, E.; Weaver, Z.R.; et al. The jet collimation profile at high resolution in BL Lacertae. *A&A* **2021**, *649*, A153. [CrossRef]
48. Nair, D.G.; Lobanov, A.P.; Krichbaum, T.P.; Ros, E.; Zensus, J.A.; Kovalev, Y.Y.; Lee, S.S.; Mertens, F.; Hagiwara, Y.; Bremer, M.; et al. Global millimeter VLBI array survey of ultracompact extragalactic radio sources at 86 GHz. *A&A* **2019**, *622*, A92. [CrossRef]
49. Rani, B.; Krichbaum, T.P.; Marscher, A.P.; Jorstad, S.G.; Hodgson, J.A.; Fuhrmann, L.; Zensus, J.A. Jet outflow and gamma-ray emission correlations in S5 0716+714. *A&A* **2014**, *571*, L2. [CrossRef]
50. Rani, B.; Krichbaum, T.P.; Marscher, A.P.; Hodgson, J.A.; Fuhrmann, L.; Angelakis, E.; Britzen, S.; Zensus, J.A. Connection between inner jet kinematics and broadband flux variability in the BL Lacertae object S5 0716+714. *A&A* **2015**, *578*, A123. [CrossRef]
51. Casadio, C.; Marscher, A.P.; Jorstad, S.G.; Blinov, D.A.; MacDonald, N.R.; Krichbaum, T.P.; Boccardi, B.; Traianou, E.; Gómez, J.L.; Agudo, I.; et al. The magnetic field structure in CTA 102 from high-resolution mm-VLBI observations during the flaring state in 2016–2017. *A&A* **2019**, *622*, A158. [CrossRef]
52. Schulz, R.; Kadler, M.; Ros, E.; Perucho, M.; Krichbaum, T.P.; Agudo, I.; Beuchert, T.; Lindqvist, M.; Mannheim, K.; Wilms, J.; et al. Sub-milliarcsecond imaging of a bright flare and ejection event in the extragalactic jet 3C 111. *A&A* **2020**, *644*, A85. [CrossRef]

53. Traianou, E.; Krichbaum, T.P.; Boccardi, B.; Angioni, R.; Rani, B.; Liu, J.; Ros, E.; Bach, U.; Sokolovsky, K.V.; Lisakov, M.M.; et al. Localizing the γ-ray emitting region in the blazar TXS 2013+370. *A&A* **2020**, *634*, A112. [CrossRef]

54. Tetarenko, A.J.; Sivakoff, G.R.; Miller-Jones, J.C.A.; Rosolowsky, E.W.; Petitpas, G.; Gurwell, M.; Wouterloot, J.; Fender, R.; Heinz, S.; Maitra, D.; et al. Extreme jet ejections from the black hole X-ray binary V404 Cygni. *Mon. Not. R. Astron. Soc.* **2017**, *469*, 3141–3162. [CrossRef]

55. Dodson, R.; Rioja, M.J.; Jung, T.; Goméz, J.L.; Bujarrabal, V.; Moscadelli, L.; Miller-Jones, J.C.A.; Tetarenko, A.J.; Sivakoff, G.R. The science case for simultaneous mm-wavelength receivers in radio astronomy. *New Astron. Rev.* **2017**, *79*, 85–102. [CrossRef]

56. Matthews, L.D.; Greenhill, L.J.; Goddi, C.; Chandler, C.J.; Humphreys, E.M.L.; Kunz, M.W. A Feature Movie of SiO Emission 20-100 AU from the Massive Young Stellar Object Orion Source I. *Astron. J.* **2010**, *708*, 80–92. [CrossRef]

57. Issaoun, S.; Goddi, C.; Matthews, L.D.; Greenhill, L.J.; Gray, M.D.; Humphreys, E.M.L.; Chandler, C.J.; Krumholz, M.; Falcke, H. VLBA imaging of the 3 mm SiO maser emission in the disk-wind from the massive protostellar system Orion Source I. *A&A* **2017**, *606*, A126. [CrossRef]

58. Selina, R.J.; Murphy, E.J.; McKinnon, M.; Beasley, A.; Butler, B.; Carilli, C.; Clark, B.; Durand, S.; Erickson, A.; Grammer, W.; et al. The ngVLA Reference Design. In *Proceedings of the Science with a Next Generation Very Large Array*; Murphy, E., Ed.; Astronomical Society of the Pacific Conference Series; NASA/ADS: Cambridge, MA, USA, 2018; Volume 517, p. 15.

59. McKinnon, M.; Beasley, A.; Murphy, E.; Selina, R.; Farnsworth, R.; Walter, A. ngVLA: The Next Generation Very Large Array. *Bull.,Am. Astron. Soc.* **2019**, *51*, 81.

60. Roelofs, F.; Blackburn, L.; Lindahl, G.; Doeleman, S.S.; Johnson, M.D.; Arras, P.; Chatterjee, K.; Emami, R.; Fromm, C.; Fuentes, A.; et al. The ngEHT Analysis Challenges. *Galaxies* **2022**, arXiv:2212.11355.

61. Raymond, A.W.; Palumbo, D.; Paine, S.N.; Blackburn, L.; Córdova Rosado, R.; Doeleman, S.S.; Farah, J.R.; Johnson, M.D.; Roelofs, F.; Tilanus, R.P.J.; et al. Evaluation of New Submillimeter VLBI Sites for the Event Horizon Telescope. *Astron. J. Suppl. Ser.* **2021**, *253*, 5. [CrossRef]

62. Chael, A.A.; Johnson, M.D.; Narayan, R.; Doeleman, S.S.; Wardle, J.F.C.; Bouman, K.L. High-resolution Linear Polarimetric Imaging for the Event Horizon Telescope. *Astron. J.* **2016**, *829*, 11. [CrossRef]

63. Chael, A.A.; Johnson, M.D.; Bouman, K.L.; Blackburn, L.L.; Akiyama, K.; Narayan, R. Interferometric Imaging Directly with Closure Phases and Closure Amplitudes. *Astron. J.* **2018**, *857*, 23. [CrossRef]

64. Kovac, J.M.; Barkats, D. CMB from the South Pole: Past, Present, and Future. *arXiv* **2007**, arXiv:0707.1075.

65. Palumbo, D.C.M.; Doeleman, S.S.; Johnson, M.D.; Bouman, K.L.; Chael, A.A. Metrics and Motivations for Earth-Space VLBI: Time-resolving Sgr A* with the Event Horizon Telescope. *Astron. J.* **2019**, *881*, 62. [CrossRef]

66. Dodson, R.; Rioja, M.J.; Jung, T.H.; Sohn, B.W.; Byun, D.Y.; Cho, S.H.; Lee, S.S.; Kim, J.; Kim, K.T.; Oh, C.S.; et al. Astrometrically Registered Simultaneous Observations of the 22 GHz H_2O and 43 GHz SiO Masers toward R Leonis Minoris Using KVN and Source/Frequency Phase Referencing. *Astron. J.* **2014**, *148*, 97. [CrossRef]

67. Rioja, M.J.; Dodson, R.; Jung, T.; Sohn, B.W.; Byun, D.Y.; Agudo, I.; Cho, S.H.; Lee, S.S.; Kim, J.; Kim, K.T.; et al. Verification of the Astrometric Performance of the Korean VLBI Network, Using Comparative SFPR Studies with the VLBA at 14/7 mm. *Astron. J.* **2014**, *148*, 84. [CrossRef]

68. Chael, A.; Johnson, M.D.; Lupsasca, A. Observing the Inner Shadow of a Black Hole: A Direct View of the Event Horizon. *Astron. J.* **2021**, *918*, 6. [CrossRef]

Article

The Transformational Power of Frequency Phase Transfer Methods for ngEHT

María J. Rioja [1,2,3,*], Richard Dodson [1] and Yoshiharu Asaki [4,5,6]

[1] International Centre for Radio Astronomy Research, M468, The University of Western Australia, 35 Stirling Hwy, Perth, WA 6009, Australia; richard.dodson@uwa.edu.au
[2] CSIRO Astronomy and Space Science, P.O. Box 1130, Kensington, WA 6102, Australia
[3] Observatorio Astronómico Nacional (IGN), Alfonso XII, 3 y 5, 28014 Madrid, Spain
[4] Joint ALMA Observatory, Alonso de Córdova 3107, Vitacura, Santiago 763 0355, Chile; yoshiharu.asaki@alma.cl
[5] National Astronomical Observatory of Japan, Alonso de Córdova 3788, Office 61B, Vitacura, Santiago 763 0492, Chile
[6] Department of Astronomical Science, School of Physical Sciences, The Graduate University for Advanced Studies (SOKENDAI), 2-21-1 Osawa, Mitaka, Tokyo 181-8588, Japan
* Correspondence: maria.rioja@uwa.edu.au

Abstract: (Sub) mm VLBI observations are strongly hindered by limited sensitivity, with the fast tropospheric fluctuations being the dominant culprit. We predict great benefits from applying next-generation frequency phase transfer calibration techniques for the next generation Event Horizon Telescope (ngEHT), using simultaneous multi-frequency observations. We present comparative simulation studies to characterise its performance, the optimum configurations, and highlight the benefits of including observations at 85 GHz along with the 230 and 340 GHz bands. The results show a transformational impact on the ngEHT array capabilities, with orders of magnitude improved sensitivity, observations routinely possible over the whole year, and ability to carry out micro-arcsecond astrometry measurements at the highest frequencies, amongst others. This will enable the addressing of a host of innovative open scientific questions in astrophysics. We present a solution for highly scatter-broadened sources such as SgrA*, a prime ngEHT target. We conclude that adding the 85 GHz band provides a pathway to an optimum and robust performance for ngEHT in sub-millimeter VLBI, and strongly recommmend its inclusion in the simultaneous multi-frequency receiver design.

Keywords: astronomical techniques; very long baseline interferometry

Citation: Rioja, M.J.; Dodson, R.; Asaki, Y. The Transformational Power of Frequency Phase Transfer Methods for ngEHT. *Galaxies* **2023**, *11*, 16. https://doi.org/10.3390/galaxies11010016

Academic Editor: Michael D. Johnson

Received: 27 November 2022
Revised: 20 December 2022
Accepted: 20 December 2022
Published: 12 January 2023

1. Introduction

VLBI observations at 230 GHz have delivered the first images of supermassive black holes. These results highlight the unique science accessible with high observing frequencies [1,2], and the interest for observations at even higher frequencies and higher angular resolutions. Nevertheless the relentless push of the upper frequency threshold in VLBI observations faces increasing challenges. The fast tropospheric phase fluctuations limit the length of time over which the signal can be coherently integrated (i.e., the coherence time) and prevent the application of standard phase referencing techniques applicable in the centimeter regime to extend this time. This propagation effect poses the main challenge in VLBI observations at high frequencies, which combined with intrinsically weaker source fluxes in general, higher instrumental noise and atmospheric opacity limit the observations to the strongest target sources, despite many advances in imaging algorithms (see [3]). It also prevents astrometric measurements.

Next-generation calibration methods and technologies have the potential to overcome these limitations. Using trans-frequency calibration, which relies on the non-dispersive nature of tropospheric fluctuations (i.e., that the tropospheric phase fluctuations are proportional to observing frequency) has a long history (e.g., [4–6]). It has only fairly recently begun to deliver on its promise for mm-VLBI [7] with the development of a suite of strategies hereafter grouped under the generic term 'frequency phase transfer methods', which share a common ground in requiring multi-frequency observations. At heart these rely on transferring the VLBI-observable solutions (for phase, delay and rate) measured at a lower frequency to correct the residuals in the analysis of simultaneous observations of the same source at a higher frequency, after scaling the phase by the frequency ratio. Encouraged by the observational success of these techniques that have been demonstrated at frequencies up to 130 GHz (e.g., [8] with the KVN) we propose to expand the application to ngEHT frequencies, where one would expect it to continue to function at two-hundred and three-hundred GHz.

The design specifications for the ngEHT observing frequencies are under revision in light of the possible benefits from the addition of a lower frequency. The proposed change to the system design is to add a 85 GHz band to the original dual-frequency system at 230 and 340-GHz. This paper is concerned with the prospects for ngEHT observations using frequency phase transfer methods and simultaneous multi-frequency receivers, with a focus on comparative performance. The methods are described in Section 2 and the simulation studies with accurate models for atmospheric propagation effects in Section 3. The results and discussions in Section 4 present the demonstrated effectiveness, the limitations, optimal configurations and some of the enabled scientific capabilities. These include the extension of effective coherence time at three-hundred GHz to hours, the array configuration requirements from the case study of SgrA* as a highly scatter-broadened source, and the astrometric measurements that could be made from such a configuration. Section 5 are the conclusions. The frequency phase transfer benefits are applicable to many different science goals, so this report stands alongside the partner reports on [9,10].

2. The Methods

The so-called frequency phase transfer paradigm encompasses a family of calibration methods that have in common the reliance of using observations at a lower reference frequency (ν_{low}) to infer the corrections at the higher target frequency (ν_{high}). Because observations at lower frequencies are more amenable, these make possible successful outcomes where single-frequency observations at the higher frequency alone are impossible. Moreover, they enhance the technical and scientific capabilities of the array.

Empirical demonstrations up to 130 GHz, detailed formulations of the methods, comprehensive error analysis, guidelines for scheduling and requirements have been presented in Rioja and Dodson [7], Rioja et al. [8], Rioja and Dodson [11] along with considerations for optimum performance in continuum and spectral line studies. For example, we strongly recommend an integer frequency ratio between the different frequency bands for robust and optimum use of these techniques, but see Dodson et al. [12] for an application when the target science requires otherwise. Here we include a brief description of these methods to support the comparative studies presented here, relevant to ngEHT.

The Frequency Phase Transfer (hereafter FPT) calibration method relies on transferring the VLBI-observable solutions (for phase, delay and rate) measured at a lower frequency, ν_{low}, with a temporal sampling τ_{low}, to the analysis of simultaneous observations of the same source at a higher frequency, ν_{high}, after scaling the phase by the frequency ratio. This is correct to address residual non-dispersive effects (i.e., frequency independent excess path length corrections), which are precisely calibrated out at the high frequency; these arise, for example, from unaccounted tropospheric contributions. On the other hand, residual dispersive effects are amplified; these arise mainly from unaccounted ionospheric and instrumental contributions (and if left uncorrected limit the functionality). The result of the precise tropospheric calibration is that the effective coherence time at the higher frequency is extended, allowing the detection of weaker sources than would be possible if using single-frequency observations. The remaining dispersive residual terms can impose limitations on the lengthened coherence time, depending on their magnitude, and additionally prevent precise astrometry. An FPT schedule consists of simultaneous multi-frequency observations of the target source. This corresponds to the observational set up used for the simulation studies presented in Section 3.

The Source/Frequency Phase Referencing (hereafter SFPR) calibration method provides a breakthrough for ultra precise mm-VLBI astrometry, beyond the scope of application of phase referencing methods, and unlimited effective coherence time. The SFPR calibration strategy comprises two steps, the first one being the FPT described above. The second step assumes that the remaining dispersive residuals can be eliminated using interleaving observations of a second calibration source. Since those terms are slowly varying, slow telescope source switching is possible, and the sources can be widely separated. Using SFPR is equivalent to carrying out observations from an excellent site and with a perfect instrument; it boosts the sensitivity by reducing coherence losses, and also provides a precise astrometric registration of the images at the two frequencies. The SFPR comprehensive astrometric error analysis in Rioja and Dodson [7] informs the astrometry estimates presented in Section 4.6.

Multi Frequency Phase Referencing (MFPR) [13] is a technique that builds on SFPR and delivers high precision astrometric measurements in the high frequency regime using observations of the target source only. Dedicated ionospheric calibration blocks (ICE-blocks) are interleaved with the simultaneous pair of dual-frequency observations at mm-wavelengths; its implementation requires an instrument with great frequency agility and wide frequency coverage, thus will be more relevant to the ngVLA.

FPT-square [14] is a technique that builds on the FPT method described above and allows for a further increase of coherence time. It uses observations of three frequency bands to form two pairs of simultaneous dual-frequency observations at mm-wavelengths and applies the scaled-correction twice to allow for the cancellation of the ionospheric contribution. As with FPT, FPT-square is not suitable for astrometric measurements in general.

All of these methods are widely applicable with no known upper frequency limit, if the instrument has the required simultaneous multi-frequency capability. These have been demonstrated for continuum and spectral line VLBI with ground observations up to 130 GHz, and at integer and non-integer frequency ratios. Furthermore they are potentially applicable in the space VLBI domain [15], improving the outcomes by calibrating out satellite orbit errors which have a non-dispersive nature, just as for the tropospheric errors.

3. Simulations for Coherence and Astrometric Studies at ngEHT Frequencies

We have carried out comparative simulation studies along with astrometric error propagation to quantify the benefits of FPT and SFPR multi-frequency techniques applied to ngEHT observing frequencies (ie bands around 85, 230 and 340 GHz). We focus on their power to overcome the dominant challenge imposed by the fast tropospheric phase fluctuations, which severely limit the coherence time and prevent astrometric measurements.

We have used ARIS [16] as our simulation tool to generate synthetic datasets. ARIS was designed explicitly for astrometric studies and includes realistic semi-analytical models of the so-called static and dynamic terms to mimic the troposphere and the ionosphere atmospheric propagation effects. The latter are implemented as moving phase screens assuming Kolmogorov turbulence characterized by a constant scale factor given by the C_w coefficient, with higher values indicative of stronger fluctuations, as expected from worsening weather conditions and/or lower quality sites. The output comprises two files, one for the target and another for the reference dataset. These can refer to observations along different lines of sight at the same frequency, as in phase referencing, or to simultaneous observations at different frequencies along the same line of sight, for the multi-frequency studies presented here. The output data files are in IDI-FITS format suitable for processing in most data reduction packages.

The studies presented here focus on the dominant atmospheric propagation effects. The simulations include four antenna sites, namely Owens Valley, Mauna Kea, KVN Yonsei and the Large Millimeter Telescope, which result in a range of baselines up to 6000 km. The target was selected to be a point source at a declination of +59°, allowing tracking for hours with Zenith angles less than $70°$ for all antennas. The simulations are for the strong signal regime, that is they do not include thermal noise. The atmospheric weather is imposed without regard for the nominal site, that is the actual antenna locations play an arbitrary role. The explored parameter space of atmospheric propagation effects are simulated using a dynamic turbulent tropospheric phase screen that is scaled by a range of C_w values equal to 0.5, 1 and 2. Best weather conditions are dubbed V for Very Good, with C_w of 0.5 and doubling for so called Good (dubbed G), and Tolerable (dubbed T) weather conditions. The static contributions use randomly determined values with standard deviations for the tropospheric excess path delay (i.e., $\Delta\ell$) of 3 cm and the ionospheric residual electron content (i.e., ΔTEC) of 6 TECU, at each VLBI station. The set of observing frequencies correspond to the originally planned 230 and 340 GHz bands, plus the new proposed 85 and 255 GHz bands. The latter is for testing the impact of frequency ratios being integer or not. We generated a complete set of synthetic single-frequency and simultaneous dual-frequency datasets, for all frequency and frequency pairs, and for all weather conditions.

The analysis of all the datasets was carried out with AIPS, using FPT calibration techniques for the dual-frequency datasets. It comprised of a self-calibration at the reference frequency (ν_{low}) with a range of solution intervals given by τ_{low} equal to 8, 15, 30 and 60 s, which are used to stabilise the phases at the target frequency ν_{high}. This was followed up with a second calibration at the target frequency with a much longer solution interval, given by τ_{high} equal to 1, 3, 10 and 60 min. These are very different scales as τ_{low} is related to the atmospheric-coherence time and τ_{high} corresponds to the FPT-coherence time, that is, after conditioning with FPT calibration. For the single-frequency datasets we use standard calibration. In all cases, following calibration, we measured the peak flux density in the images. Our preferred approach to calculate the coherence is to use the fractional peak flux recovery (FFR) quantity, calculated from the peak flux density value divided by the model point source intensity, as indicative of the coherence losses resulting from the calibration strategy. Another approach consists of measuring the baseline coherence over various timescales (e.g., using the AIPS task UVRMS). The former is closer, we feel, to what is required to characterize the quality of the image recovery. For example, the baseline coherence can be 100% whilst the FFR is zero, if the baselines are out of phase. Thus, all figures on coherence measurements use FFR, except for Figure 1.

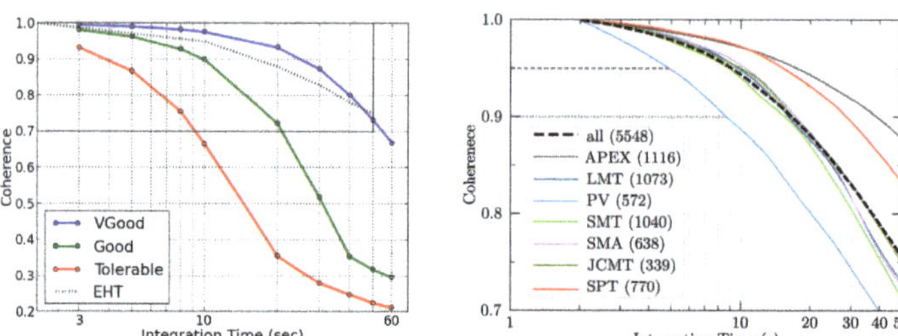

Figure 1. Comparison between atmospheric coherence measured from ARIS simulations and the empirical EHT datasets for a range of integration times up to 50 s, at 230 GHz. Plotted are first quartile values averaged over all baselines. **Left**: Solid lines are for the synthetic datasets under three weather conditions: Very Good (blue, $C_w = 0.5$), Good (green, $C_w = 1$) and Tolerable (green, $C_w = 2$). The black dotted line is for empirical EHT observations. The simulations for Very Good weather conditions (i.e., typical ALMA conditions) reproduce well the empirical measurements. **Right**: Figure 2 from Event Horizon Telescope Collaboration et al. [17] shows EHT empirical measurements. Solid lines are for individual baselines to ALMA and the dotted line is the baseline averaged coherence; the latter is overplotted in left. The axis limits are shown on the left as light black lines.

Finally, we use the SFPR astrometric propagation error formulation in Rioja and Dodson [7] to calculate the accuracy in astrometric measurements enabled with the simultaneous dual-frequency observations at ngEHT frequencies.

4. Results and Discussion

4.1. Verification of Weather Models in ARIS

We use the comparison between outcomes from the empirical EHT observations at 230 GHz and our synthetic datasets, to confirm the correspondence of the ARIS tropospheric models with the real weather conditions at EHT sites. Figure 1 (left) shows the 1st quartile baseline atmospheric coherence calculated with the AIPS task UVRMS as a function of integration time for our simulations under a range of weather conditions, shown with solid lines with different colors. Overplotted with a dotted line are the results measured from EHT observations as shown in the EHT data paper [17]. We conclude that the EHT empirical results match those of Very Good weather in our simulations, which corresponds to typical weather conditions at the ALMA site (without WVR corrections). This step is a fundamental check of the validity of the simulation studies and the conclusions extracted.

4.2. FPT Coherence at 340 GHz Using 230 GHz as the Reference Frequency

Here we present the prospects for the original system design which encompasses a dual-frequency receiver covering the 230 and 340-GHz bands.

Figure 2 plots the FPT coherence at 340 GHz (ν_{high}) as a function of calibration timescales (τ_{high}), for a range of values between 1 and 60 min, after FPT calibration using simultaneous 230 GHz (ν_{low}) observations. Shown are performances under Very Good (V, *left*) and Good (G, *right*) weather conditions. Different colors correspond to different calibration timescales at 230 GHz (τ_{low}), ranging between 8 and 60 s. The coherence corresponds to the FFR quantity obtained as described in Section 3. We find similar results for the 255-GHz and 340-GHz frequency pair.

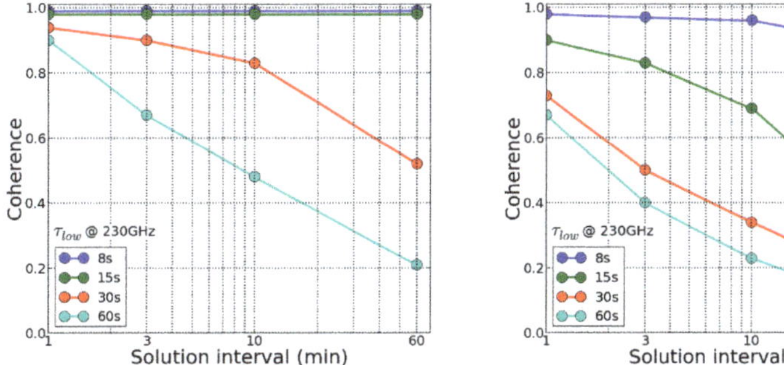

Figure 2. Expectations for FPT Coherence for the 340 GHz band under Very Good (**Left**) and Good (**Right**) weather conditions, after calibration with simultaneous 230 GHz band observations, for a range of integration times up to 60 min. Different colors are for different calibration timescales at 230 GHz (i.e., τ_{low}) as listed in the inset: 8 s (dark blue), 15 s (green), 30 s (red) and 60 s (cyan). In Very Good weather $\tau_{low} \leq 15$ s results in sustained high coherence ($\gg$0.9); longer τ_{low} timescales leads to degradation of the coherence, dropping to a coherence of 0.9 at 340 GHz for solution intervals up to 3 min or 0.8 for 10 min solutions intervals. This degradation is caused by the non integer frequency ratio in these simulations. Under G weather conditions with $\tau_{low} \sim$8 s we achieve coherence of 0.9 at 340 GHz up to an hour.

Under V weather conditions the results show very high coherence $\gg$0.9 at 340 GHz, for solution intervals up to an hour and beyond, using $\tau_{low} \leq 15$ s. Longer τ_{low}, $\sim$30 s, achieves a coherence of 0.9 at 340 GHz for solution intervals τ_{high} up to 3 min; longer solution intervals result in larger coherence losses (e.g., 10 min results in 20% loss). The performance deteriorates under G weather, but high coherence of $\gg$0.9 at 340 GHz can be obtained for up to 10 min with a sufficiently short τ_{low} of $\sim$8 s. The degradation of performance is mainly due to the non-integer frequency ratio ($\frac{\nu_{high}}{\nu_{low}}$), as discussed in Section 4.4.

We conclude that using 230-GHz as the reference frequency and under Very Good weather conditions leads to acceptable results at 340-GHz. Furthermore acceptable results are possible under Good weather conditions, with the more stringent constraint that τ_{low} is sufficiently fast. We remind the reader that for FPT a direct detection at ν_{low} within τ_{low} is a fundamental requirement. With ν_{low} equal to 230 GHz this will limit the number of possible targets. For example, only approximately 20% of the ALMA calibrator source list would be detectable by ngEHT with a τ_{low} of 8 s. This is, of course, significantly better than the percentage of that list which would be directly detected with single frequency observations at 340-GHz, which is about 8%.

4.3. FPT Coherence at 340 GHz Using 85 GHz as the Reference Frequency

With the reference frequency at 85 GHz, the target frequencies at 255- and 340-GHz can be integer frequency ratios (3 and 4, respectively), thus the solutions are more robust and coherence is not lost. Figure 3 plots the FPT coherence at 340 GHz (ν_{high}) as a function of calibration timescales (τ_{high}), for a range of values between 1 and 60 min, under Very Good (V, *left*), Good (G, *right*) and Tolerable (T, *bottom*) weather conditions, after FPT calibration using simultaneous 85 GHz (ν_{low}) observations. Different colors correspond to different calibration timescales (τ_{low}) at 85 GHz, ranging between 8 and 180 s.

Under V, G and T weather conditions we obtain very high coherence $\gg 0.9$ at 340 GHz, for solution intervals up to an hour and beyond, using $\tau_{low} \leq 30$, 15 and 8 s, respectively. Even with τ_{low} double these limits, long term coherence greater than 0.7 is maintained for an hour and beyond, as the frequency ratio is integer and phase ambiguity issues have no impact.

We conclude that using 85-GHz as the reference frequency and in Very Good and Good weather conditions, superior results are possible at 340-GHz, and with longer calibration time scales compared to using 230-GHz as the reference. Furthermore, acceptable results are possible under Tolerable weather conditions, with the more stringent constraint that τ_{low} is sufficiently fast. Approximately 90% of the ALMA calibrator source list would be detectable by ngEHT at 85 GHz with τ_{low} of 15 s. Thus, using 85-GHz as the reference frequency will dramatically increase the number of viable targets, allowing for demographic studies at 340-GHz.

Figure 4 shows a compendia of the results on FPT coherence for τ_{high} of 10 min for three frequency pairs (85→340 GHz, 85→255 GHz and 255→340 GHz) and two τ_{low} values (15 s and 30 s), as a function of the weather conditions, given by the C_w parameter (i.e., values 0.5, 1 and 2 for V, G and T, respectively). Figure 4 allows us to draw some general conclusions: (*a*) worsening weather conditions lead to degradation in the phase coherence, (*b*) that, for a given pair of frequencies, the coherence gets worse faster as τ_{low} is longer, (*c*) that, for a given ν_{low}, the higher ν_{high} the faster the degradation and (*d*) that fastest degradation is for a non-integer frequency ratio.

FPT can achieve increased effective sensitivity at (sub)mm-VLBI because it enables extended integration times, which allows the detection of sources that were otherwise too weak to be detected within the atmospheric coherence time with single-frequency observations. Our results suggest that the best performance at 340-GHz is achieved using 85-GHz as the reference frequency, compared to 230-GHz. We conclude this based on the following reasons. This approach provides superior calibration at 340-GHz, that is higher coherence and for longer integration times. Additionally, observations at 340- referenced to 85 GHz will be more robust, as they will work under a wider range of weather conditions massively expanding the observational window and/or increasing the number of suitable sites. This is partially because the analysis of data with integer frequency ratios is more straight-forward. Furthermore, direct detections at 85 GHz will be both more numerous and have higher SNR, due to the sources being intrinsically brighter, system noise being lower and τ_{low} being longer. This translates into wide applicability as more targets will be observable. Finally, 85-GHz provides a path to ultra-precise relative astrometry at 340-GHz, as discussed in Section 4.6.

Using a lower reference frequency, for example 43 GHz, results in doubling the scaling factor (given by the frequency ratio) and therefore increasing the propagation of the phase observable errors to the higher frequency. Additionally, the residual FPT ionospheric errors are larger using 43 GHz, compared to 85 GHz (see [7] for details on the impact of the frequency ratio). Lastly, 85 GHz band offers important benefits for observations of SgrA*, a main EHT target, as discussed in Section 4.5. We conclude that 85-GHz is the reference frequency of choice for ngEHT.

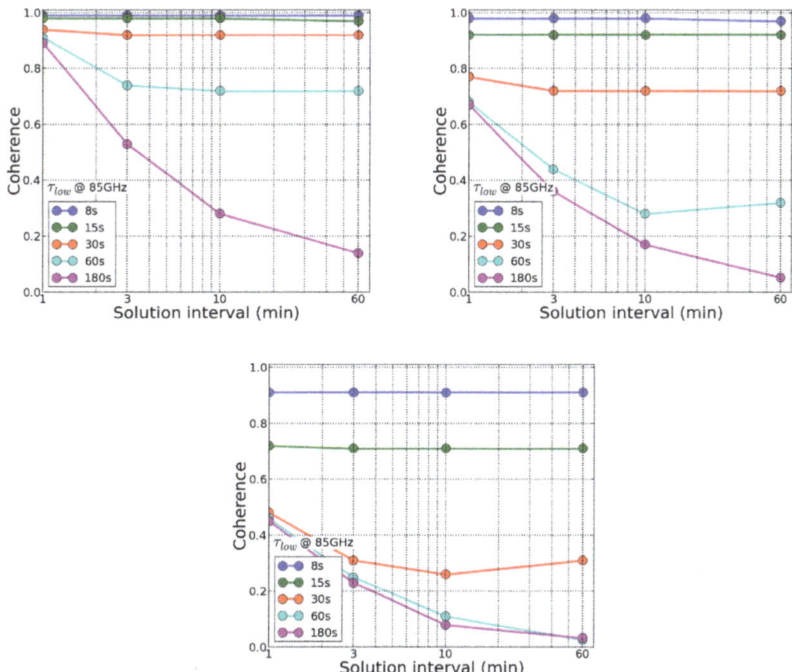

Figure 3. Expectations for FPT Coherence at 340 GHz band under Very Good (**left**) and Good (**right**) and Tolerable (**bottom**) weather conditions, after calibration with simultaneous 85 GHz band observations, for a range of integration times up to 60 min. Different colors are for different calibration timescales at 85 GHz (i.e., τ_{low}) as listed in the inset: 8 s (dark blue), 15 s (green), 30 s (red), 60 s (cyan) and 180 s (magenta). Sustained coherence of 0.9 or better can be obtained with calibration timescales of 30, 15 and 8 s, respectively. In very good weather coherence of 0.7 is obtained even with a 1 min calibration cycle, and similar levels of coherence can be achieved with a 30 s calibration cycle in good weather. Even in tolerable weather similar coherence is achievable with a fast 15 s τ_{low}.

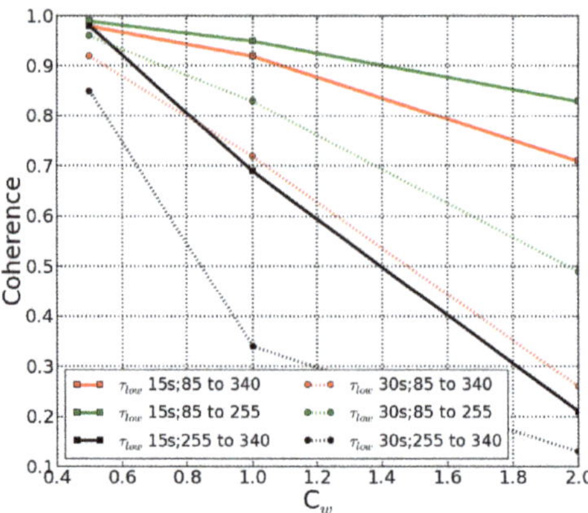

Figure 4. Expectations for FPT Coherence at 10 min integration time for a range of weather conditions (Very Good to Tolerable). Different colors are for all combinations of frequency pairs: green and red for pairs with 85 GHz as the reference frequency (i.e., 85→255 and 85→340, respectively) and black for 255 GHz as the reference frequency (i.e., 255→340). Solid and dashed lines are for calibration timescales at the low frequency equal to 15 s and 30 s, respectively. Higher frequencies and worse weather increase the degradation, but faster τ_{low} and integer frequency ratios improve the performance.

4.4. Impact of Non-Integer Frequency Ratios

In frequency phase transfer-based methods the frequency ratio is used to scale the phase measured at the low frequency. Therefore, for a non-integer frequency ratio the failure to track intrinsic 2π phase ambiguities introduces offsets and jumps in the calibrated phases at the high frequency. These jumps have a significant impact in the FPT-coherence, namely, that the phase does not change smoothly, but falls on multiple levels. Figure 5 shows the synthetic thermal-noise free residual visibility phases at 340 GHz after FPT calibration with 255 GHz simultaneous observations. Here the frequency ratio is 340/255 and thus the phase jump is MOD(340.0/255.0,1), i.e., exactly 1/3 of a turn, or $\pm120°$. Self-calibration on this data on longer timescales will introduce losses, as one would average over multiple levels. These losses are avoided if the ratio is integer, so that no phase jump is introduced, or in the non-integer case if no ambiguities are lost, which can be achieved under Very Good weather conditions (i.e., $C_w \ll 1$) or with very short solution intervals on the lower frequency (τ_{low}).

We note that observations are performed over wide bandwidths, rather than at a single frequency. The frequency reference points are selectable and these are all that need to have an integer ratio; these can in principle be outside the observed bandwidth (e.g., [12]). We would not recommend extrapolating too far from the observed frequency band to reach the reference point; we suggest no further than the spanned bandwidth. We strongly recommend using integer frequency ratios that enable a robust performance and applicability, optimizing the coherence achieved under all weather conditions, allowing for increased sensitivity because of the possibility of longer τ_{high} and enabling unambiguous bona-fide astrometry.

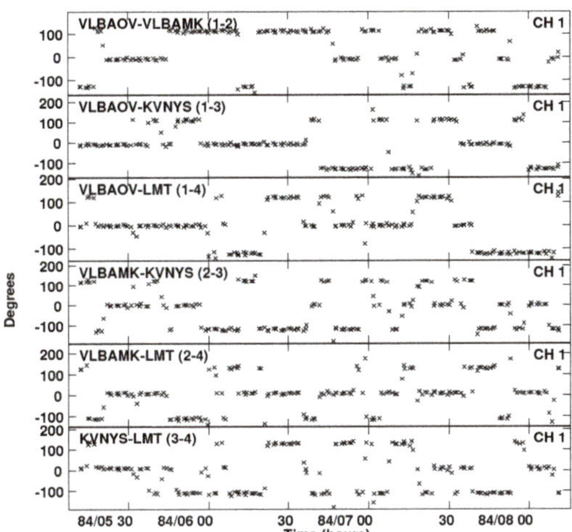

Figure 5. Impact of a non-integer frequency ratio between different observing bands. Here shown are FPT residual visibility phases at 340 GHz after calibration with simultaneous 255 GHz band observations and a calibration timescale equal to 15 s, under Good weather conditions. Phase jumps are a consequence of occasionally failing to track ambiguities. When ambiguity is lost a phase jump of $\pm 120°$ is introduced.

4.5. Simulations of SgrA* Observations, a Case Study for Scattered Sources

A successful application of frequency phase transfer techniques requires a direct detection, within τ_{low} (related to the atmospheric coherence time), of the target source at the reference or lower frequency. In the presence of strong scatter-broadening the counter-intuitive situation can arise where the source resolves at the lower frequency whilst at the higher frequency it does not, because the scattering is so much less. This is the case for SgrA*, a main target for EHT and ngEHT studies, with an apparent size of $230 \times 140\,\mu$as at 86 GHz [18]. This challenges the FPT approach, but does not make it impossible. In essence what is required is that the array configuration consists of multiple highly sensitive hubs, surrounded by 'spokes' to multiple smaller antennas that have shorter baselines, as shown in Figure 6. We thus did further ARIS simulations with Gaussian source models and a dense network of possible antenna sites. These allow for the formation of hub-and-spoke facets that, in addition to being highly sensitive, then can be phase connected as the separation between the edges of the facets are less than the uv-range limit. Given the size of the scattered source at 86 GHz, this is still difficult; baselines longer than 2250 km would be hard to use in the calibration. Therefore SgrA* is a strong driver for the requirement of the 85 GHz band to cover 115-GHz, which would still be an integer ratio to the 230- and 345-GHz frequencies. At 115-GHz the source size is $140 \times 115\,\mu$as and baselines out to 1.5 Gλ have correlated fluxes greater than 0.1 Jy. Figure 7 shows the uv-amplitude plot of the simulated data with this model at 115 GHz in the direction of SgrA*. In this case baselines out to $\sim$4000 km can be used in the calibration. Facets based in the Americas, Europe and Africa can then be connected. Identification of suitable candidate hub sites in S. America, N. America, Africa and Europe should be undertaken; the main candidates are obvious and Figure 6 provides an example.

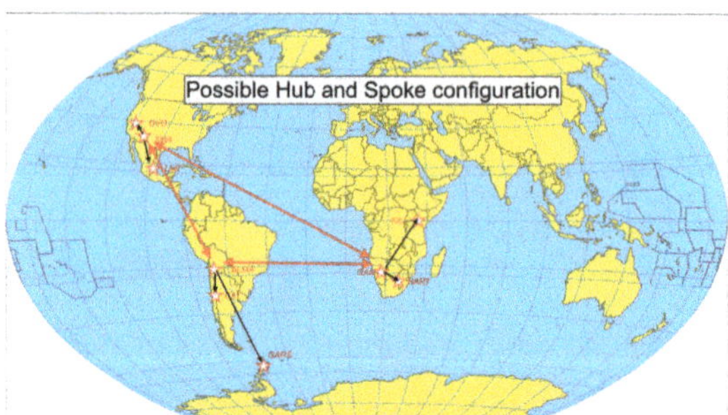

Figure 6. An example of sites that could form a hub-and-spoke configuration suitable for observing SgrA*, as used in our simulations. The black lines connect smaller antennas to a hub that provides the calibration reference site. These highly sensitive hub-based facets can then be connected together on their edges, or self-calibrated separately. Edited world map taken from Wikipedia.

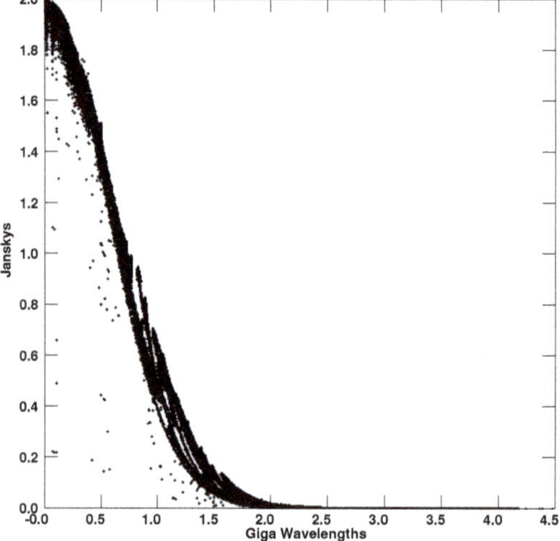

Figure 7. Visibility amplitude for the simulations as a function of uv distance for observations of SgrA* at 115 GHz. The model used in ARIS was a Gaussian based on the results of [18]. showing that significant flux (>0.1 Jy) is detectable out to 1.5 Gλ, corresponding to $\sim$4000 km. This clearly shows why the baseline lengths need to be $<$1.5 Gλ, and a two step calibration (hub-to-spoke and facet-to-global solutions) is required.

4.6. Paths to Astrometry at Sub-mm Wavelengths

A triple band receiver system and an array configuration with multiple antennas on a limited number of 'hub' sites could enable various flavours of innovative astrometric measurements with the ngEHT array that would be impossible to achieve with single frequency observations, which we will discuss here. The first flavour is 'λ-Astrometry', which provides a precise bona-fide registration of the images at different frequencies. The second flavour is 'conventional' or relative astrometry, which provides a precise bona-fide measurement of the relative angular separation between a target and a reference source(s) at a given frequency.

4.6.1. λ-Astrometry

We have estimated the astrometric accuracy at 255- or 340-GHz using the formulae for SFPR systematic astrometric error propagation in Rioja and Dodson [7], using 85-GHz as the reference frequency, with a source pair angular separation of $10°$, baseline lengths up to

6000 km and source switching times up to 10 min. In this case the astrometric precision is predicted to be about 3 µas and is dominated by the static ionospheric residuals and the large angular separation.

Source/Frequency Phase Referencing (SFPR) is a well demonstrated astrometric method and has been performed in the frequency range between 43 and to 130-GHz [8]. The performance and ease of use was excellent, even though this was a three-fold increase over the previous frequency limit for astrometry. Thus, we expect it to work at ngEHT frequencies, i.e., from 85 to 340 GHz, and SFPR should be straight forward with the current proposed array equipped with tri-band receivers. On the other hand, the proposed dual-frequency receiver (230 and 340 GHz) will struggle to provide λ-astrometry except in exceptional weather and for strong sources.

The scientific applications of λ-astrometry are extensively discussed in [10] so further comments are not needed here. However we do point out that, once the frequency position offset is measured between 85 GHz and either of the higher bands, one would be able to connect this offset to the ICRF if relative astrometry was performed at the lower frequency. This has been demonstrated at much lower frequencies in Dodson et al. [12], for example. The application of MultiView to provide this is considered below.

4.6.2. Relative Astrometry

MultiView [19] with simultaneous observations of multiple sources has the potential to provide a precise astrometry at 85 GHz using ngEHT. We note this is a regime beyond the scope of application of conventional phase referencing techniques. MultiView relies on the observations of multiple calibrators for a precise elimination of the dominant propagation errors from the observations, enabling ultra-precise relative astrometric measurements. It would provide bona-fide astrometry between sources, allowing registered dynamical measurements over time, to make movies of high impact individual sources or carry out statistical surveys of cosmological parallaxes.

MultiView has been demonstrated to provide precise corrections for the atmosphere at 1.4 [19] and 8 GHz [20], with on-going investigations at 43 GHz. The calibrator solutions can be spatially interpolated to provide the atmospheric contributions at the target line of sight, which results in an effective angular separation equal to zero degrees. The residual systematic astrometric errors are estimated to be $\sim$1 µas (see the formulae in [11]).

One would require four (or more) antennas at each astrometric site to perform Multi-View at 85 GHz, to provide simultaneous observations of the target and three calibrator sources. ALMA would be able to perform this role when sub-arraying is implemented.

This configuration seems possible for ngEHT and would enable astrometric measurements as a stand-alone instrument. Furthermore, the capability is definitely included in the ngVLA long baseline array planning [21]. A possible scenario is to perform combined measurements of the relative astrometry with the ngVLA at 85 GHz with the λ-astrometry between 85 and 340 GHz with the ngEHT; this would provide relative astrometry at 340 GHz.

4.7. Fundamental Implications on Network Specifications

Frequency phase transfer techniques result in a big boost of the sensitivity at 340 GHz, compared to single-frequency observations, by increasing the coherence time. An increase in effective coherence time from 10 s to 60 min translates to a $\sim$20-fold decrease in the flux density threshold of detectable sources and a much larger source population within reach. The FPT sensitivity is equivalent to single-frequency 340 GHz observations with an array having a 400-fold increase of bandwidth or a 4-fold increase in antenna diameter.

FPT techniques relax the constraints for suitable sites and/or weather conditions for (sub)mm-VLBI observations, with optimum outcomes having ν_{low} equal to 85-GHz. This opens the possibilities for more ngEHT sites, which would provide better uv-coverage and better image fidelity. Fewer constrains on the weather will provide extended windows of opportunity for observations, more observing time and improved temporal monitoring.

Figure 8 follows Figure 13.18 of Thompson et al. [22], showing the empirical relationship between the altitude and the RMS phase fluctuations for a number of well-known telescope sites. Based on this and the RMS path-length values for the weather models used in our simulations we overplot the estimated altitudes corresponding to V, G and T weather conditions. For the best performance with 85 GHz as the reference frequency, site altitudes greater than 2000 m are acceptable, based on the discussions in Section 4.3. For comparison, with 230 GHz as the reference frequency, site altitudes greater than 4000 m are desired, based on the discussions in Section 4.2.

Another interesting outcome from the frequency phase transfer-group of calibration methods is that they enable space VLBI at ngEHT frequencies, for highest angular resolution imaging with enhanced sensitivity and ultra precise astrometry. Orbit errors, like tropospheric errors, are non-dispersive hence FPT and SFPR methods will resolve these [15].

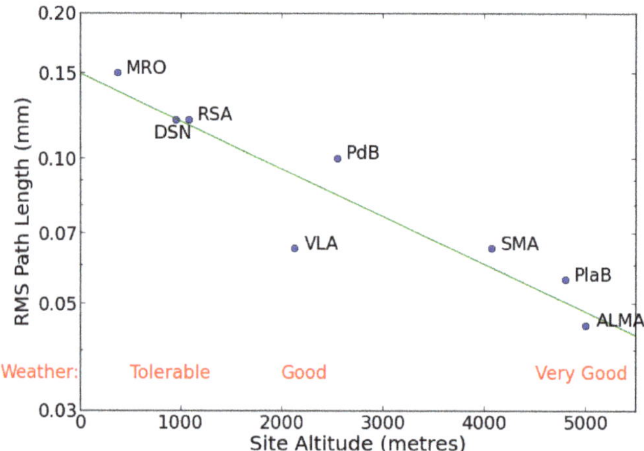

Figure 8. Implications of FPT benefits on antenna site selection. The plot follows that of Figure 13.18 using the data in Table 13.4 in Thompson et al. [22], showing the correlation between site altitude and empirical RMS pathlength measurements. Overplotted, in red , is the correspondence between altitudes and the simulation weather conditions explored in this paper, based on the RMS pathlength values from atmospheric models. The use of FPT calibration relaxes the requirements for suitable sites for (sub)mm-VLBI observations and will allow Good sites, i.e., above 2000 m, of which there are many candidates.

4.8. Requirements & Guidelines

The requirements for the application of frequency phase transfer techniques have been discussed at length in Section 4 of [7]. Here we summarise the points particularly relevant to ngEHT.

- Simultaneous observations at the multiple frequencies is an absolute key requirement, as this is the only way to fully sample the fast tropospheric fluctuations at ngEHT frequencies.
- Direct detections at the reference frequency within the relatively short calibration timescale, τ_{low}, is a fundamental requirement; the upper bound for τ_{low} sits between the atmospheric coherence time of the two frequencies (ν_{low} and ν_{high}). Note that, since phase errors are also scaled, it is important to optimise the SNR of the dectections. Having 85-GHz as the reference frequency alleviates this requirement with the lower system noise and longer τ_{low} values, and in-general larger source flux densities, compared to 230-GHz.
- We strongly recommended that the ν_{low} and ν_{high} frequencies have an integer ratio to avoid phase-ambiguity-related problems in the analysis. Note that it is sufficient that

the selected frequency reference points are within or close to the recorded frequency bands. Non-integer frequency ratios will, in general, introduce phase offsets and jumps which would have to be addressed separately.

- SgrA* requires observations at 115 GHz and a hub-and-spoke array configuration, so that all stations have sufficiently short sensitive baselines for detections.
- We recommend a tri-band receiver system with 85-, 230- and 340-GHz bands. The recorded bandwidth of the 85-GHz system should cover 115 GHz. Care should be taken to ensure that integer frequency ratios are possible within the recorded bands.
- Instrumental effects should be less than the residual atmospheric effects post frequency phase transfer analysis. For optimum performance it is required that instrumental effects, including between bands, must be measurable and stable to be better than the equivalent of 1 µas or 5 picoseconds. Larger residual instrumental effects might result in these being the dominant source of errors and limit the performance, where not thermally limited.
- MultiView will require a small number of 'hub' sites, able to carry out simultaneous observations along different lines of sight. This would be sub-arraying for ALMA and for multiple-antenna sites the requirement is for four antennas.

5. Summary and Conclusions

Frequency phase transfer techniques hold the potential to revolutionise (sub)mm-VLBI observations with ngEHT, making possible observations at 340 GHz where single-frequency observations are impossible. FPT and SFPR enable an adaptive phase-based calibration of the tropospheric fluctuations at the target frequency, using simultaneous observations at a reference frequency, of a common source. The best outcomes are expected from the proposed tri-band receiver, which encompass the addition of a 85-GHz band to the original system design of a dual-frequency receiver covering the 230 and 340-GHz bands. Our conclusions are based on the explorations of comparative ngEHT performance at 340 GHz presented in this paper.

Our metrics are primarily the quality of the FPT calibration at 340 GHz, testing the improvements of phase stability and coherence, and the lengths of time over which this is sustained. This FPT-coherence is orders of magnitude longer than the atmospheric-coherence time at the higher ngEHT frequencies. This results in reduced flux-detection limits and increased dynamic range and effective angular resolution in the images.

We show that using 85 GHz as reference in frequency phase transfer methods provides for robust operations of the ngEHT array at 340 GHz under a wider range of weather conditions and from sites that would be marginal for single-frequency 340-GHz VLBI observations. This means that observations at the highest frequencies with ngEHT will be routinely viable over the whole year, for source monitoring and movies of interesting sources. Success using 230 GHz as the reference frequency is possible, but the constraints are more severe. That is, only under Very Good weather conditions (C_w equal to 0.5, or site altitudes above ∼4000 m) would 230 GHz be a reliable reference frequency, whereas 85 GHz would be reliable even under Good weather conditions (C_w equal to 1, or site altitudes above ∼2000 m).

A fundamental requirement is for direct detections of the target source at the reference frequency within the τ_{low} timescale. This poses a more severe constraint at 230 GHz, compared to 85 GHz. Based on the ALMA calibrator list, 20% of sources are detectable by ngEHT at 230-GHz whereas 90% are detectable at 85-GHz.

Including the 85-GHz band allows for a richer range of astrometric observations. SFPR provides for μ-as bona-fide astrometry, for cross-band image registration (λ-astrometry) between the three bands, and potentially for relative astrometry up to 340 GHz, where not thermally limited. The latter requires four antennas co-observing at a subgroup of sites to perform MultiView astrometry at 85 GHz with μ-as precision.

For observing highly scatter-broaden sources such as SgrA* with frequency phase transfer methods we recommend a hub-and-spoke configuration for the array, and reference observations at a frequency of 115 GHz.

The benefits of frequency phase transfer techniques extend to the (sub)mm space VLBI regime, where the increase in spatial resolution by orders of magnitude and capacity for ultra precise astrometry will be a game changer.

Finally, a stable and well calibrated multi-frequency instrument is required, with control systems in place to reduce the level of instrumental errors to less than the residuals from the frequency phase transfer methods. We provide specifications such that they do not become the dominant source of errors and limit the performance.

Based on all of the above we conclude that the tri-band system will have a critical impact on the ngEHT scientific outcomes. Frequency phase transfer techniques will enable for (sub)mm VLBI the full range of functionalities available for the cm-VLBI regime, with μ-as resolution. That is: high sensitivity, high fidelity images, ultra-precise astrometric measurements, routinely-viable year-round imaging and using ground or space VLBI observations. These provide the basis for carrying out monitoring, making movies and performing supermassive black-hole demographic studies, which are key scientific drivers for ngEHT.

Adaptive phase-based FPT and SFPR methods overcome the challenge introduced by fast tropospheric fluctuations and the severe limitations imposed by them at ngEHT frequencies. Its application will provide a transformational impact in the ngEHT array capabilities, changing the (sub)mm-VLBI world. Therefore, we strongly recommend that establishing this capability should be an area of highest priority for the ngEHT.

Author Contributions: M.J.R. and R.D. Text and simulations. Y.A. Software and review. All authors have read and agreed to the published version of the manuscript.

Funding: This research received no external funding.

Data Availability Statement: The software ARIS is on github. github.com/yoshi-asaki/aris, (accessed on 28 July 2022)

Conflicts of Interest: The authors declare no conflict of interest.

References

1. Event Horizon Telescope Collaboration. First M87 Event Horizon Telescope Results. I. The Shadow of the Supermassive Black Hole. *Astrophys. J. Lett.* **2019**, *875*, L1.
2. Akiyama, K. et al. [Event Horizon Telescope Collaboration]. First Sagittarius A* Event Horizon Telescope Results. I. The Shadow of the Supermassive Black Hole in the Center of the Milky Way. *Astrophys. J. Lett.* **2022**, *930*, L12. [CrossRef]
3. Akiyama, K. et al. [Event Horizon Telescope Collaboration]. First M87 Event Horizon Telescope Results. IV. Imaging the Central Supermassive Black Hole. *Astrophys. J. Lett.* **2019**, *875*, L4.
4. Asaki, Y.; Saito, M.; Kawabe, R.; Morita, K.I.; Sasao, T. Phase compensation experiments with the paired antennas method. *Radio Sci.* **1996**, *31*, 1615–1626. [CrossRef]
5. Carilli, C.L.; Holdaway, M.A. Tropospheric phase calibration in millimeter interferometry. *Radio Sci.* **1999**, *34*, 817–840.
6. Middelberg, E.; Roy, A.L.; Walker, R.C.; Falcke, H. VLBI observations of weak sources using fast frequency switching. *Astron. Astrophys.* **2005**, *433*, 897–909. [CrossRef]
7. Rioja, M.; Dodson, R. High-precision Astrometric Millimeter Very Long Baseline Interferometry Using a New Method for Atmospheric Calibration. *Astron. J.* **2011**, *141*, 114.
8. Rioja, M.J.; Dodson, R.; Jung, T.; Sohn, B.W. The Power of Simultaneous Multi-Frequency Observations for mm-VLBI: Astrometry up to 130 GHz with the KVN. *Astron. J.* **2015**, *150*, 202.
9. Issaoun, S.; Pesce, D.W.; Roelofs, F.; Chael, A.A.; Dodson, R.; Rioja, M.J.; Akiyama, K.; Aran, R.; Blackburn, L.; Doeleman, S.; et al. Enabling transformational ngEHT science via the inclusion of 86 GHz capabilities. *Galaxies* **2023**, *unpublished*.
10. Jiang, W.; Zhao, G.-Y.; Shen, Z.-Q.; Rioja, M.; Dodson, R.; Cho, I.; Zhao, S.-S.; Eubanks, M.; Lu, R.-S. Applications of the source-frequency phase-referencing technique for ngEHT observations. *Galaxies* **2023**, *11*, 3. [CrossRef]
11. Rioja, M.J.; Dodson, R. Precise radio astrometry and new developments for the next-generation of instruments. *Astron. Astrophys. Rev.* **2020**, *28*, 6.

12. Dodson, R.; Rioja, M.J.; Jung, T.H.; Sohn, B.W.; Byun, D.Y.; Cho, S.H.; Lee, S.S.; Kim, J.; Kim, K.T.; Oh, C.S.; et al. Astrometrically Registered Simultaneous Observations of the 22 GHz H_2O and 43 GHz SiO Masers toward R Leonis Minoris Using KVN and Source/Frequency Phase Referencing. *Astron. J.* **2014**, *148*, 97.
13. Dodson, R.; Rioja, M.J.; Molina, S.N.; Gómez, J.L. High-precision Astrometric Millimeter Very Long Baseline Interferometry Using a New Method for Multi-frequency Calibration. *Astrophys. J.* **2017**, *834*, 177.
14. Zhao, G.Y.; Algaba, J.C.; Lee, S.S.; Jung, T.; Dodson, R.; Rioja, M.; Byun, D.Y.; Hodgson, J.; Kang, S.; Kim, D.W.; et al. The Power of Simultaneous Multi-frequency Observations for mm-VLBI: Beyond Frequency Phase Transfer. *Astron. J.* **2018**, *155*, 26.
15. Rioja, M.; Dodson, R.; Malarecki, J.; Asaki, Y. Exploration of Source Frequency Phase Referencing Techniques for Astrometry and Observations of Weak Sources with High Frequency Space Very Long Baseline Interferometry. *Astron. J.* **2011**, *142*, 157.
16. Asaki, Y.; Sudou, H.; Kono, Y.; Doi, A.; Dodson, R.; Pradel, N.; Murata, Y.; Mochizuki, N.; Edwards, P.G.; Sasao, T.; et al. Verification of the Effectiveness of VSOP-2 Phase Referencing with a Newly Developed Simulation Tool, ARIS. *Publ. Astron. Soc. Jpn.* **2007**, *59*, 397–418.
17. Akiyama, K. et al. [Event Horizon Telescope Collaboration]. First M87 Event Horizon Telescope Results. II. Array and Instrumentation. *Astrophys. J. Lett.* **2019**, *875*, L2.
18. Shen, Z.Q.; Lo, K.Y.; Liang, M.C.; Ho, P.T.P.; Zhao, J.H. A size of ∼1 AU for the radio source Sgr A* at the centre of the Milky Way. *Nature* **2005**, *438*, 62–64.
19. Rioja, M.J.; Dodson, R.; Orosz, G.; Imai, H.; Frey, S. MultiView High Precision VLBI Astrometry at Low Frequencies. *Astron. J.* **2017**, *153*, 105.
20. Hyland, L.J.; Reid, M.J.; Ellingsen, S.P.; Rioja, M.J.; Dodson, R.; Orosz, G.; Masson, C.R.; McCallum, J.M. Inverse Multiview. I. Multicalibrator Inverse Phase Referencing for Microarcsecond Very Long Baseline Interferometry Astrometry. *Astrophys. J.* **2022**, *932*, 52.
21. Reid, M.; Loinard, L.; Maccarone, T. Astrometry and Long Baseline Science. In *Proceedings of the Science with a Next Generation Very Large Array, 15 October 2018*; Murphy, E., Ed.; Astronomical Society of the Pacific: San Francisco, CA, USA, 2018; Volume 517, pp. 523–531.
22. Thompson, A.R.; Moran, J.M.; Swenson, George W.J. *Interferometry and Synthesis in Radio Astronomy*, 3rd ed.; Springer: Cham, Switzerland, 2017. [CrossRef]

Article

Applications of the Source-Frequency Phase-Referencing Technique for ngEHT Observations

Wu Jiang [1,2,*], Guang-Yao Zhao [3,*], Zhi-Qiang Shen [1,2], María J. Rioja [4,5,6], Richard Dodson [4], Ilje Cho [3], Shan-Shan Zhao [1,2], Marshall Eubanks [7] and Ru-Sen Lu [1,2,8]

1   Shanghai Astronomical Observatory, Chinese Academy of Sciences, Shanghai 200030, China
2   Key Laboratory of Radio Astronomy, Chinese Academy of Sciences, Nanjing 210008, China
3   Instituto de Astrofísica de Andalucía-CSIC, Glorieta de la Astronomía s/n, 18008 Granada, Spain
4   ICRAR, M468, The University of Western Australia, 35 Stirling Hwy, Crawley, WA 6009, Australia
5   CSIRO Astronomy and Space Science, P.O. Box 1130, Bentley, WA 6102, Australia
6   Observatorio Astronómico Nacional (IGN), Alfonso XII, 3 y 5, 28014 Madrid, Spain
7   Space Initiatives Inc., Newport, VA 24128, USA
8   Max-Planck-Institut für Radioastronomie, Auf dem Hügel 69, D-53121 Bonn, Germany
*   Correspondence: jiangwu@shao.ac.cn (W.J.); gyzhao@iaa.es (G.-Y.Z.)

Abstract: The source-frequency phase-referencing (SFPR) technique has been demonstrated to have great advantages for mm-VLBI observations. By implementing simultaneous multi-frequency receiving systems on the next-generation Event Horizon Telescope (ngEHT) antennas, it is feasible to carry out a frequency phase transfer (FPT) which could calibrate the non-dispersive propagation errors and significantly increase the phase coherence in the visibility data. Such an increase offers an efficient approach for a weak source or structure detection. The SFPR also makes it possible for high-precision astrometry, including the core-shift measurements up to sub-mm wavelengths for Sgr A*, M 87*, etc. We also briefly discuss the technical and scheduling considerations for future SFPR observations with the ngEHT.

Keywords: black hole; VLBI; ngEHT; astrometry; SFPR

Citation: Jiang, W.; Zhao, G.-Y.; Shen, Z.-Q.; Rioja, M.J.; Dodson, R.; Cho, I.; Zhao, S.-S.; Eubanks, M.; Lu, R.-S. Applications of the Source-Frequency Phase-Referencing Technique for ngEHT Observations. *Galaxies* **2023**, *11*, 3. https://doi.org/10.3390/galaxies11010003

Academic Editor: Luigina Feretti

Received: 15 November 2022
Revised: 10 December 2022
Accepted: 16 December 2022
Published: 21 December 2022

1. Introduction

The Very Long Baseline Interferometry (VLBI) technology can achieve the highest spatial angular resolution by linking intercontinental telescopes to form a virtual telescope, whose aperture size is equal to the longest baseline in the array. However, the wavefront arriving at each telescope suffers from various phase fluctuations when propagating through the atmosphere. This is even more severe at the millimeter and sub-millimeter (sub-mm) wavelengths as the phase dispersion is in proportion to the observing frequency. A novel technique called the frequency phase transfer (FPT) [1] or source-frequency phase referencing (SFPR) [2] is proposed to mitigate the fast phase fluctuations at the shorter wavelengths by referring to the phases at the longer wavelength observed close in time. The phases could be purified by two-step calibrations. The first step is the FPT calibration, where the non-dispersive phase errors, such as the tropospheric phase errors and the geometric antenna position errors, are removed. Furthermore, the unmodeled ionospheric delay and the instrumental phase offsets between the two wavelengths can be further eliminated by observations of a nearby calibrator. After the SFPR calibrations, the remaining phases just reflect the true high-frequency visibilities and the frequency-dependent shift in the positions, e.g., the frequency-dependent location of the jet cores (the core shift) [3]. The SFPR could also help to reliably align the molecular line emission seen at different frequency bands (e.g., [4]). It has great advantages in probing weak sources and high-precision astrometric measuring for the (sub-)mm-VLBI.

The capability of simultaneously receiving at four frequency bands (K/Q/W/D) makes the Korea VLBI Network (KVN) a unique prototype and instrument for the FPT/SFPR observations [5,6]. The capability of fast switching among receivers at the Very Long Baseline Array (VLBA) also makes it possible to carry out FPT/SFPR observations up to the 3 mm band [2,7], although the switching cycle time introduces coherence losses (see Figure 6 in [8]).

- ngEHT and the necessity of SFPR

  Based on the success of capturing the first images of two nearby supermassive black holes with the original Event Horizon Telescope (EHT), one at the center of the distant Messier 87 galaxy (M87*) [9] and the other at our Milky Way galaxy center (Sgr A*) [10], the next-generation Event Horizon Telescope (ngEHT) will expand the existing array (new sites) [11] and upgrade the technological deployments (receiving capabilities) significantly [12]. It aims to sharpen our view of the black holes and address fundamental questions about the accretion and jet-launching process, together with more black hole shadows captured and even making black hole "movies". Although the sensitivity of the ngEHT would be greatly improved with an ultra-wide bandwidth, the baseline sensitivity will still be limited due to the short coherent integration time at sub-mm wavelengths (a typical coherence time is $\sim$10 s at 230 GHz [13,14] and even shorter at 345 GHz) and the small dish size of most antennas. The SFPR can overcome the coherence time limitation at sub-mm wavelengths. As demonstrated in a separate technical paper in this Special Issue, the coherence time of the high frequency by referring to the low-frequency band could be increased more than 100 folds and extended to hour(s) in the simulations. See Rioja et al. in the same issue for more details. The detection threshold relies on the lower frequency rather than the higher one. Using a typical value of 10–15 s at 85 GHz, the flux density threshold for targets would become one magnitude lower ($\sim$10 mJy) and the number of targets would be hundreds under the array sensitivity. We have estimated the SFPR errors that would be introduced when referencing the 255 or 340 GHz data to 85 GHz, with an angular separation of $10°$ between sources. With simultaneous multi-frequency observations and intra-source switching times between 0 and 10 min, the astrometric precision is about 3 µas and dominated by the static ionospheric residuals. These would make the ngEHT more powerful for both astrophysical and astrometric applications.

2. Scientific Applications

2.1. Sgr A* and M 87*

Sgr A* and M 87* are the prime targets for demonstrating the application of the SFPR to observational studies of black holes and jets. The SFPR can help reduce the phase error budgets from the atmosphere and instruments, while increasing the coherence time, and thus improving the dynamical range of imaging. Furthermore, the SFPR will provide precise measurements to understand the event-horizon-scale structure adjacent to the supermassive black holes.

- Possible core-shift detection of Sgr A*

  The mm/sub-mm radio emission from Sgr A* can be produced by two generic models: an accretion flow itself [15,16] and/or an outflow [17]. To discriminate the dominant emission models of Sgr A* , the core shift, e.g., [3,18], can be used without resolving its structure. As for the jet model, based on GRMHD simulations, Mościbrodzka et al. [19] suggested the core shift of $\sim$130 µas at 22–43 GHz and $\sim$60 µas at 86–230 GHz. In a recent study by Fraga-Encinas et al. (in prep.), the core shift of Sgr A* is predicted from both the accretion disk and the jet model with different inclination angles. According to their results, a clear difference in the core shift between the two scenarios is shown. Especially at a small inclination angle, as has been suggested in recent studies [20–22], the expected core shift at 22–43 GHz is $\lesssim$10 µas in the accretion disk model while it is $\gtrsim$100 µas in the jet model. Our preliminary core-shift measurements with the Korean VLBI Network (KVN) and the Very Long Baseline Array (VLBA) at the same frequencies show $\sim$100 µas (I. Cho et al. in prep). However, the robustness has been relatively less due to large astrometric uncertainties which are mainly originated from (1) the large beam size (for the KVN) and (2) the frequency switching mode (for the VLBA). Each difficulty can be perfectly overcome through the ngEHT with the dual/triple band receiving capability.

- Connecting the jet and the black hole for M87*
  The EHT 2017 image of M87* has revealed the shadow of the central SMBH [9]. The EHT observations, however, were unable to reliably detect and image the inner jet, likely due to sensitivity limitations and the lack of short baselines in the UV coverage. At longer wavelengths, we see a well-collimated jet, but the emission is optically thick and we are only able to see the $\tau = 1$ surface and the downstream optically thin jet [23]. Furthermore, the resolution at longer wavelengths is not enough to resolve the shadow [24]. It remains uncertain how exactly the SMBH and the jet are connected. The ngEHT will improve the dynamic range of the 1.3 mm images which could enable the detection of the extended jet emission. However, it could be still challenging due to the steep spectrum of the jet. The SFPR covering 86–345 GHz bands offers an alternative way to reliably determine the relative location of the SMBH we see at 1.3 mm and the jet core at longer wavelengths. This is critical in understanding how black holes launch powerful, collimated jets (e.g., [25]).

2.2. Detection of Weak Sources and Structures

- Toward more supermassive black hole shadows
  With the increased coherent integration time, black holes, whose radio emissions are weak but shadow sizes are relatively large, can be detected by the ngEHT. According to the prediction of a semi-analytic spectral energy distribution model [26], there should be a dozen additional sources that with their horizon-scale structure resolved the ngEHT observing at 345 GHz [27]. M 84, M 104, and IC 1459 are the prominent candidates on the priority list. These targets have a correlated flux density of several tens mJy [28] and a shadow size of $\sim$10 µas. The sources could be directly fringed with a short solution interval and a relatively high signal-to-noise ratio at 85 GHz that guarantees the quality of the phases to be transferred to higher frequencies. The predicted sizes of the black hole shadows are comparable to the resolution achievable by the ngEHT at 345 GHz. It provides further test samples of black holes, whether or not described by the Kerr metric, besides M87* and Sgr A*. Vice versa, combining the diameter measurements of black hole shadows with GRMHD simulations, plus an independent distance measurement, can be used to determine the physical parameters of black holes (e.g., mass, orientation, spin, etc.).
  Toward understanding black holes, we are still on the road of pursuing precise measurements and conclusive evidence. In the case of M84 ($z = 0.00339$, D = 18.4 Mpc), the mass of the central supermassive black hole is $8.5 \times 10^8 \, M_\odot$ measured by the gas kinematics [29], or $1.8 \times 10^9 \, M_\odot$ estimated from the velocity dispersion [30]. Therefore, the diameter d of the black hole shadow would be about 5 or 10 µas, respectively. M 84 has a correlated flux of about 80 mJy at 86 GHz (Wang et al. in press), while the baseline sensitivity of the ngEHT at 86 GHz would achieve several mJy, which would guarantee the phase solutions with a signal-to-noise ratio high enough to be transferred to 345 GHz. As shown in Figure 1, the black hole mass could be independently constrained by the angular size of the shadow. It also indicates that the ngEHT with SFPR could image a batch of black hole shadows whose diameters are $\sim$10 µas. The SFPR could increase the coherent integration time that promises a firm fringe detection at 345 GHz, as well as high dynamic range imaging with a sub-diffraction-limited resolution [31].

- Detection of cosmic sources at 1 mm
  Based on the radio luminosity function, the number of AGNs detectable to the millimeter is almost inversely proportional to the array sensitivity. Besides detecting the horizon structure of faint nearby SMBHs, the SFPR could be used to increase the

detection of cosmic sources at short wavelengths. The flux threshold of the SFPR detection will be ~10 mJy through simulations. According to the ALMA calibrator catalog (https://almascience.eso.org/sc/, accessed on 1 June 2022), there would be more than nine hundred sources observable. These sources have a correlated flux (considering a resolving factor of ~0.16 with a baseline length of 5000 km) higher than 10 mJy and a flat spectrum from 85 to 345 GHz. With the increased sensitivity of the ngEHT, which is further enhanced by SFPR, it provides more diverse samples approachable at the upstream of jets for physical parameter statistics, such as the brightness temperature of the mm-core and the collimation profile of the jet base [32,33], as well as sub-structures in the core region [34].

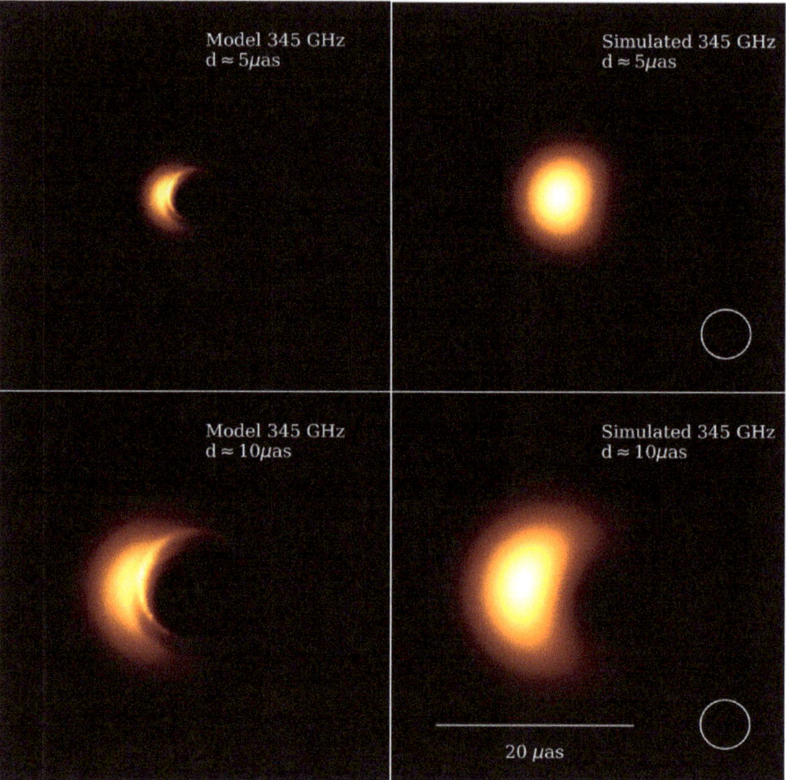

Figure 1. Model images of M84 with two different black hole masses. The images (right column) are reconstructed based on simulated ngEHT observations at 345 GHz. The empty white ring at the right bottom corner of the right panel plots is the synthesized beam of ngEHT at 345 GHz.

2.3. Microarcsecond Astrometry to the Black Holes

SFPR enables the VLBI astrometry at millimeter/sub-millimeter wavelengths with a precision of several μas. That means 0.01 pc motions of targets can be measured within a distance of Gpc. By source-frequency phase referencing, the location of a black hole could be pinpointed [28]. It enables the microarcsecond astrometry to the black hole itself in the ngEHT era.

- Orbit tracking of supermassive black hole binaries
  The merger of galaxies with central black holes can lead to the formation of a compact supermassive black hole binary (SMBHB) at the new galaxy center [35]. The early dynamical friction-driven and late gravitational radiation-driven phases of the SMBHB evolution are separated by the sub-pc orbital separation regime. How does the SMBHB

overcome this regime is known as the final-parsec problem [36]. For the ngEHT with SFPR, the propagation delays caused by the troposphere could be canceled out; we can still rely on a signal-to-noise-ratio-dependent resolution. The astrometric tracking of a black hole from an SMBHB system can reach 1 µas precision or better [37,38]. In the calculation of a population of detectable SMBHBs, we adopt the fiducial parameters of the model with a larger maximum observed binary period $P_{base} = 30$ yr (see Table 1 in [38]) and plot the number of SMBHBs as a function of the resolution θ_{min} and the sensitivity F_{min} (Figure 2). The ngEHT would provide an opportunity to track several observable sub-pc SMBHBs with a threshold of $\theta_{min} = 15$ µas and $F_{min} = 10$ mJy. While considering tracking the orbit motions of an SMBHB with respect to a background source in the same field as the upper limit, the minimum threshold of θ_{min} and F_{min} is 1 µas (the static ionospheric residuals could be minimized in the in-beam scenario) and 1 mJy, respectively, as shown in Figure 2.

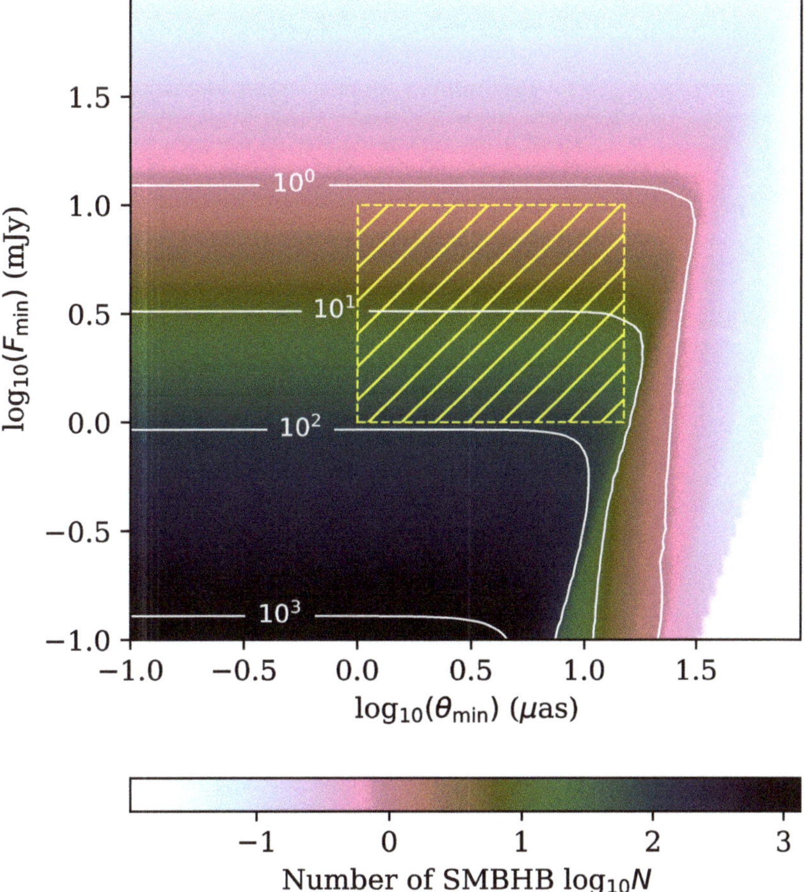

Figure 2. Number of detectable SMBHB systems (redshift z < 0.5) for the orbital tracking as a function of two main array parameters: the resolution θ_{min} and the sensitivity F_{min}. The hatched area is the target region by ngEHT, where it uses the baseline sensitivity of 10 mJy and the resolution of 15 microarcsec as the lower limit of the detection number of supermassive black hole binary systems, while the upper limit of the number is roughly corresponding to the array sensitivity and the precision of proper motion measurement by ngEHT, considering a background calibrator in the same field.

- Relative and absolute astrometric measurements

  The direct astrometric output of the SFPR is the core shift. It can be the relative positions between the 85 GHz core and the photon ring of the black hole when the 340 GHz already reaches the horizon scale. Otherwise, the core shift can be used to estimate the magnetic field and the particle density of the innermost jet [39], as well as predicting the jet apex up to the infinite frequency [3]. This provides a capability to position the black hole and track its motions by synergy with the lower-frequency VLBI, where the absolute astrometry is possible. Meanwhile, the absolute astrometry at short wavelengths needs cluster/paired antennas in each site [37]. The current proper motions of SgrA* still suffer from the scattering as measured at 43 GHz [40,41]; if one can go to a higher frequency, this effect can be largely reduced at the ngEHT frequencies. This is also very important to understand the head–tail sources (e.g., IC 310 and NGC 1265) whose hosting galaxies are infalling into the cluster at a high speed [42].

3. Requirements

3.1. Instrumentation Requirement

The capability of simultaneous observations at a lower-frequency band (85 or 110 GHz, 3 mm) and at one or two higher-frequency bands (255 or 220 GHz, 1.2 mm, and 340 or 330 GHz, 0.88 mm) is required for the frequency phase transfer. This can be accomplished with a quasi-optics tri-band receiving system [43] or a wide-band receiver [44]. In the case of a large interferometry array or co-site antennas working as a single VLBI station, the capability of forming sub-arrays corresponding to the lower and the higher observing frequency bands is feasible compared to installing trip-band receivers for each antenna. A co-located GPS will give accurate site positions for the geometric model, and the root mean square of the tropospheric path length fluctuations should be monitored for the co-site antennae. These have been found to greatly reduce the residual ionospheric, positional, and tropospheric contributions. Fuller descriptions of their impact can be found in [45]. The planned recording data rate as high as 256 Gbps would be able to incorporate the multi-band data stream simultaneously because the available bandwidth will be shared across all bands. The baseline sensitivity should be high enough to guarantee the fringe detection and minimize the phase errors on a correlated flux of a ~10 mJy source at the lower frequency, as well as to achieve a super/over-resolution power [31,46]. A detailed technical demand on the instruments is presented by Rioja et al. in the same Special Issue.

3.2. Strategy of Observation and Calibration

SFPR allows a phase calibrator within $10°$ apart in the sky and a switching cycle of more than 10 min [2]. SFPR expects a calibrator of a correlated flux higher enough at both the low- and high-frequency bands that could be fringed. A higher flux is better so as to mainly reduce the thermal noise. Meanwhile, a relatively large separation, i.e., $10°$, makes it much less restrictive to find a suitable calibrator even at the high frequencies. The core shift of the phase calibrator would be incorporated into the final core-shift measurement [6,47]. A prior core shift of a calibrator or a negligible core shift at the RA or DEC direction would be helpful to extract the true core shift of the target [7,28]. A synergy with the lower-frequency VLBI networks observing simultaneously can obtain more core-shift measurements to fit the power law scheme and perform the absolute astrometry observations.

4. Summary

With the aid of a simultaneous multi-frequency receiving system and more new stations available [6,8], the ngEHT with SFPR technique will be a very powerful tool to investigate the accretion disk and the jet/outflow connection in Sgr A* and M 87*, or other interesting targets at sub-mm wavelengths. With a dramatically increased coherence time and more feasible observational requirements (e.g., a long switching cycle time and large angular separation of calibrators), it will help to capture more images of black hole shadows and detect black hole motions in a binary system or a galaxy cluster.

Author Contributions: Conceptualization, W.J., G.-Y.Z. and Z.-Q.S.; methodology, W.J., G.-Y.Z., M.-J.R. and R.D.; software, S.-S.Z. and W.J.; writing—original draft preparation, W.J., G.-Y.Z. and I.C.; writing—review and editing, W.J., G.-Y.Z., Z.-Q.S., M.J.R., R.D., I.C., R.-S.L. and M.E. All authors have read and agreed to the published version of the manuscript.

Funding: This work was supported in part by the National Natural Science Foundation of China (grant Nos. 12173074, 11803071, and 11933007), the Key Research Program of Frontier Sciences, CAS (grant Nos. QYZDJ-SSW-SLH057 and ZDBS-LY-SLH011), the Shanghai Pilot Program for Basic Research—Chinese Academy of Science, Shanghai Branch (JCYJ-SHFY-2022-013), the Spanish Ministerio de Economía y Competitividad (grants AYA2016-80889-P and PID2019-108995GB-C21), the Consejería de Economía, Conocimiento, Empresas y Universidad of the Junta de Andalucía (grant P18-FR-1769), the Consejo Superior de Investigaciones Científicas (grant 2019AEP112), and the State Agency for Research of the Spanish MCIU through the "Center of Excellence Severo Ochoa" award to the Instituto de Astrofísica de Andalucía (SEV-2017-0709).

Institutional Review Board Statement: Not applicable.

Informed Consent Statement: Not applicable.

Data Availability Statement: Not applicable.

Acknowledgments: We thank the referees for their constructive comments and suggestions.

Conflicts of Interest: The authors declare no conflict of interest.

References

1. Middelberg, E.; Roy, A.L.; Walker, R.C.; Falcke, H. VLBI observations of weak sources using fast frequency switching. *A&A* **2005**, *433*, 897–909. [CrossRef]
2. Rioja, M.; Dodson, R. High-precision Astrometric Millimeter Very Long Baseline Interferometry Using a New Method for Atmospheric Calibration. *AJ* **2011**, *141*, 114. [CrossRef]
3. Lobanov, A.P. Ultracompact jets in active galactic nuclei. *A&A* **1998**, *330*, 79–89.
4. Yoon, D.H.; Cho, S.H.; Yun, Y.; Choi, Y.K.; Dodson, R.; Rioja, M.; Kim, J.; Imai, H.; Kim, D.; Yang, H.; et al. Astrometrically registered maps of H_2O and SiO masers toward VX Sagittarii. *Nat. Commun.* **2018**, *9*, 2534. [CrossRef]
5. Rioja, M.J.; Dodson, R.; Jung, T.; Sohn, B.W. The Power of Simultaneous Multifrequency Observations for mm-VLBI: Astrometry up to 130 GHz with the KVN. *AJ* **2015**, *150*, 202. [CrossRef]
6. Zhao, G.Y.; Jung, T.; Sohn, B.W.; Kino, M.; Honma, M.; Dodson, R.; Rioja, M.; Han, S.T.; Shibata, K.; Byun, D.Y.; et al. Source-Frequency Phase-Referencing Observation of AGNS with KAVA Using Simultaneous Dual-Frequency Receiving. *J. Korean Astron. Soc.* **2019**, *52*, 23–30. [CrossRef]
7. Jiang, W.; Shen, Z.; Jiang, D.; Martí-Vidal, I.; Kawaguchi, N. VLBI Imaging of M81* at λ = 3.4 mm with Source-frequency Phase-referencing. *ApJL* **2018**, *853*, L14. [CrossRef]
8. Rioja, M.J.; Dodson, R. Precise radio astrometry and new developments for the next-generation of instruments. *AApR* **2020**, *28*, 6. [CrossRef]
9. Event Horizon Telescope Collaboration; Akiyama, K.; Alberdi, A.; Alef, W.; Asada, K.; Azulay, R.; Baczko, A.K.; Ball, D.; Baloković, M.; Barrett, J.; et al. First M87 Event Horizon Telescope Results. I. The Shadow of the Supermassive Black Hole. *Astrophys. J. Lett.* **2019**, *875*, L1. [CrossRef]
10. Event Horizon Telescope Collaboration; Akiyama, K.; Alberdi, A.; Alef, W.; Algaba, J.C.; Anantua, R.; Asada, K.; Azulay, R.; Bach, U.; Baczko, A.K.; et al. First Sagittarius A* Event Horizon Telescope Results. I. The Shadow of the Supermassive Black Hole in the Center of the Milky Way. *ApJL* **2022**, *930*, L12. [CrossRef]
11. Raymond, A.W.; Palumbo, D.; Paine, S.N.; Blackburn, L.; Córdova Rosado, R.; Doeleman, S.S.; Farah, J.R.; Johnson, M.D.; Roelofs, F.; Tilanus, R.P.J.; et al. Evaluation of New Submillimeter VLBI Sites for the Event Horizon Telescope. *ApJS* **2021**, *253*, 5. [CrossRef]
12. Doeleman, S.; Blackburn, L.; Doeleman, S.; Dexter, J.; Gomez, J.L.; Johnson, M.D.; Palumbo, D.C.; Weintroub, J.; Farah, J.R.; Fish, V.; et al. Studying Black Holes on Horizon Scales with VLBI Ground Arrays. In Proceedings of the Bulletin of the American Astronomical Society, 2019; Volume 51, p. 256.
13. Blackburn, L.; Chan, C.k.; Crew, G.B.; Fish, V.L.; Issaoun, S.; Johnson, M.D.; Wielgus, M.; Akiyama, K.; Barrett, J.; Bouman, K.L.; et al. EHT-HOPS Pipeline for Millimeter VLBI Data Reduction. *ApJ* **2019**, *882*, 23. [CrossRef]
14. Event Horizon Telescope Collaboration; Akiyama, K.; Alberdi, A.; Alef, W.; Asada, K.; Azulay, R.; Baczko, A.K.; Ball, D.; Baloković, M.; Barrett, J.; et al. First M87 Event Horizon Telescope Results. III. Data Processing and Calibration. *ApJL* **2019**, *875*, L3. [CrossRef]

15. Narayan, R.; Yi, I.; Mahadevan, R. Explaining the spectrum of Sagittarius A* with a model of an accreting black hole. *Nature* **1995**, *374*, 623–625. [CrossRef]

16. Yuan, F.; Quataert, E.; Narayan, R. Nonthermal Electrons in Radiatively Inefficient Accretion Flow Models of Sagittarius A*. *ApJ* **2003**, *598*, 301–312. [CrossRef]

17. Falcke, H.; Markoff, S. The jet model for Sgr A*: Radio and X-ray spectrum. *A&A* **2000**, *362*, 113–118.

18. Blandford, R.D.; Königl, A. Relativistic jets as compact radio sources. *ApJ* **1979**, *232*, 34–48. [CrossRef]

19. Mościbrodzka, M.; Falcke, H.; Shiokawa, H.; Gammie, C.F. Observational appearance of inefficient accretion flows and jets in 3D GRMHD simulations: Application to Sagittarius A*. *A&A* **2014**, *570*, A7. [CrossRef]

20. Gravity Collaboration; Abuter, R.; Amorim, A.; Bauböck, M.; Berger, J.P.; Bonnet, H.; Brand ner, W.; Clénet, Y.; Coudé Du Foresto, V.; de Zeeuw, P.T.; et al. Detection of orbital motions near the last stable circular orbit of the massive black hole SgrA*. *A&A* **2018**, *618*, L10. [CrossRef]

21. Issaoun, S.; Johnson, M.D.; Blackburn, L.; Brinkerink, C.D.; Mościbrodzka, M.; Chael, A.; Goddi, C.; Martí-Vidal, I.; Wagner, J.; Doeleman, S.S.; et al. The Size, Shape, and Scattering of Sagittarius A* at 86 GHz: First VLBI with ALMA. *ApJ* **2019**, *871*, 30. [CrossRef]

22. Cho, I.; Zhao, G.Y.; Kawashima, T.; Kino, M.; Akiyama, K.; Johnson, M.D.; Issaoun, S.; Moriyama, K.; Cheng, X.; Algaba, J.C.; et al. The Intrinsic Structure of Sagittarius A* at 1.3 cm and 7 mm. *ApJ* **2022**, *926*, 108. [CrossRef]

23. Hada, K.; Doi, A.; Kino, M.; Nagai, H.; Hagiwara, Y.; Kawaguchi, N. An origin of the radio jet in M87 at the location of the central black hole. *Nature* **2011**, *477*, 185–187. [CrossRef]

24. EHT MWL Science Working Group; Algaba, J.C.; Anczarski, J.; Asada, K.; Baloković, M.; Chandra, S.; Cui, Y.Z.; Falcone, A.D.; Giroletti, M.; Goddi, C.; et al. Broadband Multi-wavelength Properties of M87 during the 2017 Event Horizon Telescope Campaign. *ApJL* **2021**, *911*, L11. [CrossRef]

25. Blandford, R.; Meier, D.; Readhead, A. Relativistic Jets from Active Galactic Nuclei. *Annu. Rev. Astron. Astrophys.* **2019**, *57*, 467–509. [CrossRef]

26. Huang, L.; Takahashi, R.; Shen, Z.Q. Testing the Accretion Flow with Plasma Wave Heating Mechanism for Sagittarius A* by the 1.3 mm VLBI Measurements. *ApJ* **2009**, *706*, 960–969. [CrossRef]

27. Pesce, D.W.; Palumbo, D.C.M.; Narayan, R.; Blackburn, L.; Doeleman, S.S.; Johnson, M.D.; Ma, C.P.; Nagar, N.M.; Natarajan, P.; Ricarte, A. Toward Determining the Number of Observable Supermassive Black Hole Shadows. *ApJ* **2021**, *923*, 260. [CrossRef]

28. Jiang, W.; Shen, Z.; Martí-Vidal, I.; Wang, X.; Jiang, D.; Kawaguchi, N. Millimeter-VLBI Observations of Low-luminosity Active Galactic Nuclei with Source-frequency Phase Referencing. *ApJL* **2021**, *922*, L16. [CrossRef]

29. Walsh, J.L.; Barth, A.J.; Sarzi, M. The Supermassive Black Hole in M84 Revisited. *ApJ* **2010**, *721*, 762–776. [CrossRef]

30. Ly, C.; Walker, R.C.; Wrobel, J.M. An Attempt to Probe the Radio Jet Collimation Regions in NGC 4278, NGC 4374 (M84), and NGC 6166. *AJ* **2004**, *127*, 119–124. [CrossRef]

31. Akiyama, K.; Kuramochi, K.; Ikeda, S.; Fish, V.L.; Tazaki, F.; Honma, M.; Doeleman, S.S.; Broderick, A.E.; Dexter, J.; Mościbrodzka, M.; et al. Imaging the Schwarzschild-radius-scale Structure of M87 with the Event Horizon Telescope Using Sparse Modeling. *ApJ* **2017**, *838*, 1. [CrossRef]

32. Asada, K.; Nakamura, M.; Pu, H.Y. Indication of the Black Hole Powered Jet in M87 by VSOP Observations. *ApJ* **2016**, *833*, 56. [CrossRef]

33. Janssen, M.; Falcke, H.; Kadler, M.; Ros, E.; Wielgus, M.; Akiyama, K.; Baloković, M.; Blackburn, L.; Bouman, K.L.; Chael, A.; et al. Event Horizon Telescope observations of the jet launching and collimation in Centaurus A. *Nat. Astron.* **2021**, *5*, 1017–1028. [CrossRef]

34. Giovannini, G.; Savolainen, T.; Orienti, M.; Nakamura, M.; Nagai, H.; Kino, M.; Giroletti, M.; Hada, K.; Bruni, G.; Kovalev, Y.Y.; et al. A wide and collimated radio jet in 3C84 on the scale of a few hundred gravitational radii. *Nat. Astron.* **2018**, *2*, 472–477. [CrossRef]

35. Kormendy, J.; Ho, L.C. Coevolution (Or Not) of Supermassive Black Holes and Host Galaxies. *ARAA* **2013**, *51*, 511–653. [CrossRef]

36. Begelman, M.C.; Blandford, R.D.; Rees, M.J. Massive black hole binaries in active galactic nuclei. *Nature* **1980**, *287*, 307–309. [CrossRef]

37. Broderick, A.E.; Loeb, A.; Reid, M.J. Localizing Sagittarius A* and M87 on Microarcsecond Scales with Millimeter Very Long Baseline Interferometry. *ApJ* **2011**, *735*, 57. [CrossRef]

38. D'Orazio, D.J.; Loeb, A. Repeated Imaging of Massive Black Hole Binary Orbits with Millimeter Interferometry: Measuring Black Hole Masses and the Hubble Constant. *ApJ* **2018**, *863*, 185. [CrossRef]

39. Zamaninasab, M.; Clausen-Brown, E.; Savolainen, T.; Tchekhovskoy, A. Dynamically important magnetic fields near accreting supermassive black holes. *Nature* **2014**, *510*, 126–128. [CrossRef]

40. Reid, M.J.; Brunthaler, A. The Proper Motion of Sagittarius A*. II. The Mass of Sagittarius A*. *ApJ* **2004**, *616*, 872–884. [CrossRef]

41. Xu, S.J.; Zhang, B.; Reid, M.J.; Zheng, X.W.; Wang, G.L.; Jung, T. A Milliarcsecond Accurate Position for Sagittarius A*. *ApJ* **2022**. [CrossRef]

42. Gendron-Marsolais, M.; Hlavacek-Larrondo, J.; van Weeren, R.J.; Rudnick, L.; Clarke, T.E.; Sebastian, B.; Mroczkowski, T.; Fabian, A.C.; Blundell, K.M.; Sheldahl, E.; et al. High-resolution VLA low radio frequency observations of the Perseus cluster: Radio lobes, mini-halo, and bent-jet radio galaxies. *Mon. Not. R. Astron. Soc.* **2020**, *499*, 5791–5805. [CrossRef]

43. Han, S.T.; Lee, J.W.; Kang, J.; Oh, C.S.; Byun, D.Y.; Je, D.H.; Chung, M.H.; Wi, S.O.; Song, M.; Kang, Y.W.; et al. Korean VLBI Network Receiver Optics for Simultaneous Multifrequency Observation: Evaluation. *Publ. Astron. Soc. Pac.* **2013**, *125*, 539. [CrossRef]
44. Yamasaki, Y.; Masui, S.; Ogawa, H.; Kondo, H.; Matsumoto, T.; Okawa, M.; Yokoyama, K.; Minami, T.; Konishi, R.; Kawashita, S.; et al. Development of a new wideband heterodyne receiver system for the Osaka 1.85 m mm-submm telescope: Corrugated horn and optics covering the 210-375 GHz band. *PASJ* **2021**, *73*, 1116–1127. [CrossRef]
45. Thompson, A.R.; Moran, J.M.; Swenson, G.W.J. *Interferometry and Synthesis in Radio Astronomy*, 2nd ed.; Springer International Publishing: Cham, Switzerland, 2001.
46. Martí-Vidal, I.; Pérez-Torres, M.A.; Lobanov, A.P. Over-resolution of compact sources in interferometric observations. *A&A* **2012**, *541*, A135. [CrossRef]
47. Jung, T.; Dodson, R.; Han, S.T.; Rioja, M.J.; Byun, D.Y.; Honma, M.; Stevens, J.; de Vincente, P.; Sohn, B.W. Measuring the Core Shift Effect in AGN Jets with the Extended Korean VLBI Network. *J. Korean Astron. Soc.* **2015**, *48*, 277–284. [CrossRef]

 galaxies

 MDPI

Article

The Role of the Large Millimeter Telescope in Black Hole Science with the Next-Generation Event Horizon Telescope

Sandra Bustamante [1,*], Lindy Blackburn [2,3], Gopal Narayanan [1], F. Peter Schloerb [1] and David Hughes [4]

1   Department of Astronomy, University of Massachusetts, Amherst, MA 01003, USA
2   Black Hole Initiative, Harvard University, 20 Garden Street, Cambridge, MA 02138, USA
3   Center for Astrophysics | Harvard & Smithsonian, 60 Garden Street, Cambridge, MA 02138, USA
4   Instituto Nacional de Astrofísica, Óptica y Electrónica, Apartado Postal 51 y 216, Puebla Pue 72000, Mexico
*   Correspondence: sbustamanteg@umass.edu

Abstract: The landmark black hole images recently taken by the Event Horizon Telescope (EHT) have allowed the detailed study of the immediate surroundings of supermassive black holes (SMBHs) via direct imaging. These tantalizing early results motivate an expansion of the array, its instrumental capabilities, and dedicated long-term observations to resolve and track faint dynamical features in the black hole jet and accretion flow. The next-generation Event Horizon Telescope (ngEHT) is a project that plans to double the number of telescopes in the VLBI array and extend observations to dual-frequency 230 + 345 GHz, improving total and snapshot coverage, as well as observational agility. The Large Millimeter Telescope (LMT) is the largest sub-mm single dish telescope in the world at 50 m in diameter, and both its sensitivity and central location within the EHT array make it a key anchor station for the other telescopes. In this work, we detail current and planned future upgrades to the LMT that will directly impact its Very Large Baseline Interferometry (VLBI) performance for the EHT and ngEHT. These include the commissioning of a simultaneous 230 + 345 GHz dual-frequency, dual-polarization heterodyne receiver, improved real-time surface measurement and setting, and improvements to thermal stability, which should enable expanded daytime operation. We test and characterize the performance of an improved LMT joining future ngEHT observations through simulated observations of Sgr A* and M 87.

Keywords: black holes; AGN; radio interferometry

Citation: Bustamante, S.; Blackburn, L.; Narayanan, G.; Schloerb, F.P.; Hughes, D. The Role of the Large Millimeter Telescope in Black Hole Science with the Next-Generation Event Horizon Telescope. *Galaxies* **2023**, *11*, 2. https://doi.org/10.3390/galaxies11010002

Academic Editor: Fulai Guo

Received: 16 November 2022
Revised: 13 December 2022
Accepted: 15 December 2022
Published: 21 December 2022

1. Introduction

The EHT Collaboration has presented the first direct images of the event horizon of the supermassive black holes (SMBHs) at the center of M 87 (M 87*) [1–8] and at the Galactic Center (Sgr A*) [9–14]. The 230 GHz radio image of M 87* appears as a linearly polarized asymmetric ring-like feature of 42 ± 3 µas with a brightness temperature of $\sim 10^{10}$ K and a derived BH mass of $6.5 \times 10^9 M_\odot$ [15]. Image features are broadly consistent with those predicted by general relativistic magneto-hydrodynamical (GRMHD) simulations including synchrotron radiation from relativistic electrons interacting with the magnetic field within a magnetically arrested accretion flow (MAD) [5,6].

Reconstructing a representative image of Sgr A* from EHT data proved more challenging due to the intrinsic variability of the source on timescales considerably faster than the baseline rotation over a night. However, characteristic images reflected most likely a compact emission region with an asymmetric thick ring of diameter 51.8 ± 2.3 µas, consistent with the known mass of $4.1 \times 10^6 M_\odot$ derived from dynamical studies of orbiting stars [16].

The EHT 2017 array viewed M 87 from only five unique geographic locations spanning the Globe, and the sparseness of the array provided the ability to reconstruct only a simple, relatively compact structure at a limited dynamic range $\sim 10{:}1$ for M 87* [4]. In particular, the faint extended features of the M 87 jet, predicted by simulations and seen at nearby wavelengths [17], were inaccessible due to the lack of dense short-baseline coverage. For Sgr A*, source variability, which is expected on the sub-hour timescale, is also evident in the EHT time-domain data [10,18], but the 2017 array coverage was too spare to confidently track these dynamical features in the image itself [11]. A much denser instantaneous coverage would enable snapshot imaging of Sgr A*, allowing hot spots and other rapid features to be followed over the course of a single night.

The primary science targets of the EHT are M 87* and Sgr A* as these are the two black holes on the sky with the largest apparent shadow diameter; however, other active galactic nuclei (AGN) sources have also been observed such as Cen A [19], 3C 279 [20], and J1924-2914 [21]. The extreme resolution of the EHT provides a view into the inner regions of the black hole systems, allowing the highest resolution imaging of SMBH jets. However these images are also at a low dynamic range, which limits the detail available for jet modeling and calls for improved u–v coverage. Since the observations of April 2017, the EHT array has been steadily improved through the addition of new telescopes and a doubling of the bandwidth (to 64 Gbps in 2018). The Greenland Telescope (GLT) joined in 2018, while the Kitt Peak 12 m Telescope (KP) and the Northern Extended Millimeter Array (NOEMA) joined in 2020. The first 345 GHz science observations with the EHT + the Atacama Large Millimeter/submillimeter Array (ALMA) are scheduled for 2023.

The ngEHT is currently undergoing an NSF-funded design and architecture study (AST-1935980). It envisions almost doubling the number of stations of the EHT with dedicated antennas and locating them at strategic points around the world to optimize Very Large Baseline Interferometry (VLBI) coverage [22,23]. Its three main goals are to: (1) increase the size of the EHT array to improve the imaging dynamic range by two to three orders of magnitude; (2) extend observations to new frequencies in particular to simultaneous 230/345 GHz (0.87 mm) as it increases the resolution to 15 µas and enables multi-frequency calibration and source modeling; and (3) use the improved snapshot coverage (for Sgr A*) and telescope availability and operational agility (for M 87 and AGN targets) to enable time-domain imaging (movies), allowing the direct study of source dynamics.

2. The Large Millimeter Telescope

The Large Millimeter Telescope (LMT) is located on Sierra Negra in central Mexico at an altitude of 4600 m. It is a 50 m-diameter single-dish radio telescope designed to operate between 0.85 mm and 4 mm with a field of view of 4 arcmins and an offset pointing accuracy better than 1 arcsec RMS [24]. It is the largest science project ever undertaken jointly by Mexico and the United States and the most expensive scientific facility built in Mexico. The telescope joined the North American Very Large Baseline Array (VLBA) for the first LMT VLBI science at 3 mm in 2014 [25], and the first 1 mm VLBI fringes to the LMT were obtained the following year. In 2017, the LMT joined as one of the telescopes in the EHT-2017 array used to obtain the first images of the black hole shadow of M 87* [1,2].

The LMT is the largest single-dish aperture in the EHT global VLBI network and provides essential short and mid-baselines to the rest of the array. In addition to filling in critical u–v coverage, the large collecting area and central location of the LMT make it particularly suitable to act as an "anchor" station for the network—where strong detections between the LMT and other EHT antennas could be used to bring the entire network to a common phase and clock reference.

The LMT is actively undergoing a series of upgrades to improve both its current performance and to prepare it as a key dual-frequency anchor station in the future ngEHT. Table 1 shows the upgrades which include: (1) an increase in the dish surface accuracy for 230 and 345 GHz efficiency, (2) an increase in the thermal stability to enable day-time operation and longer u–v tracks, and (3) a new dual-frequency receiver capable of simultaneous dual-polarization observations at 230 and 345 GHz over an intermediate frequency (IF) bandwidth of 8 GHz per sideband.

Figure 1 shows that the performance increase due to the thermal stability and the dual-frequency receiver (denoted as LMT-II) will allow the LMT to act as an anchor station for small dishes (<6 m in diameter) in the ngEHT array. This serves a particularly critical role in ngEHT observations when other high-sensitivity stations such as ALMA or NOEMA are not able to participate (such as the extended maintenance performed in February to ALMA when the weather across the Northern Hemisphere sites is often excellent).

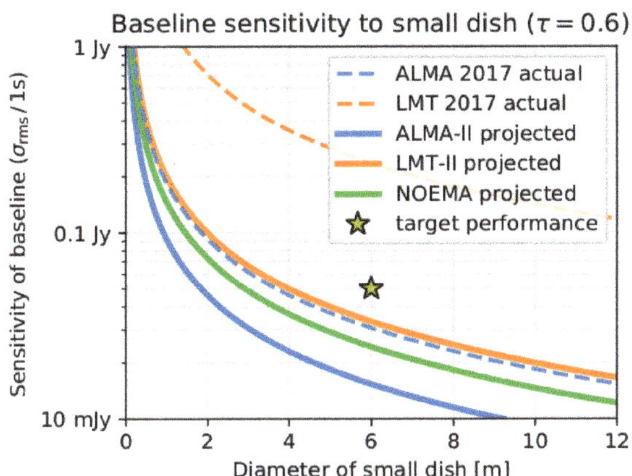

Figure 1. Figure 7 from [22], showing the expected sensitivity at 230 GHz on a VLBI baseline between LMT (and other anchor stations) and a small dish with a 0.5 aperture efficiency in moderate weather (line-of-sight $\tau = 0.6$). The curves demonstrate that baselines to LMT-II should have a sensitivity at 230 GHz to track atmospheric phase on sufficiently short timescales to phase anchor other stations. LMT-II sensitivity includes the full 50 m dish with an assumed aperture efficiency of 0.37. Simultaneous observations at 230 and 345 GHz enabled by a dual-frequency receiver (Section 2.1) at both sites allow for atmospheric phase transfer from 230 GHz to the higher frequency.

Table 1. Development Roadmap for LMT upgrades. Key upgrades target improved surface efficiency and stabilization under more varied conditions.

Development	Status (Tentative Date)	Features and Improvements
Dual-Frequency VLBI Receiver	In development (End 2023)	• Simultaneous observations at 230 and 345 GHz • dual-sideband (2xSB), dual-polarization • IF covering 4–12 GHz • 8 GHz bandwidth at 230 GHz and 12 GHz bandwidth at 345 GHz • Receiver temperature 60–80 K
Thermal stabilization	In development (End 2025)	• Ventilation system for backup structure to reduce thermal gradients • Deployment of temperature monitoring across the dish for the application of improved thermal models and compensation • Extends observing time
LASERS	In development	• Surface accuracy goal of 75 µm • Rapid active correction of dish • Improved duty cycle for VLBI

2.1. Dual-Frequency Receiver

As part of ngEHT, the University of Massachusetts (UMass) was commissioned to design and build a dual-frequency receiver capable of simultaneous observations at 1 mm (230 GHz) and 850 µm (345 GHz). In Figure 2, we present a block diagram of a dual-frequency receiver being constructed for ngEHT and to be deployed at the LMT by the end of 2023. A single cryostat will hold two different receivers, and two different frequency bands are sent to each receiver through a frequency diplexer. Each receiver is dual-polarized and features sideband separation mixers. Both bands illuminate a single beam on the sky, and the overall dual-frequency receiver has eight IF outputs, each of which is 4–12 GHz wide.

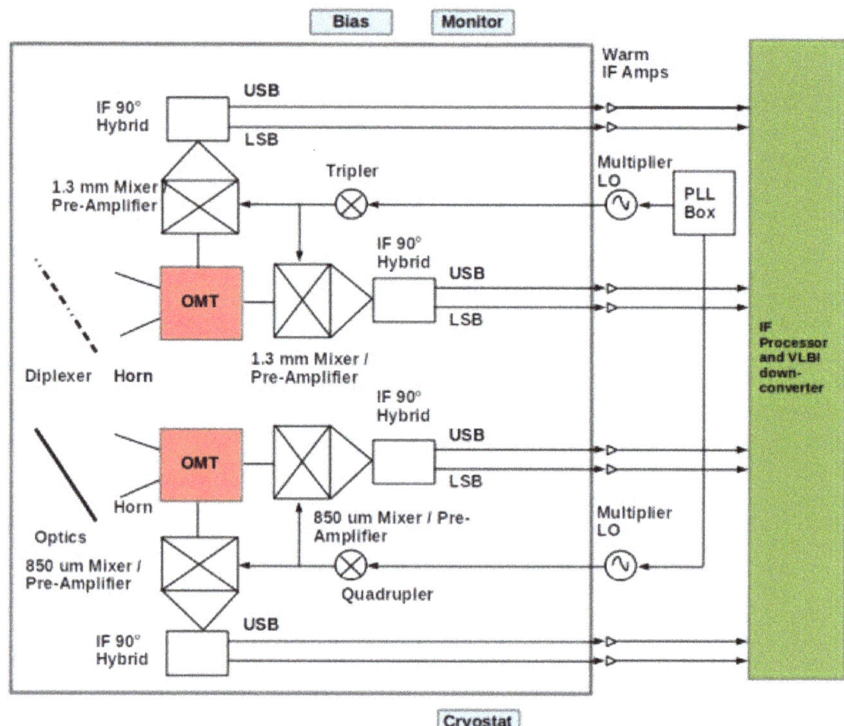

Figure 2. Block diagram for the dual-frequency receiver.

In an effort to make the design highly modular and scalable to reproduce for additional new telescopes of the ngEHT array, considerable effort has been invested into making the mixer block compact and highly integrated. In Figure 3, we show the components of this highly integrated block. Shown is the bottom block of a split-block mixer. A similar design will be employed for the 345 GHz receiver as well. The four IF outputs from each of the mixer blocks are amplified cryogenically using low-noise amplifiers from Low Noise Factory.

Each of the receiver bands is equipped with independent local-oscillator (LO) systems. YIG oscillators at lower frequencies (in the 18–30 GHz) range are multiplied up to the 95 GHz band and subsequently amplified using W-band (75 to 110 GHz) power amplifiers. This is then fed through cryogenic triplers to produce the required LO signal. The drain currents of the last stage of the W-band power amplifiers can be adjusted to set the appropriate LO power for the mixers. The whole LO system is phase locked and fully computer controlled with no mechanical moving parts.

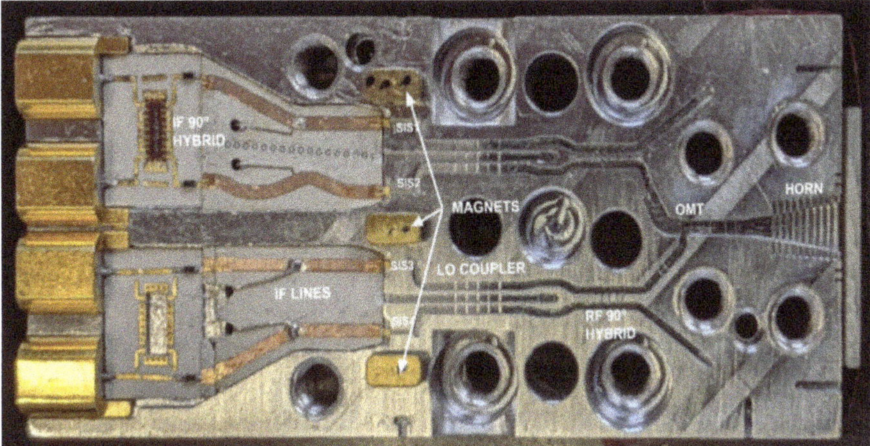

Figure 3. A view of the integrated mixer block for the 1 mm receiver. A square section of the corrugated feedhorn feeds into an orthomode transducer (OMT), which splits the two polarizations. Each polarization then enters an RF 90° hybrid and an LO coupler before going to the SIS mixers. The 4 IF outputs are fed through a IF 90° hybrid to produce 4 IF outputs, 2 polarizations, and 2 sidebands.

A prototype 230 GHz mixer block has been assembled and tested in the lab with both polarizations and sidebands with cryogenic IF amplifiers. Figure 4 shows the results of noise temperature tests of the 230 GHz receiver component. The figure shows the noise temperature is within five-times the quantum limit at all RF frequencies tested.

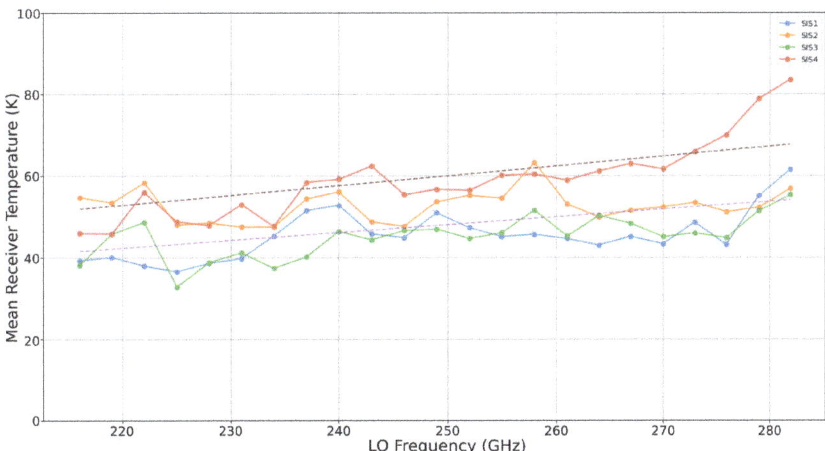

Figure 4. Mean noise temperature across an IF frequency range of 3–10 GHz, over an RF frequency range of 216–283 GHz for the 1mm receiver component of the dual-frequency receiver. The two dashed lines show the $4\times$ and $5\times$ theoretical quantum noise limits at each corresponding RF frequency.

Through a separate project, the Smithsonian Observatory and UMass will be acquiring 3 GHz superconductor–isolator–superconductor (SIS) junctions from Jet Propulsion Laboratory (JPL). Mixer blocks scaled from the design at the 230 GHz complement will be constructed at UMass and used with the JPL junctions. Each receiver has a specification of a single-sideband noise temperature of 5–6-times the quantum limit in each band.

2.2. The LMT Thermal Stabilization Program

The ultimate performance of a large telescope like the LMT is often limited by its response to thermal gradients that develop within the antenna structure. These gradients lead to deformations of the primary reflector surface, which in turn cause a loss of antenna gain. Thermally induced deformations may also change the relative position of the secondary and primary mirrors, leading to antenna focus and pointing errors. The performance of the LMT is currently limited by such effects, and as a result, scientific work at the telescope is limited to ten hours at night when the telescope is relatively thermally stable. Moreover, even during night-time conditions, the antenna still develops internal temperature gradients as it cools, and the resulting deformations require corrections every few hours to achieve the best performance. The LMT's active primary surface system provides the means to realign the surface continuously in response to any thermal deformation. Thus, it is only for the lack of the ability to measure the deformations that we are not already able to maintain the surface shape in the presence of structural temperature gradients. The LMT has undertaken a major technical program to develop systems and methods that will reduce the magnitude of temperature gradients within the structure and provide measurements to estimate the surface deformations so that the active surface can correct them. These problems are being pursued in four basic ways: (1) mitigation of thermal effects, through the installation of a ventilation system in the antenna backup structure; (2) measurement of structural temperatures, combined with finite-element modeling, to predict and correct surface deformations; (3) active measurement of the surface shape during observations using the out-of-focus (OOF) holography technique; and (4) real-time measurements of the surface shape, using laser metrology techniques, to allow the surface shape to be updated and maintained in real-time by the active surface.

2.2.1. Mitigation of Thermal Effects

Most large high-frequency antennas are equipped with systems that are designed to maintain the antenna structure at a uniform temperature. The original design of the LMT called for two measures to be installed to reduce temperature gradients within the structure. The first mitigating measure was the enclosure of the antenna in thermally insulating cladding. This was performed at the time of antenna construction. The cladding prevents solar radiation from striking the antenna backup structure and alidade directly. However, temperature gradients of a few degrees still build up within the backup structure due to the uneven distribution of solar energy and differences in the thermal mass of different parts of the structure. The second mitigation measure proposed with the original LMT design was the installation of a ventilation system. The ventilation system is a system of fans arranged to move air through the backup structure in order to minimize temperature gradients between different structural elements. The system to be installed is similar in nature to that of the IRAM 30 m antenna, but unlike that system, there is no attempt to set the structural temperatures to match a single desired value. Rather, the goal is to achieve the uniformity of the temperatures within the backup structure. The backup structure ventilation system is about to begin its detailed design work, which will be followed by the procurement and installation of the ventilation fans and ducts for the backup structure. The plan calls for the installation to take place during the summer months of 2023.

2.2.2. Measurement of Structural Temperatures

The LMT was originally equipped with a modest set of 64 temperature sensors distributed over the entire structure. These sensors have been used to provide a better understanding of thermal effects on the telescope, but the coverage of the structure with sensors is not adequate to provide the resolution needed for structural thermal modeling. Therefore, an additional 256 sensors are being procured for installation, primarily for measurements within the antenna backup structure. The goal of the sensor system is to achieve 0.1 degrees C accuracy. This can be realized with four-wire Pt100 Resistance Temperature Detector (RTD) sensors if care is taken in the calibration of both the sensor and the readout system. Sensor procurement is now underway with the expectation of the delivery of the sensors in 2023. Sensor installation will then be able to begin in early 2023 and should be completed by the spring.

Structural temperatures may be used with finite-element models to predict the deformation of the antenna structure. The LMT's original finite-element model has been reviewed and updated to allow more accuracy in predictions of the structure's thermal behavior. The model is being used to compute a set of thermal cases that can be used to predict the surface deformation given an arbitrary pattern of temperatures within the structure. After the installation of the new temperature sensors on the antenna, the system will be commissioned against direct observation of surface deformations made using photogrammetry.

2.2.3. Astronomical Measurements of Surface Shape

The out-of-focus (OOF) holography technique has been used successfully at other telescopes to measure the shape of the primary surface. At the LMT, past studies of the thermal deformation of the surface shape have shown that the dominant mode of deformation is an astigmatism. The magnitude of the effect varies from night to night and within a night. We have successfully used astronomical measurements to estimate the amount of astigmatism by using the active surface to introduce intentional deformations, following the Zernike vertical astigmatism polynomial, to find the optimum shape. The technique works well and must be used at high frequencies to obtain the best results. With OOF, we can, in principle, solve for a greater number of Zernike modes to achieve better surface accuracy. The LMT has installed OOF software on the antenna and began testing during the first half of 2022 (see Figure 5). System commissioning will be completed during the fall of 2022.

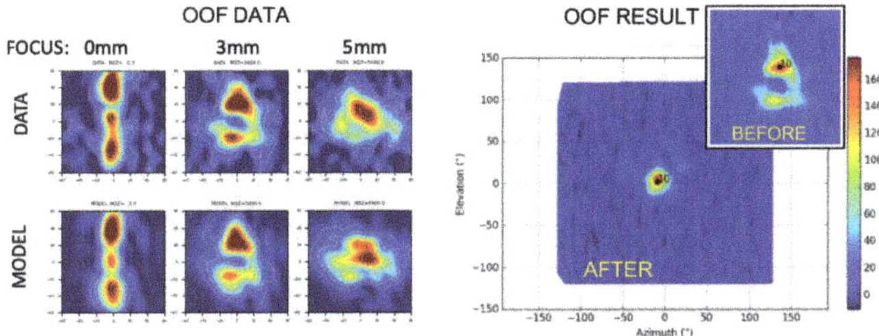

Figure 5. Example of OOF at the LMT during the daytime when surface deformations are significant. The panels on the left show the OOF data, consisting of beam maps at three positions of the secondary mirror. The upper row shows the actual data, and the lower row shows the OOF model fit. The right-hand panel shows the beam pattern after correction with the original beam shown in the inset figure.

2.2.4. Large Aperture Surface Error Recovery System

The techniques for determining the primary surface deformations of the LMT mentioned above have limitations. Astronomical measurement of surface shape can take fifteen minutes or more to complete, and this means that they can struggle to converge to a solution during times when the structure changes rapidly. Inferences about structural changes based on the temperature of the structural members require significant "calibration" of the technique and ultimately will only be as good as the underlying model. Clearly, the best solution is to directly measure, in real-time, the shape of the antenna and the relative positions of its optical elements. Thus, the LMT collaboration has embarked on the development of a new system: the Large Aperture Surface Error Recovery System (LASERS). The purpose of the LASERS project is to develop a system that will: (1) measure the LMT's low-order primary surface deformation so that the active surface may be used to remove transient deformations imposed by thermal gradients within the structure; and (2) measure the location of the LMT's secondary mirror with respect to the primary so that the relative position may be maintained to high accuracy. The system under development relies on an instrumental capability originally developed for monitoring alignments in large particle accelerators and is currently being built into large optical telescopes for the alignment of optical elements in the telescope structure. Our proposed alignment system is based on the Hexagon Absolute Multiline Technology (AMT) System, which provides a means to make highly accurate measurements of the distances between points on the antenna structure. Simulation studies (see Figure 6) suggest that LASERS will be able to remove low-spatial-order thermal deformations of the primary to a level of about 20 microns RMS over the surface.

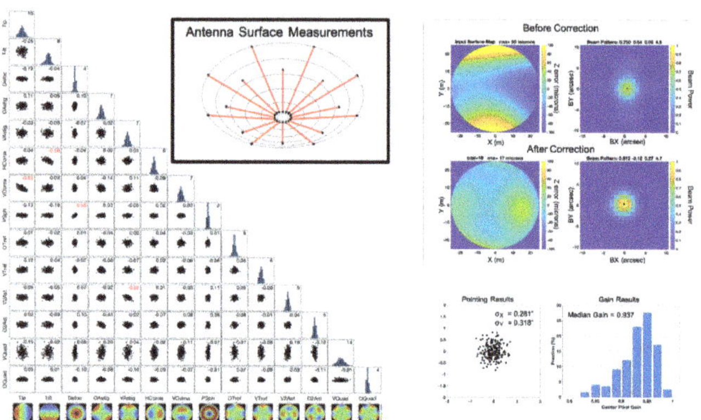

Figure 6. Monte Carlo simulation of LASERS, assuming that the surface is deformed from its parabolic shape along the optical axis using Zernike polynomials. The left panel shows the recovery of 14 Zernike coefficients from simulated distance measurements obtained according to the inset figure. The right-hand panel (upper) shows the improvement in the beam pattern after correction. The right panel (lower) shows the ability to recover possible pointing errors due to the misalignments within the primary and the distribution of errors in the recovered antenna gain.

Secondary position measurements will be made to an accuracy of about 10 microns RMS, which corresponds to a pointing error due to subreflector motions of 0.25 arcsec. The LASERS program has acquired the laser metrology device, which is currently undergoing lab tests. It is anticipated that the instrument will be sent to the LMT site for further testing by the end of 2022. Full system commissioning is expected to begin in the spring of 2023.

3. Performance Simulations

To evaluate the impact of LMT upgrades on the overall performance of both near-term EHT observations and the ngEHT, we utilize end-to-end simulations of actual observations with the VLBI array and reconstructions of ground-truth images and models. The simulations span a range of projected telescope performance parameters, (ng)EHT array configurations, and observing scenarios.

3.1. Synthetic Data

As of the most recent April 2022 observations, the EHT consists of 11 telescopes in nine distinct geographical locations: LMT on Volcan Sierra Negra in central Mexico, ALMA and the Atacama Pathfinder Experiment telescope (APEX) separated by $\sim$2 km in the Atacama Desert in Chile, the South Pole Telescope (SPT) at the South Pole, GLT in Thule, Greenland, the IRAM 30 m telescope on Pico Veleta (PV) in Spain, NOEMA in France, the Submillimeter Telescope Observatory in Arizona (SMT) on Mt. Graham and the 12 m radio telescope at Kitt Peak (KP) both in Arizona, USA, and the Submillimeter Array (SMA) and the James Clerk Maxwell Telescope (JCMT) co-located on Maunakea, Hawaii, USA.

The current EHT observes 4 GHz in dual-polarization in each of two sidebands of about 220.1 GHz and is currently commissioning non-simultaneous observations at 345 GHz observations for qualified sites for 2023. Details of the configuration of the EHT and nominal performance during the 2017 campaign are provided in [2,3].

The ngEHT concept is to supplement the current EHT by adding several new telescopes in key geographic locations, doubling the bandwidth at each observing frequency, and supporting simultaneous 230 and 345 GHz observations across the array [22]. For a hypothetical ngEHT array configuration, we adopted a selection of 10 additional candidate sites from [23], which were also used in the first ngEHT data analysis challenge [26]. The candidates sites are listed in Table 2.

Table 2. Candidates sites used in the hypothetical ngEHT array.

Site	Location
BAJA	Baja California, MX
BAR	California, USA
CNI	La Palma, ES-CN
CAT	Río Negro, AR
GAM	Khomas, NA
GARS	Antarctica
HAY	Massachusetts, USA
NZ	Canterbury, NZ
OVRO	California, USA
SGO	Santiago, CL

Station performance is based on an assumption of new 6 meter dishes installed at sites with historical records of valid local site weather and without an existing dish.

Simulated VLBI observations were generated using EHT-imaging [27,28] assuming 5 min scans were taken every 10 min. It is assumed that individual telescopes are amplitude-calibrated, but without known station phases, which is appropriate for high-frequency VLBI, where absolute phase calibration (e.g., from a nearby phase calibrator) is generally not possible.

3.2. Coverage and Daytime Operation

To test the contribution of LMT daytime observations to M 87 coverage with the ngEHT, we simulated a full-track (24-h) observation on the first day of each month of 2021. To illustrate this, Figures 7 and 8 show (left) a full-track observation and (right) the corresponding u–v plane during April and November of 2021, respectively. The orange markers show the baselines using the EHT array, while the blue markers show the baselines added to the EHT to create the ngEHT. It is clear from these figures that the ngEHT improved both the resolution and u–v coverage compared to the EHT for both months. Note that the November coverage illustrates how the baselines with the LMT almost doubled in time when daytime operations are possible after 16 UT. Thus, the ability to operate outside of a stable night-time window is important if the LMT is to serve as an anchor station throughout a multi-month monitoring campaign on M 87 during the winter months in the Northern Hemisphere.

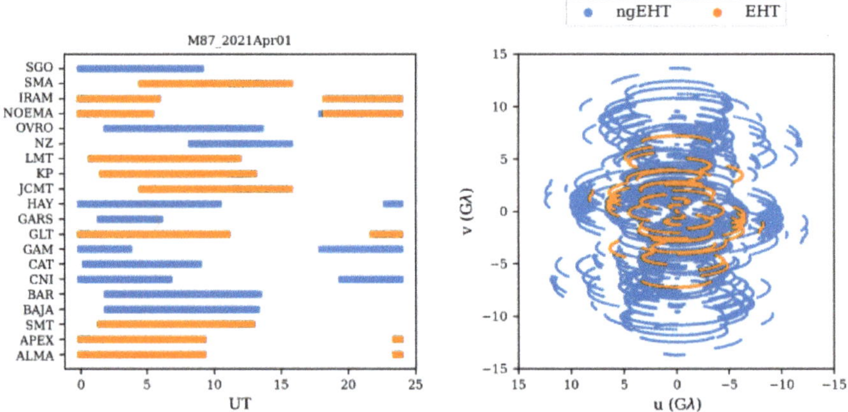

Figure 7. (**Left**) Stations that participated during a full track of observations and (**right**) u–v coverage on 1 April 2021, for M 87*. The orange markers show the observations and u–v points corresponding to the EHT. The blue markers are the ones that correspond to the added stations to form the ngEHT.

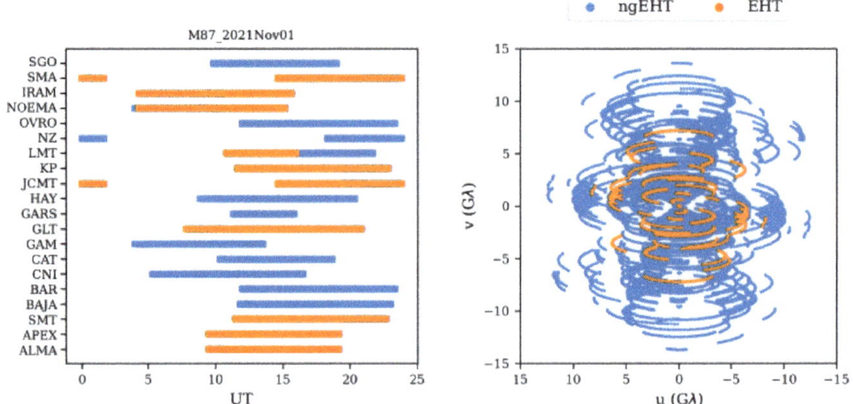

Figure 8. Same as in Figure 7, but for 1 November 2021. Note that, for the LMT, the observation finished at 16:00 UT, which corresponds to the end of night-time at 10:00 a.m. local time.

For Sgr A*, we are interested in the snapshot ability to reconstruct the source given that Sgr A* has characteristic dynamical timescales of minutes. In this case, we cannot rely on Earth's rotation over a full track to build coverage for static imaging. We determined the UT hour with the maximum number of baselines to the LMT for a given day to evaluate snapshot imaging capability with the LMT.

Figure 9 shows the active stations and the u–v coverage for a single-hour snapshot. With respect to the EHT 2022 coverage, the ngEHT is much better at filling the u–v plane through the combined benefit of double the number of stations and simultaneous multi-frequency coverage.

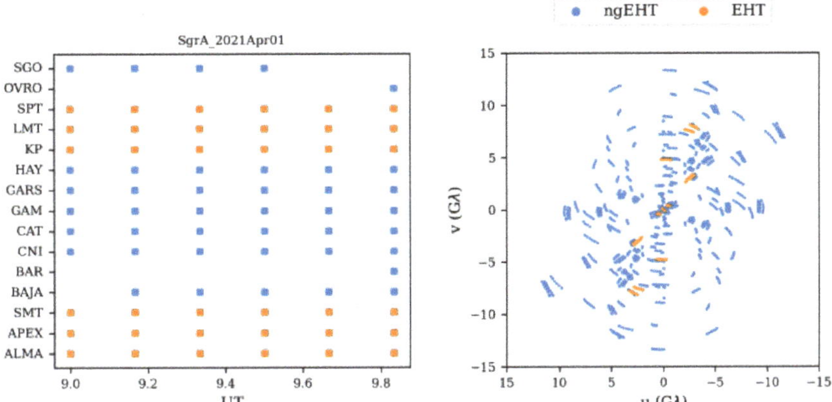

Figure 9. (**Left**) Stations that participate during an hour of observations with the best (**right**) u–v coverage on 1 April 2021, for Sgr A*. The orange markers show the observations and u–v points corresponding to the EHT, while the blue markers are the ones that correspond to the added stations and frequency coverage from the hypothetical ngEHT configuration.

3.3. Source Models

For our reconstructions, we used two GRMHD models, one for each source: M 87 and Sgr A*. Both models are ray-traced at 230 GHz and 345 GHz.

The M 87 model is a GRMHD simulation from [29]. It assumes a mass of $6.2 \times 10^9 M_\odot$ at a distance of 16.7 Mpc and a dimensionless spin $a_* = 0.9375$. This simulation allows the ions and electrons to evolve self-consistently in a MAD disk, giving, as a result, a two-temperature evolution. This simulation can reproduce a wide opening angle of the jet consistent with VLBI observations at 43 and 86 GHz [30]. In addition, it is able to produce a jet with a power of $\sim 10^{43}$ erg s^{-1} close to the correct estimated value of M 87.

The Sgr A* model, prepared by Christian Fromm, is a semi-analytic stationary RIAF model at an inclination of $i = 130°$ with $a_* = 0$. The model is scaled to have a mass of $4.14 \times 10^6 M_\odot$ at a distance of 8.178 kpc, consistent with [16].

3.4. Image Reconstruction

Once the source model is selected and a synthesized observation is created, we start the image reconstruction. We use the software package EHT-imaging [27,31] for imaging the synthetic data. EHT-imaging uses forward modeling and gradient descent to fit a regularized maximum likelihood image model to the u–v data or derived data products. In this work, we fit a combination of the closure phases, the log-closure amplitudes, and the visibility amplitudes accounting for a fractional gain error. EHT-imaging makes use of regularization to find a solution to the generally underconstrained sparse imaging problem. We use a combination of a simple entropy regularizer, total variation and total squared variation (smoothness), an image sparsity constraint (l1), and a preference for overall spatial compactness.

For M 87 reconstructions, the starting image for optimization, as well as the prior for the simple entropy regularizer are composed of two concentric Gaussians: (1) a central Gaussian with FWHM = 50 µas with an integrated flux equal to 90% of the model's total flux of the model and (2) a diffuse Gaussian with FWHM = 50 µas with an integrated flux equal to 25% of the model's total flux. The large diffuse Gaussian simulates the diffuse jet emission we expect from this source without assuming any particular direction. For Sgr A* case, the prior is a simple Gaussian with FWHM = 50 µas and an integrated flux equal to the model's integrated flux.

EHT-imaging supports simultaneous multi-frequency reconstruction by solving for the image at a reference frequency (in our case, 230 GHz) and solving for a corresponding map of the spectral index [32]. We utilize this unique capability in the case of ngEHT dual-frequency observations, where the models are typically slightly different at both frequencies since there is some shallow, but non-zero spectral index.

3.5. Image Fidelity

To evaluate the quality of our image reconstructions, we introduce a figure-of-merit for image dynamic range, which is responsive to the non-uniform image noise that is characteristic of sparse image reconstruction under various image priors. The metric is based on the pixel distribution of residual noise after subtracting the model ground truth from the reconstruction:

$$\delta I = \left\| \frac{I_{recon} - I_{model}}{I_{max}} \right\| \tag{1}$$

where I_{recon} is the intensity of the reconstruction, I_{model} is the intensity of the model, and I_{max} is the peak intensity of the model. To apply this metric to our reconstructions, we align the reconstructions with the model and blur them with a circular beam of 3 µas before calculating the model residual.

The top row of Figure 10 shows the M 87 model (left) and the reconstructions using the EHT (middle) and ngEHT (right). The ngEHT reconstruction shows a more detailed and faithful image than the EHT 2022 reconstruction. The bottom row of Figure 10 shows the residual noise of the prior (left), the EHT reconstruction (middle), and the ngEHT reconstruction (right). The prior residual noise highlights the quality of the initial guess for the model and can be compared with improvements from the reconstructions. The EHT residual noise traces the fine features in the jet that are not observed in the reconstruction, while the ngEHT residual noise looks more noise-like in the jet region. The latter shows the extent to which the ngEHT array can distinguish fine structure within the jet and the photon ring at a higher fidelity given the improved bandwidth, coverage, and frequency information.

Likewise, Figure 11 shows an improvement with the ngEHT array for the case of Sgr A* snapshot imaging. The improvement is more evident than in the M 87 case by showing a ring-like feature for a single snapshot using the ngEHT. This improvement is an encouraging step toward tracking rapid dynamical features in Sgr A* with an expanded array.

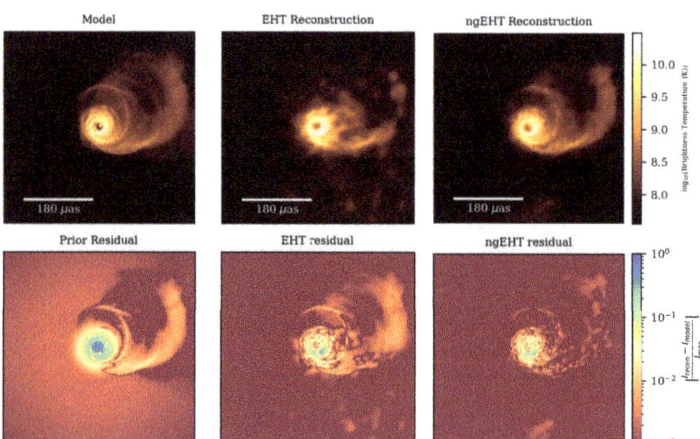

Figure 10. Top row: The model (**left**), EHT reconstruction (**middle**), and ngEHT reconstruction (**right**) 230 GHz images for the 1 April 2021 observations shown in Figure 7 of M 87*. Bottom row: Plot of the normalized residual noise of the model with the prior (**left**), the EHT reconstruction (**middle**), and the ngEHT reconstruction. For these reconstructions, the ngEHT residuals in the faint jet region approach $\sim 10^{-3}$ of the maximum brightness.

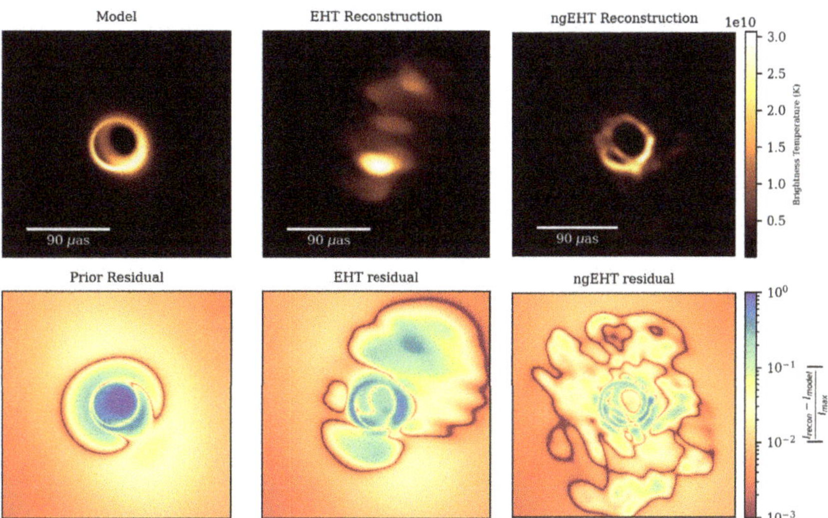

Figure 11. Same as in Figure 10, but using the snapshot reconstructions of Sgr A* on April for a 1-h period with maximum coverage. Note that the reconstructions are in a linear scale, while the residuals are in a log scale.

The distribution of residual noise can be compared via its cumulative function, as shown in Figures 12 and 13. In these cumulative distributions, we begin with the largest values of the residual error (which usually correspond to the brightest regions of the image) and record the residual error as a function of the fraction of the image field-of-view (FOV). For these plots, we also include a "model residual" (blue line) when compared to a blank image ($I_{recon} = 0$), as well as the model residual to the original starting prior for imaging ($I_{recon} = I_{prior}$, orange line) to serve as useful reference curves. When the residual curve falls below the "model residual", it shows that the reconstruction is accurately reflecting the brightness distribution of the model, and when it falls below the "prior residual", it is making a more accurate representation than the starting guess.

Figure 12 shows that the prior residual (orange) is limited in detecting the fine structure, as expected. The EHT (green) and ngEHT (red) residuals are below the model; however, we can see an ~ 1 dex improvement between both cases. The distance between the ngEHT residual noise curve and the EHT curve on the logarithmic scale is roughly the same as the distance between the EHT curve and the original model, showing a progressive improvement to the simulated reconstruction fidelity when going from EHT's 2022 configuration to ngEHT.

Figure 13 shows that the snapshot EHT residual is dominated by noise and cannot generally reproduce the model, as is clear from the reconstruction image itself. This implies that a single instantaneous observation by the EHT lacks the coverage necessary to image our simple model reliably. The ngEHT snapshot residual can follow the model's shape given the much-expanded coverage provided by the ngEHT.

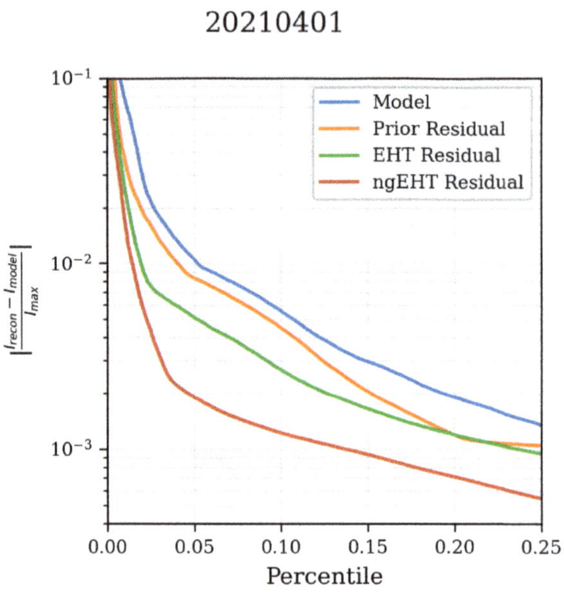

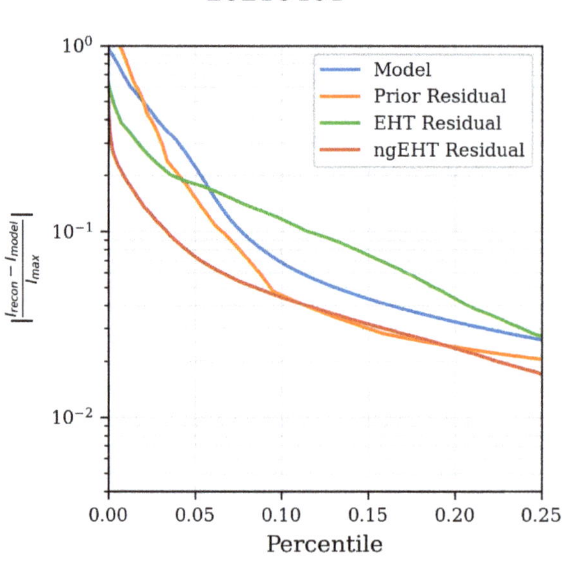

Figure 12. Cumulative distribution of the residual noise of M 87's prior (orange), EHT reconstruction (green), and ngEHT reconstruction (red). The x-axis is the percentile of the residual image intensity. The model (blue) corresponds to a normalized distribution of the model image itself, i.e., $I_{recon} = 0$ in Equation (1).

Figure 13. Same as Figure 12, but for Sgr A* snapshot imaging. Note that the y-axis (residual noise) has a different range than the M 87 one to highlight the range of interest for the different imaging scenarios (full track vs. snapshot).

4. Conclusions

In this white paper, we outlined a program of upgrades to the LMT, which are scheduled to come online before or on the timescale of the ngEHT. In addition to the simultaneous 230 + 345 GHz capabilities and quadrupling of the total on-sky bandwidth with respect to current observations, the series of upgrades include enhanced thermal stabilization through improved ventilation, measurements, and modeling, as well as a laser metrology system that can be used for real-time measurement and active correction of the surface. The improvements will allow the LMT to operate near its theoretical performance over a wide range of conditions, relaxing various current constraints on observing (such as night-time) and allowing the 50-meter dish to serve as a critical anchor station for ngEHT observations throughout much of the year.

Using the expected performance of an improved LMT as part of a hypothetical ngEHT configuration comprised of 10 additional stations alongside the current EHT 2022 array, we investigated the imaging performance of an ngEHT array using full-track simulated observations of M 87, as well as snapshot observations of Sgr A*, meant to reflect the ability of the array to form movies of both targets on their respective dynamical timescales. We introduced a metric to evaluate image fidelity based on the distribution of residual noise of simulated reconstructions, which can be used as one representation of the image dynamic range. Within a reasonable region of interest, the ngEHT is able to reconstruct structures of $\sim 10^{-3}$ peak brightness in our full-track M 87 simulations and approximately a few 10^{-2} peak brightness in snapshot reconstructions of Sgr A*.

For future work, we plan to include seasonal weather variation and limited daytime operation for the full-track observations of M 87* to investigate weather-related effects relevant to a multi-month observing campaign. Furthermore, we intend to test the effect on reconstruction fidelity when various other anchor stations (such as ALMA and NOEMA) are not able to participate, and we plan to characterize the performance of subsets of the array, which may be available during portions of an observing cycle.

Author Contributions: Conceptualization, S.B., L.B., G.N., F.P.S. and D.H.; methodology, S.B., L.B., G.N. and F.P.S.; writing—original draft preparation, S.B., L.B., G.N. and F.P.S.; writing—review and editing, S.B., L.B., G.N., F.P.S. and D.H. All authors have read and agreed to the published version of the manuscript.

Funding: This work was supported by the Black Hole Initiative at Harvard University, which is funded by grants from the John Templeton Foundation and the Gordon and Betty Moore Foundation to Harvard University. This work was also supported by National Science Foundation Grants AST-1935980 and AST-2034306 and the Gordon and Betty Moore Foundation (GBMF-10423). The LMT LASERS project is supported by National Science Foundation Grant AST-2117422.

Institutional Review Board Statement: Not applicable.

Informed Consent Statement: Not applicable.

Data Availability Statement: Not applicable.

Acknowledgments: We thank David Sanchez, Freek Roelofs, Michael Johnson, Andrew Chael, Richard Anantua, Paul Tiede, Dominic Pesce, Angelo Ricarte, Jose Gomez, and the EHT and ngEHT working groups for stimulating discussion on topics related to this work.

Conflicts of Interest: The authors declare no conflict of interest.

Abbreviations

The following abbreviations are used in this manuscript:

AGN	active galactic nuclei
ALMA	Atacama Large Millimeter Array
APEX	Atacama Pathfinder Experiment
EHT	Event Horizon Telescope
EVPA	electric-vector position angle
FOV	field-of-view
GLT	Greenland Telescope
GRMHD	general relativity and magneto-hydrodynamical
ISCO	innermost stable circular orbit
JCMT	James Clerk Maxwell Telescope
JPL	Jet Propulsion Laboratory
KP	Kitt Peak 12 m Telescope
LMT	Large Millimeter Telescope
MAD	magnetically arrested accretion flow
ngEHT	next-generation Event Horizon Telescope
OMAyA	One Millimeter Array for Astronomy
PdB	NOEMA on Plateau de Bure
PV	IRAM 30 m Telescope
RIAF	radiatively inefficient accretion flow
SMBH	super massive black hole
SMA	Submillimeter Array
SMT	Submillimeter Telescope
SPT	South Pole Telescope
STM	science traceability matrix
VLBI	Very Long Baseline Interferometer

References

1. Akiyama, K. et al. [The Event Horizon Telescope Collaboration]. First M87 Event Horizon Telescope Results. I. The Shadow of the Supermassive Black Hole. *Astrophys. J. Lett.* **2019**, *875*, L1. [CrossRef]
2. Akiyama, K. et al. [The Event Horizon Telescope Collaboration]. First M87 Event Horizon Telescope Results. II. Array and Instrumentation. *Astrophys. J. Lett.* **2019**, *875*, L2. [CrossRef]
3. Akiyama, K. et al. [The Event Horizon Telescope Collaboration]. First M87 Event Horizon Telescope Results. III. Data Processing and Calibration. *Astrophys. J. Lett.* **2019**, *875*, L3. [CrossRef]
4. Akiyama, K. et al. [The Event Horizon Telescope Collaboration]. First M87 Event Horizon Telescope Results. IV. Imaging the Central Supermassive Black Hole. *Astrophys. J. Lett.* **2019**, *875*, L4. [CrossRef]
5. Akiyama, K. et al. [The Event Horizon Telescope Collaboration]. First M87 Event Horizon Telescope Results. V. Physical Origin of the Asymmetric Ring. *Astrophys. J. Lett.* **2019**, *875*, L5. [CrossRef]
6. Akiyama, K. et al. [The Event Horizon Telescope Collaboration]. First M87 Event Horizon Telescope Results. VI. The Shadow and Mass of the Central Black Hole. *Astrophys. J. Lett.* **2019**, *875*, L6. [CrossRef]
7. Akiyama, K. et al. [The Event Horizon Telescope Collaboration]. First M87 Event Horizon Telescope Results. VII. Polarization of the Ring. *Astrophys. J. Lett.* **2021**, *910*, L12. [CrossRef]
8. Akiyama, K. et al. [The Event Horizon Telescope Collaboration]. First M87 Event Horizon Telescope Results. VIII. Magnetic Field Structure near The Event Horizon. *Astrophys. J. Lett.* **2021**, *910*, L13. [CrossRef]
9. Akiyama, K. et al. [The Event Horizon Telescope Collaboration]. First Sagittarius A* Event Horizon Telescope Results. I. The Shadow of the Supermassive Black Hole in the Center of the Milky Way. *Astrophys. J. Lett.* **2022**, *930*, L12. [CrossRef]
10. Akiyama, K. et al. [The Event Horizon Telescope Collaboration]. First Sagittarius A* Event Horizon Telescope Results. II. EHT and Multiwavelength Observations, Data Processing, and Calibration. *Astrophys. J. Lett.* **2022**, *930*, L13. [CrossRef]
11. Akiyama, K. et al. [The Event Horizon Telescope Collaboration]. First Sagittarius A* Event Horizon Telescope Results. III. Imaging of the Galactic Center Supermassive Black Hole. *Astrophys. J. Lett.* **2022**, *930*, L14. [CrossRef]
12. Akiyama, K. et al. [The Event Horizon Telescope Collaboration]. First Sagittarius A* Event Horizon Telescope Results. IV. Variability, Morphology, and Black Hole Mass. *Astrophys. J. Lett.* **2022**, *930*, L15. [CrossRef]
13. Akiyama, K. et al. [The Event Horizon Telescope Collaboration]. First Sagittarius A* Event Horizon Telescope Results. V. Testing Astrophysical Models of the Galactic Center Black Hole. *Astrophys. J. Lett.* **2022**, *930*, L16. [CrossRef]
14. Akiyama, K. et al. [The Event Horizon Telescope Collaboration]. First Sagittarius A* Event Horizon Telescope Results. VI. Testing the Black Hole Metric. *Astrophys. J. Lett.* **2022**, *930*, L17. [CrossRef]

15. Gebhardt, K.; Adams, J.; Richstone, D.; Lauer, T.R.; Faber, S.M.; Gültekin, K.; Murphy, J.; Tremaine, S. The Black Hole Mass in M87 from Gemini/NIFS Adaptive Optics Observations. *Astrophys. J.* **2011**, *729*, 119. [CrossRef]
16. GRAVITY Collaboration. A Geometric Distance Measurement to the Galactic Center Black Hole with 0.3% Uncertainty. *Astron. Astrophys.* **2019**, *625*, L10. [CrossRef]
17. Kim, J.Y.; Lu, R.S.; Krichbaum, T.; Bremer, M.; Zensus, J.; Walker, R. Resolving the Base of the Relativistic Jet in M87 at 6Rsch Resolution with Global mm-VLBI. *Galaxies* **2016**, *4*, 39. [CrossRef]
18. Wielgus, M.; Marchili, N.; Martí-Vidal, I.; Keating, G.K.; Ramakrishnan, V.; Tiede, P.; Fomalont, E.; Issaoun, S.; Neilsen, J.; Nowak, M.A.; et al. Millimeter Light Curves of Sagittarius A* Observed during the 2017 Event Horizon Telescope Campaign. *Astrophys. J. Lett.* **2022**, *930*, L19. [CrossRef]
19. Janssen, M.; Falcke, H.; Kadler, M.; Ros, E.; Wielgus, M.; Akiyama, K.; Baloković, M.; Blackburn, L.; Bouman, K.L.; Chael, A.; et al. Event Horizon Telescope Observations of the Jet Launching and Collimation in Centaurus A. *Nat. Astron.* **2021**, *5*, 1017–1028. [CrossRef]
20. Kim, J.Y.; Krichbaum, T.P.; Broderick, A.E.; Wielgus, M.; Blackburn, L.; Gómez, J.L.; Johnson, M.D.; Bouman, K.L.; Chael, A.; Akiyama, K.; et al. Event Horizon Telescope Imaging of the Archetypal Blazar 3C 279 at an Extreme 20 Microarcsecond Resolution. *Astron. Astrophys.* **2020**, *640*, A69. [CrossRef]
21. Issaoun, S.; Wielgus, M.; Jorstad, S.; Krichbaum, T.P.; Blackburn, L.; Janssen, M.; Chan, C.k.; Pesce, D.W.; Gómez, J.L.; Akiyama, K.; et al. Resolving the Inner Parsec of the Blazar J1924-2914 with the Event Horizon Telescope. *Astrophys. J.* **2022**, *934*, 145. [CrossRef]
22. Doeleman, S.; Blackburn, L.; Dexter, J.; Gomez, J.L.; Johnson, M.D.; Palumbo, D.C.; Weintroub, J.; Farah, J.R.; Fish, V.; Loinard, L.; et al. Studying Black Holes on Horizon Scales with VLBI Ground Arrays. *Bull. Am. Astron. Soc.* **2019**, *51*, 256. [CrossRef]
23. Raymond, A.W.; Palumbo, D.; Paine, S.N.; Blackburn, L.; Rosado, R.C.; Doeleman, S.S.; Farah, J.R.; Johnson, M.D.; Roelofs, F.; Tilanus, R.P.J.; et al. Evaluation of New Submillimeter VLBI Sites for the Event Horizon Telescope. *Astrophys. J. Suppl. Ser.* **2021**, *253*, 5. [CrossRef]
24. Hughes, D.H.; Schloerb, F.P.; Aretxaga, I.; Castillo-Domínguez, E.; Chávez Dagostino, M.; Colín, E.; Erickson, N.; Ferrusca Rodriguez, D.; Gale, D.M.; Gómez-Ruiz, A.; et al. The Large Millimeter Telescope (LMT) Alfonso Serrano: Current Status and Telescope Performance. In *Proc. SPIE 11445, Ground-based and Airborne Telescopes VIII*; 2020, *online*. [CrossRef]
25. Ortiz-León, G.N.; Johnson, M.D.; Doeleman, S.S.; Blackburn, L.; Fish, V.L.; Loinard, L.; Reid, M.J.; Castillo, E.; Chael, A.A.; Hernández-Gómez, A.; et al. The Intrinsic Shape of Sagittarius A* at 3.5 mm Wavelength. *Astrophys. J.* **2016**, *824*, 40. [CrossRef]
26. Roelofs, F.; Blackburn, L.;, Lindahl, G.; Doeleman, S.S.; Johnson, M.D.; Arras, P.; Chatterjee, K.; Emami, R.; Fromm, C.; Fuentes, A.; et al. The ngEHT Analysis Challenges *Galaxies* **2023**, *submitted*.
27. Chael, A.A.; Johnson, M.D.; Narayan, R.; Doeleman, S.S.; Wardle, J.F.C.; Bouman, K.L. High-resolution Linear Polarimetric Imaging for the Event Horizon Telescope. *Astrophys. J.* **2016**, *829*, 11. [CrossRef]
28. Chael, A.; Rowan, M.; Narayan, R.; Johnson, M.; Sironi, L. The Role of Electron Heating Physics in Images and Variability of the Galactic Centre Black Hole Sagittarius A*. *Mon. Not. R. Astron. Soc.* **2018**, *478*, 5209–5229. [CrossRef]
29. Chael, A.; Narayan, R.; Johnson, M.D. Two-temperature, Magnetically Arrested Disc simulations of the jet from the supermassive black hole in M87. *Mon. Not. R. Astron. Soc.* **2019**, *486*, 2873–2895. Available online: http://xxx.lanl.gov/abs/1810.01983 (accessed on 14 December 2022). [CrossRef]
30. Walker, R.C.; Hardee, P.E.; Davies, F.B.; Ly, C.; Junor, W. The Structure and Dynamics of the Subparsec Jet in M87 Based on 50 VLBA Observations over 17 Years at 43 GHz. *ApJ* **2018**, *855*, 128. [CrossRef]
31. Chael, A.A.; Johnson, M.D.; Bouman, K.L.; Blackburn, L.L.; Akiyama, K.; Narayan, R. Interferometric Imaging Directly with Closure Phases and Closure Amplitudes. *Astrophys. J.* **2018**, *857*, 23. [CrossRef]
32. Chael, A.; Issaoun, S.; Pesce, D.w.; Johnson, M.D.; Ricarte, A.; Fromm, C.M.; Mizuno, Y. Multi-frequency Black Hole Imaging for the Next-Generation Event Horizon Telescope. *Astrophys. J.* **2022**, *submitted*. [CrossRef]

Article

The Haystack Telescope as an Astronomical Instrument

Jens Kauffmann, Ganesh Rajagopalan, Kazunori Akiyama, Vincent Fish, Colin Lonsdale, Lynn D. Matthews and Thushara G.S. Pillai

Haystack Observatory, Massachusetts Institute of Technology, 99 Millstone Rd., Westford, MA 01886, USA
* Correspondence: jens.kauffmann@mit.edu

Abstract: The Haystack Telescope is an antenna with a diameter of 37 m and an elevation-dependent surface accuracy of $\leq$100 μm that is capable of millimeter-wave observations. The radome-enclosed instrument serves as a radar sensor for space situational awareness, with about one-third of the time available for research by MIT Haystack Observatory. Ongoing testing with the K-band (18–26 GHz) and W-band receivers (currently 85–93 GHz) is preparing the inclusion of the telescope into the Event Horizon Telescope (EHT) array and the use as a single-dish research telescope. Given its geographic location, the addition of the Haystack Telescope to current and future versions of the EHT array would substantially improve the image quality.

Keywords: Very Long Baseline Interferometry; radio astronomy; millimeter astronomy; radio telescopes; high angular resolution; astronomical instrumentation

Citation: Kauffmann, J.; Rajagopalan, G.; Akiyama, K.; Fish, V.; Lonsdale, C.; Matthews, L.D.; Pillai, T.G.S. The Haystack Telescope as an Astronomical Instrument. *Galaxies* **2023**, *11*, 9. https://doi.org/10.3390/galaxies11010009

Academic Editor: Lorenzo Amati

Received: 16 November 2022
Revised: 16 December 2022
Accepted: 20 December 2022
Published: 4 January 2023

1. Introduction: Astronomy Observations with the Haystack Telescope

MIT Haystack Observatory has been a home to a radome-enclosed telescope of 37 m diameter since 1964 [1][1]. Figure 1 illustrates the siting of the instrument, while Figure 2 presents an overview of the dish. The original system was primarily conceived as a space radar and as a platform for telecommunications experiments to support work by MIT Lincoln Laboratory. Ownership was transferred to the Northeast Radio Observatory Corporation[2] (NEROC) in 1970, with the goal to enable millimeter-wave observations for the astronomy community in the Northeast US, while still being available as a radar sensor to MIT Lincoln Laboratory. The site is known as MIT Haystack Observatory since that time. The telescope has undergone several upgrades since its original dedication. Some of this work focused on improving the surface accuracy of the dish, which was improved from an initial root-mean-square (RMS) error of $\sim$900 μm to $\sim$200 μm after 1992 [2].

The Haystack Telescope has enabled key scientific discoveries, as summarized by Whitney et al. [3]. Radar observations delivered key intelligence on the Apollo landing sites, and joint observations with the Westford Telescope—a dish of 18 m diameter located about a mile from the Haystack Telescope—produced the first radar maps of Venus' surface that cleanly separated radar echoes from the planet's northern and southern hemispheres. Radar observations of Venus and Mercury also delivered stringent tests of General Relativity by constraining the gravitational time delay caused by the presence of the Sun (i.e., Shapiro delay, [4]). Single-dish spectroscopy observations with the Haystack Telescope were essential in establishing "dense molecular cores" as the key star-forming sites in molecular clouds [5], and they showed how dense cores build up density as they contract out of more diffuse cloud material [6]. MIT Haystack Observatory led the way during the inception of Very Long Baseline Interferometry (VLBI), and 8 of the 22 awardees of the 1971 Rumford Prize of the American Academy of Arts and Sciences for the inception of VLBI were working at the observatory. The Haystack Telescope critically supported this work. VLBI observations with the instrument, such as the discovery of apparent superluminal motion in quasars [7], shaped our understanding of the universe at high angular resolution.

Figure 1. Aerial view of MIT Haystack Observatory. The Haystack Telescope, a dish of 37 m diameter, is located in the large radome dominating the foreground. (Used with permission, courtesy of Mark Derome).

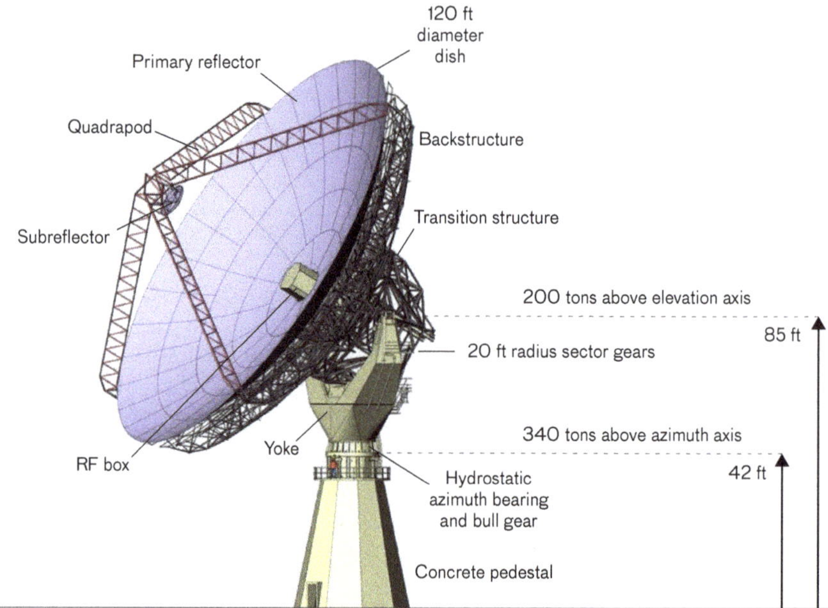

Figure 2. Overview of the Haystack Telescope, with the radome removed [8]. The receiver equipment is installed in the "RF box", a container that is brought down to ground level during "box-down" periods to enable major engineering activities. (Reprinted with permission, courtesy of MIT Lincoln Laboratory, Lexington, MA, USA).

A major upgrade, completed in 2014, improved the surface accuracy to ≤ 100 μm, depending on elevation. This work, executed by MIT Lincoln Laboratory under sponsorship by the Defense Advanced Research Projects Agency (DARPA) and the U.S. Air Force, was part of the upgrade delivering the Haystack Ultrawideband Satellite Imaging Radar (HUSIR). The HUSIR system is designed around a W-band radar covering a substantial bandwidth of 92–100 GHz, and it also includes an X-band radar operating at 9.5–10.5 GHz. The outstanding bandwidth of $\Delta \nu = 8$ GHz enables the W-band radar to directly resolve structures of $c/(2 \cdot \Delta \nu) = 1.9$ cm size in range [9], with advanced image processing techniques delivering an effective resolution well below this scale. In 2014, HUSIR's W-band radar delivered the finest spatial resolution of any imaging radar, while the X-band radar constituted the only system for imaging out to geosynchronous orbits [9]. The systems have been upgraded since, and HUSIR continues to be an essential contributing sensor for space situational awareness.

Today, NEROC has access to about one-third of the time available on the Haystack Telescope. This time can be used to conduct experiments in astronomy and other fields of fundamental research. The primary access windows are weekends, and night hours at 23:00–07:00 local time on Mon.–Fri. Access to other periods, as for example needed for time-critical experiments in astronomy, is coordinated with MIT Lincoln Laboratory. Such work can currently use K-band (18–26 GHz) and W-band receivers (85–93 GHz) dedicated to astronomical observations that are separate from the HUSIR systems. An existing Q-band system covering 36–50 GHz will be brought online in the future.

MIT Haystack Observatory currently studies the expected performance of the telescope at ∼230 GHz. This is done as part of the ngEHT project (as described elsewhere in this special issue; also see https://www.ngeht.org, accessed on 2022 Dec. 15), which seeks to deliver a "next generation EHT" by adding new stations and other capabilities to the Event Horizon Telescope (EHT; see [10] for a recent description of the system). Inclusion of the Haystack Telescope into the EHT would enhance the imaging capabilities of the array, as described below.

The upgraded dish provides exciting opportunities for astronomy. Unfortunately, until recently it was not possible to make use of the telescope's capabilities, given the lack of substantial and systematic funding for astronomical experiments. This has changed in the past few years, thanks to a private donation and a grant from the National Science Foundation supporting the ngEHT project (AST-1935980). The telescope is currently regularly used to conduct experiments in support of system commissioning and initial experiments into astrophysical research and education. This includes three VLBI runs at 86 GHz that have delivered fringe detections on intercontinental baselines.

This paper is organized as follows. The telescope, its current and future instrumentation, and the characteristics of the site are described in Section 2. The discussion in Section 3 outlines the case for research, education, and technology development on the Haystack Telescope. The connection of the telescope to the EHT and ngEHT projects is described in Section 4. The material is summarized in Section 5.

2. Telescope, Instrumentation, and Site

2.1. Telescope and Site

Figure 2 summarizes the characteristics of the dish. The reflector of the Haystack Telescope has a diameter of 120 ft, equivalent to 36.57 m. It is formed by 432 panels that each have an RMS surface accuracy of about 28 μm [8]. The main reflector itself is rigged to achieve am RMS surface accuracy of 75 μm at an elevation of 25°, with larger deformations occurring at higher or lower elevations [11]. The moving sections have a mass of 340 t, with 200 t of mass moving in elevation. The dish is capable of slewing at speeds of $5° \, \mathrm{s}^{-1}$ in azimuth and $2° \, \mathrm{s}^{-1}$ in elevation, and it achieves accelerations of $1°\!.5 \, \mathrm{s}^{-2}$ and $2° \, \mathrm{s}^{-2}$, respectively. By requirement, the pointing accuracy is $< 3''\!.6$, with a tracking accuracy $< 1''\!.8$ [11].

The telescope is housed in a radome of 150 ft diameter that was originally designed for use in extreme arctic environments and is capable of withstanding 130 mph winds (i.e., $210 \, \mathrm{km \, h}^{-1}$ or $60 \, \mathrm{m \, s}^{-1}$) [8]. The radome is skinned with three-ply ESSCOLAM 10 membranes with a hydrophobic coating, which are characterized by a transmission of about 95% at 90 GHz [8].

The receivers are housed in a "box" that is installed about 85 ft above ground. It can be brought down to the floor of the telescope building during dedicated "box-down" periods. The box houses radar equipment as well as the astronomy receivers, and it is very tightly packed with systems. As a consequence, major engineering actives can only be performed during a box-down window, during which the interior of the box can be accessed easily from all sides. The number and duration of box-down periods is minimized in support of high-priority radar observations.

The observatory's land is distributed over the Massachusetts towns of Groton, Tyngsborough, and Westford, with Westford being the administrative home of MIT Haystack Observatory. This thickly forested community is about an hour's drive away from downtown Boston (MA). The telescope itself is located at $42°62$ N vs. $71°49$ W, at an altitude of 130 m.

Haystack Observatory experiences extended periods of cold and dry weather during the winter, thus providing the weather conditions needed for observations at millimeter wavelengths. Historical measurements of the precipitable water vapor (PWV) column are available from the Suominet[3] network for atmospheric research. Archived data give a median PWV column of 8.3 mm for the period November 1 to April 30. Assuming an outside temperature of 0 °C, modeling of the atmosphere with the AM[4] radiative transfer code gives a corresponding optical depth of 0.12 at ∼86 GHz under such conditions, equivalent to an atmospheric transmission of 84% at 45° elevation. More realistically, observations by systems like the EHT are triggered in better-than-median atmospheric conditions. To give an example, the PWV column is below 5.3 mm for 25% of the winter period. Rich additional documentation about the telescope and the site can be found in Brown and Pensa [1], Whitney et al. [3], Waggener [8], Czerwinski and Usoff [9], Usoff et al. [11], MacDonald et al. [12], and Eshbaugh et al. [13].

2.2. Current Instrumentation

The telescope is equipped with receivers operating in the K (18–26 GHz), Q (36–50 GHz), and W bands (70–115 GHz). The cryogenic frontends operate at around 20 K in independent dewars. These are arranged roughly on a vertical line that is offset from the central focal point. MIT Lincoln Laboratory operates on-axis X-band and W-band radars, so that the three astronomy receivers are offset from boresight. The sub-reflector on a hexapod is remotely controlled to choose between the three K, Q and W-band receivers. Current observing projects make use of the K and W bands, and these receivers are therefore currently kept operational by engineering activities.

The K-band frontend is shared between MIT Haystack Observatory and MIT Lincoln Laboratory. One polarization is available for astronomical observations, while the other polarization is used for holography observations. Astronomical observations can be conducted anywhere in the frequency range of 18–26 GHz.

The W-band frontend is currently configured as a single-sideband receiver that senses horizontal and vertical polarization. Data are taken in a sideband of 8 GHz width that is set by an analog bandpass filter. The system is currently set up to observe at frequencies of 85–93 GHz. Modest upgrades to the hardware would allow to access the full frequency range of 70–115 GHz. The receiver was recently improved via the installation of a new wideband low -noise amplifier (LNA) and components for the sideband rejection scheme. These investments were made possible by an NSF MSRI-1 grant to the ngEHT project (AST-1935980).

The backends are located at the ground level of the telescope building. A radiofrequency-over-fiber (RFoF) system is used to transport the signals into this room. The RFoF infrastructure is currently being upgraded for transport bandwidths of up to 20 GHz for two polarizations. An up-down converter (UDC) is used to condition the signals for the backends. The single-dish backend currently processes up to 500 MHz in one polarization. Further investments in hardware and software would enable processing of larger bandwidths and of a second polarization. The backend measures continuum signals, and it currently also produces spectra of up to 500 Hz resolution. VLBI data are acquired using a ROACH2 digital backend (R2DBE) connected to a Mark 6 VLBI recorder. A Rakon Oven Controlled Crystal Oscillator (OCXO) is used as a frequency standard for ongoing engineering experiments in VLBI. The acquisition of the RFoF infrastructure, and the ongoing acquisition of a new OCXO, are supported by an NSF MSRI-1 grant to the ngEHT project (AST-1935980).

2.3. Current and Future Instrument Development

Current work on the Haystack Telescope focuses on evaluation of the newly upgraded system (i.e., after installation of the W-band LNA, RFoF system, and VLBI equipment). While characterization of the W-band system is the main activity, the K-band receiver is occasionally used to deliver complementary data on telescope performance under less ideal weather conditions. This program consists of single-dish observations of calibrators like planets, as well as participation in observations by VLBI networks. In the area of interferometry the goal is to enable future VLBI observations at $\lesssim$90 GHz, and to assess the feasibility of VLBI observations at $\sim$230 GHz in support of the EHT. More generally, the observations seek to demonstrate the capability of the Haystack Telescope to deliver exciting astrophysical research as a single-dish telescope and VLBI station.

Current funding from an NSF MSRI-1 grant (AST-1935980) supports the design of a receiver for VLBI observations with the Haystack Telescope at $\sim$230 GHz in the context of the ngEHT project. This undertaking might evolve into the design for a multi-band receiver enabling parallel observations at $\sim$86 GHz and $\sim$230 GHz. This depends on future decisions by the EHT and ngEHT projects concerning the need for multi-band observations in support of "frequency phase transfer" (FPT, [14]), i.e., the transfer of VLBI calibration information obtained at one frequency to other bands.

The long-term plan for the telescope foresees to support single-dish and VLBI observations at K, Q, and W band, as well as at $\sim$230 GHz. Ongoing investigations will clarify whether some or all of these receivers need to be able to observe in parallel (e.g., to support FPT). The installation of wideband (i.e., $\geq$8 GHz) receivers and backends for single-dish and VLBI observations, as well as of a maser clock as a frequency standard for VLBI experiments, form part of this long-term plan. The development of the telescope must be supported via dedicated grants from funding agencies, as the Haystack Telescope receives no general-purpose funding to advance the capabilities of the facility.

3. Astrophysical Research, Education, and Technology Development

Section 4 explains how the Haystack Telescope can add new, sensitive, and important baselines to the EHT. In the same way, the Haystack Telescope can complement the Global Millimeterwave VLBI Array (GMVA). The increased availability of multi-band receivers on radio observatories, as pioneered by the Korean VLBI Network (KVN) [15], raises the exciting prospect for the Haystack Telescope to join an intercontinental network building on FPT.

The outstanding scientific capabilities of large single-dish instruments at millimeter wavelengths are demonstrated by the high impact of current research on the IRAM 30m-telescope. Recent work with that telescope includes the study of star formation physics in nearby galaxies [16], and investigations of molecular cloud structure and evolution in the Milky Way that support the aforementioned extragalactic work [17,18]. Many of these studies employ the EMIR receiver system that samples 16 GHz of bandwidth per polarization in the 70–115 GHz range [19]. Installation of receiver and backend systems matching or exceeding this capability would open up new and exciting capabilities for the US-based community.

The Haystack Telescope offers unique, important, and currently missing educational opportunities in the US. The instrument is associated with the NEROC community of 13 research-intensive educational institutions in the Northeast US (see footnote 5 on page 8). Several of these are closely involved in the EHT, the Large Millimeter Telescope (LMT), and the Submillimeter Array (SMA). The Haystack Telescope is within easy driving distance from all these institutions, providing outstanding hands-on experiences for junior researchers within NEROC. More generally, the telescope can serve as a destination for educational astronomical daytrips that are independent from daytime, and only modestly dependent on the weather, by schools, colleges, and universities in six states of the US (all of MA, RI, CT; most of NH; parts of ME and NY)[5].

Operations of the Haystack Telescope are exclusively funded by grants with specific objectives and deliverables. MIT Haystack Observatory receives no general long-term funding that could make the instrument broadly accessible by the community. Future operations are expected to be supported via a mix of funding streams. This will include observations for specific user groups that will pay for having their data taken on the Haystack Telescope. The current engineering work undertaken on an NSF MSRI-1 grant supporting the ngEHT project constitutes one example for such observations. Similarly, NSF AAG grants could fund data collection for specific astrophysical research projects. It is the ambition of MIT Haystack Observatory to also make the telescope broadly available to the entire US community. The feasibility of such a program depends on the grants acquired by the observatory.

4. Connection to the Event Horizon Telescope: Current Work and Future Roles

Ongoing work on the Haystack Telescope is in part funded by an NSF MSRI-1 grant (AST-1935980). This award forms part part of the ngEHT project, and its goal is to evaluate the telescope for inclusion into the EHT at $\sim$230 GHz via quantitative modeling and VLBI test observations at $\sim$86 GHz. A private donation to MIT Haystack Observatory enables parallel activities that enhance the overall capabilities of the instrument.

Figure 3 shows that addition of the Haystack Telescope to future versions of the EHT network would produce new and critical baselines. This is specifically demonstrated here for a network that includes the telescopes that are now available for future EHT observations, but that also includes a set of additional antennas enhancing the EHT. This can be seen in Figure 3 (top right), where dishes enhancing the current EHT constellation are marked by green stars. The resulting significant improvements to the uv-coverage of the array can result in reductions of the inner sidelobes of the synthesized beam by a factor 1.4 (K. Akiyama, priv. comm.). This is indicated by imaging simulations assuming the reference array summarised in Figure 3 (top left). The simulation in particular demonstrates that the addition of the Haystack Telescope to the EHT array would substantially improve the sampling of the uv-domain at baselines of $\lesssim 4 \times 10^9 \, \lambda$ (Figure 3 bottom), resulting in the aforementioned reduction of sidelobes. Data from the Haystack Telescope can also improve other practical aspects of VLBI observations. For example, the dish can deliver an important connection between dishes in Europe and the Americas. The telescope can also add substantial sensitivity to VLBI networks, in particular at low frequencies where the telescope efficiencies are higher. All these factors aid in the overall calibration of VLBI networks, delivering advantages beyond the fundamental improvement in uv-coverage. Importantly, the specific model shown here demonstates that that the Haystack Telescope would still add substantial value to the EHT array when considering network configurations that include more telescopes than those used today. The ngEHT project is developing reference arrays for future quantitative array assessments by the collaboration. Such evaluations will in future also include the Haystack Telescope.

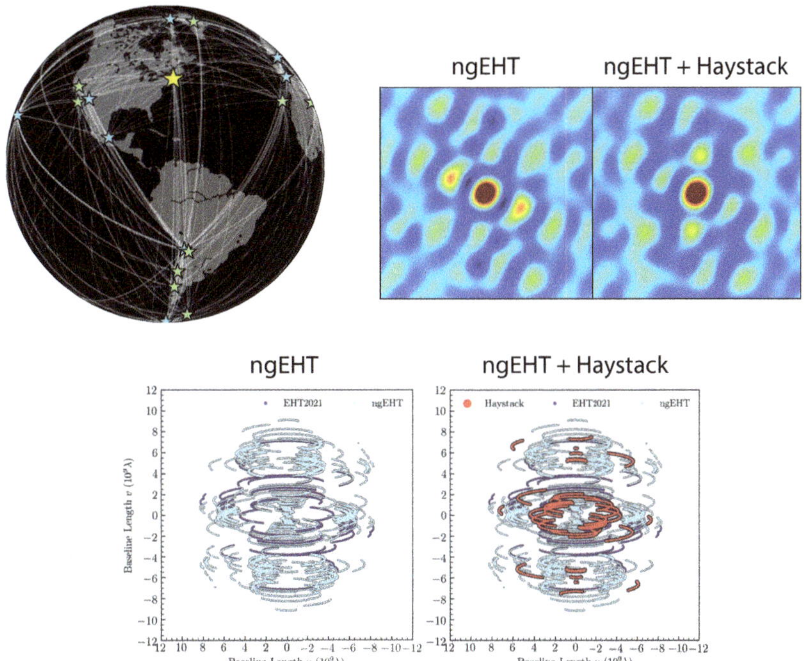

Figure 3. Outline of the impact of observations with the Haystack Telescope on VLBI arrays (K. Akiyama, priv. comm.). The **upper left panel** illustrates an ngEHT reference array used for imaging simulations, as appropriate for a target at a declination of $+10°$. Specifically, current EHT stations are marked by blue stars, while green stars indicate potential new sites. The Haystack Telescope is marked by a yellow star. This highlights the important role the Haystack Telescope can play in linking VLBI stations across the Atlantic Ocean. The **upper right panel** characterizes how inclusion of the Haystack Telescope into the adopted ngEHT array improves the synthesized beam. Inner sidelobes are reduced by a factor 1.4. The **bottom panel** illustrates that adding the Haystack Telescope would in particular help to populate the inner area of the uv-plane.

The Haystack Telescope would add substantial sensitivity to the EHT array. Consider observations at $0\,°C$ outside temperature and a better-than-median winter PWV column of 5.3 mm (Section 2.1). In that case the atmospheric transmission is 64% at 230 GHz and $45°$ elevation. Preliminary performance modeling (J. Kauffmann, priv. comm.) further indicates an aperture efficiency of 35% and a radome transmission of 73%. Multiplication of all these factors shows that the telescope's effective combined efficiency is 16%. This number is small—but the effective aperture of this telescope is still equivalent to an *ideal* dish of $0.16^{1/2} \cdot 37\,\text{m} = 15\,\text{m}$ diameter above Earth's atmosphere. This equal to the median dish size of current EHT stations[6], and smaller telescope diameters are considered for some future EHT stations. This underlines the role which the Haystack Telescope can play within the future EHT network. That said, this calculation *purely* considers the transmission losses of the system: the impact of ground-pickup and sky brightness on the system temperature are not included, given insufficient modeling at this time, while impact of the atmospheric transmission in the calibration to the T_A^*-scale *is* taken into account. For reference, repetition of the analysis for 86 GHz frequency and $45°$ elevation yields an equivalent diameter of 31 m for an ideal telescope above Earth's atmosphere. At this frequency the Haystack Telescope could serve as an "anchor station" that can be used to improve the overall calibration of the network. In particular, the instrument could serve this purpose at $\sim$90 GHz in support of FPT to smaller dishes (see Issaoun et al., this volume).

A key component of ongoing work is to validate the telescope's abilities via participation in VLBI runs conducted at ∼86 GHz frequency. The Haystack Telescope has joined three such experiments since April 2022. This has already resulted in the detection of fringes on intercontinental baselines. Ongoing analysis will quantitatively characterize the value of the Haystack Telescope in VLBI arrays.

5. Summary

The reflector of the Haystack Telescope has been upgraded to a dish of 37 m diameter that has a surface accuracy of ≤100 μm, depending on elevation (Section 1). The instrument serves as a radar sensor for space situational awareness, with about one-third of the time available for research by MIT Haystack Observatory. Current work funded by an NSF MSRI-1 grant conducts astronomical single-dish and VLBI observations at frequencies of ∼20 GHz and ∼90 GHz to study the inclusion of the telescope into the EHT array. Parallel work enabled via a private donation generally enhances the capabilities of the instrument for research and education. The telescope is housed in a radome of 150 ft diameter that is designed to support radar observations at high frequency (Section 2). Current data indicate a median precipitable water vapor (PWV) column of about 8 mm during winter months (i.e., November 1 to April 30). These characteristics enable the Haystack Telescope to provide the US-based community with new and important capabilities for research, education, and technology development in radio astronomy (Section 3). In particular, the instrument can add new transatlantic baselines to the EHT network that would drastically improve the image quality with a frequency-dependent dish sensitivity equivalent to an ideal telescope of 15 m to 31 m above Earth's atmosphere (Section 4). Initial VLBI experiments conducted in April 2022 have resulted in fringe detections on intercontinental baselines.

Author Contributions: Conceptualization, J.K. and G.R.; writing—original draft preparation, J.K. and G.R.; writing—review and editing, K.A., V.F., C.L., L.D.M. and T.G.S.P.; visualization, K.A.; funding acquisition, L.D.M. and V.F. All authors have read and agreed to the published version of the manuscript.

Funding: This work was in part enabled by grants from the National Science Foundation, including DUE-1503793 and AST-1935980. The development of the Haystack Telescope is further supported by a private donation.

Data Availability Statement: Not applicable.

Conflicts of Interest: The authors declare no conflict of interest.

Abbreviations

The following abbreviations are used in this manuscript:

EHT	Event Horizon Telescope
LMT	Large Millimeter Telescope
ngEHT	next generation Event Horizon Telescope
SMA	Submillimeter Array
VLBI	Very Long Baseline Interferometry

Notes

1. This article includes numerous references to the "Celebrating 50 Years of Haystack" Special Issue of the Lincoln Laboratory Journal, which is available at https://www.ll.mit.edu/about/lincoln-laboratory-publications/lincoln-laboratory-journal/lincoln-laboratory-journal-0 (accessed on 15 December 2022).

2. The current NEROC members are Boston College, Boston University, Brandeis University, Dartmouth College, Harvard University, Harvard-Smithsonian Center for Astrophysics, Massachusetts Institute of Technology, Merrimack College, University of Massachusetts at Amherst, University of Massachusetts at Lowell, University of New Hampshire, and Wellesley College. NEROC's mission is to further research, education, and scientific collaboration in the field of radio science. NEROC is headquartered at MIT Haystack Observatory. Also see https://www.haystack.mit.edu/about/northeast-radio-observatory-corporation-neroc/ (accessed on 15 December 2022).

³ https://www.cosmic.ucar.edu/what-we-do/suominet-weather-precipitation-data (accessed on 15 December 2022)

⁴ https://lweb.cfa.harvard.edu/~spaine/am/ (accessed on 15 December 2022)

⁵ Permitting a one-way drive time of $\leq$3 h, following https://www.smappen.com/app/ (accessed on 15 December 2022).

⁶ The median dish diameter of the EHT array available for future observation cycles is 15 m. This characterizes an array formed from the phased ALMA dishes, with a collection area equivalent to an antenna of 91 m diameter, the phased NOEMA dishes, equivalent to an antenna of 52 m, and the phased dishes of the SMA, equivalent to an antenna of 17 m. The array also includes the LMT of 50 m diameter, the IRAM 30m-telescope, the JCMT of 15 m diameter, the APEX, Kitt Peak, and GLT dishes of 12 m diameter, and the SMT and SPT dishes of 10 m diameter.

References

1. Brown, W.M.; Pensa, A.F. History of Haystack. *Linc. Lab. J.* **2014**, *21*, 4–7.
2. Barvainis, R.; Ball, J.A.; Ingalls, R.P.; Salah, J.E. The Haystack observatory lambda 3-mm upgrade. *Publ. Astron. Soc. Pac.* **1993**, *105*, 1334. [CrossRef]
3. Whitney, A.R.; Lonsdale, C.J.; Fish, V.L. Insights into the Universe: Astronomy with Haystack's Radio Telescope. *Linc. Lab. J.* **2014**, *21*, 8–27.
4. Shapiro, I.I.; Ash, M.E.; Ingalls, R.P.; Smith, W.B.; Campbell, D.B.; Dyce, R.B.; Jurgens, R.F.; Pettengill, G.H. Fourth test of general relativity: New radar result. *Phys. Rev. Lett.* **1971**, *26*, 1132–1135. [CrossRef]
5. Myers, P.C.; Benson, P.J. Dense cores in dark clouds. II - NH3 observations and star formation. *Astrophys. J.* **1983**, *266*, 309. [CrossRef]
6. Lee, C.; Myers, P.; Tafalla, M. A Survey of Infall Motions toward Starless Cores. I. CS (2-1) and N2H+ (1-0) Observations. *Astrophys. J.* **1999**, *526*, 788–805. [CrossRef]
7. Whitney, A.R.; Shapiro, I.I.; Rogers, A.E.; Robertson, D.S.; Knight, C.A.; Clark, T.A.; Goldstein, R.M.; Marandino, G.E.; Vandenberg, N.R. Quasars revisited: Rapid time variations observed via very-long-baseline interferometry. *Science* **1971**, *173*, 225–230. [CrossRef] [PubMed]
8. Waggener, N.T. Construction of the HUSIR Antenna. *Linc. Lab. J.* **2014**, *21*, 45–82.
9. Czerwinski, M.G.; Usoff, J.M. Development of the Haystack Ultrawideband Satellite Imaging Radar. *Linc. Lab. J.* **2014**, *21*, 28–44.
10. Akiyama, K. et al. [Event Horizon Telescope Collaboration]. First M87 Event Horizon Telescope Results. II. Array and Instrumentation. *Astrophys. J. Lett.* **2019**, *875*, L2,
11. Usoff, J.M.; Clarke, M.T.; Liu, C.; Silver, M.J. Optimizing the HUSIR Antenna Surface. *Linc. Lab. J.* **2014**, *21*, 83–105.
12. MacDonald, M.E.; Anderson, J.P.; Lee, R.K.; Gordon, D.A.; McGrew, G.N. The HUSIR W-Band Transmitter. *Linc. Lab. J.* **2014**, *21*, 106–114.
13. Eshbaugh, J.V.; Morrison, R.L.; Hoen, E.W.; Hiett, T.C.; Benitz, G.R. HUSIR Signal Processing. *Linc. Lab. J.* **2014**, *21*, 115–134.
14. Rioja, M.J.; Dodson, R.; Jung, T.; Sohn, B.W. THE power of simultaneous multifrequency observations for mm-VLBI: Astrometry UP to 130 GHz with the kvn. *Astron. J.* **2015**, *150*, 202.
15. Han, S.T.; Lee, J.W.; Kang, J.; Oh, C.S.; Byun, D.Y.; Je, D.H.; Chung, M.H.; Wi, S.O.; Song, M.; Kang, Y.W.; et al. Korean VLBI Network Receiver Optics for Simultaneous Multifrequency Observation: Evaluation. *Publ. Astron. Soc. Pac.* **2013**, *125*, 539–547. [CrossRef]
16. Jiménez-Donaire, M.J.; Bigiel, F.; Leroy, A.K.; Usero, A.; Cormier, D.; Puschnig, J.; Gallagher, M.; Kepley, A.; Bolatto, A.; García-Burillo, S.; et al. EMPIRE: The IRAM 30 m Dense Gas Survey of Nearby Galaxies. *Astrophys. J.* **2019**, *880*, 127,
17. Kauffmann, J.; Goldsmith, P.F.; Melnick, G.; Tolls, V.; Guzman, A.; Menten, K.M. Molecular Line Emission as a Tool for Galaxy Observations (LEGO). I. HCN as a tracer of moderate gas densities in molecular clouds and galaxies. *Astron. Astrophys.* **2017**, *605*, L4,
18. Barnes, A.T.; Kauffmann, J.; Bigiel, F.; Brinkmann, N.; Colombo, D.; Guzmán, A.E.; Kim, W.J.; Szűcs, L.; Wakelam, V.; Aalto, S.; et al. LEGO – II. A 3 mm molecular line study covering 100 pc of one of the most actively star-forming portions within the Milky Way disc. *Mon. Not. R. Astron. Soc.* **2020**, *497*, 1972–2001.
19. Carter, M.; Lazareff, B.; Maier, D.; Chenu, J.Y.; Fontana, A.L.; Bortolotti, Y.; Boucher, C.; Navarrini, A.; Blanchet, S.; Greve, A.; et al. The EMIR multi-band mm-wave receiver for the IRAM 30-m telescope. *Astron. Astrophys.* **2012**, *538*, A89. [CrossRef]

Article

Millimeter/Submillimeter VLBI with a Next Generation Large Radio Telescope in the Atacama Desert

Kazunori Akiyama [1,2,3,*], Jens Kauffmann [1], Lynn D. Matthews [1], Kotaro Moriyama [2,4], Shoko Koyama [5,6] and Kazuhiro Hada [2,7]

1 Massachusetts Institute of Technology Haystack Observatory, 99 Millstone Rd, Westford, MA 01886, USA
2 Mizusawa VLBI Observatory, National Astronomical Observatory of Japan, Iwate 023-0861, Japan
3 Black Hole Initiative at Harvard University, 20 Garden Street, Cambridge, MA 02138, USA
4 Institut für Theoretische Physik, Goethe-Universität Frankfurt, Max-von-Laue-Straße 1, D-60438 Frankfurt am Main, Germany
5 Graduate School of Science and Technology, Niigata University, 8050 Ikarashi-nino-cho, Nishi-ku, Niigata 950-2181, Japan
6 Institute of Astronomy and Astrophysics, Academia Sinica, 11F of Astronomy-Mathematics Building, AS/NTU No. 1, Sec. 4, Roosevelt Rd, Taipei 10617, Taiwan
7 Department of Astronomical Science, The Graduate University for Advanced Studies, SOKENDAI, Tokyo 181-8588, Japan
* Correspondence: kakiyama@mit.edu

Abstract: The proposed next generation Event Horizon Telescope (ngEHT) concept envisions the imaging of various astronomical sources on scales of microarcseconds in unprecedented detail with at least two orders of magnitude improvement in the image dynamic ranges by extending the Event Horizon Telescope (EHT). A key technical component of ngEHT is the utilization of large aperture telescopes to anchor the entire array, allowing the connection of less sensitive stations through highly sensitive fringe detections to form a dense network across the planet. Here, we introduce two projects for planned next generation large radio telescopes in the 2030s on the Chajnantor Plateau in the Atacama desert in northern Chile, the Large Submillimeter Telescope (LST) and the Atacama Large Aperture Submillimeter Telescope (AtLAST). Both are designed to have a 50-meter diameter and operate at the planned ngEHT frequency bands of 86, 230 and 345 GHz. A large aperture of 50 m that is co-located with two existing EHT stations, the Atacama Large Millimeter/Submillimeter Array (ALMA) and the Atacama Pathfinder Experiment (APEX) Telescope in the excellent observing site of the Chajnantor Plateau, will offer excellent capabilities for highly sensitive, multi-frequency, and time-agile millimeter very long baseline interferometry (VLBI) observations with accurate data calibration relevant to key science cases of ngEHT. In addition to ngEHT, its unique location in Chile will substantially improve angular resolutions of the planned Next Generation Very Large Array in North America or any future global millimeter VLBI arrays if combined. LST and AtLAST will be a key element enabling transformative science cases with next-generation millimeter/submillimeter VLBI arrays.

Keywords: very long baseline interferometry (1769); radio astronomy (1338); millimeter astronomy (1061); submillimeter astronomy (1647); radio telescopes (1360); high angular resolution (2167); astronomical instrumentation (799)

Citation: Akiyama, K.; Kauffmann, J.; Matthews, L.D.; Moriyama, K.; Koyama, S.; Hada, K. Millimeter/ Submillimeter VLBI with a Next Generation Large Radio Telescope in the Atacama Desert. *Galaxies* 2023, 11, 1. https://doi.org/10.3390/ galaxies11010001

Academic Editor: Bidzina Kapanadze

Received: 15 November 2022
Revised: 12 December 2022
Accepted: 13 December 2022
Published: 20 December 2022

1. Introduction

With the success of the Event Horizon Telescope[1] (EHT) [1–14], the next generation Event Horizon Telescope[2] (ngEHT) has been proposed as a development concept for the extension of EHT in the 2030s [15–17]. ngEHT aims to extend EHT by adding ∼10 new stations to its network of the very long baseline interferometry (VLBI) array together with overall upgrades in the receiving system, including a significant increase in bandwidth, comparable to the planned Next Generation Very Large Array (ngVLA; [18,19]), along with the capability to perform simultaneous dual- or tri-bands observations [16,20]. Given substantial upgrades in the instruments and VLBI network, ngEHT is anticipated to provide simultaneous multi-frequency imaging at dynamic ranges at least two orders of magnitude better than the current EHT. For instance, for M87*, ngEHT is expected to achieve an image dynamic range of >1000 enough to capture a detailed shape of the extended jet emission on scales of thousands Schwarzschild radii, e.g., [16,17], which was not possible with the EHT 2017 array, which achieved a dynamic range of only ∼10 due to its sparse baseline coverage [4]. These unprecedented capabilities allow transformative science cases at extreme high angular resolutions of a few tens of microarcseconds, not only for horizon-scale black hole astrophysics in sources such as M87* and Sgr A*, but also in potential other targets for which the array may resolve the horizon-scale emission [15,21] and various compact objects on the sky (see other articles in this special issue).

A key design aspect of the ngEHT array is the use of small and large aperture telescopes to form a dense interferometric network [15–17]. The large-aperture telescopes will work as sensitive anchor stations that will facilitate robust fringe detections across the entire array, whereas the small telescopes will fill up the Fourier coverage of the array and enable high-dynamic-range imaging. The anticipated anchor stations include existing EHT stations such as the Large Millimeter Telescope (LMT), the Atacama Large Millimeter/submillimeter Array (ALMA), and the Northern Extended Millimeter Array (NOEMA), as well as planned additional stations such as the Haystack 37 m Telescope [22]. As the baseline sensitivity is proportional to the geometric mean of the collecting area of the apertures on both ends, the participation of such sensitive facilities is essential for various science cases requiring high-sensitivity observations.

Here, we describe two international projects for planned next generation large radio telescopes in the 2030s, the Large Submillimeter Telescope[3] (LST) [23,24] and the Atacama Large Aperture Submillimeter Telescope[4] (AtLAST) [25,26], as a potential anchor stations of next-generation global VLBI arrays. Both projects aim to construct a 50-meter-class radio telescope on the Chajnantor Plateau in the Atacama desert in northern Chile operating at millimeter/submillimeter wavelengths including the planned ngEHT observing frequency bands. The remainder of the paper is constructed as follows. We first describe each project briefly in Section 2, and then discuss the prospects for having such a large-aperture dish in Atacama for the next generation global submillimeter/millimeter VLBI arrays in Section 3. Finally, we will make a brief summary and conclusion in Section 4.

2. Planned Large Submillimeter/Millimeter Radio Telescopes in the Atacama Desert

2.1. Large Submillimeter Telescope (LST)

LST is a planned 50-meter-class single-dish telescope operating at submillimeter and millimeter wavelengths to be constructed on the Chajnantor Plateau in Chile at the same site as ALMA. This project is driven by an international collaboration led by the Japanese radio astronomy community [23,24]. The LST concept was originally developed as a next-generation successor of the Nobeyama Radio Observatory (NRO) 45 m telescope [27] and the Atacama Submillimeter Telescope Experiment (ASTE) 10 m telescope [28]. LST aims to inherit two major key strengths from these predecessors: a large collection area from the NRO 45 m telescope and the submillimeter capabilities from the ASTE 10-meter telescope.

The current key conceptual design and major specifications are described in Kawabe et al. [23]. LST is planned to have a 50-meter-diameter dish (see Figure 1a) with a high surface precision (45 μm rms) designed for wide-area imaging and spectroscopic surveys with a field-of-view of ∼1° primarily focusing on the 70–420 GHz frequency range.

In the current LST design, the targeted wide field of view is enabled by adopting a Ritchey–Chrétien (RC) system for optics [23]. The project further aims to have a capability for frequencies up to 1 THz using an inner high-precision surface. To establish a high-precision surface with a large collecting area, LST plans to implement a millimetric adaptive optics system based on real-time sensing of the surface with a dedicated millimeter wavefront sensor, e.g., [29]. Key science cases enabled by the highly-sensitive large-aperture of LST, briefly summarized in Kawabe et al. [23], include black hole astrophysics with high-sensitive millimeter/submillimeter VLBI involving the LST as an anchor station for global arrays.

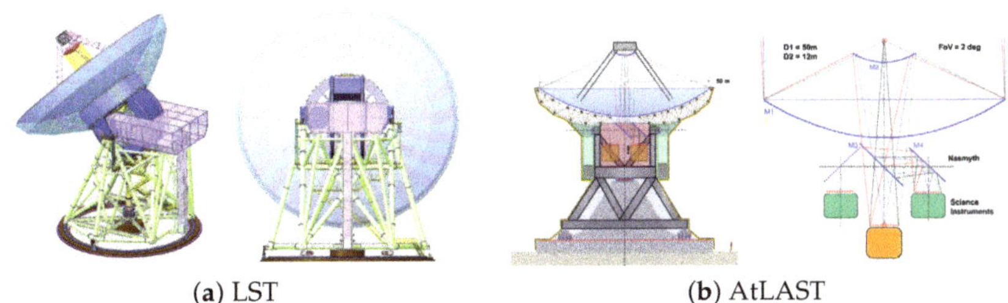

(a) LST   **(b)** AtLAST

Figure 1. Conceptual designs of **(a)** LST and **(b)** AtLAST. Both projects aim a large-aperture telescope with the diameter of ∼50 m in Atacama, Chile operating at millimeter and submillimeter wavelengths. The AtLAST concept images are designed before the ongoing European Union funded Horizon 2020 research and innovation programme. The current optics design has evolved significantly under the programme (see Section 2.2). The LST concept images are in courtesy of Mitsubishi Electric Corporation (MELCO), and adapted with permission from Ref. [23]; Copyright 2016, Society of Photo-Optical Instrumentation Engineers (SPIE). The AtLAST images are adapted with permission from Ref. [25]; Copyright 2020, SPIE.

The recent progress and the near-future prospects of the project are summarized in Kohno et al. [24]. The LST project is formally listed as a large academic project in the astronomy and astrophysics division in the Master Plan 2020 led by the Science Council of Japan, the latest series of surveys designed to review and maintain the list of high-priority large academic research projects in Japan every three or four years. The LST project anticipates the merger of the project into the AtLAST project in the mid-2020s. Collaborative studies would be undertaken under the ongoing AtLAST design study program after resolving inconsistencies in telescope specifications between the two projects.

2.2. Atacama Large Aperture Submillimeter Telescope (AtLAST)

Just like LST, AtLAST is planned to be built on the Chajnantor Plateau, in close proximity to ALMA. An ongoing design study for the observatory, supported by the European Union's Horizon 2020 research and innovation program, seeks to further refine the details in the 2021–2024 time frame. This section primarily summarizes the specifications as known at the beginning of the design study [25] and the current key science drivers [26].

The telescope will be sited at an altitude between 5100 and 5500 m, depending on the specific location chosen for construction. The project seeks to deliver a dish with a diameter of 50 m and a very large field of view ($\approx 2°$) achieved by an optics system involving a large secondary mirror (see Figure 1b for its early concept design). The latter will enable a new generation of astrophysical experiments that cannot be pursued otherwise. The goal is to provide access to frequencies $\gtrsim 850\,$GHz, resulting in a desired dish surface accuracy of 20–25 μm. An active surface would be employed to achieve this precision. We note that the AtLAST design shown in Figure 1b, adapted from Klaassen et al. [25], has matured through the European Union funded Horizon 2020 research and innovation program into a 3-mirror, hybrid Nasmyth-like design that will be able to host two 2 degree wide field-of-view instruments located along the elevation axis, as well as several additional smaller instruments with up to 1 degree diameter fields-of-view, located off axis, without requiring additional external re-imaging optics.

AtLAST would enable breakthroughs in several domains of astrophysics. These include studies of molecular clouds in the Milky Way, galaxies and their formation over cosmic time, as well as the evolution of galaxy clusters. These objects can be investigated using line emission from molecules and atoms, continuum emission from dust, and by employing the broadband spectral signature of the Sunyaev–Zeldovich (SZ) effect that probes the ionized medium in and between galaxies. AtLAST will be optimized for wide-field surveys, but it is conceived to be a multi-role observatory that would be open to PI-driven research projects.

3. Prospects for Millimeter/Submillimeter VLBI with LST and AtLAST

A next-generation large millimeter/submillimeter telescope in the Atacama desert, if realized, could play a vital role in global millimeter/submillimeter VLBI observations in multiple ways. An obvious strength is sensitivity, for instance, a 50 m diameter telescope with an aperture efficiency of $\sim$50 % and a system noise temperature of $\sim$100 K at 230 GHz anticipated for the site, e.g., [2] will achieve the system-equivalent flux density (SEFD) of $\sim$300 Jy. It is better than the anticipated sensitivity of the phased array of the Northern Extended Millimeter Array (NOEMA) with the SEFD of $\sim$700 Jy, e.g., [2], and orders of magnitude better than the typical sensitivities of the existing and anticipated EHT or ngEHT stations with SEFDs of $\sim$1000–20,000 Jy [2,3,10]. The telescope is expected to have a competitive sensitivity of $\sim$30–40 % of ALMA with the SEFD of $\sim$100 Jy[5] [2,3,10] which is the highest sensitivity among the submillimeter VLBI stations anticipated in 2030s.

The location of the planned site in the Atacama provides additional benefits to global VLBI observations. First, the telescopes in the Atacama provide very long, intercontinental baselines, especially in the north-south directions to North American, European, and Pacific stations. The Atacama baselines have been providing substantial improvements in the angular resolution of the Global Millimeter VLBI Array (GMVA), e.g., [30–33] and were essential to resolving the shadows of the supermassive black holes M87* and Sgr A* with the EHT, e.g., [4]. The 50-meter telescope in the Atacama desert will be a key anchor station to secure the detection of fringes on intercontinental baselines and enhance the sensitivity of the overall array. Second, the telescope shares its site with other submillimeter/millimeter facilities such as ALMA and APEX, providing redundant baselines as well as the intra-site baselines to the entire ALMA array. Inteferometric measurements on baselines from redundant stations allow accurate and precise absolute calibrations of interferometric data, e.g., [3,7,34,35] which are critical for both total-intensity and polarimetric imaging.

With its competitive sensitivity, a large-aperture single-dish telescope has potential strengths for global VLBI observations over phased stations such as the colocated ALMA. These strengths benefit from simpler instrumentation and observing logistics required for a single dish telescope to be a VLBI station.

3.1. Observations of Fainter Sources

With its competitive sensitivity, a large aperture single-dish telescope has a unique strength that can broaden the number of faint target sources that may not be observable sorely with phased array stations. Phased arrays often need the target sources to be bright enough for active phasing, or alternatively to have bright and compact phase calibrators nearby for passive phasing, for instance, the current ALMA phasing system has a limit on the total flux density of $\sim$500 mJy for the target sources or phase calibrators with a separation only within several degrees[6]. Its competitive sensitivity without the need for phasing will be critical for faint science targets that may not necessarily have appropriate phase calibrators.

3.2. Simultaneous Multi-Frequency Observations

Although the anticipated sensitivity is a few times lower than that of ALMA, a 50 m single dish in Atacama may have unique strengths compared to ALMA in frequency agility while maintaining its high sensitivity. An important specification of ngEHT is a capability of simultaneous multi-frequency observations [16], allowing the order-of-magnitude improvements in sensitivity with frequency phase transfer (FPT) techniques [36–39], as well as enabling various science cases based on simultaneous measurements of full-polarization spectra and Faraday-rotations of various sources. Although ngEHT originally aimed to have simultaneous dual-band receiving at 230 and 340 GHz [16], the project is now pursuing enabling simultaneous tri-band observations at 86, 230 and 345 GHz [20].

While a capability of simultaneous multi-frequency observations can be realized by the implementation of a dedicated receiving system for single-dish telescopes, it is more logistically and operationally complicated for phased ALMA. To keep its sensitivity at each frequency, it will need installations of such receiving systems across the array. Alternatively, simultaneous multi-band observations may be enabled by splitting the entire array into subarrays operating at single frequencies, which will compensate the sensitivity at each frequency due to the reduced synthesized aperture. The current 2030 roadmap of ALMA is targeting the latter approach by implementing subarraying capabilities instead of simultaneous dual- or tri-frequency operations with a single array [40], indicating that the next-generation large telescope in the Atacama has a strong potential to have a comparable or even better sensitivity than ALMA if a dedicated receiving system is implemented for simultaneous multi-band observations.

The benefits of having a large single dish for simultaneous multi-band observations are not limited to high sensitivity. A strong advantage of simultaneous multi-frequency observations with a single dish over a subarray is that the signals at multiple frequencies natively share the same atmospheric line of sight. This is important for the application of FPT techniques that allow much longer phase coherence and integration of interferometric fringes at higher frequencies by solving the atmospheric phase delays at lower frequencies. Although the application of the FPT techniques to subarrays is possible by computationally aligning the phase center of subarrays to the same location, it will complicate the signal processing and may cause the additional loss in the phase coherence and/or systematic errors in FPT-applied data.

3.3. Time-Domain Science

Another potential strength of a single-dish sensitive telescope in the Atacama is time agility, which will allow various time-domain science requiring monitoring observations over weeks and months proposed for ngEHT and also other global arrays such as ngVLA and a next-generation GMVA. Phased arrays are known to be more impacted by windy, turbulent weather, and the resulting worse atmospheric phase coherence, since the atmospheric delay in the radio signal received in each antenna needs to be corrected accurately in real time, e.g., [41]. For example, the phasing efficiency of ALMA is reported to be consistently poor at 345 GHz when wind speeds exceed $\sim$10 m/s regardless of the amount of the precipitable water vapor (PWV) [42]. While a single-dish telescope may be affected by windy weather through, for instance, losses in focus and pointing efficiency, it may provide more robust observing capabilities during periods of atmospheric stability, and ultimately enable expansion of suitable observing windows.

4. Summary

In this article, we have introduced two planned projects for a next-generation sub-millimeter/millimeter single-dish telescope, LST and AtLAST, both aiming a 50 m-class telescope in the Atacama Desert of Chile, sharing the site with two existing submillimeter/millimeter facilities, APEX and ALMA. The two projects are currently anticipated to be merged in the next several years as a result of ongoing study programs. We further discussed the strong and unique benefits of having such a large aperture dish in the Atacama as part of the planned next-generation of millimeter/sub-millimeter VLBI arrays including ngEHT and ngVLA. A large-aperture single-dish telescope at this location, with its excellent observing conditions, will play a vital role as a high-sensitivity key anchor station capable of significantly improving the sensitivity and angular resolution of planned global VLBI arrays will potentially have better time and frequency agility than phased ALMA. LST, AtLAST, or a future merged telescope thus have strong potential to be a key element for next generation millimeter/submillimeter VLBI arrays and allow them to achieve a range of new science goals requiring highly sensitive, multi-frequency, time-agile observations.

Author Contributions: Writing—original draft preparation, K.A. and J.K.; writing—review and editing, all authors (K.A., J.K., L.D.M., K.M., S.K. and K.H.). All authors contributed equally to this work. All authors have read and agreed to the published version of the manuscript.

Funding: This work is financially supported by grants from the National Science Foundation (NSF; AST-1440254, AST-1614868, AST-1935980, AST-2034306). K.A. has been also financially supported by other NSF grants (OMA-2029670, AST-2107681, AST-2132700). AtLAST has received funding from the European Union's Horizon 2020 research and innovation programme under grant agreement No. 951815.

Institutional Review Board Statement: Not applicable.

Informed Consent Statement: Not applicable.

Data Availability Statement: Not applicable.

Acknowledgments: We thank Pamela Klaassen, Tony Mroczkowski, Claudia Cicone, and Colin J. Lonsdale for helpful feedback to improve the article. We appreciate the internal reviewer of this article from the EHT collaboration, Lindy Blackburn for helpful and constructive suggestions on the article.

Conflicts of Interest: The authors declare no conflict of interest.

Abbreviations

The following abbreviations are used in this manuscript:

ALMA	Atacama Large Millimeter/submillimeter Array
AtLAST	Atacama Large Aperture Submillimeter Telescope
APEX	Atacama Pathfinder Experiment
EHT	Event Horizon Telescope
GMVA	Global Millimeter VLBI Array
LMT	Large Millimeter Telescope
LST	Large Submillimeter Telescope
ngEHT	next generation Event Horizon Telescope
ngVLA	next generation Very Large Array
NOEMA	Nothern Extended Millimeter Array
VLBI	Very Long Baseline Interferometry

Notes

[1] https://eventhorizontelescope.org/ (accessed on 15 November 2022)

[2] https://www.ngeht.org/ (accessed on 15 November 2022)

[3] https://en.lstobservatory.org/ (accessed on 15 November 2022)

[4] https://www.atlast.uio.no/ (accessed on 15 November 2022)

[5] The current ALMA 2030 roadmap aims to improve the sensitivity by the overall upgrade in the frontend and also the correlator, which would provide a better SEFD.

[6] See e.g., ALMA Cycle 8 Proposer's Guide: https://almascience.nrao.edu/documents-and-tools/cycle8/alma-proposers-guide (accessed on 15 November 2022)

References

1. Event Horizon Telescope Collaboration; Akiyama, K.; Alberdi, A.; Alef, W.; Asada, K.; Azulay, R.; Baczko, A.K.; Ball, D.; Baloković, M.; Barrett, J.; et al. First M87 Event Horizon Telescope Results. I. The Shadow of the Supermassive Black Hole. *Astrophys. J. Lett.* **2019**, *875*, L1. [CrossRef]
2. Event Horizon Telescope Collaboration; Akiyama, K.; Alberdi, A.; Alef, W.; Asada, K.; Azulay, R.; Baczko, A.K.; Ball, D.; Baloković, M.; Barrett, J.; et al. First M87 Event Horizon Telescope Results. II. Array and Instrumentation. *Astrophys. J. Lett.* **2019**, *875*, L2. [CrossRef]

3. Event Horizon Telescope Collaboration; Akiyama, K.; Alberdi, A.; Alef, W.; Asada, K.; Azulay, R.; Baczko, A.K.; Ball, D.; Baloković, M.; Barrett, J.; et al. First M87 Event Horizon Telescope Results. III. Data Processing and Calibration. *Astrophys. J. Lett.* **2019**, *875*, L3. [CrossRef]

4. Event Horizon Telescope Collaboration; Akiyama, K.; Alberdi, A.; Alef, W.; Asada, K.; Azulay, R.; Baczko, A.K.; Ball, D.; Baloković, M.; Barrett, J.; et al. First M87 Event Horizon Telescope Results. IV. Imaging the Central Supermassive Black Hole. *Astrophys. J. Lett.* **2019**, *875*, L4. [CrossRef]

5. Event Horizon Telescope Collaboration; Akiyama, K.; Alberdi, A.; Alef, W.; Asada, K.; Azulay, R.; Baczko, A.K.; Ball, D.; Baloković, M.; Barrett, J.; et al. First M87 Event Horizon Telescope Results. V. Physical Origin of the Asymmetric Ring. *Astrophys. J. Lett.* **2019**, *875*, L5. [CrossRef]

6. Event Horizon Telescope Collaboration; Akiyama, K.; Alberdi, A.; Alef, W.; Asada, K.; Azulay, R.; Baczko, A.K.; Ball, D.; Baloković, M.; Barrett, J.; et al. First M87 Event Horizon Telescope Results. VI. The Shadow and Mass of the Central Black Hole. *Astrophys. J. Lett.* **2019**, *875*, L6. [CrossRef]

7. Event Horizon Telescope Collaboration; Akiyama, K.; Algaba, J.C.; Alberdi, A.; Alef, W.; Anantua, R.; Asada, K.; Azulay, R.; Baczko, A.K.; Ball, D.; et al. First M87 Event Horizon Telescope Results. VII. Polarization of the Ring. *Astrophys. J. Lett.* **2021**, *910*, L12. [CrossRef]

8. Event Horizon Telescope Collaboration; Akiyama, K.; Algaba, J.C.; Alberdi, A.; Alef, W.; Anantua, R.; Asada, K.; Azulay, R.; Baczko, A.K.; Ball, D.; et al. First M87 Event Horizon Telescope Results. VIII. Magnetic Field Structure near The Event Horizon. *Astrophys. J. Lett.* **2021**, *910*, L13. [CrossRef]

9. Event Horizon Telescope Collaboration; Akiyama, K.; Alberdi, A.; Alef, W.; Algaba, J.C.; Anantua, R.; Asada, K.; Azulay, R.; Bach, U.; Baczko, A.K.; et al. First Sagittarius A* Event Horizon Telescope Results. I. The Shadow of the Supermassive Black Hole in the Center of the Milky Way. *Astrophys. J. Lett.* **2022**, *930*, L12. [CrossRef]

10. Event Horizon Telescope Collaboration; Akiyama, K.; Alberdi, A.; Alef, W.; Algaba, J.C.; Anantua, R.; Asada, K.; Azulay, R.; Bach, U.; Baczko, A.K.; et al. First Sagittarius A* Event Horizon Telescope Results. II. EHT and Multiwavelength Observations, Data Processing, and Calibration. *Astrophys. J. Lett.* **2022**, *930*, L13. [CrossRef]

11. Event Horizon Telescope Collaboration; Akiyama, K.; Alberdi, A.; Alef, W.; Algaba, J.C.; Anantua, R.; Asada, K.; Azulay, R.; Bach, U.; Baczko, A.K.; et al. First Sagittarius A* Event Horizon Telescope Results. III. Imaging of the Galactic Center Supermassive Black Hole. *Astrophys. J. Lett.* **2022**, *930*, L14. [CrossRef]

12. Event Horizon Telescope Collaboration; Akiyama, K.; Alberdi, A.; Alef, W.; Algaba, J.C.; Anantua, R.; Asada, K.; Azulay, R.; Bach, U.; Baczko, A.K.; et al. First Sagittarius A* Event Horizon Telescope Results. IV. Variability, Morphology, and Black Hole Mass. *Astrophys. J. Lett.* **2022**, *930*, L15. [CrossRef]

13. Event Horizon Telescope Collaboration; Akiyama, K.; Alberdi, A.; Alef, W.; Algaba, J.C.; Anantua, R.; Asada, K.; Azulay, R.; Bach, U.; Baczko, A.K.; et al. First Sagittarius A* Event Horizon Telescope Results. V. Testing Astrophysical Models of the Galactic Center Black Hole. *Astrophys. J. Lett.* **2022**, *930*, L16. [CrossRef]

14. Event Horizon Telescope Collaboration; Akiyama, K.; Alberdi, A.; Alef, W.; Algaba, J.C.; Anantua, R.; Asada, K.; Azulay, R.; Bach, U.; Baczko, A.K.; et al. First Sagittarius A* Event Horizon Telescope Results. VI. Testing the Black Hole Metric. *Astrophys. J. Lett.* **2022**, *930*, L17. [CrossRef]

15. Doeleman, S.; Akiyama, K.; Blackburn, L.; Bouman, K.L.; Bower, G.C.; Broderick, A.E.; Chael, A.; Fish, V.L.; Johnson, M.D.; Lonsdale, C.J.; et al. Black Hole Physics on Horizon Scales. *Bull. Am. Astron. Soc.* **2019**, *51*, 537.

16. Doeleman, S.; Blackburn, L.; Dexter, J.; Gomez, J.L.; Johnson, M.D.; Palumbo, D.C.; Weintroub, J.; Farah, J.R.; Fish, V.; Loinard, L.; et al. Studying Black Holes on Horizon Scales with VLBI Ground Arrays. *arXiv* **2019**, arXiv:1909.01411.

17. Raymond, A.W.; Palumbo, D.; Paine, S.N.; Blackburn, L.; Córdova Rosado, R.; Doeleman, S.S.; Farah, J.R.; Johnson, M.D.; Roelofs, F.; Tilanus, R.P.J.; et al. Evaluation of New Submillimeter VLBI Sites for the Event Horizon Telescope. *Astrophys. J. Suppl. Ser.* **2021**, *253*, 5. [CrossRef]

18. Selina, R.J.; Murphy, E.J.; McKinnon, M.; Beasley, A.; Butler, B.; Carilli, C.; Clark, B.; Durand, S.; Erickson, A.; Grammer, W.; et al. The ngVLA Reference Design. In *Science with a Next Generation Very Large Array*; Astronomical Society of the Pacific Conference Series; Murphy, E., Ed.; ASP Monograph 7; ASP: San Francisco, CA, USA, 2018; Volume 517, p. 15.

19. Murphy, E.J.; Bolatto, A.; Chatterjee, S.; Casey, C.M.; Chomiuk, L.; Dale, D.; de Pater, I.; Dickinson, M.; Francesco, J.D.; Hallinan, G.; et al. The ngVLA Science Case and Associated Science Requirements. In *Science with a Next Generation Very Large Array*; Astronomical Society of the Pacific Conference Series; Murphy, E., Ed.; ASP Monograph 7; ASP: San Francisco, CA, USA, 2018; Volume 517, p. 3.

20. Issaoun, S.; Pesce, D.W.; Roelofs, F.; Chael, A.; Dodson, R.; Rioja, M.J.; Akiyama, K.; Aran, R.; Blackburn, L.; Doeleman, S.S.; et al. Enabling transformational ngEHT science via the inclusion of 86 GHz capabilities. *Galaxies* 2022, *submitted*.

21. Pesce, D.W.; Palumbo, D.C.M.; Narayan, R.; Blackburn, L.; Doeleman, S.S.; Johnson, M.D.; Ma, C.P.; Nagar, N.M.; Natarajan, P.; Ricarte, A. Toward Determining the Number of Observable Supermassive Black Hole Shadows. *Astrophys. J.* **2021**, *923*, 260. [CrossRef]

22. Kauffmann, J.; Rajagopalan, G.; Akiyama, K.; Fish, V.L.; Lonsdale, C.J.; Matthews, L.D.; Pillai, T. The Haystack Telescope as an Astronomical Instrument. *Galaxies* 2022, *submitted*.

23. Kawabe, R.; Kohno, K.; Tamura, Y.; Takekoshi, T.; Oshima, T.; Ishii, S. New 50-m-class single-dish telescope: Large Submillimeter Telescope (LST). In *Ground-Based and Airborne Telescopes VI*; Hall, H.J., Gilmozzi, R., Marshall, H.K., Eds.; Society of Photo-Optical Instrumentation Engineers (SPIE) Conference Series; SPIE: Bellingham, WA, USA, 2016; Volume 9906, pp. 779–790. [CrossRef]

24. Kohno, K.; Kawabe, R.; Tamura, Y.; Endo, A.; Baselmans, J.J.A.; Karatsu, K.; Inoue, A.K.; Moriwaki, K.; Hayatsu, N.H.; Yoshida, N.; et al. Large format imaging spectrograph for the Large Submillimeter Telescope (LST). In *Millimeter, Submillimeter, and Far-Infrared Detectors and Instrumentation for Astronomy X*; Society of Photo-Optical Instrumentation Engineers (SPIE) Conference Series; SPIE: Bellingham, WA, USA, 2020; Volume 11453, pp. 128–138. [CrossRef]

25. Klaassen, P.D.; Mroczkowski, T.K.; Cicone, C.; Hatziminaoglou, E.; Sartori, S.; De Breuck, C.; Bryan, S.; Dicker, S.R.; Duran, C.; Groppi, C.; et al. The Atacama Large Aperture Submillimeter Telescope (AtLAST). In *Ground-Based and Airborne Telescopes VIII*; Society of Photo-Optical Instrumentation Engineers (SPIE) Conference Series; SPIE: Bellingham, WA, USA, 2020; Volume 11445, pp. 544–563. [CrossRef]

26. Ramasawmy, J.; Klaassen, P.D.; Cicone, C.; Mroczkowski, T.K.; Chen, C.C.; Cornish, T.; da Cunha, E.; Hatziminaoglou, E.; Johnstone, D.; Liu, D.; et al. The Atacama Large Aperture Submillimetre Telescope: Key science drivers. In *Millimeter, Submillimeter, and Far-Infrared Detectors and Instrumentation for Astronomy XI*; Zmuidzinas, J., Gao, J.R., Eds.; Society of Photo-Optical Instrumentation Engineers (SPIE) Conference Series; SPIE: Bellingham, WA, USA, 2022; Volume 12190, pp. 112–130. [CrossRef]

27. Ukita, N.; Tsuboi, M. A 45-meter telescope with a surface accuracy of 65 μm. *IEEE Proc.* **1994**, *82*, 725–733. [CrossRef]

28. Ezawa, H.; Kohno, K.; Kawabe, R.; Yamamoto, S.; Inoue, H.; Iwashita, H.; Matsuo, H.; Okuda, T.; Oshima, T.; Sakai, T.; et al. New achievements of ASTE: The Atacama Submillimeter Telescope Experiment. In *Ground-Based and Airborne Telescopes II*; Stepp, L.M., Gilmozzi, R., Eds.; Society of Photo-Optical Instrumentation Engineers (SPIE) Conference Series; SPIE: Bellingham, WA, USA, 2008; Volume 7012, pp. 88–96. [CrossRef]

29. Tamura, Y.; Kawabe, R.; Fukasaku, Y.; Kimura, K.; Ueda, T.; Taniguchi, A.; Okada, N.; Ogawa, H.; Hashimoto, I.; Minamidani, T.; et al. Wavefront sensor for millimeter/submillimeter-wave adaptive optics based on aperture-plane interferometry. In *Ground-Based and Airborne Telescopes VIII*; Society of Photo-Optical Instrumentation Engineers (SPIE) Conference Series; SPIE: Bellingham, WA, USA, 2020; Volume 11445, pp. 350–358. [CrossRef]

30. Issaoun, S.; Johnson, M.D.; Blackburn, L.; Brinkerink, C.D.; Mościbrodzka, M.; Chael, A.; Goddi, C.; Martí-Vidal, I.; Wagner, J.; Doeleman, S.S.; et al. The Size, Shape, and Scattering of Sagittarius A* at 86 GHz: First VLBI with ALMA. *Astrophys. J.* **2019**, *871*, 30. [CrossRef]

31. Issaoun, S.; Johnson, M.D.; Blackburn, L.; Broderick, A.; Tiede, P.; Wielgus, M.; Doeleman, S.S.; Falcke, H.; Akiyama, K.; Bower, G.C.; et al. Persistent Non-Gaussian Structure in the Image of Sagittarius A* at 86 GHz. *Astrophys. J.* **2021**, *915*, 99. [CrossRef]

32. Okino, H.; Akiyama, K.; Asada, K.; Gómez, J.L.; Hada, K.; Honma, M.; Krichbaum, T.P.; Kino, M.; Nagai, H.; Nakamura, M.; et al. Collimation of the relativistic jet in the quasar 3C 273. *arXiv* **2021**, arXiv:2112.12233.

33. Zhao, G.Y.; Gómez, J.L.; Fuentes, A.; Krichbaum, T.P.; Traianou, E.; Lico, R.; Cho, I.; Ros, E.; Komossa, S.; Akiyama, K.; et al. Unraveling the Innermost Jet Structure of OJ 287 with the First GMVA + ALMA Observations. *Astrophys. J.* **2022**, *932*, 72. [CrossRef]

34. Fish, V.L.; Doeleman, S.S.; Beaudoin, C.; Blundell, R.; Bolin, D.E.; Bower, G.C.; Chamberlin, R.; Freund, R.; Friberg, P.; Gurwell, M.A.; et al. 1.3 mm Wavelength VLBI of Sagittarius A*: Detection of Time-variable Emission on Event Horizon Scales. *Astrophys. J. Lett.* **2011**, *727*, L36. [CrossRef]

35. Johnson, M.D.; Fish, V.L.; Doeleman, S.S.; Marrone, D.P.; Plambeck, R.L.; Wardle, J.F.C.; Akiyama, K.; Asada, K.; Beaudoin, C.; Blackburn, L.; et al. Resolved magnetic-field structure and variability near the event horizon of Sagittarius A*. *Science* **2015**, *350*, 1242–1245. [CrossRef]

36. Asaki, Y.; Saito, M.; Kawabe, R.; Morita, K.; Sasao, T. A Phase Compensation Experiment with the Paired Antenna Method. *Radio Sci.* **1996**, *31*, 464. [CrossRef]

37. Dodson, R.; Rioja, M.J. VLBA Scientific Memorandum n. 31: Astrometric calibration of mm-VLBI using "Source/Frequency Phase Referenced" observations. *arXiv* **2009**, arXiv:0910.1159.

38. Rioja, M.; Dodson, R. High-precision Astrometric Millimeter Very Long Baseline Interferometry Using a New Method for Atmospheric Calibration. *Astrophys. J.* **2011**, *141*, 114. [CrossRef]

39. Rioja, M.J.; Dodson, R.; Asaki, Y. The Transformational Power of Frequency Phase Transfer Methods for ngEHT. *Galaxies* 2022, *submitted*.

40. Carlson, B.; Pleasance, M.; Gunaratne, T.; Vrcic, S. *Study Report: NRC TALON Frequency Slice Architecture Correlator/Beamformer (AT.CBF) for ALMA*; ALMA Memo 617; ALMA: New York, NY, USA, 2020.

41. Matthews, L.D.; Crew, G.B.; Doeleman, S.S.; Lacasse, R.; Saez, A.F.; Alef, W.; Akiyama, K.; Amestica, R.; Anderson, J.M.; Barkats, D.A.; et al. The ALMA Phasing System: A Beamforming Capability for Ultra-high-resolution Science at (Sub)Millimeter Wavelengths. *Publ. Astron. Soc. Pac.* **2018**, *130*, 15002. [CrossRef]

42. Crew, G.B.; Goddi, C.; Matthews, L.D.; Rottmann, H.; Saez, A.; Martí-Vidal, I. A Characterization of the ALMA Phasing System at 345 GHz. *Publ. Astron. Soc. Pac.* 2022, *submitted*.

Article

Evaluation of a Candidate Site in the Tibetan Plateau towards the Next Generation Event Horizon Telescope

Wei Yu [1,†], Ru-Sen Lu [1,*], Zhi-Qiang Shen [1] and Jonathan Weintroub [2]

1   Shanghai Astronomical Observatory, Chinese Academy of Sciences, 80 Nandan Road, Shanghai 200030, China
2   Center for Astrophysics | Harvard & Smithsonian, 60 Garden Street, Cambridge, MA 02138, USA
*   Correspondence: rslu@shao.ac.cn
†   Current Affiliation: Center for Astrophysics | Harvard & Smithsonian, 60 Garden Street, Cambridge, MA 02138, USA.

Abstract: In order to enhance the imaging capabilities of the Event Horizon Telescope (EHT) and capture the first black hole movies, the next-generation EHT (ngEHT) team is building new stations. Most stations of the EHT and ngEHT project are located in the Western Hemisphere, leaving a large vacancy in the Eastern Hemisphere. Located in the center of the Eastern Hemisphere, the Tibetan Plateau is believed to have excellent sites for (sub)millimeter astronomical radio observations. Building a telescope here could help to fill this vacancy. In this study, we evaluated the meteorological conditions of a candidate site (Shigatse, hereafter SG) with good astronomical infrastructure for this telescope. The evaluation results show that the precipitable water vapor (PWV) values of the SG site are lower than 4 mm during winter and spring, comparable to those of some existing EHT stations, and the zenith transmittances at 230 GHz and 345 GHz during March and April are excellent. We simulated VLBI observations of Sgr A* and M87 based on the conditions of the SG site and those of other existing/planned (sub)millimeter telescopes with mutual visibility at 230 GHz. The results demonstrated that images of Sgr A* and M87 could be well reconstructed, indicating that the SG site is a good candidate for future EHT/ngEHT observations.

Keywords: EHT; ngEHT; black holes; precipitable water vapor

Citation: Yu, W.; Lu, R.-S.; Shen, Z.-Q.; Weintroub, J. Evaluation of a Candidate Site in the Tibetan Plateau towards the Next Generation Event Horizon Telescope. *Galaxies* **2023**, *11*, 7. https://doi.org/10.3390/galaxies11010007

Academic Editor: Rafael Barrena

Received: 14 November 2022
Revised: 9 December 2022
Accepted: 22 December 2022
Published: 26 December 2022

1. Introduction

By combining multiple radio telescopes over a long distance, very-long-baseline interferometry (VLBI) technology can achieve an extremely high angular resolution. As a global VLBI array, the Event Horizon Telescope (EHT) has been used to image the shadow regions of the supermassive black holes Messier 87 (M87) and Sagittarius A* (Sgr A*) [1,2]. However, the baseline coverage of the current EHT is very sparse. To obtain better black hole images and high-quality movies, the next-generation EHT (ngEHT) (https://www.ngeht.org/) (accessed on 8 December 2022) [3] team is actively working to add ~10 new stations, which, together with the existing EHT stations, will greatly increase their static and dynamic imaging capabilities. In addition, several existing/planned (sub)millimeter telescopes in other regions will also be able to contribute to future EHT/ngEHT observations. These include the planned Africa Millimetre Telescope (AMT) in Namibia [4], the 10 m Solar Planetary Atmosphere Research Telescope (SPART) in Japan, and the KVN Yonsei telescope in South Korea [5]. SPART and the KVN Yonsei are located in the easternmost part of Asia, and have no common visibility with the AMT.

The Tibetan Plateau is the hinterland of Asia, and is believed to have many excellent sites for (sub)millimeter radio astronomical observations [6]. A telescope in this region could serve to connect the stations in east Asia (e.g., SPART and KVN Yonsei) with the AMT, and the EHT/ngEHT coverage in the eastern hemisphere will be well supplemented. For this reason, Shanghai Astronomical Observatory is currently leading preparations for the construction of the 15 m Sub-Millimeter Astronomical Research Telescope (SMART) in this region. Considering the geographical location and infrastructure conditions, Shigatse (hereafter SG; 29.2° N, 88.63° E, altitude of 4080 m) which is located in the south central area of the Tibetan Plateau, has been proposed as a candidate site for the telescope.

An analysis conducted using the dataset from the Modern-Era Retrospective Analysis for Research and Applications project, version 2 (MERRA-2) [7], showed that the meteorological conditions of the SG site are comparable to those of some existing EHT stations. We simulated VLBI observations of Sgr A* and M87 at 230 GHz based on the SG, SPART, KVN Yonsei, AMT, and the existing EHT sites with mutual visibility. Then, we performed imaging with the synthetic data. The results demonstrated that VLBI observations centered on the SG site could be used to reconstruct the images of Sgr A* and M87 well.

2. Evaluation of Meteorological Conditions

Water vapor in the atmosphere is the main factor related to absorbing (sub)millimeter waves. Therefore, we evaluated two related meteorological metrics, precipitable water vapor (PWV) and zenith transmittance spectra. In addition to the SG site, two representative EHT stations (GLT and SMT) were also evaluated for comparison.

The MERRA-2 data product used is M2I3NPASM, which is an instantaneous 3-dimensional 3-hourly data collection system, with a spatial resolution of 0.5° in latitude and 0.625° in longitude and 42 vertical pressure layers. The time span is from 1 January 2018 to 31 December 2021. am software [8,9] was used to calculate the average meteorological values. Before this calculation, the gridded data were interpolated to the precise location of each site [10]. The 42 pressure layers were truncated according to the actual altitude of each site, and then the values from the first layer to the truncated layer were integrated to calculate the statistical values. The truncated pressures at the SG, GLT, and SMT sites are 580 mbar, 940 mbar, and 700 mbar, respectively (All data and scripts are available at https://github.com/nomadyuwei/Site_Evaluation.git, accessed on 8 December 2022.).

The monthly averaged PWV values of the three sites are shown in Figure 1. GLT exhibits the lowest values in most months. SG and SMT are roughly equivalent, but during the winter months, the values of SG are better than those of SMT. For the months of January, February, March, April, November, and December, the PWV values of all the three sites are lower than 4 mm. This is very important, because EHT observations are often performed in March or April, when all of the existing EHT sites have median PWV values that are less than 5 mm [10]. We also compared the PWV values of GLT and SMT shown in Figure 1 with those reported in other studies from the literature (e.g., Raymond et al. [10] and references therein), and the results were found to be consistent.

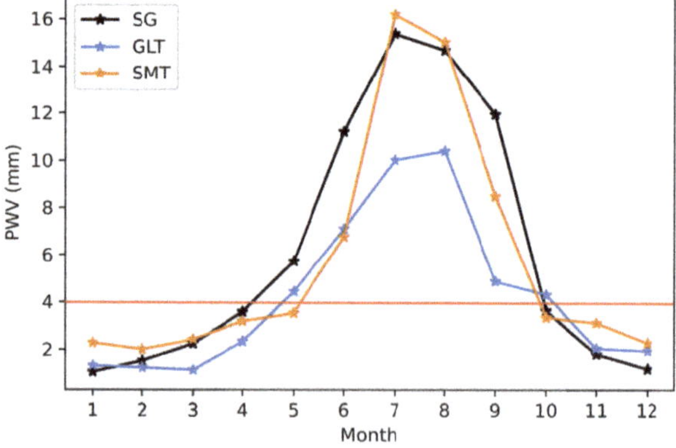

Figure 1. Monthly average PWV values of the three sites from 1 January 2018 to 31 December 2021. Values were derived from the analysis of MERRA-2 model data.

Figure 2 shows the four-year averaged zenith transmittance spectra of the three sites in March and April. For the SG site, the median values at 230 GHz and 345 GHz are about 0.9 and 0.7 in March, and about 0.85 and 0.55 in April. These values are comparable to those of GLT and SMT.

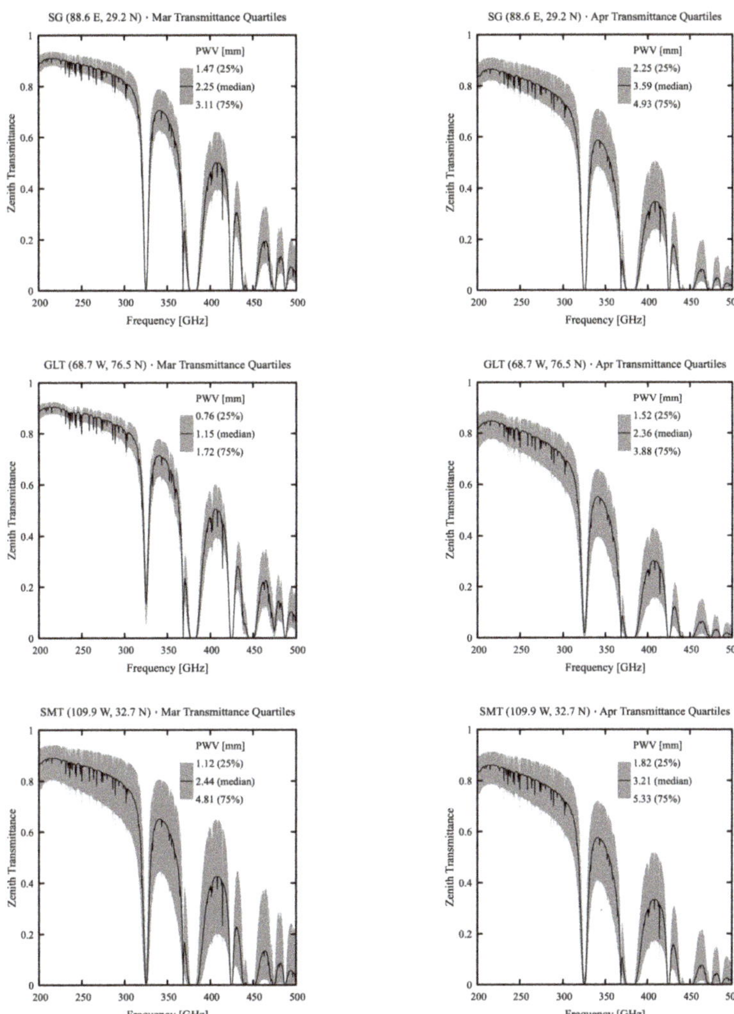

Figure 2. Four-year (2018–2021) averaged zenith transmittance spectra of the three sites in March (**left**) and April (**right**). Rows from top to bottom correspond to SG, GLT, and SMT. Values were derived from the analysis of MERRA-2 model data.

The wind speeds throughout the year of 2019 were also evaluated (Figure 3). The values of these three sites are within 20 m/s most of the time, and the SG site exhibits the smallest peak values.

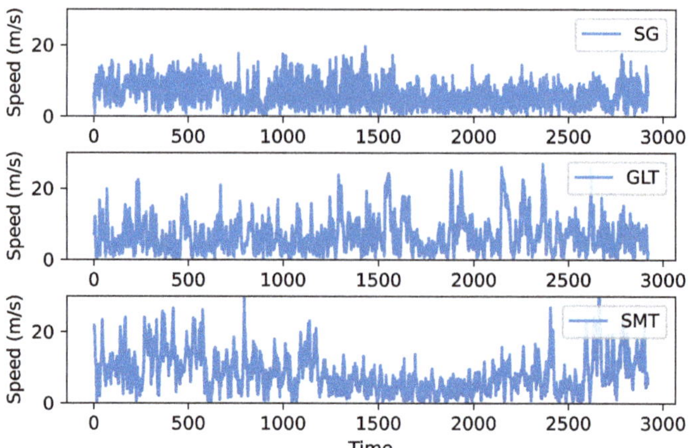

Figure 3. Wind speeds throughout the year of 2019. Values were derived from the analysis of MERRA-2 model data.

3. Synthetic Data Generation and Imaging Method

We considered an array consisting of the 11 existing EHT stations, the SG station proposed in this paper, and a further three new stations under development (SPART, KVN Yonsei, and AMT). This gives a total of 15 stations located at 13 sites. We simulated VLBI observations during the time interval that the SG site could cover. In practice, only the four new stations and a portion of the existing EHT stations with mutual visibility in relation to the new stations were used. A detailed description of the EHT stations can be found in [11] and the references therein.

The locations and system equivalent flux density (SEFD) values of the four new stations are shown in Table 1. SEFD is the most important metric of the sensitivity of a radio telescope, and it is calculated as follows:

$$SEFD = \frac{8k_B T_{sys}}{\eta_t \pi D^2}. \tag{1}$$

where k_B is the Boltzmann constant (1.38×10^3 Jy/K), T_{sys} is the effective system noise temperature [12] and D and η_t are the diameter and efficiency of the telescope. We estimate the SEFD values of SG and AMT (which are in the pre-construction stage) in Table 1 as follows. We assume that the values of T_{sys} and η_t for SG/AMT are 200 K and 0.6, which are the average values of the existing EHT stations. The diameter of the SG telescope is assumed to be 15 m, so the corresponding SEFD is about 5200. According to the literature [4], the diameter of the planned AMT is from 12 m to 16 m; we used the median value of 14 m, so the corresponding SEFD is about 6000.

Table 1. Locations and SEFD values of the four new stations. SEFD values were calculated for 230 GHz.

Telescope	Location	SEFD (Jy)	X (m)	Y (m)	Z (m)
SPART	Japan	10,000	−3,871,061.02	3,428,327.24	3,723,784.27
Yonsei	Korea	4428	−3,042,280.91	4,045,902.72	3,867,374.35
SG	China	5200	92,309.877	5,566,822.809	3,109,831.733
AMT	Namibia	6000	5,627,857.426	1,638,239.676	−2,512,266.994

The black holes Sgr A* and M87 are the main simulated observation targets. The model image of Sgr A* is a semi-analytic radiatively inefficient accretion flow (RIAF) model [13]. For M87, we used a snapshot of a GRMHD simulation [14]. The frequency is set to 230 GHz with a total bandwidth of 8 GHz. The integration time is 10 s with scans taken every 10 min. We adopted `eht-imaging` software for the creation of synthetic data and imaging.

Ideally, the VLBI imaging process uses inverse Fourier transformation to reconstruct the celestial image $I(x, y)$ based on visibility data $V(u, v)$. However, due to the sparse coverage of $V(u, v)$, the imaging process is an ill-posed problem. `eht-imaging` solves this problem by adding regularization terms based on prior information. The imaging process defines a regularized maximum likelihood (RML) algorithm through the use of a dataset, an initial image, and an objective function with given data and regularization term weights [15], reconstructing celestial image $I(x, y)$ as

$$I = \arg\min_{I}\left(\sum \alpha_D \chi_D^2(I) - \sum \beta_R S_R(I)\right). \tag{2}$$

The above equation includes two types of terms, data terms ($\chi_D^2(I)$) and regularization terms ($S_R(I)$). In this study, we used visibility amplitudes, closure phases, and log closure amplitudes as data terms. Closure phases are triple products of complex visibilities around closed triangles [16–18], which can cancel out station-based phase errors. Closure amplitudes were calculated from four telescopes forming a quadrangle. Closure amplitudes cancel out station-based gain errors.

The regularization terms used in this study are entropy, total variation, centroid position constraint, and total flux density. For a detailed description of these, we refer the reader to [19–22] and references therein. α_D and β_R are hyperparameters that were adjusted during the imaging process.

The pipeline of the imaging process is shown in Figure 4. We solved iteratively over four rounds to determine the optimal image and self-calibration. This pipeline is very similar to the one used to generate the fiducial images of M87 [1,23]. At the beginning, we used a Gaussian function to model the source structure. The full width at half maximum (FWHM) of the Gaussian function was set to 50 μas and 40 μas for Sgr A* and M87, respectively. The size of the image was set to 32×32 pixels. The iteration number of each round was set to 200. In different rounds, the weights of the data terms are different, but the weights of the regularization terms remain the same. We used the same set of weights for all simulated observations. The finely-tuned weights of each round are shown in Tables 2 and 3 (All data and Python pipelines are available at https://github.com/nomadyuwei/EHT_Asia_Simulation_Imaging.git, accessed on 8 December 2022.).

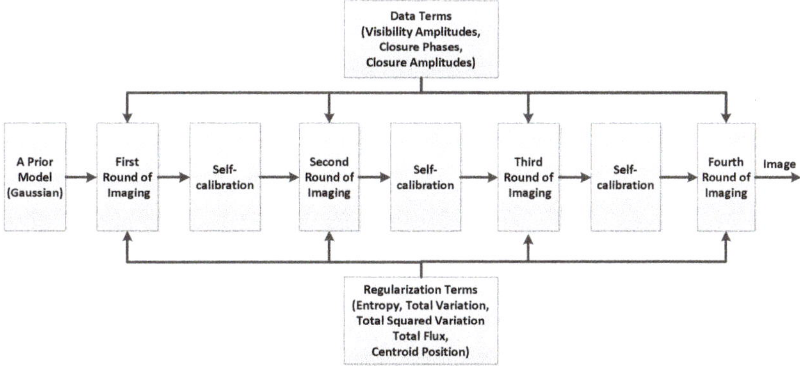

Figure 4. Pipeline of the imaging process.

Table 2. Weights of data terms.

Data/Regularizer Terms	Weight (Round 1)	Weight (Round 2)	Weight (Rounds 3 and 4)
Visibility Amplitude (α_{amp})	1	1	5
Closure phase (α_{CP})	10	1	2
Log Closure Amplitude (α_{lnCA})	10	1	2

Table 3. Weights of regularization terms.

Entropy ($\beta_{entropy}$)	10	10	10
Total Variation (β_{TV})	1	1	1
Total squared variation (β_{TSV})	1	1	1
Centroid Position (β_{cen})	20	20	20
Total Flux Density (β_{flux})	100	100	100

Normalized root-mean-square error (NRMSE) was used as a metric to evaluate the quality of the reconstructed images. The calculation of NRMSE is as follows:

$$NRMSE = \sqrt{\frac{\sum_{i=1}^{n^2} |I'_i - I_i|^2}{\sum_{i=1}^{n^2} |I_i|^2}}. \tag{3}$$

where I'_i and I_i are pixels of the reconstructed and original images and n^2 is the size of the images.

4. Simulation Results and Discussion

The simulation results are presented in this section, first in terms of (u, v) coverage as a function of source elevation, and second through a comparison of images simulated with and without the inclusion of the SG site.

4.1. Elevation and Visibility Coverage

Figure 5 shows the elevation of 13 stations above 10 degrees as a function of Greenwich mean sidereal time (GMST) when observing Sgr A* and M87. As we can see, for Sgr A*, the observable time of the SG site is in the interval of [8:00, 16:00]. With the exception of the South Pole Telescope (SPT), none of the existing EHT stations could fully cover this 8-h interval. The other two eastern telescopes (SPART and Yonsei) have observation times of about 4 h during this interval. Therefore, we conducted simulated observations of Sgr A* with this 8-h interval, and stations outside this interval did not participate in the observations.

The (u, v) coverage of the simulated observation of Sgr A* is shown in Figure 6 (top-right) with the baselines contributed by the SG site highlighted in orange. For comparison, Figure 6 (top-left) also shows the (u, v) coverage without the SG site. It can be seen that the SG site could contribute both long and middle baselines, which are very important for calibration and imaging.

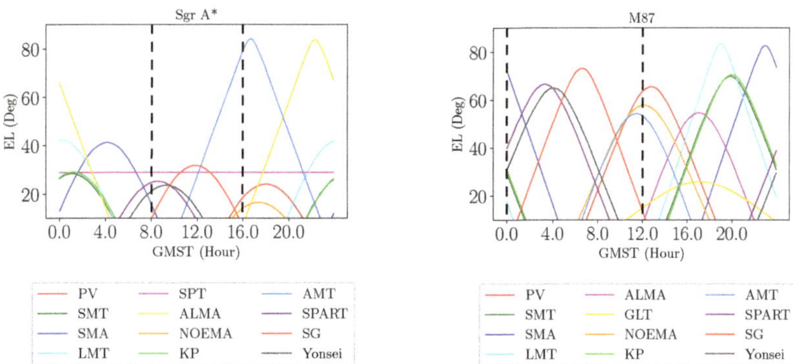

Figure 5. (**Left**) Elevation of telescopes vs GMST when observing Sgr A*. GLT is absent because it cannot observe Sgr A*. (**Right**) Elevation of telescopes vs GMST when observing M87. SPT is absent because it cannot observe M87. Since JCMT and APEX had the same sites as SMA and ALMA, we omitted them from the figure. The simulated observations were only conducted in the time intervals between the black vertical dashed lines. Stations outside these intervals did not participate in the observations.

Figure 7 (top-left and top-right) compares the dirty beams of the simulated observations of Sgr A* without and with the SG site. It can be seen that with the SG site, the side lobes are reduced. However, the dirty beams are only used to illustrate the impact of the SG site. `eht-imaging` does not use the dirty map or dirty beam for imaging, and the reconstructed images also do not need to be convolved with these beams.

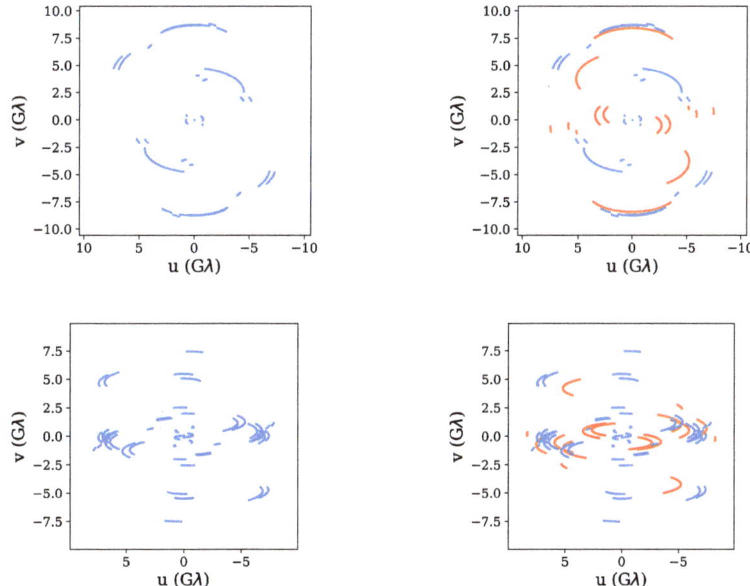

Figure 6. (u, v) coverage of Sgr A* (**top**) and M87 (**bottom**) without and with the SG site (SG site in orange). An elevation limit of 10 degrees is applied for each station.

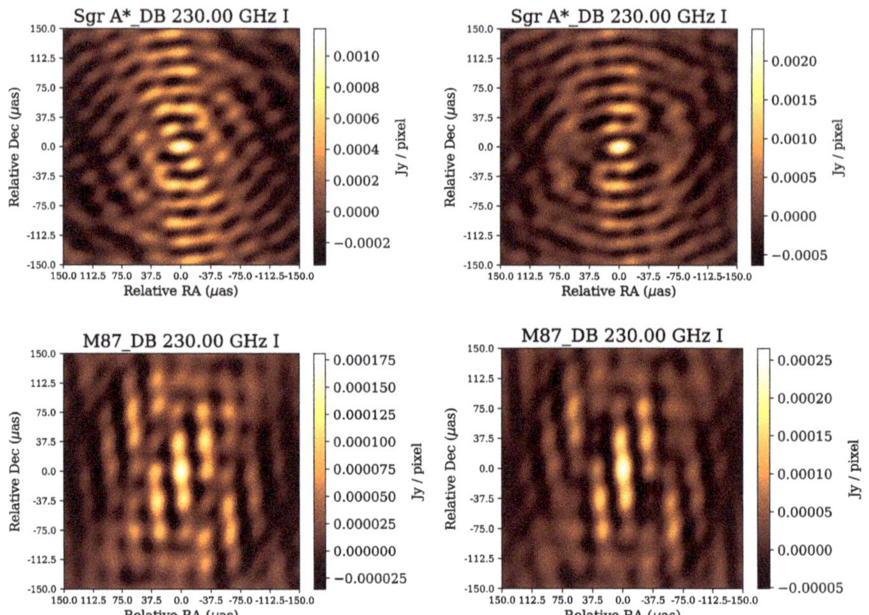

Figure 7. (**Top-left**) Dirty beam of Sgr A* without the SG site. (**Top-right**) Dirty beam of Sgr A* with the SG site. (**Bottom-left**) Dirty beam of M87 without the SG site. (**Bottom-right**) Dirty beam of M87 with the SG site.

Since M87 is in the northern celestial hemisphere, the elevation curves (Figure 5, right) of eastern telescopes (SG, SPART and Yonsei) are much better than those for Sgr A*. It can be seen that the observable time of the SG site is approximately in the interval of [0:00, 12:00]. Most of the observable time of Yonsei and SPART is also within this interval. Therefore, we conducted simulated observations for this 12-h interval. Similarly to the case of Sgr A*, stations outside this interval did not participate in the observations. Figure 6 (bottom-left and bottom-right) shows the (u, v) coverages of the observations of M87 without and with the SG site. It can be seen that, similarly to that of Sgr A*, the SG site could also contribute both long and middle baselines. Figure 7 (bottom-left and bottom-right) shows the dirty beams for M87. It can be seen that that the beam is also somewhat improved with the SG site.

4.2. Imaging

The data terms used for imaging are visibility amplitudes and closure quantities. These are shown in Figure 8 for the eight-hour simulated observations of Sgr A*. It can be seen in the left sub-figure that without the SG site, there are two large gaps in the u–v distance, which reflects the importance of the SG site in baseline coverage. In the right sub-figure, we can see that all of the representative triangles with the SG site have nonzero closure phases. This indicates that the source structure on the 50 μas scale could be resolved with these triangles. Even a constant non-zero closure phase over time indicates a spatially resolved asymmetric structure.

Figure 9 shows the model image and reconstructed images of Sgr A*. Without the SG site, the visual quality of the reconstructed image is not ideal. For example, the shadow area is not obvious, and the NRMSE is large. With the SG site, the reconstructed image shows a clear crescent shape. The shadow and the bright area are clearly visible. The NRMSE value decreases from 0.537 to 0.225.

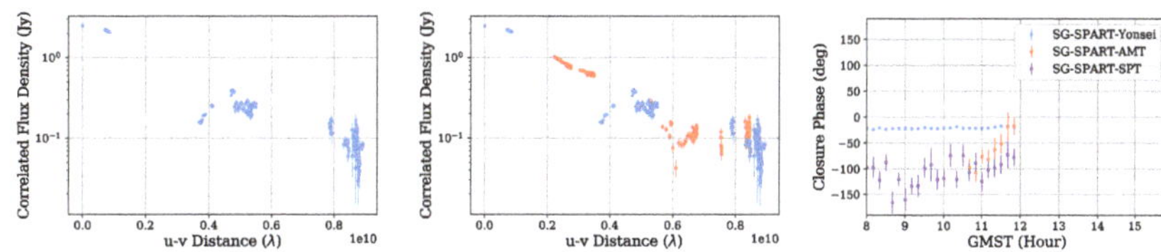

Figure 8. (**Left**) Visibility amplitudes as a function of u–v distance for Sgr A* without the SG site. (**Middle**) Visibility amplitudes as a function of u–v distance for Sgr A* with the SG site (SG site in orange). (**Right**) Closure phases as a function of GMST for Sgr A*.

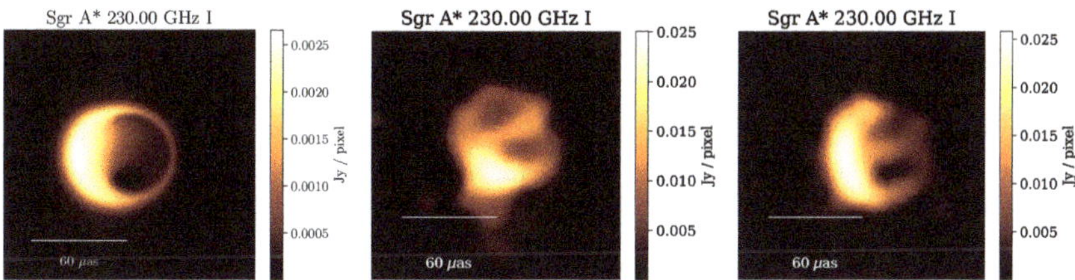

Figure 9. (**Left**) Model image of Sgr A*. (**Middle**) Reconstructed image of Sgr A* without the SG site (NRMSE = 0.537). (**Right**) Reconstructed image of Sgr A* with the SG site (NRMSE = 0.225).

For M87, the visibility amplitudes and closure phases of the twelve-hour simulated observations are shown in Figure 10. Comparing the left and middle sub-figures, we can see that without the SG site, the baseline coverage is still good. Even so, the SG site provides more baselines, which are beneficial in terms of improving the quality of the reconstructed images.

A null could be observed at the position of about 3.4 Gλ of the visibility amplitudes, which is associated with the ring-like structure in the image domain [23]. The Fourier transform of a ring structure shows the first minimum in the visibility amplitude at a baseline length b_1 for which the zero-order Bessel function is zero [24]. This allows us to estimate the source size as follows:

$$d_0 \approx 45\left(\frac{3.5G\lambda}{b_1}\right)\mu\text{as}. \tag{4}$$

The null in the position of 3.4 Gλ (b_1) corresponds to a source size of 42 μas (d_0), which matches the model image. With the non-zero closure phases shown in Figure 10 (right), we could resolve the source structure on a 40 μas scale.

Figure 11 shows the model image and reconstructed images of M87. As we can see, the twelve-hour simulated observation without the SG site is still able to reconstruct the shadow and bright areas of the model image. With the SG site, the quality of the reconstructed image can be further improved, although the improvement is not as great as for Sgr A*. This is indicated by the smaller NRMSE value, which is 0.224 without SG, and 0.175 with the extra site.

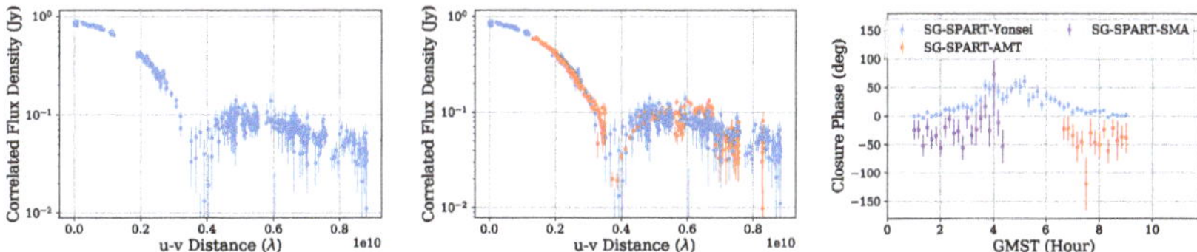

Figure 10. (**Left**) Visibility amplitude as a function of u–v distance for M87 without the SG site. (**Middle**) Visibility amplitude as a function of UV-distance for M87 with the SG site (SG site in orange). (**Right**) Closure phases as a function of GMST for M87.

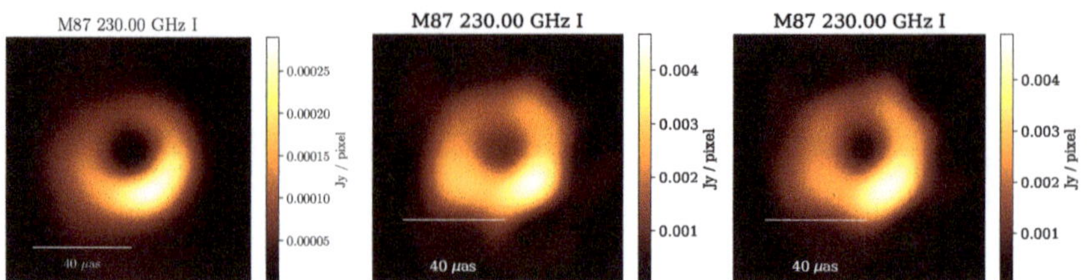

Figure 11. (**Left**) Model image of M87 (blurred with 11 μas circle Gauss beam). (**Middle**) Reconstructed image of M87 without the SG site (NRMSE = 0.224). (**Right**) Reconstructed image of M87 with the SG site (NRMSE = 0.175).

5. Conclusions

In this study we used four-year MERRA-2 data to evaluate the meteorological conditions of a candidate EHT site at Shigatse, or SG, located in the Tibetan Plateau. The PWV values were found to be very low during winter and spring, and the transmittance values, at 230 GHz and 345 GHz, are excellent in March and April. Thus, SG is a suitable site for (sub)millimeter observations.

We combined the SG site with other (sub)millimeter stations to carry out simulated observations. The other sites include three proposed new VLBI sites and a portion of the current EHT sites having mutual visibility of Sgr A* and M87 with SG. Our simulated results include both (u, v) coverage analysis and imaging comparisons with and without SG. For both sources, SG can provide important baseline coverage. The imaging results showed that with the SG site, the "Eastern ngEHT Array" plus a portion of the existing EHT stations could reconstruct good images for both Sgr A* and M87. In the imaging results, the improvement in the Sgr A* simulation is clearer than for M87; however, there is a measurable improvement for both sources.

In future works, we plan to evaluate the improvement in images obtained from the Eastern ngEHT Array for sources other than SgrA* and M87. With this limited analysis we have shown that the SG site can make a promising contribution to the ngEHT for global observations with continuous time coverage. The SG site will make progressively more significant improvements to imaging as more sites are added to the Eastern ngEHT Array.

Author Contributions: Conceptualization, R.-S.L., Z.-Q.S. and W.Y.; methodology, W.Y. and R.-S.L.; software, W.Y.; writing—original draft, W.Y. and R.-S.L.; writing—review and editing, W.Y., R.-S.L., Z.-Q.S. and J.W.; supervision, R.-S.L., Z.-Q.S. and J.W. All authors have read and agreed to the published version of the manuscript.

Funding: R.-S. Lu was supported by the National Natural Science Foundation of China (1193307), the Key Research Program of Frontier Sciences, CAS (ZDBS-LY-SLH011), and the Shanghai Pilot Program for Basic Research - Chinese Academy of Science, Shanghai Branch: JCYJ-SHFY-2022-013.

Institutional Review Board Statement: Not applicable.

Informed Consent Statement: Not applicable.

Data Availability Statement: Publicly available datasets were analyzed in this study. All data can be found here: https://github.com/nomadyuwei/Site_Evaluation.git and https://github.com/nomadyuwei/EHT_Asia_Simulation_Imaging.git, all accessed on 8 December 2022.

Acknowledgments: We thank Scott Paine, Mark Freeman, Geoffrey Bower, Iniyan Natarajan, Razieh Emami, Freek Roelofs, Daniel Palumbo, and Sheperd Doeleman for their helpful conversations about station meteorological analysis and VLBI imaging.

Conflicts of Interest: The authors declare no conflict of interest.

References

1. Event Horizon Telescope Collaboration; Akiyama, K.; Alberdi, A.; Alef, W.; Asada, K.; Azulay, R.; Baczko, A.K.; Ball, D.; Baloković, M.; Barrett, J.; et al. First M87 Event Horizon Telescope Results. I. The Shadow of the Supermassive Black Hole. *Astrophys. J. Lett.* **2019**, *875*, L1. [CrossRef]
2. Event Horizon Telescope Collaboration; Akiyama, K.; Alberdi, A.; Alef, W.; Algaba, J.C.; Anantua, R.; Asada, K.; Azulay, R.; Bach, U.; Baczko, A.K.; et al. First Sagittarius A* Event Horizon Telescope Results. I. The Shadow of the Supermassive Black Hole in the Center of the Milky Way. *Astrophys. J. Lett.* **2022**, *930*, L12. [CrossRef]
3. Doeleman, S.; Blackburn, L.; Dexter, J.; Gomez, J.L.; Johnson, M.D.; Palumbo, D.C.; Weintroub, J.; Farah, J.R.; Fish, V.; Loinard, L.; et al. Studying Black Holes on Horizon Scales with VLBI Ground Arrays. *Bull. Am. Astron. Soc.* **2019**, *51*, 256.
4. Backes, M.; Müller, C.; Conway, J.E.; Deane, R.; Evans, R.; Falcke, H.; Fraga-Encinas, R.; Goddi, C.; Klein Wolt, M.; Krichbaum, T.P.; et al. The Africa Millimetre Telescope. In Proceedings of the 4th Annual Conference on High Energy Astrophysics in Southern Africa (HEASA 2016), Cape Town, South Africa, 25–26 August 2016; p. 29. [CrossRef]
5. Asada, K.; Kino, M.; Honma, M.; Hirota, T.; Lu, R.S.; Inoue, M.; Sohn, B.W.; Shen, Z.Q.; Ho, P.T.P.; Akiyama, K.; et al. White Paper on East Asian Vision for mm/submm VLBI: Toward Black Hole Astrophysics down to Angular Resolution of $1{\sim}R_S$. *arXiv* **2017**, arXiv:1705.04776.
6. Ye, Q.Z.; Su, M.; Li, H.; Zhang, X. Tibet's Ali: Asia's Atacama? *Mon. Not. R. Astron. Soc. Lett.* **2016**, *457*, L1–L4. [CrossRef]
7. Gelaro, R.; McCarty, W.; Suárez, M.J.; Todling, R.; Molod, A.; Takacs, L.; Randles, C.A.; Darmenov, A.; Bosilovich, M.G.; Reichle, R.; et al. The Modern-Era Retrospective Analysis for Research and Applications, Version 2 (MERRA-2). *J. Clim.* **2017**, *30*, 5419–5454. [CrossRef] [PubMed]
8. Paine, S. The Am Atmospheric Model. Zenodo. 2019. Available online: https://doi.org/10.5281/zenodo.3406483 (accessed on 8 December 2022).
9. Paine, S. BH PIRE Winter School 2018 Tutorial: Radio Climatology. 2018. Available online: https://bhpire.arizona.edu/education/tutorials/ (accessed on 8 December 2022).
10. Raymond, A.W.; Palumbo, D.; Paine, S.N.; Blackburn, L.; Córdova Rosado, R.; Doeleman, S.S.; Farah, J.R.; Johnson, M.D.; Roelofs, F.; Tilanus, R.P.J.; et al. Evaluation of New Submillimeter VLBI Sites for the Event Horizon Telescope. *Astrophys. J. Suppl. Ser.* **2021**, *253*, 5. [CrossRef]
11. Event Horizon Telescope Collaboration; Akiyama, K.; Alberdi, A.; Alef, W.; Asada, K.; Azulay, R.; Baczko, A.K.; Ball, D.; Baloković, M.; Barrett, J.; et al. First M87 Event Horizon Telescope Results. II. Array and Instrumentation. *Astrophys. J. Lett.* **2019**, *875*, L2. [CrossRef]
12. Event Horizon Telescope Collaboration; Akiyama, K.; Alberdi, A.; Alef, W.; Asada, K.; Azulay, R.; Baczko, A.K.; Ball, D.; Baloković, M.; Barrett, J.; et al. First M87 Event Horizon Telescope Results. III. Data Processing and Calibration. *Astrophys. J. Lett.* **2019**, *875*, L3. [CrossRef]
13. Broderick, A.E.; Fish, V.L.; Doeleman, S.S.; Loeb, A. Evidence for Low Black Hole Spin and Physically Motivated Accretion Models from Millimeter-VLBI Observations of Sagittarius A*. *Astrophys. J.* **2011**, *735*, 110. [CrossRef]
14. Chael, A.; Narayan, R.; Johnson, M.D. Two-temperature, Magnetically Arrested Disc simulations of the jet from the supermassive black hole in M87. *Mon. Not. R. Astron. Soc.* **2019**, *486*, 2873–2895. [CrossRef]
15. Chael, A.A. Simulating and Imaging Supermassive Black Hole Accretion Flows. Ph.D. Thesis, Harvard University, Cambridge, MA, USA, 2019.
16. Jennison, R.C. A phase sensitive interferometer technique for the measurement of the Fourier transforms of spatial brightness distributions of small angular extent. *Mon. Not. R. Astron. Soc.* **1958**, *118*, 276. [CrossRef]
17. Rogers, A.E.E.; Hinteregger, H.F.; Whitney, A.R.; Counselman, C.C.; Shapiro, I.I.; Wittels, J.J.; Klemperer, W.K.; Warnock, W.W.; Clark, T.A.; Hutton, L.K.; et al. The structure of radio sources 3C 273B and 3C 84 deduced from the "closure" phases and visibility amplitudes observed with three-element interferometers. *Astrophys. J.* **1974**, *193*, 293–301. [CrossRef]
18. Lu, R.S.; Roelofs, F.; Fish, V.L.; Shiokawa, H.; Doeleman, S.S.; Gammie, C.F.; Falcke, H.; Krichbaum, T.P.; Zensus, J.A. Imaging an Event Horizon: Mitigation of Source Variability of Sagittarius A*. *Astrophys. J.* **2016**, *817*, 173. [CrossRef]
19. Chael, A.A.; Johnson, M.D.; Narayan, R.; Doeleman, S.S.; Wardle, J.F.C.; Bouman, K.L. High-resolution Linear Polarimetric Imaging for the Event Horizon Telescope. *Astrophys. J.* **2016**, *829*, 11. [CrossRef]
20. Akiyama, K.; Kuramochi, K.; Ikeda, S.; Fish, V.L.; Tazaki, F.; Honma, M.; Doeleman, S.S.; Broderick, A.E.; Dexter, J.; Mościbrodzka, M.; et al. Imaging the Schwarzschild-radius-scale Structure of M87 with the Event Horizon Telescope Using Sparse Modeling. *Astrophys. J.* **2017**, *838*, 1. [CrossRef]
21. Akiyama, K.; Ikeda, S.; Pleau, M.; Fish, V.L.; Tazaki, F.; Kuramochi, K.; Broderick, A.E.; Dexter, J.; Mościbrodzka, M.; Gowanlock, M.; et al. Superresolution Full-polarimetric Imaging for Radio Interferometry with Sparse Modeling. *Astrophys. J.* **2017**, *153*, 159. [CrossRef]
22. Chael, A.A.; Johnson, M.D.; Bouman, K.L.; Blackburn, L.L.; Akiyama, K.; Narayan, R. Interferometric Imaging Directly with Closure Phases and Closure Amplitudes. *Astrophys. J.* **2018**, *857*, 23. [CrossRef]

23. Event Horizon Telescope Collaboration; Akiyama, K.; Alberdi, A.; Alef, W.; Asada, K.; Azulay, R.; Baczko, A.K.; Ball, D.; Baloković, M.; Barrett, J.; et al. First M87 Event Horizon Telescope Results. IV. Imaging the Central Supermassive Black Hole. *Astrophys. J. Lett.* **2019**, *875*, L4. [CrossRef]
24. Event Horizon Telescope Collaboration; Akiyama, K.; Alberdi, A.; Alef, W.; Asada, K.; Azulay, R.; Baczko, A.K.; Ball, D.; Baloković, M.; Barrett, J.; et al. First M87 Event Horizon Telescope Results. VI. The Shadow and Mass of the Central Black Hole. *Astrophys. J. Lett.* **2019**, *875*, L6. [CrossRef]

Article

Prospects of GPU Tensor Core Correlation for the SMA and the ngEHT

Wei Yu [1,*], John W. Romein [2], L. Jonathan Dursi [3], Ru-Sen Lu [4], Adrian Pope [5], Gareth Callanan [6], Dominic W. Pesce [1,7], Lindy Blackburn [1,7], Bruce Merry [8], Ranjani Srinivasan [1], Jongsoo Kim [9] and Jonathan Weintroub [1]

[1] Center for Astrophysics | Harvard & Smithsonian, 60 Garden Street, Cambridge, MA 02138, USA
[2] ASTRON, Netherlands Institute for Radio Astronomy, Oude Hoogeveensedijk 4, 7991 PD Dwingeloo, The Netherlands
[3] NVIDIA Canada, 431 King St W, Toronto, ON M5V 1K4, Canada
[4] Shanghai Astronomical Observatory, Chinese Academy of Sciences, 80 Nandan Road, Shanghai 200030, China
[5] Argonne National Laboratory, 9700 S. Cass Avenue, Lemont, IL 60439, USA
[6] Department of Computer Science, Lund University, Ole Römers väg 3, 223 63 Lund, Sweden
[7] Black Hole Initiative, Harvard University, 20 Garden Street, Cambridge, MA 02138, USA
[8] South African Radio Astronomy Observatory, 2 Fir Street, Black River Park, Observatory 7925, South Africa
[9] Korea Astronomy and Space Science Insitute, 776 Daedeok-daero, Yuseong-gu, Daejeon 34055, Republic of Korea
* Correspondence: wei.yu@cfa.harvard.edu

Abstract: Building on the base of the existing telescopes of the Event Horizon Telescope (EHT) and ALMA, the next-generation EHT (ngEHT) aspires to deploy ∼10 more stations. The ngEHT targets an angular resolution of ∼15 microarcseconds. This resolution is achieved using Very Long Baseline Interferometry (VLBI) at the shortest radio wavelengths ∼1 mm. The Submillimeter Array (SMA) is both a standalone radio interferometer and a station of the EHT and will conduct observations together with the new ngEHT stations. The future EHT + ngEHT array requires a dedicated correlator to process massive amounts of data. The current correlator-beamformer (CBF) of the SMA would also benefit from an upgrade, to expand the SMA's bandwidth and also match the EHT + ngEHT observations. The two correlators share the same basic architecture, so that the development time can be reduced using common technology for both applications. This paper explores the prospects of using Tensor Core Graphics Processing Units (TC GPU) as the primary digital signal processing (DSP) engine. This paper describes the architecture, aspects of the detailed design, and approaches to performance optimization of a CBF using the "FX" approach. We describe some of the benefits and challenges of the TC GPU approach.

Keywords: ngEHT; VLBI; SMA; correlation; GPU; Tensor Core

Citation: Yu, W.; Romein, J.W.; Dursi, L.J.; Lu, R-S.; Pope, A.; Callanan, G.; Pesce, D.W.; Blackburn, L.; Merry, B.; Srinivasan, R.; et al. Prospects of GPU Tensor Core Correlation for the SMA and the ngEHT. *Galaxies* **2023**, *11*, 13. https://doi.org/10.3390/galaxies11010013

Academic Editor: Phil Edwards

Received: 15 November 2022
Revised: 5 January 2023
Accepted: 9 January 2023
Published: 11 January 2023

1. Introduction

The Event Horizon Telescope (EHT) is a globe-spanning Very Long Baseline Interferometry (VLBI) array that has captured images of the shadow region of black holes at the center of the galaxy M87 and the Milky Way [1,2], attracting worldwide attention. The next-generation EHT (ngEHT)[1] will push this scientific frontier even further, building ∼10 more stations and doubling the observation bandwidth to improve the imaging capabilities and capture the first black hole movies.

The future EHT + ngEHT array will include ∼21 stations. For each station, the sideband bandwidth will increase from the current 4 GHz to 8 GHz. Dual polarization, two sidebands, and simultaneous dual-frequency (230 GHz and 345 GHz) observations correspond to a bandwidth of 64 GHz and recording data rate of 256 Gbps for two-bit recording and Nyquist sampling [3]. Current EHT observations are 7 days per year, while ngEHT observations are expected to run 60 days per year. Thus, compared with the current EHT, the future array will have more stations, a wider bandwidth, and more observation days each year. The amount of observation data is anticipated to be tens of petabytes in size, which brings great challenges to the correlation. This is the motivation to build a dedicated high-performance correlator[2].

The Submillimeter Array (SMA) is a standalone radio interferometer consisting of eight antennas. A digital correlator-beamformer called SWARM [6] functions as two distinct instruments: an FX correlator that computes fringe visibilities across frequencies for every pair of antennas, and a beamformer that forms the coherent phased array sum of the eight antennas, aggregating the SMA collecting area to an equivalent single large-aperture telescope. The beamformer mode was designed in SWARM to enable its participation in EHT observations.

In order to match the observational capabilities of the ngEHT and the Wideband SMA Project (wSMA), it is desirable to upgrade the correlator-beamformer of the SMA. Likewise, the VLBI correlator now supporting EHT is limited in bandwidth, which can be processed in a reasonable computing time scale, and also the number of stations that it can handle. We propose a common GPU Tensor Core-based architecture for both of these wideband applications.

This paper reports on the prospects for a correlator-beamformer system having common features that can potentially benefit wSMA and ngEHT. The primary focus of the work described here is the design of an X-engine prototype based on an open-source Tensor Core library [7]. We note that we have already built a small prototype two-server four-GPU system, on which we have micro-benchmarked and tuned some of the codes and subsystems described in this work. We have made extensive use of the NVIDIA profiling tool Nsight to optimize subsystem performance. We have not yet built the full system described in this paper; however, we continue to actively develop the experimental counterpart to this work.

2. Related Work

The correlator-beamformer (CBF) is a key data processing system for radio interferometers and VLBI arrays. The correlator cross-correlates the baseband data of each antenna/station pair and outputs visibility data for imaging, while the beamformer aggregates the collecting area of the array for VLBI operations.

The correlator function of the CBF is divided into two broad classes. The XF type computes a direct cross-correlation between time series data from pairs of telescopes before transforming to frequency with Fast Fourier Transforms (FFT). The FX type computes the Fourier transform of time series data from each station first, followed by a cross-correlation. The two processing stages are called F-engine and X-engine, respectively. In the F-engine, data from each antenna/station are divided into multiple frequency channels. In the X-engine, a subset of frequency channels from all antenna/station pairs are bin-wise multiplied and the result integrated.

The trade-off between the XF and FX correlation architectures is multidimensional. For high spectral resolutions, the use of FFTs reduces the number of required multiplications in aggregate, although, because of bit growth in the FFTs, these multipliers need to be wider. There are other trade-offs in respect to memory utilization and the number of baselines. Reference [6] discusses multiplier utilization quantitatively and recommends the FX approach. Whereas XF correlators had been favored historically because wide multipliers were expensive, we observe that almost all new correlator designs (in the last decade, for example) use the FX architecture. We attribute this to the wider availability and lower resource cost of wide multipliers. In addition, FX correlators naturally provide more parallelism through early spectral decomposition. Modern correlator platforms mainly use Field-Programmable Gate Arrays (FPGAs), Graphics Processing Units (GPUs), or CPUs, which are both well equipped with sufficiently wide or even floating point multiplication and are designed at the root for a very high degree of parallelism.

FPGAs have the advantage of hardware programmability. Almost all of the existing hardware correlators are developed with FPGAs. Among them, the Collaboration for Astronomy Signal Processing and Electronics Research (CASPER)[3] open-source FPGA platform is the most popular solution; for example, the SWARM correlator for the SMA [6], the correlator for MeerKAT [8], and the correlator for the Arcminute Microkelvin Imager (AMI) array [9] are all developed with CASPER FPGA boards.

The powerful parallel computing resources of GPUs present an opportunity, and correlators are increasingly adopting GPUs for accelerated computing, especially in X-engines. Hybrid FPGA+GPU architectures have become a very popular solution for correlators. For example, the Canadian Hydrogen Intensity Mapping Experiment (CHIME) [10], the Large Aperture Experiment to Detect the Dark Ages (LEDA) project of the Long Wavelength Array (LWA) [11,12], the Donald C. Backer Precision Array for Probing the Epoch of Reionization (PAPER) [13], and the Murchison Widefield Array (MWA) [14] all adopt the hybrid FPGA+GPU architecture for their correlators, where FPGA boards (including ADCs) are used for data acquisition and channelization (F-engines), and GPU boards are used for cross-correlation computing (X-engines). There are also some correlators developed with pure GPUs, such as the early MWA 32-antenna prototype correlator [15] and the Cobalt correlator of the Low-Frequency Array (LOFAR) [16]. In these systems, the GPUs are used for the fine channelization of F-engines and cross-correlation of X-engines. ADCs and FPGAs are still required for data acquisition and coarse channelization.

For VLBI arrays, data acquisition and correlation are separated due to the remote station locations. Raw station data are recorded and shipped to a central location for correlation. VLBI correlators are typically CPU-based (software-based). Many VLBI institutions have developed a variety of software correlators. Among them, DiFX [4,5] is the most widely used one. Facilities such as the VLBI Global Observing System (VGOS), the Very Long Baseline Array (VLBA), the Australian Long Baseline Array (LBA), and the EHT [17] all use DiFX to correlate data. Recently, the EHT has been investigating a Cloud-based correlation scheme [18], which is also based on DiFX.

For VLBI and ngEHT, we sometimes use the term "near-real-time" to describe the desired performance of a wideband correlator. VLBI correlators use recorded data and therefore need not be strictly real-time. "Near-real-time" under these conditions signals that it is desirable to improve the efficiency of media recycling, reducing media costs, which can be very high for the wide bandwidths, high cadences, and increased number of stations of the ngEHT. As a side benefit, faster correlation reduces the time to science and improves fringe search iteration times and efficiency, and feedback on prior campaign results can improve the planning of future ones.

While various architectures have been introduced above, in this paper, we propose a GPU Tensor Core-based correlator for the wSMA and the ngEHT for the following reasons.

(1) GPUs have powerful computing resources, which are far more efficient than CPUs for parallel computing. In particular, the newly embedded Tensor Cores in NVIDIA GPUs are much more efficient than regular CUDA cores as well as FPGAs in terms of matrix multiplication [7].

(2) There is a wealth of open-source software libraries (both GPU-based and CPU-based) available for radio astronomical instruments. Although CASPER also has various open-source libraries, users still sometimes need to develop new firmware modules with the hardware description languages (HDL). Compared with CASPER libraries and HDL, the GPU source code is easier to upgrade and modify. The GPU development ecosystem includes sophisticated debugging and optimization tools and uses standard languages familiar to a wide swath of software engineers.

(3) The architecture of GPU-based correlators is flexible, maintainable, and scalable. High-performance hardware such as CPUs, GPUs, network interface cards (NICs), etc., can be integrated together easily.

3. System Design and Selection of Open-Source Libraries

We propose two schemes for the architecture of the correlator. In the first "expanded" scheme, the F-engines and X-engines are implemented in different GPU servers, which is shown in Figure 1. The corner-turner transpose between the two types of engines is implemented through the network switch. Each network packet from the F-engines contains data containing a subset of frequency channels. Different subsets of packets have different destination IP addresses and are transmitted to different X-engines through the switch. Thus, each X-engine processes a subset of frequency channels from all of the F-engines to form all of the array baselines.

In the second "compact" scheme, the core functions of the F-engine and the X-engine are arranged to execute in the same GPU server. This compact scheme reduces hardware resources such as GPUs, NICs, and CPUs. Thus, it is the preferred architecture for scaled deployment for smaller arrays. Since both wSMA and ngEHT have relatively few antennas and baselines, the compact scheme is potentially attractive. The expanded first scheme is simpler to build and debug, is the better starting point for development and laboratory debug, and is preferred for larger wideband arrays, such as ALMA.

We discuss the compact scheme in more detail in Section 7.

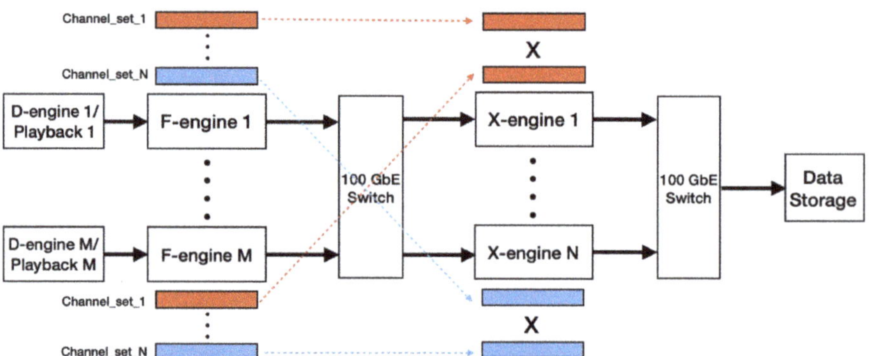

Figure 1. Level architecture of the proposed packetized FX-type correlator pioneered by CASPER. From left to right, note the D-engine samplers or VLBI playback recorders, the F-engines, a fast network switch implementing the corner-turner transpose, the X-engine, and archival visibility data storage. Not shown is the beamformer B-engine typically co-located with the X-engine.

The main difference between the wSMA real-time correlator and the ngEHT VLBI correlator is that the former uses D-engines to generate data, while the latter uses playback servers. A D-engine is a hardware system with ADCs and FPGAs to digitize the analog signals from each antenna, and it then transmits the data to the corresponding F-engine. The D-engine FPGA firmware optionally includes data processing modules that implement coarse channelization, slope and ripple equalization, and requantization[4].

Based on this same ADC-FPGA hardware, we are developing a wideband digital backend for the ngEHT, as shown in Figure 2, which includes a Xilinx VCU128 FPGA evaluation motherboard and a sub-board with 4×16 Gsps ADCs. This backend can be used as D-engines for the wSMA correlator with only slight modifications of the FPGA firmware. For the ngEHT correlator, the playback servers play back the recorded data from the digital backends to the corresponding F-engines.

Figure 2. The FPGA evaluation and custom ADC board set proof-of-concept platform. This is a general-purpose open-source platform, which, at SAO, will find application as the ngEHT digital backend and wSMA D-engine. The left board in the set is a commercial-off-the-shelf (COTS) Xilinx VCU128 FPGA evaluation motherboard. The right board is a custom plug-in to the VCU128's "FPGA Mezzanine Card +" (FMC+ or VITA 57.4) standard high-speed connector. This custom ADC board has four analog inputs leading to four Adsantec ASNT7123A ADC chips, each proven to run at 16.384 gigasamples-per-second (GS/s). The 4 × 100 GbE QSFP28 network ports on the FPGA motherboard can connect to the F-engines of the wSMA correlator or ngEHT recorders.

For the F-engine, a GPU-based library called katgpucbf[5] developed by the MeerKAT team is being considered. We have installed and tested the F-engine of the library on our experimental server. The next step is to modify it according to the specific requirements. A geometric delay model is required by both applications. The ngEHT VLBI correlator will have much more rapid fringe (or delay rate) correction. It also needs fringe search features, an across delay and delay rate using a geographic model of station placement, and a VLBI Data Interchange Format (VDIF) deformatter. The present focus of this work is on the correlation engine with delay and fringe correction development, and VLBI fringe search features, planned for future development.

Another library under consideration is the new cuFFTDx library developed by NVIDIA. As it generates FFTs that are GPU-callable, it is possible to merge FIR filters, FFTs, and phase corrections in a single GPU kernel instead of three. As a result, only one pass over the data is made, and as each of these functions is limited by the GPU memory bandwidth, one can expect an almost three-fold speedup.

For the X-engine, we developed a pipeline based on open-source libraries. Key components include a network transmission/receiving module, a format conversion module, a GPU-based correlation module, a long-term accumulation (LTA) module, and a pipeline framework to manage the above modules. We review and evaluate the existing open-source libraries and select the appropriate ones in the following sections.

The open-source libraries for network transmission/receiving that were under consideration include PF_ring[6], jive5ab[7], and SPEAD2[8]. PF_ring has been adopted by Mark6 equipments to capture network packets from VLBI digital backends. jive5ab is mainly used for data transfer and recording in e-VLBI systems, which has been used in Mark5/Mark6 and FlexBuffer by the European VLBI network. SPEAD is a data format for radio astronomy. SPEAD2 is a python/C++ library with the functions of the SPEAD formatter/deformatter and the network system. It supports both the traditional networking stack and the verbs API [19]. In order to match the MeerKAT katgpucbf F-engines, we select SPEAD2 as the basic library for the network transmission/receiving module.

The original and most popular GPU correlation library is xGPU [20], which has been adopted by many radio arrays, such as the LWA, MWA, PAPER, etc. When xGPU was developed 10 years ago, Tensor Cores had not appeared yet, so regular CUDA cores were used for correlation computing, which are not as efficient as Tensor Cores. The CHIME team developed a correlation library based on the AMD GPU. The AMD offerings feature higher computational throughput per unit cost than the comparable NVIDIA GPUs [21]; however, the former is more difficult to program than the latter. The recently developed Tensor Core Correlator (TCC) library adopts Tensor Cores as computing resources; the library resolves the complexity of using Tensor Cores and addresses several optimization challenges, such as the missing hardware support for complex numbers [7], which makes it much more efficient than xGPU. Considering the computational efficiency and development threshold, we select TCC as the correlation library.

The format conversion module is used to implement data conversion between F-engines and X-engines. For the implementation platform, we have two options, CPU-based and GPU-based. The input data of the xGPU library must be in the host memory, so the format conversion module must be implemented on the CPU. TCC does not have this restriction, so we implement this module on the GPU, which is much faster than on the CPU. Some observations require long-time integration, and the GPU memory may not be able to cache the pre-integrated data for such a long time. Thus, the TCC library is only used for short-term integration, and an LTA module is required after TCC, which can also be executed on the CPU or GPU. Currently, we implement this module on the CPU.

In order to improve the throughput, each of the above modules should be executed on an individual CPU thread under a framework[9]. Some open-source frameworks for radio astronomy include kotekan [22], PSRDADA[10], bifrost [23], and HASHPIPE[11]. Among them, kotekan, PSRDADA, and bifrost already include the network function inside; kotekan even includes the GPU correlation function. Our computing and network requirements are different from these functions. If we use them in our X-engine, these extra functions will need to be removed, which adds difficulty to the development. The HASHPIPE framework is a very general and convenient multi-thread management pipeline, which does not contain any specific function other than a framework. Since we have already selected modules with specific functions, HASHPIPE is an ideal management framework.

4. Key Hardware Technologies

As the correlator will be a cluster composed of multiple GPU servers, high-throughput network connectivity is key to harnessing the power of CPUs and GPUs. For this reason, we choose the NVIDIA ConnectX-5 2×100 Gbps NICs for the network system. Other key hardware includes the NVIDIA A5000 GPUs with Tensor Cores inside and the Intel Xeon Silver 4314 CPUs, shown in Figure 3. In this section, we provide a description of the advantages of the key hardware technologies of the proposed correlator.

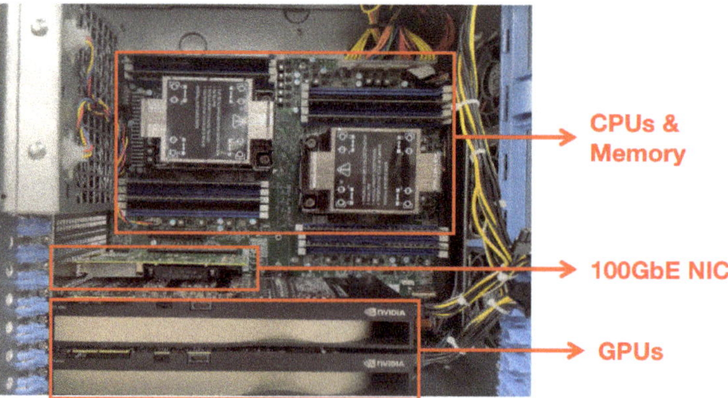

Figure 3. Key hardware inside the GPU server.

4.1. The NVIDIA ConnectX NICs

In the traditional socket-based network scheme, the CPU copies network packets from the NIC to the networking stack; after processing, the payload data are copied to the user space. This approach results in many memory copies and lots of processing for the CPUs, which limits the throughput. The SPEAD2 library is able to use hardware support in the NVIDIA ConnectX NICs to bypass the kernel's networking stack and directly access Ethernet frames with minimal copying[12]. The wire protocol is standard UDP. A potential alternative would be to use an RDMA protocol (such as RoCE) to have the NIC place data exactly where they are needed, but this is not currently supported by SPEAD2. In addition to being used for data communications between F-engines and X-engines, the NVIDIA ConnectX NICs will also be used between the D-engines and F-engines of the wSMA correlator, and between the digital backends and recorders of the ngEHT stations.

4.2. GPU Tensor Cores

Tensor Cores are mixed-precision computing units of NVIDIA GPUs. A Tensor Core can perform the matrix-multiply-and-accumulate operation ($D = A \times B + C$) in one GPU clock cycle, where A, B, C, and D are fixed-size matrices (typically 16×16) [7]. The A5000 GPU that we are using contains 256 third-generation Tensor Cores. Currently, NVIDIA provides three different ways to program Tensor Cores: the WMMA API, CUTLASS, and cuBLAS GEMM [24]. Here, TCC uses the lowest-level interface, the WMMA API, which mainly includes 3 functions. The first function is *load_matrix_sync()*, which loads matrices from the GPU memory to the registers of GPU threads. The second function is *mma_sync()*, which implements the actual matrix-multiply-and-accumulate operation. The third function is *store_matrix_sync()*, which copies the calculated results from the GPU registers to GPU memory. Currently, Tensor Cores are only used in the X-engines, and the F-engines use the regular CUDA cores.

5. Introduction of the Katgpucbf F-Engine

As mentioned previously, we are considering to use and modify the F-engine of the katgpucbf library, so we describe it briefly here. A detailed description can be found in the online documentation of https://katgpucbf.readthedocs.io/en/latest/index.html, accessed on 20 December 2022. The framework of the F-engine is developed in Python and mainly includes three functions: the *run_receive()* function, which receives network packets from a D-engine or a playback server; the *run_processing()* function, which calls the GPU to process the received network packets; and the *run_transmit()* function, which transmits the processed data to the X-engines through the network. The three functions run in parallel with the framework of Asyncio, which is an asynchronous programming framework of Python.

The *run_processing()* function is the core of this library. After a chunk of data is received, coarse delay compensation is first performed, and then the GPU kernel functions process the data to achieve channelization. The main signal processing algorithm is the polyphase filter bank (PFB), which consists of an FIR filter and an FFT. The FIR filter has branches equal to the FFT size, and each branch is executed on an individual GPU thread. The output of PFB is a frequency-domain spectrum. However, the X-engines expect time-domain samples of each channel. Thus, a transpose operation is required to convert the data from the frequency domain to time domain. Other functions, such as fine delay compensation, fringe rotation, and quantization, are also required; these functions are integrated into one GPU kernel function called the PostProc function. All of the above kernel functions (FIR Filter, FFT, and PostProc) are implemented with floating point arithmetic.

6. The GPU Tensor Core X-Engine

6.1. The X-Engine Pipeline and Key Modules

The diagram of the proposed X-engine pipeline is shown in Figure 4, which includes 4 main modules working in parallel (each module is an independent thread) within the framework of HASHPIPE.

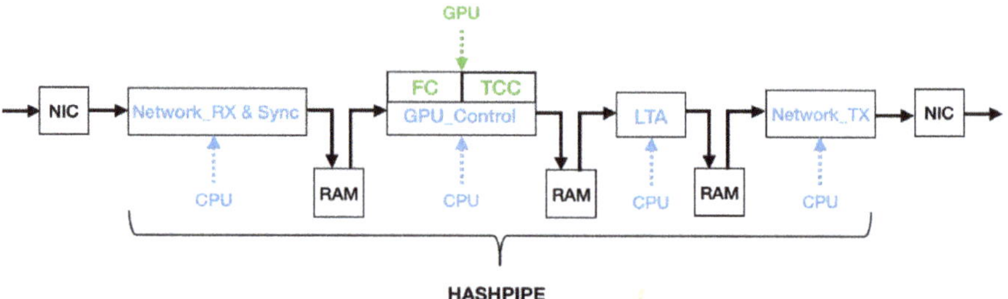

Figure 4. Diagram of the proposed X-engine pipeline.

The first module of the pipeline is called *Network_RX & Sync*, which is based on the SPEAD2 library. The network packets output from the F-engines are captured and reconstructed into heaps by the SPEAD2 library. A heap consists of a header containing a timestamp and an array of time-domain samples of one F-engine. Timestamps of multiple heaps may be out of order and thus reordered. Then, heaps with the same timestamp from different F-engines are synchronized into a heap array. Some heaps may be lost due to various reasons. If this happens, the software replaces them with zero values. After synchronization, all K^{13} heap arrays are assembled into one large data block, which will be transmitted to the subsequent *GPU_Control* module.

The throughput of the entire X-engine pipeline depends on the module that consumes the longest time—that is, the *GPU_Control* module. This module includes 4 processes: memory copy from CPU to GPU, the format conversion (FC) kernel function, the TCC kernel function, and memory copy from GPU to CPU. In order to improve the throughput of this module, the four processes are executed with 3 streams to form a pipeline; that is, the two kernel functions use a common stream, and each memory copy uses an individual stream. Through profiling with the software of Nsight, the throughput of the pipeline is around 140 Gbps, and the main time is spent on the memory copy from CPU to GPU. After TCC operation, the data have been integrated greatly, so the time consumption of the memory copy from GPU to CPU is not significant.

The input data format of TCC and output data format of katgpucbf F-engines are as follows:

$$Input\_TCC[Channel\_ID][Block\_ID][Fengine\_ID][Pol\_ID][TimePerBlock\_ID].$$
$$Output\_Fengine[Fengine\_ID][Channel\_ID][Time\_ID][Pol\_ID]. \tag{1}$$

The dimensions from left to right of the above equation change from slow to fast. The format conversion (FC) kernel function is used to rearrange the high-dimensional matrices between these two formats.

Before using the WMMA API to implement the matrix-multiply-and-accumulate operation ($D = A \times B + C$) for correlation, TCC will construct two matrices of A and B at first. For matrix A, the first axis represents time-domain samples of each channel for integration, and the second axis represents antennas/stations. Matrix B is the transpose of matrix A. Values of matrix C should be set to be all zero. After the multiplication of A and B, we can obtain the visibility data of each channel between any two antennas/stations.

After TCC, the data rate has been greatly reduced, and the LTA module will further reduce the rate by integrating the data over a long period of time. The *Network_TX* module is also based on the SPEAD2 library to transmit the visibility data to the data storage servers. Since the visibility data have been integrated by the TCC kernel function and the LTA module, the pressure of data transmission is far less than the *Network_RX & Sync* module.

6.2. Hardware Requirements

As mentioned previously, the key hardware of the proposed correlator includes GPUs, NICs, and CPUs. For the F-engines, the required hardware resources depend on the number of antennas/stations. For the X-engines, the required hardware resources depend on the computing capability and IO bandwidth. We analyze the hardware requirements and try to find the bottleneck in the following content.

For the correlation computing of X-engines, the required performance of multiply-and-accumulate operations per second (OPS) can be calculated with the following equation:

$$N\_corr = B \times 2N(N+1) \times 8. \tag{2}$$

where B is the bandwidth of each sideband, which is 8 GHz. N is the number of antennas/stations, which is 8/21 for the wSMA/ngEHT. We consider full-stokes correlation with dual polarization—for each antenna pair, the vertical polarization and horizontal polarization must be correlated [25]. Auto-correlation of each antenna/station per polarization is also considered, so the number of correlation operations is $2N(N+1)$. The factor 8 arises from the complex-valued multiply–accumulate operation [20]. The result of Equation (2) is 9.216/59.136 TOPS for the wSMA/ngEHT correlator. We evaluated the the performance of TCC on an A5000 GPU at different bit widths; the results are shown in Figure 5. For the 4-bit situation, when the number of antennas is 8, the performance is approximately 6.52 TOPS, which means that two A5000 GPUs can meet the computing requirements of all X-engines of the wSMA correlator. When the number of stations is 21, the performance is approximately 31.36 TOPS, which means that two A5000 GPUs are also sufficient for the computing requirements of all X-engines of the ngEHT correlator. Through the above analysis, we can see that due to the extremely high performance of Tensor Cores, the correlation computation is no longer a bottleneck in the correlator.

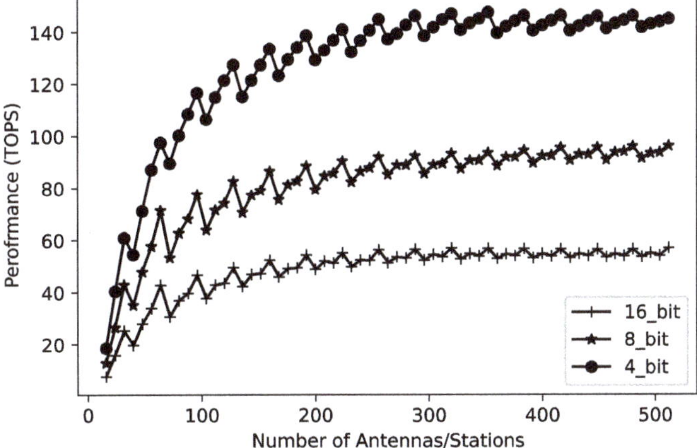

Figure 5. Performance of TCC on an A5000 GPU. The sawtooth shapes are caused by work distribution imbalances and redundant computations for non-multiples of 48 or 64 antennas/stations [7]. For smaller numbers of antennas or stations, the TCC is memory-I/O-bound, so it cannot achieve peak performance. For larger numbers of stations, the TCC converges to a plateau as it becomes compute-bound, even though the memory bandwidth use remains high. The GPU is eventually limited by the power use, as the driver slows down the clock to keep the GPU within its power limit.

For the IO bandwidth, the total data rate between the F-engines and X-engines can be calculated with the following equation:

$$B \times 2N \times 2 \times 4bits. \tag{3}$$

The result of the above equation is 1024/2688 Gbps for the wSMA/ngEHT correlator. The A5000 GPU uses $PCIe4.0 \times 16$ with an actual throughput of around 200 Gbps as the data path. As mentioned previously, the throughput of the $GPU\_Control$ module is around 140 Gbps, and the main time is spent on memory copying, so the utilization of $PCIe4.0$ effective bandwidth is approximately 70%. Therefore, due to the bandwidth limitation, theoretically, the number of GPUs required by the X-engines of the wSMA correlator and ngEHT correlator is $\lceil \frac{1024Gbps}{140Gbps} \rceil = 8$ and $\lceil \frac{2688Gbps}{140Gbps} \rceil = 20$, respectively. The NVIDIA ConnectX-5 NIC also uses $PCIe4.0 \times 16$ to connect to the host server, and the throughput of the network interface is 200 Gbps, which is almost the same as the throughput of PCIe, so the required number of NVIDIA ConnectX-5 NICs is the same as the number of GPUs.

Although we use GPUs to implement correlation computing, CPU resources are also required. The reason is that the HASHPIPE framework and the multiple modules in Figure 4 require multiple CPU cores. The specific number required cannot be calculated theoretically, but depends on the actual situation during the development and testing stage. Currently, each server has 2 powerful Intel Xeon Silver 4314 CPUs with a total of 32 cores, which are sufficient for our proof-of-concept.

7. The Full Compact Architecture

The full compact architecture is shown as Figure 6. Compared with the previous architecture in Figure 1, the changes are as follows. First, the D-engines of the wSMA correlator need to support the functions of coarse channelization and corner turning. For the ngEHT correlator, since the recorded data have been coarsely channelized into sub-bands by the digital backends, the playback servers simply need to corner turn these sub-bands to different GPU servers for further processing. Second, the $Network\_RX$ & $Sync$ module needs to perform coarse delay compensation for each sub-band. Third, the main functions of the F-engine and X-engine are integrated into the same GPU to form a new engine called the FX-engine.

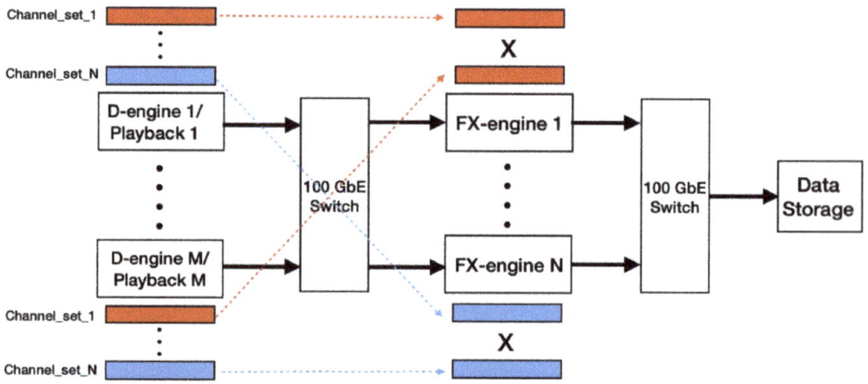

Figure 6. Full compact architecture of the proposed correlator.

The pipeline of the FX-engine is shown in Figure 7, which is based on the previous X-engine pipeline in Figure 4, but with the addition of new kernel functions of FIR filters (FIR_Pol0/FIR_Pol1[14]), FFTs (FFT_Pol0/FFT_Pol1), and PostProc to achieve fine channelization. These new kernel functions are very similar to that of the katgpucbf F-engine library, but the difference is that the latter only processes data from one antenna/station, while the former needs to process data from all antennas/stations. The other two kernel functions, Format Conversion (FC) and TCC, are the same as in Figure 7. At present, we have developed an initial design with a throughput of around 86 Gbps for the whole pipeline, but there is still much room for optimization in the future.

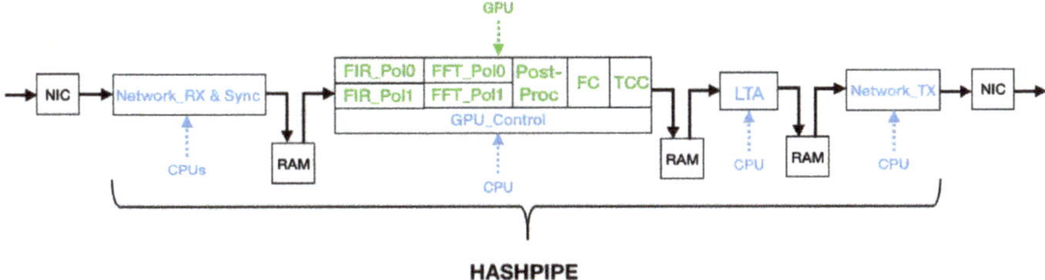

Figure 7. Diagram of the proposed full compact FX-engine pipeline.

8. Acceleration with New Hardware

The data path from the NIC to the GPU in Figures 4 and 7 can be divided into 3 steps. At first, data are transmitted from the NIC to the host server's memory via PCIe. Then, the host server caches and synchronizes these data. Finally, the synchronized data are sent to the GPU via PCIe for correlation. Multiple passes over bandwidth-limited PCIe are very inefficient and represent the bottleneck of the entire pipeline.

The NVIDIA ConnectX NIC that we are using can accelerate the first step through bypassing the kernel's networking stack. Going a step further, NVIDIA's new product of BlueField Data Processing Units (DPUs) can provide an acceleration of the first and second steps of the data path. By combining a ConnectX NIC with an array of ARM CPU cores, BlueField DPUs can be fully software-programmable [26], which can improve the performance of the *Network_RX & Sync* module in Figures 4 and 7. The embedded NIC captures network packets at first, and then forwards them to the subsequent embedded ARM CPUs, which can replace the host's CPUs to maintain network connections and execute data crossover and time synchronization operations in parallel multi-threads. Since the data receiving, caching, and synchronization are combined and implemented on the same DPU, the whole system can be more efficient and has a lower cost due to the reduced demand on the host CPUs [27].

The NVIDIA converged accelerator is an upcoming product, which integrates a Blue-filed DPU and a GPU on one board[15]. If the X-engine is equipped with this board, the entire data path from the NIC to the GPU no longer needs to go through the host server's memory. Data can be sent from the DPU to the GPU directly, so all three steps can be accelerated, and the end-to-end throughput can be further improved.

9. Conclusions

We have reviewed the architectures of existing correlator-beamformers. With the benefit of this context, and using codes shared by collaborating institutions including SARAO and ASTRON, we built a proof-of-concept GPU Tensor Core-based correlator for the ngEHT and the wSMA. The requirement is to provide a solution to the transmission and computing challenges brought about by the massive data rates required by wideband instruments.

It is notable that this architecture is a purely software- as opposed to firmware-driven design. Compared to the currently favored approach using FPGAs, the development and ongoing maintenance of a software machine is much easier compared to the very difficult hardware description language (HDL) FPGA firmware design. The GPU's floating point arithmetic yields improved digital efficiency with lower quantization and clipping losses.[16] In summary, GPUs are an order of magnitude faster to code for complex applications, while being both flexible and maintainable after deployment.

To start the proof-of-concept, we evaluated open-source libraries and selected the appropriate ones according to the requirements. Then, we described the proposed X-engine pipeline and analyzed the hardware resource requirements. We also introduced a full compact architecture, which integrates the main functions of the F-engine and X-engine into one GPU. Finally, we discussed how new hardware can accelerate future correlator-beamformers.

The paper presents proof-of-concept design work and associated benchmarks, which lead us to an optimistic assessment of the prospects for true- or near-real-time computation for connected and VLBI interferometry. We anticipate that this new development will support the improved imaging capability of the ngEHT and wSMA upgrades, thus allowing both instruments to achieve their respective transformative scientific goals.

Author Contributions: Conceptualization, J.W., J.W.R. and W.Y.; methodology, J.W., W.Y., J.W.R., L.J.D., R.-S.L., A.P., G.C., D.W.P., L.B., B.M., R.S. and J.K.; software, W.Y., J.W.R., L.J.D., G.C. and B.M.; writing and editing, W.Y., J.W., J.W.R., G.C., D.W.P., L.B. and B.M. All authors have read and agreed to the published version of the manuscript.

Funding: This work was supported by the Black Hole Initiative at Harvard University, which is funded by grants from the John Templeton Foundation and the Gordon and Betty Moore Foundation to Harvard University. This work was also supported by National Science Foundation grants AST-1726637, AST-1935980, AST-2034306; the Gordon and Betty Moore Foundation (GBMF-10423); the European Commission (DEEP-EST grant agreement 754304); the Netherlands eScience Center (PADRE, RECRUIT), and the European Southern Observatory. We acknowledge donations of hardware and software from NVIDIA and from AMD/Xilinx. The SMA is a joint project between the SAO and ASIAA and is funded by the Smithsonian Institution and the Academia Sinica.

Data Availability Statement: Not applicable.

Acknowledgments: This research has made use of the NASA Astrophysics Data System (ADS). The authors have benefited from technology shared under open-source license by the Collaboration for Astronomy Signal Processing and Electronics Research (CASPER). We thank Sheperd Doeleman, Robert Wilson, Garrett Karto Keating and Ram Rao for their helpful conversations. We acknowledge the significance that Maunakea, where the SMA is located, has for the indigenous Hawaiian people.

Conflicts of Interest: L. Jonathan Dursi discloses that he is an employee of NVIDIA. The authors declare no other conflicts of interest.

Notes

[1] https://www.ngeht.org/ (accessed on 20 December 2022).

[2] The current EHT data are correlated by two DiFX [4,5] clusters located at the Bonn and MIT Haystack observatories. These two correlators are not specifically deployed for the EHT; they have other routine correlation tasks.

[3] https://casper.berkeley.edu/ (accessed on 20 December 2022).

[4] These functions are also available in the CBF. In a full system, it is not decided yet whether they will be implemented in the D-engine, CBF, or split across both.

[5] https://katgpucbf.readthedocs.io/en/latest/index.html (accessed on 20 December 2022).

[6] https://www.ntop.org/guides/pf_ring/ (accessed on 20 December 2022).

[7] https://github.com/jive-vlbi/jive5ab (accessed on 20 December 2022).

[8] https://spead2.readthedocs.io/en/latest/ (accessed on 20 December 2022).

[9] The format conversion module and the TCC module are two kernel functions executed on the same GPU, so they will be executed on the same CPU thread.

[10] http://psrdada.sourceforge.net/ (accessed on 20 December 2022).

[11] https://github.com/david-macmahon/hashpipe (accessed on 20 December 2022).

[12] It uses the ibverbs library with IBV_QPT_RAW_PACKET queue pairs.

[13] At present, we fix K to be 128.

[14] The two polarization data paths are processed separately.

[15] https://www.nvidia.com/en-us/data-center/products/converged-accelerator/ (accessed on 20 December 2022).

[16] VLBI correlation is limited in most practical applications to 88% digital efficiency because samples are typically quantized to 2-bit width for recording. Greater efficiency using floating point arithmetic is achievable for real-time tied array correlators, such as

for wSMA. For the VLBI case, starting with 2-bit samples, data widths grow in the correlation processing, so the floating point arithmetic can still be beneficial to actually achieve the 88% efficiency, which is possible in principle.

References

1. Event Horizon Telescope Collaboration; Akiyama, K.; Alberdi, A.; Alef, W.; Asada, K.; Azulay, R.; Baczko, A.K.; Ball, D.; Baloković, M.; Barrett, J.; et al. First M87 Event Horizon Telescope Results. I. The Shadow of the Supermassive Black Hole. *ApJL* **2019**, *875*, L1. [CrossRef]
2. Event Horizon Telescope Collaboration; Akiyama, K.; Alberdi, A.; Alef, W.; Algaba, J.C.; Anantua, R.; Asada, K.; Azulay, R.; Bach, U.; Baczko, A.K.; et al. First Sagittarius A* Event Horizon Telescope Results. I. The Shadow of the Supermassive Black Hole in the Center of the Milky Way. *ApJL* **2022**, *930*, L12. [CrossRef]
3. Doeleman, S.; Blackburn, L.; Dexter, J.; Gomez, J.L.; Johnson, M.D.; Palumbo, D.C.; Weintroub, J.; Farah, J.R.; Fish, V.; Loinard, L.; et al. Studying Black Holes on Horizon Scales with VLBI Ground Arrays. *arXiv* **2019**, arXiv:1909.01411.
4. Deller, A.T.; Tingay, S.J.; Bailes, M.; West, C. DiFX: A Software Correlator for Very Long Baseline Interferometry Using Multiprocessor Computing Environments. *Publ. Astron. Soc. Pac.* **2007**, *119*, 318–336.
5. Deller, A.T.; Brisken, W.F.; Phillips, C.J.; Morgan, J.; Alef, W.; Cappallo, R.; Middelberg, E.; Romney, J.; Rottmann, H.; Tingay, S.J.; et al. DiFX-2: A More Flexible, Efficient, Robust, and Powerful Software Correlator. *Publ. Astron. Soc. Pac.* **2011**, *123*, 275–287. [CrossRef]
6. Primiani, R.A.; Young, K.H.; Young, A.; Patel, N.; Wilson, R.W.; Vertatschitsch, L.; Chitwood, B.B.; Srinivasan, R.; MacMahon, D.; Weintroub, J. SWARM: A 32 GHz Correlator and VLBI Beamformer for the Submillimeter Array. *J. Astron. Instrum.* **2016**, *5*, 1641006. [CrossRef]
7. Romein, J.W. The Tensor-Core Correlator. *Astron. Astrophys.* **2021**, *656*, A52. [CrossRef]
8. van der Byl, A.; Smith, J.; Martens, A.; Manley, J.; van Balla, T.; Rust, A.; Patel, A.; Callanan, G.; Isaacson, A.; New, W.; et al. MeerKAT correlator-beamformer: A real-time processing back-end for astronomical observations. *J. Astron. Telesc. Instruments Syst.* **2022**, *8*, 011006. [CrossRef]
9. Hickish, J.; Razavi-Ghods, N.; Perrott, Y.C.; Titterington, D.J.; Carey, S.H.; Scott, P.F.; Grainge, K.J.B.; Scaife, A.M.M.; Alexander, P.; Saunders, R.D.E.; et al. A digital correlator upgrade for the Arcminute MicroKelvin Imager. *Mon. Not. R. Astron. Soc.* **2018**, *475*, 5677–5687. [CrossRef]
10. Denman, N.; Renard, A.; Vanderlinde, K.; Berger, P.; Masui, K.; Tretyakov, I. A GPU Spatial Processing System for CHIME. *J. Astron. Instrum.* **2020**, *9*, 2050014. [CrossRef]
11. Kocz, J.; Greenhill, L.J.; Barsdell, B.R.; Bernardi, G.; Jameson, A.; Clark, M.A.; Craig, J.; Price, D.; Taylor, G.B.; Schinzel, F.; et al. A Scalable Hybrid Fpga/gpu FX Correlator. *J. Astron. Instrum.* **2014**, *3*, 1450002. [CrossRef]
12. Kocz, J.; Greenhill, L.J.; Barsdell, B.R.; Price, D.; Bernardi, G.; Bourke, S.; Clark, M.A.; Craig, J.; Dexter, M.; Dowell, J.; et al. Digital Signal Processing Using Stream High Performance Computing: A 512-Input Broadband Correlator for Radio Astronomy. *J. Astron. Instrum.* **2015**, *4*, 1550003. [CrossRef]
13. Ali, Z.S.; Parsons, A.R.; Zheng, H.; Pober, J.C.; Liu, A.; Aguirre, J.E.; Bradley, R.F.; Bernardi, G.; Carilli, C.L.; Cheng, C.; et al. PAPER-64 Constraints on Reionization: The 21 cm Power Spectrum at z = 8.4. *Astrophys. J.* **2015**, *809*, 61. [CrossRef]
14. Ord, S.M.; Crosse, B.; Emrich, D.; Pallot, D.; Wayth, R.B.; Clark, M.A.; Tremblay, S.E.; Arcus, W.; Barnes, D.; Bell, M.; et al. The Murchison Widefield Array Correlator. *Publ. Astron. Soc. Aust.* **2015**, *32*, e006. [CrossRef]
15. Wayth, R.B.; Greenhill, L.J.; Briggs, F.H. A GPU-based Real-time Software Correlation System for the Murchison Widefield Array Prototype. *Publ. Astron. Soc. Pac.* **2009**, *121*, 857. [CrossRef]
16. Broekema, P.C.; Mol, J.J.D.; Nijboer, R.; van Amesfoort, A.S.; Brentjens, M.A.; Loose, G.M.; Klijn, W.F.A.; Romein, J.W. Cobalt: A GPU-based correlator and beamformer for LOFAR. *Astron. Comput.* **2018**, *23*, 180. [CrossRef]
17. Event Horizon Telescope Collaboration; Akiyama, K.; Alberdi, A.; Alef, W.; Asada, K.; Azulay, R.; Baczko, A.K.; Ball, D.; Baloković, M.; Barrett, J.; et al. First M87 Event Horizon Telescope Results. III. Data Processing and Calibration. *ApJL* **2019**, *875*, L3. [CrossRef]
18. Gill, A.; Blackburn, L.; Roshanineshat, A.; Chan, C.K.; Doeleman, S.S.; Johnson, M.D.; Raymond, A.W.; Weintroub, J. Prospects for Wideband VLBI Correlation in the Cloud. *Publ. Astron. Soc. Pac.* **2019**, *131*, 124501. [CrossRef]
19. Kalia, A.; Kaminsky, M.; Andersen, D.G. Design Guidelines for High Performance RDMA Systems. In Proceedings of the 2016 USENIX Annual Technical Conference (USENIX ATC 16), Denver, CO, USA, 22–24 June 2016.
20. Clark, M.A.; LaPlante, P.C.; Greenhill, L.J. Accelerating radio astronomy cross-correlation with graphics processing units. *Int. J. High Perform. Comput. Appl.* **2013**, *27*, 178–192. [CrossRef]
21. Denman, N.; Amiri, M.; Bandura, K.; Cliche, J.F.; Connor, L.; Dobbs, M.; Fandino, M.; Halpern, M.; Hincks, A.; Hinshaw, G.; et al. A GPU-based Correlator X-engine Implemented on the CHIME Pathfinder. *arXiv* **2015**, arXiv:1503.06202.
22. Recnik, A.; Bandura, K.; Denman, N.; Hincks, A.D.; Hinshaw, G.; Klages, P.; Pen, U.L.; Vanderlinde, K. An efficient real-time data pipeline for the CHIME Pathfinder radio telescope X-engine. In Proceedings of the 2015 IEEE 26th International Conference on Application-Specific Systems, Architectures and Processors (ASAP), Toronto, ON, Canada, 27–29 July 2015.
23. Cranmer, M.D.; Barsdell, B.R.; Price, D.C.; Dowell, J.; Garsden, H.; Dike, V.; Eftekhari, T.; Hegedus, A.M.; Malins, J.; Obenberger, K.S.; et al. Bifrost: A Python/C++ Framework for High-Throughput Stream Processing in Astronomy. *J. Astron. Instrum.* **2017**, *6*, 1750007. [CrossRef]

24. Markidis, S.; Chien, S.W.D.; Laure, E.; Peng, I.B.; Vetter, J.S. NVIDIA Tensor Core Programmability, Performance & Precision. In Proceedings of the 2018 IEEE International Parallel and Distributed Processing Symposium Workshops (IPDPSW), Vancouver, BC, Canada, 21–25 May 2018.
25. Callanan, G.M. A GPU based X-Engine for the MeerKAT Radio Telescope. Master's Thesis, University of Cape Town, Cape Town, South Africa, 2020.
26. NVIDIA BLUEFIELD-2 DPU Data Center Infrastructure on a Chip. Available online: https://www.nvidia.com/content/dam/en-zz/Solutions/Data-Center/documents/datasheet-nvidia-bluefield-2-dpu.pdf (accessed on 20 December 2022).
27. Deierling, K. Achieving a Cloud-Scale Architecture with DPUs. Available online: https://developer.nvidia.com/blog/achieving-a-cloud-scale-architecture-with-dpus/ (accessed on 20 December 2022).

Technical Note

Progress on the Haystack Observatory Postprocessing System

Daniel Hoak *,†, John Barrett †, Geoffrey Crew † and Violet Pfeiffer †

Massachusetts Institute of Technology, Haystack Observatory, 99 Millstone Rd, Westford, MA 01886, USA
* Correspondence: dhoak@mit.edu
† These authors contributed equally to this work.

Abstract: The Haystack Observatory Postprocessing System (HOPS) is a multipurpose tool for post-correlation calibration and data analysis in Very-Long Baseline Interferometry experiments. The requirements on stations, baselines, and bandwidth for the Next Generation Event Horizon Telescope (ngEHT) have motivated a significant refactoring of the HOPS codebase. In this paper, we present the requirements, specifications, and design of HOPS 4.0 and the current state of the refactoring, and we discuss future work.

Keywords: VLBI; black holes; signal processing

Citation: Hoak, D.; Barrett, J.; Crew, G.; Pfeiffer, V. Progress on the Haystack Observatory Postprocessing System. *Galaxies* **2022**, *10*, 119. https://doi.org/10.3390/galaxies10060119

Academic Editors: Michael D. Johnson, Shep Doeleman and Jose L. Gómez

Received: 15 November 2022
Accepted: 10 December 2022
Published: 17 December 2022

Publisher's Note: MDPI stays neutral with regard to jurisdictional claims in published maps and institutional affiliations.

1. Introduction

In Very-Long Baseline Interferometry (VLBI), the signal from widely separated radio observatories is correlated between each pair of antennas, known as a *baseline*, to generate complex time-averaged quantities, known as *visibilities*. A critical step between correlation and further data analysis (e.g., imaging) is solving for corrections to the relative difference in arrival times of the wavefront at the antennas, which can be caused by geometric path-length differences, atmospheric effects, and instrumental effects that are not accounted for in the timing model used for correlation. The procedure that solves for these residual delay and delay-rate solutions that maximize the visibility amplitude on each baseline is known as *fringe fitting*.

The Haystack Observatory Postprocessing System, or HOPS, is a multipurpose software package designed to facilitate fringe fitting, phase calibration/correction, and data analysis for VLBI experiments. HOPS has a multi-decade history as a VLBI tool, beginning with work by Alan Rogers in FORTRAN in the 1970s, followed by a complete rewrite into C by Colin Lonsdale and Roger Cappallo in the 1990s. There have been incremental improvements since then by a large cast of contributors. The primary tool in HOPS is the fringe-fitting program `fourfit`; separate tools provide data summary and visualization methods (`aedit`, `alist`), and there are functions to segment, merge, average, export, and incoherently search data (`fringex`, `fourmer`, `average`, `CorAsc2`, and `search`, respectively).

The Event Horizon Telescope (EHT) Collaboration has recently used VLBI techniques to image the horizon scale structure of two supermassive black holes, M87* [1] and Sagittarius A* [2]. The HOPS software was a critical component of the EHT data-processing pipeline [3,4], along with CASA [5–7] and AIPS [8]. The next generation EHT (ngEHT) is currently being designed and is expected to dramatically expand the EHT network [9]. The ngEHT is planned to include up to 30 stations recording four, 2-bit, dual-frequency, dual-polarization channels at 8 GHz sampling frequency. These design goals amount to a 10x increase in the number of baselines and a 4x increase in the total bandwidth compared to the EHT, which exceeds the capabilities of the current HOPS software[1]. Thus, as part of the ngEHT design effort, the HOPS software is being significantly refactored to support the expansion of the network.

2. Goals of the Refactoring

The current HOPS software (major version 3, or HOPS3[2]) is written in C and dates from the 1990s. While HOPS3 has had tremendous success as a tool for the VLBI community, the design of the existing codebase has several limitations. We will address these by refactoring the existing functions and methods into C/C++ for HOPS4.

The memory allocation of the pipeline is controlled by hard-coded parameters that place a limit on the number of stations, channels/sub-bands/IFs[3], accumulation periods (APs), and other dimensions of the data. HOPS4 will use dynamic memory allocation and will have no practical limit on the number of stations, baselines, channels, and APs in a fringe search.

Currently, HOPS has the ability to perform phase corrections on a per-channel basis, but it cannot perform fully complex corrections (amplitude and phase) at the sub-channel level. HOPS4 will support amplitude, phase, and delay corrections at each frequency bin.

The existing code relies on legacy software packages that are no longer supported, such as the plotting utility PGPLOT [10], which is deeply integrated into the HOPS code. HOPS4 will decouple plotting and analysis routines and provide new plotting tools using modern packages such as `matplotlib` [11]. HOPS4 will also provide hooks for user-defined plotting packages.

HOPS requires a user-generated configuration file to set basic analysis parameters. The syntax for this file is complex and does not support operations such as flagging or vetoing data beyond a rudimentary manual selection in time or sub-band (channel). Furthermore, the code base as a whole is monolithic and difficult to extend or use in a modular way for either debugging or analysis. HOPS4 will use modern wrappers (e.g., Python) for initializing the configuration parameters and refactor the code into modular, independent libraries.

Most importantly, HOPS4 will be capable of performing any operation that HOPS3 is capable of and will continue to support current data formats.

The following sections describe particular design choices for HOPS4.

2.1. Data Format

HOPS processes VLBI data that have been correlated using the DiFX software correlator [12,13]. DiFX output (which contains the complex visibilities) is in the so-called "Swinburne" format. Currently, HOPS requires users to convert the DiFX output files into the legacy "Mark4" format using the `difx2mark4` utility. The Mark4 data format is based on I/O methods from the tape-drive era, and the format of the in-memory data structures in HOPS3 is tightly coupled to this disk storage format. As a result, modifications to the in-memory structures require similar modifications to the disk storage format and vice versa. HOPS4 has replaced the Mark4 disk storage format with a binary data format that allows increased flexibility and improved file I/O performance. A new utility, `difx2hops`, converts the visibilities and metadata from the Swinburne format into the HOPS4 format.

The binary file format is well-suited for large homogeneous data files such as visibilities. These can be several gigabytes in size for a single (few minute) EHT scan. Heterogeneous data types, such as experiment metadata, are currently stored in memory as C structures with hard-coded sizes. HOPS4 stores these data types as JSON key-value pairs. The JSON format supports lists and is not restricted by compile-time size definitions.

Finally, the HOPS4 team has implemented support for the `vex` [14] and `ovex` metadata formats, including the new `Vex 2.0` [15] format. The metadata from these files are stored as key-value pairs in a JSON object. HOPS4 will support output of fringe data to other formats such as `uvfits` or `hdf5`.

2.2. Data Structures

Currently, HOPS imports data from the Mark4 data format into C-type structures, whose parameters are hard-coded and are difficult to modify. HOPS4 will utilize C++ template classes to construct multidimensional arrays of any trivially constructable data type. This feature provides a method for augmenting the code with a wide variety of possible data types that all share the same unified array-like access interface. The base class has methods to handle memory allocation, resizing, data access via indices or iterators, and data operators for streaming, and so new data types can be defined with minimal effort to meet new use-cases without requiring extensive changes to the code base. While the dimensionality and element types composing a data array in HOPS4 must be fixed at compile time (for example, channelized visibility data with four dimensions: time, frequency, channel, and polarization), the size of each dimension is not fixed or limited. This allows HOPS4 to analyze datasets of arbitrary size. In addition, these template classes support labeling each axis with intervals defined by key-value pairs to facilitate data flagging.

The data operations that act upon multidimensional data in HOPS4 follow a common interface that only requires methods to set the data objects that are used in the operation, initialize the internal state, and execute the operation. This allows users to chain operations in an ordered list for execution, while allowing each operation to be self-contained and unit-testable.

Furthermore, we are implementing bindings and plugins to expose the data structures to external methods. We are using the `pybind11` and `SWIG` libraries to construct bindings to the channelized visibility containers, which allows users to manipulate the data with Python code. We also verify that the operator interface works with OpenCL extensions, which allows the most time-consuming data operations to be parallelized on GPUs.

2.3. Plotting and Data Summary

HOPS currently generates plots using the PGPLOT graphics library, which has been unsupported for over a decade and has become difficult to install on contemporary operating systems. Unfortunately, the PGPLOT functions are deeply embedded in the HOPS code. Significant refactoring is required to separate the data processing from plotting and file output.

The HOPS4 analysis methods will be completely independent of plotting and visualization tools. The analysis code will export results to standard file formats, and the plotting functions will read the data from the disk. Our default plotting routines will be written in `matplotlib`, but the modularity of the plotting functions will enable users to implement their own plotting methods (and allow HOPS to be compiled and executed without linking to any plotting libraries).

The basic plotting result from HOPS is the *fringe plot* (see Figure 1), which has been widely recognized by VLBI users for many years. The fringe plot will be replicated in HOPS4, but options will be available for different use-cases, for example replacing the cross-power spectrum with an alternative plot, or simplifying the dense text and metadata at the bottom of the figure.

The HOPS `alist` utility summarizes the results of a VLBI experiment in an "A-file", a text file that records the useful parameters for each scan, baseline, and polarization. The utility `aedit` provides a number of command-line tools that manipulate A-list data, such as plotting, filtering, editing and sorting. Currently, `aedit` includes both command-line and GUI interfaces. The GUI capability currently requires PGPLOT, which will be replaced in HOPS4. We have prototyped a PyQt-based GUI that replicates the graphical interface of `aedit` and supports the command-line features.

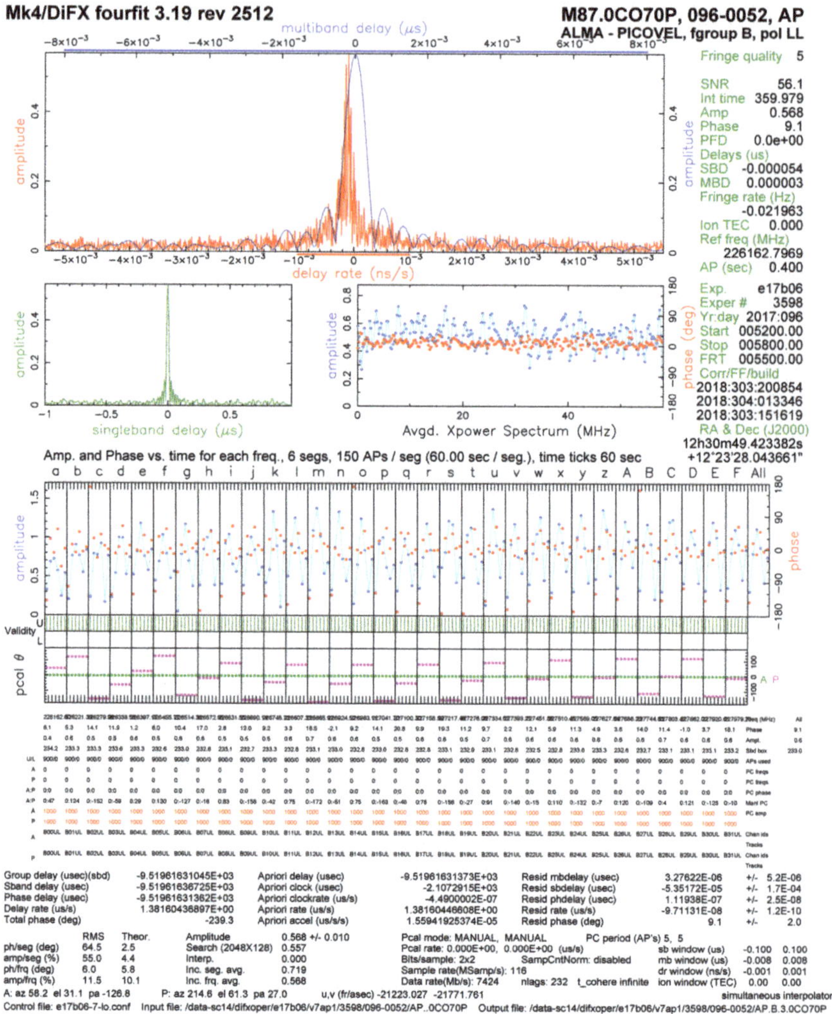

Figure 1. An example fringe plot from the 2017 EHT observations of M87.

3. Testing

The HOPS4 team is following best practices with regards to test-driven development. These include regression, unit, component, and integration tests. We plan to use HOPS3 as a test oracle and currently support test-coverage reports. The suite of tests is intended to provide developers with up-to-date verification of the required HOPS functionality across the supported platforms and distributions. It also provides end users with build-time checks to indicate that the installation was successful. We use coverage tools to demonstrate that the test plan executes a satisfactory fraction of the code, and that all required functions and cases are exercised by the tests. Additionally, we collect performance assessments and benchmarks on a regular basis using captured datasets.

4. Future Work

HOPS currently fringe-fits VLBI data on a per-polarization, per-baseline basis; each pair of polarizations and stations are treated independently. Closure quantities are not considered during the fringe-finding stage, so non-closing errors can occasionally be introduced with this method unless care is taken to iteratively generate a global fringe solution by repeated re-fringing. A number of global fringe-fitting algorithms exist (e.g., [16,17]) that minimize the residual phase, delay, and delay-rate solutions for arrays with three or more stations and can aid in recovering fringes on baselines with low signal-to-noise. One of the goals of the refactoring effort of HOPS is to modify the software so it may accommodate a choice of alternative algorithms for fringe fitting in addition to the native baseline-based algorithm.

Sources with weak continuum emission but bright spectral lines can be imaged using spectral-line VLBI techniques. In principle, the current version of HOPS can perform spectral-line VLBI, but the procedure is quite technical and requires a significant amount of hand-tuning. Implementing a fringe-fitting algorithm that supports spectral-line VLBI by searching for the fringe maximum over delay-rate and frequency space is a highly desirable goal for HOPS4.

HOPS3 calculates per-baseline, per-scan fringe solutions in a one-shot execution of `fourfit` from the command line; crude parallelization can be made by running multiple `fourfit` jobs over multiple datasets, but this must be orchestrated by the user. While this sort of simple parallelization is crucial for processing large swaths of independent data needed for VLBI-imaging, it is not particularly useful for acceleration on data with granularity below that of a single baseline/scan. The basic fringe-fitting algorithm in HOPS3 computes the Fast-Fourier Transform (FFT) and searches for the fringe maximum in the three-dimensional space defined by the single-band delay, the multi-band delay, and the delay rate; the computationally intensive portion of this method can be parallelized relatively easily either with multi-threading or via single-instruction–multiple-data (SIMD) techniques. Given the raw computational power of modern GPUs, the SIMD avenue is particularly attractive as it is well-tailored for the repeated calculation of simple mathematical functions (e.g., the delay/delay-rate phase rotation) as well as array manipulation, which dominates the fringe-fitting computation. OpenCL extensions of simple data operations (array scaling/multiplication) have been demonstrated and will be applied to additional operations as computational bottlenecks in the existing and future algorithms are identified.

5. Conclusions

The refactoring of the HOPS VLBI analysis software for the ngEHT is well underway. Approximately 23k lines of code have been written and define new data structures and I/O routines, import/export to legacy data formats, and perform data analysis on the new structures. The goal of the refactoring is to maintain the current functionality and performance of HOPS while supporting the increased number of stations, baselines, and frequency bands for the ngEHT. HOPS4 will rely on standard software packages (C/C++, Python, and minimal associated tools) that are readily available on common Linux-based operating systems such as Ubuntu, Debian, and CentOS. HOPS4 will have improved modularity and extensibility compared to HOPS3, allowing users to export data to common formats, inject code to test new methods, and ease debugging.

The HOPS development team[4] welcomes user feedback, questions, and feature requests. The team looks forward to releasing a beta version of `fourfit` in early 2023 and a beta version of the full HOPS4 software package in 2024.

Author Contributions: Authors D.H., J.B., G.C. and V.P. contributed equally to this work. All authors have read and agreed to the published version of the manuscript.

Funding: This research was supported by the National Science Foundation through grant numbers AST-1935980 and AST-2034306.

Data Availability Statement: Not applicable.

Acknowledgments: The authors gratefully acknowledge useful discussions from their MIT Haystack, EHT, and ngEHT colleagues.

Conflicts of Interest: The authors declare no conflict of interest.

Notes

[1] HOPS3 has hard-coded limits on the number of stations (16), baselines (120), channels/sub-bands (64), and accumulation periods (8192) that are challenging to modify in the current architecture.

[2] The latest release as of this article is version 3.24: ftp://gemini.haystack.mit.edu/pub/hops (accessed on 15 December 2022)

[3] In VLBI, each spectral band measured at the antenna is typically divided into several smaller bands for averaging and analysis. In HOPS, these sub-bands are called *channels*, while in other fringe-fitting packages (e.g., AIPS) they are called intermediate-frequency bands or "IFs". For example, the EHT collects data in 2 GHz bands, which in HOPS are divided into 32 channels, each 58 MHz wide; see Figure 3 in [4]. One or more individual frequency bins ("channels" in AIPS) are referred to as a "sub-channel" in HOPS.

[4] hops-dev@mit.edu

References

1. The Event Horizon Telescope Collaboration. First M87 Event Horizon Telescope Results. I. The Shadow of the Supermassive Black Hole. *Astrophys. J. Lett.* **2019**, *875*, L1. [CrossRef]
2. The Event Horizon Telescope Collaboration. First Sagittarius A* Event Horizon Telescope Results. I. The Shadow of the Supermassive Black Hole in the Center of the Milky Way. *Astrophys. J. Lett.* **2022**, *930*, L12. [CrossRef]
3. The Event Horizon Telescope Collaboration. First M87 Event Horizon Telescope Results. III. Data Processing and Calibration. *Astrophys. J. Lett.* **2019**, *875*, L3. [CrossRef]
4. Blackburn, L.; Chan, C.; Crew, G.B.; Fish, V.L.; Issaoun, S.; Johnson, M.D.; Wielgus, M.; Akiyama, K.; Barrett, J.; Bouman, K.L.; et al. EHT-HOPS Pipeline for Millimeter VLBI Data Reduction. *Astrophys. J.* **2019**, *882*, 23. [CrossRef]
5. The CASA Team; Bean, B.; Bhatnagar, S.; Castro, S.; Donovan Meyer, J.; Emonts, B.; Garcia, E.; Garwood, R.; Golap, K.; Gonzalez Villalba, J.; et al. CASA, the Common Astronomy Software Applications for Radio Astronomy. *arXiv* **2022**, arXiv:2210.02276.
6. van Bemmel, I.M.; Kettenis, M.; Small, D.; Janssen, M.; Moellenbrock, G.A.; Petry, D.; Goddi, C.; Linford, J.D.; Rygl, K.L.J.; Liuzzo, E.; et al. CASA on the fringe—Development of VLBI processing capabilities for CASA. *arXiv* **2022**, arXiv:2210.02275.
7. Janssen, M.; Goddi, C.; van Bemmel, I.M.; Kettenis, M.; Small, D.; Liuzzo, E.; Rygl, K.; Martí-Vidal, I.; Blackburn, L.; Wielgus, M.; et al. rPICARD: A CASA-based calibration pipeline for VLBI data—Calibration and imaging of 7 mm VLBA observations of the AGN jet in M 87. *Astron. Astrophys.* **2019**, *626*, A75. [CrossRef]
8. Greisen, E.W. AIPS, the VLA, and the VLBA. In *Information Handling in Astronomy—Historical Vistas*; Heck, A., Ed.; Astrophysics and Space Science Library; Springer: Dordrecht, The Netherlands, 2003; Volume 285, p. 109. _7. [CrossRef]
9. Blackburn, L.; Doeleman, S.; Dexter, J.; Gómez, J.L.; Johnson, M.D.; Palumbo, D.C.; Weintroub, J.; Bouman, K.L.; Chael, A.A.; Farah, J.R.; et al. Studying Black Holes on Horizon Scales with VLBI Ground Arrays. *arXiv* **2019**, arXiv:1909.01411.
10. Pearson, T. PGPLOT: Device-independent Graphics Package for Simple Scientific Graphs. Astrophysics Source Code Library: 2011; p. 03002. Available online: https://ui.adsabs.harvard.edu/abs/2011ascl.soft03002P/abstract (accessed on 9 December 2022).
11. Hunter, J.D. Matplotlib: A 2D graphics environment. *Comput. Sci. Eng.* **2007**, *9*, 90–95. [CrossRef]
12. Deller, A.T.; Tingay, S.J.; Bailes, M.; West, C. DiFX: A Software Correlator for Very Long Baseline Interferometry Using Multiprocessor Computing Environments. *Publ. Astron. Soc. Pac.* **2007**, *119*, 318–336. [CrossRef]
13. Deller, A.T.; Brisken, W.F.; Phillips, C.J.; Morgan, J.; Alef, W.; Cappallo, R.; Middelberg, E.; Romney, J.; Rottmann, H.; Tingay, S.J.; et al. DiFX-2: A More Flexible, Efficient, Robust, and Powerful Software Correlator. *Publ. Astron. Soc. Pac.* **2011**, *123*, 275–287. [CrossRef]
14. VEX 1.5 File Definition. 2002. Available online: https://vlbi.org/wp-content/uploads/2019/03/vex-definition-15b1.pdf (accessed on 9 December 2022).
15. VEX 2.0 File Definition. 2021. Available online: https://safe.nrao.edu/wiki/bin/view/VLBA/Vex2doc (accessed on 9 December 2022).
16. Schwab, F.R.; Cotton, W.D. Global fringe search techniques for VLBI. *Astron. J.* **1983**, *88*, 688–694. [CrossRef]
17. Alef, W.; Porcas, R. VLBI fringe-fitting with antenna-based residuals. *Astron. Astrophys.* **1986**, *168*, 365–368.

Article

The ngEHT's Role in Measuring Supermassive Black Hole Spins

Angelo Ricarte [1,2,*], **Paul Tiede** [1,2], **Razieh Emami** [1], **Aditya Tamar** [3] and **Priyamvada Natarajan** [2,4,5]

[1] Center for Astrophysics | Harvard & Smithsonian, 60 Garden Street, Cambridge, MA 02138, USA; paul.tiede@cfa.harvard.edu (P.T.); razieh.emami_meibody@cfa.harvard.edu (R.E.)

[2] Black Hole Initiative, 20 Garden Street, Cambridge, MA 02138, USA; priyamvada.natarajan@yale.edu

[3] Independent Researcher, Delhi 110092, India; adityatamar@gmail.com

[4] Department of Astronomy, Yale University, New Haven, CT 06511, USA

[5] Department of Physics, Yale University, New Haven, CT 06520, USA

* Correspondence: angelo.ricarte@cfa.harvard.edu

Abstract: While supermassive black-hole masses have been cataloged across cosmic time, only a few dozen of them have robust spin measurements. By extending and improving the existing Event Horizon Telescope (EHT) array, the next-generation Event Horizon Telescope (ngEHT) will enable multifrequency, polarimetric movies on event-horizon scales, which will place new constraints on the space-time and accretion flow. By combining this information, it is anticipated that the ngEHT may be able to measure tens of supermassive black-hole masses and spins. In this white paper, we discuss existing spin measurements and many proposed techniques with which the ngEHT could potentially measure spins of target supermassive black holes. Spins measured by the ngEHT would represent a completely new sample of sources that, unlike pre-existing samples, would not be biased towards objects with high accretion rates. Such a sample would provide new insights into the accretion, feedback, and cosmic assembly of supermassive black holes.

Keywords: supermassive black holes; accretion; general relativity; very long baseline interferometry; Messier 87; Sagittarius A*

Citation: Ricarte, A.; Tiede, P; Emami, R; Tamar, A; Natarajan, P The ngEHT's Role in Measuring Supermassive Black-Hole Spins. *Galaxies* **2023**, *11*, 6. https://doi.org/10.3390/galaxies11010006

Academic Editor: Bidzina Kapanadze

Received: 7 November 2022
Revised: 19 December 2022
Accepted: 20 December 2022
Published: 26 December 2022

1. Introduction

Astrophysical supermassive black holes (SMBHs) can be fully described by just two parameters: their mass (which we denote as $M_\bullet$) and their dimensionless spin parameter (which we denote as $a_\bullet$) [1]. A variety of techniques ranging from dynamical modeling to calibrated scaling relations to broad emission lines have been developed to estimate SMBH masses across the Universe, e.g., [2–4], as far out to redshifts of $z \sim 6$–7, e.g., [5–7]. These investigations reveal that SMBH masses correlate well with several properties of their host galaxies, most famously leading to tight empirical relationships between SMBH mass and bulge mass, as well as velocity dispersion [8–12]. This suggests growth of SMBHs and their hosts occurs in tandem, as gas is transported to galactic nuclei to form stars and grow SMBHs, and SMBHs inject energy into their hosts in the form of radiation, winds, and jets [13,14]. Indeed, virtually all current models of cosmic galaxy evolution include SMBH growth and active galactic nucleus (AGN) feedback as a necessary ingredient for suppressing star formation in the most massive host galaxies to a level consistent with observations, e.g., [15–22].

Compared to their masses, much less is known observationally about SMBH spins. For actively accreting SMBHs with geometrically thin and optically thick accretion disks (those with Eddington ratios roughly in the range $0.01 \lesssim f_{\rm edd} \lesssim 0.3$), spin can be estimated from their spectral properties. A black hole's innermost stable circular orbit shrinks as a function of spin, which leads to higher temperatures and stronger Doppler effects, the latter of which is seen most clearly in the shape of the Iron K-alpha line in X-ray spectra [23,24]. Dozens of

SMBH spins have been measured using the X-ray reflection spectroscopy technique, which involves modeling X-ray spectra by convolving a rest-frame spectrum with relativistic broadening and redshift effects. These investigations have found that most SMBHs to which this technique has been applied are highly spinning [25], with hints of decreases at both high and low masses [25,26]. However, since accretion directly affects a SMBH's spin, these high spin measurements may be biased and not representative of the spin distribution of the overall SMBH population. Intriguingly, although this does not apply to most AGN samples, radio-selected AGN tend to be found in galaxy mergers. A possible explanation is that SMBH mergers occur along with galaxy mergers, and the spin acquired from these mergers helps power jets to allow them to be detected more easily in the radio [27]. There also exist indirect spin estimates of stellar mass BHs detected via gravitational waves, which unlike AGN probed by X-ray reflection spectroscopy seem not to be maximally spinning, e.g., [28], although perhaps not exactly zero either [29]. Since the formation of these objects has little in common with SMBHs and are not accessible to the ngEHT, we restrict ourselves to their more massive counterparts.

Little is known observationally about the spins of more typical and ubiquitous, low Eddington rate black holes. In the case of our own galaxy, Fragione and Loeb [30] argue that the co-existence of two stellar disks at the galactic center places an upper limit on the spin of Sagittarius A* of $a_\bullet \lesssim 0.1$ needed to prevent Lense–Thirring precession from disrupting them. EHT observations indirectly rule out certain spin values via near-horizon mapping of the accretion flow [31–33]. Non-spinning SMBH models also fail to produce high enough jet powers for M87*, as expected [31]. Spin information may also be encoded on jet scales in their their multi-frequency images and jet power, which is active field of research, e.g., [34–36]. In this white paper, we restrict ourselves to horizon scales and review ongoing theoretical work to determine how reliably spin maps onto high spatial resolution images that could be constructed by the ngEHT.

Quasi-periodic Oscillations (QPOs) are sometimes detected when analyzing light curves of AGN or stellar mass black holes, and their characteristic timescales can also be related to the orbital timescale of matter orbiting in the innermost region of the accretion disc [37–39]. However, the origin of QPOs is poorly understood, and spin measurements originating from QPOs rely on uncertain origin models that range from disk oscillations [40] to the orbital motion of hotspots (e.g., [37,41–44]). For Very Long Baseline Interferometry (VLBI) observations, tracking the motion of hotspots produced during flares [45] is a promising approach. Closure quantities alone have been shown to be sensitive to periodicities in orbiting hotspots [46,47]. By strategically placing additional antennas around the globe to produce denser *uv* coverage [48], the next-generation Event Horizon Telescope (ngEHT) will be able to resolve orbital motion. ngEHT will probe the largest SMBHs on the sky with unprecedented spatial and temporal resolution. By directly observing regions where general relativistic effects are strong, it may recover exquisite constraints on the space-time of these SMBHs and their properties, including their spins. It is anticipated that the Phase I ngEHT array will enable dozens of mass and spin measurements [49]. In the more distant future (outside the present scope of the ngEHT), space-based extensions to the array should be able to place even stronger constraints due to higher spatial resolution [50].

In this white paper, we discuss several proposed methods by which the ngEHT could provide novel measures of the spin, and then discuss our current understanding of SMBH spin in the context of the cosmic co-evolution of SMBHs and their host galaxies. Since these new methodologies are completely independent of existing techniques, and because they would be subject to very different selection effects, we argue therefore that even a handful of spin measurements would be greatly impactful for understanding both SMBH accretion flows and their cosmic co-evolution with their host galaxies. In particular, the nearby ngEHT accessible source SMBHs are expected to be accreting preferentially at lower rates than AGN and therefore these measurements will provide a new window into the overall spin distribution of the more characteristic SMBHs.

2. Novel Techniques to Infer Spin with ngEHT

Inferring spin from spatially resolved polarimetric observables and movies of EHT/ngEHT sources is the goal of several recent and ongoing investigations. As discussed above, any spin constraints derived by ngEHT would probe an entirely new sample of objects, a population of more typical low-Eddington ratio sources. The most direct probes of spin are based on "sub-images" of the accretion flow within the photon ring, which are determined directly by the space-time geometry. More indirect but more easily achievable approaches involve properties of the accretion flow itself, which is affected not only by the space-time geometry, but also magneto-hydrodynamic (MHD) forces.

2.1. Spin from Sub-Images: Theoretically Cleaner, Observationally Harder

The most direct impact of spin on black hole images is its effect on the trajectories of photon geodesics, particularly the properties of "sub-images" within the photon ring. A sub-image of a given order is assigned an integer n, referring to the number of half-orbits a photon makes around the SMBH on the way to the observer. The direct image is denoted as $n = 0$, while all sub-images with $n \geq 1$ produce "photon rings" in the SMBH image. Each sub-image is ≈4–13% the width and the flux of the sub-image preceding it, depending on the spin, viewing angle, and position on the ring [51].

As $n \to \infty$, the shape of the sub-image approaches a "critical curve" that is independent of the direct $n = 0$ image and completely determined by the space-time [51,52]. In Figure 1, we plot the shape of the critical curve for SMBHs of different spins for both a face-on and an edge-on viewing angle in geometrized units of $M = GM_\bullet/c^2 D$), where G is the gravitational constant, $M_\bullet$ is the SMBH mass, and c is the speed of light, and D is the distance. For $a_\bullet = 0$, this is simply a circle with radius $\sqrt{27}\,M$ since the Schwarzschild metric is spherically symmetric. For $a_\bullet > 0$ SMBHs, the effect of spin is slight for pole-on viewing angles, decreasing the radius of the critical curve by only ≈6%. The effect of spin is more noticeable for an edge-on viewing angle, where the critical curve shifts and grows more asymmetric as a function of spin.

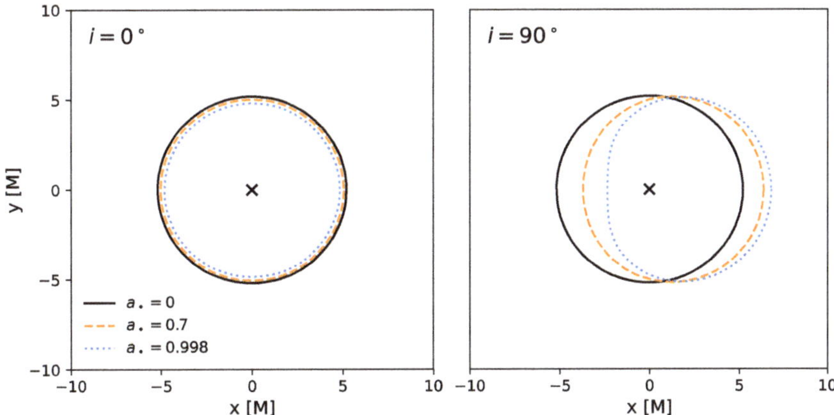

Figure 1. Shape of the $n = \infty$ photon ring or "critical curve" as a function of spin and inclination, using the analytic formulae provided in Chael et al. [53]. For face-on viewing angles (**left**), the critical curve remains circular and shrinks only by about 7% between 0 and maximal spin. For edge-on viewing angles (**right**), the critical curve becomes horizontally displaced and asymmetric as spin increases.

Directly resolving the width of even the $n = 1$ photon ring already requires much finer spatial resolution than accessible to a millimeter array restricted to the size of the Earth, motivating Very Long Baseline Interferometry (VLBI) experiments in space [51]. Unfortunately, we view M87* [54] and seemingly also Sgr A* at pole-on viewing angles [33,45] for which the spin signature requires the most precise imaging measurements. Finally, although the shape of the $n = \infty$ sub-image is fully determined by the space-time, low-order photon rings exhibit a non-negligible dependence on the emission geometry. It is still possible to break degeneracies and place constraints on spin (as well as mass) with precision measurements of the diameters of the first few sub-images [55]. With high spatial resolution and dynamic range, a measurement of the "inner shadow", the lensed image of the equatorial horizon, can also be used to break degeneracies [53]. In the near future, these methods may rely on "super-resolution" techniques, imposing strong constraints on the image in an approach between direct imaging and modeling to outperform the nominal spatial resolution of an array [56,57].

2.2. Spin from Accretion Flows: Theoretically Dirtier, Observationally Easier

Although sub-images offer the cleanest constraints on the space-time, the signal is weak for face-on viewing angles, and sub-images are extremely narrow and faint. Alternative approaches for inferring spin are emerging that may utilize properties of the plasma embedded in the space-time. Inferring spin from the plasma structure and dynamics is less clean than through photon trajectories, since (i) plasma is affected by MHD forces and therefore is not restricted to flow along geodesics and (ii) the emitting region does not necessarily trace the bulk dynamics of the plasma. Nevertheless, several recent works suggest that this may be a promising avenue, at least in magnetically arrested disk (MAD) systems, requiring much lower spatial resolution than would be necessary to resolve the photon ring. Through frame-dragging, spin can govern the average magnetic field structure, the morphology of infalling streams, and the orbital motion and appearance of hotspots.

General relativistic magnetohydrodynamic (GRMHD) simulations, wherein magnetized plasma is allowed to evolve in a Kerr space-time, are key numerical tools used to interpret EHT data by evolving plasma and integrating polarized radiative transfer self-consistently [31,33]. Palumbo et al. [58] studied the morphology of linearly polarized images of GRMHD models of M87* and found that the twisty morphology of these ticks, quantified by a parameter β_2, could be used to discriminate between strongly and weakly magnetized accretion disks. Moreover, as we illustrate in Figure 2, the pitch angle of these ticks demonstrates a clear spin dependence. Here we plot time-averaged (over 5000 $GM_\bullet/c^3$) polarimetric from GRMHD simulations for spins $a_\bullet \in \{-0.94, -0.5, 0, 0.5, 0.94\}$, where a negative sign denotes a retrograde or counter-rotating accretion flow. Recent detailed work delving into the origin of this signal has found that it originates directly from the changing magnetic field structure of the plasma as a function of spin [59]. In this picture, frame dragging pulls along plasma which advects magnetic fields along with it. Larger spins result in a magnetic field in the mid-plane which is more toroidal and wrapped in the direction of the SMBH's spin. Then, since synchrotron emission is linearly polarized perpendicular to the direction of the magnetic field, the magnetic field geometry is imprinted onto linear polarization ticks. Indeed, machine learning algorithms point towards this twisty linear polarization morphology as the most important feature for inferring spin [60]. Image asymmetry also emerges a spin indicator, reflecting increased Doppler beaming in systems with higher spin [61].

It may also be possible to observe frame dragging directly. In systems where the disk and black hole angular momenta are misaligned, frame dragging can impart a characteristic "S"-shaped signature onto infalling streams. Due to magnetic flux freezing in ideal GRMHD, a similar signature can also be imparted onto the linear polarization [62]. An example retrograde accretion flow model of M87* is shown in Figure 3, including a simulated image reconstruction using the Phase I ngEHT array. The ngEHT could observe the sign flip in $\angle\beta_2$ across the photon ring corresponding to a turnaround in the accretion flow. At larger radii, misalignment is generally expected between the SMBH angular momentum axis and that of inflowing gas. As a result, accretion disks can warp, tear, and potentially undergo Lense–Thirring precession, e.g., [63,64]. With sufficient dynamic range in both intensity and spatial scale, this may result in a visible transition at some radius, or potentially impart a QPO signal in the time domain, meriting further study. Frame dragging may also allow spin measurements if a pulsar is discovered in close proximity to Sgr A*, e.g., [65–68], although transient searches have not yet uncovered one [69,70].

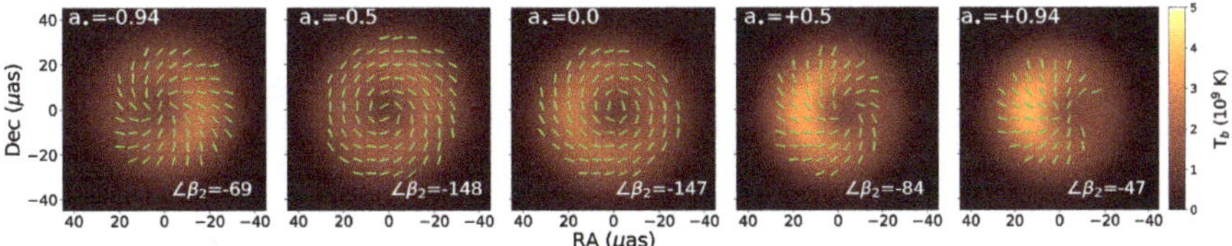

Figure 2. Time-averaged polarimetric images of M87* GRMHD models with increasing spin from left to right, $a_\bullet \in \{-0.94, -0.5, 0, 0.5, 0.94\}$ reproduced from Emami et al. [59], using models presented in Event Horizon Telescope Collaboration et al. [32]. The morphology of these linear polarization ticks originates from an evolving magnetic field structure as a function of spin. Models with higher spin have more toroidal magnetic fields due to frame dragging, which leads to a more radial polarization pattern, while the opposite is true for $a_\bullet \sim 0$. This is reflected in $\angle\beta_2$, written in degrees at the bottom of each panel, which is closer to $-180°$ for toroidal patterns like the $a_\bullet = 0$ model, but moves towards $0°$ for more radial patterns like the $a_\bullet = 0.94$ model [58].

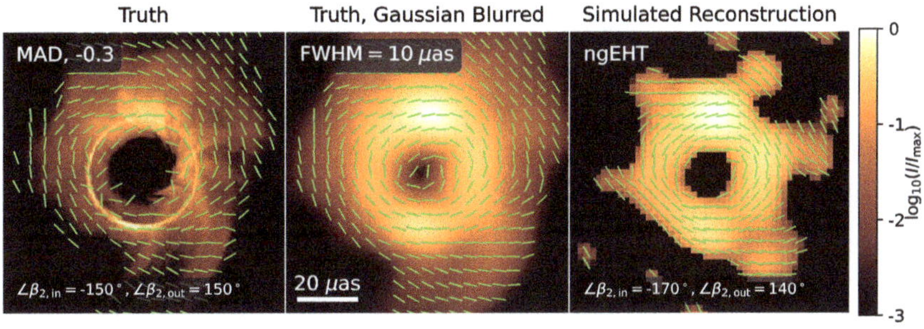

Figure 3. Example total intensity and linear polarization map of a model of M87* that exhibits direct signatures of frame dragging, adapted from Ricarte et al. [62]. Inflowing streams approaching the horizon must turn around due to the existence of an ergosphere in the Kerr space-time, leading to characteristic "S"-shaped streams and flips in the sign of $\angle\beta_2$ as a function of radius. The central panel shows the same simulation blurred with a Gaussian with a full-width at half-maximum of 10 micro-arcseconds, which resembles the spatial resolution obtained when simulating the image reconstruction process. The right-most panel shows a simulated image reconstruction using the phase I ngEHT. Although the spatial resolution is not sufficient to observe the turnaround of individual streams, the ngEHT can observe the flip in the sign of $\angle\beta_2$ as a function of radius.

2.3. Spin from the Time Domain: Movies and Motion

Lacking spatial resolution, it may also be possible to extract spin with excellent temporal resolution. Due to lensing, light that travels from the black hole to an observer travels along multiple paths. Spatially, this leads to multiple sub-images or light echoes. However, these echoes appear around the black hole's photon ring at different times. Wong [71] demonstrated that measuring the light echoes or glimmer location in time and angle around the black hole precisely encodes the spin and inclination of the central black hole. The utility of temporal measurements was noticed earlier by Broderick and Loeb [42], who demonstrated that even the fractional polarization light curve during a flare was very sensitive to the inclination of the black holes accretion disk. The autocorrelation structure of glimmer was further analyzed in [72,73]. Hadar et al. [72] and Wong [71] demonstrated that measuring the angular location and arrival time of glimmer is feasible for finite-resolution images like those produced from the EHT. Conceptually, excellent snapshot imaging substitutes for exquisite spatial resolution for measuring glimmer. This means that one must quickly construct SMBH images with high dynamic range (more than an order of magnitude to measure the sub-image) on timescales similar to the light-crossing time of the black hole, which is on the order of minutes for Sgr A*. So far, studies of SMBH glimmer have been limited to simple toy models, motivating additional study to better understand observational requirements in the presence of realistic stochastic accretion flow.

Finally, movies capturing the dynamics of the plasma in the vicinity of the event horizon made by ngEHT will also allow us to access dynamics more directly in the time domain. Tiede et al. [74], Moriyama et al. [75] demonstrated that by directly modeling the appearance and evolution of hotspots seen by GRAVITY [45] around Sgr A*, the EHT/ngEHT could measure both the black-hole spin and accretion dynamics. One illustrative example is shown in Figure 4. Follow-up work by Levis et al. [76] demonstrated that the ngEHT could potentially measure an arbitrarily complicated emissivity profile using similar methods, although they assumed a fixed spacetime and hotspot velocity field. However, these direct modeling methods are still quite restrictive in terms of types of magnetic fields and velocity fields used to describe the hotspot. Additionally, both Tiede et al. [74] and Levis et al. [76] ignored the surrounding stochastic accretion flow. More research into the expected velocity field of hotspots and how the surrounding accretion flow impacts measurements of plasma dynamics and black-hole spin is needed.

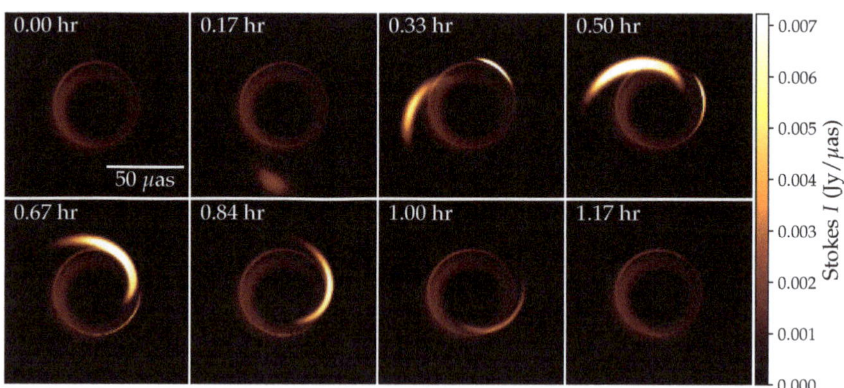

Figure 4. Frames of a hotspot simulation using the model from Tiede et al. [74], where a shearing hotspot is superimposed on top of a static accretion flow. By tracking the motion of the hotspot, the ngEHT could constrain the dynamics of the accretion flow. Additionally, while the motion of the primary hotspot is highly dependent on the accretion flow, the appearance of the secondary image probes the black hole spacetime [71,72] allowing for a direct measurement of the black holes spin and inclination.

Another avenue to measuring plasma dynamics is to extract the motion of the on-sky image instead of around the black hole. The direct image domain approach tends to be computationally simpler, and several different dynamical models [77–79] exist. Additionally, these models make fewer assumptions about the nature of the motion compared to the direct modeling approach described above and are agnostic about the underlying physics. Emami et al. [80] presents an initial exploration demonstrating the feasibility of tracking hotspots around Sgr A* using the ngEHT. The downside of the image domain approach is that relating the on-sky motion to the dynamics of the plasma and surrounding spacetime is poorly understood and requires additional research. In the end for actual observations, both the more direct but restricted parametric modeling and the flexible but less specific image domain modeling will be necessary to test of the robustness of any measurements to different modeling choices.

At present, inferring spin from accretion flow properties is limited not only by observational limitations, but also in large part by theoretical uncertainties. A GRMHD-based analysis of the polarized EHT image of M87* already suggestively rules out certain spin values, but uncertainties regarding electron heating and cooling currently limit our conclusions [32]. These uncertainties propagate into the geometry of the emitting region and Faraday rotation, both of which are integral for interpreting polarized data. Theoretical developments in this area could significantly improve spin constraints by reducing the allowable parameter space.

3. Implications of SMBH Spin

A SMBH grows via both accretion (which may include multiple triggering mechanisms and accretion modes) and mergers with other SMBHs. Its spin encodes recent gas dynamical activity determined by its mode of accretion as well as its merger history. Accretion via a thin disk imparts angular momentum in the direction of the disk's angular momentum. Retrograde accretion therefore spins a SMBH down, while prograde accretion for a SMBH surrounded by a thin disk can spin up a SMBH up to the theoretical maximum of $a_\bullet = 0.998$ [81]. This may be related to the tendency for AGN spins probed by X-ray reflection spectroscopy to be large, since they are necessarily high-Eddington rate systems [25]. For geometrically thick disks, on the other hand, energy extracted to power jets via the Blandford and Znajek [82] process can cause spin down even in the prograde case, which may have interesting implications for SMBHs imparting "maintenance-mode" feedback for Gyrs [83]. SMBHs may also accrete chaotically, for example from the stochastic scattering of molecular clouds with random angular momenta, which would decrease spin on average, e.g., [84,85]. If this is the case, then accretion may on average spin SMBHs down over cosmic time. The relative alignment between disk and SMBH angular momentum vectors over cosmic timescales remains an open question, since accretion disk scales are much smaller than the scale height of the galactic disk. Our own galactic center exhibits a complex environment with substructures that change angular momentum direction across spatial scales [86]. A picture where the relative alignment is chaotic and time-varying is supported by sub-pc resolution zoom-in simulations of gas from stellar winds fueling Sgr A* [87] as well as zoom-in simulations of a generic quasar [88]. Finally, SMBH-SMBH mergers also impact the remnant's spin, depending on the mass ratio and the relative alignments between the orbital angular momentum vector and the spin vectors of the two SMBHs [89]. An equal mass merger of SMBHs without pre-existing spins will tend to produce a remnant with a spin of $a_\bullet \sim 0.7$, but as with accretion, many low-mass mergers on random orbits will decrease the spin [90,91]. All of these complexities can now be modeled self-consistently in semi-analytic models, and dramatically different results can be obtained depending upon one's assumptions about the alignment of accretion disks and mergers.

As a demonstration, we compute spin probability distributions for SMBHs hosted in 100 different $10^{15} \, M_\odot$ halos (like M87*) using the simple semi-analytic model for SMBH evolution developed in Ricarte and Natarajan [92,93], Ricarte et al. [94]. We build upon the Ricarte et al. [94] model by including spin evolution by accretion and mergers self-consistently with mass assembly as in previous works, e.g., [90,91]. In this model, accretion is triggered by halo mergers with a mass ratio of 1:10 or larger. When this occurs, an Eddington ratio is drawn from a distribution appropriate for Sloan Digital Sky Survey (SDSS) broad line quasars [95], near Eddington including a super-Eddington tail. With minimal assumptions, this model reproduces the bolometric luminosity function of AGN out to $z = 6$ quite well [94]. Here, spin is evolved using analytic calculations appropriate for a thin disk [96] up to a maximum value of $a_\bullet = 0.998$ [81]. For this simple, illustrative calculation, we assume that merger-triggered accretion always occurs via prograde thin disks. Following a SMBH-SMBH merger, which we assume is randomly aligned, we use the equations of Rezzolla et al. [89] to compute the spin of the remnant.

We isolate one of the theoretical uncertainties that affects the cosmic evolution of SMBH spin: the probability that a SMBH merger occurs following a halo merger. The most massive SMBHs in the universe are especially sensitive to this astrophysics, as both cosmological simulations and semi-analytic models predict that SMBH-SMBH mergers can in fact dominate the final mass budget of these SMBHs [92,97,98]. However, the journey between halo/galaxy merger and SMBH merger involves traversing many orders of magnitude in spatial scale, and requires multiple physical mechanisms from dynamical friction on the largest scales to gravitational wave emission on the smallest scales [99,100]. When a major galaxy merger occurs, the central SMBHs may not merge for a variety of reasons, including kilo-parsec scale wandering owing to a messy and cosmologically evolving potential, e.g., [20,101–103], potential delays around one parsec when neither dynamical friction nor gravitational wave emission are efficient [104], or even multi-body scatterings in the galactic nucleus [105]. To simply and clearly illustrate our model's sensitivity to SMBH-SMBH mergers, we vary a constant SMBH merging probability following a halo merger, for which we select three values, $p_{\mathrm{merge}} \in \{0.1, 0.3, 1.0\}$. We keep this probability equal to 0 if the halo merger had a mass ratio more extreme than 1:10, in which case the satellite should be stripped, leaving the central SMBH in the outskirts of the halo.

In Figure 5, we plot three representative example SMBH assembly histories as a function of redshift for our different values of p_{merge}. In the left panel, we plot the evolution of $a_\bullet$, and in the right panel, we plot the evolution of $M_\bullet / M_{\bullet,0}$, where $M_{\bullet,0}$ is the final SMBH mass at $z = 0$. Open circles mark SMBH mergers with a mass ratio of at least 1:100. Many mergers may occur at low-redshift for these massive halos if p_{merge} is large, which can lead to sharp jumps in $a_\bullet$. On the other hand, if p_{merge} is small, $a_\bullet$ stays near its maximum value of $a_\bullet = 0.998$, since all accretion is assumed to occur via prograde thin disks. The SMBH in the model with $p_{\mathrm{merge}} = 1$ assembles its final mass latest in cosmic time, as many SMBH mergers contribute to its final mass budget.

Using all 100 different assembly histories that we have computed, we then plot distributions of spin (left; now plotted in terms of $\log(1 - a_\bullet)$) and the fraction of the final mass accumulated via mergers (right) in Figure 6. As expected, these three different values of p_{merge} yield different spin distributions, with more maximal spins in the model with the fewest mergers. In the right panel, we see that for models with $p_{\mathrm{merge}} = 1$, over 80% of the final mass is accumulated via SMBH-SMBH mergers. In this model, this is because their gas-driven accretion is occurs very early in the universe in order to produce a sufficient quantity of luminous quasars at $z = 6$. Then, the model then shuts off gas-driven growth at later times, so as not to overshoot the $M_\bullet - \sigma$ relation, but not SMBH-SMBH mergers.

Apart from cosmic evolution, a SMBH's spin also has an immediate impact on its accretion and feedback processes. The radiative efficiency of a thin disk is strongly sensitive to the SMBH's spin, reaching up to $\epsilon = 42\%$ for $a_\bullet = 1$ compared to a mere $\epsilon = 6\%$ for $a_\bullet = 0$. For geometrically thick disks, jet efficiencies also scale strongly with spin, and may even exceed 100% for prograde disks approaching $a_\bullet = 1$ that power jets via the Blandford-Znajek mechanism [83,106]. A sample of SMBHs with both spin and jet power measurements by the ngEHT could help elucidate the mechanism that powers jets. Finally, following a SMBH-SMBH merger, spin and orbital energy can be converted into a velocity kick, which may offset SMBHs from their host's centers. These kick velocities have been computed from general relativistic simulations and are found to range between 100 and 1000 km s^{-1}. Therefore, in some cases where the kick velocity exceeds the velocity dispersion of the galaxy's gravitational potential, the remnant can be ejected from the nucleus, e.g., [107].

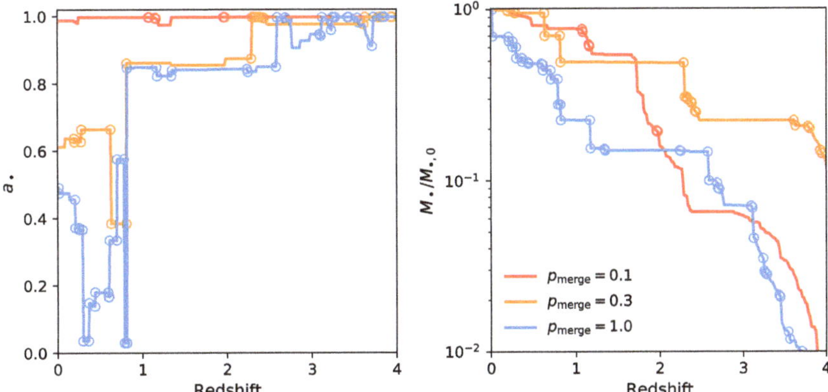

Figure 5. Example evolutionary histories of SMBH spin (**left**) and mass normalized by final mass (**right**) as a function of redshift in $10^{15}\ M_\odot$ halos using a semi-analytic model for cosmological SMBH assembly. We demonstrate sensitivity to the merger history of SMBHs by varying a free parameter p_{merge}, which sets the probability that a SMBH merger occurs following a major halo merger. Different colors encode different values of p_{merge}, and open circles mark SMBH-SMBH mergers with mass ratios of at least 1:100. Many mergers occur in these massive halos, which can cause sharp jumps in $a_\bullet$ if p_{merge} is large. If p_{merge} is small, $a_\bullet$ stays near the maximum value of $a_\bullet = 0.998$.

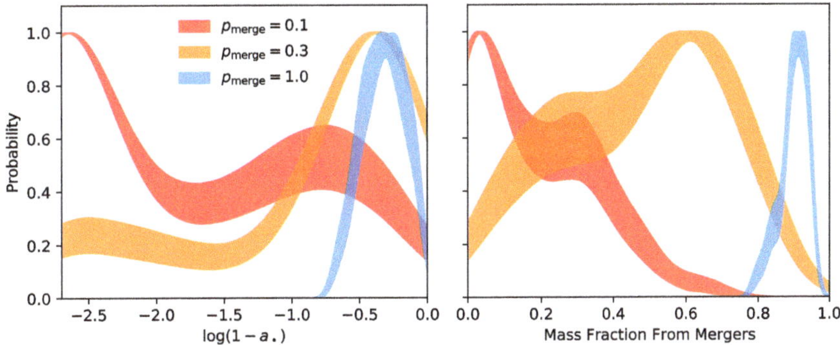

Figure 6. Distributions of spin (cast in terms of $\log(1 - a_\bullet)$) and mass fraction accumulated via mergers for SMBHs in $10^{15}\ M_\odot$ halos using a semi-analytic model. In the left panel, we see that models which allow more mergers produce SMBHs with less extreme spin values. In the right panel, we see that most SMBH mass can originate from mergers if all of them are allowed to occur.

4. Summary

Robust SMBH spin measurements have only been inferred for about twenty sources, and this sample is methodologically biased towards objects with high accretion rates. With spatially resolved polarimetry and time-domain information, the ngEHT has the potential to place spin constraints on a completely different sample of SMBHs, with very different model assumptions and selection effects. Spins measured by ngEHT would sample an ideal type of source: typical low-Eddington ratio objects with large masses, responsible for maintenance mode AGN feedback and whose growth history can be majorly impacted by SMBH-SMBH mergers. Spin constraints of such a sample would have implications for the accretion, feedback, and cosmic assembly of SMBHs.

The cleanest probes of spin for ngEHT sources involve "sub-images" of accretion flows in what is referred to as the photon ring of optically thin SMBH images. These methods are clean because the paths that photons take is determined entirely by the space-time. The signature of spin in the image of the photon ring is quite subtle except for the edge-on cases with substantial spin. With time domain information, we may also constrain spin through characteristic light echoes or "glimmer." More indirect but more observationally accessible probes of spin rely on the structure and motion of the plasma in the accretion flow, which are affected by the forces of MHD in addition to gravity. We have discussed how the frame-dragging of plasma threaded with magnetic fields as a function of spin can impart a signature in linear polarization ticks. Inflowing streams may also exhibit changes in pitch angle, even flipping handedness with radius, depending on the spin. In the time domain, the apparent motion of hotspots can also encode the underlying geometry. These indirect and model-dependent spin inferences would benefit from continued theoretical development in the upcoming years to determine sensitivity to initial conditions and model assumptions.

The prospect of measuring spin motivates event horizon-scale polarimetry with high spatial resolution, high dynamic range, and fine temporal sampling. For each of these, the *uv* coverage provided by the ngEHT will be essential.

Author Contributions: Conceptualization, A.R. and P.N.; methodology, A.R., P.T., R.E. and P.N.; software, A.R., P.T. and R.E.; investigation, A.R., P.T., R.E., A.T. and P.N.; writing—original draft preparation, A.R., P.T., R.E., A.T. and P.N.; writing—review and editing, A.R., P.T., R.E., A.T. and P.N.; visualization, A.R., P.T. and R.E. All authors have read and agreed to the published version of the manuscript.

Funding: We thank the National Science Foundation (AST-1716536, AST-1935980, AST-2034306, AST-1816420, and OISE-1743747) for financial support of this work. This work was supported in large part by the Black Hole Initiative, which is supported by grants from the Gordon and Betty Moore Foundation and the John Templeton Foundation. The opinions expressed in this publication are those of the author(s) and do not necessarily reflect the views of the Moore or Templeton Foundations. RE acknowledges the support from NASA via grant HST-GO-16173.001-A.

Data Availability Statement: The data underlying the figures of this article can be shared upon reasonable request to the corresponding author.

Conflicts of Interest: The authors declare no conflict of interest.

References

1. Kerr, R.P. Gravitational Field of a Spinning Mass as an Example of Algebraically Special Metrics. *Phys. Rev. Lett.* **1963**, *11*, 237–238. [CrossRef]
2. Peterson, B.M. Reverberation Mapping of Active Galactic Nuclei. *Publ. Astron. Soc. Pac.* **1993**, *105*, 247. [CrossRef]
3. Wandel, A.; Peterson, B.M.; Malkan, M.A. Central Masses and Broad-Line Region Sizes of Active Galactic Nuclei. I. Comparing the Photoionization and Reverberation Techniques. *Astrophys. J.* **1999**, *526*, 579–591.
4. Greene, J.E.; Peng, C.Y.; Kim, M.; Kuo, C.Y.; Braatz, J.A.; Impellizzeri, C.M.V.; Condon, J.J.; Lo, K.Y.; Henkel, C.; Reid, M.J. Precise Black Hole Masses from Megamaser Disks: Black Hole-Bulge Relations at Low Mass. *Astrophys. J.* **2010**, *721*, 26–45.
5. Fan, X.; Strauss, M.A.; Schneider, D.P.; Becker, R.H.; White, R.L.; Haiman, Z.; Gregg, M.; Pentericci, L.; Grebel, E.K.; Narayanan, V.K.; et al. A Survey of z>5.7 Quasars in the Sloan Digital Sky Survey. II. Discovery of Three Additional Quasars at z>6. *Astron. J.* **2003**, *125*, 1649–1659.
6. Bañados, E.; Venemans, B.P.; Mazzucchelli, C.; Farina, E.P.; Walter, F.; Wang, F.; Decarli, R.; Stern, D.; Fan, X.; Davies, F.B.; et al. An 800-million-solar-mass black hole in a significantly neutral Universe at a redshift of 7.5. *Nature* **2018**, *553*, 473–476.
7. Wang, F.; Yang, J.; Fan, X.; Hennawi, J.F.; Barth, A.J.; Banados, E.; Bian, F.; Boutsia, K.; Connor, T.; Davies, F.B.; et al. A Luminous Quasar at Redshift 7.642. *Astrophys. J.* **2021**, *907*, L1.
8. Magorrian, J.; Tremaine, S.; Richstone, D.; Bender, R.; Bower, G.; Dressler, A.; Faber, S.M.; Gebhardt, K.; Green, R.; Grillmair, C.; et al. The Demography of Massive Dark Objects in Galaxy Centers. *Astron. J.* **1998**, *115*, 2285–2305.
9. Ferrarese, L.; Merritt, D. A Fundamental Relation between Supermassive Black Holes and Their Host Galaxies. *Astrophys. J.* **2000**, *539*, L9–L12.
10. Gebhardt, K.; Bender, R.; Bower, G.; Dressler, A.; Faber, S.M.; Filippenko, A.V.; Green, R.; Grillmair, C.; Ho, L.C.; Kormendy, J.; et al. A Relationship between Nuclear Black Hole Mass and Galaxy Velocity Dispersion. *Astrophys. J.* **2000**, *539*, L13–L16.
11. Tremaine, S.; Gebhardt, K.; Bender, R.; Bower, G.; Dressler, A.; Faber, S.M.; Filippenko, A.V.; Green, R.; Grillmair, C.; Ho, L.C.; et al. The Slope of the Black Hole Mass versus Velocity Dispersion Correlation. *Astrophys. J.* **2002**, *574*, 740–753.

12. Gültekin, K.; Richstone, D.O.; Gebhardt, K.; Lauer, T.R.; Tremaine, S.; Aller, M.C.; Bender, R.; Dressler, A.; Faber, S.M.; Filippenko, A.V.; et al. The M-σ and M-L Relations in Galactic Bulges, and Determinations of Their Intrinsic Scatter. *Astrophys. J.* **2009**, *698*, 198–221.

13. Kormendy, J.; Ho, L.C. Coevolution (Or Not) of Supermassive Black Holes and Host Galaxies. *Annu. Rev. Astron. Astrophys.* **2013**, *51*, 511–653.

14. Heckman, T.M.; Best, P.N. The Coevolution of Galaxies and Supermassive Black Holes: Insights from Surveys of the Contemporary Universe. *Annu. Rev. Astron. Astrophys.* **2014**, *52*, 589–660.

15. Vogelsberger, M.; Genel, S.; Sijacki, D.; Torrey, P.; Springel, V.; Hernquist, L. A model for cosmological simulations of galaxy formation physics. *Mon. Not. R. Astron. Soc.* **2013**, *436*, 3031–3067.

16. Schaye, J.; Crain, R.A.; Bower, R.G.; Furlong, M.; Schaller, M.; Theuns, T.; Dalla Vecchia, C.; Frenk, C.S.; McCarthy, I.G.; Helly, J.C.; et al. The EAGLE project: Simulating the evolution and assembly of galaxies and their environments. *Mon. Not. R. Astron. Soc.* **2015**, *446*, 521–554.

17. Khandai, N.; Di Matteo, T.; Croft, R.; Wilkins, S.; Feng, Y.; Tucker, E.; DeGraf, C.; Liu, M.S. The MassiveBlack-II simulation: The evolution of haloes and galaxies to z $\sim$ 0. *Mon. Not. R. Astron. Soc.* **2015**, *450*, 1349–1374.

18. Dubois, Y.; Peirani, S.; Pichon, C.; Devriendt, J.; Gavazzi, R.; Welker, C.; Volonteri, M. The HORIZON-AGN simulation: Morphological diversity of galaxies promoted by AGN feedback. *Mon. Not. R. Astron. Soc.* **2016**, *463*, 3948–3964.

19. Steinborn, L.K.; Dolag, K.; Comerford, J.M.; Hirschmann, M.; Remus, R.S.; Teklu, A.F. Origin and properties of dual and offset active galactic nuclei in a cosmological simulation at $z = 2$. *Mon. Not. R. Astron. Soc.* **2016**, *458*, 1013–1028.

20. Tremmel, M.; Karcher, M.; Governato, F.; Volonteri, M.; Quinn, T.R.; Pontzen, A.; Anderson, L.; Bellovary, J. The Romulus cosmological simulations: A physical approach to the formation, dynamics and accretion models of SMBHs. *Mon. Not. R. Astron. Soc.* **2017**, *470*, 1121–1139.

21. Davé, R.; Anglés-Alcázar, D.; Narayanan, D.; Li, Q.; Rafieferantsoa, M.H.; Appleby, S. SIMBA: Cosmological simulations with black hole growth and feedback. *Mon. Not. R. Astron. Soc.* **2019**, *486*, 2827–2849.

22. Ni, Y.; Di Matteo, T.; Bird, S.; Croft, R.; Feng, Y.; Chen, N.; Tremmel, M.; DeGraf, C.; Li, Y. The ASTRID simulation: The evolution of supermassive black holes. *Mon. Not. R. Astron. Soc.* **2022**, *513*, 670–692.

23. Remillard, R.A.; McClintock, J.E. X-ray Properties of Black-Hole Binaries. *Annu. Rev. Astron. Astrophys.* **2006**, *44*, 49–92.

24. Brenneman, L. *Measuring the Angular Momentum of Supermassive Black Holes*; Springer: Berlin, Germany, 2013. [CrossRef]

25. Reynolds, C.S. Observational Constraints on Black Hole Spin. *Annu. Rev. Astron. Astrophys.* **2021**, *59*, 117–154.

26. Mallick, L.; Fabian, A.C.; García, J.A.; Tomsick, J.A.; Parker, M.L.; Dauser, T.; Wilkins, D.R.; De Marco, B.; Steiner, J.F.; Connors, R.M.T.; et al. High-density disc reflection spectroscopy of low-mass active galactic nuclei. *Mon. Not. R. Astron. Soc.* **2022**, *513*, 4361–4379.

27. Chiaberge, M.; Gilli, R.; Lotz, J.M.; Norman, C. Radio Loud AGNs are Mergers. *Astrophys. J.* **2015**, *806*, 147.

28. Miller, S.; Callister, T.A.; Farr, W.M. The Low Effective Spin of Binary Black Holes and Implications for Individual Gravitational-wave Events. *Astrophys. J.* **2020**, *895*, 128.

29. Callister, T.A.; Miller, S.J.; Chatziioannou, K.; Farr, W.M. No Evidence that the Majority of Black Holes in Binaries Have Zero Spin. *Astrophys. J.* **2022**, *937*, L13.

30. Fragione, G.; Loeb, A. An Upper Limit on the Spin of SgrA* Based on Stellar Orbits in Its Vicinity. *Astrophys. J.* **2020**, *901*, L32.

31. Akiyama, K. et al. [Event Horizon Telescope Collaboration]. First M87 Event Horizon Telescope Results. V. Physical Origin of the Asymmetric Ring. *Astrophys. J.* **2019**, *875*, L5.

32. Akiyama, K. et al. [Event Horizon Telescope Collaboration]. First M87 Event Horizon Telescope Results. VIII. Magnetic Field Structure near The Event Horizon. *Astrophys. J.* **2021**, *910*, L13.

33. Akiyama, K.; Alberdi, A.; Alef, W.; Algaba, J.C.; Anantua, R.; Asada, K.; Azulay, R.; Bach, U.; Baczko, A.K.; Ball, D.; et al. First Sagittarius A* Event Horizon Telescope Results. V. Testing Astrophysical Models of the Galactic Center Black Hole. *Astrophys. J.* **2022**, *930*, L16. [CrossRef]

34. Mościbrodzka, M.; Falcke, H.; Noble, S. Scale-invariant radio jets and varying black hole spin. *Astron. Astrophys.* **2016**, *596*, A13.

35. Feng, J.; Wu, Q. Constraint on the black hole spin of M87 from the accretion-jet model. *Mon. Not. R. Astron. Soc.* **2017**, *470*, 612–616.

36. Cruz-Osorio, A.; Fromm, C.M.; Mizuno, Y.; Nathanail, A.; Younsi, Z.; Porth, O.; Davelaar, J.; Falcke, H.; Kramer, M.; Rezzolla, L. State-of-the-art energetic and morphological modelling of the launching site of the M87 jet. *Nat. Astron.* **2022**, *6*, 103–108.

37. Dokuchaev, V.I. Spin and mass of the nearest supermassive black hole. *Gen. Relativ. Gravit.* **2014**, *46*, 1832.

38. Brink, J.; Geyer, M.; Hinderer, T. Astrophysics of resonant orbits in the Kerr metric. *Phys. Rev. D* **2015**, *91*, 083001.

39. Dolence, J.C.; Gammie, C.F.; Shiokawa, H.; Noble, S.C. Near-infrared and X-Ray Quasi-periodic Oscillations in Numerical Models of Sgr A*. *Astrophys. J.* **2012**, *746*, L10.

40. Kato, Y.; Miyoshi, M.; Takahashi, R.; Negoro, H.; Matsumoto, R. Measuring spin of a supermassive black hole at the Galactic centre—implications for a unique spin. *Mon. Not. R. Astron. Soc.* **2010**, *403*, L74–L78.

41. Dovčiak, M.; Karas, V.; Yaqoob, T. An Extended Scheme for Fitting X-ray Data with Accretion Disk Spectra in the Strong Gravity Regime. *Astrophys. J. Suppl. Ser.* **2004**, *153*, 205–221.

42. Broderick, A.E.; Loeb, A. Imaging bright-spots in the accretion flow near the black hole horizon of Sgr A*. *Mon. Not. R. Astron. Soc.* **2005**, *363*, 353–362.

43. Broderick, A.E.; Loeb, A. Imaging optically-thin hotspots near the black hole horizon of Sgr A* at radio and near-infrared wavelengths. *Mon. Not. R. Astron. Soc.* **2006**, *367*, 905–916.
44. Eckart, A.; Schödel, R.; Meyer, L.; Trippe, S.; Ott, T.; Genzel, R. Polarimetry of near-infrared flares from Sagittarius A*. *Astron. Astrophys.* **2006**, *455*, 1–10. [CrossRef]
45. Gravity Collaboration; Bauböck, M.; Dexter, J.; Abuter, R.; Amorim, A.; Berger, J.P.; Bonnet, H.; Brandner, W.; Clénet, Y.; Coudé Du Foresto, V.; et al. Modeling the orbital motion of Sgr A*'s near-infrared flares. *Astron. Astrophys.* **2020**, *635*, A143.
46. Doeleman, S.S.; Fish, V.L.; Broderick, A.E.; Loeb, A.; Rogers, A.E.E. Detecting Flaring Structures in Sagittarius A* with High-Frequency VLBI. *Astrophys. J.* **2009**, *695*, 59–74.
47. Fraga-Encinas, R.; Mościbrodzka, M.; Brinkerink, C.; Falcke, H. Probing spacetime around Sagittarius A* using modeled VLBI closure phases. *Astron. Astrophys.* **2016**, *588*, A57.
48. Raymond, A.W.; Palumbo, D.; Paine, S.N.; Blackburn, L.; Córdova Rosado, R.; Doeleman, S.S.; Farah, J.R.; Johnson, M.D.; Roelofs, F.; Tilanus, R.P.J.; et al. Evaluation of New Submillimeter VLBI Sites for the Event Horizon Telescope. *Astrophys. J. Suppl. Ser.* **2021**, *253*, 5.
49. Pesce, D.W.; Palumbo, D.C.M.; Ricarte, A.; Broderick, A.E.; Johnson, M.D.; Nagar, N.M.; Natarajan, P.; Gómez, J.L. Expectations for Horizon-Scale Supermassive Black Hole Population Studies with the ngEHT. *Galaxies* **2022**, *10*, 109.
50. Roelofs, F.; Fromm, C.M.; Mizuno, Y.; Davelaar, J.; Janssen, M.; Younsi, Z.; Rezzolla, L.; Falcke, H. Black hole parameter estimation with synthetic very long baseline interferometry data from the ground and from space. *Astron. Astrophys.* **2021**, *650*, A56.
51. Johnson, M.D.; Lupsasca, A.; Strominger, A.; Wong, G.N.; Hadar, S.; Kapec, D.; Narayan, R.; Chael, A.; Gammie, C.F.; Galison, P.; et al. Universal interferometric signatures of a black hole's photon ring. *Sci. Adv.* **2020**, *6*, eaaz1310.
52. Gralla, S.E.; Holz, D.E.; Wald, R.M. Black hole shadows, photon rings, and lensing rings. *Phys. Rev. D* **2019**, *100*, 024018.
53. Chael, A.; Johnson, M.D.; Lupsasca, A. Observing the Inner Shadow of a Black Hole: A Direct View of the Event Horizon. *Astrophys. J.* **2021**, *918*, 6.
54. Walker, R.C.; Hardee, P.E.; Davies, F.B.; Ly, C.; Junor, W. The Structure and Dynamics of the Subparsec Jet in M87 Based on 50 VLBA Observations over 17 Years at 43 GHz. *Astrophys. J.* **2018**, *855*, 128.
55. Broderick, A.E.; Tiede, P.; Pesce, D.W.; Gold, R. Measuring Spin from Relative Photon-ring Sizes. *Astrophys. J.* **2022**, *927*, 6.
56. Broderick, A.E.; Pesce, D.W.; Tiede, P.; Pu, H.Y.; Gold, R. Hybrid Very Long Baseline Interferometry Imaging and Modeling with THEMIS. *Astrophys. J.* **2020**, *898*, 9. [CrossRef]
57. Broderick, A.E.; Pesce, D.W.; Gold, R.; Tiede, P.; Pu, H.Y.; Anantua, R.; Britzen, S.; Ceccobello, C.; Chatterjee, K.; Chen, Y.; et al. The Photon Ring in M87*. *Astrophys. J.* **2022**, *935*, 61.
58. Palumbo, D.C.M.; Wong, G.N.; Prather, B.S. Discriminating Accretion States via Rotational Symmetry in Simulated Polarimetric Images of M87. *Astrophys. J.* **2020**, *894*, 156.
59. Emami, R.; Ricarte, A.; Wong, G.N.; Palumbo, D.; Chang, D.; Doeleman, S.S.; Broaderick, A.; Narayan, R.; Weintroub, J.; Wielgus, M.; et al. Unraveling Twisty Linear Polarization Morphologies in Black Hole Images. *arXiv* **2022**, arXiv:2210.01218.
60. Qiu, R.; Ricarte, A.; Narayan, R.; Wong, G.N.; Chael, A.; Palumbo, D. Using Machine Learning to Link Black Hole Accretion Flows with Spatially Resolved Polarimetric Observables. *arXiv* **2022**, arXiv:2212.04852.
61. Medeiros, L.; Chan, C.K.; Narayan, R.; Özel, F.; Psaltis, D. Brightness Asymmetry of Black Hole Images as a Probe of Observer Inclination. *Astrophys. J.* **2022**, *924*, 46.
62. Ricarte, A.; Palumbo, D.C.M.; Narayan, R.; Roelofs, F.; Emami, R. Observational Signatures of Frame Dragging in Strong Gravity. *arXiv* **2022**, arXiv:2211.01810.
63. Fragile, P.C.; Blaes, O.M.; Anninos, P.; Salmonson, J.D. Global General Relativistic Magnetohydrodynamic Simulation of a Tilted Black Hole Accretion Disk. *Astrophys. J.* **2007**, *668*, 417–429.
64. Liska, M.; Hesp, C.; Tchekhovskoy, A.; Ingram, A.; van der Klis, M.; Markoff, S.B.; Van Moer, M. Disc tearing and Bardeen-Petterson alignment in GRMHD simulations of highly tilted thin accretion discs. *Mon. Not. R. Astron. Soc.* **2021**, *507*, 983–990.
65. Wex, N.; Kopeikin, S.M. Frame Dragging and Other Precessional Effects in Black Hole Pulsar Binaries. *Astrophys. J.* **1999**, *514*, 388–401.
66. Pfahl, E.; Loeb, A. Probing the Spacetime around Sagittarius A* with Radio Pulsars. *Astrophys. J.* **2004**, *615*, 253–258.
67. Psaltis, D.; Wex, N.; Kramer, M. A Quantitative Test of the No-hair Theorem with Sgr A* Using Stars, Pulsars, and the Event Horizon Telescope. *Astrophys. J.* **2016**, *818*, 121.
68. De Laurentis, M.; Younsi, Z.; Porth, O.; Mizuno, Y.; Rezzolla, L. Test-particle dynamics in general spherically symmetric black hole spacetimes. *Phys. Rev. D* **2018**, *97*, 104024.
69. Liu, K.; Desvignes, G.; Eatough, R.P.; Karuppusamy, R.; Kramer, M.; Torne, P.; Wharton, R.; Chatterjee, S.; Cordes, J.M.; Crew, G.B.; et al. An 86 GHz Search for Pulsars in the Galactic Center with the Atacama Large Millimeter/submillimeter Array. *Astrophys. J.* **2021**, *914*, 30.
70. Mus, A.; Martí-Vidal, I.; Wielgus, M.; Stroud, G. A first search of transients in the Galactic center from 230 GHz ALMA observations. *Astron. Astrophys.* **2022**, *666*, A39.
71. Wong, G.N. Black Hole Glimmer Signatures of Mass, Spin, and Inclination. *Astrophys. J.* **2021**, *909*, 217.
72. Hadar, S.; Johnson, M.D.; Lupsasca, A.; Wong, G.N. Photon ring autocorrelations. *Phys. Rev. D* **2021**, *103*, 104038.
73. Chesler, P.M.; Blackburn, L.; Doeleman, S.S.; Johnson, M.D.; Moran, J.M.; Narayan, R.; Wielgus, M. Light echos and coherent autocorrelations in a black hole spacetime. *Class. Quantum Gravity* **2021**, *38*, 125006.

74. Tiede, P.; Pu, H.Y.; Broderick, A.E.; Gold, R.; Karami, M.; Preciado-López, J.A. Spacetime Tomography Using the Event Horizon Telescope. *Astrophys. J.* **2020**, *892*, 132.
75. Moriyama, K.; Mineshige, S.; Honma, M.; Akiyama, K. Black Hole Spin Measurement Based on Time-domain VLBI Observations of Infalling Gas Clouds. *Astrophys. J.* **2019**, *887*, 227.
76. Levis, A.; Srinivasan, P.P.; Chael, A.A.; Ng, R.; Bouman, K.L. Gravitationally Lensed Black Hole Emission Tomography. *arXiv* **2022**, arXiv:2204.03715.
77. Johnson, M.D.; Bouman, K.L.; Blackburn, L.; Chael, A.A.; Rosen, J.; Shiokawa, H.; Roelofs, F.; Akiyama, K.; Fish, V.L.; Doeleman, S.S. Dynamical Imaging with Interferometry. *Astrophys. J.* **2017**, *850*, 172.
78. Arras, P.; Frank, P.; Haim, P.; Knollmüller, J.; Leike, R.; Reinecke, M.; Enßlin, T. Variable structures in M87* from space, time and frequency resolved interferometry. *Nat. Astron.* **2022**, *6*, 259–269.
79. Bouman, K.L.; Johnson, M.D.; Dalca, A.V.; Chael, A.A.; Roelofs, F.; Doeleman, S.S.; Freeman, W.T. Reconstructing Video from Interferometric Measurements of Time-Varying Sources. *arXiv* **2017**, arXiv:1711.01357.
80. Emami, R.; Tiede, P.; Doeleman, S.S.; Roelofs, F.; Wielgus, M.; Blackburn, L.; Liska, M.; Chatterjee, K.; Ripperda, B.; Fuentes, A.; et al. Tracing the hot spot motion using the next generation Event Horizon Telescope (ngEHT). *arXiv* **2022**, arXiv:2211.06773.
81. Thorne, K.S. Disk-Accretion onto a Black Hole. II. Evolution of the Hole. *Astrophys. J.* **1974**, *191*, 507–520. [CrossRef]
82. Blandford, R.D.; Znajek, R.L. Electromagnetic extraction of energy from Kerr black holes. *Mon. Not. R. Astron. Soc.* **1977**, *179*, 433–456. [CrossRef]
83. Narayan, R.; Chael, A.; Chatterjee, K.; Ricarte, A.; Curd, B. Jets in magnetically arrested hot accretion flows: Geometry, power, and black hole spin-down. *Mon. Not. R. Astron. Soc.* **2022**, *511*, 3795–3813.
84. King, A.R.; Pringle, J.E.; Hofmann, J.A. The evolution of black hole mass and spin in active galactic nuclei. *Mon. Not. R. Astron. Soc.* **2008**, *385*, 1621–1627.
85. Volonteri, M.; Sikora, M.; Lasota, J.P.; Merloni, A. The Evolution of Active Galactic Nuclei and their Spins. *Astrophys. J.* **2013**, *775*, 94.
86. Murchikova, E.M.; Phinney, E.S.; Pancoast, A.; Blandford, R.D. A cool accretion disk around the Galactic Centre black hole. *Nature* **2019**, *570*, 83–86.
87. Ressler, S.M.; Quataert, E.; Stone, J.M. Hydrodynamic simulations of the inner accretion flow of Sagittarius A* fuelled by stellar winds. *Mon. Not. R. Astron. Soc.* **2018**, *478*, 3544–3563.
88. Anglés-Alcázar, D.; Quataert, E.; Hopkins, P.F.; Somerville, R.S.; Hayward, C.C.; Faucher-Giguère, C.A.; Bryan, G.L.; Kereš, D.; Hernquist, L.; Stone, J.M. Cosmological Simulations of Quasar Fueling to Subparsec Scales Using Lagrangian Hyper-refinement. *Astrophys. J.* **2021**, *917*, 53.
89. Rezzolla, L.; Barausse, E.; Dorband, E.N.; Pollney, D.; Reisswig, C.; Seiler, J.; Husa, S. Final spin from the coalescence of two black holes. *Phys. Rev. D* **2008**, *78*, 044002.
90. Volonteri, M.; Madau, P.; Quataert, E.; Rees, M.J. The Distribution and Cosmic Evolution of Massive Black Hole Spins. *Astrophys. J.* **2005**, *620*, 69–77.
91. Berti, E.; Volonteri, M. Cosmological Black Hole Spin Evolution by Mergers and Accretion. *Astrophys. J.* **2008**, *684*, 822–828.
92. Ricarte, A.; Natarajan, P. The observational signatures of supermassive black hole seeds. *Mon. Not. R. Astron. Soc.* **2018**, *481*, 3278–3292.
93. Ricarte, A.; Natarajan, P. Exploring SMBH assembly with semi-analytic modelling. *Mon. Not. R. Astron. Soc.* **2018**, *474*, 1995–2011.
94. Ricarte, A.; Pacucci, F.; Cappelluti, N.; Natarajan, P. The clustering of undetected high-redshift black holes and their signatures in cosmic backgrounds. *Mon. Not. R. Astron. Soc.* **2019**, *489*, 1006–1022.
95. Kelly, B.C.; Shen, Y. The Demographics of Broad-line Quasars in the Mass-Luminosity Plane. II. Black Hole Mass and Eddington Ratio Functions. *Astrophys. J.* **2013**, *764*, 45.
96. Bardeen, J.M. Kerr Metric Black Holes. *Nature* **1970**, *226*, 64–65. [CrossRef] [PubMed]
97. Weinberger, R.; Springel, V.; Pakmor, R.; Nelson, D.; Genel, S.; Pillepich, A.; Vogelsberger, M.; Marinacci, F.; Naiman, J.; Torrey, P.; et al. Supermassive black holes and their feedback effects in the IllustrisTNG simulation. *Mon. Not. R. Astron. Soc.* **2018**, *479*, 4056–4072.
98. Pacucci, F.; Loeb, A. Separating Accretion and Mergers in the Cosmic Growth of Black Holes with X-Ray and Gravitational-wave Observations. *Astrophys. J.* **2020**, *895*, 95.
99. Begelman, M.C.; Blandford, R.D.; Rees, M.J. Massive black hole binaries in active galactic nuclei. *Nature* **1980**, *287*, 307–309. [CrossRef]
100. Colpi, M. Massive Binary Black Holes in Galactic Nuclei and Their Path to Coalescence. *Space Sci. Rev.* **2014**, *183*, 189–221.
101. Bortolas, E.; Capelo, P.R.; Zana, T.; Mayer, L.; Bonetti, M.; Dotti, M.; Davies, M.B.; Madau, P. Global torques and stochasticity as the drivers of massive black hole pairing in the young Universe. *Mon. Not. R. Astron. Soc.* **2020**, *498*, 3601–3615.
102. Izquierdo-Villalba, D.; Bonoli, S.; Dotti, M.; Sesana, A.; Rosas-Guevara, Y.; Spinoso, D. From galactic nuclei to the halo outskirts: Tracing supermassive black holes across cosmic history and environments. *Mon. Not. R. Astron. Soc.* **2020**, *495*, 4681–4706.
103. Ricarte, A.; Tremmel, M.; Natarajan, P.; Zimmer, C.; Quinn, T. Origins and demographics of wandering black holes. *Mon. Not. R. Astron. Soc.* **2021**, *503*, 6098–6111.
104. Milosavljević, M.; Merritt, D. Formation of Galactic Nuclei. *Astrophys. J.* **2001**, *563*, 34–62.

105. Volonteri, M.; Haardt, F.; Madau, P. The Assembly and Merging History of Supermassive Black Holes in Hierarchical Models of Galaxy Formation. *Astrophys. J.* **2003**, *582*, 559–573.
106. Tchekhovskoy, A.; Narayan, R.; McKinney, J.C. Efficient generation of jets from magnetically arrested accretion on a rapidly spinning black hole. *Mon. Not. R. Astron. Soc.* **2011**, *418*, L79–L83.
107. Volonteri, M.; Perna, R. Dynamical evolution of intermediate-mass black holes and their observable signatures in the nearby Universe. *Mon. Not. R. Astron. Soc.* **2005**, *358*, 913–922.

Article

Measuring Photon Rings with the ngEHT

Paul Tiede [1,2,*], Michael D. Johnson [1,2], Dominic W. Pesce [1,2], Daniel C. M. Palumbo [1,2], Dominic O. Chang [1,2] and Peter Galison [2,3,4]

1   Center for Astrophysics | Harvard & Smithsonian, 60 Garden Street, Cambridge, MA 02138, USA
2   Black Hole Initiative, Harvard University, 20 Garden Street, Cambridge, MA 02138, USA
3   Department of Physics, Harvard University, Cambridge, MA 02138, USA
4   Department of History of Science, Harvard University, Cambridge, MA 02138, USA
*   Correspondence: paul.tiede@cfa.harvard.edu

Abstract: General relativity predicts that images of optically thin accretion flows around black holes should generically have a "photon ring", composed of a series of increasingly sharp subrings that correspond to increasingly strongly lensed emission near the black hole. Because the effects of lensing are determined by the spacetime curvature, the photon ring provides a pathway to precise measurements of the black hole properties and tests of the Kerr metric. We explore the prospects for detecting and measuring the photon ring using very long baseline interferometry (VLBI) with the Event Horizon Telescope (EHT) and the next-generation EHT (ngEHT). We present a series of tests using idealized self-fits to simple geometrical models and show that the EHT observations in 2017 and 2022 lack the angular resolution and sensitivity to detect the photon ring, while the improved coverage and angular resolution of ngEHT at 230 GHz and 345 GHz is sufficient for these models. We then analyze detection prospects using more realistic images from general relativistic magnetohydrodynamic simulations by applying "hybrid imaging", which simultaneously models two components: a flexible raster image (to capture the direct emission) and a ring component. Using the Bayesian VLBI modeling package `Comrade.jl`, we show that the results of hybrid imaging must be interpreted with extreme caution for both photon ring detection and measurement—hybrid imaging readily produces false positives for a photon ring, and its ring measurements do not directly correspond to the properties of the photon ring.

Keywords: black holes; photon rings; Radio Astronomy; VLBI

Citation: Tiede, P.; Johnson, M.D.; Pesce, D.W.; Palumbo, D.C.M.; Chang, D.O.; Galison, P. Measuring Photon Rings with the ngEHT. *Galaxies* **2022**, *10*, 111. https://doi.org/10.3390/galaxies10060111

Academic Editor: Emilio Elizalde

Received: 24 October 2022
Accepted: 2 December 2022
Published: 6 December 2022

Publisher's Note: MDPI stays neutral with regard to jurisdictional claims in published maps and institutional affiliations.

1. Introduction

Simulated images of optically thin accretion flows around supermassive black holes (SMBHs) generically exhibit a nested series of "photon rings" produced from strong gravitational lensing of photon trajectories near the black hole (e.g., [1,2]). These increasingly sharp ring-like features are exponentially demagnified as they converge on an asymptotic critical curve [3–5], and they can be indexed by the number n of half-orbits that light takes around the black hole, as shown in Figure 1 [6–8]. Because the null geodesics that define the photon ring are determined by the spacetime curvature and are negligibly affected by accreting plasma, detection of an $n > 0$ photon ring would provide striking evidence that the supermassive compact objects in galactic cores are Kerr black holes and would provide a pathway to precisely measuring their properties.

To date, measurements of the horizon-scale emission structure around black holes are only possible using millimeter-wavelength very long baseline interferometry (VLBI). The Event Horizon Telescope (EHT) is a globe-spanning network of (sub)millimeter radio telescopes that has carried out VLBI observations of the SMBHs M87* and Sgr A* on horizon scales [1,2,9–20]. The next-generation EHT (ngEHT) plans to build on the capabilities of the EHT by adding multiple new telescopes to the array, increasing the frequency coverage, and improving the sensitivity by observing with wider bandwidths [21]. Though the ngEHT will operate with an unprecedentedly fine diffraction-limited angular resolution of $\sim$15 µas, the $n = 1$ photon ring is anticipated to be finer still; the expected thickness of the $n = 1$ photon ring in M87* corresponds to an angular size of less than $\sim$4 µas [8]. Direct imaging of the $n = 1$ photon ring will thus likely remain unachievable for the foreseeable future, and studies of this feature using ground-based VLBI will require some degree of "superresolution" via judicious application of parameterized models of the source structure.

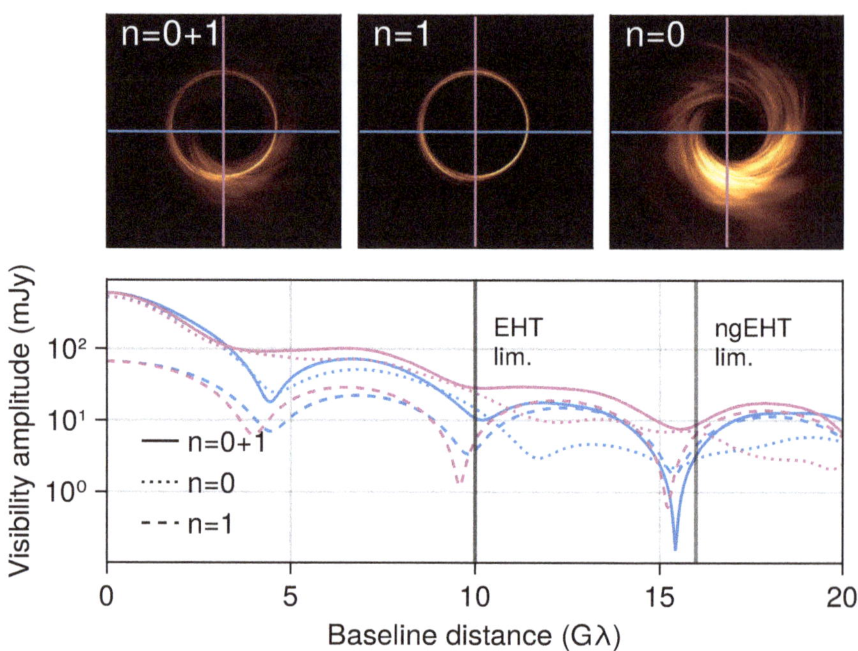

Figure 1. The image of a black hole can be decomposed into subimages that are indexed by the number of half orbits that their photons traveled around the black hole before reaching the observer. In this scheme, the $n = 0$ emission (top right panel) is the "direct" image of the accretion flow and is dominated by astrophysical emission structure. The $n = 1$ emission (top middle panel) is the "secondary" image, consisting of photons that have traveled a half orbit around the black hole before reaching the observer. The actual observed image is a sum of all n subimages (top right panel). The bottom panel shows visibility amplitudes of these (sub)images for projected baselines that are parallel (blue) and perpendicular (pink) to the black hole spin axis. The longest EHT and ngEHT baselines, indicated with vertical black lines, occur at baseline lengths for which the $n = 0$ and $n = 1$ contributions are comparable, raising the prospect of distinguishing them through modeling.

At least two classes of modeling methodology currently show promise for extracting superresolved photon ring signatures from VLBI measurements of black holes: models that parameterize the three-dimensional distribution of the material in the vicinity of the black hole (e.g., [22–24]), and models that parameterize the two-dimensional distribution of the emission morphology as seen on the sky [25,26]. In either case, because the additional information supplied by the model specification is supporting the extraction of superresolved structural information, it is important to quantify precisely what defines a photon ring "detection." For instance, the most compelling detection might not require the assumption that general relativity (GR) is true, while a somewhat weaker claim of detection might test for the presence of this feature under the assumptions of GR. Likewise, methods could utilize models that *assume* the existence of the photon ring to make measurements of black hole parameters without needing to meet potentially more stringent criteria for an unambiguous *detection* of the same feature.

A parameterized modeling approach to study the photon ring was recently developed by Broderick et al. [25] (hereafter [25]), who employ a "hybrid imaging" technique that fits a thin geometric ring component alongside a more flexible pixel-based image component, where the pixel flux densities are treated as model parameters. Broderick et al. [26] (hereafter [26]) applied this technique to the EHT observations of M87*, finding that the diameter of the thin ring component is well-constrained by the EHT data; the authors associate this component with the $n = 1$ photon ring. While the value and stability of the diameter of this component across different datasets support its identification as an image feature that is determined by the spacetime, other aspects—particularly the fraction of the total flux density that is recovered in the thin ring—challenge its association with the $n = 1$ ring. This ambiguity underscores the need to quantify exactly what constitutes a photon ring detection.

In this paper, we aim to investigate the efficacy of tools such as hybrid imaging to extract photon ring signatures from EHT- and ngEHT-like data and to determine what VLBI measurements are necessary and sufficient to reliably detect a photon ring. In Section 2, we conduct tests using simple geometric models, deriving necessary conditions to detect the $n = 1$ photon ring. Next, in Section 3, we explore the application and limitations of the hybrid imaging approach to detect and measure the photon ring, and we perform tests using more realistic synthetic data from general relativistic magnetohydrodynamic (GRMHD) simulations. In Section 4, we summarize these results and discuss their implications for the EHT, ngEHT, and other future VLBI arrays.

2. Geometric Modeling

We begin with a series of idealized tests, generating simulated data from a simple geometric on-sky model that includes a proxy for the photon ring and then fitting the same model to these data. This so-called self-fit procedure guarantees that model parameter posteriors are directly interpretable. However, the clarity of this procedure comes with the penalty of being artificially optimistic; it provides requirements for detecting the photon ring that are likely necessarily but almost certainly not sufficient. Hence, if these self-fits to simulated data cannot detect a photon ring with a given array, then we expect that photon ring detection with the same array in realistic settings will be impossible.

The structure of this section is as follows. First, we describe our geometric model (Section 2.1). Next, we outline our construction of simulated data and the fitting procedure (Section 2.2). Finally, we perform self-fits for a variety of EHT and ngEHT arrays to assess the requirements for detecting the $n = 1$ photon ring (Section 2.3).

2.1. Specifying the Geometric Model

Our simple parametric model is motivated by the expected image structure for optically thin emission near a black hole consisting of multiple ring-like structures. For each component, we use the m-ring model from Johnson et al. [8] and Event Horizon Telescope Collaboration et al. [19]. This model is an infinitesimally thin ring with azimuthal brightness modulation determined by angular Fourier coefficients, which is then convolved with a Gaussian kernel G. We restrict ourselves to a first-order Fourier expansion, giving the following intensity profile for the thin ring:

$$M(r, \theta | d_i, a_i, b_i, F_i) = \frac{F_i}{\pi d_i} \delta(r - d_i/2)(1 + a_i \cos(\theta) - b_i \sin(\theta)), \qquad (1)$$

where we parameterize a_i, b_i using a polar representation $a_i = A_i \cos \phi_i$ and $b_i = A_i \sin \phi_i$, where A_i is the amplitude and ϕ_i is the phase of the first-order Fourier coefficient. Finally, F_i and d_i are the flux density and diameter of the ring, respectively. Note that we have included a subscript, i, in anticipation of the nested photon rings. The location of the observed n photon rings relative to the emitting plasma are shifted as a function of spin and inclination. Therefore, we allow the centroid of the rings to be displaced by an amount x_i, y_i. To give the ring finite width, we convolve the m-ring with a symmetric Gaussian:

$$G(r, \theta | w_i) = \frac{4\log(2)}{\pi w_i^2} \exp\left(-\frac{4\log(2)r^2}{w_i^2}\right) \tag{2}$$

where w_i is the Gaussian's full width at half maximum (FWHM). We denote the thick m-ring model by $T(x, y) = M \star G$, where $\star$ is the convolution operator. Finally, the shape of the ring is also of interest since it encodes information about the spin and inclination of the central black hole (see, e.g., [8,27–32]). To add ring ellipticity, we modify the intensity map of the thick m-ring

$$T(x, y) \rightarrow R_\zeta T(x, (1 + \tau)y) = I(x, y) \tag{3}$$

where R_ζ rotates the image by ζ radians counter-clockwise, and $\tau > 0$ parametrizes the ring ellipticity. Formally, τ is related to the eccentricity e of the elliptical (stretched) ring via $e = \sqrt{1 - 1/(1 + \tau)^2} \approx \sqrt{2\tau}$. We denote this model by I and call it the stretched thick m-ring.

The stretched thick m-ring forms the base image for each nested photon ring. The final model that we use is a sum of multiple stretched thick m-ring components:

$$I^{0:N}(x, y) = \sum_{n=0}^{N} I(x, y | F_i, d_i, w_i, A_i, \phi_i, \tau_i, \zeta_i, x_i, y_i). \tag{4}$$

2.2. Simulated Observations and Fitting Procedure

To create simulated data, we use Equation (4) with m-ring parameters motivated by the observed structure and expected gravitational lensing of M87* [13]. Because we are focused on distinguishing the $n = 0$ and $n = 1$ structure, our model for the construction of the simulated data consists of two nested rings (i.e., $N = 1$) with equal brightness. Their diameters d_i are related by

$$d_1 = 2\rho_c + (d_0 - 2\rho_c)e^{-\gamma}. \tag{5}$$

We set $\rho_c = 19\,\mu$as and $\gamma = \pi$, which approximates the structure of the photon ring in M87* given its mass [13] and low viewing inclination [33]. Additionally, the width of the rings will be given by

$$w_1 = w_0 e^{-\gamma}, \tag{6}$$

and the flux density ratio by

$$F_{n+1}/F_n \sim e^{-\gamma}. \tag{7}$$

Note that these expressions are a rather crude approximation of the precise structure expected in black hole images. Nevertheless, given that we want to explore the potential for photon ring detections that do not explicitly assume general relativity and that our goal is only to define a minimum threshold for detection, we do not regard this as a significant limitation.

Our complete model description is as follows:

- **Flux density:** $F_0 = 0.6\,$Jy, and $F_1 = e^{-\pi}F_0 = 0.03\,$Jy (Equation (7))
- **Diameter:** $d_0 = 45\,\mu$as, and $d_1 = 2\rho_c + (d_0 - 2\rho_c)e^{-\pi} = 38.3\,\mu$as (Equation (5))
- **Width:** $w_0 = 18\,\mu$as, and $w_1 = w_0 e^{-\pi} = 0.78\,\mu$as (Equation (6))
- **Brightness Asymmetry:** $A_0 = A_1 = 0.15$, and $\phi_0 = \phi_1 = -\pi/2$
- **Ellipticity:** $\tau_0 = \tau_1 = 0.05$, and $\zeta_0 = \zeta_1 = \pi/3$
- **Ring Centers:** $x_0 = y_0 = x_1 = y_1 = 0\,\mu$as

The image structure of this model is shown in the leftmost panel of Figure 2.

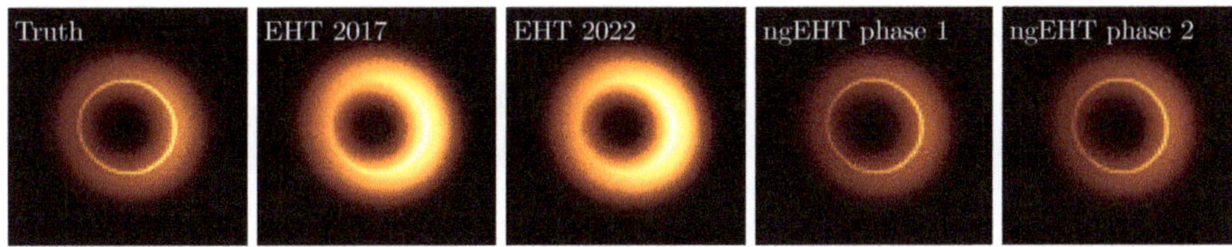

Figure 2. Results from self-fits of geometric models to synthetic data for a series of EHT and ngEHT arrays. Each fit includes two m-ring components, as described in Section 2.1. The left panel shows the ground truth model; the remaining panels show the mean image for the array noted at the top. Even for this optimistic test, the EHT coverage is insufficient to identify the $n = 1$ photon ring. However, the longer baselines of the ngEHT allow clear detection of the $n = 1$ photon ring.

We generate simulated data using `eht-imaging` [34] and the `ngehtsim`[1] package. This procedure integrates historical weather data at each site using similar methods to Raymond et al. [35]. For the EHT arrays, we use 2 GHz of recorded bandwidth with dual polarization, while for the ngEHT arrays, we use 8 GHz of recorded bandwidth with dual polarization. To approximate the practical limitations related to fringe detection, we remove baselines that are not connected to another baseline with an SNR > 5 within a 10 s integration time. Finally, we segment and coherently average the data over 5 min intervals, emulating standard on-sky scans. For each 5 min scan, we add station-based gain corruptions, which add 10% Gaussian multiplicative Gaussian noise in amplitudes and uniform $[0, 2\pi)$ noise in visibility phases. We include thermal noise but do not include any additional non-closing errors, such as polarimetric leakage. The list of arrays we use in this paper is listed in Table 1.

Table 1. Arrays used for synthetic data. For additional details on EHT sites, see [10]; for additional details on ngEHT sites, see [35]. Note that the SPT cannot observe M87* so does not contribute to the tests shown in this paper. New ngEHT phase 1 sites use specifications for existing facilities (HAY: 37-m, OVRO: 10.4-m) and are 6.1-m for new locations (BAJA, CNI, LAS); new ngEHT phase 2 sites assume 8-m diameters with the exception of the AMT, which is planned to be 15-m [36].

Array	Freq. (GHz)	Sites
EHT 2017	230	(8) ALMA, APEX, JCMT, LMT, IRAM, SMA, SMT, SPT
EHT 2022	230	(11) EHT 2017, KP, NOEMA, GLT
ngEHT phase 1	230, 345	(16) EHT 2022, BAJA, CNI, HAY, LAS, OVRO
ngEHT phase 2	230, 345	(22) ngEHT phase 1, GARS, AMT, CAT, BOL, BRZ, PIKE

To extract model parameters from the simulated data, we use Bayesian inference. Our goal is to find the posterior distribution $p(\theta|D)$ for our model parameters θ given the data products D:

$$p(\theta|D) = \frac{p(D|\theta)p(\theta)}{p(D)}. \tag{8}$$

The $p(D|\theta)$ distribution is often called the likelihood and is sometimes denoted by $\mathcal{L}(D|\theta)$, $p(\theta)$ is the prior, and $p(D)$ is the marginal likelihood or evidence.

In this work, we use log-closure amplitudes and closure phases as our data products[2] The benefit of closure products is that they are immune to station-based gain errors (such as those introduced in the synthetic data generation). However, the likelihood functions of closure quantities are non-Gaussian in the low-SNR limit [37,38]. For this paper, we used the high-SNR expression from Event Horizon Telescope Collaboration et al. [19] and flagged any closure products that had SNR < 3. For the EHT and ngEHT 230 GHz observations, this cut removed 1% of the closure products. For the ngEHT 345 GHz observations, this cut removed 25% of the closure products.

For model fitting, we use the same model prescription as above, but we parameterize the width and total flux density as follows:

- $w_1 = \epsilon_1 w_0$ (the width of the $n = 1$ ring is forced to be thinner than the first)
- $F_0 + F_1 = 1\,\text{Jy}$ (the total flux density is forced to be unity)

Both of these choices are for computational efficiency when sampling the posterior. The first prevents a trivial label-swapping degeneracy between the two m-rings, and the second resolves the trivial flux density-rescaling degeneracy that occurs when using only closure products. Additionally, we force the $x_0 = y_0 = 0\,\mu\text{as}$ to provide a phase center for the reconstructions. For multi-frequency observations with the ngEHT, we assume a flat spectral index (i.e., we use the same model to fit both 230 and 345 GHz observations and also for the ground-truth model). Table 2 lists the priors assumed for each model parameter.

Table 2. Priors used for the geometric self-fits with two m-ring components. $\mathcal{U}(a, b)$ represents the uniform distribution on the interval (a, b).

Parameter	Prior
$d_{0,1}$	$\mathcal{U}(20\,\mu\text{as}, 60\,\mu\text{as})$
w_0	$\mathcal{U}(0.1\,\mu\text{as}, 40\,\mu\text{as})$
ϵ_1	$\mathcal{U}(0.0, 1.0)$
$A_{0,1}$	$\mathcal{U}(0.0, 0.5)$
$\phi_{0,1}$	$\mathcal{U}(-\pi, \pi)$
$\tau_{0,1}$	$\mathcal{U}(0.0, 0.5)$
$\xi_{0,1}$	$\mathcal{U}(0, \pi)$
y_1	$\mathcal{U}(-10\,\mu\text{as}, 10\,\mu\text{as})$
x_1	$\mathcal{U}(-10\,\mu\text{as}, 10\,\mu\text{as})$

To fit the model to the simulated data, we use the Bayesian VLBI modeling package `Comrade.jl` [39] implemented within the Julia programming language [40]. To sample from the posterior, we first use the pathfinder algorithm [41] and its Julia implementation `Pathfinder.jl` [42] to find a Gaussian approximation of the posterior. This approximation tends to be poor, but it helps initialize MCMC sampling methods that enable more precise estimates of the posterior. To sample from the posterior, we use the no-u-turn sampler (NUTS) algorithm [43] and its Julia implementation [44]. To initialize our sampling, we draw a random sample from the pathfinder variational approximation, and we use the diagonal elements of the covariance matrix to initialize the NUTS mass matrix. We find that this greatly reduces the amount of time required for NUTS to adapt to our posterior. We run the NUTS sampler for 3000 adaptation steps and 5000 sampling steps. To check MCMC convergence, we compute the effective sample size of each chain (after removing the adaptation steps) and found > 500 effective samples for all model parameters. Note that NUTS struggles to explore multi-modal posteriors, making it possible that we are missing parts of the posterior distribution. To test for multi-modality, we ran an optimizer from many starting locations. Other than pathological cases where the fit quality is extremely poor, we find no evidence for a multi-modal posterior.

2.3. Results

Figure 2 shows the mean image estimated using synthetic data for each array in Table 1, while Figure 3 shows the marginal posteriors for the parameters of the secondary ring component. These results show that the secondary ring parameters have a hierarchy of measurement difficulty. For instance, all the arrays provide tight constraints on the ring diameter, flux density, and centroid. The ring asymmetry is more challenging to constrain; the EHT 2017 coverage is inadequate to constrain the $n = 1$ ring asymmetry[3], while the EHT 2022 coverage is sufficient to constrain asymmetry to the ~2% level, and the ngEHT coverage provides further improvement. The most challenging parameter to constrain is the width of the secondary ring, which is weakly constrained by both EHT arrays but is tightly constrained for the ngEHT arrays. Note that we used the combined 230 and 345 GHz coverage for the ngEHT arrays; if restricted to just 230 or 345 GHz, the arrays no longer constrain the width of the m-ring to be less than $M/D \approx 4\,\mu$as. Thus, these tests demonstrate the importance of ngEHT observations at both frequencies for M87*.

Interestingly, the width posterior for ngEHT coverage does not appreciably change when moving from phase 1 to phase 2. This result arises because the six additional dishes in phase 2 bring more complete baseline coverage but do not increase the maximum baseline length or the SNR on long baselines, both of which are crucial for measuring the properties of the photon ring. Similarly to the marginal posterior for the width, the ellipticity posterior also shrinks considerably when moving from the EHT to the ngEHT arrays. Both ngEHT arrays measure the ellipticity to $\lesssim 1\%$ precision.

These idealized tests show that the EHT cannot detect the photon ring. However, the ngEHT may be able to meaningfully constrain its size, width, flux density ratio, and ellipticity. Because the principal difficulty in a photon ring detection will be in distinguishing it from the direct emission, measuring each of these ring properties is important for claiming a detection of the photon ring. For instance, treating the ring thickness as a model parameter permits a crucial diagnostic of whether the data show a preference for a narrow ring, which is a key identifying property of the photon ring. Anomalous values in any of the fitted ring parameters may indicate that the direct emission affects the measured photon ring structure.

In the next section, we analyze the prospects for detecting the $n = 1$ photon ring under more realistic circumstances, using images from GRMHD simulations to generate synthetic data and fitting models to these data using the hybrid imaging approach from [25].

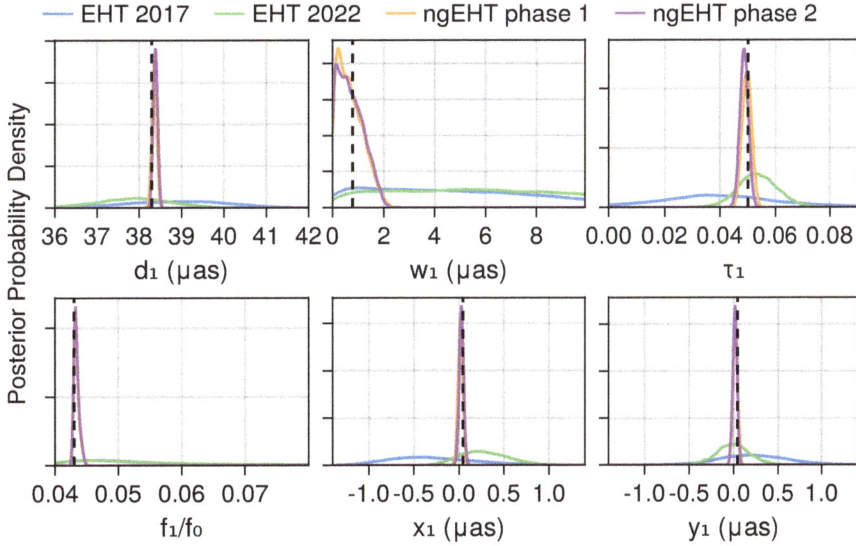

Figure 3. Geometric self-fit marginal posterior distributions for the examples shown in Figure 2. The curves show the marginal posterior distributions for $n = 1$ ring parameters of interest: diameter (d_1), width (w_1), ellipticity (τ_1), fractional flux density (f_1/f_0), and relative displacement (x_1, y_1). The black dashed line shows the true value for the $n = 1$ ring component. Overall, we find that the EHT 2017 and 2022 coverage is insufficient to fully constrain the second ring. In particular, the width of the $n = 1$ ring component is the most difficult quantity to measure.

3. Hybrid Imaging

We now explore more realistic tests of whether the EHT and ngEHT can detect and measure the photon ring, using synthetic data from GRMHD simulations and applying a flexible hybrid imaging approach from [25] that we have implemented in `Comrade.jl`. First, we review the original [25] model and describe the modifications in our implementation of it (Section 3.1). Next, we apply hybrid imaging to a series of simulated datasets using the 2017 EHT array (Section 3.2). Finally, we apply the hybrid model to simulated ngEHT data for the first time and assess the viability of photon ring measurements with the ngEHT and the hybrid imaging approach (Section 3.3).

3.1. Review of Hybrid Imaging

Ref. [25] proposed modeling a black hole image using a decomposition consisting of two components. The first component is a rasterized image model given by

$$I(x,y) = \sum_{ij} I_{ij}\kappa(x - x_i)\kappa(y - y_j), \tag{9}$$

where $\kappa(x)^4$ is the *pulse* function that converts the raster of pixel flux densities I_{ij} into a continuous image. For this work, we use a third-order B-spline kernel (Ref. [25] uses a slightly different pulse function that does not preserve image positivity.). The B-spline kernel is given by successive convolutions of the square wave pulse ($\mathrm{Sq}(x)$) with itself; e.g., the third order B-spline is given by

$$\kappa(x) = (\mathrm{Sq} \star \mathrm{Sq} \star \mathrm{Sq})(x), \tag{10}$$

where $\mathrm{Sq}(x) = 1$ when $|x| < 1/2$ and 0 otherwise.

The second model component in [25] is a ring that is forced to be thin (<2 µas), creating a natural scale separation in the model components. The authors suggested that this scale separation would allow the rasterized image model to predominantly fit the $n = 0$ emission, while the ring component would predominantly fit the $n = 1$ emission. Hence, the ring component would measure the properties of the photon ring.

To test whether the hybrid imaging hypothesis works for the EHT data, ref. [25] analyzed mock data from five GRMHD simulations, with coverage and sensitivity corresponding to the 2017 EHT observations of M87*. They found that in 4/5 cases, the correct $n = 1$ photon ring diameter was contained within the 95% highest posterior density interval (HPDI) for the fitted ring diameter, and 5/5 models had the $n = 1$ photon ring diameter within the 99% HPDI. These results suggested that the measured diameter is correlated with the true $n = 1$ photon ring diameter. However, ref. [25] found that the recovered ring flux density was a factor of 2–3 times higher than the true $n = 1$ photon ring flux density; they argue that this excess flux density matches expectations for an array with an angular resolution of 20 µas. Namely, the flux density approximately matches the integrated flux density within an annulus with a diameter of the $n = 1$ photon ring and a width of 20 µas.

The ref. [25] analysis has two notable limitations: (1) it does not answer the question of whether the hybrid imaging approach will always favor placing a thin ring feature in the image regardless of the true on-sky appearance, and (2) it does not demonstrate that hybrid imaging can distinguish the $n = 0$ and $n = 1$ emission because the diameters of these components were very similar in all five tests. In the next section, we assess hybrid imaging for 2017 EHT data, focusing on tests that address both of these limitations.

3.2. Testing Hybrid Imaging on EHT 2017 Data

We first explore the application of hybrid imaging to synthetic data matching the EHT 2017 observations of M87*. For these tests, we generate data using the GRMHD simulation snapshot of M87* shown in Figure 1. The GRMHD simulation is of a blackhole in the magnetically arrested accretion state (MAD) taken from Johnson et al. [8]. It has a blackhole with a spin of 0.9375 that is viewed at an inclination of 163° with respect to the spin axis and. The electron temperature is described through a temperature model that has a maximum proton to electron temperature ratio, $r_{\mathrm{high}} = 10$ [46]. The GRMHD fluid simulations used in this paper were run using `iharm3d` [47,48] and ray-traced using `ipole` [49]. The ray tracing decomposition into sub-images is taken from Palumbo and

Wong [50]; for additional details on the snapshot generation pipeline, see Wong et al. [51]. To assess the ability of hybrid imaging to detect and/or measure the photon ring, we analyze four separate images: the first and second images contain just the $n = 1$ and $n = 0$ emission, respectively, and the third image is the combined $n = 0 + 1$ emission, and the fourth image is similar to the $n = 0 + 1$ image, but we have artificially shrunk the $n = 1$ photon ring by 21%, corresponding to a 30 µas diameter for the $n = 1$ photon ring. The resulting images are shown in the first column of Figure 4.

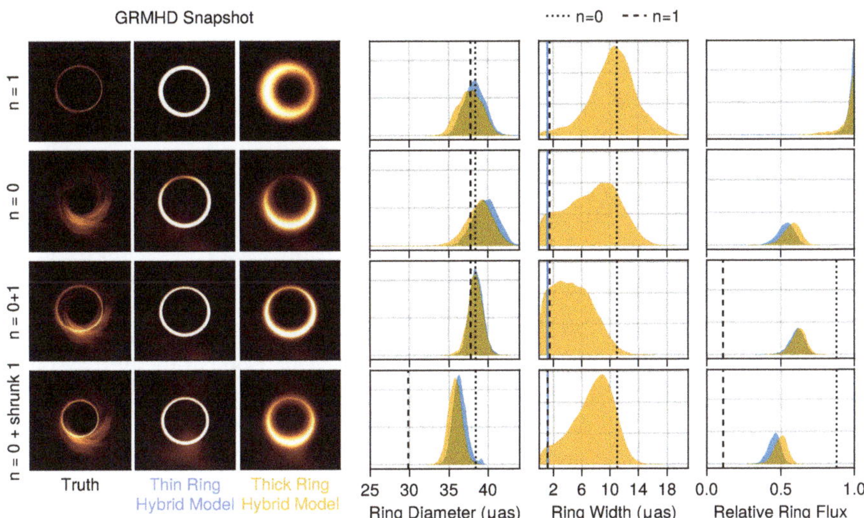

Figure 4. Results of fitting a GRMHD simulation snapshot with coverage corresponding to the EHT 2017 array. The GRMHD simulation has a spin of 0.9375, inclination of 163°, $r_{\text{high}} = 10$, and is in the MAD accretion state. The top row shows the results from fitting only the $n = 1$ emission with the hybrid model, the second row shows fits to only the $n = 0$ emission, and the third row shows fits to the combined $n = 0 + 1$ emission. The bottom row shows fits to the combined $n = 0 + 1$ emission after reducing the diameter of the $n = 1$ ring by 21%. All rows show both fits that force the ring component to be thin (blue) and fits that do not constrain the ring width (orange). The black dashed lines show true values for the $n = 1$ photon ring, while the black dotted lines show the true values for the $n = 0$ photon ring.

For each image, we generated simulated data whose properties matched the April 11 2017 EHT observations. We then fit two versions of the hybrid model described in the previous section for each dataset. Both models use an 8×8 raster (see Equation (9)) with a 90 µas field of view and also a thick m-ring model. Note that we do not include ellipticity in the ring model fit to EHT 2017 data since we found that it was poorly constrained even for the geometric self-fits using EHT data (Ref. [25] also did not include ellipticity).

The difference between the two models is the prior on the thickness of the blurred m-ring. The first model forces the FWHM of the ring to be 1 µas (similar to [25], who force the ring to be thin); we call this the *thin-ring hybrid model*. For the second model, we also fit the thickness of the thick m-ring; we call this the *thick-ring hybrid model*. The priors for the raster model are given by a Dirichlet prior with concentration parameter $\alpha = 1$, effectively placing a uniform prior on the N simplex where N is the number of pixels in the image.[5] For the ring component, we use the same priors as those for the first ring component in Table 2, except for the thin-ring hybrid model where $w_0 = 1$ µas. We force the ring component and raster to be centered on the origin, which differs from [25], which fits for the ring centroid.

In our view, the two hybrid models serve different purposes, distinguished by their ability to fit all relevant ring parameters. The thick-ring hybrid model makes fewer assumptions about the fitted ring component, making it a useful basis for *detecting* a photon ring.

The thin-ring hybrid model imposes more assumptions about the fitted ring component, making it a useful basis for *measuring* the remaining photon ring properties. Figure 3 provides motivation for forcing the ring to be thin, suggesting that the ring thickness is the most difficult parameter to constrain.

To measure the ring diameters and widths for the $n = 0$ and $n = 1$ photon rings from the GRMHD images, we used the VIDA.jl package [52]. VIDA extracts image parameters by optimizing approximate template images that are parameterized by the features of interest. We used VIDA's SlashedGaussianRing template, which provides estimates for the ring diameter, width, and brightness position angle. For our objective function, we used the Bhattacharyya divergence [53]:

$$\mathrm{Bh}(f|I) = \int \sqrt{f(x,y)I(x,y)}dxdy, \tag{11}$$

where f and I are the template and image, respectively. We use the Bhattacharyya divergence since it follows the recommendations from Tiede et al. [52]. While other divergences could be used, we found that they did not appreciably change the extracted parameters. To extract the ring parameters, we used VIDA on $n = 0$ and $n = 1$ images separately.

The results of the test are shown in Figure 4. We find that the thick and thin ring models give similar results for the measured ring diameter and relative ring flux density. This suggests that the thin ring component is not focusing on a different aspect of the image than the thick ring component. Comparing the $n = 0$ and $n = 0 + 1$ fits, we find that the results are very similar for both ring models.

Focusing on the thin ring model, we find that the measured diameter is $39.8^{+1.7}_{-0.7}$ µas for the $n = 0$ fits and $38.5^{+1.0}_{-0.3}$ µas for the $n = 1$ fits; the relative flux density of the ring is $0.52^{+0.06}_{-0.03}$ for the $n = 0$ fits and $0.61^{+0.05}_{-0.02}$ for the $n = 0 + 1$ fits. Therefore, we find that the assumption of a thin ring does not appreciably change the estimated ring parameters. These results are consistent with the conclusions of Section 2, which showed that the EHT 2017 array could not meaningfully constrain the thickness of a thin ring feature in M87*.

For the thin ring fits, we find that the model always places ~50–60% of the flux density in the thin ring component. Additionally, the thin ring diameter of the $n = 0 + 1$ image appears to be an average of the $n = 0$ and $n = 1$ fits, suggesting that its measured diameter is a combination of both. For the thick ring fits, we find that the measured flux density and diameter of the m-ring component are very similar to the thin ring fits. The width for the thick ring analysis is very uncertain—going from 0 µas to 15–20 µas.

From Figure 4, it appears that the thin ring component is modeling the combined emission of the $n = 0$ and $n = 1$ photon rings. One reason for this could be the fact that the $n = 0$ emission has substantial small-scale (< 5 µas) structure due to plasma turbulence. This structure could be causing the thin ring to fit both the $n = 0$ and $n = 1$ structures. To assess whether this is the case, we repeated the above analysis, but we replaced the GRMHD snapshot with a time-averaged GRMHD simulation. By time-averaging, we have averaged over the small-scale turbulence and created a smooth image $n = 0$, yielding a more natural scale separation between the $n = 0$ and $n = 1$ emission that may make hybrid imaging more successful. Nevertheless, we find similar results when fitting the time-averaged images as those for the snapshots (see Figure 5). Namely, the $n = 0$ and $n = 0 + 1$ fits give similar marginal posteriors for the thin ring diameter, and they are not substantially changed from the thick ring diameters. Additionally, the $n = 0 + \mathrm{shrunk1}$ fits also show that the measured ring diameter is substantially biased towards the $n = 0$ emission diameter.

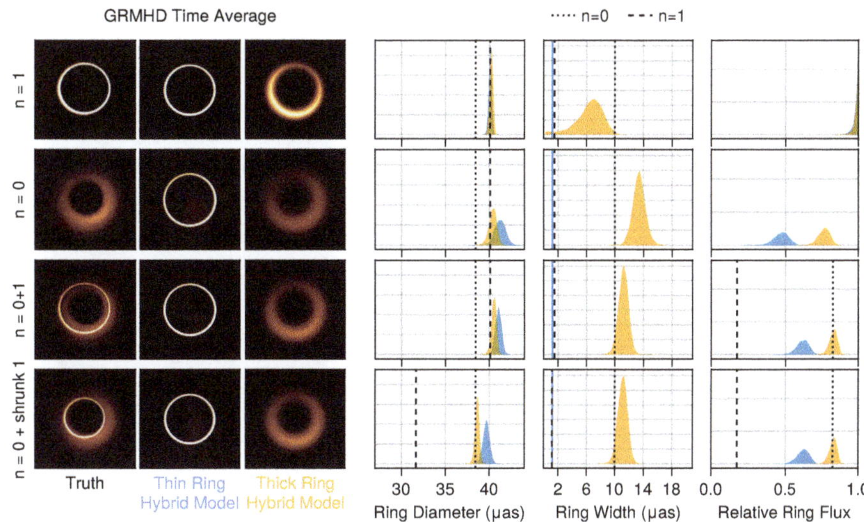

Figure 5. Results of fitting a time-averaged GRMHD simulation with coverage corresponding to the EHT 2017 array, following the same procedure and format of Figure 4. The GRMHD simulation has a spin of 0.5, inclination of 163°, $r_{\text{high}} = 20$, and is in the MAD accretion state.

3.3. Hybrid Imaging with the ngEHT

We now explore the prospects for hybrid imaging of M87* with the ngEHT, focusing on the ngEHT phase 1 array. We use a different GRMHD simulation for these tests, selecting one with no large-scale jet (see Figure 6). Because the ngEHT has many short baselines, using an image with a prominent jet would significantly increase the necessary field of view and, hence, the number of raster elements required. Additionally, we only use a time-averaged simulation, since we expect hybrid imaging to perform best after time-averaging (see Section 3.2). Finally, as for the geometric models, we assume a flat spectral index between 230 GHz and 345 GHz so that these bands can be easily combined in the modeling.

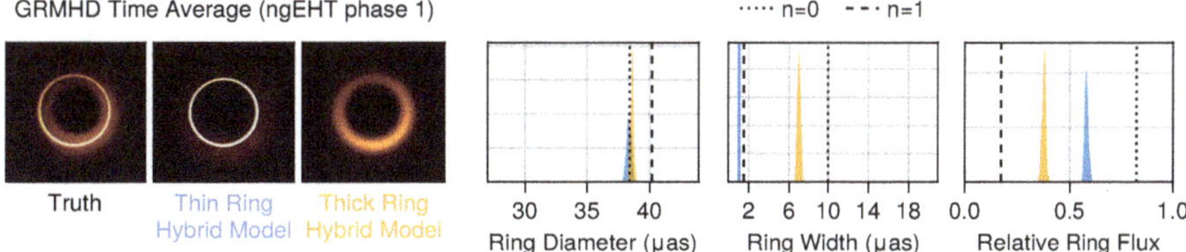

Figure 6. Results of fitting a time-averaged GRMHD simulation with no large-scale jet using a thick ring hybrid model and baseline coverage corresponding to the ngEHT phase 1 array.

The simulated dataset was created using the `ngehtsim` package using the same settings described in Section 2.2. The model we fit matches Section 3.2, except that we use a 13×13 raster with a 110 μas field of view, which corresponds to an 8 μas pixel size. We again fit the data using only closure data products, flagging any that have SNR < 3, and we use a similar sampling strategy as for the geometric models in Section 2. Because of the computational expense of this test, we only fit data for the full GRMHD image rather than examining the four decompositions into specific subimages (Figure 5). As in Section 3.2, we explore the prospects for both detection and measurement of the photon ring by fitting both the thick ring hybrid model and the thin ring hybrid model. Note that we include the ring ellipticity τ as a model parameter when fitting ngEHT synthetic data due to the improved baseline coverage.

The results are shown in Figure 6. Unfortunately, the additional coverage of the ngEHT does not improve the biases seen in Section 3.2. Nearly all ring parameters indicate that they are a combination of the $n = 0$ and $n = 1$ photon ring properties; the single exception is the measured ring diameter, which is consistent with the $n = 0$ photon ring diameter for both the thin and thick ring fits. Unlike the EHT fits, the ring width is tightly constrained; however, it is much larger than the $n = 1$ ring width and smaller than the $n = 0$ ring width. The recovered fractional flux density in the ring component is 0.36 ± 0.01, while the true value for the $n = 0$ and $n = 1$ photon rings are 0.83 and 0.17, respectively.

Comparing the ngEHT hybrid imaging results to the $n = 0 + 1$ 2017 EHT results in Figure 2, we find that the results are somewhat similar, with the major difference being the posterior concentration for both the raster and ring parameters. In particular, rather than identifying a new solution mode, the posterior of the ring diameter and width for the ngEHT phase 1 array results appear to be subsets of the EHT 2017 array results. Thus, even with the significant improvements of the ngEHT resolution, baseline coverage, and sensitivity, the hybrid imaging methodology does not successfully isolate the $n = 1$ photon ring.

Figures 7 and 8 show the mean reconstructions and horizontal cross sections for the thick ring hybrid model and the thin ring hybrid model, respectively. The cross sections demonstrate the significant improvement in image fidelity and dynamic range for the ngEHT relative to the EHT; all ngEHT reconstructions robustly identify a central brightness depression, which is predominantly caused by the deep "inner shadow" in the simulated image [54]. However, no reconstructions for either model indicate a ring component that corresponds directly to the $n = 1$ photon ring.

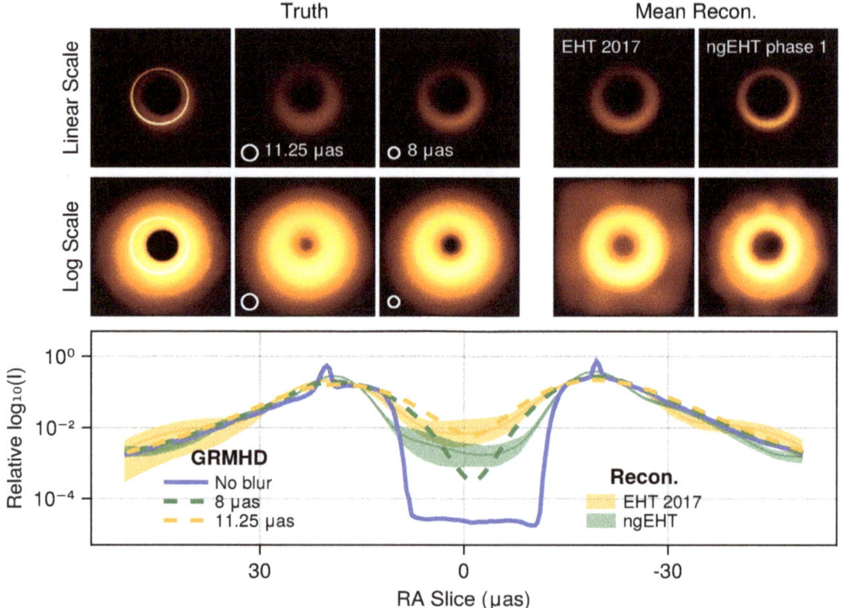

Figure 7. Summary of applying the hybrid model to fit both EHT 2017 and ngEHT phase 1 synthetic data for the time-averaged GRMHD simulation shown in Figure 5. The top row shows the ground truth image with different degrees of blurring (left group) and mean reconstructions using the thick ring hybrid model (right group) on a linear scale. The second row is the same set of images but plotted on a logarithmic scale. The 11 µas and 8 µas blurring kernels applied to the GRMHD simulation were chosen to match the raster resolutions of the hybrid models used for the EHT and ngEHT reconstructions, respectively. The bottom row shows the emission profile along the x-axis for the various images in the top two rows. Bands for the reconstructions denote 95% credible intervals (including both the raster and ring components).

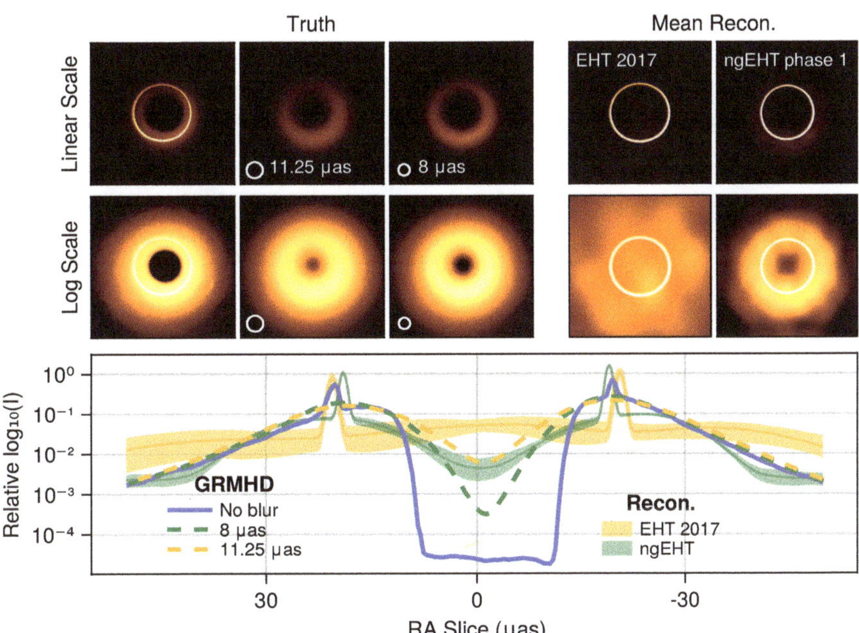

Figure 8. Similar to Figure 7 except fitting the thin ring hybrid model. In this case, the profiles show localized peaks associated with the thin ring, but they are displaced from the profile peaks of the $n = 1$ photon ring.

4. Discussion

We have explored the prospects for detecting and measuring the properties of the photon ring in M87* using VLBI. Specifically, we have performed two types of tests using synthetic VLBI data, both within a Bayesian modeling framework implemented in the open-source library `Comrade.jl`. The first type is simple geometric "self-fits", which are idealized but easily interpreted. These geometric fits can be used to define firm requirements to detect the photon ring and to quantify how different array design choices affect the accuracy of photon ring parameter measurements. The second type generates synthetic data from more realistic GRMHD models and fits them in a "hybrid imaging" framework that simultaneously models a raster grid (similar to conventional VLBI imaging) and a geometric ring component.

For the geometric tests, we find that the EHT baseline coverage and sensitivity cannot distinguish the direct ($n = 0$) and secondary ($n = 1$) emission (see also [55]). In particular, the width of the $n = 1$ component is weakly constrained. However, adding longer baselines and higher observing frequencies in the simulated ngEHT coverage allows a firm detection of the $n = 1$ photon ring. Thus, this test successfully provides minimal requirements for a photon ring detection with the ngEHT.

For the more realistic GRMHD tests, we have demonstrated that the Bayesian VLBI modeling package `Comrade.jl` can readily support posterior estimation using the hybrid imaging methodology with large rasters, even with ngEHT baseline coverage. Note that our results are more nuanced and are strongly tied to limitations of the hybrid imaging methodology and of the VLBI data fitted. For our tests with EHT and ngEHT phase 1 and phase 2 coverage, we find that

- **Hybrid imaging is prone to false positive detections of the photon ring.** Tests using images that only have direct ($n = 0$) emission still show a strong preference for a ring component, even if the ring is restricted to be narrow.

- **Assuming a thin ring does not appreciably affect the other inferred ring parameters.** While the physically motivated assumption that the $n = 1$ ring is narrow could plausibly affect the success of hybrid imaging, our fits are only weakly affected by this assumption.
- **The fitted ring parameters in hybrid imaging do not correspond to the $n = 1$ photon ring in the presence of confounding $n = 0$ emission.** In our tests, the ring flux density, width, and diameter are all affected by both the $n = 0$ and $n = 1$ emission and are generally most consistent with properties of the direct emission.

In short, our tests indicate that estimates of black hole properties that rely on a rigid association of the ring component in hybrid imaging with the $n = 1$ photon ring, such as those presented in [26], should be regarded with caution. A statistically significant detection of a thin ring component in a hybrid imaging model fit does not by itself demonstrate the existence of a photon ring in the source, because of the false positive tendencies of the method. Moreover, at minimum, mass-to-distance posteriors derived using hybrid imaging require an additional systematic error budget to account for the unknown bias from $n = 0$ emission.

Our results highlight the challenge of estimating photon ring parameters in the superresolution regime, even when the modeling is informed by knowledge of the true image. Future studies, including blind testing within frameworks such as the ngEHT analysis challenge [56][6], will provide additional guidance on what inferences are reliable. To convincingly make photon ring detections and measurements with real data, it will be imperative to demonstrate frequency and temporal independence of the inferred black hole parameters. Black hole images have strong dependence on frequency because of changing optical depth and synchrotron emissivity, so we expect that independent but consistent inferences across the full ngEHT frequency range (86–345 GHz) will provide the most compelling empirical tests.

We have explored a narrow range of possible implementations of hybrid imaging, examining only the difference between assuming a thick or thin ring component. Additional studies should explore the role of ring ellipticity, diameter, and relative flux density, all of which could use physically informed priors or information from complementary observations (e.g., from resolved stellar orbits of Sgr A*).

Future studies that explore other observational signatures of the photon ring, such as those in linear polarization (e.g., [50,57]), in circular polarization (e.g., [58,59]), and in the time domain (e.g., [23,60–62]), can provide important pathways to detection and measurement, as well as creating additional validation opportunities across data products and analysis methods. Finally, while we have focused our tests on M87*, Sgr A* is another target for millimeter VLBI for which photon ring detection may soon be possible. Sgr A* has a somewhat larger angular gravitational radius than M87*, so its photon ring is likely to be larger as well. Unlike M87*, Sgr A* has an exquisitely measured mass from resolved stellar orbits (e.g., [63,64]) which can either be integrated as an informative prior or can be used as a powerful consistency test on photon ring inferences. However, Sgr A* has the additional challenges of strong interstellar scattering (e.g., [65–68]) and rapid variability (e.g., [19,69]), and we expect that the requirements to detect the photon ring in Sgr A* may be more stringent than those for M87*.

Author Contributions: Conceptualization, P.T., M.D.J. and D.W.P.; methodology, P.T. and M.D.J.; software, P.T.; writing—original draft preparation, P.T., M.D.J. and D.W.P.; writing—review and editing, P.T., M.D.J., D.W.P., D.C.M.P., D.O.C. and P.G. All authors have read and agreed to the published version of the manuscript.

Funding: This work was supported by the Black Hole Initiative at Harvard University, which is funded by grants from the John Templeton Foundation and the Gordon and Betty Moore Foundation to Harvard University. This work was also supported by National Science Foundation grants AST-1935980 and AST-2034306 and the Gordon and Betty Moore Foundation (GBMF-10423).

Institutional Review Board Statement: Not applicable.

Informed Consent Statement: Not applicable.

Data Availability Statement: Not applicable.

Acknowledgments: We thank Avery Broderick, Boris Georgiev, and Britt Jeter for many helpful conversations about hybrid imaging.

Conflicts of Interest: The authors declare no conflict of interest.

Notes

1. https://github.com/Smithsonian/ngehtsim, accessed on 19 June 2022.
2. We have also explored fitting other data products, including visibility amplitudes and complex visibilities, and we find that our conclusions are unchanged.
3. Tiede et al. [45] find that the EHT 2017 coverage cannot even constrain the asymmetry of the $n = 0$ image, τ_0.
4. For this work we assume that $\kappa(x)$ has units of sr^{-1}.
5. Our raster prior differs from the [25] model which uses a log-uniform prior on the pixel intensity. Ref. [26] also fits for the raster field of view and orientation.
6. The future ngEHT analysis challenges will be hosted on https://challenge.ngeht.org/.

References

1. Akiyama, K. et al. [Event Horizon Telescope Collaboration] First M87 Event Horizon Telescope Results. V. Physical Origin of the Asymmetric Ring. *Astrophys. J. Lett.* **2019**, *875*, L5. [CrossRef]
2. Akiyama, K. et al. [Event Horizon Telescope Collaboration] First Sagittarius A* Event Horizon Telescope Results. V. Testing Astrophysical Models of the Galactic Center Black Hole. *Astrophys. J. Lett.* **2022**, *930*, L16. [CrossRef]
3. Bardeen, J.M. Timelike and null geodesics in the Kerr metric. In *Black Holes (Les Astres Occlus)*; Gordon and Breach: New York, NY, USA, 1973; pp. 215–239.
4. Cunningham, C.T.; Bardeen, J.M. The Optical Appearance of a Star Orbiting an Extreme Kerr Black Hole. *Astrophys. J.* **1973**, *183*, 237–264. [CrossRef]
5. Luminet, J.P. Image of a spherical black hole with thin accretion disk. *A&A* **1979**, *75*, 228–235.
6. Darwin, C. The Gravity Field of a Particle. *Proc. R. Soc. Lond. Ser. A* **1959**, *249*, 180–194. [CrossRef]
7. Gralla, S.E.; Holz, D.E.; Wald, R.M. Black hole shadows, photon rings, and lensing rings. *Phys. Rev. D* **2019**, *100*, 024018. [CrossRef]
8. Johnson, M.D.; Lupsasca, A.; Strominger, A.; Wong, G.N.; Hadar, S.; Kapec, D.; Narayan, R.; Chael, A.; Gammie, C.F.; Galison, P.; et al. Universal interferometric signatures of a black hole's photon ring. *Sci. Adv.* **2020**, *6*, eaaz1310. [CrossRef]
9. Akiyama, K. et al. [Event Horizon Telescope Collaboration] First M87 Event Horizon Telescope Results. I. The Shadow of the Supermassive Black Hole. *Astrophys. J. Lett.* **2019**, *875*, L1. [CrossRef]
10. Akiyama, K. et al. [Event Horizon Telescope Collaboration] First M87 Event Horizon Telescope Results. II. Array and Instrumentation. *Astrophys. J. Lett.* **2019**, *875*, L2. [CrossRef]
11. Akiyama, K. et al. [Event Horizon Telescope Collaboration] First M87 Event Horizon Telescope Results. III. Data Processing and Calibration. *Astrophys. J. Lett.* **2019**, *875*, L3. [CrossRef]
12. Akiyama, K. et al. [Event Horizon Telescope Collaboration] First M87 Event Horizon Telescope Results. IV. Imaging the Central Supermassive Black Hole. *Astrophys. J. Lett.* **2019**, *875*, L4. [CrossRef]
13. Akiyama, K. et al. [Event Horizon Telescope Collaboration] First M87 Event Horizon Telescope Results. VI. The Shadow and Mass of the Central Black Hole. *Astrophys. J. Lett.* **2019**, *875*, L6. [CrossRef]
14. Akiyama, K. et al. [Event Horizon Telescope Collaboration] First M87 Event Horizon Telescope Results. VII. Polarization of the Ring. *Astrophys. J. Lett.* **2021**, *910*, L12. [CrossRef]
15. Akiyama, K. et al. [Event Horizon Telescope Collaboration] First M87 Event Horizon Telescope Results. VIII. Magnetic Field Structure near The Event Horizon. *Astrophys. J. Lett.* **2021**, *910*, L13. [CrossRef]
16. Akiyama, K. et al. [Event Horizon Telescope Collaboration] First Sagittarius A* Event Horizon Telescope Results. I. The Shadow of the Supermassive Black Hole in the Center of the Milky Way. *Astrophys. J. Lett.* **2022**, *930*, L12. [CrossRef]
17. Akiyama, K. et al. [Event Horizon Telescope Collaboration] First Sagittarius A* Event Horizon Telescope Results. II. EHT and Multiwavelength Observations, Data Processing, and Calibration. *Astrophys. J. Lett.* **2022**, *930*, L13. [CrossRef]
18. Akiyama, K. et al. [Event Horizon Telescope Collaboration] First Sagittarius A* Event Horizon Telescope Results. III. Imaging of the Galactic Center Supermassive Black Hole. *Astrophys. J. Lett.* **2022**, *930*, L14. [CrossRef]
19. Akiyama, K. et al. [Event Horizon Telescope Collaboration] First Sagittarius A* Event Horizon Telescope Results. IV. Variability, Morphology, and Black Hole Mass. *Astrophys. J. Lett.* **2022**, *930*, L15. [CrossRef]
20. Akiyama, K. et al. [Event Horizon Telescope Collaboration] First Sagittarius A* Event Horizon Telescope Results. VI. Testing the Black Hole Metric. *Astrophys. J. Lett.* **2022**, *930*, L17. [CrossRef]

21. Doeleman, S.; Blackburn, L.; Dexter, J.; Gomez, J.L.; Johnson, M.D.; Palumbo, D.C.; Weintroub, J.; Farah, J.R.; Fish, V.; Loinard, L.; et al. Studying Black Holes on Horizon Scales with VLBI Ground Arrays. *Bull. Am. Astron. Soc.* **2019**, *51*, 256.

22. Broderick, A.E.; Fish, V.L.; Doeleman, S.S.; Loeb, A. Estimating the Parameters of Sagittarius A*'s Accretion Flow Via Millimeter VLBI. *Astrophys. J.* **2009**, *697*, 45–54. [CrossRef]

23. Tiede, P.; Pu, H.Y.; Broderick, A.E.; Gold, R.; Karami, M.; Preciado-López, J.A. Spacetime Tomography Using the Event Horizon Telescope. *Astrophys. J.* **2020**, *892*, 132. [CrossRef]

24. Palumbo, D.C.; Gelles, Z.; Tiede, P.; Chang, D.O.; Pesce, D.W.; Chael, A.; Johnson, M.D. Bayesian Accretion Modeling: Axisymmetric Equatorial Emission in the Kerr Spacetime. *Astrophys. J.* **2022**, *939*, 107. [CrossRef]

25. Broderick, A.E.; Pesce, D.W.; Tiede, P.; Pu, H.Y.; Gold, R. Hybrid Very Long Baseline Interferometry Imaging and Modeling with themis. *Astrophys. J.* **2020**, *898*, 9. [CrossRef]

26. Broderick, A.E.; Pesce, D.W.; Gold, R.; Tiede, P.; Pu, H.Y.; Anantua, R.; Britzen, S.; Ceccobello, C.; Chatterjee, K.; Chen, Y.; et al. The Photon Ring in M87*. *Astrophys. J.* **2022**, *935*, 61. [CrossRef]

27. Takahashi, R. Shapes and Positions of Black Hole Shadows in Accretion Disks and Spin Parameters of Black Holes. *Astrophys. J.* **2004**, *611*, 996–1004. [CrossRef]

28. Johannsen, T.; Psaltis, D. Testing the no-hair theorem with observations in the electromagnetic spectrum. II. BLACK HOLE IMAGES. *Astrophys. J.* **2010**, *718*, 446–454. [CrossRef]

29. Broderick, A.E.; Johannsen, T.; Loeb, A.; Psaltis, D. Testing the no-hair theorem with event horizon telescope observations of sagittarius A*. *Astrophys. J.* **2014**, *784*, 7. [CrossRef]

30. Medeiros, L.; Psaltis, D.; Özel, F. A Parametric Model for the Shapes of Black Hole Shadows in Non-Kerr Spacetimes. *Astrophys. J.* **2020**, *896*, 7. [CrossRef]

31. Farah, J.R.; Pesce, D.W.; Johnson, M.D.; Blackburn, L. On the Approximation of the Black Hole Shadow with a Simple Polar Curve. *Astrophys. J.* **2020**, *900*, 77. [CrossRef]

32. Gralla, S.E.; Lupsasca, A.; Marrone, D.P. The shape of the black hole photon ring: A precise test of strong-field general relativity. *Phys. Rev. D* **2020**, *102*, 124004. [CrossRef]

33. Walker, R.C.; Hardee, P.E.; Davies, F.B.; Ly, C.; Junor, W. The Structure and Dynamics of the Subparsec Jet in M87 Based on 50 VLBA Observations over 17 Years at 43 GHz. *Astrophys. J.* **2018**, *855*, 128. [CrossRef]

34. Chael, A.A.; Johnson, M.D.; Bouman, K.L.; Blackburn, L.L.; Akiyama, K.; Narayan, R. Interferometric Imaging Directly with Closure Phases and Closure Amplitudes. *Astrophys. J.* **2018**, *857*, 23. [CrossRef]

35. Raymond, A.W.; Palumbo, D.; Paine, S.N.; Blackburn, L.; Córdova Rosado, R.; Doeleman, S.S.; Farah, J.R.; Johnson, M.D.; Roelofs, F.; Tilanus, R.P.J.; et al. Evaluation of New Submillimeter VLBI Sites for the Event Horizon Telescope. *Astrophys. J. Suppl. Ser.* **2021**, *253*, 5. [CrossRef]

36. Backes, M.; Müller, C.; Conway, J.E.; Deane, R.; Evans, R.; Falcke, H.; Fraga-Encinas, R.; Goddi, C.; Klein Wolt, M.; Krichbaum, T.P.; et al. The Africa Millimetre Telescope. In Proceedings of the 4th Annual Conference on High Energy Astrophysics in Southern Africa (HEASA 2016), Cape Town, South Africa, 25–26 August 2016; p. 29. [CrossRef]

37. Thompson, A.R.; Moran, J.M.; Swenson, G.W., Jr. *Interferometry and Synthesis in Radio Astronomy*, 3rd ed.; Springer International Publishing: Berlin/Heidelberg, Germany, 2017. [CrossRef]

38. Blackburn, L.; Pesce, D.W.; Johnson, M.D.; Wielgus, M.; Chael, A.A.; Christian, P.; Doeleman, S.S. Closure Statistics in Interferometric Data. *Astrophys. J.* **2020**, *894*, 31. [CrossRef]

39. Tiede, P. Comrade: Composable Modeling of Radio Emission. *J. Open Source Softw.* **2022**, *7*, 4457. [CrossRef]

40. Bezanson, J.; Edelman, A.; Karpinski, S.; Shah, V.B. Julia: A fresh approach to numerical computing. *SIAM Rev.* **2017**, *59*, 65–98. [CrossRef]

41. Zhang, L.; Carpenter, B.; Gelman, A.; Vehtari, A. Pathfinder: Parallel quasi-Newton variational inference. *arXiv* **2021**, arXiv:2108.03782.

42. Axen, S.; Karrasch, D.; Burton, J. sethaxen/Pathfinder.jl: V0.4.11. 2022. Available online: https://zenodo.org/record/7037652 (accessed on 2 October 2022).

43. Hoffman, M.D.; Gelman, A. The No-U-Turn sampler: Adaptively setting path lengths in Hamiltonian Monte Carlo. *J. Mach. Learn. Res.* **2014**, *15*, 1593–1623.

44. Xu, K.; Ge, H.; Tebbutt, W.; Tarek, M.; Trapp, M.; Ghahramani, Z. AdvancedHMC. jl: A robust, modular and efficient implementation of advanced HMC algorithms. In Proceedings of the Symposium on Advances in Approximate Bayesian Inference (PMLR), Vancouver, BC, Canada, 8 December 2019; pp. 1–10.

45. Tiede, P.; Broderick, A.E.; Palumbo, D.C.M.; Chael, A. Measuring the Ellipticity of M87* Images. *Astrophys. J.* **2022**, *940*, 182. [CrossRef]

46. Mościbrodzka, M.; Falcke, H.; Shiokawa, H. General relativistic magnetohydrodynamical simulations of the jet in M 87. *A&A* **2016**, *586*, A38. [CrossRef]

47. Gammie, C.F.; McKinney, J.C.; Tóth, G. HARM: A Numerical Scheme for General Relativistic Magnetohydrodynamics. *Astrophys. J.* **2003**, *589*, 444–457. [CrossRef]

48. Prather, B.; Wong, G.; Dhruv, V.; Ryan, B.; Dolence, J.; Ressler, S.; Gammie, C. iharm3D: Vectorized General Relativistic Magnetohydrodynamics. *J. Open Source Softw.* **2021**, *6*, 3336. [CrossRef]

49. Mościbrodzka, M.; Gammie, C.F. IPOLE - semi-analytic scheme for relativistic polarized radiative transport. *Mon. Not. R. Astron. Soc.* **2018**, *475*, 43–54. [CrossRef]

50. Palumbo, D.C.M.; Wong, G.N. Photon Ring Symmetries in Simulated Linear Polarization Images of Messier 87*. *Astrophys. J.* **2022**, *929*, 49. [CrossRef]

51. Wong, G.N.; Prather, B.S.; Dhruv, V.; Ryan, B.R.; Moscibrodzka, M.; Chan, C.k.; Joshi, A.V.; Yarza, R.; Ricarte, A.; Shiokawa, H.; et al. PATOKA: Simulating Electromagnetic Observables of Black Hole Accretion. *arXiv* **2022**, arXiv:2202.11721.

52. Tiede, P.; Broderick, A.E.; Palumbo, D.C.M. Variational Image Feature Extraction for the Event Horizon Telescope. *Astrophys. J.* **2022**, *925*, 122. [CrossRef]

53. Bhattacharyya, A. On a measure of divergence between two statistical populations defined by their probability distributions. *Bull. Calcutta Math. Soc.* **1943**, *35*, 99–109.

54. Chael, A.; Johnson, M.D.; Lupsasca, A. Observing the Inner Shadow of a Black Hole: A Direct View of the Event Horizon. *Astrophys. J.* **2021**, *918*, 6. [CrossRef]

55. Lockhart, W.; Gralla, S.E. How narrow is the M87* ring? II. A new geometric model. *arXiv* **2022**, arXiv:2208.09989.

56. Roelofs, F.; Blackburn, L.; Lindahl, G.; Doeleman, S.S.; Johnson, M.D.; Arras, P.; Chatterjee, K.; Emami, R.; Fromm, C.; Fuentes, A.; et al. The ngEHT Analysis Challenges. 2022, *in preparation*.

57. Himwich, E.; Johnson, M.D.; Lupsasca, A.; Strominger, A. Universal polarimetric signatures of the black hole photon ring. *Phys. Rev. D* **2020**, *101*, 084020. [CrossRef]

58. Mościbrodzka, M.; Janiuk, A.; De Laurentis, M. Unraveling circular polarimetric images of magnetically arrested accretion flows near event horizon of a black hole. *Mon. Not. R. Astron. Soc.* **2021**, *508*, 4282–4296. [CrossRef]

59. Ricarte, A.; Qiu, R.; Narayan, R. Black hole magnetic fields and their imprint on circular polarization images. *Mon. Not. R. Astron. Soc.* **2021**, *505*, 523–539. [CrossRef]

60. Moriyama, K.; Mineshige, S.; Honma, M.; Akiyama, K. Black Hole Spin Measurement Based on Time-domain VLBI Observations of Infalling Gas Clouds. *Astrophys. J.* **2019**, *887*, 227. [CrossRef]

61. Hadar, S.; Johnson, M.D.; Lupsasca, A.; Wong, G.N. Photon ring autocorrelations. *Phys. Rev. D* **2021**, *103*, 104038. [CrossRef]

62. Wong, G.N. Black Hole Glimmer Signatures of Mass, Spin, and Inclination. *Astrophys. J.* **2021**, *909*, 217. [CrossRef]

63. Do, T.; Hees, A.; Ghez, A.; Martinez, G.D.; Chu, D.S.; Jia, S.; Sakai, S.; Lu, J.R.; Gautam, A.K.; O'Neil, K.K.; et al. Relativistic redshift of the star S0-2 orbiting the Galactic Center supermassive black hole. *Science* **2019**, *365*, 664–668. [CrossRef]

64. Gravity Collaboration; Abuter, R.; Amorim, A.; Bauböck, M.; Berger, J.P.; Bonnet, H.; Brandner, W.; Clénet, Y.; Coudé Du Foresto, V.; de Zeeuw, P.T.; et al. A geometric distance measurement to the Galactic center black hole with 0.3% uncertainty. *A&A* **2019**, *625*, L10. [CrossRef]

65. Bower, G.C.; Goss, W.M.; Falcke, H.; Backer, D.C.; Lithwick, Y. The Intrinsic Size of Sagittarius A* from 0.35 to 6 cm. *Astrophys. J.* **2006**, *648*, L127–L130. [CrossRef]

66. Psaltis, D.; Johnson, M.; Narayan, R.; Medeiros, L.; Blackburn, L.; Bower, G. A Model for Anisotropic Interstellar Scattering and its Application to Sgr A*. *arXiv* **2018**, arXiv:1805.01242.

67. Johnson, M.D.; Narayan, R.; Psaltis, D.; Blackburn, L.; Kovalev, Y.Y.; Gwinn, C.R.; Zhao, G.Y.; Bower, G.C.; Moran, J.M.; Kino, M.; et al. The Scattering and Intrinsic Structure of Sagittarius A* at Radio Wavelengths. *Astrophys. J.* **2018**, *865*, 104. [CrossRef]

68. Issaoun, S.; Johnson, M.D.; Blackburn, L.; Brinkerink, C.D.; Mościbrodzka, M.; Chael, A.; Goddi, C.; Martí-Vidal, I.; Wagner, J.; Doeleman, S.S.; et al. The Size, Shape, and Scattering of Sagittarius A* at 86 GHz: First VLBI with ALMA. *Astrophys. J.* **2019**, *871*, 30. [CrossRef]

69. Wielgus, M.; Marchili, N.; Martí-Vidal, I.; Keating, G.K.; Ramakrishnan, V.; Tiede, P.; Fomalont, E.; Issaoun, S.; Neilsen, J.; Nowak, M.A.; et al. Millimeter Light Curves of Sagittarius A* Observed during the 2017 Event Horizon Telescope Campaign. *Astrophys. J. Lett.* **2022**, *930*, L19. [CrossRef]

Article

Relativistic Signatures of Flux Eruption Events near Black Holes

Zachary Gelles [1,2,3,*], **Koushik Chatterjee** [1,2,*], **Michael Johnson** [1,2], **Bart Ripperda** [4,5] **and Matthew Liska** [1,6,†]

1. Center for Astrophysics, Harvard & Smithsonian, 60 Garden Street, Cambridge, MA 02138, USA
2. Black Hole Initiative, Harvard University, 20 Garden Street, Cambridge, MA 02138, USA
3. Department of Physics, Princeton University, Princeton, NJ 08544, USA
4. Department of Astrophysical Sciences, Peyton Hall, Princeton University, Princeton, NJ 08544, USA
5. Center for Computational Astrophysics, Flatiron Institute, 162 Fifth Avenue, New York, NY 10010, USA
6. Institute for Theory and Computation (ITC), Harvard University, 60 Garden Street, Cambridge, MA 02138, USA
* Correspondence: zgelles@princeton.edu (Z.G.); koushik.chatterjee@cfa.harvard.edu (K.C.)
† John Harvard Distinguished Science and ITC Fellow.

Abstract: Images of supermassive black holes produced using very long baseline interferometry provide a pathway to directly observing effects of a highly curved spacetime, such as a bright "photon ring" that arises from strongly lensed emission. In addition, the emission near supermassive black holes is highly variable, with bright high-energy flares regularly observed. We demonstrate that intrinsic variability can introduce prominent associated changes in the relative brightness of the photon ring. We analyze both semianalytic toy models and GRMHD simulations with magnetic flux eruption events, showing that they each exhibit a characteristic "loop" in the space of relative photon ring brightness versus total flux density. For black holes viewed at high inclination, the relative photon ring brightness can change by an order of magnitude, even with variations in total flux density that are comparatively mild. We show that gravitational lensing, Doppler boosting, and magnetic field structure all significantly affect this feature, and we discuss the prospects for observing it in observations of M87* and Sgr A* with the next-generation Event Horizon Telescope.

Keywords: black holes; general relativity; accretion; relativistic jets; very-long-baseline interferometry

Citation: Gelles, Z.; Chatterjee, K.; Johnson, M.; Ripperda, B.; Liska, M. Relativistic Signatures of Flux Eruption Events near Black Holes. *Galaxies* **2022**, *10*, 107. https://doi.org/10.3390/galaxies10060107

Academic Editor: Bidzina Kapanadze

Received: 12 October 2022
Accepted: 10 November 2022
Published: 24 November 2022

Publisher's Note: MDPI stays neutral with regard to jurisdictional claims in published maps and institutional affiliations.

1. Introduction

The Event Horizon Telescope released the first images of M87* in 2019 and the first images of Sgr A* in 2022 [1,2], enabling new measurements of black hole accretion flow properties directly from VLBI data and event-horizon-scale images. A theorized component of all black hole images is the photon ring: a thin annulus of light composed of photons traveling on nearly-bound geodesics. This photon ring splits into a series of self-similar subrings, each of which reflects a different degree of light-bending around the hole see, e.g., [3–9].

In particular, each subring is labeled by an index n, which counts the number of half-orbits that a photon completes on its trajectory from emitter to observer. In the optically thin limit, each successive subring has similar brightness but is exponentially demagnified, with the demagnification related to Lyapunov exponents that are governed by the properties of unstable spherical orbits of null geodesics see, e.g., [3,9–11].

One consequence of the demagnification is that the direct image ($n = 0$) of a black hole tends to be the dominant source of observed flux, with indirect images ($n \geq 1$) appearing exponentially suppressed. Hence, an important quantity is the Photon ring Flux Ratio (PFR), defined as the fraction of total flux contained in a particular subimage. We focus on the $n = 1$ ratio, which we denote as f_1:

$$f_1 \equiv \frac{F_1}{F_{\text{tot}}}. \tag{1}$$

This quantity depends both on the spacetime (which entirely determines the relative demagnification of the subring), the emission geometry, and the magnetic field structure. Moreover, it can potentially be measured with the next-generation Event Horizon Telescope (ngEHT) by using modeling methods that isolate the contribution of the photon ring e.g., [12,13].

General relativistic magneto-hydrodynamic (GRMHD) simulations and general relativistic radiative transfer (GRRT) are important tools for connecting observations of the photon ring to the underlying plasma and emission physics of the accretion disk e.g., [14,15]. GRMHD simulations used to interpret the 2017 EHT observations favor strongly magnetized gas accreting onto Sgr A* and M87* [16,17]. In the magnetically arrested disk MAD [18,19] limit, the magnetic field near the BH becomes strong enough to vertically squeeze the accreting gas. These fields ultimately undergo magnetic reconnection, allowing a bundle of vertical fields to escape from the vicinity of the BH, and in the process, eject out a large portion of the disk e.g., [20]. Observable signatures of such MAD system behavior in Sgr A* were recently reported by Wielgus et al. [21]. These "magnetic flux eruption events" exhibit large gas temperatures, strong vertical fields and occur quasi-periodically in the MAD state e.g., [22], and hence, are a prime candidate for the origin of high-energy flux eruptions, e.g., in Sgr A* and M87*. An example of such a flux eruption is depicted in Figure 1, and a more detailed description of the accretion flow's response to these events is outlined in Appendix A.

Figure 1. Magnetic flux eruptions can remove over half of the disk from near the black hole, significantly changing the resultant horizon-scale image. **Top**: Snapshots of simulation in quiescent state (**left**) and flux eruption event (**right**), ray-traced with equal mass scale units at a viewing inclination of 17°. **Bottom**: 3D rendering of gas density (in GRMHD code units) for the quiescent state (**left**) and flux eruption event (**right**) within the inner 15 M, viewed at 17°. The white region in the density plot shows the evacuation of the disk and the formation of a low density magnetospheric region near the black hole.

During these eruptions, the flux density in the image can drop by a factor of ~10 (see Jia et al. in prep for more details about the lightcurve). Observations of Sagittarius A* indicate a flux density drop of a factor of 2 at millimeter wavelengths following a high energy flaring event [23,24]. Furthermore, simulations indicate that individual eruptions may last for $\sim$100–300 M,[1] with the flux bundle eventually dissipating in the disk. This timescale corresponds to a few weeks to months for M87* or an $\sim$hour for Sgr A*, and is reasonably consistent with the analysis of Sgr A* X-ray flares population [25]. Such a long evolution period allows us to possibly capture eruption events with instruments such as the ngEHT and to resolve the flaring state of a SMBH. The presence of strong vertical fields and the ejection of gas present an exceptional opportunity to probe the detailed structure of the flux eruption through its observable features in polarization and variability e.g., [26,27].

In this work, we explore how the dynamics of flux eruptions are manifest in image morphology, with specific attention to the underlying factors that directly control the relative brightness of the photon ring during these events. In Section 2, we introduce a toy model to simplify the depiction of a flux eruption event, allowing us to separate the individual influences of Doppler boosting and gravitational lensing on the photon ring's appearance. In Section 3, we repeat a similar analysis on a set of GRMHD data, showing that the results of the toy model are recovered provided that the emission profile is not changed. In Section 4, we show that the introduction of a magnetic field (with a thermal emission profile) dramatically alters the photon ring brightness. Finally, in Section 5, we discuss our results in the context of observable targets for future EHT and ngEHT science.

2. Toy Model

In this section, we develop a toy model to illuminate the effects of Doppler boosting on the PFR during flux eruption events.

2.1. Description of the Model

During a flux eruption event, a substantial portion of the high-density material in the disk is ejected, leaving a low-density magnetosphere (with an equatorial current sheet) in the inner region. Our toy model therefore consists of a half-wedge (i.e., a wedge that spans 180° in ϕ) surrounding a Kerr black hole, with emission extending down to the horizon. In practice, one could experiment with smaller wedges that span a narrower range of ϕ values, but we restrict our focus to the half-wedge for simplicity. The emission from the wedge represents the near-horizon sub-millimeter emission seen during a flux eruption event.

From here, we build off of the semianalytic models of Gold et al. [15], modifying the prescriptions so that emission is confined to exactly one half of the spacetime. Among these semianalytic test models, we adapt "Test 5", which entails a simple, isotropic emissivity function (i.e., independent of magnetic field direction) and a thin scale height for the disk:

$$n(\vec{r}) \propto \exp\left\{-\frac{1}{2}\left[\left(\frac{r}{10}\right)^2 + \left(\frac{100}{3}\cos\theta\right)^2\right]\right\}, \tag{2}$$

$$j_\nu(\vec{r}) \propto n(\vec{r}),$$

$$\alpha_\nu(\vec{r}) = 10^6\left(\frac{\nu}{230\,\text{GHz}}\right)^{-2.5} \times j_\nu(\vec{r}).$$

Here, n is the number density, j_ν is the emissivity, and α_ν is the absorptivity, with θ and r taking on their Boyer-Lindquist coordinate values. In addition to cutting out half of the emitting region, we also modify the original prescription of Gold et al. [15] so that distance of the black hole matches that of M87*, the spin of the black hole is $a = +0.94$, and the overall density is rescaled so $F_{230} \sim 0.5$ Jy for one of the snapshots.

2.2. Photon Ring Flux Ratios

During a flux eruption event, there is a disruption of the approximate axisymmetry in the accretion structure. One expects the reconnection layer powering the eruption to emit strongly in the X-ray, leaving the cool gas in the disk to emit in the sub-millimeter and radio. Hence, we expect the PFR to depend strongly on the position of the observer relative to the flux eruption itself. This orientation is encoded through both the observer's polar inclination angle (θ_{cam}), which is measured from the black hole's spin axis, as well as the observer's azimuthal angle (ϕ_{cam}). We orient the azimuthal coordinate so that $\phi_{\text{cam}} = 0$ when the region of highest density (and hence highest emissivity) is positioned directly in front of the hole. In a physical black hole, such a scenario would take place when the flux tube is positioned directly *behind* the black hole.

We ray-trace the toy model for a range of values of ϕ_{cam} as a proxy for tracking the flux eruption over time, as increasing ϕ_{cam} is equivalent to rotating the accretion disk with respect to the observer. We ray-trace the disk for both a "static" fluid rotation profile ($u_\phi = 0$) as well as an asymptotically Keplerian rotation profile ($u_\phi \to (r \sin \theta)^{1/2}$ as $r \to \infty$), corresponding to $\ell_0 = 0$ and $\ell_0 = 1$ in Gold et al. [15], respectively. This allows us to isolate the effects of Doppler boosting, which is relevant only when objects are in motion. We further repeat this procedure for $\theta_{\text{cam}} = 17°$ and $\theta_{\text{cam}} = 80°$, representing both the low-inclination and high-inclination limits of the viewing geometry.

The ray-tracing is performed using the adaptive sampling scheme of `ipole` [28,29]. The field of view is taken to be 160 µas with an effective pixel size of ~ 0.15 µas, sufficient to fully resolve the $n = 1$ subring. Each image is decomposed into its individual subrings using the procedure described by Gelles et al. [29]. In this scheme, the intensity of a pixel in the nth subimage is computed by performing radiative transfer along the corresponding geodesic's n^{th} pass around the black hole.

In Figure 2, we plot the ratio f_1 for each of these configurations and rotation profiles as a function of the total image flux. We normalize the total image flux F_{tot} with a mean flux $\overline{F}_{\text{tot}}$ taken over all values of ϕ_{cam}, which effectively represents a time-averaged flux. Since we may be able to measure the ratio f_1 along with the total compact flux using the ngEHT, we choose to plot f_1 against $F_{\text{tot}}/\overline{F}_{\text{tot}}$ for each camera azimuthal position ϕ_{cam}. The flux eruption induces a strong correlation between the total image flux and the fractional flux contained in the subring. As the half-disk orbits around the black hole, the PFR changes dramatically, more than doubling over the course of one revolution in the edge-on limit.

The specific shape of the curves traced out in Figure 2 is due to a confluence of numerous factors. In the low inclination limit ($\theta_{\text{cam}} = 17°$), the PFR remains relatively constant over azimuth, regardless of whether or not the disk is rotating. This is because the fluid velocity is confined to the midplane, so with a low-inclination viewing geometry, the half-disk cannot acquire a large velocity tangent to null geodesics that reach the observer.

The specific value of the PFR is consistent with theoretical predictions as well. For an $a = +0.94$ Kerr black hole viewed at $\theta_{\text{cam}} = 17°$, the Lyapunov exponent ranges from $\gamma = 2.36$ to $\gamma = 2.76$, depending on the azimuthal screen coordinate. Following Johnson et al. [9], in the asymptotic limit of large n (as well as optical transparency and axisymmetry), one expects

$$\frac{F_{n+1}}{F_n} \sim e^{-\gamma} \tag{3}$$

$$\implies \frac{F_1}{F_{\text{tot}}} = \frac{F_1}{F_0} \frac{1}{1 + \frac{F_1}{F_0} + \dots}$$

$$\sim e^{-\gamma} \frac{1}{\sum\limits_{n=0}^{\infty} e^{-\gamma n}}$$

$$= e^{-\gamma} - e^{-2\gamma},$$

which ranges from 0.059 to 0.085 for this particular black hole. This range falls not far from the PFR of the low inclination snapshots in Figure 2.

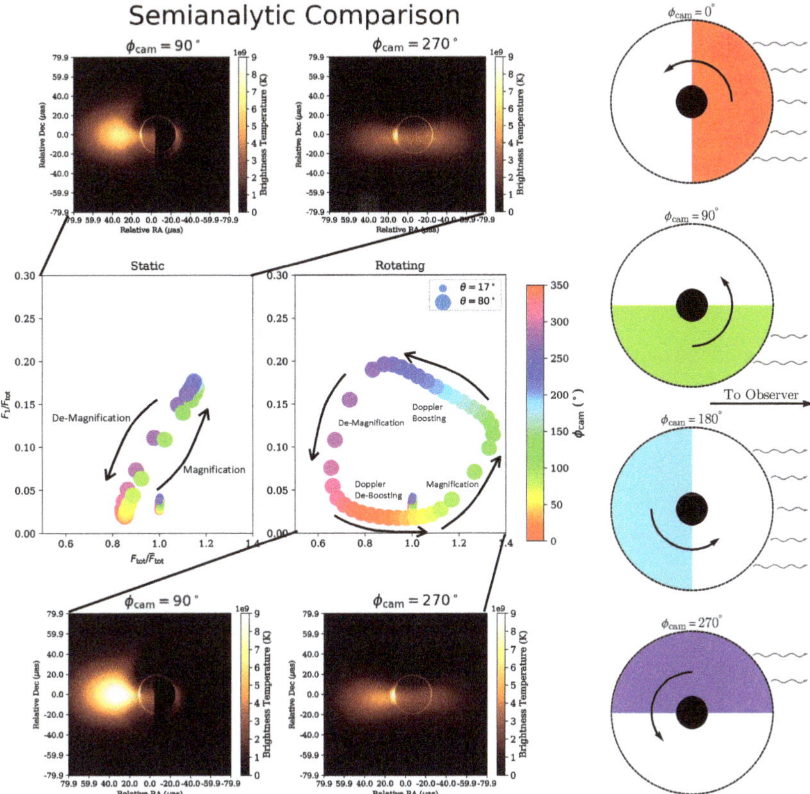

Figure 2. PFR in half-disk toy model, along with a schematic depicting the disk orientation on the right. Snapshots are for "static" disk (**top**) and rotating disk (**bottom**), both with $\theta_{cam} = 80°$. Both curves show a steep change in the PFR with ϕ_{cam} because of gravitational lensing, but the addition of rotation introduces Doppler effects that modulate the relative flux, spreading the curve horizontally. The Doppler effects are also seen to a small degree for the "static" case because of the angular velocity of the zero-angular momentum frame.

On the other hand, in the high-inclination case ($\theta_{cam} = 80°$), the brightness of the photon ring depends heavily on the location of the observer with respect to the flux eruption. In the case of a static rotation profile (Figure 2; left panel), the PFR is primarily determined by the relative magnification from gravitational lensing. As the half-disk passes behind the black hole, more of the received flux is bent around the hole, leading to a magnification of the $n = 1$ image and a resultant brightening of the photon ring. Indeed, the PFR is largest when $\phi_{cam} \sim 180°$ and the emission wedge is located behind the black hole. The phases of increasing magnification and demagnification are demarcated with arrows in Figure 2.

However, when the disk is rotating (Figure 2; right panel), Doppler effects become important. The maximum and minimum PFR's still occur when $\phi_{cam} \sim 180°$ and $\phi_{cam} \sim 0°$ respectively, as the gravitational lensing is identical to the non-rotating case. However, Doppler boosting stretches the PFR curve out horizontally, as the recessional speed of the eruption is largest when $\phi_{cam} \sim 90°$ and $\phi_{cam} \sim 270°$. Indeed, in the bottom left snapshot of Figure 2, one can see the Doppler boosted direct image of the flux eruption on the left, leading to a Doppler deboosted indirect image on the right. In the bottom right snapshot, one sees the Doppler deboosted direct image of the flux eruption on the right, leading to a Doppler boosted indirect image on the left. The Doppler boosting and deboosting phases (which refer to the indirect image) are also demarcated with arrows in Figure 2. [2]

We next return to the flaring GRMHD simulations to evaluate whether the gross behavior of the toy model is seen under more physically plausible circumstances.

3. GRMHD

In this section, we describe the GRMHD simulation and ray-tracing techniques that we use to measure subring fluxes in various accreting environments.

3.1. Procedure

To further investigate the connection between photon ring brightness and flux eruption events, we ray-trace the ideal GRMHD simulation from Ripperda et al. [20], performed using the h-amr code [30]. The simulation shows MAD accretion cycles, separated by prominent plasmoid-mediated magnetic reconnection events through which magnetic flux is expelled from the event horizon. The dimensionless black hole spin parameter is $a = 15/16$ and the effective grid resolution is $N_r \times N_\theta \times N_\phi = 5376 \times 2304 \times 2304$ defined for logarithmic Kerr–Schild spherical polar coordinates. The simulation was evolved to $t = 10,000$ M.

We ray-trace the simulation using the adaptive sampling scheme of ipole. We used the mass and distance of M87*, as for the toy model discussed in Section 2, but we reduce the FOV to 80 μas. The GRMHD scale factor is calibrated so that the average flux density is $F_{230} \sim 0.5$ Jy. In generating the images, we rotated our azimuthal coordinates clockwise by $150°$ from the ipole default to align the region of highest synchrotron emissivity with the observer at $\phi_{\rm cam} = 0$, following the conventions of the toy model described in Section 2.

Over the course of the simulation, the flow exhibits a quiescent state (wherein material accretes steadily onto the black hole) and a flaring state (wherein magnetic flux is expelled outward as described in Section 1). We isolate two GRMHD time slices of the simulation that directly showcase these different states: $t = 8858$ M is identified with quiescent accretion, and $t = 9553$ M is identified with the flux eruption event. During the latter state, roughly half the fluid cells have radially inward velocities (hence accreting) while half have radially outward velocities (hence ejecting). The structure of the flux eruption in the GRMHD simulation is thus broadly consistent with our choice of toy model in Section 2.

We note that not only are there many smaller-scale eruption events that occur during the GRMHD simulation, but the size of the ejection region for any one specific eruption event also changes over time. We used our toy model to represent the peak of a particularly large flux eruption event from this simulation so as to capture the near-horizon image structure during a potentially bright high-energy flare.

3.2. Photon Ring Flux Ratios

As with the toy model, to a leading order approximation, we can ray-trace a single time-slice of data for a range of values of $\phi_{\rm cam}$ as a proxy for tracking the eruption over time. In particular, this eliminates the need to account for the time-dependence of the eruption shape, which would introduce additional non-linearities into our analysis.

Unlike the toy model, however, the GRMHD data contains information about magnetic fields. We expect the orientation of these fields in the accretion flow to directly control the brightness of the photon ring; for synchrotron processes, the plasma emissivity depends on $\vec{k} \times \vec{B}$, where $\vec{k}$ is the spatial 3-vector of the null geodesic and $\vec{B}$ is the magnetic field, both of which are measured in the local Minkowski frame of the source [31]. Indeed, for M87* and Sgr A* at millimeter wavelengths, the specific intensity will be roughly proportional to $\sim \sin^2 \zeta$, where ζ is the "pitch angle" between the emitted wavevector and the magnetic field see, e.g., [31].

To compare the GRMHD case to Figure 2, we first ray-trace the data for a range of $\phi_{\rm cam}$ values with $\sin \zeta \equiv 1$ everywhere, thus eliminating magnetic field directional dependence and causing the GRMHD emissivity prescription to resemble that of the toy model. We then construct curves of the GRMHD PFR as a function of $\phi_{\rm cam}$ for both $\theta_{\rm cam} = 17°$ and $\theta_{\rm cam} = 80°$, and the results are plotted in Figure 3.

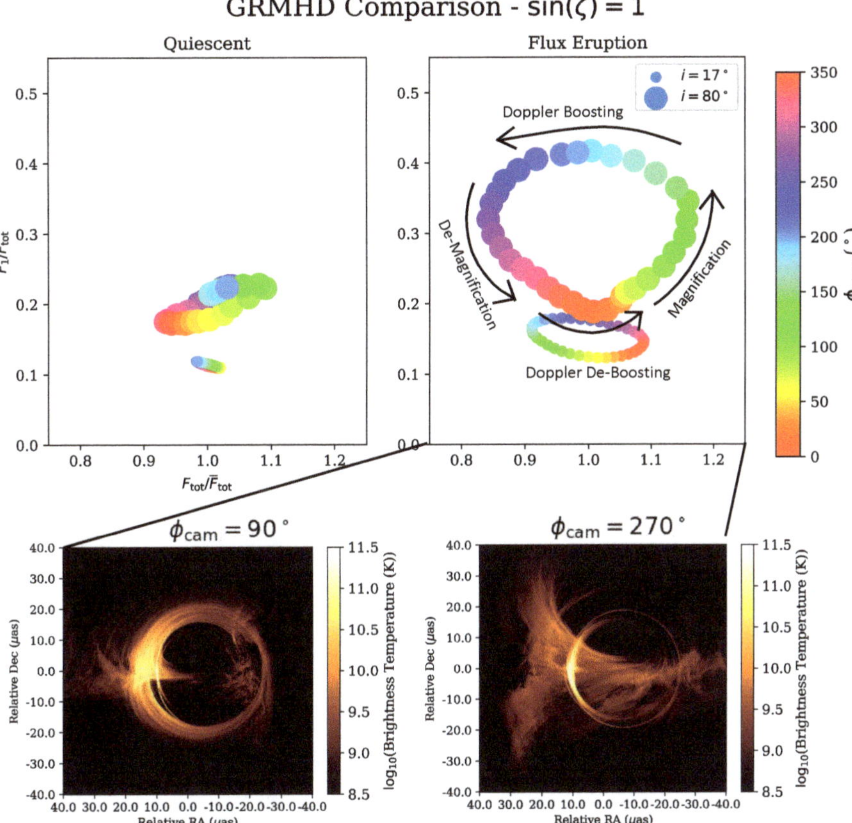

Figure 3. GRMHD PFR's after removing magnetic field directional dependence. Snapshots (bottom) are for $\theta_{cam} = 80°$. Here, "quiescent" and "flux eruption" refer to time slices $t = 8858$ M and $t = 9553$ M respectively. During the flux eruption, the shape of the high-inclination ($\theta_{cam} = 80°$) PFR curve closely resembles that of the toy model, driven by the effects of Doppler boosting and magnification. Furthermore, the direct and indirect images of the flux eruption can each be seen in the snapshots below.

Qualitatively, the high-inclination curves of Figure 3 match those of semianalytic case in Figure 2. All four phases (magnification, boosting, demagnification, and deboosting) are evident. Also evident are the Doppler boosted direct image of the eruption in the lower left panel of Figure 3, as well as the Doppler boosted indirect image of the eruption in the lower right panel Figure 3.

However, the low-inclination curve now rotates in the opposite direction (i.e., clockwise corresponds to *increasing* ϕ_{cam}) and has been stretched out dramatically, indicating a break in axisymmetry; the region emitting in the sub-millimeter has been restricted to a narrower range of ϕ values. The effects of magnetic field magnitude are primarily responsible for this change.

Next, we will investigate the role of the magnetic field direction by re-introducing the $\sin \zeta$ dependence in the synchrotron emissivity.

4. Magnetic Fields

In this section, we describe the specific role that magnetic field direction plays in altering the relative intensity of the direct vs. indirect images.

4.1. Background

The pitch angle ζ, which encodes the directional dependence of synchrotron emissivity, can dramatically alter the brightness of the image in a way that cannot be predicted from properties of the spacetime alone. In particular, ζ is *different* for the direct and indirect images and therefore directly influences the PFR. For many field configurations, the pitch angle will be larger for the strongly lensed geodesics and can hence artificially inflate the relative brightness of the photon ring.

For M87*, it is believed that the accretion flow has a strongly poloidal (i.e., a mix of vertical and radial) magnetic field and that the hole is viewed from Earth at a relatively low inclination [1,32]. For instance, for a stationary emitter at the ISCO ($r = 6$ M) of a Schwarzschild black hole viewed at face-on inclination in a purely vertical magnetic field, the pitch angle is $\zeta = 19.4°$ for the $n = 0$ image and $\zeta = 41.8°$ for the $n = 1$ image. In this case, $\sin^2 \zeta$ increases by more than a factor of 4 in the indirect image, leading to a corresponding increase in the relative brightness of the photon ring.

Such differences in field configurations are particularly relevant to our discussion of flux eruption events. During the quiescent state, the magnetic field is less ordered and accretes together with the infalling gas. However, during the flaring state, the magnetic field in the remaining magnetosphere (after material has been ejected) is connected to the event horizon and the jet, rather than the disk. This causes the magnetic field to be nearly equatorial and contain a magnetic null (i.e., a current sheet where reconnection takes place) separating the northern and southern jets. These contrasting magnetic field configurations may result in observable differences in direct emission that may be more prominent in polarized emission.

4.2. Effects of Magnetic Field Direction on GRMHD PFR's

To demonstrate the effects of magnetic field direction on the observed PFR's, we ray-trace the GRMHD simulation anew with a synchrotron emissivity profile that depends appropriately on $\sin \zeta$ (as would normally be done). The resultant PFR's are shown in Figure 4. Qualitatively, Figure 4 matches Figure 3. All four phases are once again identifiable in the high-inclination PFR curve, although the Doppler deboosting phase has shrunk to a smaller region of the phase space.

The low-inclination curve, while similar in orientation to Figure 3, has been dilated and translated upward, indicating the presence of a brighter photon ring. This transformation is consistent with a predominantly vertical magnetic field configuration in the accretion flow. As explained in Section 4.1, the contrast in pitch angle between direct/indirect images is particularly strong when the magnetic field threading the disk points directly at (or directly away from) the observer, as is the case for a vertical field viewed nearly face-on. In this case, we expect the PFR to increase, and we subsequently expect the behavior seen in Figure 4.

The correspondence between Figures 2–4 suggest that the half-disk is an appropriate (albeit simplistic) toy model for broadly representing the shape of a flux eruption event, as well as its signatures of gravitational lensing and Doppler effects. However, a dynamically important magnetic field is necessary to fully capture the correct emissivity profile and resultant PFR.

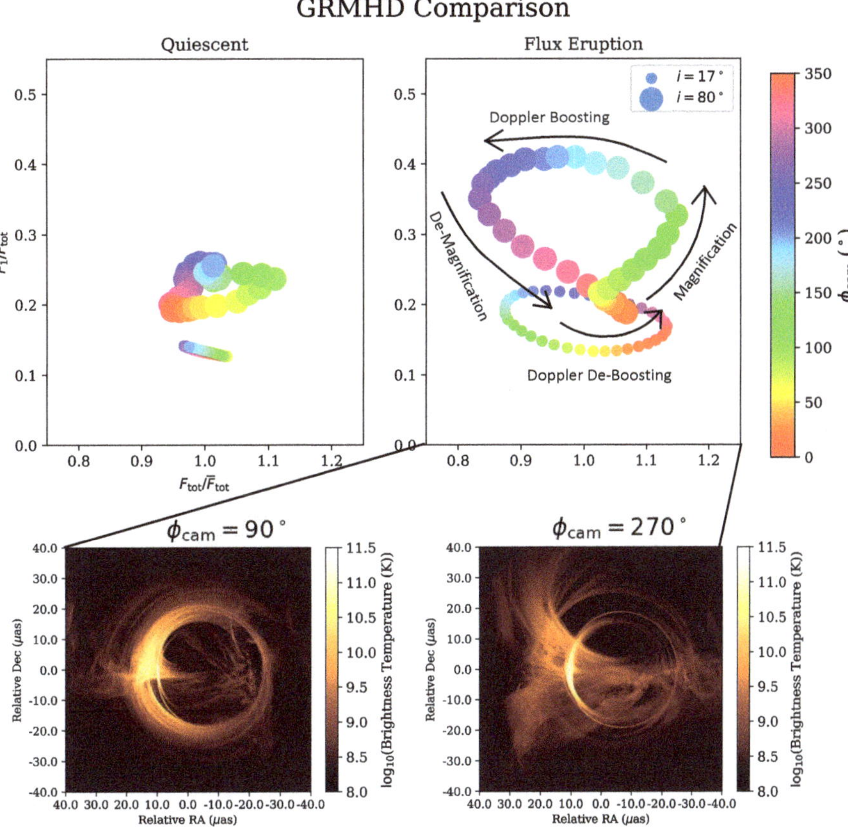

Figure 4. PFR curves for GRMHD ray-traced with full synchrotron emissivities (i.e., including magnetic field dependence). The results are similar to Figure 3, showing that the effects of magnetic field direction are insignificant in shaping the PFR for this example both in the quiescent state and during the flux eruption.

5. Discussion

We have analyzed sub-millimeter images of black holes from semi-analytic toy models and GRMHD simulations with plasmoid-mediated magnetic reconnection events. We have shown that these flux eruption events illuminate clear signatures of light bending near a black hole, and that these specific signatures depend on a multitude of factors stemming from both the spacetime and astrophysics at hand. In particular, at sub-millimeter wavelengths, the relative brightness of the $n = 1$ photon ring, f_1, is sensitive to the orbital phase of the flux eruption and can exceed 40% in systems viewed at high inclination. Furthermore, we have shown that during these flux eruption events, the PFR traces out a generic looping pattern over time, which is due to a combination of gravitational lensing, Doppler boosting, and changing synchrotron emissivity. Each of these factors is significant.

These striking features in the time-variable images of black holes are an exceptional opportunity for studies with the next generation EHT (ngEHT). In particular, the ngEHT will enable time-resolved images of both Sgr A* and M87* over hundreds-to-thousands of gravitational timescales, sufficient to catch rare events and to monitor their evolution. The enhanced angular resolution, baseline coverage, and sensitivity of the ngEHT will enable detailed studies of the photon ring see, e.g., [13], allowing measurements of the PFR during flux eruptions and flares [3]. Finally, simultaneous multi-frequency capabilities of the ngEHT will allow for spectral index and polarization measurements during flares see, e.g., [33], providing additional information that can resolve the degeneracies between effects from the curved spacetime and emitting plasma. Along with the ngEHT, simultaneous multiwavelength coverage at higher energies such as near-infrared and X-rays for Sgr A* e.g., [26,34] or TeV for M87* e.g., [35] could constrain both the flow structure and particle acceleration mechanisms during eruption events.

We emphasize that our model does not capture several intricacies of the problem at hand. While our moving camera does serve as an effective time-proxy for the rotation of the accretion flow, we employ a "fast-light" ray tracing algorithm that does not take into account the time delay between direct and indirect images. While "slow-light" simulations are significantly more computationally expensive, time delays can be integral to our study of hotspot lightcurves and images e.g., [36,37], so they should be employed in future studies of the PFR. Qualitatively, we suspect that the effects of time delay on the PFR curves are two-fold. First, the curve should rotate, as the point of maximal/minimal lensing will be shifted. Second, the curve should shrink, as direct/indirect images will no longer be diametrically opposed on the observer's screen, leading to a smaller range of possible values for f_1. In any case, we do not expect the time delay to modify our conclusions, as the generic shape of the PFR curve will remain consistent.

This study has provided a first glimpse of the scientific opportunities that may be possible with time-resolved studies of the photon ring during flux eruptions of magnetically arrested disks. Additional crucial topics for future studies include the accessible signatures in polarization, the effects of optical depth, and reconstructed movies with the ngEHT.

Author Contributions: Conceptualization, Z.G., K.C. and M.J.; methodology, Z.G. and K.C.; software, Z.G., K.C. and M.L.; formal analysis, Z.G. and K.C.; writing—original draft preparation, Z.G., K.C. and M.J.; writing—review and editing, Z.G., K.C., M.J., B.R. and M.L.; visualization, Z.G., K.C. and M.J.; supervision, K.C. and M.J. All authors have read and agreed to the published version of the manuscript.

Funding: We thank the National Science Foundation (AST-1716536 and AST-1935980) and the Gordon and Betty Moore Foundation (GBMF-10423) for financial support of this work. This work was supported in part by the Black Hole Initiative, which is funded by grants from the John Templeton Foundation and the Gordon and Betty Moore Foundation to Harvard University. K.C. is also supported in part by the Black Hole PIRE program (NSF grant OISE-1743747). B.R. is supported by a Joint Princeton/Flatiron Postdoctoral Fellowship. Research at the Flatiron Institute is supported by the Simons Foundation. M.L. is supported by John Harvard Distinguished Science Fellowship and ITC Fellowship.

Data Availability Statement: Simulation and raytraced data available on request to authors. Software: eht-imaging library [38], h-amr [30], ipole [28], Numpy [39], Matplotlib [40].

Acknowledgments: We thank Maciek Wielgus and Dominic Chang for useful discussions. This research was enabled by support provided by grant no. NSF PHY-1125915 along with a INCITE program award PHY129, using resources from the Oak Ridge Leadership Computing Facility, Summit, which is a US Department of Energy office of Science User Facility supported under contract DE-AC05- 00OR22725, as well as Calcul Quebec (http://www.calculquebec.ca, accessed on 9 November 2022) and Compute Canada (http://www.computecanada.ca, accessed on 9 November 2022). The computational resources and services used in this work were partially provided by facilities supported by the Scientific Computing Core at the Flatiron Institute, a division of the Simons Foundation. This research is part of the Frontera (Stanzione et al. [41]) computing project at the Texas Advanced Computing Center (LRAC-AST20008). Frontera is made possible by National Science Foundation award OAC-1818253.

Conflicts of Interest: The authors declare no conflict of interest.

Abbreviations

The following abbreviations are used in this manuscript:

BH	Black Hole
EHT	Event Horizon Telescope
FOV	Field Of View
GRMHD	General Relativistic Magneto Hydro Dynamic
GRRT	General Relativistic Ray Tracing
ISCO	Inner Stable Circular Orbit
MAD	Magnetically Arrested Disk
ngEHT	next generation Event Horizon Telescope
PFR	Photon ring Flux Ratio
SMBH	Super Massive Black Hole
VLBI	Very Long Baseline Interferometry

Appendix A. Magnetic Flux Eruptions in GRMHD

From Figure 1, we see that the gas inspirals towards the black hole during the quiescent state ($t = 8858$ M), while roughly half of the disk is ejected during the flux eruption event ($t = 9553$ M). Here we briefly discuss the changes in other properties of the accretion flow brought about by magnetic flux eruptions. Figure A1 shows the midplane cross-section of the gas density ρ, plasma-β (i.e., the ratio of the thermal and magnetic pressures, p_{gas}/p_{mag}), and gas temperature $T_{gas} = p_{gas}/\rho$. We also show the proxy for the 230 GHz thermal synchrotron emissivity j_{syn} given by the EHT code comparison project Porth et al. [14]:

$$j_{syn} = \frac{\rho^3}{p_{gas}^2} \exp\left[-C\left(\frac{\rho^2}{B p_{gas}^2} \right)^{1/3} \right]. \tag{A1}$$

This emissivity prescription is designed to resemble the true synchrotron emissivity fitting function given in Leung et al. [42]. Following Porth et al. [14], we assume $C = 0.2$ such that the emission drops exponentially beyond a few gravitational radii. We further normalize j_{syn} such that the volume-integrated total synchrotron emissivity is 1.

The quiescent state is characterized by gas-rich spiral features that interact with strong magnetic fields (indicated by $\beta < 1$) near the black hole. There are occasional reconnecting sites where magnetic dissipation leads to relativistic gas temperatures (i.e., $T_{gas} > 1$). overall, we see a roughly axisymmetric disk structure that produces an azimuthally-uniform emissivity profile.

On the other hand, the flux eruption state exhibits highly relativistic temperatures in the evacuated region. This occurs due to the formation of a long thin current sheet that ultimately destabilizes and undergoes reconnection, allowing the escape of a magnetic flux-tube [20]. The large gas temperatures could potentially produce high-energy emission that may be able to explain X-ray/$\gamma-$ray flares seen in low-luminosity super-massive black holes such as Sgr A* and M87*. For the 230 GHz image, the synchrotron emissivity map predicts that the bulk of the emission originates in the highly-dense relatively low-temperature accreting region. The evacuated region, despite the high gas temperatures, produces little to no flux, leading to the unique azimuthally-dependent image morphologies shown in Figure 4.

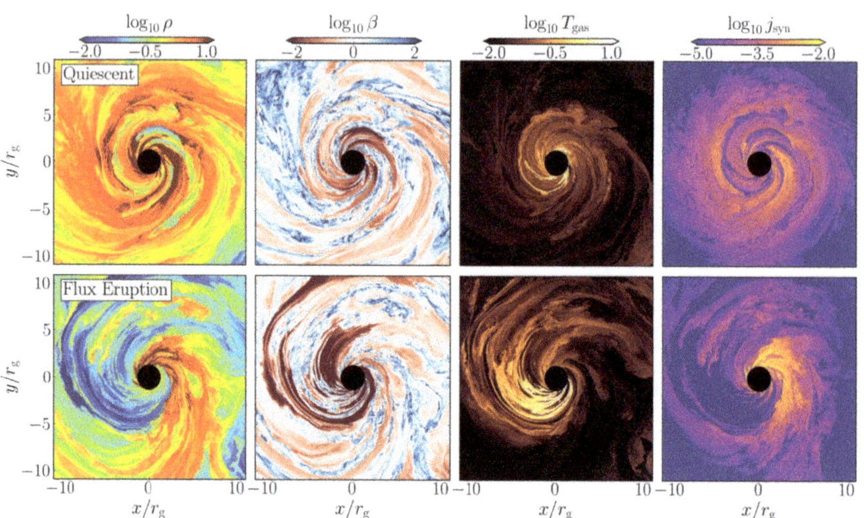

Figure A1. Magnetic flux eruptions produce regions of relativistically hot gas that could potentially produce high-energy flares in Sgr A* and M87*. This figure shows midplane cross-sections of the quiescent (**top**) and flux eruption (**bottom**) states from the GRMHD simulation. From left-to-right, panels show the gas density ρ, plasma-β, gas temperature T_{gas} (in relativistic units), and a proxy for the 230 GHz synchrotron emissivity j_{syn}.

Notes

[1]   We use natural units, i.e., $G = c = 1$, and the length and time units are both defined only in terms of the black hole mass M.

[2]   The effects of black hole spin on these conclusions are minimal, and these results are very similar for a Schwarzschild ($a = 0$) black hole. In the Schwarzschild case, Doppler effects vanish completely, as the zero-angular momentum frame is motionless everywhere.

[3]   By flares, we refer to transient bright emission usually observed at wavelengths much smaller than sub-millimeter, e.g., X-ray flares in Sgr A* [25].

References

1.   Akiyama, K. et al. [Event Horizon Telescope Collaboration].   First M87 Event Horizon Telescope Results. I. The Shadow of the Supermassive Black Hole. *Astrophys. J. Lett.* **2019**, *875*, L1. [CrossRef]
2.   Akiyama, K. et al. [Event Horizon Telescope Collaboration].   First Sagittarius A* Event Horizon Telescope Results. I. The Shadow of the Supermassive Black Hole in the Center of the Milky Way. *Astrophys. J. Lett.* **2022**, *930*, L12. [CrossRef]
3.   Luminet, J.P. Image of a spherical black hole with thin accretion disk. *Astron. Astrophys.* **1979**, *75*, 228–235.
4.   de Vries, A. The apparent shape of a rotating charged black hole, closed photon orbits and the bifurcation set A_4. *Class. Quantum Gravity* **2000**, *17*, 123–144. [CrossRef]
5.   Takahashi, R. Shapes and Positions of Black Hole Shadows in Accretion Disks and Spin Parameters of Black Holes. *Astrophys. J.* **2004**, *611*, 996–1004. [CrossRef]
6.   Beckwith, K.; Done, C. Extreme gravitational lensing near rotating black holes. *Mon. Not. R. Astron. Soc.* **2005**, *359*, 1217–1228. [CrossRef]
7.   Johannsen, T.; Psaltis, D. Testing the No-hair Theorem with Observations in the Electromagnetic Spectrum. II. Black Hole Images. *Astrophys. J.* **2010**, *718*, 446–454. [CrossRef]
8.   Gralla, S.E.; Holz, D.E.; Wald, R.M. Black hole shadows, photon rings, and lensing rings. *Phys. Rev. D* **2019**, *100*, 024018. [CrossRef]
9.   Johnson, M.D.; Lupsasca, A.; Strominger, A.; Wong, G.N.; Hadar, S.; Kapec, D.; Narayan, R.; Chael, A.; Gammie, C.F.; Galison, P.; et al. Universal interferometric signatures of a black hole's photon ring. *Sci. Adv.* **2020**, *6*, eaaz1310. [CrossRef]
10.   Darwin, C. The Gravity Field of a Particle. *Proc. R. Soc. Lond. Ser. A* **1959**, *249*, 180–194. [CrossRef]
11.   Ohanian, H.C. The black hole as a gravitational "lens". *Am. J. Phys.* **1987**, *55*, 428–432. [CrossRef]
12.   Broderick, A.E.; Pesce, D.W.; Tiede, P.; Pu, H.Y.; Gold, R. Hybrid Very Long Baseline Interferometry Imaging and Modeling with THEMIS. *Astrophys. J.* **2020**, *898*, 9. [CrossRef]
13.   Tiede, P.; Broderick, A.E.; Palumbo, D.C.M.; Chael, A. Measuring the Ellipticity of M 87* Images. *arXiv* **2022**, arXiv:2210.13499.

14. Porth, O.; Chatterjee, K.; Narayan, R.; Gammie, C.F.; Mizuno, Y.; Anninos, P.; Baker, J.G.; Bugli, M.; Chan, C.k.; Davelaar, J.; et al. The Event Horizon General Relativistic Magnetohydrodynamic Code Comparison Project. *Astrophys. J. Suppl. Ser.* **2019**, *243*, 26. [CrossRef]
15. Gold, R.; Broderick, A.E.; Younsi, Z.; Fromm, C.M.; Gammie, C.F.; Mościbrodzka, M.; Pu, H.Y.; Bronzwaer, T.; Davelaar, J.; Dexter, J.; et al. Verification of Radiative Transfer Schemes for the EHT. *Astrophys. J.* **2020**, *897*, 148. [CrossRef]
16. Akiyama, K. et al. [Event Horizon Telescope Collaboration]. First M87 Event Horizon Telescope Results. V. Physical Origin of the Asymmetric Ring. *Astrophys. J. Lett.* **2019**, *875*, L5. [CrossRef]
17. Akiyama, K. et al. [Event Horizon Telescope Collaboration]. First Sagittarius A* Event Horizon Telescope Results. V. Testing Astrophysical Models of the Galactic Center Black Hole. *Astrophys. J. Lett.* **2022**, *930*, L16. [CrossRef]
18. Narayan, R.; Igumenshchev, I.V.; Abramowicz, M.A. Magnetically Arrested Disk: An Energetically Efficient Accretion Flow. *Publ. Astron. Soc. Jpn.* **2003**, *55*, L69–L72. [CrossRef]
19. Igumenshchev, I.V.; Narayan, R.; Abramowicz, M.A. Three-dimensional Magnetohydrodynamic Simulations of Radiatively Inefficient Accretion Flows. *Astrophys. J.* **2003**, *592*, 1042–1059. [CrossRef]
20. Ripperda, B.; Liska, M.; Chatterjee, K.; Musoke, G.; Philippov, A.A.; Markoff, S.B.; Tchekhovskoy, A.; Younsi, Z. Black Hole Flares: Ejection of Accreted Magnetic Flux through 3D Plasmoid-mediated Reconnection. *Astrophys. J. Lett.* **2022**, *924*, L32. [CrossRef]
21. Wielgus, M.; Moscibrodzka, M.; Vos, J.; Gelles, Z.; Martí-Vidal, I.; Farah, J.; Marchili, N.; Goddi, C.; Messias, H. Orbital motion near Sagittarius A*. Constraints from polarimetric ALMA observations. *Astron. Astrophys.* **2022**, *665*, L6. [CrossRef]
22. Narayan, R.; Chael, A.; Chatterjee, K.; Ricarte, A.; Curd, B. Jets in magnetically arrested hot accretion flows: Geometry, power, and black hole spin-down. *Mon. Not. R. Astron. Soc.* **2022**, *511*, 3795–3813. [CrossRef]
23. Akiyama, K. et al. [Event Horizon Telescope Collaboration]. First Sagittarius A* Event Horizon Telescope Results. II. EHT and Multiwavelength Observations, Data Processing, and Calibration. *Astrophys. J. Lett.* **2022**, *930*, L13. [CrossRef]
24. Wielgus, M.; Marchili, N.; Martí-Vidal, I.; Keating, G.K.; Ramakrishnan, V.; Tiede, P.; Fomalont, E.; Issaoun, S.; Neilsen, J.; Nowak, M.A.; et al. Millimeter Light Curves of Sagittarius A* Observed during the 2017 Event Horizon Telescope Campaign. *Astrophys. J. Lett.* **2022**, *930*, L19. doi: 10.3847/2041-8213/ac6428. [CrossRef]
25. Haggard, D.; Nynka, M.; Mon, B.; de la Cruz Hernandez, N.; Nowak, M.; Heinke, C.; Neilsen, J.; Dexter, J.; Fragile, P.C.; Baganoff, F.; et al. Chandra Spectral and Timing Analysis of Sgr A*'s Brightest X-ray Flares. *Astrophys. J.* **2019**, *886*, 96. [CrossRef]
26. Abuter, R. et al. [GRAVITY Collaboration]. Constraining particle acceleration in Sgr A* with simultaneous GRAVITY, Spitzer, NuSTAR, and Chandra observations. *Astron. Astrophys.* **2021**, *654*, A22. [CrossRef]
27. Gelles, Z.; Himwich, E.; Johnson, M.D.; Palumbo, D.C.M. Polarized image of equatorial emission in the Kerr geometry. *Phys. Rev. D* **2021**, *104*, 044060. [CrossRef]
28. Mościbrodzka, M.; Gammie, C.F. IPOLE—Semi-analytic scheme for relativistic polarized radiative transport. *Mon. Not. R. Astron. Soc.* **2018**, *475*, 43–54. [CrossRef]
29. Gelles, Z.; Prather, B.S.; Palumbo, D.C.M.; Johnson, M.D.; Wong, G.N.; Georgiev, B. The Role of Adaptive Ray Tracing in Analyzing Black Hole Structure. *Astrophys. J.* **2021**, *912*, 39. [CrossRef]
30. Liska, M.; Chatterjee, K.; Tchekhovskoy, A.; Yoon, D.; van Eijnatten, D.; Hesp, C.; Markoff, S.; Ingram, A.; van der Klis, M. H-AMR: A New GPU-accelerated GRMHD Code for Exascale Computing With 3D Adaptive Mesh Refinement and Local Adaptive Time-stepping. *arXiv* **2019**, arXiv:1912.10192 .
31. Narayan, R.; Palumbo, D.C.M.; Johnson, M.D.; Gelles, Z.; Himwich, E.; Chang, D.O.; Ricarte, A.; Dexter, J.; Gammie, C.F.; Chael, A.A.; et al. The Polarized Image of a Synchrotron-emitting Ring of Gas Orbiting a Black Hole. *Astrophys. J.* **2021**, *912*, 35. [CrossRef]
32. Akiyama, K. et al. [Event Horizon Telescope Collaboration]. First M87 Event Horizon Telescope Results. VIII. Magnetic Field Structure near The Event Horizon. *Astrophys. J. Lett.* **2021**, *910*, L13. [CrossRef]
33. Ricarte, A.; Gammie, C.; Narayan, R.; Prather, B.S. Probing Plasma Physics with Spectral Index Maps of Accreting Black Holes on Event Horizon Scales. *arXiv* **2022**, arXiv:2202.02408.
34. Boyce, H.; Haggard, D.; Witzel, G.; Fellenberg, S.v.; Willner, S.P.; Becklin, E.E.; Do, T.; Eckart, A.; Fazio, G.G.; Gurwell, M.A.; et al. Multiwavelength Variability of Sagittarius A* in 2019 July. *Astrophys. J.* **2022**, *931*, 7. [CrossRef]
35. Aharonian, F.; Akhperjanian, A.G.; Bazer-Bachi, A.R.; Beilicke, M.; Benbow, W.; Berge, D.; Bernlöhr, K.; Boisson, C.; Bolz, O.; Borrel, V.; et al. Fast Variability of Tera-Electron Volt γ Rays from the Radio Galaxy M87. *Science* **2006**, *314*, 1424–1427. [CrossRef]
36. Broderick, A.E.; Loeb, A. Imaging bright-spots in the accretion flow near the black hole horizon of Sgr A*. *Mon. Not. R. Astron. Soc.* **2005**, *363*, 353–362. [CrossRef]
37. Broderick, A.E.; Loeb, A. Imaging optically-thin hotspots near the black hole horizon of Sgr A* at radio and near-infrared wavelengths. *Mon. Not. R. Astron. Soc.* **2006**, *367*, 905–916. [CrossRef]
38. Chael, A.A.; Johnson, M.D.; Narayan, R.; Doeleman, S.S.; Wardle, J.F.C.; Bouman, K.L. High-resolution Linear Polarimetric Imaging for the Event Horizon Telescope. *Astrophys. J.* **2016**, *829*, 11. [CrossRef]
39. Harris, C.R.; Millman, K.J.; van der Walt, S.J.; Gommers, R.; Virtanen, P.; Cournapeau, D.; Wieser, E.; Taylor, J.; Berg, S.; Smith, N.J.; et al. Array programming with NumPy. *Nature* **2020**, *585*, 357–362. [CrossRef]
40. Hunter, J.D. Matplotlib: A 2D graphics environment. *Comput. Sci. Eng.* **2007**, *9*, 90–95. [CrossRef]

41. Stanzione, D.; West, J.; Evans, R.T.; Minyard, T.; Ghattas, O.; Panda, D.K. Frontera: The Evolution of Leadership Computing at the National Science Foundation. In Proceedings of the Practice and Experience in Advanced Research Computing, PEARC'20, Portland, OR, USA, 26–30 July 2020; Association for Computing Machinery: New York, NY, USA, 2020; pp. 106–111.
42. Leung, P.K.; Gammie, C.F.; Noble, S.C. Numerical Calculation of Magnetobremsstrahlung Emission and Absorption Coefficients. *Astrophys. J.* **2011**, *737*, 21. [CrossRef]

Article

Event Horizon and Environs (ETHER): A Curated Database for EHT and ngEHT Targets and Science

Venkatessh Ramakrishnan [1,2,*], Neil Nagar [2,*], Vicente Arratia [2], Joaquín Hernández-Yévenes [2], Dominic W. Pesce [3,4], Dhanya G. Nair [2], Bidisha Bandyopadhyay [2], Catalina Medina-Porcile [2], Thomas P. Krichbaum [5], Sheperd Doeleman [3,4], Angelo Ricarte [3,4], Vincent L. Fish [6], Lindy Blackburn [3], Heino Falcke [7], Geoffrey Bower [8] and Priyamvada Natarajan [3,4,9]

[1] Finnish Centre for Astronomy with ESO, University of Turku, 20014 Turku, Finland
[2] Astronomy Department, Universidad de Concepción, Barrio Universitario S/N, Concepción 4030000, Chile
[3] Center for Astrophysics, Harvard & Smithsonian, 60 Garden Street, Cambridge, MA 02138, USA
[4] Black Hole Initiative, Harvard University, 20 Garden Street, Cambridge, MA 02138, USA
[5] Max Planck Institute for Radioastronomy, Auf dem Hugel 69, 53121 Bonn, Germany
[6] Massachusetts Institute of Technology, Haystack Observatory, 99 Millstone Hill Road, Westford, MA 01886, USA
[7] Department of Astronomy, Radboud Universiteit Nijmegen, 6525 AJ Nijmegen, The Netherlands
[8] Academia Sinica Institute of Astronomy and Astrophysics, 645 N. A'ohoku Place, Hilo, HI 96720, USA
[9] Department of Astronomy, Yale University, 52 Hillhouse Avenue, New Haven, CT 06511, USA
* Correspondence: venkatesshr@gmail.com (V.R.); nagar@astro-udec.cl (N.N.)

Abstract: The next generation Event Horizon Telescope (ngEHT) will observe multiple supermassive black hole (SMBH) candidates down to a few tens of mJy, and profoundly transform our understanding of the local SMBH population. Given the impossibility of large-area high-resolution millimeter surveys, multi-frequency spectral energy densities (SEDs), and models are required to both identify source samples tailored to specific science goals, and to predict the feasibility of detection of individual interesting sources. Here, we present the Event Horizon and Environs (ETHER) source and SED model database whose primary use is to enable the selection and optimization of targets for EHT and ngEHT science. The living ETHER database currently consolidates 1.6 million black hole mass estimates, ~15,500 milliarcsec-scale radio fluxes, ~14,000 hard X-ray fluxes (expected to grow by factor $\gtrsim$40 with the eROSITA data release) and SED information as obtained from catalogs and database queries, the literature, and our own new observations. Jet and accretion flow models are fit to individual SEDs in an automated way in order to predict the ngEHT observable fluxes from the jet base and accretion inflow. The database can be filtered by parameters or cross matched to a user source list, with the automated SED fitting models optionally fine tuned by the user. We have identified an initial ngEHT 'gold sample' for jet base studies and potentially black hole shadows; this sample will grow significantly in the coming years. While the ngEHT requires and will best exploit the ETHER database, six (eleven) ETHER sources have already been observed (scheduled) with the EHT in 2022 (2023), and the database has wide ranging applications in galaxy and black hole mass evolution studies.

Keywords: supermassive black holes; accretion inflows; jet launching; event horizon telescope; next-generation event horizon telescope; active galactic nuclei

Citation: Ramakrishnan, V.; Nagar, N.; Arratia, V.; Hernández-Yévenes, J.; Pesce, D.W.; Nair, D.G.; Bandyopadhyay, B.; Medina-Porcile, C.; Krichbaum, T.P.; Doeleman, S.; et al. Event Horizon and Environs (ETHER): A Curated Database for EHT and ngEHT Targets and Science. *Galaxies* 2023, 11, 15. https://doi.org/10.3390/galaxies11010015

Academic Editor: Bidzina Kapanadze

Received: 21 November 2022
Revised: 23 December 2022
Accepted: 26 December 2022
Published: 12 January 2023

1. Introduction

Supermassive black holes are believed to transition through 'hard' and 'soft' states [1]. Current theoretical models describe these states as arising from differences in the inner flow geometries. The latter are characterized by high accretion rates relative to Eddington, with emission in the inner 1000 gravitational radii; $R_g = GM/c^2$ where M is the BH mass dominated by UV to optical emission from an optically-thick geometrically-thin accretion disk (10 s to 1000 s of gravitational radii) and X-rays from a 'corona' around the accretion disk. The 'hard' states reveal low Eddington accretion rates (low-luminosity active galactic nuclei or LLAGNs) and their radio to gamma-ray emission in the inner 1000 R_g comes primarily from a quasi-spherical accretion inflow in the innermost tens of R_g e.g., [2], and potentially from jets launched within 10 s of R_g. The innermost LLAGN accretion flows have been modelled with radiatively inefficient accretion flows (RIAFs), jets, or both e.g., [2–6]. The SMBH, when embedded in an accretion flow, or back-lit by a receding jet, produces a gravitationally-lensed bright ring with diameter 10.4 R_g [7], within which sits the shadow produced by its event horizon.

The next-generation event horizon telescope (ngEHT) will provide a spectacular increase in both sensitivity (detecting sources at ∼10 s of mJy), and native resolution (15 μarcsec) with super-resolution techniques potentially enabling resolutions of a few μarcsec. Additionally, the decreasing electron scattering with increasing frequencies and the often steeply increasing flux with frequency of the mm-wave emission from the inner accretion inflow, allows one to better resolve, or at least pinpoint accurately, the innermost region around the supermassive black hole as compared to cm-wave very long baseline interferometry (VLBI).

While the superlative sensitivity and resolution of the ngEHT will leverage wide ranging and transformational results in explicitly selected target samples, or in serendipitous sources of interest, it also creates a significant challenge. How do we identify and optimally select these few tens to thousands of ngEHT targets? Current predictions posit several to several thousand ngEHT detectable sources [8,9] with the range of predicted numbers depending on the still to be frozen technical specifications of the ngEHT. If we limit ourselves to SMBH with large black hole rings, e.g., within factor 5 of those in M87 and Sgr A*, we have only a few hundred potential targets, predominantly in the local universe. However, there are compelling reasons to include SMBHs with a large range of ring sizes. In terms of 'resolved' science, super-resolution techniques, future earth-space VLBI, and studies of jet launching at 10 s to 1000 s R_g, motivate the inclusion of SMBHs with ring sizes down to ∼1 μarcsec. Further, transformational science can be leveraged by using ngEHT-detected jets and accretion flows as 'signposts' independent of the linear resolution in scales of R_g, e.g., tracking orbits of binary SMBH. Direct selection of ngEHT targets is difficult for several reasons: (a) a sub-arcsec large-area survey at 230–345 GHz is curently impossible: for reference, a sub-arcsec ALMA 230 GHz survey of GOODS-S required ∼30 h to cover a 69 arcmin^2 area to a depth of 0.2 mJy/beam [10]. Observing the same area distributed over the sky is significantly more expensive due to calibration overheads; (b) all-sky mm-surveys have detection limits of a few hundred mJy, and resolutions of several arcminutes, e.g., Planck Collaboration et al. [11]; (c) multi-year large-area mm-surveys, e.g., SPT-3G [12] and JCMT-SCUBA2 [13], can only cover a few thousand of square degrees, at sensitivities of a few mJy to micro Jy, and resolutions of several tens of arcseconds to arcmin; and (d) dust is a strong contributor to 230–345 GHz emission, so that flux measurements at ≥ arcsec-scale are often dominated by galactic dust rather than the SMBH environment.

The few thousand ngEHT targets—which require to be bright (≥few tens of mJy) and compact (≲1000 R_g) at 230–345 GHz—are, thus, best pre-identified using SED fits on as large a sample of AGN and LLAGN possible. Posterior lower frequency VLBI and/or lower resolution 230 GHz imaging can then be used to confirm feasibility of ngEHT detection or imaging. For the first step, given the 'contamination' from host galaxy emission, one requires high resolution (ideally milliarcsec or mas scale, i.e., VLBI) flux measurements, and/or arcsec-scale fluxes at energies where the host galaxy is relatively faint, e.g., hard X-ray to γ-ray.

In this work, we present and describe the Event Horizon and Environment (ETHER) sample and database, which, independent of any specific science goal, aims to provide the definitive SMBH parent sample from which to select EHT and ngEHT targets in the coming decades. Additionally, ETHER will offer an easy to use 'pipeline' to evaluate the observational feasibility of any given new source.

2. Materials and Methods

In this section, we describe the compilation of observational data which forms the basis of the ETHER database, the immediately derived quantities from these data, and the modelling procedure used to predict the ngEHT observing feasibility of any given source.

2.1. Catalogs and Data

Since the mass of a SMBH is one of its key physical and observable properties, ETHER compiles a mass estimate for every source in the database. This permits an estimate of the angular size of the black hole shadow and ring, constrains the scales (in terms of R_g) on which we can resolve the jet base, and is a necessary ingredient for the SED modelling. Only ~230 galaxies have SMBH measurements, and we, thus, primarily depend on estimates. SMBH measurements, from stellar dynamics and ionized, molecular, and maser, gas kinematics, where available are taken from large compilations e.g., [14–16] individual publications, e.g., [17,18] and references therein, and online databases e.g., the Reverberation Mass (RM) database at http://www.astro.gsu.edu/AGNmass/ (accessed on 1 April 2020).

Our more reliable SMBH mass estimates come from the multiple black hole mass—host galaxy empirical scaling relationships, including that between SMBH mass and velocity dispersion (M-σ), or luminosity of the galaxy bulge (M-L_{bulge}), e.g., [15], 'single-epoch' reverberation mapping e.g., [19], and so-called 'fundamental planes' between SMBH mass and the pairs [effective radius of the bulge, central surface brightness]; e.g., Woo and Urry [20], [arcsec-scale radio and hard X-ray flux]; e.g., Gültekin et al. [21]. These scaling relationships are best calibrated at low redshifts and for SMBH masses $\gtrsim 10^7$ $M_\odot$ and we expect typical uncertainties of 0.3–0.6 dex in these. Here, velocity dispersions are from the HET massive galaxy survey [22], SDSS DR17 via astroqueries, and the Hyperleda database [23]. SMBH mass estimates from M-L_{bulge} are taken from Caramete and Biermann [24] and Mezcua et al. [25]. SMBH mass estimates from the following references—which use multiple methods—are also incorporated: Woo and Urry [20], Shaw et al. [26], Chen et al. [27].

If no black hole mass estimate is available from any of the above sources, we use WISE W1, W2, and W3 magnitudes from the AllWISE catalog [28] to derive a SMBH mass estimate (for details see Hernández-Yévenes, J., et al., in prep.). Briefly, the process is as follows: (a) WISE W1 and W2 band magnitudes are used to estimate the total stellar mass of the galaxy via Equation (2) of Cluver et al. [29]; (b) the WISE W2−W3 color is used to estimate the morphological type (T) of the galaxy. This color-morphological type relationship is trained on ~25,000 galaxies with morphological classifications from the 2MRS catalog [30] and from NED[1] or SIMBAD; (c) the morphological type is used to estimate the bulge to total mass ratio following Figure 1 of Caramete and Biermann [24], and, thus, the bulge stellar mass, and (d) the SMBH mass estimate is derived using the $M_{BH} - M_{Bulge}$ relationship of Schutte et al. [31].

Two further corrections are applied to the WISE derived mass estimations. First, the WISE-derived mass shows small but systematic offsets to the values from the ~230 ETHER galaxies with SMBH measurements and ~400 ETHER galaxies with high-quality stellar velocity dispersion (thus high-quality SMBH estimation via the M-σ relationship). After empirically removing this offset the WISE-derived SMBH masses and the above mentioned ETHER SMBH masses (all in relatively nearby galaxies) agree within 0.6 dex. Second, powerful AGNs and ULIRGs are identified by their W2−W3 colors. In these, AGN emission could contaminate the W1 and W2 magnitudes, thus overestimating the SMBH mass. In these cases, we compare AGN bolometric luminosities with W3-derived total IR luminosities in order to flag those in which the AGN potentially contaminate the SMBH mass estimate from W1 and W2.

The details of the WISE based SMBH mass estimations will be presented in Hernández-Yévenes, J., et al., currently in preparation. While there are potentially large errors in individual values, statistically, these estimations are highly useful in the absence of any other. For example, this additional method provides us with SMBH mass estimates for a third of the VLBI-detected sources with no previous mass estimate, and 70% of the $\sim$520,000 ETHER galaxies at D $\leq$ 400 Mpc.

Redshifts and distances are incorporated from the original SMBH mass reference, or are taken, when necessary, from the Roma BZCAT 5th Ed. [32], the Veron-Cetty and Veron AGN catalog 13th Ed. [33], the Milliquas catalog [34], and via astroqueries to SIMBAD and NED. Luminosity and angular distances are derived from the EDD Database [35] for nearby ($\lesssim$50 Mpc) galaxies, or a standard flat cosmology with H_0 = 70 km/s, Omega$_m$ = 0.30, and Lambda$_0$ = 0.70.

High-resolution (mas-scale) radio fluxes come from the 2022b version of the RFC catalog[2], the VLBA calibrator list[3] , supplemented by 230 GHz EHT fluxes [5,6,36–38], 86 GHz GMVA fluxes [39,40], 43 GHz VLBA fluxes [41], 15 GHz VLBA fluxes [42], 5 GHz VLBA fluxes [43], NED and SIMBAD (via astroqueries), diverse literature results, and our ongoing VLBA and LBA programs (Ramakrishnan et al., in prep.). When mas-scale fluxes are not available, we use arcsec-scale fluxes from the 14 April 2022 version of the ALMA Calibrator catalog[4], from the literature for individual galaxies of interest, and from our own survey programs with ALMA (Nair et al., in prep.).

Hard X-ray fluxes are incorporated from NED photometry tables (via astroqueries) and the Chandra Source Catalogue (CSC) through the CSCview application [44]. Currently we have a total of 19,158 hard X-ray flux values, of which, 14,242 are high-resolution ($\sim$1 arcsec) measurements.

Astroqueries to NED, SIMBAD, and SDSS are used to supplement information on position, redshift, SED fluxes, galaxy and AGN type, and galaxy size.

2.2. Models

The SED coverage, when considering only relatively high-resolution ($\lesssim$1 arcsec) measurements, is highly variable from source to source, is often sparse, and almost always non-simultaneous. We thus employ relatively simple analytical Advection Dominated Accretion Flow (ADAF) [8] and jet SED models [4,45] to predict fluxes at multiple frequencies (43, 86, 230, and 345 GHz) directly relevant to the EHT and ngEHT, or to our preliminary surveys aimed at selecting ngEHT targets. Details on these ADAF and/or jet model fits will be published in Arratia et al., in prep.

The analytic ADAF model is particularly useful for a large sample such as ETHER. While it internally uses many parameters, the total (synchrotron, bremsstrahlung, plus inverse Compton) ADAF SED is obtained by varying only the two most important variables—the SMBH mass and the Eddington ratio (l_{Edd}; between 10^{-7} and $10^{-1.7}$). For the jet component, we currently use multiple ($\sim$980) templates built from the models used in Bandyopadhyay et al. [4]. While this jet model has 15 free parameters, we only iterate over those that result in the most significant impact to the model SED, namely black hole mass, jet outflow rate, fraction of electrons accelerated in the shock, and the fraction of shock energy in these electrons. We are currently implementing a similar approach using the *BHJet* model [45], which has 26 free parameters, of which those that most effect the cm to hard X-ray SED include the black hole mass, viewing angle, injected power, and jet maximum extent.

To model the SED of an ETHER source (see, e.g., Figure 1), we require at least a distance or redshift, an SMBH mass estimate, and a hard X-ray flux. The Eddington rate is then varied until the the ADAF SED model fits the hard X-ray point. Errors in this estimate are based on errors of the black hole mass and hard X-ray flux.

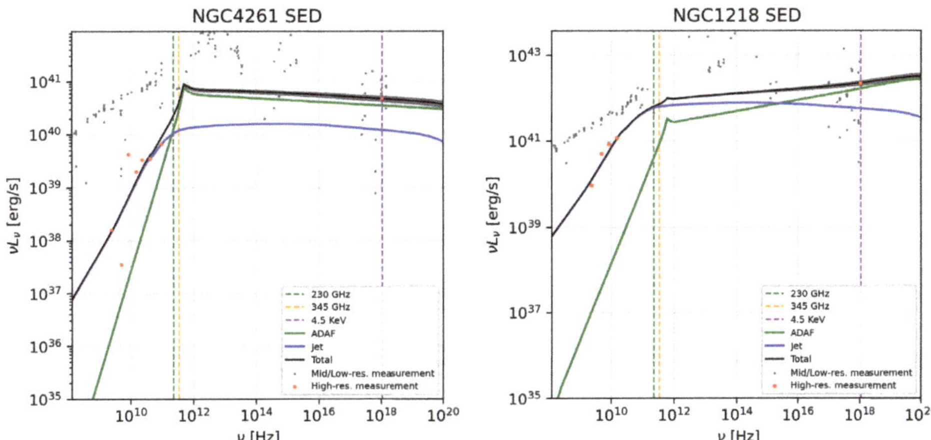

Figure 1. Spectral energy density plots contrasting two ETHER galaxies with known cm-wave nuclear jets: NGC 4261 (**left**) and NGC 1218 (3C 78; **right**). Red (grey) points indicate high (low) resolution flux measurements. For illustration, the cm-wave and hard X-ray data are fit with a scaled jet [4] and analytic ADAF model [8]. Vertical dotted lines indicate the current EHT observing frequency (230 GHz), and the hard X-ray band. For NGC 4261, the fitted jet component (blue) dominates at cm-waves while the ADAF component (green) is expected to dominate at 230 GHz. For NGC 1218, the jet is expected to dominate the 230 GHz EHT flux. We aim to systematically predict the EHT fluxes from the ADAF and jet base in ETHER galaxies using a SMBH mass estimate, hard X-ray flux (e.g., eROSITA), and, if a jet is present, high frequency VLBI observations. For SMBH with ring size ≥ 3 μarcsec, direct 230 GHz imaging with ALMA and SMA is also being pursued.

If one or more high resolution cm-wave fluxes are available, and these are, on average, above the ADAF-only model predictions, then a combined jet plus ADAF model is attempted. Here we attempt jet + ADAF fits with all of our current jet templates, and if multiple high resolution radio fluxes exist, we choose the fit with the lowest root-mean square (RMS) value between predicted and observed radio data. This combined jet plus ADAF fit to a given source thus primarily constrains the full range of expected fluxes from the jet and ADAF, rather than necessarily providing a single best-fit flux estimate plus its error.

Given the above caveats, and the fact that a hard X-ray flux is not currently available for many cm-wave VLBI sources in ETHER, Figures 2, 3 and 6 do not use fluxes derived from these model fits for VLBI-detected sources. Instead, for consistency and for illustration, the expected 230 GHz jet flux for the EHT is extrapolated from the the flux measured at the highest available VLBI frequency using the median spectral slope seen in our sample ($S_\nu \propto \nu^{-0.4}$), plus the median (from our sample) resolution factor which accounts for the smaller field of view as frequency increases.

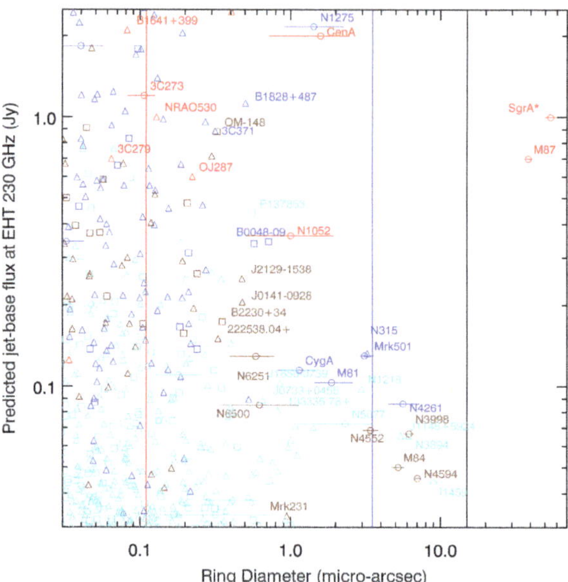

Predicted jet-base flux at EHT 230 GHz (Jy)

Ring Diameter (micro-arcsec)

Figure 2. The estimated 0.1-mas-scale flux from the jet base when observed at 230 GHz with the EHT as a function of ring size (i.e., 10.4 R_g) for brighter VLBI-detected SMBH in ETHER. The intrinsic resolution of ground EHT, geostationary orbit to EHT, and L2 orbit to EHT will resolve the rings to the right of the black, blue, and red lines, respectively. Red, blue, and cyan symbols denote galaxies already observed with the EHT, with 43–86 GHz VLBI, and with <40 GHz VLBI, respectively. Circles (triangles) are used for SMBH mass measurements (estimations), and as a special case, our WISE-based SMBH mass estimations are shown with squares (see text).

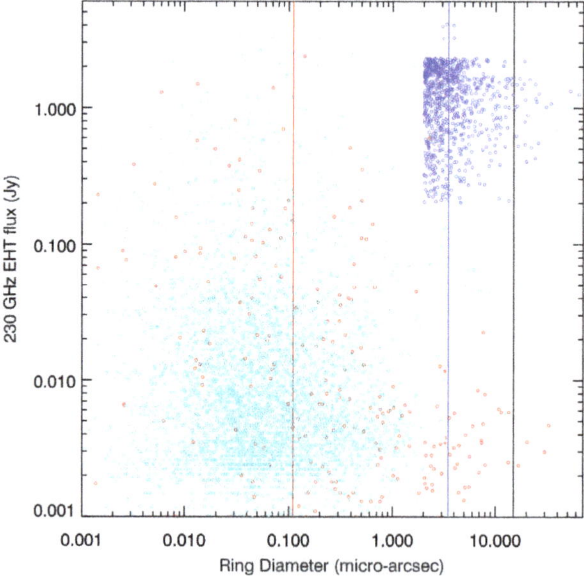

230 GHz EHT flux (Jy)

Ring Diameter (micro-arcsec)

Figure 3. As in Figure 2 but this time with a larger axis range, and plotting additional SMBHs. The points of Figure 2 (VLBI-detected) are now all plotted as cyan circles: these represent the expected flux from the jet base; the EHT flux will be higher if a significant ADAF component is present. For other SMBH, we plot the following fluxes, in order of availability: (a) SMBH with observed 230 GHz fluxes at arcsec-scales are plotted in red. No flux corrections for resolution are applied; (b) SMBH with observed-frame 230 GHz flux estimated from the ADAF-only models of Section 3.1.2 are plotted in green; (c) galaxies with an SMBH estimate, but none of the above, are plotted in blue *at an arbitrary y axis value.*

3. Results

3.1. ETHER: Overall Statistics

The ETHER sample (left panel of Figure 4) merges our collection of SMBH mass estimations, radio fluxes (VLBI, ALMA Calibrator database, literature, and our own work), of AGN and quasars identified in large catalogs (Section 2.1), and all ~580 K galaxies in HyperLeda with Vrecc $\leq$ 25,000 km/s. Currently ETHER contains 1.9 million galaxies, of which 230 have a measured SMBH mass (from resolved stellar dynamics or gas kinematics), and ~1.6 million have an indirectly derived estimate of the SMBH mass. The remaining galaxies are part of the sample due to being listed as an AGN or quasar in redshift catalogs, or for being a bright radio source, but are missing a redshift and/or a SMBH mass estimate. Of the 1.1 million SMBH estimations, 294 K are from M$-\sigma$, 290 K from WISE M-L$_{bulge}$, 5 K from other (non-WISE) M-L$_{bulge}$, and 500 K from single-epoch RM (primarily from [19]). Focusing on galaxies at D $\leq$ 300 Mpc, ETHER currently includes 77,000 galaxies with an SMBH estimate. The mass function of these, in shells of 20 Mpc width out to 290 Mpc is shown in the right panel of Figure 4. At D $\leq$ 100 Mpc our mass function is in agreement with the 'low' theoretical mass function adopted by Pesce et al. [8] except at the lowest masses. However, at larger distances we are increasingly incomplete at the low and higher mass ends. Ongoing SDSS-V (and future 4MOST) programs for spectroscopic observations of eROSITA sources, refinements of our WISE-derived SMBH mass estimates (Section 2.2) to lower signal to noise and/or W1-only detections, and the ingestion of further galaxy catalogs will help fill in this shortfall in the coming years.

Among the 15.4 K AGN detected with cm-wave VLBI, only 8.6 K have an SMBH mass measurement or estimate in ETHER. Of the remaining 6.8 K AGN, we do not currently have a redshift for 91%. Some fraction of these could be recovered with a more exhaustive ingestion of current literature and databases. Determining the missing redshifts and SMBH mass estimates, especially in the brighter VLBI sources, is a high priority in our future work.

Given an SMBH estimate, and a distance, we can estimate the diameter of the ring around the black hole: 10.4R$_g$ [7].

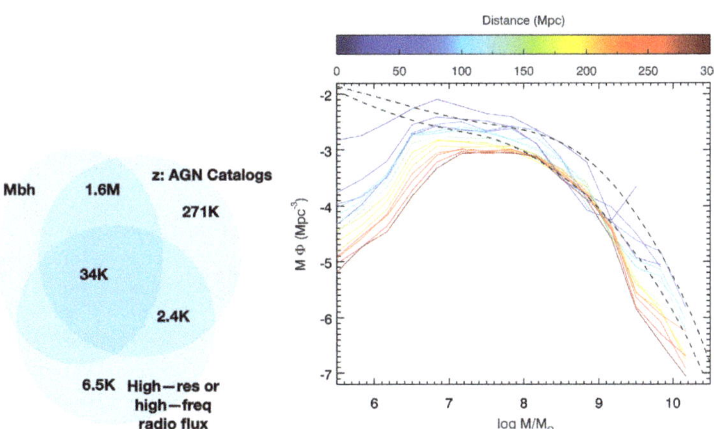

Figure 4. Left: summary statistics of the current ETHER sample. The number of SMBHs with black hole mass estimations, redshifts in AGN catalogs, and high (mas) resolution or high ($\geq$230 GHz with

few to 20″ resolution) frequency radio fluxes are shown, together with their intersections. **Right**: the black hole mass function in ETHER for shells of width 20 Mpc at distances of 10 to 290 Mpc, following the color bar on top. For reference the z = 0 'low' and 'high' theoretical mass functions adopted by [8] are shown in dashed lines.

3.1.1. Jet Bases and Shadows

A major goal of the ngEHT is to study the physics of jet launching and collimation across the parameter space of, e.g., black hole mass, accretion rate, spin, and jet morphology at larger scales: here the ngEHT can uniquely constrain theoretical models in the innermost 10 s to 1000 s R_g e.g., [46]. For such goals, the optimal targets can be selected among the ∼15,400 AGN with already detected cm-wave VLBI cores. If an accretion inflow is present, or if the black hole is 'back-lit' by the receding jet base, then we can image, or constrain, the size of the black hole shadow and gravitationally lensed ring, in a subset of these jetted SMBH.

Figure 2 shows the expected EHT 230 GHz flux from the jet base (inner few 100 s of R_g) as a function of the linear resolution in R_g. Here, the expected EHT flux is illustrative. As we discuss in Section 2.2, we do model the combined jet and ADAF spectrum for AGN with cm-wave VLBI and hard X-ray fluxes. However, the bulk of the ETHER sample have one to few, non-simultaneous, VLBI flux measurements so that SED modelling is not an option for the overall sample. In Figure 2, for uniformity, we simplistically extrapolate the highest frequency VLBI flux to 230 GHz using the median (in our sample) spectral index of −0.4, plus a median (in our sample) flux dilution factor due to the increasing resolution with frequency (0.4–0.8 for 5–86 GHz)). Extrapolations from previous VLBI observations at ≥86 GHz (blue points in the figure) are thus more reliable than others (brown for 43 GHz measurements and cyan for lower frequencies—predominantly 8 GHz). Additionally, for illustration, in the case of Sgr A*—which currently has no detected jet emission—we use the flux of the accretion inflow.

Figure 2 clearly shows that Sgr A* and M87 are, unfortunately, relatively isolated in terms of large shadow size and bright EHT 230 GHz fluxes. While the blue and red lines in the figure delineate the limits for resolving the black hole ring for ngEHT to Geostationary orbit (ngEHT-Geo) satellite(s) and ngEHT to L2 orbit (ngEHT-L2) satellite(s), they also roughly represent the limits for resolving jet bases at 100 R_g and 1000 R_g, respectively, with a ground-based ngEHT. We, thus, already have ∼10 (few hundred) candidates in which the jet base can be resolved at better than 100 R_g (1000 R_g) with the ground-based EHT and ngEHT.

The 230 GHz EHT image of M87 Event Horizon Telescope Collaboration et al. [5] revealed that the flux in the innermost 100 s of R_g is dominated by the accretion inflow, rather than jet base (∼700 mJy in the bright ring produced by the gravitationally lensed accretion inflow, leaving only ∼300 mJy in the still-to-be-EHT-imaged jet base). This is also the scenario we expect from model fits to NGC 4261 (Figure 1); strong 230 GHz fringes detected on a ∼1000 km baseline in our March 2022 EHT observations of this source further supports this. Our current SED fits to other AGN to the right of the blue line in the figure favor the dominance of the accretion flow over the jet in most cases. Thus, as we pursue EHT imaging the the jet base in large ring, relatively low luminosity (as compared to Cen A, or even M87) jets, there is a large possibility that flux from the accretion inflow is detected in, and perhaps dominates, the EHT image in many cases. Super-resolution techniques could constrain black hole ring sizes in targets between the blue and black lines for a ground only ngEHT, and future ground to space VLBI can potentially resolve 10 s (100 s) of black hole rings if the accretion flow, or the receding jet-base, is bright enough.

Our current 'gold sample' of VLBI-detected jet bases includes 15 SMBHs (beyond Sgr A* and M87) with ring size ≥5 μarcsec, plus 39 SMBHs with ring size in the range 2–5 μarcsec. Based on current estimates and SED modelling, there is no reason to rule out an ngEHT goal of constraining $\sim$5 to 10 black hole shadows, and imaging $\sim$10–20 jet bases at resolutions of a few tens of R_g.

While a true exploitation of this sample's potential requires the ngEHT, we have already commenced observations with the current EHT and GMVA+ALMA. All targets to the right of the blue line in Figure 2 were observed (imaging or fringe tests) in March 2022 with the EHT. All are scheduled be imaged with the GMVA+ALMA in March 2023, and an additional 11 SMBHs with ring diameter $\geq$1 μarcsec are scheduled for fringe-test observations (i.e., quick flux determinations) with the EHT in April 2023. Finally, since the March 2022 EHT observations of NGC 4261 showed strong fringes on a $\sim$1000 km baseline, a full track EHT observation of this is scheduled for April 2023.

It is relevant to mention the future growth of this jet-base sample. Currently, a third of cm-wave VLBI detected sources in ETHER lack redshifts. An additional 20% lack a SMBH mass estimate: assuming a M87-like SMBH for these latter yield shadow sizes primarily in the range $\sim$0.1–3 μarcsec. Finally, several VLBI compilations in the literature remain to be ingested into ETHER. The number of potential jet base targets appearing in this figure is thus expected to increase by factor up to two in the coming years, through black hole mass estimates from large surveys, e.g., SDSS-V and 4MOST, and with our own programs. Our programs to obtain higher frequency VLBI (including EHT) observations of the best targets here will allow improvements in our SED model procedure and, thus, 230 GHz flux estimations, and lead to a larger and more reliably identified 'gold' sample for the ngEHT.

As our parent sample increases, we require more efficient pathways to confirm model-selected ngEHT targets. Sources with large ring size and promising jet + ADAF fits to observed fluxes at ν < 43 GHz, are currently being followed up with 43 GHz VLBA snapshot imaging. The combination of 43 GHz VLA and 230 GHz ALMA (or ALMA-ACA) can efficiently survey several hundreds of sources and identify those with 'sub-mm' bumps from the ADAF. Snapshot imaging with the (phased-reference) VLBA at 43 GHz or with GMVA+ALMA can then further filter this sample before ngEHT fringe tests are attempted.

3.1.2. ETHER: Accretion Flows

The absence of a bright radio jet does not necessarily imply the absence of a EHT-detectable accretion inflow (Section 1). Pesce et al. [8] posit that massive ($\geq 10^{9.5}$ M$_\odot$), high accretion rate ($\log l_{\mathrm{Edd}} \sim -1.7$ to -3) SMBHs have the most luminous accretion inflows at 230 GHz. These extreme objects are not in the 'low-hard' state, and may not launch powerful jets (Section 1). We thus require to identify 230 GHz bright accretion flow sources independent of the presence of a radio jet or a previous cm-wave VLBI detection of this. It would not be at all surprising if such accretion-inflow-only ngEHT targets eventually outnumber ngEHT targets selected via the presence of VLBI jets (Section 3.1.1). Resolving accretion inflows allow us to access resolved shadows and rings, to obtain constraints on competing accretion models (e.g., MAD or SANE), and detect orbiting hotspots. Detected unresolved accretion inflows are highly useful signposts for other science cases (Section 3.1.3).

We are following two pathways towards identifying accretion-inflow-only ngEHT targets. First, the largest known black hole rings. ETHER currently includes 623 (85) SMBHs with estimated ring diameters $\geq$3 μarcsec ($\geq$10 μarcsec), independent of the presence of a VLBI-detected jet. These 623 SMBHs are either in relatively nearby (D $\leq$ 100 Mpc) galaxies, in central galaxies of galaxy clusters at z $\lesssim$ 0.3, or massive broad-line SMBHs at higher redshifts from Rakshit et al. [19]. We are in the process of directly measuring the arcsec-scale 230 GHz flux in all of these. Between the literature and calibrator databases ($\sim$30 SMBHs), our ALMA-ACA programs in the last two years ($\sim$130 SMBHs), and programs proposed to the SMA (currently $\sim$100 SMBHs), we are less than half-way through this process, though new candidates are added as ETHER grows. Leaving aside VLBI detected galaxies (addressed in Section 3.1.1), we have thus far detected 35 targets with 230 GHz fluxes between $\sim$1 and 22 mJy at resolutions of 1" to 5". The remaining are not detected at 3σ upper limits of $\sim$1–2 mJy. The detected sources require to be observed at higher resolution (e.g., $\sim$50 mas with the ALMA 12 m array), before clearer statistics, and a 'gold sample' emerge. Further, we are still highly incomplete in large diameter rings (Figure 4) even in the local universe, so this sample of large rings without jets is expected to grow significantly in the coming years (see below).

Second, for future larger samples of large ring SMBHs, and ngEHT-bright accretion flow SMBHs across all ring diameters, direct arcsec-resolution all-sky mm surveys to identify the few hundreds of EHT science targets are impossible (Section 1). While detailed 1-D models, e.g., [4] and full GRMHD models, e.g., [47] of the accretion inflow exist, applying these to millions of SMBH with sparse data is currently not feasible. We, thus, commence with the relatively simple analytical model of Pesce et al. [8] which requires only two user parameters—SMBH mass and accretion rate—to predict the combined synchrotron, inverse Compton, and thermal emission from the accretion inflow (see Section 2). High-resolution data points at frequencies higher than mm-wave are best used to constrain the ADAF SED, since cm-wave fluxes are dominantly from the jet and other galaxy components. Hard X-ray fluxes are an ideal solution since even at resolutions of 1–10″ the bulk of the emission is expected to come from the SMBH environs, and for $\log l_{\mathrm{Edd}} \lesssim -2$ it is likely that a significant fraction of this originates in the accretion inflow rather than a corona.

Currently, we have hard X-ray fluxes and SMBH mass estimates for $\sim$16,800 ETHER galaxies. We fit analytical ADAF models to all of these in order to predict the 43 to 345 GHz fluxes from the ADAF. Figure 5 shows preliminary results of our observed 230 GHz fluxes (red and blue points) and predicted 230 GHz fluxes from the ADAF fits (green points) and compare them with the ADAF fluxes predicted from the analytic models for three specific (and one illustrative) SMBHs over a range of redshifts. We should immediately note that most high redshift 230 GHz fluxes are ALMA Calibrator Database flux measurements of known cm-wave VLBI sources (blue points in the figure). In most of these we expect that the jet, rather than accretion inflow, dominates the 230 GHz flux, and this is borne out by our SED model fits. Even when we assume that all the hard X-ray flux originates in the ADAF the predicted 230 GHz ADAF fluxes are significantly lower than the measured arcsec-scale fluxes. These bright jet sources are thus best modelled with a combined jet plus ADAF model, though an additional complication is contamination from dust emission in the 1 to 5″ beam of the 230 GHz flux measurements.

At D $\leq$ 400 Mpc, we see a larger overlap between arcsec-scale measured and mas-scale ADAF-predicted 230 GHz fluxes. In the few test cases where the mm-wave flux is measured at sub-arcsec resolution, and there is no indication of a cm-wave jet, we obtain a factor $\sim$5 agreement. We clearly require a larger sample of jet-free, or weak-jet, cases to best determine the precision of our ADAF-only estimates. Our ongoing ALMA (and proposed SMA) programs to observe large-ring galaxies without evidence for cm-wave jets (see above), together with hard X-ray fluxes from eROSITA will increase our test sample size by factor $\gtrsim$10–100 and allow refinements and consistency tests in these model predictions. Future improvements also include the use of SEDs from full (scaled) GRMHD modelling of accretion inflows, and/or fine tuning of the Pesce et al. [8] models.

Our identification of accretion-inflow-only bright targets for the ngEHT is very much at an initial stage. Current results favor the requirement of a ngEHT detection threshold of a few mJy, even for studies of unresolved accretion inflows. However, there are several reasons we expect the sample to expand by factor $\sim$10 to 50. First, eROSITA expects to detect $\sim$1 million hard X-ray sources in their all sky survey (first data release scheduled for March 2023), with ongoing SDSS-V and future 4MOST spectroscopic follow-ups to determine black hole masses of eROSITA detections. Massive SMBHs with relatively high accretion rates are expected to have ADAFs bright in both the hard X-ray and at 230 GHz (see above and [8]), so that the eROSITA selection, and posterior SED modelling with ADAFs, is an especially promising pathway to select ngEHT targets in the coming years. Further, our current ETHER SMBH mass function (Figure 4) remains incomplete at the high mass end, even in the local universe. Filling in this deficit, via additional WISE- and

2MASS-derived SMBH estimates, plus other (and ours) spectroscopic and photometric programs, will allow an increase in our target samples.

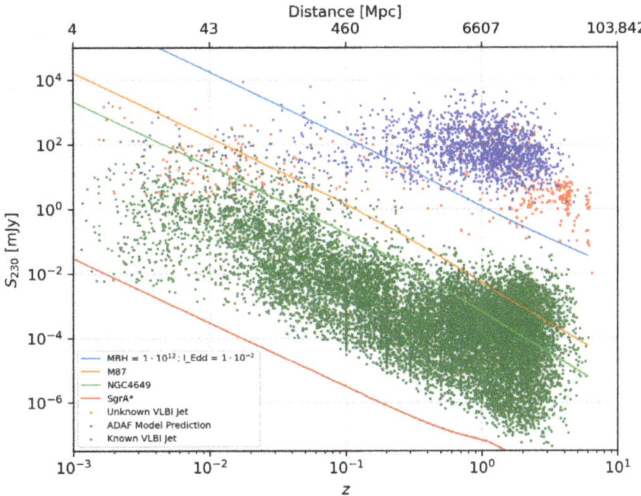

Figure 5. Measured (red and blue points) and model-predicted (green points) 230 GHz fluxes as a function of redshift (lower axis) and distance (upper axis). Blue points are used for SMBH with known cm-wave VLBI detections, i.e., AGN with compact jets. Green points show the expected flux from the analytic model of Pesce et al. [8] given our SMBH mass estimate and a measured hard X-ray flux (presumed to come only from the ADAF). The evolution of the ADAF flux with redshift—using the Pesce et al. [8] models—for SgrA*, M87, and NGC4649, and an illustrative extreme high-mass high-accretion-rate SMBH—are shown in colored lines following the legend on the panel. Details can be found in Arratia et al., in prep.

3.1.3. ETHER: Other Science Cases

The ngEHT will pursue several science cases in varied targets independent of their linear resolution in R_g scales, e.g., using the emission from the jet base and/or accretion inflow as signposts (e.g., the search for resolved binary supermassive black holes in their gravitational emitting phase) or to better constrain multi-wavelength and multi-messenger phenomena.

Figure 6 shows the predicted 230 GHz EHT flux (the same extrapolation used in Figure 2) but this time as a function of linear resolution at the redshift of the SMBH. The EHT has a sub-0.2 pc resolution across the universe, and can uniquely resolve binary SMBHs during their gravitation wave emitting inspiral. Here we once more only plot VLBI-detected SMBHs and use an estimated flux of the jet base. The future addition of additional VLBI-detected SMBHs and also bright ADAF-only sources will significantly increase the number of SMBHs in this figure.

There are currently $\sim$160 binary SMBH candidates in ETHER, as identified by diverse methods for a review, see [48]. Only about 10, e.g., [49–51] come from direct VLBI imaging of multiple potential SMBHs. The majority of posited binary SMBHs come from the interpretation of periodic flux variability, e.g., [52,53], double velocity components in the Narrow, e.g., [54] or Broad, e.g., [55] line region. X-shaped radio sources are also candidates [56]: while the X-shape can be explained by the characteristics of the spin axis of a single SMBH, they could also potentially be produced by binary SMBH. Additionally, Ref. [57] have identified the nearby galaxies most likely to host binary SMBHs based on simulating their merger history.

An analysis similar to that of single SMBHs in ETHER, but including a statistical treatment of the SMBH mass ratio, is being used to select the best candidates for ngEHT monitoring surveys. The Vera C. Rubin Observatory will enable extensive studies of periodic flux variability, and is thus expected to significantly enlarge the sample of binary black hole candidates. The ETHER database will provide a pathway towards identifying the subset of these which are best observable with the ngEHT.

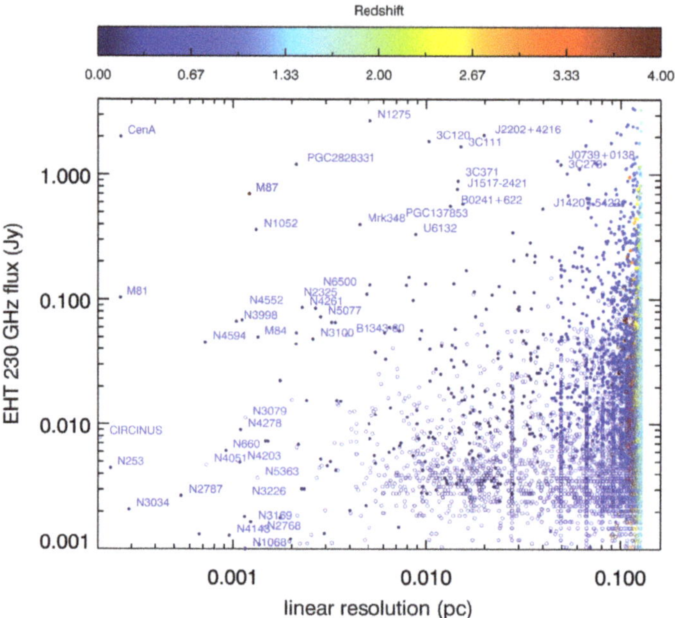

Figure 6. The estimated 230 GHz flux from the jet base—for VLBI detected sources only—as a function of linear resolution of the EHT assuming a 15 μarcsec angular resolution: the EHT has a sub 0.2 pc resolution across the universe. EHT-Geo and EHT-L2 baselines would increase this by factor ∼4 and ∼130, respectively. Symbols are colored by redshift following the color bar, and some individual galaxies are named in uncrowded regions of the plot. Vertical striations are primarily due to redshifts with few significant digits in Véron-Cetty and Véron [33]. We expect to add a significant number of additional sources on this plot, both from new cm-VLBI detections, and for AGN with a bright accretion inflow.

3.2. ETHER: Querying and Usage

The ETHER database and server is currently being prepared for release on a dedicated server via a *Python* interface. The tabulated searchable data will include: (a) basic source data including position, distance, redshift, AGN type, galaxy morphological type; (b) the best available (and the WISE-based) black hole mass estimate, plus errors and quality flags, the estimated shadow size and linear resolution at the SMBH angular distance, the estimated stellar mass and star formation rate, and the AGN bolometric luminosity; (c) fluxes at all typical VLBI frequencies, and at hard X-ray and γ-ray; and (d) model predicted ADAF-only and jet and ADAF fluxes at ngEHT frequencies. For proper attribution of credit, all user outputs will contain easy-to-use references (in 'bibtex' and equivalent formats) for data ingested from the literature or other databases. Each source will additionally link to files containing its full SED data, and the model parameters and SEDs of the best-fit ADAF and jet+ADAF models.

The database interface will support two types of usage: (a) cross-matching to a list of names or positions. This interface will match targets to the existing database and return tabulated data, model results, and figures. In case input positions are not matched to the database, the server will generate *Python*-based 'astro-queries' to relevant databases and virtual observatories, run relevant models, and incorporate the new data into the database and the returned results; and (b) exploration of the current database, using a GUI with multi-parameter filters in order to obtain specific sub-samples of science targets.

In the future, the database will be interfaced to other ngEHT simulation tools currently in development, so that observing constraints and optimizations can be taken into account when choosing a final sample.

4. Discussion

Figure 3 combines the results of our previous sections to show a more comprehensive view of (the current status of) the ngEHT target pool. We emphasize that this figure is illustrative and preliminary: many data-points in the figure come from low-resolution fluxes, model predictions, and extrapolations still undergoing refinement. On the positive side, our parent sample will grow significantly in the coming years (we expect a factor at least two for jet sources and factor $\sim$10–50 for accretion-only inflow sources, see Sections 3.1.1 and 3.1.2).

Jet-base targets for the EHT (cyan circles)—whose brighter subset are seen in more detail in Figure 2—primarily cluster at expected EHT 230 GHz fluxes of 2 to 30 mJy, and expected ring diameters of 0.005 to 0.5 μarcsec. While these are identified via their cm-wave VLBI jet emission, note that in M87 the EHT showed that the 230 GHz nuclear emission is dominated by accretion inflow rather than jet base [5]. Remaining SMBHs with observed arcsec-scale resolution 230 GHz fluxes are shown in red. Here, those at smaller ring sizes are primarily bright sources from the ALMA Calibrator database (and likely dominated by jet emission at 230 GHz) but also include weaker sources: those at the largest ring sizes come primarily from our ongoing ALMA-ACA programs to directly measure 230 GHz fluxes. In the latter, the $\sim$5 arcsec resolution of ALMA-ACA means that the fluxes are potentially contaminated by dust from the host galaxy, so future $\sim$50 mas resolution ALMA observations of the brighter detections will be required to confirm their ADAF origin. For several hundred SMBH with large ring sizes, we currently do not have an accurate path towards estimating 230 GHz fluxes. Those at estimated ring sizes ≥ 2 μarcsec are shown as blue circles (with an arbitrary flux value in the y axis in Figure 2): the number of these SMBH increases rapidly below this ring size but we do not show them to avoid crowding on the plot. These SMBH—divided roughly equally between nearby galaxies and high-z galaxies with large SMBH mass estimates from Rakshit et al. [19]—are the targets of our ongoing ALMA/ACA and SMA (proposed) programs. Finally, SMBH with 230 GHz fluxes predicted from our ADAF-only models are shown with green. Currently, these cluster at predicted fluxes of $\lesssim$10 mJy.

The cumulative histograms of ring sizes in the current ETHER sample are shown in the right panel of Figure 7. While the cumulative distribution (red) shows >10 rings already resolvable, in principle, by the EHT, and rises quickly to smaller shadows, the subset previously detected at mas-scales at $\nu > 40$ GHz (primarily jet-base candidates) significantly lags behind. Two factors contribute to this effect: (a) intrinsically weak radio emission at all frequencies from the innermost SMBH environs—as evidenced by across the board low fluxes at lower spatial resolutions, and (b) the lack of high resolution—and in many cases any resolution—mm-wave observations of many sources. In these sources, weak cm-wave fluxes (from a jet base) does not necessarily imply weak mm-wave fluxes (from the accretion inflow). As described in the previous sections, we are actively attempting to mitigate the second factor. Further, flux monitoring of SMBH in the former sub-class may reveal new targets due to variability, which is not uncommon in the mm [58].

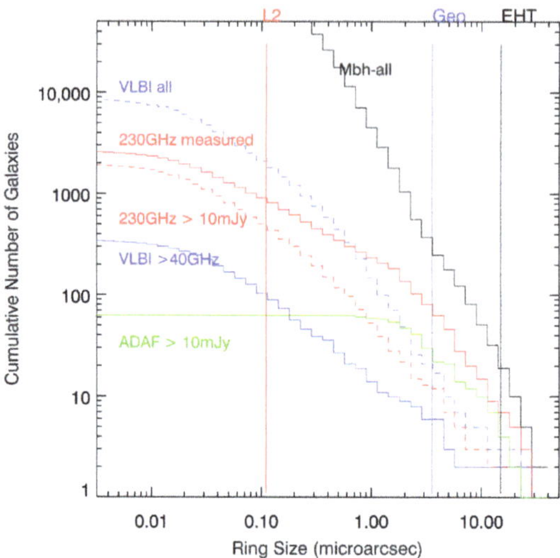

Figure 7. Cumulative histograms of 'ring' sizes in the ETHER sample. The intrinsic 15 μarcsec resolution of ground EHT, geostationary orbit to EHT, and L2 orbit to EHT at 345 GHz will resolve the rings to the right of the black, blue, and red lines, respectively. The cumulative histogram of all 230 GHz flux measurements, and all 230 GHz flux measurements greater than 10 mJy, are shown with the red solid and dashed lines, respectively.

The cumulative distribution of detectable accretion inflows (ADAFs) remains to be determined, as we are in the initial phase of this pathway (see above, and Section 3.1.2). The cumulative distribution of sources with measured or expected flux $\geq$ 10 mJy are shown in red. These should be treated with caution: the number of AGN with 230 GHz fluxes predicted from hard X-ray fluxes and ADAF models will increase by factor $\sim$100 with eROSITA and SDSS-V, while the current measured 230 GHz fluxes will likely decrease as higher resolution observations are obtained. The cumulative histogram of sources with measured 230 GHz flux smaller than 10 mJy is shown in green. Clearly, even at large black hole ring sizes, the many targets lack a 230 GHz measured flux, at any resolution, or a flux prediction via ADAF models. Our ongoing ALMA, and proposed SMA, programs aim to obtain fluxes for all with estimated ring size $\geq$ 2 μarcsec.

The millions of black hole mass estimates used in ETHER come from diverse methods, including our automated WISE-based estimations, with varying random and systematic errors. Detailed manual analysis of existing data, and new data, will be (and is being) used to refine these black hole mass estimates for specific targets which enter our ngEHT 'Gold Samples'.

In conclusion, the ETHER sample and database consolidates data and model predictions of all potential ngEHT candidates independent of science case. It also provides pathways towards determining the factability of ngEHT observations of any given new source. While we expect the sample to expand significantly (factor $\sim$10–50 for accretion-inflow-only targets, and $\sim$2 for known cm-wave VLBI jets) we have already identified a 'gold sample' for ngEHT jet base and potentially black hole shadow studies, and have commenced to observe these with the existing EHT.

Author Contributions: Conceptualization, V.R., N.N., D.W.P., T.P.K., S.D., H.F., G.B. and P.N.; Data curation, V.R., N.N., V.A., J.H.-Y., D.G.N., V.L.F., L.B. and G.B.; Funding acquisition, N.N., S.D. and H.F.; Investigation, V.R., N.N., V.A., J.H.-Y., D.G.N., B.B., C.M.-P., V.L.F. and P.N.; Methodology, V.R., N.N., D.W.P., B.B., T.P.K., S.D., A.R., V.L.F., L.B., H.F., G.B. and P.N.; Project administration, V.R. and N.N.; Resources, V.R., S.D., V.L.F., L.B. and H.F.; Software, N.N., V.A., J.H.-Y., D.W.P., B.B. and L.B.; Supervision, N.N.; Validation, N.N., V.A. and C.M.-P.; Writing—original draft, V.R. and N.N.; Writing—review & editing, V.A., J.H.-Y., D.W.P., T.P.K., A.R. and P.N. All authors have read and agreed to the published version of the manuscript.

Funding: We acknowledge funding from ANID Chile via Nucleo Milenio TITANs (Project NCN19-058), Fondecyt Regular (Project 1221421), Fondecyt Postdoctorado (Project 3220195) and BASAL (Projects AFB-170002 and FB210003).

Data Availability Statement: The ETHER database will be available at www.titans.cl/ether (accessed on 8 March 2020).

Acknowledgments: This paper makes use of ALMA data. ALMA is a partnership of ESO (representing its member states), NSF (USA) and NINS (Japan), together with NRC (Canada), MOST and ASIAA (Taiwan), and KASI (Republic of Korea), in cooperation with the Republic of Chile. The Joint ALMA Observatory is operated by ESO, AUI/NRAO and NAOJ. We acknowledge the usage of the HyperLeda database (http://leda.univ-lyon1.fr; accessed on 8 March 2020 for all galaxy velocity dispersions, and on 29 November 2022 for the list of all $D \leq 350$ Mpc galaxies), and congratulate its managers for the easy to use interface provided. This research has made use of the NASA/IPAC Extragalactic Database (NED), which is funded by the National Aeronautics and Space Administration and operated by the California Institute of Technology. This research has made use of the SIMBAD database, operated at CDS, Strasbourg, France. This research has made use of the VizieR catalogue access tool, CDS, Strasbourg, France. This research was supported in part by the Black Hole Initiative, which is supported by grants by the Gordon and Betty Moore Foundation and John Templeton Foundation. The opinions expressed in this publication are those of the author(s) and do not necessarily reflect the views of the Moore or Templeton Foundations.

Conflicts of Interest: The authors declare no conflict of interest.

Abbreviations

The following abbreviations are used in this manuscript:

ADAF	Advection Dominated Accretion Flow
EHT	Event Horizon Telescope
ETHER	Event Horizon and Environs Sample
ngEHT	next-generation Event Horizon Telescope
SMBH	supermassive black hole(s)

Notes

[1] The NASA/IPAC Extragalactic Database (NED) is funded by the National Aeronautics and Space Administration and operated by the California Institute of Technology.

[2] http://astrogeo.org/rfc/ (accessed on 5 February 2021).

[3] http://www.vlba.nrao.edu/astro/calib/ (accessed on 5 February 2021).

[4] https://almascience.eso.org/alma-data/calibrator-catalogue (accessed on 5 February 2021).

References

1. Yuan, F.; Narayan, R. Hot Accretion Flows around Black Holes. *Annu. Rev. Astron. Astrophys.* **2014**, *52*, 529–588.
2. Narayan, R.; Yi, I.; Mahadevan, R. Advection-Dominated Accretion Model of Sagittarius A* and Other Accreting Black Holes. *Astron. Astrophys. Suppl.* **1996**, *120*, 287–290.
3. Markoff, S.; Bower, G.C.; Falcke, H. How to hide Large-Scale Outflows: Size Constraints on the Jets of Sgr A*. *Mon. Not. R. Astron. Soc. Lett.* **2007**, *379*, 1519–1532.
4. Bandyopadhyay, B.; Xie, F.G.; Nagar, N.M.; Schleicher, D.R.G.; Ramakrishnan, V.; Arévalo, P.; López, E.; Diaz, Y. Resolving Accretion Flows in Nearby Active Galactic Nuclei with the Event Horizon Telescope. *Mon. Not. R. Astron. Soc. Lett.* **2019**, *490*, 4606–4621.

5. Event Horizon Telescope Collaboration; Akiyama, K.; Alberdi, A.; Alef, W.; Asada, K.; Azulay, R.; Baczko, A.K.; Ball, D.; Baloković, M.; Barrett, J.; et al. First M87 Event Horizon Telescope Results. I. The Shadow of the Supermassive Black Hole. *Astrophys. J. Lett.* **2019**, *875*, L1.

6. Event Horizon Telescope Collaboration; Akiyama, K.; Alberdi, A.; Alef, W.; Algaba, J.C.; Anantua, R.; Asada, K.; Azulay, R.; Bach, U.; Baczko, A.K.; et al. First Sagittarius A* Event Horizon Telescope Results. I. The Shadow of the Supermassive Black Hole in the Center of the Milky Way. *Astrophys. J. Lett.* **2022**, *930*, L12. [CrossRef]

7. Johannsen, T.; Psaltis, D. Testing the No-hair Theorem with Observations in the Electromagnetic Spectrum. II. Black Hole Images. *Astrophys. J.* **2010**, *718*, 446–454.

8. Pesce, D.W.; Palumbo, D.C.M.; Narayan, R.; Blackburn, L.; Doeleman, S.S.; Johnson, M.D.; Ma, C.P.; Nagar, N.M.; Natarajan, P.; Ricarte, A. Toward Determining the Number of Observable Supermassive Black Hole Shadows. *Astrophys. J.* **2021**, *923*, 260.

9. Pesce, D.W.; Palumbo, D.C.M.; Ricarte, A.; Broderick, A.E.; Johnson, M.D.; Nagar, N.M.; Natarajan, P.; Gómez, J.L. Expectations for Horizon-Scale Supermassive Black Hole Population Studies with the ngEHT. *Galaxies* **2022**, *10*, 109. [CrossRef]

10. Franco, M.; Elbaz, D.; Béthermin, M.; Magnelli, B.; Schreiber, C.; Ciesla, L.; Dickinson, M.; Nagar, N.; Silverman, J.; Daddi, E.; et al. GOODS-ALMA: 1.1 mm Galaxy Survey. I. Source Catalog and Optically Dark Galaxies. *Astron. Astrophys.* **2018**, *620*, A152.

11. Planck Collaboration; Ade, P.A.R.; Aghanim, N.; Argüeso, F.; Arnaud, M.; Ashdown, M.; Aumont, J.; Baccigalupi, C.; Banday, A.J.; Barreiro, R.B.; et al. Planck 2015 results. XXVI. The Second Planck Catalogue of Compact Sources. *Astron. Astrophys.* **2016**, *594*, A26.

12. Sobrin, J.A.; Anderson, A.J.; Bender, A.N.; Benson, B.A.; Dutcher, D.; Foster, A.; Goeckner-Wald, N.; Montgomery, J.; Nadolski, A.; Rahlin, A.; et al. The Design and Integrated Performance of SPT-3G. *Astrophys. J. Suppl. Ser.* **2022**, *258*, 42.

13. Simpson, J.M.; Smail, I.; Swinbank, A.M.; Chapman, S.C.; Chen, C.C.; Geach, J.E.; Matsuda, Y.; Wang, R.; Wang, W.H.; Yang, Y.; et al. The East Asian Observatory SCUBA-2 Survey of the COSMOS Field: Unveiling 1147 Bright Sub-Millimeter Sources across 2.6 Square Degrees. *Astrophys. J.* **2019**, *880*, 43.

14. van den Bosch, R.C.E. Unification of the Fundamental Plane and Super Massive Black Hole Masses. *Astrophys. J.* **2016**, *831*, 134.

15. Saglia, R.P.; Opitsch, M.; Erwin, P.; Thomas, J.; Beifiori, A.; Fabricius, M.; Mazzalay, X.; Nowak, N.; Rusli, S.P.; Bender, R. The SINFONI Black Hole Survey: The Black Hole Fundamental Plane Revisited and the Paths of (Co)evolution of Supermassive Black Holes and Bulges. *Astrophys. J.* **2016**, *818*, 47.

16. Thater, S.; Krajnović, D.; Cappellari, M.; Davis, T.A.; de Zeeuw, P.T.; McDermid, R.M.; Sarzi, M. Six New Supermassive Black Hole Mass Determinations from Adaptive-Optics Assisted SINFONI Observations. *Astron. Astrophys.* **2019**, *625*, A62.

17. Boizelle, B.D.; Walsh, J.L.; Barth, A.J.; Buote, D.A.; Baker, A.J.; Darling, J.; Ho, L.C.; Cohn, J.; Kabasares, K.M. Black Hole Mass Measurements of Radio Galaxies NGC 315 and NGC 4261 Using ALMA CO Observations. *Astrophys. J.* **2021**, *908*, 19.

18. North, E.V.; Davis, T.A.; Bureau, M.; Cappellari, M.; Iguchi, S.; Liu, L.; Onishi, K.; Sarzi, M.; Smith, M.D.; Williams, T.G. WISDOM project—V. Resolving Molecular Gas in Keplerian Rotation around the Supermassive Black Hole in NGC 0383. *Mon. Not. R. Astron. Soc. Lett.* **2019**, *490*, 319–330.

19. Rakshit, S.; Stalin, C.S.; Kotilainen, J.; Shin, J. High-Redshift Narrow-Line Seyfert 1 Galaxies: A Candidate Sample. *Astrophys. J. Suppl. Ser.* **2021**, *253*, 28.

20. Woo, J.H.; Urry, C.M. Active Galactic Nucleus Black Hole Masses and Bolometric Luminosities. *Astrophys. J.* **2002**, *579*, 530–544.

21. Gültekin, K.; King, A.L.; Cackett, E.M.; Nyland, K.; Miller, J.M.; Di Matteo, T.; Markoff, S.; Rupen, M.P. The Fundamental Plane of Black Hole Accretion and Its Use as a Black Hole-Mass Estimator. *Astrophys. J.* **2019**, *871*, 80.

22. van den Bosch, R.C.E.; Gebhardt, K.; Gültekin, K.; Yıldırım, A.; Walsh, J.L. Hunting for Supermassive Black Holes in Nearby Galaxies With the Hobby-Eberly Telescope. *Astrophys. J. Suppl. Ser.* **2015**, *218*, 10.

23. Makarov, D.; Prugniel, P.; Terekhova, N.; Courtois, H.; Vauglin, I. HyperLEDA. III. The catalogue of extragalactic distances. *Astron. Astrophys.* **2014**, *570*, A13. [CrossRef]

24. Caramete, L.I.; Biermann, P.L. The Mass Function of Nearby Black HOLE Candidates. *Astron. Astrophys.* **2010**, *521*, A55.

25. Mezcua, M.; Hlavacek-Larrondo, J.; Lucey, J.R.; Hogan, M.T.; Edge, A.C.; McNamara, B.R. The Most Massive Black Holes on the Fundamental Plane of Black Hole Accretion. *Mon. Not. R. Astron. Soc. Lett.* **2018**, *474*, 1342–1360.

26. Shaw, M.S.; Romani, R.W.; Cotter, G.; Healey, S.E.; Michelson, P.F.; Readhead, A.C.S.; Richards, J.L.; Max-Moerbeck, W.; King, O.G.; Potter, W.J. Spectroscopy of Broad-line Blazars from 1LAC. *Astrophys. J.* **2012**, *748*, 49.

27. Chen, Y.Y.; Zhang, X.; Zhang, H.J.; Yu, X.L. Core-Dominance Parameter, Black Hole Mass and Jet-Disc Connection for Fermi blazars. *Mon. Not. R. Astron. Soc. Lett.* **2015**, *451*, 4193–4206.

28. Cutri, R.M.; Wright, E.L.; Conrow, T.; Fowler, J.W.; Eisenhardt, P.R.M.; Grillmair, C.; Kirkpatrick, J.D.; Masci, F.; McCallon, H.L.; Wheelock, S.L.; et al. VizieR Online Data Catalog: AllWISE Data Release (Cutri+ 2013). *Vizier Online Data Cat.* **2021**, *II*, 328.

29. Cluver, M.E.; Jarrett, T.H.; Hopkins, A.M.; Driver, S.P.; Liske, J.; Gunawardhana, M.L.P.; Taylor, E.N.; Robotham, A.S.G.; Alpaslan, M.; Baldry, I.; et al. Galaxy and Mass Assembly (GAMA): Mid-infrared Properties and Empirical Relations from WISE. *Astrophys. J.* **2014**, *782*, 90.

30. Huchra, J.P.; Macri, L.M.; Masters, K.L.; Jarrett, T.H.; Berlind, P.; Calkins, M.; Crook, A.C.; Cutri, R.; Erdoğdu, P.; Falco, E.; et al. The 2MASS Redshift Survey—Description and Data Release. *Astrophys. J. Suppl. Ser.* **2012**, *199*, 26.

31. Schutte, Z.; Reines, A.E.; Greene, J.E. The Black Hole-Bulge Mass Relation Including Dwarf Galaxies Hosting Active Galactic Nuclei. *Astrophys. J.* **2019**, *887*, 245.

32. Massaro, E.; Maselli, A.; Leto, C.; Marchegiani, P.; Perri, M.; Giommi, P.; Piranomonte, S. The 5th Edition of the Roma-BZCAT. A Short Presentation. *Astrophys. Space Sci.* **2015**, *357*, 75.

33. Véron-Cetty, M.P.; Véron, P. A catalogue of quasars and active nuclei: 13th edition. *Astron. Astrophys.* **2010**, *518*, A10. [CrossRef]

34. Flesch, E.W. The Half Million Quasars (HMQ) Catalogue. *PASA* **2015**, *32*, e010.

35. Tully, R.B.; Rizzi, L.; Shaya, E.J.; Courtois, H.M.; Makarov, D.I.; Jacobs, B.A. The Extragalactic Distance Database. *Astron. J.* **2009**, *138*, 323–331. [CrossRef]

36. Kim, J.Y.; Krichbaum, T.P.; Broderick, A.E.; Wielgus, M.; Blackburn, L.; Gómez, J.L.; Johnson, M.D.; Bouman, K.L.; Chael, A.; Akiyama, K.; et al. Event Horizon Telescope imaging of the archetypal blazar 3C 279 at an extreme 20 microarcsecond resolution. *Astron. Astrophys.* **2020**, *640*, A69. [CrossRef]

37. Janssen, M.; Falcke, H.; Kadler, M.; Ros, E.; Wielgus, M.; Akiyama, K.; Baloković, M.; Blackburn, L.; Bouman, K.L.; Chael, A.; et al. Event Horizon Telescope Observations of the Jet Launching and Collimation in Centaurus A. *Nat. Astron.* **2021**, *5*, 1017–1028.

38. Issaoun, S.; Wielgus, M.; Jorstad, S.; Krichbaum, T.P.; Blackburn, L.; Janssen, M.; Chan, C.k.; Pesce, D.W.; Gómez, J.L.; Akiyama, K.; et al. Resolving the Inner Parsec of the Blazar J1924-2914 with the Event Horizon Telescope. *Astrophys. J.* **2022**, *934*, 145.

39. Nair, D.G.; Lobanov, A.P.; Krichbaum, T.P.; Ros, E.; Zensus, J.A.; Kovalev, Y.Y.; Lee, S.S.; Mertens, F.; Hagiwara, Y.; Bremer, M.; et al. Global Millimeter VLBI Array Survey of Ultracompact Extragalactic Radio Sources at 86 GHz. *Astron. Astrophys.* **2019**, *622*, A92.

40. Lee, S.S.; Lobanov, A.P.; Krichbaum, T.P.; Witzel, A.; Zensus, A.; Bremer, M.; Greve, A.; Grewing, M. A Global 86 GHz VLBI Survey of Compact Radio Sources. *Astron. J.* **2008**, *136*, 159–180.

41. Cheng, X.P.; An, T.; Frey, S.; Hong, X.Y.; He, X.; Kellermann, K.I.; Lister, M.L.; Lao, B.Q.; Li, X.F.; Mohan, P.; et al. Compact Bright Radio-loud AGNs. III. A Large VLBA Survey at 43 GHz. *Astrophys. J. Suppl. Ser.* **2020**, *247*, 57.

42. Lister, M.L.; Aller, M.F.; Aller, H.D.; Hodge, M.A.; Homan, D.C.; Kovalev, Y.Y.; Pushkarev, A.B.; Savolainen, T. MOJAVE. XV. VLBA 15 GHz Total Intensity and Polarization Maps of 437 Parsec-scale AGN Jets from 1996 to 2017. *Astrophys. J. Suppl. Ser.* **2018**, *234*, 12.

43. Helmboldt, J.F.; Taylor, G.B.; Tremblay, S.; Fassnacht, C.D.; Walker, R.C.; Myers, S.T.; Sjouwerman, L.O.; Pearson, T.J.; Readhead, A.C.S.; Weintraub, L.; et al. The VLBA Imaging and Polarimetry Survey at 5 GHz. *Astrophys. J.* **2007**, *658*, 203–216.

44. Evans, I.N.; Primini, F.A.; Glotfelty, K.J.; Anderson, C.S.; Bonaventura, N.R.; Chen, J.C.; Davis, J.E.; Doe, S.M.; Evans, J.D.; Fabbiano, G.; et al. The Chandra Source Catalog. *Astrophys. J. Suppl. Ser.* **2010**, *189*, 37–82.

45. Lucchini, M.; Ceccobello, C.; Markoff, S.; Kini, Y.; Chhotray, A.; Connors, R.M.T.; Crumley, P.; Falcke, H.; Kantzas, D.; Maitra, D. Bhjet: A public multi-zone, steady state jet plus thermal corona spectral model. *Mon. Not. R. Astron. Soc.* **2022**, *517*, 5853–5881. [CrossRef]

46. Ricarte, A.; Gammie, C.; Narayan, R.; Prather, B.S. Probing Plasma Phys. Spectr. Index Maps Accreting Black Holes Event Horiz. Scales. *arXiv* **2022**, arXiv:202.02408.

47. Porth, O.; Chatterjee, K.; Narayan, R.; Gammie, C.F.; Mizuno, Y.; Anninos, P.; Baker, J.G.; Bugli, M.; Chan, C.k.; Davelaar, J.; et al. The Event Horizon General Relativistic Magnetohydrodynamic Code Comparison Project. *Astrophys. J. Suppl. Ser.* **2019**, *243*, 26.

48. De Rosa, A.; Vignali, C.; Bogdanović, T.; Capelo, P.R.; Charisi, M.; Dotti, M.; Husemann, B.; Lusso, E.; Mayer, L.; Paragi, Z.; et al. The Quest for Dual and Binary Supermassive Black Holes: A Multi-Messenger View. *New Astron. Rev.* **2019**, *86*, 101525.

49. Gitti, M.; Giroletti, M.; Giovannini, G.; Feretti, L.; Liuzzo, E. A Candidate Supermassive Binary Black Hole System in the Brightest Cluster Galaxy of RBS 797. *Astron. Astrophys.* **2013**, *557*, L14.

50. Bansal, K.; Taylor, G.B.; Peck, A.B.; Zavala, R.T.; Romani, R.W. Constraining the Orbit of the Supermassive Black Hole Binary 0402+379. *Astrophys. J.* **2017**, *843*, 14.

51. An, T.; Zhang, Y.; Wang, A.; Shu, X.; Yang, H.; Jiang, N.; Dou, L.; Pan, Z.; Wang, T.; Zheng, Z. VLBI Imaging of the Pre-Coalescence SMBHB Candidate SDSS J143016.05+230344.4. *Astron. Astrophys.* **2022**, *663*, A139.

52. Graham, M.J.; Djorgovski, S.G.; Stern, D.; Drake, A.J.; Mahabal, A.A.; Donalek, C.; Glikman, E.; Larson, S.; Christensen, E. A Systematic Search for Close Supermassive Black Hole Binaries in the Catalina Real-Time Transient Survey. *Mon. Not. R. Astron. Soc. Lett.* **2015**, *453*, 1562–1576.

53. Ren, H.X.; Cerruti, M.; Sahakyan, N. Quasi-Periodic Oscillations in the γ-ray Light Curves of Bright Active Galactic Nuclei. *arXiv* **2022**, arXiv:2204.13051.

54. Severgnini, P.; Braito, V.; Cicone, C.; Saracco, P.; Vignali, C.; Serafinelli, R.; Della Ceca, R.; Dotti, M.; Cusano, F.; Paris, D.; et al. A possible sub-kiloparsec dual AGN buried behind the galaxy curtain. *Astron. Astrophys.* **2021**, *646*, A153.

55. Terwel, J.H.; Jonker, P.G. Discovery of a Quasar with Double-Peaked Broad Balmer Emission lines. *Mon. Not. R. Astron. Soc. Lett.* **2022**, *512*, L80–L84.

56. Saripalli, L.; Roberts, D.H. What Are "X-shaped" Radio Sources Telling Us? II. Properties of a Sample of 87. *Astrophys. J.* **2018**, *852*, 48. [CrossRef]

57. Mingarelli, C.M.F.; Lazio, T.J.W.; Sesana, A.; Greene, J.E.; Ellis, J.A.; Ma, C.P.; Croft, S.; Burke-Spolaor, S.; Taylor, S.R. The Local Nanohertz Gravitational-Wave Landscape from Supermassive Black Hole Binaries. *Nat. Astron.* **2017**, *1*, 886–892.
58. Bower, G.C.; Dexter, J.; Markoff, S.; Gurwell, M.A.; Rao, R.; McHardy, I. A Black Hole Mass-Variability Timescale Correlation at Submillimeter Wavelengths. *Astrophys. J. Lett.* **2015**, *811*, L6.

Article

Expectations for Horizon-Scale Supermassive Black Hole Population Studies with the ngEHT

Dominic W. Pesce [1,2,*], Daniel C. M. Palumbo [1,2], Angelo Ricarte [1,2], Avery E. Broderick [3,4,5], Michael D. Johnson [1,2], Neil M. Nagar [6], Priyamvada Natarajan [2,7,8] and José L. Gómez [9]

1. Center for Astrophysics | Harvard & Smithsonian, 60 Garden Street, Cambridge, MA 02138, USA
2. Black Hole Initiative, Harvard University, 20 Garden Street, Cambridge, MA 02138, USA
3. Perimeter Institute for Theoretical Physics, 31 Caroline Street North, Waterloo, ON N2L 2Y5, Canada
4. Department of Physics and Astronomy, University of Waterloo, 200 University Avenue West, Waterloo, ON N2L 3G1, Canada
5. Waterloo Centre for Astrophysics, University of Waterloo, Waterloo, ON N2L 3G1, Canada
6. Astronomy Department, Universidad de Concepción, Casilla 160-C, Concepción 4030000, Chile
7. Department of Astronomy, Yale University, 52 Hillhouse Avenue, New Haven, CT 06511, USA
8. Department of Physics, Yale University, P.O. Box 208121, New Haven, CT 06520, USA
9. Instituto de Astrofísica de Andalucía-CSIC, Glorieta de la Astronomía s/n, E-18008 Granada, Spain
* Correspondence: dpesce@cfa.harvard.edu

Abstract: We present estimates for the number of supermassive black holes (SMBHs) for which the next-generation Event Horizon Telescope (ngEHT) can identify the black hole "shadow", along with estimates for how many black hole masses and spins the ngEHT can expect to constrain using measurements of horizon-resolved emission structure. Building on prior theoretical studies of SMBH accretion flows and analyses carried out by the Event Horizon Telescope (EHT) collaboration, we construct a simple geometric model for the polarized emission structure around a black hole, and we associate parameters of this model with the three physical quantities of interest. We generate a large number of realistic synthetic ngEHT datasets across different assumed source sizes and flux densities, and we estimate the precision with which our defined proxies for physical parameters could be measured from these datasets. Under April weather conditions and using an observing frequency of 230 GHz, we predict that a "Phase 1" ngEHT can potentially measure $\sim$50 black hole masses, $\sim$30 black hole spins, and $\sim$7 black hole shadows across the entire sky.

Keywords: SMBHs; VLBI; ngEHT

Citation: Pesce, D.W.; Palumbo, D.C.M.; Ricarte, A.; Broderick, A.E.; Johnson, M.D.; Nagar, N.M.; Natarajan, P.; Gómez, J.L. Expectations for Horizon-Scale Supermassive Black Hole Population Studies with the ngEHT. *Galaxies* **2022**, *10*, 109. https://doi.org/10.3390/galaxies10060109

Academic Editor: Luigina Feretti

Received: 6 November 2022
Accepted: 25 November 2022
Published: 2 December 2022

Publisher's Note: MDPI stays neutral with regard to jurisdictional claims in published maps and institutional affiliations.

1. Introduction

The Event Horizon Telescope (EHT) observations of the supermassive black holes (SMBHs) in M87 [1–8] and Sgr A* [9–14] are the first in a new era of horizon-scale studies of black holes. The primary observational signature on horizon scales is the black hole "shadow", a ring-like emission structure surrounding a darker central region [15,16]. Simulations of accretion flows around SMBHs generically produce images that exhibit such shadows [5,13], which typically have a size comparable to that of the theoretical curve bounding the locus of impact parameters for photon trajectories that escape the black hole (i.e., the "apparent shape" of the black hole, from Bardeen [17]). A driving motivation for the EHT to pursue observations of M87* and Sgr A* was because these sources were anticipated to have the largest shadow sizes of all black holes on the sky [1].

The next-generation Event Horizon Telescope (ngEHT) will build on the capabilities of the EHT by improving (u, v)-coverage through the addition of more stations to the array, increasing baseline sensitivities by using wider observing bandwidths, and accessing finer angular resolution by observing at higher frequencies [18]. A natural question to ask is whether these improved capabilities will yield access to a larger pool of shadow-resolved

SMBHs. The horizon-scale emission structure around a black hole encodes spacetime properties such as its mass and spin, and the detection of a shadow is a distinct and relatively unambiguous identifier of the observed object's black hole nature. Access to a population of shadow-resolved SMBHs would thus provide an opportunity to make uniquely direct and self-consistent measurements of such spacetime properties, with attendant implications for studies of SMBH formation, growth, and co-evolution with host galaxies.

The suitability of any particular SMBH for shadow-resolving ngEHT observations depends primarily on three properties [19]:

1. the angular size of the SMBH shadow (θ);
2. the total horizon-scale flux density emitted by the source (S_ν); and
3. the optical depth of the emitting material.

The first of the above properties is set primarily by the mass of and distance to the black hole, while the latter two are more complex and depend also on the mass accretion rate and other physical conditions in the accretion flow. However, the detectability of horizon-scale structure from a SMBH does not guarantee the measurability of any particular quantity of interest; additional conditions must be met to ensure that, e.g., a black hole mass can be measured, or that the ring-like structure associated with the black hole shadow can be distinguished from other possible emission morphologies.

In this paper, we provide estimates for the number of SMBHs for which the ngEHT could plausibly make mass, spin, and shadow measurements. In Section 2, we define observational proxies for each of these quantities of interest that can be accessed from the horizon-scale emission structure. Section 3 describes our synthetic data generation procedure and our approach to estimating parameter measurement precision from ngEHT data. Our conditions for the measurability of each proxy are defined in Section 4, where we also report the number of objects expected to satisfy these conditions for each quantity of interest. We summarize and conclude in Section 5. Throughout this paper, we use the results from Pesce et al. [19] as our baseline for how many SMBHs satisfy the above three detection criteria as a function of θ and S_ν.[1]

2. Measurable Proxies for Quantities of Interest

For a given SMBH, the two primary quantities of scientific interest are its mass and spin, neither of which is directly observable by the ngEHT. Instead, analyses of ngEHT observations will need to identify and measure features of the emission structure that serve as proxies for the desired quantities, or else they will need to carry out some form of physical modeling to infer the SMBH mass and/or spin from the ngEHT data. For the proof-of-concept analyses presented in this paper, we pursue the former strategy.

2.1. Proxy for SMBH Shadows

One of the most generic predictions from simulated images of SMBHs is that the observed emission structure on event horizon scales should exhibit a ring-like morphology associated with the black hole shadow (e.g., [5,13]). Though it is possible for other processes to give rise to ring-like emission structures—e.g., the Einstein ring from a bright, compact emitter passing behind the black hole—in such cases the ring-like structure is expected to be transient. For the purposes of this paper, we thus consider the observation of a ring-like emission morphology to be a proxy for verifying the object's black hole nature. If we can determine from ngEHT observations that the emission structure from a particular object is ring-like—i.e., if we can discern the shadow—then we can identify that object as a black hole.

2.2. Proxy for SMBH Masses

The mass of a SMBH sets the physical scale for its associated spacetime metric, and all spacetime-sensitive length scales in the system should thus exhibit a proportionality with the gravitational radius,

$$\theta_g = \frac{GM}{c^2 D},$$

(1)

with M the black hole mass and D its distance from Earth. The most observationally accessible length scale is the overall size of the emission region, which for a ring-like emission structure corresponds to the ring diameter, d. The EHT has demonstrated that black hole mass measurements for both M87* and Sgr A* can be made by calibrating the scaling relationship between d and θ_g using a large number of simulated images of the emission structure [6,12]. In this paper, we thus take d to be a proxy for M^2; i.e., we assume that if d can be measured for a particular SMBH, then M can also be determined.

2.3. Proxy for SMBH Spins

The spin, a, of a SMBH has historically proven to be more difficult to measure than the mass; e.g., the EHT observations of M87* and Sgr A* have not yet yielded strong constraints on the spin of either SMBH [5,13]. There are a number of possible avenues for measuring a from horizon-scale images of SMBH systems (e.g., [23]), but the most observationally accessible of these approaches target the imprint of the SMBH spin on the horizon-scale magnetic field topology, which in turn can be accessed through observations of the linear polarization structure around the ring (e.g., [7,8]). Palumbo et al. [24] have developed a useful decomposition of the polarization structure in terms of a basis that captures the azimuthal behavior of the electric vector position angle (EVPA, i.e., the orientation of the linear polarization around the ring). This decomposition takes the form

$$\beta_m = \frac{1}{S_0} \iint P(r,\phi) e^{-im\phi} r \, dr \, d\phi,$$

(2)

where (r,ϕ) are polar coordinates in the image, $P(r,\phi) = Q(r,\phi) + iU(r,\phi)$ is the complex-valued linear polarization field (with Q and U the standard Stokes intensities), and S_0 is a flux normalization factor. When studying images of M87* from GRMHD simulations, Palumbo et al. [24] found that the "twisted" morphology of the linear polarization pattern, quantified by the (complex-valued) β_2 coefficient, is correlated with the spin of the black hole. It is now believed that this relation arises from a magnetic field geometry that evolves with the black hole spin: black holes with larger spins exhibit more frame dragging, and produce more strongly toroidal magnetic fields than lower-spin black holes [25]. Qiu et al. [26] further explored the connection between polarized image morphology and SMBH spin, finding that the asymmetry (A) of the Stokes I emission, the polarimetric β_1 mode, and the modulus of the polarimetric β_2 mode also encode spin information (though β_2 continues to stand out as the most discriminating measurable parameter). In this paper, we thus take a joint measurement of β_1, β_2, and A to be our proxy for a.

3. Synthetic Data Generation and Fitting Procedure

To determine the region of the (θ, S_ν) parameter space—and thus the number of SMBHs—for which the quantities of interest described in the previous section could be measured by the ngEHT, we carry out a series of model-fitting exercises using synthetic data. We use a model for the SMBH emission structure that captures the salient features relevant for measuring the physical quantities of interest. Per Section 2, these salient features include the diameter and thickness of the emitting ring, as well as the structure of the linear polarization pattern. As our parameterization of the SMBH emission structure, we thus use a polarized "m-ring" model [12,27] convolved with a circular Gaussian blurring kernel. This model is restricted to describing ring-like morphologies, but it can flexibly distribute both the total intensity and the linearly polarized flux about the ring using a relatively small number of parameters. The emission structures produced by this model qualitatively match those expected from both simple analytic treatments (e.g., [28]) as well as numerical GRMHD simulations (e.g., [8,12]).

In our polarized source model, the Stokes I image structure is given by

$$I(r,\phi) = \left[\frac{S_0}{\pi d}\delta\left(r - \frac{d}{2}\right)\sum_{k=-m}^{m}\alpha_k e^{ik\phi}\right] * \left[\frac{4\ln(2)}{\pi W^2}\exp\left(-\frac{4\ln(2)r^2}{W^2}\right)\right], \qquad (3)$$

where $*$ denotes the convolution operation, d is the ring diameter, W is the FWHM ring width, and δ denotes the Dirac delta function. We enforce $\alpha_0 = 1$ so that S_0 is the total flux density, and we also enforce $\alpha_{-k} = \alpha_k^*$ so that the image intensity is real-valued. We define $A = |\alpha_1|$ to be the asymmetry parameter mentioned in Section 2 as potentially relevant for spin constraints.[3] The linear polarization structure is similarly given by

$$P(r,\phi) = \left[\frac{1}{\pi d}\delta\left(r - \frac{d}{2}\right)\sum_{k=-m}^{m}\beta_k e^{ik\phi}\right] * \left[\frac{4\ln(2)}{\pi W^2}\exp\left(-\frac{4\ln(2)r^2}{W^2}\right)\right], \qquad (4)$$

where we now allow both β_{-k} and β_k to be free parameters because P is complex-valued in general.

We generate a number of synthetic SMBH images by gridding the (d, S_0) parameter space, spanning $[0.1, 100]$ μas in d and $[10^{-3}, 1]$ Jy in S_0, with both dimensions uniformly gridded on a logarithmic scale. We set $m = 1$ for the Stokes I emission, with both the real and imaginary parts of α_1 uniformly sampled within $[-0.5, 0.5]$. For the polarized emission we set $m = 2$, with the real and imaginary parts of β_0 and β_{-2} uniformly sampled within $[-0.1, 0.1]$, the real and imaginary parts of β_1 and β_{-1} uniformly sampled within $[-0.05, 0.05]$, and the real and imaginary parts of β_2 uniformly sampled within $[-0.3, 0.3]$. For all synthetic images, we enforce $W = d/3$. Though these choices are not unique, they cover a range of parameter values similar to that seen in the GRMHD simulations developed by the EHT collaboration [5,13]. An example polarized m-ring image generated using these specifications is shown in Figure 1.

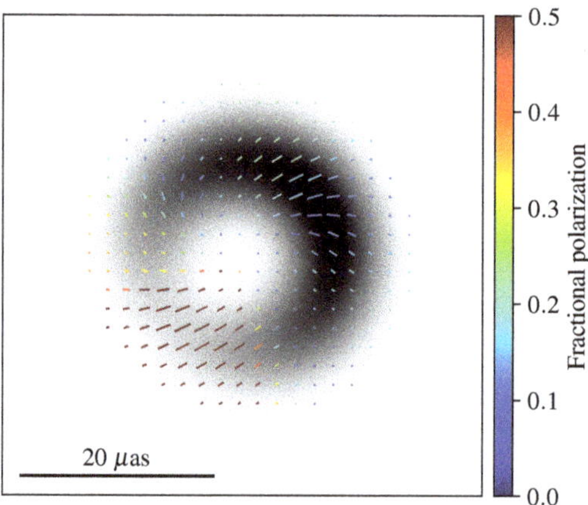

Figure 1. Example polarized source model used for generating the synthetic data described in Section 3. The grayscale image shows the Stokes I emission, while the colored ticks mark the EVPA of the linear polarization structure. The length of each tick is proportional to the intensity of the linear polarization (i.e., $|P|$), while the color of each tick reflects the fractional polarization (i.e., $|P|/I$).

To generate synthetic ngEHT observations corresponding to the synthetic images, we use the ngehtsim[4] package, which expands on the synthetic data generating functionality of the ehtim library [30,31]. We assume the observations are carried out at an observing frequency of 230 GHz and with 8 GHz of bandwidth using the "full" ngEHT Phase 1 array configuration from [32], which consists of the 2022 EHT array plus the OVRO 10.4 m dish, the Haystack 37 m dish, and three 6.1 m dishes located in Baja California (Mexico), Las Campanas Observatory (Chile), and the Canary Islands (Spain). We use historical weather data to determine appropriate system equivalent flux densities at each site following a procedure similar to that in Raymond et al. [33]. To emulate fringe-finding signal-to-noise ratio (SNR) thresholds, we flag any visibilities from baselines that contain a station not participating in at least one other baseline that achieves an SNR of 5 in a 10-s integration time. We add complex station gain corruptions at the level of 10% in amplitude and uniformly sampled within $[0, 2\pi]$ in phase for all stations on every 300-s time interval, to emulate scans, and we assume that the data have been calibrated to remove polarimetric leakage effects.

We generate synthetic datasets across a grid in right ascension and declination, with spacings between grid points of 1 h in right ascension and 10 degrees in declination. To gather information on the performance of the array in different weather conditions and for different black hole structure realizations, we generate 100 instantiations of synthetic data at each grid location. We assume weather conditions typical for the month of April.

For each synthetic dataset, we estimate the precision with which the parameters of a polarized m-ring model fit to the data could be recovered. We compute these estimates using a Fisher matrix approach implemented within the ngEHTforecast[5] package. This approach does not explicitly carry out fits of the model to the data; instead, it assumes that a "good" fit to the data has already been achieved, and it then provides an estimate of the uncertainty in each of the fitted parameters via a second-order expansion of the logarithmic probability density around the best-fit location. We compute parameter precision estimates assuming that the fits have been carried out using complex visibilities as the input data products, with broad priors on the station gain amplitudes and phases at every scan.

4. Results: The Expected Number of Measurable SMBH Masses, Spins, and Shadows

The results of the modeling exercises described in the previous section are summarized in Figure 2, which shows the sky density of sources expected to have measurable masses (top panel), spins (middle panel), and shadows (bottom panel). At each sky location, the plotted density corresponds to an average over 100 instantiations of weather conditions and source structure. Our criteria for determining whether a particular mass, spin, or shadow is deemed "measurable" are as follows:

1. Our condition for whether a SMBH has a measurable mass is that the fractional uncertainty in the measurement of the ring diameter d must be at the level of 20% or lower (i.e., it is measured with a statistical significance $\gtrsim 5\sigma$). Values of (θ, S_ν) for which this condition is satisfied fall to the upper right of the red dashed curve in Figure 3.
2. Our condition for whether a SMBH has a measurable spin is that the uncertainty in the measurement of all spin-relevant parameters (as determined by Qiu et al. [26]; see also Section 2.3) must be at the level of 20% or lower. Specifically, we require the fractional uncertainty in $|\alpha_1|$, $|\beta_1|$, and $|\beta_2|$ and the uncertainty in $\arg(\beta_1)$ and $\arg(\beta_2)$ to all be less than 0.2 (i.e., 20%). Values of (θ, S_ν) for which this condition is satisfied fall to the upper right of the green dashed curve in Figure 3.
3. Our condition for whether a SMBH has a measurable shadow is that the fractional width W/d deviates from unity with an uncertainty of 20% or smaller; i.e., we require that $W < d$ with a statistical significance $\gtrsim 5\sigma$. Values of (θ, S_ν) for which this condition is satisfied fall to the upper right of the blue dashed curve in Figure 3.

Given the above measurability thresholds, we can see from Figure 3 that there is a hierarchy of measurement difficulty with increasing S_ν and θ. The "easiest" quantity to measure is d (and thus the black hole mass), which can be recovered for ~50 sources after integrating over the whole sky. The next most well-constrained quantities are those pertaining to the black hole spin, which we find can be recovered for ∼30 sources. The most difficult quantity to measure is W (and thus the black hole shadow), which can be recovered for ∼7 sources. The measurements are cumulative within this hierarchy: for all sources for which spin is measurable, mass is also measurable; for all sources for which the shadow is measurable, both spin and mass are also measurable.

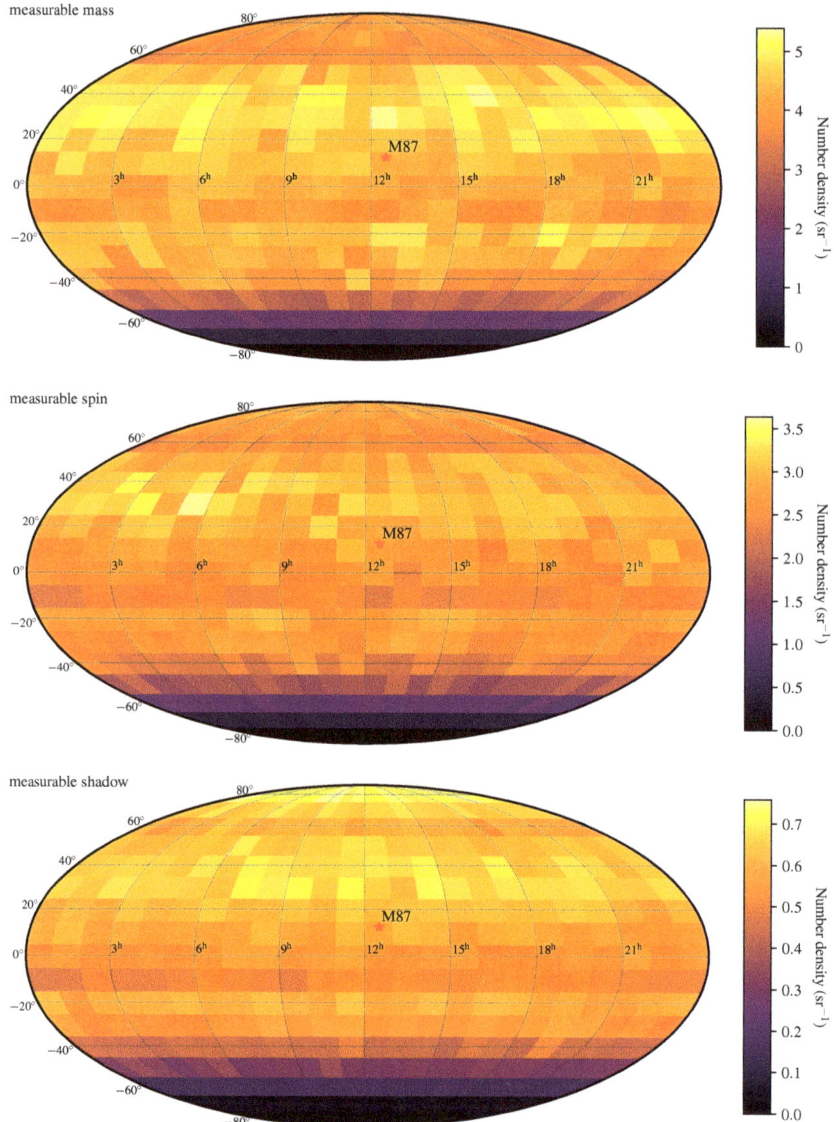

Figure 2. Estimated sky density of SMBHs with measurable masses (top), spins (middle), and shadows (bottom), as a function of right ascension and declination. These estimates have been determined according to the criteria outlined in Section 4, and they assume an underlying distribution of observable SMBHs from Pesce et al. [19]. The stochastic variations seen from pixel to pixel are primarily the result of sampling noise. The location of M87* is marked with a red star.

All three quantities of interest exhibit two regimes of non-measurability in Figure 3. For sources with flux densities below $S_\nu \lesssim 10\,\text{mJy}$, the source is too weak to be detected on most baselines, and there are thus simply insufficient data to enable significant constraints on the parameters of interest. For sources that are stronger than $\sim 10\,\text{mJy}$ but smaller than several μas, there can be many detected data points, but the source is insufficiently resolved to enable significant constraints on morphological parameters. In both regimes, all three quantities of interest exhibit a measurability tradeoff between θ and S_ν. In the second regime, this tradeoff is such that it is possible to make a measurement for sources with smaller θ so long as they have sufficiently larger S_ν (because increasing signal-to-noise ratio permits subtler features to be recovered), while in the first regime the tradeoff is reversed (because compact sources yield more detections—particularly on long baselines—than extended sources).

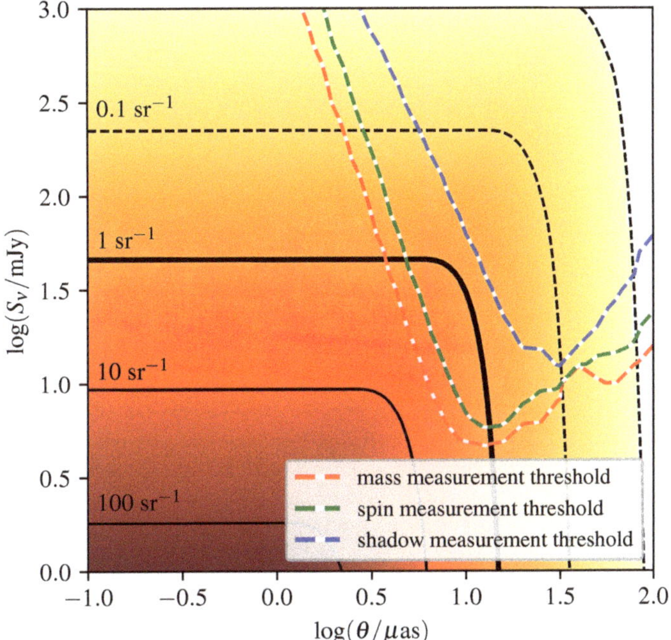

Figure 3. Approximate number density of SMBHs that are expected to satisfy different thresholds of measurability, assuming an observing frequency of 230 GHz. The background colorscale and contours mark the number density (per unit solid angle) of SMBHs that have flux densities greater than S_ν and shadow diameters larger than θ, as a function of S_ν and θ and assuming that sources are distributed isotropically on the sky [19]. The solid contours start with the thick contour indicating a count of 1 and then increase by factors of 10 towards the lower left, while the dashed contours each decrease by a factor of ten towards the upper right. The overplotted colored dashed contours indicate where various parameters of interest could be measurable for different combinations of (θ, S_ν), assuming observations appropriate for the "full" ngEHT Phase 1 array observing at a declination of 10 degrees (i.e., averaged over right ascension). The red dashed contour marks the lower boundary of the region in which black hole mass can be measured, the green dashed contour marks the lower boundary of the region in which black hole spin can be measured, and the blue dashed contour marks the lower boundary of the region in which black hole shadow can be measured.

Figure 2 shows the sky distribution of objects with measurable masses, spins, and shadows, after averaging over weather and source structure instantiations. We find that the distribution is quite uniform, and that accessible objects can be located almost anywhere in the sky; there is no strong dependence on right ascension. The only major gaps in accessibility are for sources having declinations within $\sim$30 degrees of the southern celestial pole, for which the (u, v)-coverage of the array is particularly poor. A modest increase in source density is seen around declinations of $\sim$30–40 degrees, where the (u, v)-coverage of the array is densest. For the shadow measurements, we also see a modest increase in source density around the northern celestial pole; northern polar observations provide the most complete long-baseline coverage, so this bump in density may indicate that the long baselines are the most constraining for the width parameter.

5. Summary and Conclusions

To date, the EHT has observed the horizon-scale emission structure around two SMBHs. The ngEHT aims to improve on the capabilities of the EHT by adding new dishes to the array, increasing the observing bandwidth, and expanding the frequency coverage, all of which will improve the sensitivity and fidelity of reconstructed images.

Motivated by the promise of the ngEHT for population studies of SMBHs, we have identified three scientific quantities of interest that the ngEHT can expect to measure for a number of SMBHs: the black hole mass, the black hole spin, and the black hole shadow. We construct a geometric ring model for the polarized emission structure around a SMBH, and we identify parameters of this model as observable proxies for the scientific quantities of interest. Specifically, we associate the diameter of the ring with measurements of the black hole mass, the thickness of the ring with measurements of the black hole shadow, and the linear polarization structure with measurements of the black hole spin.

Assuming a Phase 1 ngEHT array configuration observing in April conditions at a frequency of 230 GHz, we generate a large number of realistic synthetic observations spanning a range of source structure (i.e., flux density S_ν and angular size θ) and site weather (i.e., opacity and atmospheric temperature) instantiations. For each synthetic dataset, we use a Fisher matrix formalism to estimate the precision with which each of the geometric ring model parameters of interest could be measured. We use the statistics of these measurement precision estimates (across all weather instantiations) to determine the corresponding number of SMBHs on the sky whose properties could be well-constrained as a function of S_ν and θ. We carry out this procedure for synthetic observations covering a grid in right ascension and declination, finding that the sky density of measurable sources (in each parameter of interest) is approximately uniform for declinations above roughly $-60°$.

Associating these measurable parameters with their corresponding physical quantities of interest, we present estimates for the number of SMBHs for which the Phase 1 ngEHT can expect to make measurements of these quantities. Integrating over the whole sky, we find that the Phase 1 ngEHT should be able to measure $\sim$50 black hole masses, $\sim$30 black hole spins, and $\sim$7 black hole shadows. The measurable SMBHs have characteristic observed flux densities of $\sim$30 mJy and angular sizes of $\sim$10 μas; per Pesce et al. [19], we expect the bulk of these SMBHs to lie in the redshift range between $z \approx 0.01$ and $z \approx 0.1$. Our estimate for the number of measurable shadows is consistent with the predictions from Pesce et al. [19].

We note that our detection criteria for mass and spin are likely optimistic. A primary analysis limitation is that our model for the appearance of an SMBH does not include emission that extends much beyond the near-horizon region. Mass estimates for SMBHs of interest to the ngEHT may be complicated by additional image features—such as, e.g., AGN jets (e.g., [34]—that could limit the ability to accurately estimate the ring diameter when it is only marginally resolved or weakly detected. For spin, the situation is even more uncertain: the EHT has already produced tight estimates for the ring parameters β_1, β_2, and A for M87*, but it has not yet claimed a corresponding measurement of the black hole spin [6–8]. A secure association between these ring parameters and spin will require a combination of continued observational and theoretical studies.

Galaxies **2022**, *10*, 109

On the other hand, we have employed a simplified analysis that likely underestimates the number of accessible sources, given any particular set of detection criteria. For instance, the synthetic datasets used in this paper are currently limited to April weather conditions and an observing frequency of 230 GHz; a more comprehensive exploration of year-round weather conditions and the addition of a 345 GHz observing band would likely increase the number of accessible sources. Furthermore, the synthetic datasets generated for the analyses in this paper have assumed an EHT-like calibration procedure; more advanced calibration strategies that can bootstrap phase information across frequency bands (e.g., [35]) are also expected to increase the number of accessible sources. Addressing these shortcomings will be the focus of future work.

Observationally, the most critical next step is to identify a list of credible targets and start surveying them to determine flux densities and compactness for ngEHT followup. Ramakrishnan et al. [36] are compiling a comprehensive sample of all plausible ngEHT AGN targets, which is expected serve as a source catalog for pursuing SMBH population studies with the ngEHT.

Author Contributions: Conceptualization, J.L.G., P.N., D.W.P.; methodology, M.D.J., D.C.M.P., D.W.P., A.R.; software, A.E.B., M.D.J., D.W.P.; writing—original draft preparation, D.W.P.; writing—review and editing, A.E.B., J.L.G., M.D.J., N.M.N., P.N., D.C.M.P., D.W.P., A.R.; visualization, D.W.P.; supervision, J.L.G., P.N. All authors have read and agreed to the published version of the manuscript.

Funding: Support for this work was provided by the NSF through grants AST-1440254, AST-1935980, and AST-2034306, and by the Gordon and Betty Moore Foundation through grant GBMF-10423. This work has been supported in part by the Black Hole Initiative at Harvard University, which is funded by grants from the John Templeton Foundation and the Gordon and Betty Moore Foundation to Harvard University. NN acknowledges funding from TITANs NCN19-058 and Fondecyt 1221421.

Conflicts of Interest: The authors declare no conflict of interest.

Notes

[1] The procedure Pesce et al. [19] used to determine the number of observable SMBHs involves integrating the supermassive black hole mass function (BHMF) to determine how many objects have shadow diameters larger than θ, while also using a semi-analytic spectral energy distribution model and adopting an empirically motivated prescription for the SMBH Eddington ratio distribution function to restrict the objects under consideration to those that have flux densities greater than S_ν and accretion flows that are optically thin. The distribution of sources used in this paper assumes an observing frequency of 230 GHz and a BHMF determined using the stellar mass function from Behroozi et al. [20] scaled according to the relation determined by Kormendy and Ho [21] (i.e., the "upper BHMF" from Pesce et al. [19]).

[2] We note that the spin of a black hole also has an effect on the shadow size, but the impact of spin is small ($\sim$4%; Takahashi [22]) and is dominated by the $\gtrsim$10% systematic uncertainty associated with the unknown accretion flow morphology [6,12].

[3] Note that this definition for A differs from that in Qiu et al. [26], who instead adopt the asymmetry definition used in Medeiros et al. [29].

[4] https://github.com/Smithsonian/ngehtsim, accessed on 5 November 2022.

[5] https://github.com/aeb/ngEHTforecast, accessed on 5 November 2022.

References

1. Akiyama, K. et al. [Event Horizon Telescope Collaboration] First M87 Event Horizon Telescope Results. I. The Shadow of the Supermassive Black Hole. *Astrophys. J. Lett.* **2019**, *875*, L1.
2. Akiyama, K. et al. [Event Horizon Telescope Collaboration] First M87 Event Horizon Telescope Results. II. Array and Instrumentation. *Astrophys. J. Lett.* **2019**, *875*, L2.
3. Akiyama, K. et al. [Event Horizon Telescope Collaboration] First M87 Event Horizon Telescope Results. III. Data Processing and Calibration. *Astrophys. J. Lett.* **2019**, *875*, L3.
4. Akiyama, K. et al. [Event Horizon Telescope Collaboration] First M87 Event Horizon Telescope Results. IV. Imaging the Central Supermassive Black Hole. *Astrophys. J. Lett.* **2019**, *875*, L4.
5. Akiyama, K. et al. [Event Horizon Telescope Collaboration] First M87 Event Horizon Telescope Results. V. Physical Origin of the Asymmetric Ring. *Astrophys. J. Lett.* **2019**, *875*, L5.
6. Akiyama, K. et al. [Event Horizon Telescope Collaboration] First M87 Event Horizon Telescope Results. VI. The Shadow and Mass of the Central Black Hole. *Astrophys. J. Lett.* **2019**, *875*, L6.

7. Akiyama, K. et al. [Event Horizon Telescope Collaboration] First M87 Event Horizon Telescope Results. VII. Polarization of the Ring. *Astrophys. J. Lett.* **2021**, *910*, L12.

8. Event Horizon Telescope Collaboration.; Akiyama, K.; Algaba, J.C.; Alberdi, A.; Alef, W.; Anantua, R.; Asada, K.; Azulay, R.; Baczko, A.K.; Ball, D.; et al. First M87 Event Horizon Telescope Results. VIII. Magnetic Field Structure near The Event Horizon. *Astrophys. J. Lett.* **2021**, *910*, L13.

9. Akiyama, K. et al. [Event Horizon Telescope Collaboration] First Sagittarius A* Event Horizon Telescope Results. I. The Shadow of the Supermassive Black Hole in the Center of the Milky Way. *Astrophys. J. Lett.* **2022**, *930*, L12. [CrossRef]

10. Akiyama, K. et al. [Event Horizon Telescope Collaboration] First Sagittarius A* Event Horizon Telescope Results. II. EHT and Multiwavelength Observations, Data Processing, and Calibration. *Astrophys. J. Lett.* **2022**, *930*, L13. [CrossRef]

11. Akiyama, K. et al. [Event Horizon Telescope Collaboration] First Sagittarius A* Event Horizon Telescope Results. III. Imaging of the Galactic Center Supermassive Black Hole. *Astrophys. J. Lett.* **2022**, *930*, L14. [CrossRef]

12. Akiyama, K. et al. [Event Horizon Telescope Collaboration] First Sagittarius A* Event Horizon Telescope Results. IV. Variability, Morphology, and Black Hole Mass. *Astrophys. J. Lett.* **2022**, *930*, L15. [CrossRef]

13. Akiyama, K. et al. [Event Horizon Telescope Collaboration] First Sagittarius A* Event Horizon Telescope Results. V. Testing Astrophysical Models of the Galactic Center Black Hole. *Astrophys. J. Lett.* **2022**, *930*, L16. [CrossRef]

14. Akiyama, K. et al. [Event Horizon Telescope Collaboration] First Sagittarius A* Event Horizon Telescope Results. VI. Testing the Black Hole Metric. *Astrophys. J. Lett.* **2022**, *930*, L17. [CrossRef]

15. Falcke, H.; Melia, F.; Agol, E. Viewing the Shadow of the Black Hole at the Galactic Center. *Astrophys. J. Lett.* **2000**, *528*, L13–L16.

16. Narayan, R.; Johnson, M.D.; Gammie, C.F. The Shadow of a Spherically Accreting Black Hole. *Astrophys. J. Lett.* **2019**, *885*, L33.

17. Bardeen, J.M. Timelike and null geodesics in the Kerr metric. In *Proceedings of the Black Holes (Les Astres Occlus)*; Gordon and Breach: New York, NY, USA, 1973; pp. 215–239.

18. Doeleman, S.; Blackburn, L.; Dexter, J.; Gomez, J.L.; Johnson, M.D.; Palumbo, D.C.; Weintroub, J.; Farah, J.R.; Fish, V.; Loinard, L.; et al. Studying Black Holes on Horizon Scales with VLBI Ground Arrays. *Bull. Am. Astron. Soc.* **2019**, *51*, 256.

19. Pesce, D.W.; Palumbo, D.C.M.; Narayan, R.; Blackburn, L.; Doeleman, S.S.; Johnson, M.D.; Ma, C.P.; Nagar, N.M.; Natarajan, P.; Ricarte, A. Toward Determining the Number of Observable Supermassive Black Hole Shadows. *Astrophys. J.* **2021**, *923*, 260.

20. Behroozi, P.; Wechsler, R.H.; Hearin, A.P.; Conroy, C. UNIVERSEMACHINE: The correlation between galaxy growth and dark matter halo assembly from z = 0-10. *Mon. Not. R. Astron. Soc.* **2019**, *488*, 3143–3194.

21. Kormendy, J.; Ho, L.C. Coevolution (Or Not) of Supermassive Black Holes and Host Galaxies. *Annu. Rev. Astron. Astrophys.* **2013**, *51*, 511–653.

22. Takahashi, R. Shapes and Positions of Black Hole Shadows in Accretion Disks and Spin Parameters of Black Holes. *Astrophys. J.* **2004**, *611*, 996–1004.

23. Ricarte, A.; Tiede, P.; Emami, R.; Tamar, A.; Natarajan, P. The ngEHT's Role in Measuring Supermassive Black Hole Spins. *arXiv* **2022**, arXiv:2211.03910.

24. Palumbo, D.C.M.; Wong, G.N.; Prather, B.S. Discriminating Accretion States via Rotational Symmetry in Simulated Polarimetric Images of M87. *Astrophys. J.* **2020**, *894*, 156.

25. Emami, R.; Ricarte, A.; Wong, G.N.; Palumbo, D.; Chang, D.; Doeleman, S.S.; Broaderick, A.; Narayan, R.; Weintroub, J.; Wielgus, M.; et al. Unraveling Twisty Linear Polarization Morphologies in Black Hole Images. *arXiv* **2022**, arXiv:2210.01218.

26. Qiu, R.; Ricarte, A.; Narayan, R.; Wong, G.N.; Chael, A.; Palumbo, D.C.M. Using Machine Learning to Link Black Hole Accretion Flows with Spatially Resolved Polarimetric Observables. 2022, *in preparation*.

27. Johnson, M.D.; Lupsasca, A.; Strominger, A.; Wong, G.N.; Hadar, S.; Kapec, D.; Narayan, R.; Chael, A.; Gammie, C.F.; Galison, P.; et al. Universal interferometric signatures of a black hole's photon ring. *Sci. Adv.* **2020**, *6*, eaaz1310.

28. Gelles, Z.; Himwich, E.; Johnson, M.D.; Palumbo, D.C.M. Polarized image of equatorial emission in the Kerr geometry. *Phys. Rev. D* **2021**, *104*, 044060.

29. Medeiros, L.; Chan, C.K.; Narayan, R.; Özel, F.; Psaltis, D. Brightness Asymmetry of Black Hole Images as a Probe of Observer Inclination. *Astrophys. J.* **2022**, *924*, 46,

30. Chael, A.A.; Johnson, M.D.; Narayan, R.; Doeleman, S.S.; Wardle, J.F.C.; Bouman, K.L. High-resolution Linear Polarimetric Imaging for the Event Horizon Telescope. *Astrophys. J.* **2016**, *829*, 11,

31. Chael, A.A.; Johnson, M.D.; Bouman, K.L.; Blackburn, L.L.; Akiyama, K.; Narayan, R. Interferometric Imaging Directly with Closure Phases and Closure Amplitudes. *Astrophys. J.* **2018**, *857*, 23.

32. Doeleman, S.S. et al. [Event Horizon Telescope Collaboration] Reference Array and Design Consideration for the next-generation Event Horizon Telescope. 2022. *in preparation*.

33. Raymond, A.W.; Palumbo, D.; Paine, S.N.; Blackburn, L.; Córdova Rosado, R.; Doeleman, S.S.; Farah, J.R.; Johnson, M.D.; Roelofs, F.; Tilanus, R.P.J.; et al. Evaluation of New Submillimeter VLBI Sites for the Event Horizon Telescope. *Astrophys. J.* **2021**, *253*, 5.

34. Janssen, M.; Falcke, H.; Kadler, M.; Ros, E.; Wielgus, M.; Akiyama, K.; Baloković, M.; Blackburn, L.; Bouman, K.L.; Chael, A.; et al. Event Horizon Telescope observations of the jet launching and collimation in Centaurus A. *Nat. Astron.* **2021**, *5*, 1017–1028.

35. Rioja, M.J.; Dodson, R. Precise radio astrometry and new developments for the next-generation of instruments. *Astron. Astrophys. Rev.* **2020**, *28*, 6.

36. Ramakrishnan, V. et al. [Event Horizon Telescope Collaboration] ETHER. 2022. *in preparation*.

Article

Multi-Wavelength and Multi-Messenger Studies Using the Next-Generation Event Horizon Telescope

Rocco Lico [1,2,*], Svetlana G. Jorstad [3,4,*], Alan P. Marscher [3], Jose L. Gómez [1], Ioannis Liodakis [5], Rohan Dahale [1], Antxon Alberdi [1], Roman Gold [6], Efthalia Traianou [1], Teresa Toscano [1] and Marianna Foschi [1]

[1] Instituto de Astrofísica de Andalucía-CSIC, Glorieta de la Astronomía s/n, 18008 Granada, Spain
[2] INAF Istituto di Radioastronomia, Via Gobetti 101, 40129 Bologna, Italy
[3] Institute for Astrophysical Research, Boston University, 725 Commonwealth Avenue, Boston, MA 02215, USA
[4] Astronomical Institute, St. Petersburg State University, Universitetskij Pr. 28, Petrodvorets, St. Petersburg 198504, Russia
[5] Finnish Centre for Astronomy with ESO, University of Turku, FI-20014 Turku, Finland
[6] CP3-Origins, University of Southern Denmark, Campusvej 55, DK-5230 Odense M, Denmark
[*] Correspondence: rlico@iaa.es (R.L.); jorstad@bu.edu (S.G.J.)

Abstract: The next-generation Event Horizon Telescope (ngEHT) will provide us with the best opportunity to investigate supermassive black holes (SMBHs) at the highest possible resolution and sensitivity. With respect to the existing Event Horizon Telescope (EHT) array, the ngEHT will provide increased sensitivity and uv-coverage (with the addition of new stations), wider frequency coverage (from 86 GHz to 345 GHz and higher), finer resolution ($<$15 micro-arcseconds), and better monitoring capabilities. The ngEHT will offer a unique opportunity to deeply investigate the physics around SMBHs, such as the disk-jet connection, the mechanisms responsible for high-energy photon and neutrino events, and the role of magnetic fields in shaping relativistic jets, as well as the nature of binary SMBH systems. In this white paper we describe some ngEHT science cases in the context of multi-wavelength studies and synergies.

Keywords: very long baseline interferometry (VLBI); supermassive black holes; active galactic nuclei; multi-wavelength studies; relativistic jets

Citation: Lico, R.; Jorstad, S.G.; Marscher, A.P.; Gómez, J.L.; Liodakis, I.; Dahale, R.; Alberdi, A.; Gold, R.; Traianou, E.; Toscano, T.; et al. Multi-Wavelength and Multi-Messenger Studies Using the Next-Generation Event Horizon Telescope. *Galaxies* **2023**, *11*, 17. https://doi.org/10.3390/galaxies11010017

Academic Editor: Massimiliano De Pasquale

Received: 15 November 2022
Revised: 10 January 2023
Accepted: 12 January 2023
Published: 14 January 2023

1. Introduction

Cosmic systems powered by accretion onto supermassive black holes (SMBHs) are the most intense sources of energy in the universe. The tremendous power they generate manifests in a variety of forms: electromagnetic waves from radio to γ-ray wavelengths, ultra-energetic particles, strong magnetic fields, and plasma jets propelled outwards at near-light speeds. There is now firm evidence that SMBHs lie at the hearts of nearly all galaxies. The first direct measurements of the sizes and masses of SMBHs have been obtained with the Event Horizon Telescope (EHT) at 230 GHz for the radio galaxy, M87, ($M_{BH} = 6.5 \pm 0.7 \times 10^9\,M_{Sun}$, [1]) and the Milky Way black hole, SgrA*, ($M_{BH} \approx 4 \times 10^6\,M_{Sun}$, [2]). These measurements show consistency with the predictions of general relativity spanning over three orders of magnitude in central mass. Systems where matter falls into the black hole at a high rate create the phenomenon of active galactic nuclei (AGNs). The masses of black holes in AGNs range from less than 1 million to roughly 10 billion solar masses (e.g., [3,4]). The luminosity of the jets can be as high as $\sim 10^{15}$ times that of the sun.

AGNs with powerful jets are rare in the universe, so most lie at distances of hundreds of millions or billions of light-years from the Earth. Much of their light comes from regions of a similar size as our solar system. Their extremely compact nature, combined with the great distances involved, causes these regions to appear extremely small in the sky; therefore, observations designed to study AGNs require telescopes with extraordinary resolving powers. Furthermore, the extremely luminous emission, and thus the physical conditions that create it, are highly variable with respect to time. Due to these two factors, instruments that probe the nature of AGNs and their jets must provide ultra-fine resolution imaging, as well as the ability to monitor rapid changes in the images.

Such opportunities and capabilities will be offered in the near future by the next-generation Event Horizon Telescope (ngEHT). The main goal of the ngEHT is to upgrade and enhance the current capabilities of the EHT array by adding more than 10 new stations, improving the image dynamic range by $\sim$2 orders of magnitude, enabling simultaneous multi-frequency observations up to 345 GHz and possibly higher frequencies, including polarization capabilities at all frequencies, and quadrupling the observing bandwidth [5]. The ngEHT will have an angular resolution of $\sim$10–15 micro-arcseconds (μas) and reach linear scales that are a fraction of a light-year. Moreover, the development of new super-resolution algorithms (e.g., [6–8]) will allow the smallest features resolved by the ngEHT to be further improved by a factor of up to $\sim$3. Observations at 86 GHz will also be included, which are expected to be important for understanding the connection between the photon ring and inner jet. The ngEHT project development and implementation will consist of two phases. The first phase is based on the instrument design and site selection, and it is expected to be concluded by the end of 2023. The second phase will consist of the actual construction and commissioning of the ngEHT array, which is expected to be fully operational by 2030, with a gradual deployment in the period 2026–2030.

2. Science Cases

The EHT has already demonstrated the power of multi-wavelength (MWL) synergies for characterizing the underlying physics and emission mechanisms in AGNs. This was by means of an intensive multi-wavelength observation campaign of M87 conducted in 2017 with 19 different observing facilities (Figure 1, EHT MWL Science Working Group et al. [9]). In this section, we describe some potential science cases for the ngEHT in the context of MWL studies.

2.1. Disk-Jet Connection

The ngEHT will allow us to determine a jet's profile starting at its origin in the vicinity of an SMBH, the location and its effect on the jet and length of the jet flow's acceleration and collimation zone (ACZ), and the location of the Bondi radius (inside of which the SMBH's gravitational influence is important). This will provide crucial insights toward understanding the connection and interplay between the jet, the accretion disk, and the black hole. Currently, the jet-disk connection has only been reported for two radio galaxies: 3C 120 [10–12] and 3C 111 [13,14]. Both these radio galaxies possess relativistic jets with apparent speeds of $\sim$5c [15]. They exhibit X-ray properties similar to Seyfert galaxies, which implies that their X-ray emissions are produced in the accretion disk (plus its "corona" of hot electrons) in the vicinity of the SMBH. Monitoring with the Very Long Baseline Array (VLBA) at 43 GHz and X-ray satellites (*RXTE, Swift, and Suzaku*), a connection was established between X-ray flux variations and the emergence of new superluminal emission features ("knots") in the jet: appearances of new knots are preceded by dips in the X-ray luminosity. This X-ray/radio connection conclusively demonstrates a link between changes in accretion disk structure and powerful ejection events. Moreover, there is a similarity with the behavior of X-ray binary systems ("microquasars"), allowing comparisons that are important for generalizing insights obtained regarding the origin of relativistic jets in both types of systems [16].

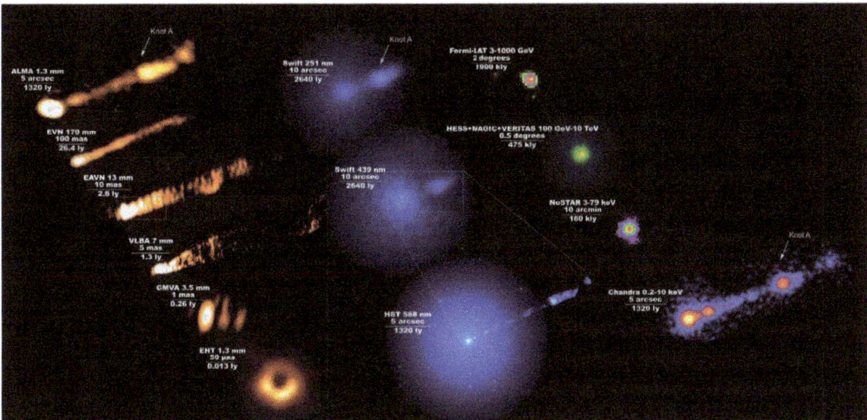

Figure 1. Composite image of M87 as seen across the entire electromagnetic spectrum during the 2017 EHT observation campaign, adapted from [9] under the terms of the CC BY 3.0 licence. For each image, the instrument, the observing wavelength, and the scale are shown next to it. The frequency–wavelength scale at the bottom indicates the location of each instrument in the electromagnetic spectrum. Image credit: the EHT Multi-wavelength Science Working Group; the EHT Collaboration; ALMA (ESO/NAOJ/NRAO); the EVN; the EAVN Collaboration; VLBA (NRAO); the GMVA; the Hubble Space Telescope; the Neil Gehrels Swift Observatory; the Chandra X-ray Observatory; the Nuclear Spectroscopic Telescope Array; the Fermi-LAT Collaboration; the MAGIC collaboration; the VERITAS collaboration; NASA and ESA.

The innermost jet profile and its evolution with respect to distance from the SMBH are key factors in understanding a jet's collimation and acceleration. Recently, a number of studies have been devoted to this topic. Kovalev et al. [17] performed a search for transitions in the jet shape in a sample of radio-loud AGNs (367 sources) at 15 and 1.4 GHz. The authors found that, for nearby AGNs (redshift $z < 0.07$), 10 out of 29 jets exhibited a transition from a parabolic to a conical shape. They concluded that the transition in geometry is perhaps a common feature of AGNs jets, but they can be observed only when sufficient linear resolution is achieved. The authors also indicated that a jet transition occurs when the particle kinetic energy becomes equal to the energy of the magnetic field, while the Bondi radius position is governed by ambient medium pressure and may not coincide with the transition break. Casadio et al. [18] have analyzed stacked VLBI images at 86 and 43 GHz of BL Lacertae (BL Lac, $z = 0.069$). They found that its jet expanded with a conical geometry, with a higher expansion rate between $\sim$5 and 10 pc (de-projected) from the BH. They proposed that the transition in the profile is connected with a change in the external pressure at the location of the Bondi radius ($\sim 3.3 \times 10^5$ R$_g$).

The capabilities of the high-resolution ngEHT will allow direct imaging of these regions, tracing the jet down to the expected accretion disk and the areas where the jet is ejected and collimated in several AGNs. Such images and the studies they allow in combination with MWL observations, will provide decisive information for constructing and testing theoretical models.

2.2. Nature of the VLBI Core

VLBI images of blazar jets, the most extreme objects in the family of AGNs, are dominated by a compact feature known as the "core". The nature of the core remains an open question, mostly because of optical depth effects and the limited angular resolution of VLBI arrays. At wavelengths longer than $\sim$7 mm, the core is optically thick and essentially unresolved, probably representing the area in the jet where the optical depth is $\sim$1. At 0.8–3 mm, where the resolution is higher and the core is optically thin, it should represent a physical structure in the jet's acceleration/collimation zone, such as a standing shock [19], a kink in the jet flow, a site of magnetic reconnection (Figure 2), or some other phenomenon. The EHT Collaboration imaged the core region of the blazar 3C 279, finding it to be dynamic with a puzzling multi-component structure [20]. The ngEHT will provide the data needed to make MWL movies to follow the motions, and perhaps the formation, of the components. This will lead to an understanding of the core, the brightest feature seen in relativistic jets, and how it is connected to the SMBH and its accretion disk.

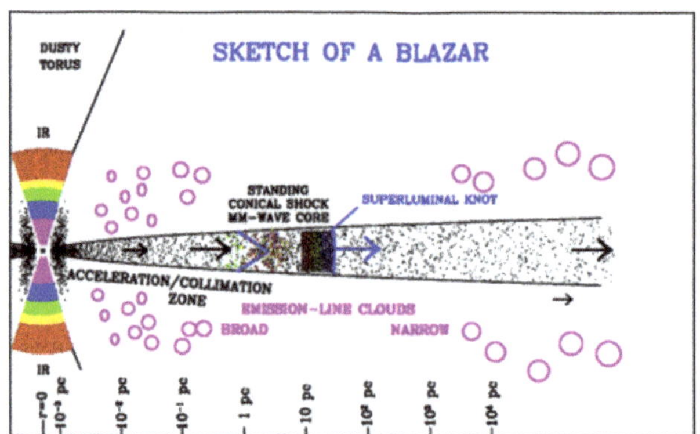

Figure 2. Sketch of the innermost jet region in blazars (image credit: A.Marscher) [21].

A particularly important diagnostic of the nature of the core is its polarization. For example, a standing conical shock is expected to display a radial linear polarization pattern for viewing angles very close to the jet axis [22], a pattern that can be distinguished by the ngEHT. A kink can develop from an instability in a helical field, which has the signature of a rotation measure gradient across the jet (e.g., [23]) that can be detected at the resolution of the ngEHT. Magnetic reconnection can occur where a jet's magnetic field is highly turbulent, with low polarization and chaotic position angles of polarization across the jet. They can also originate where the magnetic field lines form loops that are stretched parallel to the jet, in which case the linear polarization vectors should lie transversely to the jet.

2.3. Spine-Sheath Structure

High-frequency synchrotron peaked (HSP) blazars, such as Mrk 421 and Mrk 501, are associated with a long-standing problem: how to reconcile extreme luminosities of tera-electronvolt (TeV) gamma rays, which require high relativistic Doppler beaming factors, with low apparent speeds observed in microwave VLBI images (known as the "Doppler crisis"; e.g., [15,24]). One of the most favored models to resolve the Doppler crisis involves the presence of a transverse structure in the jet, with an extremely fast (Lorentz factor > 20) inner spine surrounded by a slower outer sheath (e.g., [25]). This produces distinctive observable signatures in VLBI images, such as limb brightening, especially in linearly polarized intensities, with electric-vector position angles (EVPAs) that are transverse to the jet. All of these signatures have already been observed in the jets of both Mrk 421 and Mrk 501 (e.g., [26–28]), as well as in other blazars (e.g., [23,29,30]). However, as it was previously mentioned, very-high angular resolutions and sensitivities are needed to resolve the transverse structure in the innermost jet regions; these will become routinely available with the ngEHT. In fact, for the case of Centaurus A, EHT observations at 228 GHz have revealed a highly collimated, asymmeytrically edge-brightened jet together with a fainter counterjet [31]. In addition, a spine-sheath structure was recently detected in the bright quasar 3C 273 with the fine resolution of space-VLBI *RadioAstron* observations at wavelengths of 18 and 6 cm [32]. At 18 cm, the jet exhibited limb-brightening that had not previously been seen in this source. At 6 cm, an emission from the central stream of plasma was imaged, with a spatial distribution that was complementary to the limb-brightened emission, which indicated the presence of a spine. Regular monitoring with high resolution and sensitivity is needed to determine the jet stratification structure and its evolution. Such observations do not yet exist, but will be possible with the ngEHT, which will allow measurements of apparent speeds across the jets of several HSP blazars.

2.4. Supermassive Binary Black Holes

Binary SMBH systems represent a natural product of hierarchical galaxy formation and merger events. Currently, one of the best candidates for hosting a binary SMBH system is OJ 287, a bright blazar showing quasi-periodic variability patterns in its optical light-curve that could be explained as the result of the secondary SMBH impacting the accretion disk of a primary one [33,34]. However, alternative scenarios to the binary SMBH system can also explain the variability pattern (e.g., [35]).

Moreover, according to current theoretical models, when the system reaches an orbital separation of 0.1–10 parsec the dynamical processes that drive the coalescence of the two SMBHs cease to be effective and inhibit the merging process. This is known as "the final parsec problem". The ngEHT, with a resolution finer than $\sim$15 μas, will represent a unique opportunity to detect and spatially resolve from tens to hundreds of sub-parsec binary SMBH systems (see Pesce et al. [36] for more details about the statistics and the expected detections). Spatially resolving such binary SMBH systems would provide the ultimate proof of their existence. With its monitoring capabilities, an ngEHT will allow us to determine and characterize the orbital parameters of such tight binary systems, whose orbital periods are expected to range from months to a few years, and fully resolve their orbital motion. This is currently being investigated in a set of numerical simulations by using the Black Hole Accretion Code (BHAC, [37]), see Figure 3.

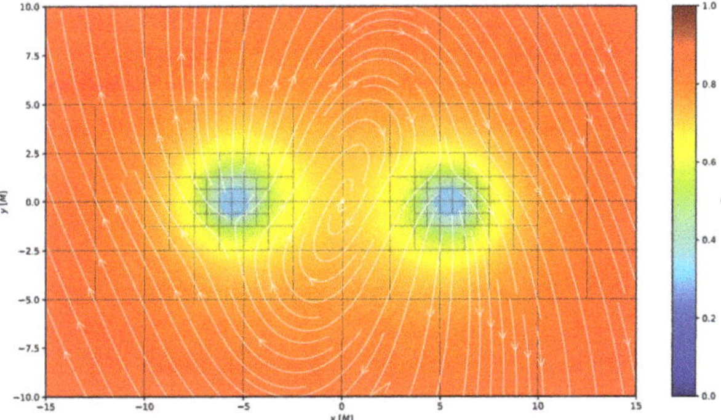

Figure 3. Equatorial slice of a binary SMBH system in the corotating frame interpolated on the Black Hole Accretion Code (BHAC) adaptive mesh grid. The colorscale shows the spacetime lapse function and the arrows show the shift vector.

Ongoing and upcoming large surveys in the optical to X-ray bands (e.g., the European-Chinese X-ray mission Einstein Probe) will provide prime lists of new binary SMBH candidates, among which the best-suited ones can be followed up with the ngEHT to check the source association and measure the orbital parameters. Moreover, the ngEHT high-resolution observations would also represent an important resource for future gravitational wave detectors, both ground-based ones similar to the pulsar timing arrays (PTAs) and space-based ones similar to the Laser Interferometer Space Antenna (LISA), which is sensitive to the low-frequency band of the gravitational wave spectrum expected from binary supermassive black holes.

2.5. High-Energy and Neutrino Events

Blazars are very efficient cosmic accelerators and have been proposed as a possible source of several past high-energy neutrino events, although the production mechanisms and the precise region(s) where they form are still unknown [38]. The identification of the astrophysical counterparts of high-energy neutrinos is not straightforward (see e.g., [39]), and, in this context, the ngEHT could be helpful in two ways. On one hand, the high-resolution mm-wavelength ngEHT observations will allow us to identify and constrain the region(s) responsible for neutrino production by tracing the variations and delay between the radio emission from the innermost jet regions and the neutrino detection event. On the other hand, owing to its monitoring and observing agility capabilities, an ngEHT will help to more accurately localize neutrino emission sources by triggering prompt observations when a detection event occurs (see Y. Y. Kovalev et al., this issue of *Galaxies*).

The ngEHT will also provide essential details about the innermost regions of blazar jets, allowing us to disentangle the different regions where visible X-ray and γ-ray emissions originate (e.g., near the SMBH, in the jet's acceleration zone, in a standing shock, in a moving knot, in a kink, in regions of magnetic reconnection, etc.), and to better understand the mechanisms and physical processes responsible for the observed emission. The time-domain imaging capabilities of the ngEHT will allow events to be matched in the jet with outbursts of high-energy photons with high accuracy. Moreover, multi-epoch and multi-frequency ngEHT polarization observations will allow investigations into the possible connection of magnetic field variations to the high-energy emission events (e.g., [40]). This can be achieved by determining the time evolution of fundamental polarization parameters, such as the fractional polarization, rotation of the polarization plane, and Faraday rotation measure.

2.6. Polarization and Magnetic Field Studies

The ngEHT will allow the characterization of a polarized radio emission in the near vicinity of the SMBH and constrain the magnetic field geometry, the plasma temperature, and the underlying particle density. This will determine whether the magnetic field is helical, as predicted by most theories, turbulent, or compressed by shocks (e.g., [41–43]). The presence of an ionized medium with a magnetic field causes the direction of linear polarization to rotate as radio waves pass through it; this is also known as Faraday rotation. MWL observations with the ngEHT can measure this effect to explore the medium both inside and surrounding the jet near its vertex. This can determine whether the jet is confined by a wind, an accretion flow, or its own magnetic field. Moreover, with the ngEHT it will be possible to measure the velocity field at the base of a jet, allowing us to test the different magneto-hydrodynamic jet formation models on scales of hundreds to thousands of Schwarzschild radii, and to conclusively determine whether the jets are powered by the extraction of rotational energy from the spinning black hole (Blandford–Znajek process, [44]) or from the accretion flow (Blandford–Payne process, [45]).

Synergies with current (Imaging X-ray Polarimetry Explorer—IXPE [46]) and future (e.g., e-XTP [47], COSI [48]) high-energy polarimeters could also prove invaluable. The combination of broadband radio polarization from the ngEHT and high-energy polarization observations will allow us to distinguish between shock and magnetic reconnection models for particle energization in the jets (e.g., [19,49]), as well as accurately pinpoint the particle acceleration regions. This could have significant implications for models of jet propagation as well as the nature of the blazar core (see Section 2.2). It will also bring a unique view of the spatial evolution of the magnetic field configuration on diverse scales from the vicinity of the black hole to pc and kpc away.

3. Technical Requirements

In order to measure Faraday rotation and gradients in the magnetic field of the most compact regions of jets, and other μas-scale emission regions, simultaneous multi-frequency ngEHT polarization observations are required (see A. Ricarte et al., this issue of *Galaxies*). This can be implemented in several ways: (1) by using antennas whose frequency can be switched quickly, as is the case, for example, for a VLBA; (2) by combining ngEHT observations with other VLBI arrays (e.g., VLBA or next-generation VLA, at 43 and 86 GHz, respectively), or with other future space observatories (e.g., Millimetron); (3) by splitting the ngEHT into sub-arrays operating at different frequencies.

To achieve the science goals presented in this paper it will be essential to access the innermost jet regions, with typical 230 GHz flux densities ranging from a fraction to a few tens of mJy. For this reason, a sub-mJy sensitivity will be an essential requirement for the ngEHT, which is achievable thanks to next-generation receivers and the addition of new telescopes to the array. More details about the sensitivity requirements will be presented in M. Johnson et al. in preparation. In order to unambiguously connect MWL flaring and neutrino emission events to changes in the jets, monitoring capabilities are required for the ngEHT at a cadence shorter than the lifetime of the above-mentioned multi-messenger events. This could be possible during 1–2 week campaigns in the fortunate case that we are indeed observing a source with such an event. Either year-long or multiple weeks-long sessions per year would produce a high chance of success. Such multi-epoch ngEHT observations will need to be coordinated with other existing facilities, as well as with the new generation of instruments, such as the Cherenkov Telescope Array (CTA), LISA, and Athena. Moreover, the Russian-led next generation millimeter-band Space VLBI mission Millimetron, when launched, will provide a space arm to the ngEHT. This will enhance the resolution of the ngEHT to provide both space VLBI imaging, although with somewhat limited uv-coverage, and uv-domain studies of photon rings.

Author Contributions: Conceptualization and writing—original draft preparation, R.L., S.G.J. and A.P.M.; writing—review and editing, all authors. All authors have read and agreed to the published version of the manuscript.

Funding: R.L., J.L.G., and R.D. acknowledge financial support from the State Agency for Research of the Spanish MCIU through the "Center of Excellence Severo Ochoa" award for the Instituto de Astrofísica de Andalucía (SEV-2017-0709), from the Spanish Ministerio de Economía y Competitividad and Ministerio de Ciencia e Innovación (grants AYA2016-80889-P, PID2019-108995GB-C21), the Consejería de Economía, Conocimiento, Empresas y Universidad of the Junta de Andalucía (grant P18-FR-1769), the Consejo Superior de Investigaciones Científicas (grant 2019AEP112). A.P.M. and S.G.J. acknowledge financial support from NASA Fermi Guest Investigator grants 80NSSC20K1567 and 80NSSC22K1571.

Data Availability Statement: Not applicable.

References

1. Akiyama, K.; et al. [Event Horizon Telescope Collaboration]. First M87 Event Horizon Telescope Results. I. The Shadow of the Supermassive Black Hole. *Astrophys. J. Lett.* **2019**, *875*, L1.
2. Akiyama, K.; et al. [Event Horizon Telescope Collaboration]. First Sagittarius A* Event Horizon Telescope Results. I. The Shadow of the Supermassive Black Hole in the Center of the Milky Way. *Astrophys. J. Lett.* **2022**, *930*, L12. [CrossRef]
3. Woo, J.H.; Urry, C.M. Active Galactic Nucleus Black Hole Masses and Bolometric Luminosities. *Astrophys. J.* **2002**, *579*, 530–544.
4. Ghez, A.M.; Duchêne, G.; Matthews, K.; Hornstein, S.D.; Tanner, A.; Larkin, J.; Morris, M.; Becklin, E.E.; Salim, S.; Kremenek, T.; et al. The First Measurement of Spectral Lines in a Short-Period Star Bound to the Galaxy's Central Black Hole: A Paradox of Youth. *Astrophys. J. Lett.* **2003**, *586*, L127–L131.
5. Doeleman, S.; Blackburn, L.; Dexter, J.; Gomez, J.L.; Johnson, M.D.; Palumbo, D.C.; Weintroub, J.; Farah, J.R.; Fish, V.; Loinard, L.; et al. Studying Black Holes on Horizon Scales with VLBI Ground Arrays. *arXiv* **2019**, arXiv:1909.01411.
6. Akiyama, K.; Kuramochi, K.; Ikeda, S.; Fish, V.L.; Tazaki, F.; Honma, M.; Doeleman, S.S.; Broderick, A.E.; Dexter, J.; Mościbrodzka, M.; et al. Imaging the Schwarzschild-radius-scale Structure of M87 with the Event Horizon Telescope Using Sparse Modeling. *Astrophys. J.* **2017**, *838*, 1.
7. Chael, A.A.; Johnson, M.D.; Bouman, K.L.; Blackburn, L.L.; Akiyama, K.; Narayan, R. Interferometric Imaging Directly with Closure Phases and Closure Amplitudes. *Astrophys. J.* **2018**, *857*, 23.
8. Broderick, A.E.; Pesce, D.W.; Tiede, P.; Pu, H.Y.; Gold, R. Hybrid Very Long Baseline Interferometry Imaging and Modeling with THEMIS. *Astrophys. J.* **2020**, *898*, 9.
9. Algaba, J.C.; et al. [EHT MWL Science Working Group]. Broadband Multi-wavelength Properties of M87 during the 2017 Event Horizon Telescope Campaign. *Astrophys. J. Lett.* **2021**, *911*, L11.
10. Marscher, A.P.; Jorstad, S.G.; Gómez, J.L.; Aller, M.F.; Teräsranta, H.; Lister, M.L.; Stirling, A.M. Observational evidence for the accretion-disk origin for a radio jet in an active galaxy. *Nature* **2002**, *417*, 625–627. [CrossRef] [PubMed]

11. Chatterjee, R.; Marscher, A.P.; Jorstad, S.G.; Olmstead, A.R.; McHardy, I.M.; Aller, M.F.; Aller, H.D.; Lähteenmäki, A.; Tornikoski, M.; Hovatta, T.; et al. Disk-Jet Connection in the Radio Galaxy 3C 120. *Astrophys. J.* **2009**, *704*, 1689–1703.
12. Lohfink, A.M.; Reynolds, C.S.; Jorstad, S.G.; Marscher, A.P.; Miller, E.D.; Aller, H.; Aller, M.F.; Brenneman, L.W.; Fabian, A.C.; Miller, J.M.; et al. An X-Ray View of the Jet Cycle in the Radio-loud AGN 3C120. *Astrophys. J.* **2013**, *772*, 83.
13. Chatterjee, R.; Marscher, A.P.; Jorstad, S.G.; Markowitz, A.; Rivers, E.; Rothschild, R.E.; McHardy, I.M.; Aller, M.F.; Aller, H.D.; Lähteenmäki, A.; et al. Connection Between the Accretion Disk and Jet in the Radio Galaxy 3C 111. *Astrophys. J.* **2011**, *734*, 43.
14. Tombesi, F.; Sambruna, R.M.; Marscher, A.P.; Jorstad, S.G.; Reynolds, C.S.; Markowitz, A. Comparison of ejection events in the jet and accretion disc outflows in 3C 111. *Mon. Not. R. Astron. Soc.* **2012**, *424*, 754–761.
15. Weaver, Z.R.; Jorstad, S.G.; Marscher, A.P.; Morozova, D.A.; Troitsky, I.S.; Agudo, I.; Gómez, J.L.; Lähteenmäki, A.; Tammi, J.; Tornikoski, M. Kinematics of Parsec-scale Jets of Gamma-Ray Blazars at 43 GHz during 10 yr of the VLBA-BU-BLAZAR Program. *Astrophys. J. Suppl. Ser.* **2022**, *260*, 12.
16. Fender, R.P.; Belloni, T.M.; Gallo, E. Towards a unified model for black hole X-ray binary jets. *Mon. Not. R. Astron. Soc.* **2004**, *355*, 1105–1118.
17. Kovalev, Y.Y.; Pushkarev, A.B.; Nokhrina, E.E.; Plavin, A.V.; Beskin, V.S.; Chernoglazov, A.V.; Lister, M.L.; Savolainen, T. A transition from parabolic to conical shape as a common effect in nearby AGN jets. *Mon. Not. R. Astron. Soc.* **2020**, *495*, 3576–3591.
18. Casadio, C.; MacDonald, N.R.; Boccardi, B.; Jorstad, S.G.; Marscher, A.P.; Krichbaum, T.P.; Hodgson, J.A.; Kim, J.Y.; Traianou, E.; Weaver, Z.R.; et al. The jet collimation profile at high resolution in BL Lacertae. *Astron. Astrophys.* **2021**, *649*, A153.
19. Liodakis, I.; Marscher, A.P.; Agudo, I.; Berdyugin, A.V.; Bernardos, M.I.; Bonnoli, G.; Borman, G.A.; Casadio, C.; Casanova, V.; Cavazzuti, E.; et al. Polarized Blazar X-rays imply particle acceleration in shocks. *arXiv* **2022**, arXiv:2209.06227.
20. Kim, J.Y.; Krichbaum, T.P.; Broderick, A.E.; Wielgus, M.; Blackburn, L.; Gómez, J.L.; Johnson, M.D.; Bouman, K.L.; Chael, A.; Akiyama, K.; et al. Event Horizon Telescope imaging of the archetypal blazar 3C 279 at an extreme 20 microarcsecond resolution. *Astron. Astrophys.* **2020**, *640*, A69. [CrossRef]
21. Marscher, A.P.; Jorstad, S.G.; D'Arcangelo, F.D.; Smith, P.S.; Williams, G.G.; Larionov, V.M.; Oh, H.; Olmstead, A.R.; Aller, M.F.; Aller, H.D.; et al. The inner jet of an active galactic nucleus as revealed by a radio-to-γ-ray outburst. *Nature* **2008**, *452*, 966–969. [CrossRef]
22. Cawthorne, T.V.; Jorstad, S.G.; Marscher, A.P. Polarization Structure in the Core of 1803+784: A Signature of Recollimation Shocks? *Astrophys. J.* **2013**, *772*, 14.
23. Gabuzda, D.C.; Knuettel, S.; Reardon, B. Transverse Faraday-rotation gradients across the jets of 15 active galactic nuclei. *Mon. Not. R. Astron. Soc.* **2015**, *450*, 2441–2450.
24. Lico, R.; Giroletti, M.; Orienti, M.; Giovannini, G.; Cotton, W.; Edwards, P.G.; Fuhrmann, L.; Krichbaum, T.P.; Sokolovsky, K.V.; Kovalev, Y.Y.; et al. VLBA monitoring of Mrk 421 at 15 GHz and 24 GHz during 2011. *Astron. Astrophys.* **2012**, *545*, A117.
25. Ghisellini, G.; Tavecchio, F.; Chiaberge, M. Structured jets in TeV BL Lac objects and radiogalaxies. Implications for the observed properties. *Astron. Astrophys.* **2005**, *432*, 401–410. [CrossRef]
26. Giroletti, M.; Giovannini, G.; Cotton, W.D.; Taylor, G.B.; Pérez-Torres, M.A.; Chiaberge, M.; Edwards, P.G. The jet of Markarian 501 from millions of Schwarzschild radii down to a few hundreds. *Astron. Astrophys.* **2008**, *488*, 905–914. [CrossRef]
27. Lico, R.; Giroletti, M.; Orienti, M.; Gómez, J.L.; Casadio, C.; D'Ammando, F.; Blasi, M.G.; Cotton, W.; Edwards, P.G.; Fuhrmann, L.; et al. Very Long Baseline polarimetry and the γ-ray connection in Markarian 421 during the broadband campaign in 2011. *Astron. Astrophys.* **2014**, *571*, A54.
28. Koyama, S.; Kino, M.; Doi, A.; Niinuma, K.; Giroletti, M.; Paneque, D.; Akiyama, K.; Giovannini, G.; Zhao, G.Y.; Ros, E.; et al. Stable Radio Core of the Blazar Mrk 501 during High-energy Active State in 2012. *Astrophys. J.* **2019**, *884*, 132. [CrossRef]
29. Hovatta, T.; Lister, M.L.; Aller, M.F.; Aller, H.D.; Homan, D.C.; Kovalev, Y.Y.; Pushkarev, A.B.; Savolainen, T. MOJAVE: Monitoring of Jets in Active Galactic Nuclei with VLBA Experiments. VIII. Faraday Rotation in Parsec-scale AGN Jets. *Astron. J.* **2012**, *144*, 105.
30. Nagai, H.; Haga, T.; Giovannini, G.; Doi, A.; Orienti, M.; D'Ammando, F.; Kino, M.; Nakamura, M.; Asada, K.; Hada, K.; et al. Limb-brightened Jet of 3C 84 Revealed by the 43 GHz Very-Long-Baseline-Array Observation. *Astrophys. J.* **2014**, *785*, 53.
31. Janssen, M.; Falcke, H.; Kadler, M.; Ros, E.; Wielgus, M.; Akiyama, K.; Baloković, M.; Blackburn, L.; Bouman, K.L.; Chael, A.; et al. Event Horizon Telescope observations of the jet launching and collimation in Centaurus A. *Nat. Astron.* **2021**, *5*, 1017–1028.
32. Bruni, G.; Gómez, J.L.; Vega-García, L.; Lobanov, A.P.; Fuentes, A.; Savolainen, T.; Kovalev, Y.Y.; Perucho, M.; Martí, J.M.; Anderson, J.M.; et al. RadioAstron reveals a spine-sheath jet structure in 3C 273. *Astron. Astrophys.* **2021**, *654*, A27.
33. Valtonen, M.J.; Lehto, H.J.; Nilsson, K.; Heidt, J.; Takalo, L.O.; Sillanpää, A.; Villforth, C.; Kidger, M.; Poyner, G.; Pursimo, T.; et al. A massive binary black-hole system in OJ287 and a test of general relativity. *Nature* **2008**, *452*, 851–853.
34. Gómez, J.L.; Traianou, E.; Krichbaum, T.P.; Lobanov, A.P.; Fuentes, A.; Lico, R.; Zhao, G.Y.; Bruni, G.; Kovalev, Y.Y.; Lähteenmäki, A.; et al. Probing the Innermost Regions of AGN Jets and Their Magnetic Fields with RadioAstron. V. Space and Ground Millimeter-VLBI Imaging of OJ 287. *Astrophys. J.* **2022**, *924*, 122.
35. Liska, M.; Hesp, C.; Tchekhovskoy, A.; Ingram, A.; van der Klis, M.; Markoff, S. Formation of precessing jets by tilted black hole discs in 3D general relativistic MHD simulations. *Mon. Not. R. Astron. Soc.* **2018**, *474*, L81–L85.
36. Pesce, D.W.; Palumbo, D.C.M.; Narayan, R.; Blackburn, L.; Doeleman, S.S.; Johnson, M.D.; Ma, C.P.; Nagar, N.M.; Natarajan, P.; Ricarte, A. Toward Determining the Number of Observable Supermassive Black Hole Shadows. *Astrophys. J.* **2021**, *923*, 260.

37. Porth, O.; Olivares, H.; Mizuno, Y.; Younsi, Z.; Rezzolla, L.; Moscibrodzka, M.; Falcke, H.; Kramer, M. The black hole accretion code. *Comput. Astrophys. Cosmol.* **2017**, *4*, 1.
38. Giommi, P.; Padovani, P. Astrophysical Neutrinos and Blazars. *Universe* **2021**, *7*, 492.
39. Nanci, C.; Giroletti, M.; Orienti, M.; Migliori, G.; Moldón, J.; Garrappa, S.; Kadler, M.; Ros, E.; Buson, S.; An, T.; et al. Observing the inner parsec-scale region of candidate neutrino-emitting blazars. *Astron. Astrophys.* **2022**, *663*, A129.
40. Blinov, D.; Pavlidou, V.; Papadakis, I.E.; Hovatta, T.; Pearson, T.J.; Liodakis, I.; Panopoulou, G.V.; Angelakis, E.; Baloković, M.; Das, H.; et al. RoboPol: Optical polarization-plane rotations and flaring activity in blazars. *Mon. Not. R. Astron. Soc.* **2016**, *457*, 2252–2262.
41. Gómez, J.L.; Lobanov, A.P.; Bruni, G.; Kovalev, Y.Y.; Marscher, A.P.; Jorstad, S.G.; Mizuno, Y.; Bach, U.; Sokolovsky, K.V.; Anderson, J.M.; et al. Probing the Innermost Regions of AGN Jets and Their Magnetic Fields with RadioAstron. I. Imaging BL Lacertae at 21 Microarcsecond Resolution. *Astrophys. J.* **2016**, *817*, 96.
42. Issaoun, S.; Wielgus, M.; Jorstad, S.; Krichbaum, T.P.; Blackburn, L.; Janssen, M.; Chan, C.k.; Pesce, D.W.; Gómez, J.L.; Akiyama, K.; et al. Resolving the Inner Parsec of the Blazar J1924-2914 with the Event Horizon Telescope. *Astrophys. J.* **2022**, *934*, 145.
43. Zhao, G.Y.; Gómez, J.L.; Fuentes, A.; Krichbaum, T.P.; Traianou, E.; Lico, R.; Cho, I.; Ros, E.; Komossa, S.; Akiyama, K.; et al. Unraveling the Innermost Jet Structure of OJ 287 with the First GMVA + ALMA Observations. *Astrophys. J.* **2022**, *932*, 72.
44. Blandford, R.D.; Znajek, R.L. Electromagnetic extraction of energy from Kerr black holes. *Mon. Not. R. Astron. Soc.* **1977**, *179*, 433–456. [CrossRef]
45. Blandford, R.D.; Payne, D.G. Hydromagnetic flows from accretion disks and the production of radio jets. *Mon. Not. R. Astron. Soc.* **1982**, *199*, 883–903. [CrossRef]
46. Weisskopf, M.C.; Soffitta, P.; Baldini, L.; Ramsey, B.D.; O'Dell, S.L.; Romani, R.W.; Matt, G.; Deininger, W.D.; Baumgartner, W.H.; Bellazzini, R.; et al. Imaging X-ray Polarimetry Explorer: Prelaunch. *J. Astron. Telesc. Instrum. Syst.* **2022**, *8*, 026002. [CrossRef]
47. Zhang, S.; Santangelo, A.; Feroci, M.; Xu, Y.; Lu, F.; Chen, Y.; Feng, H.; Zhang, S.; Brandt, S.; Hernanz, M.; et al. The enhanced X-ray Timing and Polarimetry mission—eXTP. *Sci. China Phys. Mech. Astron.* **2019**, *62*, 29502.
48. Beechert, J.; Lazar, H.; Boggs, S.E.; Brandt, T.J.; Chang, Y.C.; Chu, C.Y.; Gulick, H.; Kierans, C.; Lowell, A.; Pellegrini, N.; et al. Calibrations of the Compton Spectrometer and Imager. *Nucl. Instrum. Methods Phys. Res. A* **2022**, *1031*, 166510.
49. Di Gesu, L.; Donnarumma, I.; Tavecchio, F.; Agudo, I.; Barnounin, T.; Cibrario, N.; Di Lalla, N.; Di Marco, A.; Escudero, J.; Errando, M.; et al. The X-ray Polarization View of Mrk~421 in an Average Flux State as Observed by the Imaging X-ray Polarimetry Explorer. *arXiv* **2022**, arXiv:2209.07184.

Communication

Probing Neutrino Production in Blazars by Millimeter VLBI

Yuri Y. Kovalev [1,2,*], **Alexander V. Plavin** [3], **Alexander B. Pushkarev** [3,4] and **Sergey V. Troitsky** [5,6]

1   Max–Planck–Institut für Radioastronomie, Auf dem Hügel 69, 53121 Bonn, Germany
2   Black Hole Initiative, Harvard University, 20 Garden Street, Cambridge, MA 02138, USA
3   Lebedev Physical Institute of the Russian Academy of Sciences, Leninsky Prospekt 53, 119991 Moscow, Russia; alexander@plav.in (A.V.P.); pushkarev.alexander@gmail.com (A.B.P.)
4   Crimean Astrophysical Observatory, Crimea, 298409 Nauchny, Russia
5   Institute for Nuclear Research of the Russian Academy of Sciences, 60th October Anniversary Prospect 7a, 117312 Moscow, Russia; st@inr.ac.ru
6   Faculty of Physics, M.V. Lomonosov Moscow State University, 1-2 Leninskie Gory, 119991 Moscow, Russia
*   Correspondence: yykovalev@gmail.com

Abstract: The advancement of neutrino observatories has sparked a surge in multi-messenger astronomy. Multiple neutrino associations among blazars are reported while neutrino production sites are located within their central (sub)parsecs. Yet, many questions remain on the nature of those processes. The next generation Event Horizon Telescope (ngEHT) is uniquely positioned for these studies, as its high frequency and resolution can probe both the accretion disk region and the parsec-scale jet. This opens up new opportunities for connecting the two regions and unraveling the proton acceleration and neutrino production in blazars. We outline observational strategies for ngEHT and highlight what it can contribute to the multi-messenger study of blazars.

Keywords: neutrinos; active galaxies; galaxy jets; quasars; radio continuum; interferometric techniques

Citation: Kovalev, Y.Y.; Plavin, A.V.; Pushkarev, A.B.; Troitsky, S.V. Probing Neutrino Production in Blazars by Millimeter VLBI. *Galaxies* **2023**, *11*, 84. https://doi.org/10.3390/galaxies11040084

Academic Editor: Luigina Feretti

Received: 17 April 2023
Revised: 1 July 2023
Accepted: 5 July 2023
Published: 10 July 2023

1. Introduction: Current Status of High-Energy Neutrino Studies, Blazar–Neutrino Connections

Neutrino observatories, such as IceCube, ANTARES (Astronomy with a Neutrino Telescope and Abyss environmental RESearch project), and Baikal-GVD (Gigaton Volume Detector) have been convincingly detecting astrophysical neutrinos at TeV to PeV energies [1–4]. Despite these observations, little was known about the origin of energetic astrophysical neutrinos until recently.

Blazars, a class of active galactic nuclei (AGN), have been considered as potential neutrino sources since the very early days of multi-messenger astronomy [5]. Observational evidence for a blazar–neutrino connection has started to emerge in recent years. First, the blazar TXS 0506+056 was associated with a high-energy neutrino, which coincided with a gamma-ray flare in 2017 [6]. This association was in contrast with a lack of systematic connection between gamma-ray-loud blazars and neutrinos (see, e.g., [7,8]). Then, numerous radio-bright blazars were shown to emit neutrinos with energies from TeVs to PeVs [9–16]. The detection of this correlation is driven by the unique capabilities of very-long-baseline interferometry (VLBI): the only technique able to directly probe and resolve central (sub)parsecs in AGNs at cosmological distances. Blazars emit neutrinos preferentially at the times of their flares (Figure 1), visible in radio bands [10,12,15,17]. Still, the neutrino production mechanism and the physical regions where it occurs remain unclear. The observed connection of neutrinos with radio emission from compact jet regions emphasizes the importance of high-resolution studies in answering these questions. VLBI is the best direct visual evidence we can obtain in astronomy.

For a general discussion of multi-wavelength and multi-messenger studies with the ngEHT, see Lico et al. [18]. In this paper, we present the progress in multi-messenger astronomy studies of cosmic neutrinos, their probable association with blazars, challenges and a critical role to be played by ngEHT [19–21] in addressing exciting open questions of high-energy neutrino production.

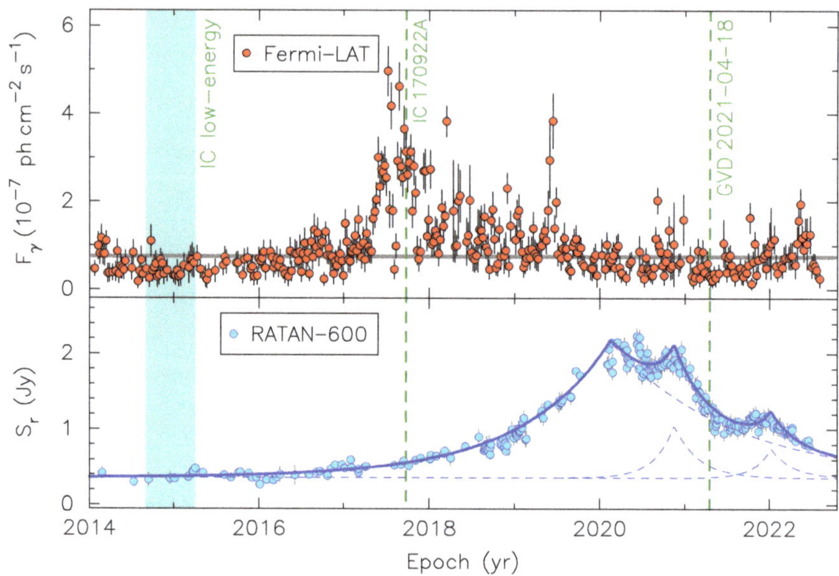

Figure 1. Radio and gamma-ray light curves of TXS 0506+056. Top: *Fermi* LAT weekly binning light curve [22] of the gamma-ray source 4FGLJ0509.4+0542 positionally associated with the quasar TXS 0506+056. The gray horizontal line denotes the median gamma-ray flux. Bottom: RATAN-600 light curve at 11 GHz. The radio light curve is decomposed of three radio flares depicted by the blue dashed lines, the sum of which is represented by a blue thick line [23]. The vertical lines denote the times of neutrino arrival [6,23].

2. Neutrino Production in Blazars: Open Questions

Assuming no particle physics beyond the standard model, astrophysical neutrinos with energies above TeV can only be produced in interactions of relativistic hadrons—protons or nuclei—with ambient matter or radiation, see, e.g., [24] for a recent review. This fits the observational evidence discussed in Section 1 because the non-thermal radiation of blazars gives a clear signal that particles are accelerated there. However, both the amount of relativistic hadrons in AGN, and the degree to which these hadrons contribute to the observed electromagnetic radiation, are uncertain. Population studies suggest [24–26] that their contribution is small, and neutrino luminosities of blazars are orders of magnitude lower than photon luminosities. Consequently, one may imagine neutrino production in various places in a blazar and by means of different mechanisms.

The main challenge is to explain the production of neutrinos of very different energies, from a few TeV [11,27] to sub-PeV [6,10], in sources of the same class. For the $p\gamma$ mechanism, expected to dominate in blazars [28], the wide neutrino energy range requires the presence of target photons with a very broad distribution of energies. Conventional models of high-energy neutrino production in AGN, known for decades, e.g., [29–31], as well as their modern versions, e.g., [32–34], often experience problems in explaining the lower-energy part of the observed neutrino flux, particularly because the target photons from the accretion disk are expected to have energies $\sim(10\ldots100)$ eV, while ~10 keV are required for the intense production of ~10 TeV neutrinos.

While neutrinos have already been associated with VLBI-bright blazars [11,15] and with their radio flares [10,12], these results were based on observations at centimeter wavelengths. There, synchrotron self-absorption prevents one from obtaining detailed spatio-temporal studies of the AGN central sub-parsec parts, e.g., [35]. To summarize, the open questions of the blazar–neutrino astrophysics are the following: (i) how are protons accelerated; (ii) what is the neutrino production process, $p\gamma$ or pp; (iii) from where do seed (X-ray) photons originate from in case of $p\gamma$; (iv) where are neutrinos produced? Note that (ii) and (iv) can be different, and multi-zone models may be required to explain all observations consistently.

3. Neutrino Astronomy in the ngEHT Era

Currently, studies of high-energy astrophysical neutrinos and their sources are limited by the sensitivity and resolution of neutrino observatories. The situation is rapidly changing, as their capabilities are increasingly improving. The next-generation IceCube-Gen2 will grow the telescope volume ten-fold, from 1 to 10 km^3, aiming at a corresponding increase in detection rates by 2033 [36]. The Baikal-GVD detector has already reached the effective volume of 0.5 km^3 and continues to grow and improve event reconstruction algorithms [37]. KM3Net (Cubic Kilometer Neutrino Telescope), a neutrino observatory in the Mediterranean, is being constructed and has already started yielding its first results [38]. Together, these instruments will provide a qualitative leap in both the number of detected astrophysical neutrinos and their precise localization.

An increasing number of well-localized neutrinos will lead to reliable identification of individual blazars as neutrino sources. Moreover, it should be possible to highlight specific time periods with more prominent neutrino emission. This brings new challenges and opportunities to the EM counterpart of such multi-messenger studies.

The planned ngEHT array [19–21] will provide superior angular resolution, dynamic range and sensitivity in Stokes I and polarization at 3, 1.3, and 0.7 mm. This will allow scientists to observe and monitor the central (sub)parsecs of neutrino-emitting blazars at the highest resolution and frequency possible, significantly alleviating the synchrotron opacity problem of the current centimeter-wavelength VLBI. The ngEHT will be able to probe both the accretion disk region and the parsec-scale jet base, opening new opportunities for connecting the two regions and unraveling the proton acceleration and neutrino production in blazars.

4. Planning ngEHT Experiments

Below, we discuss several approaches to study and understand the physics behind the connection between neutrino production and EM activity from the jet upstream of the central engine—a possibility which will be realized by ngEHT. Before elaborating on observational campaigns, we note the following important complications of neutrino astrophysics that affect the suggested scenarios below. A typical probability of a neutrino with an energy above 100 TeV to be of an astrophysical origin is around 50%, and it drops significantly for lower energies [39,40]. A typical 90% error region of a highly probable high-energy neutrino is several square degrees [40,41]. Some neutrinos might arrive from nearby non-jetted AGNs [42] or even from our Galaxy and its relativistic objects [43–48]. On top of this, we know very little about the mechanisms of neutrino production in blazars; therefore, there is no streetlight under which we can plan our search.

We expect that a variety of blazars could be associated with neutrinos, allowing us to select optimal ngEHT targets by accounting for both their physical properties and technical or observational limitations. Within our current understanding and the experience accumulated from observational searches for high-energy neutrino counterparts, the following three scenarios for monitoring observations are suggested.

Scenario 1: Observation of blazars associated with selected new high-energy neutrino alerts immediately after neutrino arrival. Several blazar-associated high-energy alerts per year are expected. When two or three neutrino telescopes become fully operational, one might conservatively require two alerts for a given target to arrive within several days.

Pros: The most efficient strategy since it is linked to a specific event.
Cons: It will only be able to probe the state of an associated object after neutrino arrival.

Scenario 2: Observation of a sample of selected blazars reliably identified previously as neutrino sources. See Table 1 for the current most probable neutrino candidates.

Pros: This strategy is optimal in terms of the observed sample and complete temporal coverage of events.

Cons: so far, a very limited number of cases are known with repeated neutrino detection from the same source (Table 1, column 5), but this list could grow.

Scenario 3: Observations of a complete VLBI-flux-density limited sample of 50–100 of the brightest blazars with a 3 mm VLBI flux density above 1 Jy [49,50].

Pros: Offers full temporal coverage of the expected events, with the possibility to compare neutrino-emitting and neutrino-non-emitting blazars to calculate the robust significance of a coincidence [15,51]. Furthermore, the strategy provides the option to combine such observations with other ngEHT cases [20].

Cons: Observationally expensive.

Table 1. Most probable neutrino candidates among the VLBI-selected bright blazars.

| Blazar Name | | z | $S^{\text{VLBI}}_{86\,\text{GHz}}$ (Jy) | Number of High-Energy | References |
B1950	Alias			Neutrinos (and Dates)	
0506+056		0.34	0.6 [†]	2 (2017-09-22, 2021-04-18)	[6,23]
0735+178	OI 158	0.45	0.6	1–4 (2021-12-04&08)	[15,49]
1253−055	3C 279	0.54	22.7	1 (2015-09-26)	[10,50]
1502+106	OR 103	1.84	0.6	1 (2019-07-30)	[10,49]
1730−130	NRAO 530	0.90	1.9 [†]	1 (2016-01-28)	[10,52]
1741−038		1.05	3.2	2 (2011-09-30, 2022-02-05)	[15,49]
1749+096	OT 081	0.32	2.4	1 (2022-03-03)	[15,49]
2145+067	4C +06.69	1.00	3.6 [†]	1 (2015-08-12)	[10,52]

Notes: Publications that selected each blazar as a highly probable neutrino-emitter and measured their flux densities are referenced in the last column. The 0506+056 blazar was the first and only blazar distinguished by the IceCube, while the others were found by statistical analysis of complete VLBI-selected samples. The dates for high-energy neutrinos are shown in the format YYYY-MM-DD. [†] Estimated from nearby VLBI measurements at 15 and 43 GHz of MOJAVE and Boston University VLBA programs.

Tracing changes in the compact structure of blazars during and around periods of increased neutrino emission requires multi-epoch monitoring at the high resolution provided by ngEHT. To roughly estimate the required observation time, we expect that one imaging epoch per target will take 4–8 h. The observations should happen with a cadence between two weeks and one month (an estimate based on experience gained by the 7 mm blazar VLBA-monitoring program [53]) and produce polarization images with Stokes I dynamic range or better than 1000:1, preferably multi-frequency with a possibility for Faraday rotation measurements (RMs) and spectral analysis. From this, we will be able to constrain the following source properties.

1. Jet kinematic measurements will allow us to better estimate Doppler boosting and jet viewing angle following, e.g., [53,54], constrain plasma acceleration, e.g., [53,55]. Jet geometry profile studies will constrain jet formation and collimation [56,57].
2. Jet kinematics will also deliver information about newborn jet features, e.g., [53,58], measure ejection epochs of features possibly associated with neutrino events, compare these with neutrino arrival times and locate the neutrino production zone from the measured delay. Comparison with similar analyses for VLBI-γ-ray studies [59,60].
3. Faraday RM, reconstructed EVPAs and analysis of radio spectra together with core-shift measurements will deliver information on the magnetic field structure, its strength and changes, e.g., [61–63], which might be related to the physical conditions required for neutrino production.
4. Monitoring the overall changes in the millimeter parsec- and sub-parsec-scale structure of blazars at the extreme resolution of ngEHT will allow us to distinguish between

flares in disks and in jets, e.g., [64,65] related to neutrino production if the resolution, sensitivity, and opacity permit. Observing in this regime, we will be able to overcome significant delays related to synchrotron self-absorption at lower radio frequencies (see Figure 1 and [60]).

We underline that studying a complete sample of AGN with understandable properties will allow us to not only relate the observed changes to detected neutrinos but also set a robust significance on that association, following the approach suggested by Plavin et al. [15].

5. Synergy with Other Facilities

The Square Kilometer Array, SKA [66] and especially the next-generation Very Large Array, ngVLA [67,68] going as high as 100 GHz will allow the monitoring of much larger samples of VLBI-selected AGNs as well as the faster imaging of neutrino arrival fields, and pre-selecting probable neutrino candidates for targeted ngEHT studies. Wide-field telescopes such as the optical Legacy Survey of Space and Time, LSST [69] will allow scientists to better associate blazars with neutrinos in cases where flaring activity is confirmed as a valid indicator, e.g., [6,15,70]. Moreover, optical and UV telescopes can separate flares occurring in jets and accretion disks, analyzing the optical color and polarization. Seed photons are expected from X-rays [11,64,71], where current and new-generation space X-ray telescopes will be very helpful. High energies, e.g., the Cherenkov Telescope Array, CTA [72,73], will continue to support gamma-ray/TeV–neutrino analyses and identify whether neutrino production zones are opaque to gamma-rays.

6. Summary

The ngEHT will revolutionize VLBI-imaging capabilities by bringing together the power of very high resolution, advanced dynamic range, and sensitive polarization data. What makes it unique, however, is its remarkable immunity to synchrotron absorption. It will allow for regions to be extensively probed from the accretion disk to the parsec-scale jet, e.g., [74], and study the most probable sources of high-energy neutrinos.

When ngEHT is fully operational, three large high-energy neutrino telescopes will be fully functional: IceCube, KM3NeT, and Baikal-GVD. This paper formulates the case, presents eight most probable associations to date, and suggests observational strategies to address very exciting and open questions concerning proton acceleration and neutrino production.

Author Contributions: Conceptualization and methodology, Y.Y.K.; writing—original draft preparation as well as review and editing, Y.Y.K., A.V.P., A.B.P. and S.V.T.; visualization, A.B.P. All authors have read and agreed to the published version of the manuscript.

Funding: This publication was funded in part by the Gordon and Betty Moore Foundation. It was also made possible through the support of a grant from the John Templeton Foundation. The opinions expressed in this publication are those of the author(s) and do not necessarily reflect the views of these foundations.

Data Availability Statement: Not applicable.

Acknowledgments: We thank the ngEHT team for discussions, Eduardo Ros as well as anonymous referees for useful comments on the manuscript, and Elena Bazanova for language editing. The VLBA is an instrument of the National Radio Astronomy Observatory. The National Radio Astronomy Observatory is a facility of the National Science Foundation operated by Associated Universities, Inc.

Conflicts of Interest: The authors declare no conflict of interest.

References

1. Kopper, C.; Whitehorn, N.; Kurahashi Neilson, N.; IceCube Collaboration. Evidence for High-Energy Extraterrestrial Neutrinos at the IceCube Detector. *Science* **2013**, *342*, 1242856. [CrossRef]
2. Aartsen, M.G.; Abraham, K.; Ackermann, M.; Adams, J.; Aguilar, J.A.; Ahlers, M.; Ahrens, M.; Altmann, D.; Andeen, K.; Anderson, T.; et al. Observation and Characterization of a Cosmic Muon Neutrino Flux from the Northern Hemisphere Using Six Years of IceCube Data. *Astrophys. J.* **2016**, *833*, 3. [CrossRef]

3. Albert, A.; André, M.; Anghinolfi, M.; Anton, G.; Ardid, M.; Aubert, J.J.; Aublin, J.; Avgitas, T.; Baret, B.; Barrios-Martí, J.; et al. All-flavor Search for a Diffuse Flux of Cosmic Neutrinos with Nine Years of ANTARES Data. *Astrophys. J. Lett.* **2018**, *853*, L7. [CrossRef]

4. Allakhverdyan, V.A.; Avrorin, A.D.; Avrorin, A.V.; Aynutdinov, V.M.; Bardačová, Z.; Belolaptikov, I.A.; Borina, I.V.; Budnev, N.M.; Dik, V.Y.; Domogatsky, G.V.; et al. Diffuse neutrino flux measurements with the Baikal-GVD neutrino telescope. *Phys. Rev. D* **2023**, *107*, 042005. [CrossRef]

5. Berezinsky, V. Extraterrestrial neutrino sources and high energy neutrino astrophysics. In Proceedings of the Neutrino-77 Conference, Moscow, Russia, 18–24 June 1977; p. 177.

6. Aartsen, M.G.; Ackermann, M.; Adams, J.; Aguilar, J.A.; Ahlers, M.; Ahrens, M.; Al Samarai, I.; Altmann, D.; Andeen, K.; Anderson, T.; et al. Multimessenger observations of a flaring blazar coincident with high-energy neutrino IceCube-170922A. *Science* **2018**, *361*, aat1378. [CrossRef]

7. Neronov, A.; Semikoz, D.V.; Ptitsyna, K. Strong constraints on hadronic models of blazar activity from Fermi and IceCube stacking analysis. *Astron. Astrophys.* **2017**, *603*, A135. [CrossRef]

8. Murase, K.; Oikonomou, F.; Petropoulou, M. Blazar Flares as an Origin of High-energy Cosmic Neutrinos? *Astrophys. J.* **2018**, *865*, 124. [CrossRef]

9. Kadler, M.; Krauß, F.; Mannheim, K.; Ojha, R.; Müller, C.; Schulz, R.; Anton, G.; Baumgartner, W.; Beuchert, T.; Buson, S.; et al. Coincidence of a high-fluence blazar outburst with a PeV-energy neutrino event. *Nat. Phys.* **2016**, *12*, 807–814. [CrossRef]

10. Plavin, A.; Kovalev, Y.Y.; Kovalev, Y.A.; Troitsky, S. Observational Evidence for the Origin of High-energy Neutrinos in Parsec-scale Nuclei of Radio-bright Active Galaxies. *Astrophys. J.* **2020**, *894*, 101. [CrossRef]

11. Plavin, A.V.; Kovalev, Y.Y.; Kovalev, Y.A.; Troitsky, S.V. Directional Association of TeV to PeV Astrophysical Neutrinos with Radio Blazars. *Astrophys. J.* **2021**, *908*, 157. [CrossRef]

12. Hovatta, T.; Lindfors, E.; Kiehlmann, S.; Max-Moerbeck, W.; Hodges, M.; Liodakis, I.; Lähteemäki, A.; Pearson, T.J.; Readhead, A.C.S.; Reeves, R.A.; et al. Association of IceCube neutrinos with radio sources observed at Owens Valley and Metsähovi Radio Observatories. *Astrophys. J.* **2021**, *650*, A83. [CrossRef]

13. Illuminati, G.; ANTARES Collaboration. [ANTARES Collaboration] ANTARES search for neutrino flares from the direction of radio-bright blazars. In Proceedings of the 37th International Cosmic Ray Conference, Barcelona, Spain, 4–8 July 2022; p. 972. [CrossRef]

14. Allakhverdyan, V.A.; Avrorin, A.D.; Avrorin, A.V.; Aynutdinov, V.M.; Bannasch, R.; Bardačová, Z.; Zaborov, D.N. The Baikal-GVD neutrino telescope: Search for high-energy cascades. *arXiv* **2021**, arXiv:2108.01894.

15. Plavin, A.V.; Kovalev, Y.Y.; Kovalev, Y.A.; Troitsky, S.V. Growing evidence for high-energy neutrinos originating in radio blazars. *Mon. Not. R. Astron. Soc.* **2023**, *523*, 1799–1808. [CrossRef]

16. Buson, S.; Tramacere, A.; Oswald, L.; Barbano, E.; Fichet de Clairfontaine, G.; Pfeiffer, L.; Azzollini, A.; Baghmanyan, V.; Ajello, M. Extragalactic neutrino factories. *arXiv* **2023**, arXiv:2305.11263.

17. Britzen, S.; Zajaček, M.; Popović, L.Č.; Fendt, C.; Tramacere, A.; Pashchenko, I.N.; Jaron, F.; Pánis, R.; Petrov, L.; Aller, M.F.; et al. A ring accelerator? Unusual jet dynamics in the IceCube candidate PKS 1502+106. *Mon. Not. R. Astron. Soc.* **2021**, *503*, 3145–3178. [CrossRef]

18. Lico, R.; Jorstad, S.G.; Marscher, A.P.; Gómez, J.L.; Liodakis, I.; Dahale, R.; Alberdi, A.; Gold, R.; Traianou, E.; Toscano, T.; et al. Multi-Wavelength and Multi-Messenger Studies Using the Next-Generation Event Horizon Telescope. *Galaxies* **2023**, *11*, 17. [CrossRef]

19. Raymond, A.W.; Palumbo, D.; Paine, S.N.; Blackburn, L.; Córdova Rosado, R.; Doeleman, S.S.; Farah, J.R.; Johnson, M.D.; Roelofs, F.; Tilanus, R.P.J.; et al. Evaluation of New Submillimeter VLBI Sites for the Event Horizon Telescope. *Astrophys. J. Suppl.* **2021**, *253*, 5. [CrossRef]

20. Johnson, M.D.; Akiyama, K.; Blackburn, L.; Bouman, K.L.; Broderick, A.E.; Cardoso, V.; Fender, R.P.; Fromm, C.M.; Galison, P.; Gómez, J.L.; et al. Key Science Goals for the Next-Generation Event Horizon Telescope. *Galaxies* **2023**, *11*, 61. [CrossRef]

21. Doeleman, S.S.; Barrett, J.; Blackburn, L.; Bouman, K.; Broderick, A.E.; Chaves, R.; Fish, V.L.; Fitzpatrick, G.; Fuentes, A.; Freeman, M.; et al. Reference Array and Design Consideration for the next-generation Event Horizon Telescope. *arXiv* **2023**, arXiv:2306.08787.

22. Abdollahi, S.; Ajello, M.; Baldini, L.; Ballet, J.; Bastieri, D.; Becerra Gonzalez, J.; Bellazzini, R.; Berretta, A.; Bissaldi, E.; Bonino, R.; et al. The Fermi-LAT Lightcurve Repository. *Astrophys. J. Suppl.* **2023**, *265*, 31. [CrossRef]

23. Erkenov, A.K.; Kosogorov, N.A.; Kovalev, Y.A.; Kovalev, Y.Y.; Plavin, A.V.; Popkov, A.V.; Pushkarev, A.B.; Semikoz, D.V.; Sotnikova, Y.V.; Troitsky, S.V.; et al. [Baikal-GVD Collaboration] High-energy neutrino-induced cascade from the direction of the flaring radio blazar TXS 0506+056 observed by the Baikal Gigaton Volume Detector in 2021. *arXiv* **2022**, arXiv:2210.01650.

24. Troitsky, S. Constraints on models of the origin of high-energy astrophysical neutrinos. *Usp. Fiz. Nauk* **2021**, *191*, 1333–1360.

25. Neronov, A.; Semikoz, D. Self-consistent model of extragalactic neutrino flux from evolving blazar population. *J. Exp. Theor. Phys.* **2020**, *158*, 295. [CrossRef]

26. Capel, F.; Mortlock, D.J.; Finley, C. Bayesian constraints on the astrophysical neutrino source population from IceCube data. *Phys. Rev. D* **2020**, *101*, 123017. [CrossRef]

27. Aartsen, M.G.; Ackermann, M.; Adams, J.; Aguilar, J.A.; Ahlers, M.; Ahrens, M.; Al Samarai, I.; Altmann, D.; Andeen, K.; Anderson, T.; et al. Neutrino emission from the direction of the blazar TXS 0506+056 prior to the IceCube-170922A alert. *Science* **2018**, *361*, 147–151. [CrossRef]

28. Böttcher, M. Progress in Multi-wavelength and Multi-Messenger Observations of Blazars and Theoretical Challenges. *Galaxies* **2019**, *7*, 20. [CrossRef]

29. Berezinskii, V.S.; Ginzburg, V.L. On high-energy neutrino radiation of quasars and active galactic nuclei. *Mon. Not. R. Astron. Soc.* **1981**, *194*, 3–14. [CrossRef]

30. Eichler, D. High-energy neutrino astronomy: A probe of galactic nuclei? *Astrophys. J.* **1979**, *232*, 106–112. [CrossRef]

31. Stecker, F.W.; Done, C.; Salamon, M.H.; Sommers, P. High-energy neutrinos from active galactic nuclei. *Phys. Rev. Lett.* **1991**, *66*, 2697–2700. [CrossRef]

32. Neronov, A.Y.; Semikoz, D.V. Which blazars are neutrino loud? *Phys. Rev.* **2002**, *D66*, 123003. [CrossRef]

33. Stecker, F.W. PeV neutrinos observed by IceCube from cores of active galactic nuclei. *Phys. Rev.* **2013**, *D88*, 047301. [CrossRef]

34. Kalashev, O.; Semikoz, D.; Tkachev, I. Neutrinos in IceCube from active galactic nuclei. *J. Exp. Theor. Phys.* **2015**, *120*, 541–548. [CrossRef]

35. Boccardi, B.; Krichbaum, T.P.; Ros, E.; Zensus, J.A. Radio observations of active galactic nuclei with mm-VLBI. *Astron. Astrophys. Rev.* **2017**, *25*, 4. [CrossRef]

36. Aartsen, M.G.; Abbasi, R.; Ackermann, M.; Adams, J.; Aguilar, J.A.; Ahlers, M.; Ahrens, M.; Alispach, C.; Allison, P.; Amin, N.M.; et al. IceCube-Gen2: The window to the extreme Universe. *J. Phys. Nucl. Phys.* **2021**, *48*, 060501. [CrossRef]

37. Belolaptikov, I.; Dzhilkibaev, Z.A.M.; Allakhverdyan, V.A.; Avrorin, A.D.; Avrorin, A.V.; Aynutdinov, V.M.; Bannasch, R.; Bardacová, Z.; Belolaptikov, I.A.; Borina, I.V.; et al. Neutrino Telescope in Lake Baikal: Present and Nearest Future. In Proceedings of the 37th International Cosmic Ray Conference, Berlin, Germany, 12–23 July 2022; p. 2.

38. Aiello, S.; Akrame, S.E.; Ameli, F.; Anassontzis, E.G.; Andre, M.; Androulakis, G.; Anghinolfi, M.; Anton, G.; Ardid, M.; Aublin, J.; et al. Sensitivity of the KM3NeT/ARCA neutrino telescope to point-like neutrino sources. *Astropart. Phys.* **2019**, *111*, 100–110. [CrossRef]

39. Ahlers, M.; Halzen, F. Opening a new window onto the universe with IceCube. *Prog. Part. Nucl. Phys.* **2018**, *102*, 73–88. [CrossRef]

40. Abbasi, R.; Ackermann, M.; Adams, J.; Agarwalla, S.K.; Aguilar, J.A.; Ahlers, M.; Alameddine, J.M.; Amin, N.M.; Andeen, K.; Anton, G.; et al. IceCat-1: The IceCube Event Catalog of Alert Tracks. *arXiv* **2023**, arXiv:2304.01174.

41. Aartsen, M.G.; Ackermann, M.; Adams, J.; Aguilar, J.A.; Ahlers, M.; Ahrens, M.; Altmann, D.; Andeen, K.; Anderson, T.; Ansseau, I.; et al. The IceCube Realtime Alert System. *Astropart. Phys.* **2017**, *92*, 30–41. [CrossRef]

42. Abbasi, R.; Ackermann, M.; Adams, J.; Aguilar, J.A.; Ahlers, M.; Ahrens, M.; Alameddine, J.M.; Alispach, C.; Alves Jr, A.A.; Amin, N.M.; et al. Evidence for neutrino emission from the nearby active galaxy NGC 1068. *Science* **2022**, *378*, 538–543. [CrossRef]

43. Kovalev, Y.Y.; Plavin, A.V.; Troitsky, S.V. Galactic Contribution to the High-energy Neutrino Flux Found in Track-like IceCube Events. *Astrophys. J. Lett.* **2022**, *940*, L41. [CrossRef]

44. Troitsky, S.V. Constraints on models of the origin of high-energy astrophysical neutrinos. *Phys. Uspekhi* **2021**, *64*, 1261–1285. [CrossRef]

45. Bykov, A.M.; Petrov, A.E.; Kalyashova, M.E.; Troitsky, S.V. PeV Photon and Neutrino Flares from Galactic Gamma-Ray Binaries. *Astrophys. J. Lett.* **2021**, *921*, L10. [CrossRef]

46. Koljonen, K.I.I.; Satalecka, K.; Lindfors, E.J.; Liodakis, I. Microquasar Cyg X-3—A unique jet-wind neutrino factory? *Mon. Not. R. Astron. Soc.* **2023**, *524*, L89–L93 . [CrossRef]

47. Abbasi, R.; Ackermann, M.; Adams, J.; Agarwalla, S.K.; Aguilar, J.A.; Ahlers, M.; Alameddine, J.M.; Amin, N.M.; Andeen, K.; Anton, G.; et al. Constraining High-energy Neutrino Emission from Supernovae with IceCube. *Astrophys. J. Lett.* **2023**, *949*, L12. [CrossRef]

48. Abbasi, R.; Ackermann, M.; Adams, J.; Aguilar, J.A.; Ahlers, M.; Ahrens, M.; Alameddine, J.M.; Alves, A.A., Jr.; Amin, N.M.; Andeen, K.; et al. Observation of high-energy neutrinos from the Galactic plane. *Science* **2023**, *380*, 1338–1343. [CrossRef] [PubMed]

49. Lee, S.S.; Lobanov, A.P.; Krichbaum, T.P.; Witzel, A.; Zensus, A.; Bremer, M.; Greve, A.; Grewing, M. A Global 86 GHz VLBI Survey of Compact Radio Sources. *Astron. J.* **2008**, *136*, 159–180. [CrossRef]

50. Nair, D.G.; Lobanov, A.P.; Krichbaum, T.P.; Ros, E.; Zensus, J.A.; Kovalev, Y.Y.; Lee, S.S.; Mertens, F.; Hagiwara, Y.; Bremer, M.; et al. Global millimeter VLBI array survey of ultracompact extragalactic radio sources at 86 GHz. *Astrophys. J.* **2019**, *622*, A92. [CrossRef]

51. Liodakis, I.; Hovatta, T.; Pavlidou, V.; Readhead, A.C.S.; Blandford, R.D.; Kiehlmann, S.; Lindfors, E.; Max-Moerbeck, W.; Pearson, T.J.; Petropoulou, M. The hunt for extraterrestrial high-energy neutrino counterparts. *Astrophys. J.* **2022**, *666*, A36. [CrossRef]

52. Agudo, I.; Thum, C.; Wiesemeyer, H.; Krichbaum, T.P. A 3.5 mm Polarimetric Survey of Radio-loud Active Galactic Nuclei. *Astrophys. J. Suppl.* **2010**, *189*, 1–14. [CrossRef]

53. Weaver, Z.R.; Jorstad, S.G.; Marscher, A.P.; Morozova, D.A.; Troitsky, I.S.; Agudo, I.; Gómez, J.L.; Lähteenmäki, A.; Tammi, J.; Tornikoski, M. Kinematics of Parsec-scale Jets of Gamma-Ray Blazars at 43 GHz during 10 yr of the VLBA-BU-BLAZAR Program. *Astrophys. J. Suppl.* **2022**, *260*, 12. [CrossRef]

54. Homan, D.C.; Cohen, M.H.; Hovatta, T.; Kellermann, K.I.; Kovalev, Y.Y.; Lister, M.L.; Popkov, A.V.; Pushkarev, A.B.; Ros, E.; Savolainen, T. MOJAVE. XIX. Brightness Temperatures and Intrinsic Properties of Blazar Jets. *Astrophys. J.* **2021**, *923*, 67. [CrossRef]

55. Homan, D.C.; Lister, M.L.; Kovalev, Y.Y.; Pushkarev, A.B.; Savolainen, T.; Kellermann, K.I.; Richards, J.L.; Ros, E. MOJAVE. XII. Acceleration and Collimation of Blazar Jets on Parsec Scales. *Astrophys. J.* **2015**, *798*, 134. [CrossRef]

56. Asada, K.; Nakamura, M. The Structure of the M87 Jet: A Transition from Parabolic to Conical Streamlines. *Astrophys. J. Lett.* **2012**, *745*, L28. [CrossRef]

57. Kovalev, Y.Y.; Pushkarev, A.B.; Nokhrina, E.E.; Plavin, A.V.; Beskin, V.S.; Chernoglazov, A.V.; Lister, M.L.; Savolainen, T. A transition from parabolic to conical shape as a common effect in nearby AGN jets. *Mon. Not. R. Astron. Soc.* **2020**, *495*, 3576–3591. [CrossRef]

58. Lister, M.L.; Homan, D.C.; Kellermann, K.I.; Kovalev, Y.Y.; Pushkarev, A.B.; Ros, E.; Savolainen, T. Monitoring of Jets in Active Galactic Nuclei with VLBA Experiments. XVIII. Kinematics and Inner Jet Evolution of Bright Radio-loud Active Galaxies. *Astrophys. J.* **2021**, *923*, 30. [CrossRef]

59. Jorstad, S.G.; Marscher, A.P.; Morozova, D.A.; Troitsky, I.S.; Agudo, I.; Casadio, C.; Foord, A.; Gómez, J.L.; MacDonald, N.R.; Molina, S.N.; et al. Kinematics of Parsec-scale Jets of Gamma-Ray Blazars at 43 GHz within the VLBA-BU-BLAZAR Program. *Astrophys. J.* **2017**, *846*, 98. [CrossRef]

60. Kramarenko, I.G.; Pushkarev, A.B.; Kovalev, Y.Y.; Lister, M.L.; Hovatta, T.; Savolainen, T. A decade of joint MOJAVE-Fermi AGN monitoring: Localization of the gamma-ray emission region. *Mon. Not. R. Astron. Soc.* **2022**, *510*, 469–480. [CrossRef]

61. Lobanov, A.P. Spectral distributions in compact radio sources. I. Imaging with VLBI data. *Astron. Astrophys. Suppl. Ser.* **1998**, *132*, 261–273. [CrossRef]

62. Kravchenko, E.V.; Kovalev, Y.Y.; Sokolovsky, K.V. Parsec-scale Faraday rotation and polarization of 20 active galactic nuclei jets. *Mon. Not. R. Astron. Soc.* **2017**, *467*, 83–101. [CrossRef]

63. Martí-Vidal, I.; Muller, S.; Vlemmings, W.; Horellou, C.; Aalto, S. A strong magnetic field in the jet base of a supermassive black hole. *Science* **2015**, *348*, 311–314. [CrossRef]

64. Murase, K. Active Galactic Nuclei as High-Energy Neutrino Sources. In *Neutrino Astronomy: Current Status, Future Prospects*; Gaisser, T., Karle, A., Eds.; World Scientific Publishing Co. Pte. Ltd.: Singapore, 2017; pp. 15–31. ISBN 9789814759410. [CrossRef]

65. Kalashev, O.E.; Kivokurtseva, P.; Troitsky, S. Neutrino production in blazar radio cores. *arXiv* **2022**, arXiv:2212.03151.

66. Paragi, Z.; Godfrey, L.; Reynolds, C.; Rioja, M.J.; Deller, A.; Zhang, B.; Gurvits, L.; Bietenholz, M.; Szomoru, A.; Bignall, H.E.; et al. Very Long Baseline Interferometry with the SKA. In Proceedings of the Advancing Astrophysics with the Square Kilometre Array (AASKA14), Giardini Naxos, Italy, 9–13 June 2014; p. 143. [CrossRef]

67. Murphy, E.J.; Bolatto, A.; Chatterjee, S.; Casey, C.M.; Chomiuk, L.; Dale, D.; de Pater, I.; Dickinson, M.; Francesco, J.D.; Hallinan, G.; et al. The ngVLA Science Case and Associated Science Requirements. In *Science with a Next Generation Very Large Array*; Murphy, E., Ed.; Astronomical Society of the Pacific Conference Series; ASP: San Francisco, CA, USA, 2018; Volume 517, p. 3. [CrossRef]

68. Selina, R.J.; Murphy, E.J.; McKinnon, M.; Beasley, A.; Butler, B.; Carilli, C.; Clark, B.; Durand, S.; Erickson, A.; Grammer, W.; et al. The ngVLA Reference Design. In *Science with a Next Generation Very Large Array*; Murphy, E., Ed.; Astronomical Society of the Pacific Conference Series; ASP: San Francisco, CA, USA, 2018; Volume 517, p. 15. [CrossRef]

69. Ivezić, Ž.; Kahn, S.M.; Tyson, J.A.; Abel, B.; Acosta, E.; Allsman, R.; Alonso, D.; AlSayyad, Y.; Anderson, S.F.; Andrew, J.; et al. LSST: From Science Drivers to Reference Design and Anticipated Data Products. *Astrophys. J.* **2019**, *873*, 111. [CrossRef]

70. Lipunov, V.M.; Kornilov, V.G.; Zhirkov, K.; Gorbovskoy, E.; Budnev, N.M.; Buckley, D.A.H.; Rebolo, R.; Serra-Ricart, M.; Podesta, R.; Tyurina, N.; et al. Optical Observations Reveal Strong Evidence for High-energy Neutrino Progenitor. *Astrophys. J. Lett.* **2020**, *896*, L19. [CrossRef]

71. Plavin, A.V.; Burenin, R.A.; Kovalev, Y.Y.; Lutovinov, A.A.; Starobinsky, A.A.; Troitsky, S.V.; Zakharov, E.I. Hard X-ray emission from blazars associated with high-energy neutrinos. *arXiv* **2023**, arXiv:2306.00960.

72. Actis, M.; Agnetta, G.; Aharonian, F.; Akhperjanian, A.; Aleksić, J.; Aliu, E.; Allan, D.; Allekotte, I.; Antico, F.; Antonelli, L.A.; et al. Design concepts for the Cherenkov Telescope Array CTA: An advanced facility for ground-based high-energy gamma-ray astronomy. *Exp. Astron.* **2011**, *32*, 193–316. [CrossRef]

73. Cherenkov Telescope Array Consortium; Acharya, B.S.; Agudo, I.; Al Samarai, I.; Alfaro, R.; Alfaro, J.; Alispach, C.; Alves Batista, R.; Amans, J.P.; Amato, E.; et al. *Science with the Cherenkov Telescope Array*; World Scientific: Singapore , 2019. [CrossRef]

74. Lu, R.S.; Asada, K.; Krichbaum, T.P.; Park, J.; Tazaki, F.; Pu, H.Y.; Nakamura, M.; Lobanov, A.; Hada, K.; Akiyama, K.; et al. A ring-like accretion structure in M87 connecting its black hole and jet. *Nature* **2023**, *616*, 686–690. [CrossRef]

Article

Emission Modeling in the EHT–ngEHT Age

Richard Anantua [1,2,3,*], Joaquín Dúran [1,†], Nathan Ngata [4,†], Lani Oramas [1,*,†], Jan Röder [5,†], Razieh Emami [3], Angelo Ricarte [2,3], Brandon Curd [1,2,3], Avery E. Broderick [6,7,8], Jeremy Wayland [9,10], George N. Wong [11,12], Sean Ressler [13], Nitya Nigam [2,3,14] and Emmanuel Durodola [15]

1   Department of Physics & Astronomy, The University of Texas at San Antonio, One UTSA Circle, San Antonio, TX 78249, USA
2   Black Hole Initiative at Harvard University, 20 Garden Street, Cambridge, MA 02138, USA
3   Center for Astrophysics|Harvard & Smithsonian, 60 Garden Street, Cambridge, MA 02138, USA
4   Claudia Taylor Lady Bird Johnson High School, San Antonio, TX 78259, USA
5   Max-Planck-Institut für Radioastronomie, Auf dem Hügel 69, D-53121 Bonn, Germany
6   Perimeter Institute for Theoretical Physics, 31 Caroline Street North, Waterloo, ON N2L 2Y5, Canada
7   Department of Physics and Astronomy, University of Waterloo, 200 University Avenue West, Waterloo, ON N2L 3G1, Canada
8   Waterloo Centre for Astrophysics, University of Waterloo, Waterloo, ON N2L 3G1, Canada
9   Institute of AI for Health at Helmholtz Munich, D-85754 Munich, Germany
10   Department of Mathematics, The Technical University of Munich, D-85748 Garching, Germany
11   School of Natural Sciences, Institute for Advanced Study, 1 Einstein Drive, Princeton, NJ 08540, USA
12   Princeton Gravity Initiative, Princeton University, Princeton, NJ 08544, USA
13   Kavli Institute for Theoretical Physics, University of California Santa Barbara, Kohn Hall, Santa Barbara, CA 93107, USA
14   Department of Physics, Columbia University, 538 West 120th Street, New York, NY 10027, USA
15   Department of Physics and Astronomy, Dartmouth College, 14 North Main Street, Hanover, NH 03755, USA
*   Correspondence: richard.anantua@utsa.edu (R.A.); lani.oramas@utsa.edu (L.O.)
†   These authors contributed equally to this work.

Citation: Anantua, R.; Dúran, J.; Ngata, N.; Oramas, L.; Röder, J.; Emami, R.; Ricarte, A.; Curd, B.; Broderick, A.E.; Wayland, J.; et al. Emission Modeling in the EHT–ngEHT Age. *Galaxies* **2023**, *11*, 4. https://doi.org/10.3390/galaxies11010004

Academic Editor: Dimitris M. Christodoulou

Received: 11 November 2022
Revised: 9 December 2022
Accepted: 15 December 2022
Published: 23 December 2022

Abstract: This work proposes a methodology for testing phenomenologically motivated emission processes that account for the flux and polarization distribution and global structure of the 230 GHz sources imaged by the Event Horizon Telescope (EHT): Messier (M)87* and Sagittarius (Sgr) A*. We introduce into general relativistic magnetohydrodynamic (GRMHD) simulations some novel models to bridge the largely uncertain mechanisms by which high-energy particles in jet/accretion flow/black hole (JAB) system plasmas attain billion-degree temperatures and emit synchrotron radiation. The "Observing" JAB Systems methodology then partitions the simulation to apply different parametric models to regions governed by different plasma physics—an advance over methods in which one parametrization is used over simulation regions spanning thousands of gravitational radii from the central supermassive black hole. We present several classes of viewing-angle-dependent morphologies and highlight signatures of piecewise modeling and positron effects, including a MAD/SANE dichotomy in which polarized maps appear dominated by intrinsic polarization in the MAD case and by Faraday effects in the SANE case. The library of images thus produced spans a wide range of morphologies awaiting discovery by the groundbreaking EHT instrument and its yet more sensitive, higher-resolution next-generation counterpart, ngEHT.

Keywords: accretion disk; relativistic jet; GRMHD

1. Introduction

With some of the highest-resolution images ever obtained in astronomy, the Event Horizon Telescope has probed the horizon scale of the supermassive black holes M87* [1] and Sagittarius A* [2]. Incidentally, both sources possess ringlike morphologies with diameters of 42 and 52 µas, respectively [1,2]. Based on a seminal 2017 data collection campaign, the ring imaged around M87* was seen in 2019 to possess brightness asymmetry dominated by a Doppler-boosted Southern bright spot, indicating a black hole spin direction pointing away from the Earth [3]. After the observations in 2017 and publication of the initial total intensity in 2019, the M87 image in linearly polarized light was published in 2021, revealing a dynamically important poloidal B-field threading a plasma with a polarization pattern that spiraled azimuthally into the hole and electrons with inferred synchrotron temperatures from 10 to 120 billion K [4]. Following this milestone, this year, the EHT published an image of the supermassive black hole at our Galactic Center, likewise observed in 2017, revealing a ring with a face-on inclination of $i < 50°$ and azimuthal hotspots in most reconstructions [5], affirming work done by the very large infrared telescopes of GRAVITY four years prior [6].

Vast simulation libraries have modeled tens of thousands of parameter combinations to infer properties of such accreting black hole systems, including black hole spin, surrounding magnetic flux, and emitting particles' thermodynamics [3,7]. The M87 image libraries produced by the EHT Collaboration's Theory and Simulations Team [3] strongly favored non-zero black hole spin values in order to exceed lower limits for the relativistic jet power in concordance with the Blandford–Znajek mechanism [8]. Some of the greatest uncertainties, however, lie in the interpretation of the heating mechanisms required to produce up-to-billion-degree bright features and the overall flux distribution of not only the emitting rings of M87 and Sgr A*, but of jet/accretion flow/black hole, or JAB, systems in general. To this end, we have developed "Observing" JAB Systems to bridge state-of-the-art simulations and cutting-edge observations.

2. Methodology

2.1. "Observing" JAB Systems

The "Observing" JAB Systems pipeline can be summarized as follows:

- Start with a general relativistic magnetohydrodynamic (GRMHD) simulation or semi-analytic model of a jet (or outflow)/accretion flow/black hole (JAB) system
- Convert GRMHD variables into radiation prescriptions for emission, absorption, polarization, particle acceleration, and/or dissipation to emulate sources, using piecewise models when appropriate to assign parametrizations to each distinct JAB system region
- Add a realistic, synthetic "observer" in postprocessing—which includes all radiating species that significantly contribute to radiative transfer—in order to view sources—specifically, images, spectra, light curves, and Stokes maps.

Repeated applications of "Observing" JAB Systems to broad classes of phenomenological processes can naturally lead to model feature libraries with significant clusters in parameter space, as shown in our application to Sgr A* [9]. Note the provisos in (2) and (3) are often overlooked without adherence to this methodology.[1] We illustrate the importance of local piecewise modeling and the inclusion of significant radiating particle species, such as positrons, for our M87 application.

2.2. GRHMD

The first step in "Observing" JAB Systems makes use of the powerful simulated plasma physics laboratories produced by general relativistic magnetohydrodynamic (GRMHD) numerical methods. GRMHD methods are (typically) conservative in mass ($\nabla_\mu (\rho u^\mu) = 0$) and stress–energy–momentum ($\nabla_\mu T^\mu_\nu = 0$);[2] however, recent advances have enabled the inclusion of dissipative effects, such as viscous heating and heat conduction [11], as well as simulations of both thick torii and thin disks [12] and different implementations of the interaction between radiation and the background plasma fluid [13].

We use implementations of the HARM method [14,15] as a testbed for emission models. The simulations presented in this work were generated by using descendants of the harm code [14,15], a conservative second-order explicit shock-capturing finite-volume method for solving the equations of ideal GRMHD in arbitrary stationary spacetimes. On a coordinate basis, the governing equations are

$$\partial_t\left(\sqrt{-g}\rho u^t\right) = -\partial_i\left(\sqrt{-g}\rho u^i\right), \tag{1}$$

$$\partial_t\left(\sqrt{-g}T^t{}_\nu\right) = -\partial_i\left(\sqrt{-g}T^i{}_\nu\right) + \sqrt{-g}T^\kappa{}_\lambda\Gamma^\lambda{}_{\nu\kappa}, \tag{2}$$

$$\partial_t\left(\sqrt{-g}B^i\right) = -\partial_j\left[\sqrt{-g}\left(b^j u^i - b^i u^j\right)\right], \tag{3}$$

along with the constraint

$$\partial_i\left(\sqrt{-g}B^i\right) = 0, \tag{4}$$

where the plasma is defined by its rest mass density ρ and its four-velocity u^μ, and b^μ is the magnetic field four-vector. More details about the simulation process can be found in [16].

All simulations were initialized from a Fishbone–Moncrief (FM) torus [17] with a spin of $a = 0.5$ (Sgr A*) or $a = -0.5$ (M87) (the latter is loosely based on a well-performing model with respect to the M87 polarization constraints in [4]). The Sgr A* simulations were performed by using the `iharm3d` code [18]. The M87 simulations were produced by using `iharm3d`'s kokkos/GPU-based descendent, kharma.[3] Other physically motivated flow geometries are possible. The Bondi spherical accretion flow solution [19] is analytically tractable, but requires an exquisite degree of symmetry and non-magnetized flows to be realistic. Magnetized Bondi flows in numerical calculations display similar properties to those of torus simulations in terms of horizon-scale emission [20,21]. Stellar winds from a small population of ~ 30 Wolf–Rayet stars were theorized, e.g., in [22–25] to source the disk and inflow of Sgr A* in a model that reasonably accounted for the diffuse X-ray emissions. However, the precise knowledge of the position and orbits of this population is currently only possible in our own Galactic Center, while the FM torus model can be generalized to other AGNs. This generalization includes the possibility that the FM torus can be tilted with respect to the black hole spin (a situation that could arise often in a low-luminosity AGN), but we neglect this possibility in our work, which is focused on emission physics.

The initial magnetic field geometry is computed as the curl of a prescribed axisymmetric electromagnetic vector potential $A_\phi(r,\phi)$, which is computed at simulation zone corners. In the SANE case, the vector potential is linear in the plasma density $A_\phi \sim \rho$; in the MAD case, $A_\phi \sim \rho r^3 e^r$. The field strength is normalized so that the ratio of the maximum gas pressure to the maximum magnetic pressure over the domain is 100. The simulations are performed in a cylindrified, modified spherical Kerr–Schild coordinate system that concentrates the resolution near the midplane and close to the event horizon. The domain extends to at least $r_{\mathrm{max}} = 1000\,GM/c^2$ with a resolution of $N_r \times N_\theta \times N_\phi = 288 \times 128 \times 128$. More details about the simulation procedure and the initial conditions can be found in [16].

Magnetic flux is a key distinguishing factor among accreting plasmas. The time-averaged magnetic flux $\Phi = (1/2)\int_\theta\int_\varphi |B^r| dA_{\theta\varphi} / \sqrt{\dot{m}r_g}c$ determines two distinct regimes. For $\Phi \gtrsim 50$, the disk is magnetically arrested (MAD) by its own magnetic pressure as it plunges into the hole. A much smaller Φ governs standard and normal evolution (SANE). Analyses of EHT simulation libraries tend to prefer MAD models over SANE, e.g., with MAD $a = -0.5$ earning the highest average image score for M87 with the parametric likelihood estimation procedure THEMIS [26].

2.3. Plasma-Heating-Based Emission Models

2.3.1. $R - \beta$ Model

In general, GRMHD simulations evolve only the bulk of a fluid, i.e., the dynamically important ions.[4] Therefore, in radiative post-processing, we seek to bridge the ions to the radiating electrons by the ratio of their temperatures T_i/T_e. In a hot, low-density, collision-less plasma, electrons can radiate and, therefore, cool efficiently, while the ion cooling through Coulomb collisions is suppressed. The heating processes believed to be in action, e.g., viscous, compressional, and turbulent heating, can have similarly asymmetric effects on the temperatures and are more poorly understood [27]. Based on the tendency of plasma turbulence to preferentially heat electrons at low gas-to-magnetic pressure ratios $\beta = P_g/P_B$ and ions at high β [28,29], the $R - \beta$ model is well motivated (e.g., [10,30–33]):

$$R = \frac{T_i}{T_e} = \frac{\beta^2}{1+\beta^2}R_{\text{high}} + \frac{1}{1+\beta^2}R_{\text{low}}. \tag{5}$$

The R-β model is the primary model used by the EHT Collaboration [3,34]. The electron-to-ion temperature ratio asymptotically reaches $1/R_{\text{low}}$ in the low β regime and $1/R_{\text{high}}$ in the high β regime. In turn, this means that R_{high} controls the electron temperature in the disk or torus, and R_{low} governs T_e in the jet or wind outflow. Often, R_{low} is fixed to 1 and only R_{high} is varied, since by normalizing the flux to fit observations, the jet appears comparatively brighter upon an increase in R_{high}, i.e., upon a decrease in T_e in the disk (see also Figure 3 in [10]). The $R - \beta$ model has been extensively compared to GRMHD simulations that readily employ heating models [35,36], concluding that on scales probed by the EHT, the $R - \beta$ model approximates the influences of electron heating physics reasonably well.

2.3.2. Critical β Model

This alternative turbulent heating model has an exponential parameter β_c that controls the transition between electron- and ion-dominated heating [9]

$$\frac{T_e}{T_e + T_i} = f e^{-\beta/\beta_c}. \tag{6}$$

The exponential parametric control over the maximum β contributing to the emitting region is the basis for distinctly different near-horizon electron heating behavior compared to the R-β model.

We compare the R-β and Critical Beta models in Figure 1. For the same range of electron-to-proton temperature ratios, the Critical Beta model can have a sharper or smoother decline in the emission contributions from the highest β regions, smoothly transitioning from funnel/outflow- to near-horizon/inflow-dominant heating profiles as β_c increases.

At low β, the electron-to-ion temperature ratio approaches parametrically determined maximum values (f and $1/R_{\text{low}}$) for the Critical Beta and R-Beta models, respectively. However, at high β, the electron temperature always asymptotes to 0 in the former model and is adjustable (through R_{high}) only for the latter model. The exponential rate of electron temperature fall-off in the Critical Beta parametrization should have testably different spectral properties, such as the lowering of the bremsstrahlung contribution to the spectral energy distribution.

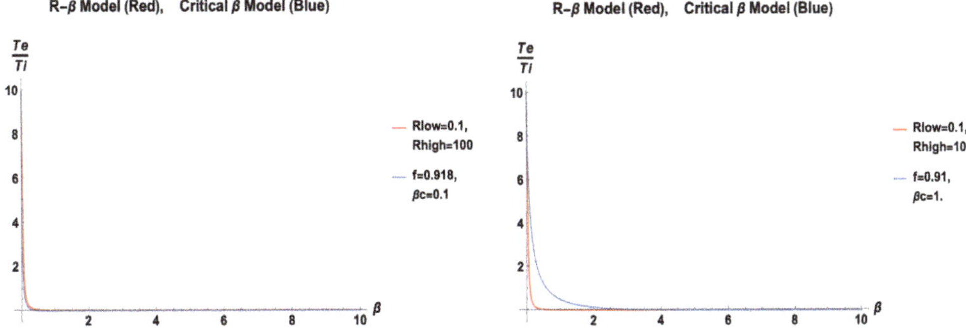

Figure 1. Electron-to-proton temperature ratio variation between $T_e/T_i = 10$ and ~ 0 for the R-Beta and Critical Beta Models as a function of β. The electron-to-ion temperature ratio varies as $T_e/T_i = \frac{1}{(1+\beta^2)^{-1}R_{\text{low}} + (1+\beta^2)^{-1}\beta^2 R_{\text{high}}}$ for the R-β Model and $\frac{fe^{-\beta/\beta_c}}{1-fe^{-\beta/\beta_c}}$ for the Critical Beta Model. For the R-Beta parameters $(R_{\text{low}}, R_{\text{high}}) = (0.1, 100)$ and Critical Beta parameters $(f, \beta_c) = (0.918, 0.1)$ (**Left**), the models nearly coincide. For the R-beta parameters $(R_{\text{low}}, R_{\text{high}}) = (0.1, 100)$ and Critical Beta parameters $(f, \beta_c) = (0.91, 1)$ (**Right**), the Critical Beta model has a softer transition.

2.4. Sub-Equipartition-Based Models

2.4.1. Constant β_e

One of the simplest yet most powerful models for understanding jet emission in JAB systems is the Constant Electron Beta Model [37,38]:

$$P_e = \beta_{e0}P_B. \tag{7}$$

This model exploits our analytic knowledge of relativistic jets as being commonly generated by the Blandford–Znajek mechanism [8], which converts energy from accreting magnetized plasmas and spinning black holes into Poynting flux relativistic electromagnetic outflows with a power of $P_{BZ} \propto a^2\Phi_B^2$. We now assume that a fixed fraction β_{e0} of jet magnetic energy is available for the emission observed at radio VLBI frequencies, where β_{e0} is related to the efficiency of conversion.

2.4.2. Magnetic Bias

This class of model generalizes the Constant β_e model by relating the pressure P_e of relativistic emitters (electrons or positrons) to the conversion of magnetic energy into particle energy through the powers n of the magnetic pressure:

$$P_e = K_n P_B^n \sim B^{2n}. \tag{8}$$

The constant K_n ensures that the right-hand side has units of pressure, and it can be estimated from a simulation by taking the average value of B^N/B^2 (where $N = 2n$) over a simulation surface enclosing the flux threading the black hole. The Bias Model parameter n helps modulate jet collimation in circumstances in which azimuthal magnetic fields are expected to scale simply with the cylindrical radius [37,39], e.g., the Blandford–Königl [40] model, where $B \sim B_\phi \sim r$ for radio jets.

2.5. Hybrid Models

We recognize the broad diversity of plasma regimes represented in a typical GRMHD field of view. Disk plasmas at large radii (a few tens or hundreds of M) are often not in inflow/outflow equilibrium [9], and thus, many of the thermal emission modeling approaches fail. As a Keplerian or sub-Keplerian disk approaches the horizon, its behavior near the plunging region beyond the innermost stable circular orbit (ISCO) is heavily dependent on SANE and MAD. For the MAD case, the disk can barely trickle into the supermassive black hole. For SANE flows, plasma can continuously flow. In both cases, the black hole spin can interact magnetohydrodynamically with disk plasma to form stable relativistic jets in simulations [41]. These jets, which are of underdense material themselves, are subjected to instabilities that are often distinct from those found in the disk, such as $m = 0$ pinch, $m = 1$ kink, and magnetic Kelvin–Helmholtz instabilities.

The inflow/outflow division in JAB systems gives an impetus for generating piecewise models. Jet plasmas are characterized by their low density and high energy, making the magnetization

$$\sigma = \frac{b^2}{\rho},\tag{9}$$

which is a natural demarcation for the transition from jet-dominated emission to accretion-flow-dominated emission. When jets are well collimated, we may also use parabolic or other geometric cuts to isolate the jet region [37,39,42,43].

2.6. Phenomenological Models

We postulate other mechanisms here (detailing their full functional form in the model compendium in Appendix A.1) by noting the synchrotron emissivity $j_\nu \propto P_e$, where the pressure of relativistic emitters can be written as $P_e \propto W t_{\mathrm{cool}}$ in terms of the dissipation rate per unit time W and the cooling time t. In the following, we proceed to specify more potential models in our arsenal, systematically carrying out the second step of "Observing" JAB Simulations by relating the pressure of relativistic emitters to energetic processes in an AGN.

The Current Density Model relates the dissipation of energy into emitting particles to the current density $W \propto j^2$. This is seen to trace a co-axial current morphology with a central outgoing current and a return current, thus creating a boundary layer in jets [39].

The Jet Alpha Model parametrizes the efficiency of linear momentum transport in jets by α_J, i.e., $W \propto \alpha_J$, in a manner analogous to that of the Shakura–Sunyaev model for angular momentum transport in accretion disks [44]. This dissipation rate is also linear in the shear stress, as seen in Appendix A.1, thus enabling the disk–jet interface to be visible [39].

Lastly, the Shear Model adopts a Newtonian framework for velocity shear $\tau = \mu S$, where μ is the dynamical viscosity and S is the shear stress ($|dv_z/ds|$ in cylindrical coordinates). Then, $W \propto \tau S \propto S^2$. This quadratic dependence on shear stress gives us an edge-brightened model relative to the Jet Alpha Model.

These jet models may naturally be glued to disk and corona models (especially those based on turbulent heating), as they capture the behavior of the inflow/outflow interface. Together, these phenomenological emission prescriptions may form the building blocks for detailed hybrid models that account for different plasma flow physics throughout the JAB system.

2.7. Electron Distribution Functions

A fixed temperature for electron thermodynamics models is often an idealization when particles are found in nearly collisionless plasmas (such as in JAB systems with mean free paths of $\mathcal{O}(10^5 M)$). The influence of a non-thermal electron population on horizon- and jet-base-scale emission remains an important subject of investigation in the EHT Collaboration [3,7]. These non-thermal distributions need not be applied globally in the GRMHD domain. Rather, determining the proper region's home for charged particle acceleration processes is an integral part of these studies. We explore some possibilities for particle acceleration and the concomitant energy distributions below.

2.7.1. Power Law

One of the simplest phenomenologically viable assumptions for the energy distribution of a population of n_e emitting particles in an astrophysical plasma is a power-law energy decay:

$$\frac{dn_e}{d\gamma} = K\gamma^{-p} \tag{10}$$

The normalization factor K depends on the synchrotron pressure. Power-law particle distributions are naturally produced in shock waves through diffusive shock acceleration, whereby particles are energized by repeated interactions with magnetic inhomogeneities as they propagate alongside the shock. This is a first-order Fermi process, as the energy gain of the particles is linear in the shock velocity. Astrophysical shocks may occur at the interface of fluids with differing velocities, such as the interface of a jet with its ambient medium.

2.7.2. The Kappa Model

It is often preferable to be able to model the full SED with a single distribution function. This can be achieved with the kappa electron energy distribution function (e.g., [45,46]), which has its theoretical foundation in non-extensive Tsallis statistics [47,48]. Looking like a thermal distribution at low energies, it smoothly transitions into a non-thermal power-law tail with index s towards high energies, so that $\kappa = 1 + s$ (see, e.g., Figure 4 in [10]). The thermal [49] and kappa [50] electron energy distribution functions read

$$\frac{dn_e}{d\gamma_e \, d\cos\zeta \, d\phi} = \begin{cases} \dfrac{n_e}{4\pi\Theta_e} \dfrac{\gamma_e\left(\gamma_e^2 - 1\right)^{1/2}}{K_2(1/\Theta_e)} \exp\left(-\dfrac{\gamma_e}{\Theta_e}\right), & \text{thermal} \\[2ex] \dfrac{N}{4\pi}\gamma_e\left(\gamma_e^2 - 1\right)^{1/2}\left(1 + \dfrac{\gamma_e - 1}{\kappa w}\right)^{-(\kappa+1)}, & \text{kappa} \end{cases}$$

where n_e is the electron number density, ϕ is the gyrophase, γ_e is the electron Lorentz factor, ζ is the electron pitch angle, and K_2 is the modified Bessel function of the second kind. From these distribution functions, emission and absorption coefficients are determined by using fit functions [51]. w is the "width" of the distribution and describes the energy in the system.

From particle-in-cell (PIC) simulations of magnetized current sheets, it is evident that the kappa index is not constant in all sub-regions of the system, requiring kappa to be variable and dependent on plasma quantities [52]. Further, since many GRMHD codes do not provide accurate values close to the jet spine due to boundary conditions, the inner spine is usually excluded from the emission by imposing a maximum in the magnetization σ_{cut} (e.g., [10,31,32]).

Additionally, the distribution function can be modified to account for a thermal and a magnetic contribution to the total energy [10,31,32,53]. That way, it is possible to control the amount of magnetically accelerated electrons and the distance of their point of injection into the jet from the central engine.

It is important to distinguish the influences of the individual dials of this non-thermal model on the image morphology. This requires extensive parameter surveys, which would be too computationally expensive in a two-temperature GRMHD simulation. Therefore, an implementation in radiative post-processing is currently the only feasible option for these surveys [10]. Increasing ε or σ_{cut} raises the SED at energies past the synchrotron turnover, with a growing influence towards the highest energies (e.g., Figure 12 in [10]).

2.8. Emission Modeling in Non-Kerr Spacetimes

For over a century, Einstein's theory of general relativity has been thoroughly tested in many different ways (e.g., [54–63]). In the era of the EHT and ngEHT, a test of GR in an imaging-based approach is on the horizon for the first time. Specifically, the shape and size of the black hole shadow and photon ring are crucial characteristic properties of horizon-scale images. Photons rings in the Kerr metric are predicted to be nested in subrings with exponentially decreasing separation, the lowest orders of which may be observed with the aid of longer baselines in the ngEHT [64]. In addition to this observational test of the Kerr solution for the spacetime around astrophysical black holes, many studies on alternative theories to general relativity make use of semi-analytical models for both plasma accretion and emission processes to compare models to observations (e.g., [65–67]). Alternative spacetime geometries are rarely investigated in full GRMHD and GRRT, and if so, the emission physics are based on a constant proton-to-electron temperature ratio and purely thermal radiation [68,69]. Only recently have advances been made to study the influence of emission models more akin to reality in an alternative spacetime [33,53]. The fundamental difficulties in finding deviations from the Kerr metric arise from the presence of greater astrophysical uncertainties. For example, the magnetic field configuration in a GRMHD simulation appears to have a much larger influence on the source morphology on horizon scales than the background spacetime [33,53]. Moreover, the differences in image features, such as the shadow size or photon sub-ring spacing caused by a deviation from GR, are often small and subject to degeneracy with accretion and emission models. Still, past EHT observations of black hole shadows have helped constrain alternative theories to GR [65,70]. With the help of the ngEHT, we aim to resolve the degeneracy between the effects of plasma physics and GR and put even more robust constraints onto alternative spacetime geometries.

3. Commencing the Computing: Emission Models in Numerical Codes

The emission models discussed above have been implemented in a variety of numerical codes (e.g., [71–77]; for an extensive comparison, see [78]). Since GRMHD simulations are already computationally expensive, radiation is commonly modeled in post-processing. While there are a handful of radiative GRMHD (GRrMHD) codes (e.g., [36]), when evolving a two-temperature plasma while accounting for ions and electrons, the computational cost only increases. In the case of pure jet simulations, however, it is possible to bring the cost down by moving to special relativity (e.g., [79–83]).

On event horizon scales, i.e., in the strong-gravity regime, we are required to take the full effects of GR into account. Usually, radiative transfer codes first calculate light rays by integrating the geodesic equation and, subsequently, solve the radiative transfer equation along those rays. The use of post-processing enables us to freely investigate a variety of emission models, with many going beyond thermal synchrotron radiation. For example, Compton/inverse Compton scattering, bremsstrahlung, and non-thermal emission processes all have their own imprints on the SED and the image morphology in both total intensity and polarized light [43,76].

Positrons' Effects on Radiative Transfer

Positrons in JAB systems can be produced through photon–photon (Breit–Wheeler) interactions in jet funnel walls [84] and spark gap processes near the magnetospheric poles of supermassive Kerr black holes [85]. Each of these processes may contribute electron–positron pair densities exceeding the Goldrich–Julien value [86], which is required in order to screen the large-scale electric fields responsible for high-energy lepton cascades. In sources such as M87 that have significant pair production through these channels, positron effects abound. These include the increase in linear and vanishing of circular polarization, as well as the higher energy fall-off of the circular polarization spectrum, all of which have been modeled as potential discriminators of JAB systems that are rich in ionic versus pair plasma [43,83].

4. Results: Adding an Observer

We present a suite of parametric emission models and illustrate them in GRMHD simulations that are scaled to the first horizons observed: M87* and Sgr A*. We start by painting a single semi-MAD Sgr A* simulation with several phenomenologically motivated prescriptions of electron thermodynamics. Thus, by using the second step of the "Observing" JAB Simulations pipeline, we illustrate the sensitivity of the interpretations of near-horizon structure in real-world observations to our models of plasma physics. We then use a SANE and MAD M87 simulation to reveal widely divergent—and observationally distinguishable—polarized emission signatures governed by Faraday and positron effects. The latter effects enable us to better constrain a fundamental, though largely uncertain, property of relativistic jets: ion- versus lepton-dominated composition. In what follows, we use perceptually uniform color maps to identify image morphology with the underlying emission mechanisms that energize relativistic particles to produce synchrotron radiation.

4.1. Sgr A*

4.1.1. Parametric Model Comparison

We explore the Critical Beta model parameter space from $f \in \{0.1, 0.5\}$ and $\beta_c \in \{0.01, 0.1, 1\}$, starting with an edge in view to highlight the lensing effects on image morphology in Figure 2. As the f parameter increases, the overall electron temperature increases, and changes in the image morphology at fixed flux are effected through the M_{unit} used in codes to scale the relative importance of inertial plasma properties, such as the density and mass accretion rate, relative to the plasma's electromagnetic properties. As the critical beta parameter increases, the locus of the greatest electron contributions moves from the low β outflow to higher β regions that are lensed around a compact crescent near the horizon due to higher values of β being found in the inflow around the black hole. Thus, β_c serves as a dial in JAB emission modeling for compactifying emitting regions and asymmetrizing them from an edge-on orientation, as shown in Figure 2.

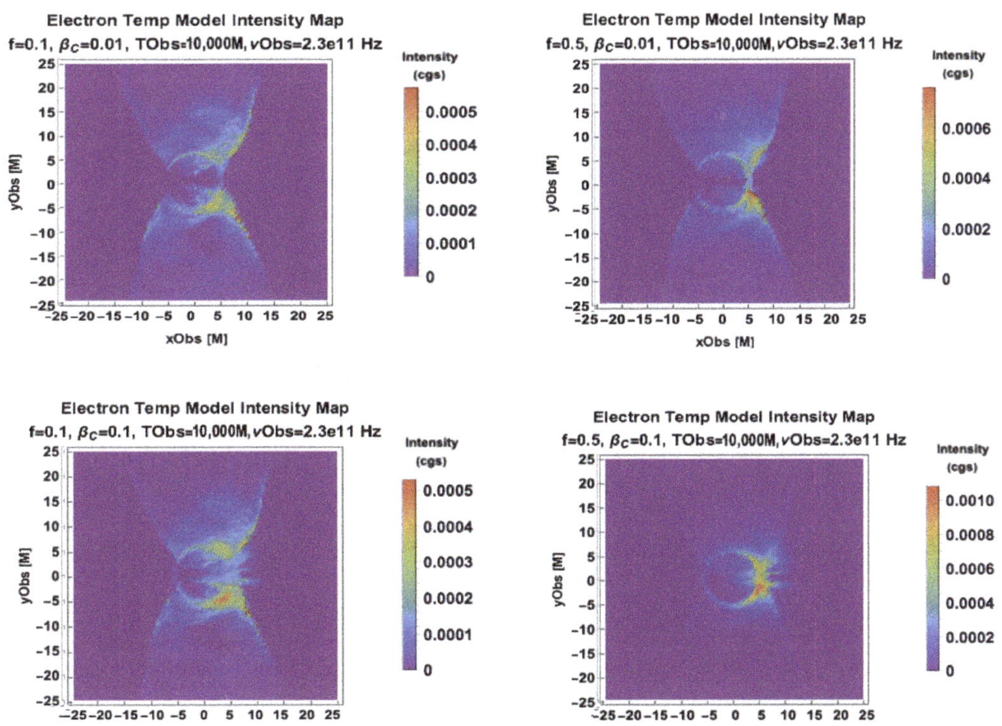

Figure 2. *Cont.*

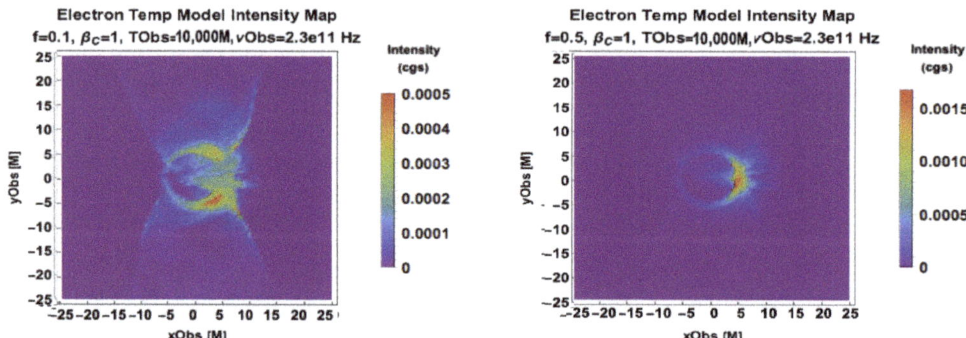

Figure 2. A parameter scan of the Critical Beta Model with (**Left**) $f = .1$, (**Right**) $f = 0.5$, $\beta_c = 0.01$ (**Top**), $\beta_c = 0.1$ (**Middle**), and $\beta_c = 1$ (**Bottom**) from an edge-on view. For Sgr A* models, the cgs conversion into Jy is found by multiplying each cgs-intensity-colored pixel value by 57.9 to get its flux density in Jy [9]. The scale is $M \equiv GM_{\text{BH}}/c^2$ when used as a length and $\equiv GM_{\text{BH}}/c^3$ when used as time. The notation 1e2 is a compact form of the scientific notation 1×10^2.

In Figure 3, we see different morphologies associated with our equipartition-inspired Constant β_e and Magnetic Bias models. The locus of emission for the Constant Electron Beta Model approaches the funnel for low β_{e0} and broadens into a thick, lensed torus for higher β_{e0}. When the bias parameter goes to 0, the pressure from relativistic electrons goes to a constant and does not sharply decline with the radius, leaving extended outflow signatures.

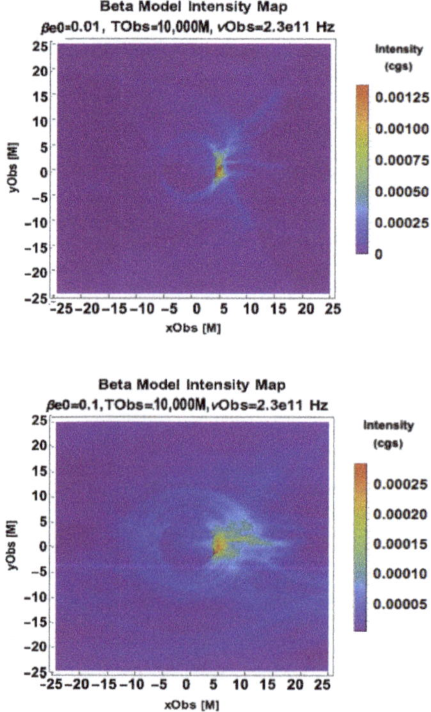

Figure 3. *Cont.*

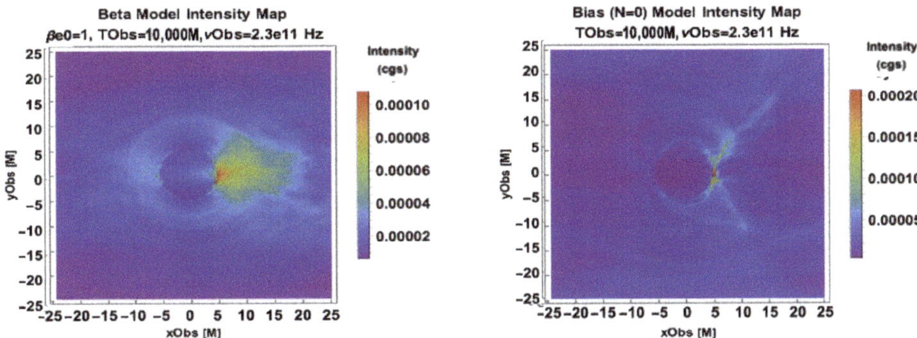

Figure 3. A parameter scan of the Constant Electron Beta Model (**Left**) with $f = 0.1$, (**Right**) $f = 0.5$, $\beta_{e0} = 0.01$ (**Top**), $\beta_{e0} = 0.01 = 0.1$ (**Middle**), and $\beta_{e0} = 1$ (**Bottom**). Magnetic Bias Model with $\beta_{e0} = 1$ and $N = 0$ (**Bottom Right**).

We now change the orientation and consider face-on models in Figure 4. An Sgr A* 20° spin axis orientation was observationally preferred by GRAVITY [6] , and this was corroborated by the EHT, disfavoring inclination angles above 50°. Here, tuning up the critical value of beta, β_c, still leads to more compact images, but now they maintain a ring symmetry even near the gravitational lensing profile of a Kerr black hole.

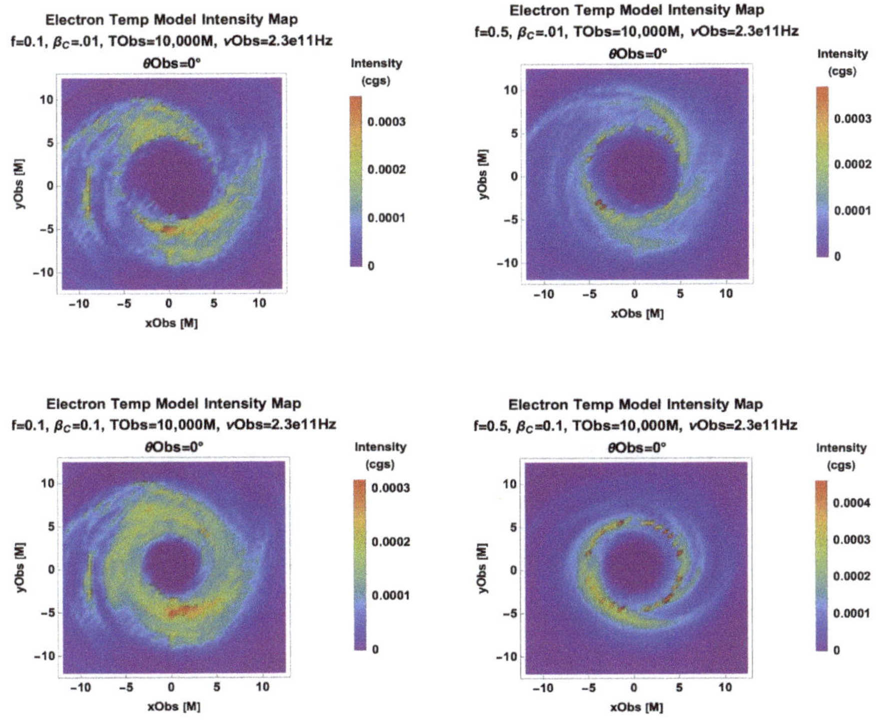

Figure 4. *Cont.*

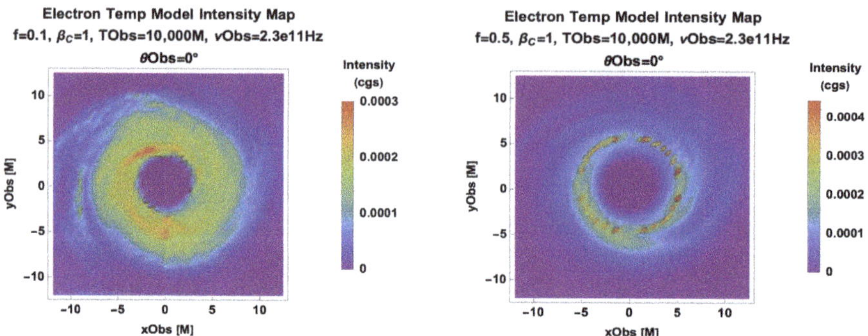

Figure 4. A parameter scan of the face-on Critical Beta Model with (**Left**) $f = 0.1$, (**Right**) $f = 0.5$, $\beta_c = 0.01$ (**Top**), $\beta_c = 0.1$ (**Middle**), and $\beta_c = 1$ (**Bottom**).

4.1.2. Morphological Classification

The ring morphology has dominated the focus of emission modeling since the advent of the first two ring-like images of the horizon scale were released by the EHT in 2019 and 2022. However, we showed examples of the strong inclination dependence of 230 GHz images above. It is worth taking stock of how fortuitous the near-face-on ($\sim 20°$) spin axis viewing angles of Sgr A* and M87* are given that their main selection criterion is the exceptionally large angular width of their horizon gravitational radii as seen from Earth, which is independent of their spin inclination angle.

We preview the edge-on morphologies that the ngEHT will see based on morphological clusters in the parameter space of the Critical Beta, Constant Electron Beta, and Magnetic Bias Models in Figure 5. These types are:

1. A thin, compact asymmetric photon ring/crescent with the best fit or flat spectrum (with the spectral energy distribution shown in [9]);
2. Inflow–outflow boundary + thin photon ring with a steep spectrum;
3. Thick photon ring with spectral excesses at high and low frequencies;
4. Extended outflow and a flat low frequency spectrum with excesses at high and low frequencies.

In Figure 6, we plot the morphological types on a common intensity scale to emulate observing with a single instrument. The EHT's dynamical range of ~ 2 orders of magnitude in intensity is used to set this scale. Morphological degeneracies, e.g., between Types I (large f, β_c) and III (large β_{e0}) and between Types II (small f, β_c) and IV (small n), are more likely to emerge as more regions fall below the flux threshold of the observing instrument. The ngEHT's dynamical range will span a few orders of magnitude. The ngEHT's improved sensitivity down to ~ 5 mJy and increased frequency range to $\gtrsim 345$ GHz [87] will enable us to resolve the currently excluded low-flux density regions in the field of view and at larger radii. This will lead to more accurate determination of the morphological type, thus breaking degeneracies. Thus, as the ngEHT increases the dynamic range of JAB system observations to lower fluxes, we are able to better classify images and pinpoint the underlying emission physics related to their morphologies.

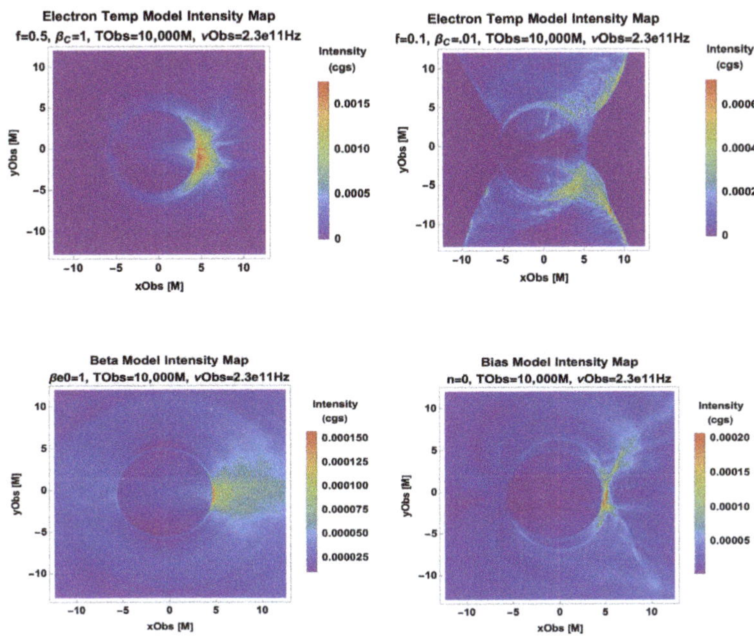

Figure 5. Semi-MAD simulation [9] models that were ray traced at 230 GHz at $T = 10{,}000\ M$: (**Top Left**) best-fit Critical Beta model (**Top Right**) R Beta with $(f, \beta_c) = (0.5, 1)$; (**Bottom Left**) Constant Electron Beta model with $\beta_{e0} = 1$; (**Bottom Right**) Magnetic Bias with $N = 0$ jet.

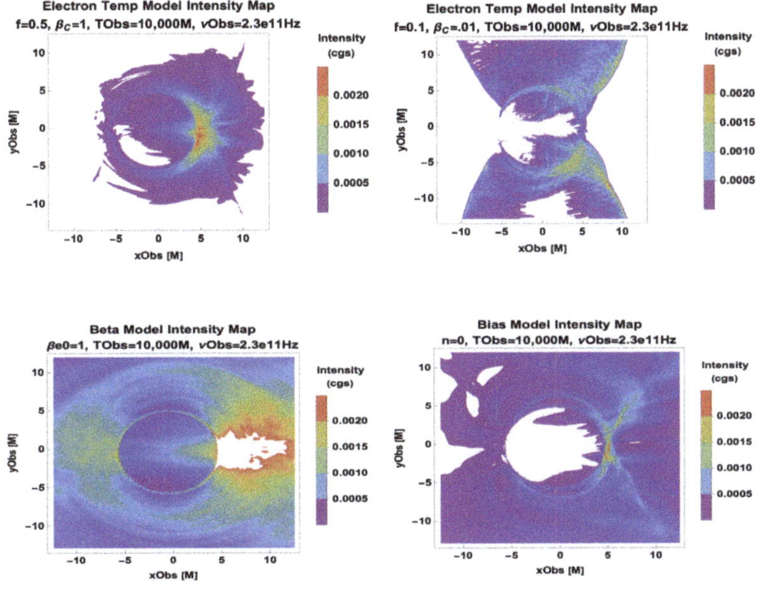

Figure 6. Semi-MAD simulation [9] model classes that were ray traced at 230 GHz at $T = 10{,}000\ M$, as shown in Figure 5, but with a common intensity scale.

4.2. M87

4.2.1. Parametric Model Comparison

We implement hybrid models that include turbulent heating and sub-equipartition emitting regions for M87. We make a natural choice to partition the simulation region according to the magnetization (i.e., the magnetic energy density to enthalpy density ratio $\sigma = \frac{b^2}{\rho}$). The value $\sigma_{\text{transition}} = 1/2$ that determines the jet regions is near the inflow/outflow interface and is less than the σ_{cut} of 2 for these simulations.

A SANE/MAD dichotomy emerges in Figures 7 and 8. The SANE intensity maps are more ring symmetric, whereas the MAD case has a prominent flux loop (lower left of all panels in Figure 8). The SANE circular polarization varies on smaller spatial scales than the MAD does. The sign of the circular polarization changes on sub-M scales for SANE. The SANE and MAD hybrid models that have added a Constant Electron Beta jet for $\sigma > \sigma_{\text{transition}}$ have broader flux distributions over the field of view (all of the right panels in Figures 7 and 8).

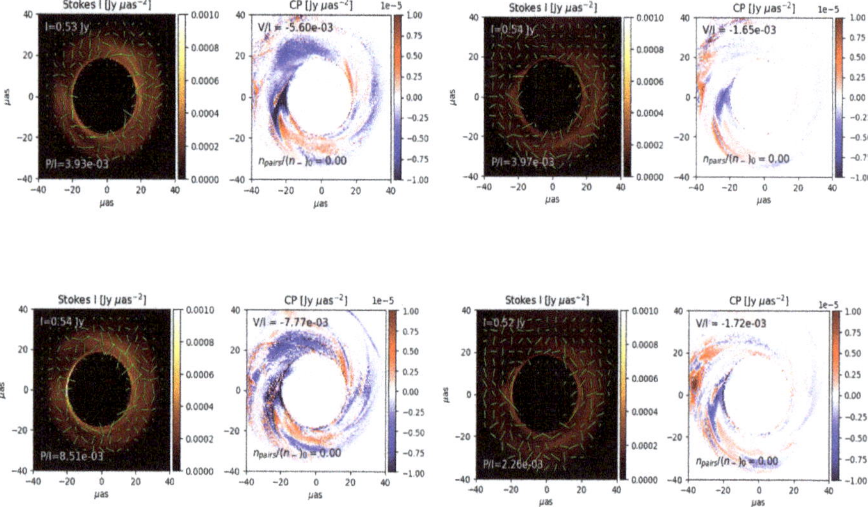

Figure 7. Synthetic intensity with the electric vector polarization angle (EVPA) and circular polarization maps for models that include positron effects and piecewise modeling. For the $a = -0.5$ SANE at 230 GHz and at $T = 25{,}000\,M$: (**Top Left**) R-Beta with a $\beta_{e0} = 0.01$ jet model; (**Top Right**) R-Beta with a $\beta_{e0} = 0.01$ jet; (**Bottom Left**) Critical Beta Model; (**Bottom Right**) Critical Beta with a $\beta_{e0} = 0.01$ jet. For each case, the intensity is overplotted with the electric vector polarization angle on the left panel, and the circular polarization degree is mapped on the right panel.

4.2.2. Positron Effects

We see the effect of the addition of positrons in Figure 9. There, we start with an ionic plasma with a number density of n_0 (see the top panels), and then add enough positron pairs n_+ so that the pair number density equals that of the original electrons (and the unit for M is adjusted to match the normalized flux (see the bottom panels)). In Figure 9, we see that in the MAD case (left), the degree of circular polarization is proportional to the unpaired emitter fraction (1/3 in the case of $f_{\text{pos}} = 1$) [43,88]. The SANE case (right), which has Faraday rotation depths that are thousands of times greater than the less dense, more highly magnetized MAD, does not have a simple linear relationship between the pair content and V/I. In the linear polarization, we also see greater positron effects for SANE than for MAD, as the addition of positrons scrambles the EVPA pattern only in the SANE case.

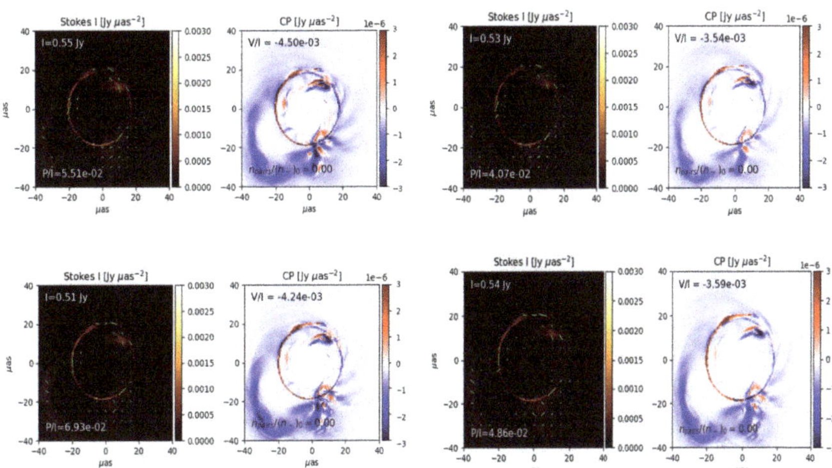

Figure 8. For the $a = -0.5$ MAD at 230 GHz and at $T = 25{,}000\ M$: (**Top Left**) R-Beta Model; (**Top Right**) R-Beta with a $\beta_{e0} = 0.01$ jet; (**Bottom Left**) Critical Beta Model; (**Bottom Right**) Critical Beta with a $\beta_{e0} = 0.01$ jet. For each case, the intensity is overplotted with the electric vector polarization angle on the left panel, and the circular polarization degree is mapped on the right panel.

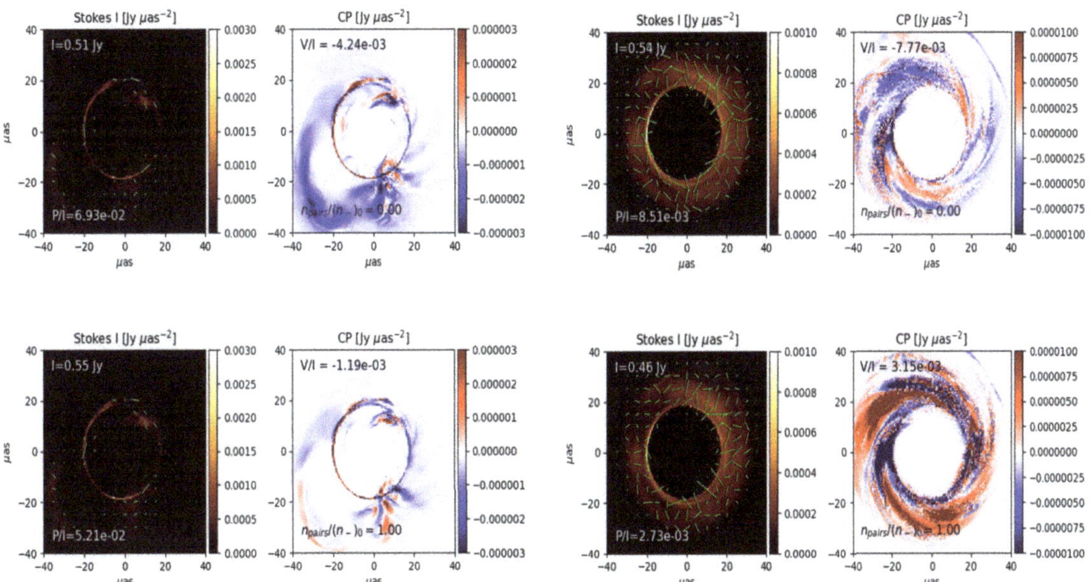

Figure 9. For $a = -0.5$ at 230 GHz and at $T = 25{,}000\ M$: (**Top**) Critical Beta Model without positrons; (**Bottom**) Critical Beta with $n_{\text{pairs}}/n_0 = 1$. We also compare MAD (**Left**) and SANE (**Right**).

5. Discussion and Conclusions

We have seen from a SANE and a MAD M87 simulation and one semi-MAD simulation for Sgr A* that the thermodynamic properties of emitting particle populations can be as important as those of the underlying plasma for predicting the properties of intensity and polarization maps near the horizon. Moreover, special relativistic effects, such as beaming and general relativistic lensing, can produce nonlinear modifications of the flux received by observers away from the JAB system. Understanding the relative contributions of these effects can be facilitated by idealized cases, such as in semi-analytic jet models, e.g., those in [42]. We considered the impacts of some of the competing effects listed here:

- The plasma β controls the emitting region size in turbulent heating models, where parameter combinations with greater emission contributions from low β tend to have more extended outflow/coronal regions, and those with contributions from high β are more compact and dominated by near-horizon inflow, as shown in Figure 2, going from top to bottom.
- Inclination has a pronounced effect on the 230 GHz observer plane image morphology due to special relativistic beaming and the focusing properties of gravitational lensing. Thus, we predict a wide variety of image morphologies beyond ring structures that may be uncovered by the ngEHT, as shown in Figure 6.
- SANE and MAD simulations have widely divergent positron effects that are modulated by the larger Faraday depth of SANE plasmas, which are constrained to achieve the same image fluxes that MADs acquire through magnetic fields, with SANEs having EVPAs that are highly sensitive to positron content and MADs having a circular polarization degree that is greatly suppressed by positrons, as shown in Figure 9.

By using a variety of turbulent heating and magnetic-to-particle-energy-ratio-based emission models, we found a number of distinct image morphologies that we expect to encounter in future observational campaigns with the ngEHT. In particular, the ngEHT will have the resolution to probe images with areas a few times smaller than those probed with the EHT, allowing the imaging of more compact ringlike structures that are viewed face-on (corresponding to high values of Critical Beta, should this turbulent heating mechanism predominate). This will enable us to better determine which processes dominate the emissions. Furthermore, we showed the strong dependence on inclination of the relationship between electron thermodynamics and image morphology. We identified at least four distinct image signatures of turbulent heating and magnetic-to-particle energy ratios, mainly visible in the edge-on view. We also showed the EHT's/ngEHT's ability to discriminate between MAD and SANE accretion states based on ordered vs. scrambled polarization signatures. This dichotomy is accentuated by the presence of positrons, allowing us to probe the plasma composition of JAB systems, which remains a long-standing question.

5.1. Limits of Instrumentation

The 20 µas angular resolution of the EHT (enough to resolve an atom at arm's length or an apple on the Moon) is exquisite according to the VLBI observational standards. More remarkably still, the ngEHT may double this angular resolution. However, the angular gravitational radius of even the largest black hole shadow ($M = 5$ µas for Sgr A*) is still smaller than the size of the synthesized beam of the ngEHT ($\lesssim 20$ µas). This leads to blurring in the image plane, washing out intricate features. Due to this instrumental limitation, numerical simulations are crucial for our understanding of the properties of accretion flows. Though simulations themselves produce grid-resolution-dependent models (cf. Appendix A.2), they remain our most reliable probe below the scale set by the gravitational radius, and will continue to motivate observing missions for the ngEHT and beyond to confirm their predictions.

5.2. Universality of Select Measures

The existence of flaring emissions in Sgr A* and extended synchrotron emissions in M87's jet provide strong evidence for the presence of a high-energy tail to the lepton distribution, i.e., non-thermal lepton distribution functions. This is generally expected as a consequence of the microphysical processes responsible for the dissipation of turbulence and the attendant injection of energy explored above. However, the detailed shape and magnitude of this additional tail depend sensitively on the specific acceleration mechanism considered. Nevertheless, for the physical parameters that are relevant for Sgr A* and M87, below 690 GHz, the synchrotron coefficients are only weakly sensitive to the particular choice of the extension of the lepton distribution function [89].

The lepton distributions are expected to be rapidly isotropized on the cyclotron scale by plasma instabilities [14,15]. For such distributions, the synchrotron emission and absorption coefficients for all Stokes parameters may be expressed as convolutions with broad kernels in frequency space [90]. As a consequence, all of these transfer coefficients are well approximated by a universal expression that is dependent only on the local spectral index, which conceptually corresponds to a measure of the relative number of "hot" non-thermal and "cold" thermal leptons [89].

At a single observation frequency, this approximation is as good as to better than 2%. Thus, for images at 230 or 345 GHz, for example, the specific nature of the acceleration mechanism may be effectively and efficiently parameterized by a single lepton distribution model (e.g., thermal, κ, power law, etc.), thus eliminating a key systematic degeneracy between the location of and microphysical processes responsible for turbulence dissipation. Across large frequency ranges, e.g., from 230 to 345 GHz, the accuracy of this approximation falls to $\sim$10%, or from 230 to 480 GHz to $\sim$40%. Therefore, multi-frequency image reconstructions remain a powerful discriminant between different acceleration mechanisms.

Author Contributions: Conceptualization, R.A.; methodology, R.A.; software, A.R., R.E.; validation, J.D., L.O. and N.N. (Nathan Ngata); formal analysis, J.W.; investigation, R.A.; resources, A.R.; data curation, B.C. and S.R.; writing—original draft preparation, R.A., J.R., A.E.B. and G.N.W.; writing—review and editing, J.R., J.D., L.O. and N.N. (Nathan Ngata).; visualization, J.D., L.O., N.N. (Nathan Ngata), N.N. (Nitya Nigam) and E.D.; supervision, R.A.; project administration, R.A.; funding acquisition, R.E. All authors have read and agreed to the published version of the manuscript.

Funding: Jan Röder received financial support for this research from the International Max Planck Research School (IMPRS) for Astronomy and Astrophysics at the Universities of Bonn and Cologne. Razieh Emami acknowledges the support from the Institute for Theory and Computation at the Center for Astrophysics, as well as grant numbers 21-atp21-0077, NSF AST-1816420, and HST-GO-16173.001-A, for their very generous support.

Data Availability Statement: The data presented in this study are available on request from the corresponding author.

Conflicts of Interest: The authors declare no conflict of interest.

Abbreviations

The following abbreviations are used in this manuscript:

(ng)EHT	(Next-Generation) Event Horizon Telescope
GRMHD	General relativistic magnetohydrodynamics
GRRT	General relativistic radiative transfer
SED	Spectral energy distribution

Appendix A

Appendix A.1. List of Emission Models

We summarize the emission models employed above (and more) in Table A1.

Table A1. JAB Emission Model List. This is a compendium of the phonologically motivated emission models mentioned in this work. The shear stress $S = \gamma^2 |dv_z/ds|$ and dimensional parameters L_j and L_S are set by the width of jet system.

Model Name	Parameters	Functional Form
R-Beta	$R_{\text{low}}, R_{\text{high}}$	$R = \frac{T_i}{T_e} = \frac{\beta^2}{1+\beta^2} R_{\text{high}} + \frac{1}{1+\beta^2} R_{\text{low}}$
Critical Beta	f, β_c	$\frac{T_e}{T_e+T_i} = f e^{-\beta/\beta_c}$

Table A1. *Cont.*

Model Name	Parameters	Functional Form
Const. β_e Jet	β_{e0}	$P_e = \beta_{e0} P_B$
Magnetic Bias Jet	β_{e0}, n	$P_e = K_n(\beta_{e0}) P_B^n$
R Beta w. Const. β_e Jet	$R_{\text{low}}, R_{\text{high}}, \beta_{e0}$	Const. β_e in Jet, $R - \beta$ o.w.
Critical Beta w. Const. β_e Jet	f, β_c, β_{e0}	Const. β_e in Jet, Crit. β o.w
Current Density	L_j	$P_e = \mu_0 c L_j j^\mu j_\mu t_{\text{cool}}$
Jet Alpha	α_j	$P_e = \frac{1}{2}\tau S t_{\text{cool}}, \tau = \alpha_j \left(\frac{B^\mu B_\mu}{2\mu_0} + \frac{u_g}{3} \right)$
Shear	L_S	$P_e = \frac{1}{2}\tau S t_{\text{cool}}, \tau = \mu S,$ $\mu = \frac{L_s}{3c} \sqrt{\left(\rho c^2 + \frac{B^\mu B_\mu}{2\mu_0} \right) + \left(\frac{u_g}{3} + \frac{B^\mu B_\mu}{2\mu_0} \right)}$

Appendix A.2. Resolution Dependence

In Figure A1, we briefly consider the robustness of image properties to resolution effects. Changing the resolution can shift the local and global maxima, as seen in these $50\,M \times 50\,M$ fields of view, where going from 100 pixels per side to 500 pixels per side makes the maximum intensity (red) pixels more concentrated in a more compact equatorial region. Increasing the resolution also increases the overall maximum intensity, as one may expect due to the breaking up of the high-intensity regions into more pixels over which to perform the maximization. In both cases shown, as discussed in Section 5.1, the sub-gravitational-radius pixel spacing in the synthetic images exceeds current and planned observational resolution limits. However, for even the modestly sized supermassive black hole Sgr A*, $M = 10^{10}$ cm, so the role of emission modeling to fill in the sizable gap left by subgrid physics will remain paramount for years to come.

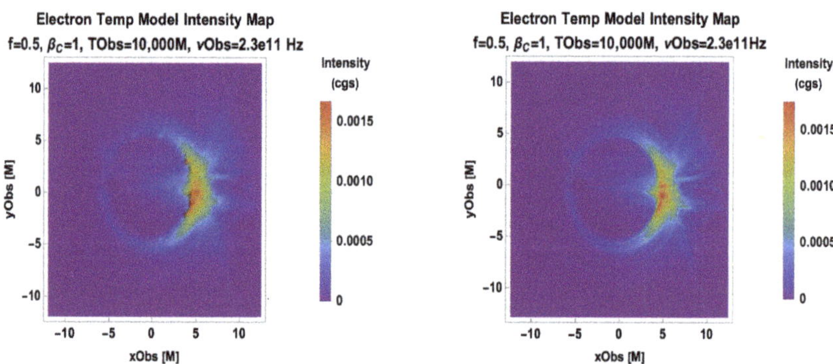

Figure A1. Resolution dependence of the Critical Beta Model. Increasing from 2 pixels/M to 10 pixels/M alters the image morphology such that the overall image is sharper and the brightest features are more localized and extreme in intensity.

Notes

[1] There have been some notable recent attempts to bridge the gap through hybrid electron distribution functions (edfs), such as the κ-model smoothly joining thermal electrons to a high-energy power-law tail [10].

[2] Here, ρ is the rest mass density, u^μ is the four-velocity, and T_ν^μ is the stress–energy–momentum tensor.

[3] This was written using Parthenon; see https://github.com/lanl/parthenon.

[4] In this treatment, ions can be treated interchangeably with protons.

References

1. Event Horizon Telescope Collaboration; Akiyama, K.; Alberdi, A.; Alef, W.; Asada, K.; Azulay, R.; Baczko, A.K.; Ball, D.; Baloković, M.; Barrett, J.; et al. First M87 Event Horizon Telescope Results. I. The Shadow of the Supermassive Black Hole. *Astrophys. J. Lett.* **2019**, *875*, L1. [CrossRef]
2. Event Horizon Telescope Collaboration; Akiyama, K.; Alberdi, A.; Alef, W.; Algaba, J.C.; Anantua, R.; Asada, K.; Azulay, R.; Bach, U.; Baczko, A.K.; et al. First Sagittarius A* Event Horizon Telescope Results. I. The Shadow of the Supermassive Black Hole in the Center of the Milky Way. *Astrophys. J. Lett.* **2022**, *930*, L12. [CrossRef]
3. Event Horizon Telescope Collaboration; Akiyama, K.; Alberdi, A.; Alef, W.; Asada, K.; Azulay, R.; Baczko, A.K.; Ball, D.; Baloković, M.; Barrett, J.; et al. First M87 Event Horizon Telescope Results. V. Physical Origin of the Asymmetric Ring. *Astrophys. J. Lett.* **2019**, *875*, L5. [CrossRef]
4. Event Horizon Telescope Collaboration; Akiyama, K.; Algaba, J.C.; Alberdi, A.; Alef, W.; Anantua, R.; Asada, K.; Azulay, R.; Baczko, A.K.; Ball, D.; et al. First M87 Event Horizon Telescope Results. VII. Polarization of the Ring. *Astrophys. J. Lett.* **2021**, *910*, L12. [CrossRef]
5. Akiyama, K.; Alberdi, A.; Alef, W.; Algaba, J.C.; Anantua, R.; Asada, K.; Azulay, R.; Bach, U.; Baczko, A.-K.; Ball, D.; et al. First Sagittarius A* Event Horizon Telescope Results. III. Imaging of the Galactic Center Supermassive Black Hole. *Astrophys. J. Lett.* **2022**, *930*, L14.
6. GRAVITY Collaboration; Abuter, R.; Amorim, A.; Anugu, N.; Bauböck, M.; Benisty, M.; Berger, J.P.; Blind, N.; Bonnet, H.; Brandner, W.; et al. Detection of the gravitational redshift in the orbit of the star S2 near the Galactic centre massive black hole. *Astron. Astrophys.* **2018**, *615*, L15. [CrossRef]
7. Akiyama, K.; Alberdi, A.; Alef, W.; Algaba, J.C.; Anantua, R.; Asada, K.; Azulay, R.; Bach, U.; Baczko, A.K.; Ball, D.; et al. First Sagittarius A* Event Horizon Telescope Results. V. Testing Astrophysical Models of the Galactic Center Black Hole. *Astrophys. J. Lett.* **2022**, *930*, L16.
8. Blandford, R.D.; Znajek, R.L. Electromagnetic extraction of energy from Kerr black holes. *Mon. Not. R. Astron. Soc.* **1977**, *179*, 433–456. [CrossRef]
9. Anantua, R.; Ressler, S.; Quataert, E. On the comparison of AGN with GRMHD simulations: I. Sgr A*. *Mon. Not. R. Astron. Soc.* **2020**, *493*, 1404–1418. [CrossRef]
10. Fromm, C.M.; Cruz-Osorio, A.; Mizuno, Y.; Nathanail, A.; Younsi, Z.; Porth, O.; Olivares, H.; Davelaar, J.; Falcke, H.; Kramer, M.; et al. Impact of non-thermal particles on the spectral and structural properties of M87. *Astron. Astrophys.* **2022**, *660*, A107. [CrossRef]
11. Ressler, S.M.; Tchekhovskoy, A.; Quataert, E.; Chandra, M.; Gammie, C.F. Electron thermodynamics in GRMHD simulations of low-luminosity black hole accretion. *Mon. Not. R. Astron. Soc.* **2015**, *454*, 1848–1870. [CrossRef]
12. Chatterjee, K.; Younsi, Z.; Liska, M.; Tchekhovskoy, A.; Markoff, S.B.; Yoon, D.; van Eijnatten, D.; Hesp, C.; Ingram, A.; van der Klis, M.B.M. Observational signatures of disc and jet misalignment in images of accreting black holes. *Mon. Not. R. Astron. Soc.* **2020**, *499*, 362–378. [CrossRef]
13. Curd, B.; Emami, R.; Anantua, R.; Palumbo, D.; Doeleman, S.; Narayan, R. Jets from SANE Super-Eddington Accretion Disks: Morphology, Spectra, and Their Potential as Targets for ngEHT. *arXiv* **2022**, arXiv:2206.06358.
14. Gammie, C.F.; McKinney, J.C.; Tóth, G. HARM: A Numerical Scheme for General Relativistic Magnetohydrodynamics. *Astrophys. J.* **2003**, *589*, 444–457. [CrossRef]
15. Gammie, C.F.; McKinney, J.C.; Tóth, G. HARM: A Numerical Scheme for General Relativistic Magnetohydrodynamics. *Astrophys. Source Code Libr.* Available online: http://ascl.net/1209.005 (accessed on 10 November 2022).
16. Wong, G.N.; Prather, B.S.; Dhruv, V.; Ryan, B.R.; Mościbrodzka, M.; Chan, C.k.; Joshi, A.V.; Yarza, R.; Ricarte, A.; Shiokawa, H.; et al. PATOKA: Simulating Electromagnetic Observables of Black Hole Accretion. *Astrophys. J. Suppl. Ser.* **2022**, *259*, 64. [CrossRef]
17. Fishbone, L.G.; Moncrief, V. Relativistic fluid disks in orbit around Kerr black holes. *Astrophys. J.* **1976**, *207*, 962–976. [CrossRef]
18. Prather, B.; Wong, G.; Dhruv, V.; Ryan, B.; Dolence, J.; Ressler, S.; Gammie, C. iharm3D: Vectorized General Relativistic Magnetohydrodynamics. *J. Open Source Softw.* **2021**, *6*, 3336. [CrossRef]
19. Bondi, H. On spherically symmetrical accretion. *Mon. Not. R. Astron. Soc.* **1952**, *112*, 195. [CrossRef]
20. Ressler, S.M.; Quataert, E.; White, C.J.; Blaes, O. Magnetically modified spherical accretion in GRMHD: Reconnection-driven convection and jet propagation. *Mon. Not. R. Astron. Soc.* **2021**, *504*, 6076–6095. [CrossRef]
21. Jia, H.; White, C.J.; Quataert, E.; Ressler, S.M. Observational signatures of black hole accretion: Rotating versus spherical flows with tilted magnetic fields. *Mon. Not. R. Astron. Soc.* **2022**, *515*, 1392–1403. [CrossRef]
22. Quataert, E. A Dynamical Model for Hot Gas in the Galactic Center. *Astrophys. J.* **2004**, *613*, 322–325. [CrossRef]
23. Cuadra, J.; Nayakshin, S.; Martins, F. Variable accretion and emission from the stellar winds in the Galactic Centre. *Mon. Not. R. Astron. Soc.* **2008**, *383*, 458–466. [CrossRef]
24. Shcherbakov, R.V.; Baganoff, F.K. Inflow-Outflow Model with Conduction and Self-consistent Feeding for Sgr A*. *Astrophys. J.* **2010**, *716*, 504–509. [CrossRef]
25. Ressler, S.M.; Quataert, E.; Stone, J.M. Hydrodynamic simulations of the inner accretion flow of Sagittarius A* fuelled by stellar winds. *Mon. Not. R. Astron. Soc.* **2018**, *478*, 3544–3563. [CrossRef]

26. Broderick, A.E.; Gold, R.; Karami, M.; Preciado-López, J.A.; Tiede, P.; Pu, H.Y.; Akiyama, K.; Alberdi, A.; Alef, W.; Asada, K.; et al. THEMIS: A Parameter Estimation Framework for the Event Horizon Telescope. *Astrophys. J.* **2020**, *897*, 139. [CrossRef]
27. Yuan, F.; Narayan, R. Hot Accretion Flows Around Black Holes. *Annu. Rev. Astron. Astrophys.* **2014**, *52*, 529–588. [CrossRef]
28. Quataert, E.; Gruzinov, A. Turbulence and Particle Heating in Advection-dominated Accretion Flows. *Astrophys. J.* **1999**, *520*, 248–255. [CrossRef]
29. Howes, G.G. A prescription for the turbulent heating of astrophysical plasmas. *Mon. Not. R. Astron. Soc.* **2010**, *409*, L104–L108. [CrossRef]
30. Mościbrodzka, M.; Falcke, H.; Shiokawa, H. General relativistic magnetohydrodynamical simulations of the jet in M 87. *Astron. Astrophys.* **2016**, *586*, A38. [CrossRef]
31. Davelaar, J.; Olivares, H.; Porth, O.; Bronzwaer, T.; Janssen, M.; Roelofs, F.; Mizuno, Y.; Fromm, C.M.; Falcke, H.; Rezzolla, L. Modeling non-thermal emission from the jet-launching region of M 87 with adaptive mesh refinement. *Astron. Astrophys.* **2019**, *632*, A2. [CrossRef]
32. Cruz-Osorio, A.; Fromm, C.M.; Mizuno, Y.; Nathanail, A.; Younsi, Z.; Porth, O.; Davelaar, J.; Falcke, H.; Kramer, M.; Rezzolla, L. State-of-the-art energetic and morphological modelling of the launching site of the M87 jet. *Nat. Astron.* **2022**, *6*, 103–108. [CrossRef]
33. Röder, J.; Cruz-Osorio, A.; Fromm, C.M.; Mizuno, Y.; Younsi, Z.; Rezzolla, L. Comparison of Kerr and dilaton black hole shadows. In Proceedings of the European VLBI Network Mini-Symposium and Users' Meeting 2021, Online, 12–14 July 2021; p. 24.
34. Event Horizon Telescope Collaboration; Akiyama, K.; Algaba, J.C.; Alberdi, A.; Alef, W.; Anantua, R.; Asada, K.; Azulay, R.; Baczko, A.K.; Ball, D.; et al. First M87 Event Horizon Telescope Results. VIII. Magnetic Field Structure near The Event Horizon. *Astrophys. J. Lett.* **2021**, *910*, L13. [CrossRef]
35. Mizuno, Y.; Fromm, C.M.; Younsi, Z.; Porth, O.; Olivares, H.; Rezzolla, L. Comparison of the ion-to-electron temperature ratio prescription: GRMHD simulations with electron thermodynamics. *Mon. Not. R. Astron. Soc.* **2021**, *506*, 741–758. [CrossRef]
36. Chael, A.; Rowan, M.; Narayan, R.; Johnson, M.; Sironi, L. The role of electron heating physics in images and variability of the Galactic Centre black hole Sagittarius A*. *Mon. Not. R. Astron. Soc.* **2018**, *478*, 5209–5229. [CrossRef]
37. Anantua, R.J. Towards Multi-Wavelength Observations of Relativistic Jets from General Relativistic Magnetohydrodynamic Simulations. Ph.D. Thesis, Stanford University, Stanford, CA, USA , 2016.
38. Blandford, R.; Anantua, R. The Future of Black Hole Astrophysics in the LIGO-VIRGO-LPF Era. *J. Phys. Conf. Ser.* **2017**, *840*, 012023. [CrossRef]
39. Anantua, R.; Blandford, R.; Tchekhovskoy, A. Multiwavelength Observations of Relativistic Jets from General Relativistic Magnetohydrodynamic Simulations. *Galaxies* **2018**, *6*, 31. [CrossRef]
40. Blandford, R.D.; Königl, A. Relativistic jets as compact radio sources. *Astrophys. J.* **1979**, *232*, 34–48. [CrossRef]
41. McKinney, J.C.; Blandford, R.D. Stability of relativistic jets from rotating, accreting black holes via fully three-dimensional magnetohydrodynamic simulations. *Mon. Not. R. Astron. Soc.* **2009**, *394*, L126–L130. [CrossRef]
42. Anantua, R.; Emami, R.; Loeb, A.; Chael, A. Determining the Composition of Relativistic Jets from Polarization Maps. *Astrophys. J.* **2020**, *896*, 27. [CrossRef]
43. Emami, R.; Anantua, R.; Chael, A.; Loeb, A. Positron Effects on Polarized Images and Spectra from Jet and Accretion Flow Models of M87* and Sgr A*. *Astrophys. J.* **2021**, *923*, 272. [CrossRef]
44. Shakura, N.I.; Sunyaev, R.A. Black holes in binary systems. Observational appearance. *Astron. Astrophys.* **1973**, *24*, 337–355.
45. Vasyliunas, V.M. Low-energy electrons on the day side of the magnetosphere. *J. Geophys. Res.* **1968**, *73*, 7519–7523. [CrossRef]
46. Livadiotis, G.; McComas, D.J. Beyond kappa distributions: Exploiting Tsallis statistical mechanics in space plasmas. *J. Geophys. Res. (Space Phys.)* **2009**, *114*, A11105. [CrossRef]
47. Tsallis, C. Possible generalization of Boltzmann-Gibbs statistics. *J. Stat. Phys.* **1988**, *52*, 479–487. [CrossRef]
48. Tsallis, C.; Mendes, R.; Plastino, A.R. The role of constraints within generalized nonextensive statistics. *Phys. Stat. Mech. Its Appl.* **1998**, *261*, 534–554. [CrossRef]
49. Leung, P.K.; Gammie, C.F.; Noble, S.C. Numerical Calculation of Magnetobremsstrahlung Emission and Absorption Coefficients. *Astrophys. J.* **2011**, *737*, 21. [CrossRef]
50. Xiao, F. Modelling energetic particles by a relativistic kappa-loss-cone distribution function in plasmas. *Plasma Phys. Control. Fusion* **2006**, *48*, 203–213. [CrossRef]
51. Pandya, A.; Zhang, Z.; Chandra, M.; Gammie, C.F. Polarized Synchrotron Emissivities and Absorptivities for Relativistic Thermal, Power-law, and Kappa Distribution Functions. *Astrophys. J.* **2016**, *822*, 34. [CrossRef]
52. Ball, D.; Sironi, L.; Özel, F. Electron and Proton Acceleration in Trans-relativistic Magnetic Reconnection: Dependence on Plasma Beta and Magnetization. *Astrophys. J.* **2018**, *862*, 80. [CrossRef]
53. Röder, J.; Cruz-Osorio, A.; Fromm, C.M.; Mizuno, Y.; Younsi, Z.; Rezzolla, L. Probing spacetime and accretion model for the Galactic Center: Comparison of Kerr and dilaton black hole shadows. 2022, *manuscript in preparation*.
54. Pound, R.V.; Rebka, G.A. Gravitational Red-Shift in Nuclear Resonance. *Phys. Rev. Lett.* **1959**, *3*, 439–441. . [CrossRef]
55. Pound, R.V.; Rebka, G.A. Apparent Weight of Photons. *Phys. Rev. Lett.* **1960**, *4*, 337–341. . [CrossRef]
56. Pound, R.V.; Snider, J.L. Effect of Gravity on Nuclear Resonance. *Phys. Rev. Lett.* **1964**, *13*, 539–540. . [CrossRef]
57. Weisberg, J.M.; Taylor, J.H.; Fowler, L.A. Gravitational waves from an orbiting pulsar. *Sci. Am.* **1981**, *245*, 74–82. [CrossRef]

58. Hafele, J.C.; Keating, R.E. Around-the-World Atomic Clocks: Predicted Relativistic Time Gains. *Science* **1972**, *177*, 166–168. [CrossRef]

59. Hafele, J.C.; Keating, R.E. Around-the-World Atomic Clocks: Observed Relativistic Time Gains. *Science* **1972**, *177*, 168–170. [CrossRef]

60. Kramer, M.; Stairs, I.H.; Manchester, R.N.; McLaughlin, M.A.; Lyne, A.G.; Ferdman, R.D.; Burgay, M.; Lorimer, D.R.; Possenti, A.; D'Amico, N.; et al. Tests of General Relativity from Timing the Double Pulsar. *Science* **2006**, *314*, 97–102. [CrossRef]

61. Stairs, I.H. Testing General Relativity with Pulsar Timing. *Living Rev. Relativ.* **2003**, *6*, 5. [CrossRef]

62. Will, C.M. Testing the General Relativistic "No-Hair" Theorems Using the Galactic Center Black Hole Sagittarius A*. *Astrophys. J. Lett.* **2008**, *674*, L25. [CrossRef]

63. Abbott, B.P.; Abbott, R.; Abbott, T.D.; Abernathy, M.R.; Acernese, F.; Ackley, K.; Adams, C.; Adams, T.; Addesso, P.; Adhikari, R.X.; et al. Observation of Gravitational Waves from a Binary Black Hole Merger. *Phys. Rev. Lett.* **2016**, *116*, 061102. [CrossRef]

64. Johnson, M.D.; Lupsasca, A.; Strominger, A.; Wong, G.N.; Hadar, S.; Kapec, D.; Narayan, R.; Chael, A.; Gammie, C.F.; Galison, P.; et al. Universal interferometric signatures of a black hole's photon ring. *Sci. Adv.* **2020**, *6*, eaaz1310. [CrossRef]

65. Kocherlakota, P.; Rezzolla, L. Distinguishing gravitational and emission physics in black hole imaging: Spherical symmetry. *Mon. Not. R. Astron. Soc.* **2022**, *513*, 1229–1243. [CrossRef]

66. Younsi, Z.; Psaltis, D.; Özel, F. Black Hole Images as Tests of General Relativity: Effects of Spacetime Geometry. *arXiv* **2021**, arXiv:2111.01752.

67. Özel, F.; Psaltis, D.; Younsi, Z. Black Hole Images as Tests of General Relativity: Effects of Plasma Physics. *arXiv* **2021**, arXiv:2111.01123.

68. Mizuno, Y.; Younsi, Z.; Fromm, C.M.; Porth, O.; De Laurentis, M.; Olivares, H.; Falcke, H.; Kramer, M.; Rezzolla, L. The current ability to test theories of gravity with black hole shadows. *Nat. Astron.* **2018**, *2*, 585–590. [CrossRef]

69. Olivares, H.; Younsi, Z.; Fromm, C.M.; De Laurentis, M.; Porth, O.; Mizuno, Y.; Falcke, H.; Kramer, M.; Rezzolla, L. How to tell an accreting boson star from a black hole. *Mon. Not. R. Astron. Soc.* **2020**, *497*, 521–535. [CrossRef]

70. Event Horizon Telescope Collaboration; Akiyama, K.; Alberdi, A.; Alef, W.; Algaba, J.C.; Anantua, R.; Asada, K.; Azulay, R.; Bach, U.; Baczko, A.K.; et al. First Sagittarius A* Event Horizon Telescope Results. VI. Testing the Black Hole Metric. *Astrophys. J. Lett.* **2022**, *930*, L17. [CrossRef]

71. Younsi, Z.; Porth, O.; Mizuno, Y.; Fromm, C.M.; Olivares, H. Modelling the polarised emission from black holes on event horizon-scales. *Proc. Int. Astron. Union* **2020**, *14*, 9–12. [CrossRef]

72. Pu, H.Y.; Yun, K.; Younsi, Z.; Yoon, S.J. Odyssey: A Public GPU-based Code for General Relativistic Radiative Transfer in Kerr Spacetime. *Astrophys. J.* **2016**, *820*, 105. [CrossRef]

73. Dexter, J. A public code for general relativistic, polarised radiative transfer around spinning black holes. *Mon. Not. R. Astron. Soc.* **2016**, *462*, 115–136. [CrossRef]

74. Mościbrodzka, M.; Gammie, C.F. IPOLE - semi-analytic scheme for relativistic polarized radiative transport. *Mon. Not. R. Astron. Soc.* **2018**, *475*, 43–54. [CrossRef]

75. Chan, C.k.; Psaltis, D.; Özel, F. GRay: A Massively Parallel GPU-based Code for Ray Tracing in Relativistic Spacetimes. *Astrophys. J.* **2013**, *777*, 13. [CrossRef]

76. Kawashima, T.; Ohsuga, K.; Takahashi, H.R. RAIKOU: A General Relativistic, Multi-wavelength Radiative Transfer Code. *arXiv* **2021**, arXiv:2108.05131.

77. Bronzwaer, T.; Davelaar, J.; Younsi, Z.; Mościbrodzka, M.; Falcke, H.; Kramer, M.; Rezzolla, L. RAPTOR. I. Time-dependent radiative transfer in arbitrary spacetimes. *Astron. Astrophys.* **2018**, *613*, A2. [CrossRef]

78. Gold, R.; Broderick, A.E.; Younsi, Z.; Fromm, C.M.; Gammie, C.F.; Mościbrodzka, M.; Pu, H.Y.; Bronzwaer, T.; Davelaar, J.; Dexter, J.; et al. Verification of Radiative Transfer Schemes for the EHT. *Astrophys. J.* **2020**, *897*, 148. [CrossRef]

79. Mignone, A.; Bodo, G.; Massaglia, S.; Matsakos, T.; Tesileanu, O.; Zanni, C.; Ferrari, A. PLUTO: A Numerical Code for Computational Astrophysics. *Astrophys. J. Suppl. Ser.* **2007**, *170*, 228–242. [CrossRef]

80. Mignone, A.; Zanni, C.; Tzeferacos, P.; van Straalen, B.; Colella, P.; Bodo, G. The PLUTO Code for Adaptive Mesh Computations in Astrophysical Fluid Dynamics. *Astrophys. J. Suppl. Ser.* **2012**, *198*, 7. [CrossRef]

81. Perucho, M.; Martí, J.M.; Cela, J.M.; Hanasz, M.; de La Cruz, R.; Rubio, F. Stability of three-dimensional relativistic jets: implications for jet collimation. *Astron. Astrophys.* **2010**, *519*, A41. [CrossRef]

82. Kramer, J.A.; MacDonald, N.R. Ray-tracing in relativistic jet simulations: A polarimetric study of magnetic field morphology and electron scaling relations. *Astron. Astrophys.* **2021**, *656*, A143. [CrossRef]

83. MacDonald, N.R.; Nishikawa, K.I. From electrons to Janskys: Full stokes polarized radiative transfer in 3D relativistic particle-in-cell jet simulations. *Astron. Astrophys.* **2021**, *653*, A10. [CrossRef]

84. Mościbrodzka, M.; Gammie, C.F.; Dolence, J.C.; Shiokawa, H. Pair Production in Low-luminosity Galactic Nuclei. *Astrophys. J.* **2011**, *735*, 9. [CrossRef]

85. Broderick, A.E.; Tchekhovskoy, A. Horizon-Scale Lepton Acceleration in jets: Explaining the Compact Radio Emission in M87. *Astrophys. J.* **2015**, *809*, 97. [CrossRef]

86. Goldreich, P.; Julian, W.H. Pulsar Electrodynamics. *Astrophys. J.* **1969**, *157*, 869. [CrossRef]

87. Raymond, A.W.; Palumbo, D.; Paine, S.N.; Blackburn, L.; Córdova Rosado, R.; Doeleman, S.S.; Farah, J.R.; Johnson, M.D.; Roelofs, F.; Tilanus, R.P.J.; et al. Evaluation of New Submillimeter VLBI Sites for the Event Horizon Telescope. *Astrophys. J. Suppl. Ser.* **2021**, *253*, 5. [CrossRef]

88. Emami, R.; Anantua, R.; Ricarte, A.; Doeleman, S.S.; Broderick, A.; Wong, G.; Blackburn, L.; Wielgus, M.; Narayan, R.; Tremblay, G.; et al. Probing plasma composition with the next generation Event Horizon Telescope (ngEHT). *arXiv* **2022**, arXiv:2211.07306.

89. Broderick, A.; Loeb, A. Local Universality of Nonthermal Synchrotron Emission from EHT Targets. 2022; *manuscript in preparation*.

90. Rybicki, G.B.; Lightman, A.P. *Radiative Processes in Astrophysics*; Wiley-VCH: Weinheim, Germany, 1986.

Article

Accretion Flow Morphology in Numerical Simulations of Black Holes from the ngEHT Model Library: The Impact of Radiation Physics

Koushik Chatterjee [1,2,*], Andrew Chael [3], Paul Tiede [1,2], Yosuke Mizuno [4,5,6], Razieh Emami [2], Christian Fromm [6,7,8], Angelo Ricarte [1,2], Lindy Blackburn [1,2], Freek Roelofs [1,2], Michael D. Johnson [1,2], Sheperd S. Doeleman [1,2], Philipp Arras [9,10], Antonio Fuentes [11], Jakob Knollmüller [10,12], Nikita Kosogorov [13,14], Greg Lindahl [2], Hendrik Müller [8], Nimesh Patel [2], Alexander Raymond [1,2], Efthalia Traianou [11] and Justin Vega [2,15]

[1] Black Hole Initiative, Harvard University, 20 Garden Street, Cambridge, MA 02138, USA
[2] Center for Astrophysics, Harvard & Smithsonian, 60 Garden Street, Cambridge, MA 02138, USA
[3] Princeton Gravity Initiative, Princeton University, Jadwin Hall, Princeton, NJ 08544, USA
[4] Tsung-Dao Lee Institute, Shanghai Jiao-Tong University, 520 Shengrong Road, Shanghai 201210, China
[5] School of Physics & Astronomy, Shanghai Jiao-Tong University, 800 Dongchuan Road, Shanghai 200240, China
[6] Institut für Theoretische Physik, Goethe-Universität Frankfurt, Max-von-Laue-Straße 1, D-60438 Frankfurt am Main, Germany
[7] Institut für Theoretische Physik und Astrophysik, Universität Würzburg, Emil-Fischer-Str. 31, D-97074 Würzburg, Germany
[8] Max–Planck–Institut für Radioastronomie, Auf dem Hügel 69, D-53121 Bonn, Germany
[9] Technical University Munich (TUM), Boltzmannstr. 3, 85748 Garching, Germany
[10] Max–Planck Institute for Astrophysics, Karl-Schwarzschild-Str. 1, 85748 Garching, Germany
[11] Instituto de Astrofísica de Andalucía-CSIC, Glorieta de la Astronomía s/n, E-18008 Granada, Spain
[12] Excellence Cluster ORIGINS, Boltzmannstr. 2, 85748 Garching, Germany
[13] Moscow Institute of Physics and Technology, Institutsky per. 9, Dolgoprudny 141700, Russia
[14] Lebedev Physical Institute of the Russian Academy of Sciences, Leninsky Prospekt 53, Moscow 119991, Russia
[15] Department of Physics, Northeastern University, 360 Huntington Ave, Boston, MA 02115, USA
* Correspondence: koushik.chatterjee@cfa.harvard.edu

Citation: Chatterjee, K.; Chael, A.; Tiede, P.; Mizuno, Y.; Emami, R.; Fromm, C.; Ricarte, A.; Blackburn, L.; Roelofs, F.; Johnson, M.D.; et al. Accretion Flow Morphology in Numerical Simulations of Black Holes from the ngEHT Model Library: The Impact of Radiation Physics. *Galaxies* **2023**, *11*, 38. https://doi.org/10.3390/galaxies11020038

Academic Editor: Dimitris M. Christodoulou

Received: 4 December 2022
Revised: 14 February 2023
Accepted: 20 February 2023
Published: 23 February 2023

Abstract: In the past few years, the Event Horizon Telescope (EHT) has provided the first-ever event horizon-scale images of the supermassive black holes (BHs) M87* and Sagittarius A* (Sgr A*). The next-generation EHT project is an extension of the EHT array that promises larger angular resolution and higher sensitivity to the dim, extended flux around the central ring-like structure, possibly connecting the accretion flow and the jet. The ngEHT Analysis Challenges aim to understand the science extractability from synthetic images and movies to inform the ngEHT array design and analysis algorithm development. In this work, we compare the accretion flow structure and dynamics in numerical fluid simulations that specifically target M87* and Sgr A*, and were used to construct the source models in the challenge set. We consider (1) a steady-state axisymmetric radiatively inefficient accretion flow model with a time-dependent shearing hotspot, (2) two time-dependent single fluid general relativistic magnetohydrodynamic (GRMHD) simulations from the H–AMR code, (3) a two-temperature GRMHD simulation from the BHAC code, and (4) a two-temperature radiative GRMHD simulation from the KORAL code. We find that the different models exhibit remarkably similar temporal and spatial properties, except for the electron temperature, since radiative losses substantially cool down electrons near the BH and the jet sheath, signaling the importance of radiative cooling even for slowly accreting BHs such as M87*. We restrict ourselves to standard torus accretion flows, and leave larger explorations of alternate accretion models to future work.

Keywords: black holes; general relativity; accretion; relativistic jets; very-long-baseline interferometry

1. Introduction

With the advent of the Event Horizon Telescope ([EHT; [1,2]), imaging the near-horizon structure of supermassive black holes (BHs) is now a reality. The primary targets of the EHT and the future next-generation EHT (or ngEHT[1]) are M87* (the supermassive BH in the elliptical galaxy M87; [1]) and Sgr A* in the Galactic Center [2], which are two of the most well-studied low-luminosity active galactic nuclei. Extracting information about the event horizon-scale accretion flows in these two sources using the EHT's enormous resolving power is an active area of research. With the ngEHT, we will achieve unprecedented levels of angular resolution and sensitivity to low-flux regions, with the dynamic range in flux expected to increase to $\sim$1000 compared to the EHT's current dynamic range of $\sim$10 (e.g., [3]). This would enable us to investigate the BH shadow shape with higher precision as well as provide a crucial connection between the accretion flow and the jet launching region. The expected advances in sensitivity require deeper investigations of feature extraction from simulated synthetic reconstructions of BH systems. Hence, we designed the ngEHT analysis challenges [2] [4] to test our ability to capture the complex dynamics of gas and magnetic fields around M87* and Sgr A* using the ngEHT reference array (e.g., [5]) with various analysis methods.

Black hole accretion and jet physics have been intensively studied over the past few decades (e.g., [6–15]). In the context of M87* and Sgr A*, we expect the accretion flow to be highly sub-Eddington, radiatively inefficient, and geometrically thick, popularly known as radiatively inefficient accretion flows (RIAFs). This accretion flow solution has been used to successfully model the multiwavelength spectrum of Sgr A* (e.g., [16]). On the other hand, semi-analytical models of jets are preferred to explain the spectrum of M87* (e.g., [17]). Thus, these two sources already provide a means to probe two different components of BH accretion, namely, the inner accretion flow structure and turbulence in Sgr A* and the prominent jet feature in M87*. The first three EHT numerical simulation papers [18–20] already provide us with important clues about the horizon-scale conditions of these BH systems based on numerical simulations: (1) these BHs probably have non-zero spin, (2) the accretion disk is expected to have colder electrons than the jet sheath, and (3) the observations favor the presence of dynamically important magnetic fields close to the BH. All of these results point us toward the magnetically arrested disk (MAD; [10,21]) state, an accretion mode where the BH magnetosphere becomes over-saturated with magnetic flux and exhibits quasi-periodic explosions of vertical magnetic field bundles. MAD flows around spinning BHs also have powerful relativistic jets, where the jet power can exceed the input accretion power [13], which is a definite signature of BH spin energy extraction via the Blandford and Znajek [22] process.

Building on the semi-analytical RIAF models, the time-dependent general relativistic magneto-hydrodynamic (GRMHD) simulations have become important tools for deciphering BH accretion physics in a variety of astrophysical systems (e.g., [11,13,23–27]). Indeed, the EHT regularly uses the libraries of GRMHD simulations to model the observed horizon-scale BH images of M87* and Sgr A* as well as larger-scale jet images (such as for Centaurus A; [28]) in order to constrain the time-variable plasma properties. In designing the ngEHT reference array, it is therefore crucial to use GRMHD simulations for understanding the attainability of specific science goals, such as resolving the photon ring and the disk-jet connection region as well as tracing out time-variable features via the ngEHT analysis challenges.

In this work, we discuss the numerical fluid simulations that were used as source models for the ngEHT analysis challenges. In particular, our objective is to compare between the models that incorporate increasingly complicated levels of accretion and electron physics, focusing on M87* and Sgr A*. Our model set consists of a time-dependent shearing hotspot stitched to a steady-state RIAF solution, two standard GRMHD simulations of MAD accretion flows, a GRMHD MAD simulation with electron heating via incorporating two-temperature physics, and a fully radiative, two-temperature, GRMHD MAD simulation.

We describe the equations and setup of our numerical models in Section 2, show our comparison results in Section 3, and, finally, conclude in Section 4.

2. Numerical Simulations

In this section, we provide a brief description of the semi-analytical stationary RIAF and shearing hotspot model as well as the (two-temperature/radiative) GRMHD simulations used for the ngEHT analysis challenges.

2.1. RIAF + Hotspot Solutions

RIAF models attempt to describe the time and azimuthally averaged appearance of accretion flows. This is performed using a set of building blocks. The specific RIAF models used in the challenges are based on Yuan et al. [16], Broderick and Loeb [29], Broderick and McKinney [30] approach. We decompose the accretion flow into a set of phenomenological models that describe the electron density, temperature, magnetic field, and velocity profile. We have a cylindrical coordinate system $x^{\mu} \equiv (R_{\mathrm{cyl}}, \varphi, z)$. The electron density profile is defined in terms of the cylindrical radius $R_{\mathrm{cyl}} = r|\sin(\theta)|$ and vertical displacement $z = r\cos(\theta)$, and is provided by

$$n_{e,X}(R_{\mathrm{cyl}}, z) = n_{e,X,0} R_{\mathrm{cyl}}^{p_X} \exp\left(-\frac{z^2}{2h^2 R_{\mathrm{cyl}}^2}\right) \tag{1}$$

where X denotes the population of electrons. The disk height h is set to unity for this work. For the challenge dataset, we included both thermal synchrotron emitting ($X \equiv$ th), and non-thermal synchrotron ($X \equiv$ nth) emitting electrons. The thermal electrons have $n_{e,\mathrm{th},0} = 1.3 \times 10^8$ and $p_{\mathrm{th}} = -1.1$, while the non-thermal electrons are provided by $n_{e,\mathrm{nth},0} = 1.3 \times 10^5$, $p_{\mathrm{nth}} = -2.02$. These numbers are from Tiede et al. [31] and are set to match the best-fit parameters for the Sgr A* spectrum from Broderick et al. [32] and Broderick et al. [33].

The temperature profile of the thermal electrons is also provided by a radial power law with a Gaussian envelope describing the height:

$$T_e(t, r, \theta, \varphi) = T_{e,0} R_{\mathrm{cyl}}^{-0.84} \exp\left(-\frac{z^2}{2h^2 R_{\mathrm{cyl}}^2}\right), \tag{2}$$

where, for the challenge data, we set $T_{e,0} = 6.3 \times 10^{10}$ K.

Following Pu et al. [34], we define the gas pressure by assuming that the ultra-relavistic protons are in roughly virial equilibrium with the gravitational force providing:

$$p_{\mathrm{gas}}(x^{\mu}) = \frac{m_p n_{e,\mathrm{th}}(x^{\mu})}{6} \frac{GM_{\mathrm{BH}}}{r}, \tag{3}$$

where m_p, G, and M_{BH} are the proton mass, gravitational constant, and the black hole mass.

For the local magnetic field, we then assume a constant $\beta = p_{\mathrm{gas}}/p_{\mathrm{mag}} = 10$ plasma, which, combined with Equation (3), provide us the magnetic field strength. The orientation of the magnetic field is then provided by a purely toroidal configuration, relative to the plasma observer. Finally, we extract the emission and absorption coefficients from the synchrotron self-absorption model in Broderick and Blandford [35] for both the thermal and non-thermal synchrotron electrons. For the non-thermal electrons, we use a power law prescription with radiation coefficients from Jones and O'Dell [36], and a photon spectral power law index of 1.25 (see [31] for more details).

We follow the velocity field prescription from Pu et al. [34] to describe the accretion flow dynamics. Using the notation from Tiede et al. [31], the velocity field $u^\mu = u^t v^\mu$ is provided by

$$v^r = v_K^r + \alpha_{\text{vel}}(v_{ff}^r - v_K^r)$$
$$v^\theta = 0 \tag{4}$$
$$v^\varphi = v_K^\varphi + (1 - \kappa_{\text{vel}})(v_{ff}^\varphi - v_K^\varphi),$$

where v_K^μ denotes the Keplerian velocity field, v_{ff}^μ is the free-fall velocity field, and $\alpha_{\text{vel}} = 0.01$ and $\kappa_{\text{vel}} = 1.0$, i.e., the hotspot rotates with Keplerian angular velocity and is weakly free-falling. For $\alpha_{\text{vel}} = 0, \kappa_{\text{vel}} = 1$, we have a Keplerian orbit, whereas, for $\alpha_{\text{vel}} = 1, \kappa_{\text{vel}} = 0$, we have free-fall. The remaining component u^t is provided by the normalization condition $u^\mu u_\mu = -1$. The radial component outside the inner stable circular orbit (ISCO) is $v_K^r = 0$ as the disk is in steady-state. However, inside the ISCO, we use plunging geodesics that are specified by matching the angular momentum and energy at the ISCO.

The hotspot evolution follows the model from Tiede et al. [31] (also see [29]), where we assume that the hotspot travels passively along the accretion flow velocity field Equation (4). This implies that the equation of motion is provided by the conservation of particle number Equation (6). For the emission, we assume a non-thermal synchrotron hotspot, with an initial Gaussian density profile

$$n_e(x^\mu) = n_0 e^{-\Delta r^\mu r_\mu + (\Delta r_\mu v^\mu)^2 / 2R_s^2}, \tag{5}$$

where we have set $n_0 = 6 \times 10^6$, hotspot size $R_s = 0.5\, r_g = 0.5 G M_{\text{BH}}/c^2$, and Δr^μ is the displacement from the hotspot center.

2.2. GRMHD Simulations

Over the previous two decades, multiple GRMHD codes have been developed and utilized to model black hole accretion and jet launching physics over long dynamical timescales. The wide usage of GRMHD simulations is particularly encouraging since this allows for verification of code-specific numerical choices that users usually have to make even while solving the same base set of GRMHD equations. Indeed, recently, there was a community-wide effort to benchmark these codes against each other for a standard problem: evolving a weakly magnetized torus of gas around a spinning black hole [26]. It was found that these codes largely provide similar solutions, though some disk quantities remain unconverged with increasing grid resolutions, suggesting more investigation is required. For this work, we employ three different GRMHD codes to probe black hole accretion, increasing the complexity of the equations solved at each step: (1) single fluid GRMHD simulations from the H-AMR code [37], (2) a two-temperature single fluid GRMHD simulation from the BHAC code [38], and (3) a two-temperature radiative GRMHD simulation from the KORAL code [39].

First, we describe the set of GRMHD Equations (e.g., from [23,26]). We have the conservation of particle number and energy-momentum:

$$\partial_t(\sqrt{-g}\rho u^t) = -\partial_i(\sqrt{-g}\rho u^i) \tag{6}$$
$$\partial_t(\sqrt{-g}T_\nu^t) = -\partial_i(\sqrt{-g}T_\nu^i) + \sqrt{-g}T_\beta^\alpha \Gamma_{\nu\alpha}^\beta. \tag{7}$$

Here, ρ is the rest-mass gas density and can also be written in the form of $\rho = mn$, where m is the mean rest–mass per particle and n is the particle number density. We also have the four-velocity u^μ, stress–energy tensor T_ν^μ, metric determinant $g \equiv det(g_{\mu\nu})$, and the metric connection $\Gamma_{\nu\alpha}^\beta$. Note that the index t refers to the temporal component of the vector or tensor and i denotes the spatial indices.

The stress-energy tensor T^μ_ν is provided as

$$T^\mu_\nu = (\rho + U_{\text{gas}} + p_{\text{gas}} + 2p_{\text{mag}})u^\mu u_\nu + (p_{\text{gas}} + p_{\text{mag}})\delta^\mu_\nu - b^\mu b_\nu, \tag{8}$$

where U_{gas} and p_{gas} are the gas internal energy and pressure, related by the ideal gas equation: $p_{\text{gas}} = (\Gamma_{\text{gas}} - 1)U_{\text{gas}}$ assuming a gas adiabatic index Γ_{gas}. We also have the magnetic pressure $p_{\text{mag}} = b^2/2$ and the magnetic field 4-vector b^μ, which can be defined in terms of the magnetic field 3-vector B^i:

$$b^t = B^i u^\mu g_{i\mu} \tag{9}$$
$$b^i = (B^i + b^t u^i)/u^t. \tag{10}$$

Here, we included a factor of $\sqrt{4\pi}$ into the definition of B^i. We evolve the magnetic field B^i using the spatial components of the induction equation,

$$\partial_t(\sqrt{-g}B^i) = -\partial_j(\sqrt{-g}(b^j u^i - b^i u^j)), \tag{11}$$

while the temporal component provides the no monopoles constraint,

$$\frac{1}{\sqrt{-g}}\partial_i(\sqrt{-g}B^i) = 0. \tag{12}$$

These equations are numerically integrated in the conservative form [23] to obtain the physically relevant quantities ρ, U_{gas}, u^μ, and B^i. We refer the reader to the corresponding code papers for more information on the numerical techniques used to evolve these equations over space and time.

In this work, we use two GRMHD simulations performed with the H-AMR code, one targeting M87* and the other Sgr A*. These simulations employ logarithmic Kerr–Schild coordinates and the grid resolutions are $N_r \times N_\theta \times N_\varphi = 580 \times 288 \times 512$ for the M87* simulation and $348 \times 192 \times 192$ for the Sgr A* simulation. All the simulations in this work adopt the geometrical unit convention, $G = c = 1$ and using $M_{\text{BH}} = 1$, normalizing the length scale to the gravitational radius $r_g = GM_{\text{BH}}/c^2$. The M87* GRMHD simulation evolves a MAD flow around a black hole with spin $a = 0.9375$. The Sgr A* model also simulates a MAD flow but around a black hole with spin $a = 1/2$. H-AMR uses outflowing radial boundary conditions (BCs), transmissive polar BCs, and periodic azimuthal BCs (for more details, see [40]).

Since GRMHD simulations are scale-free, we determine the gas density code-to-CGS units conversion factor (hereafter, "density scaling") by raytracing the simulation at 230 GHz for a target source and flux. We use the general relativistic ray-tracing codes BHOSS [41] and IPOLE [42] to compute images at 230 GHz and set the compact flux to be approximately 0.5 Jy for M87* [1] and 2.4 Jy for Sgr A* [2]. We use the black hole masses and distances of $M_{\text{BH}} = 6.2 \times 10^9 M_\odot$ and $D_{\text{BH}} = 16.9$ Mpc for M87* (e.g., [18,43] and references therein) and $M_{\text{BH}} = 4.14 \times 10^6 M_\odot$ and $D_{\text{BH}} = 8.127$ kpc for Sgr A* [20,44,45].

GRMHD simulations evolve a single temperature fluid. At the low accretion rates seen in M87* and Sgr A*, Coulomb coupling between ions and electrons is inefficient and, therefore, the two particle species are not in thermal equilibrium. Since ions are much heavier than electrons, the ions dominate the single fluid thermodynamics evolved in GRMHD simulations. Hence, to calculate the radiative output from GRMHD simulations, we calculate the electron temperature T_e using sub-grid models such as the $R - \beta$ prescription [46] based on local gas plasma-β ($\equiv p_{\text{gas}}/p_{\text{mag}}$):

$$T_e = \frac{2m_p U_{gas}}{3k_B \rho (2 + R)},\tag{13}$$

$$\text{where,}\quad R = \frac{1 + R_{high}\beta^2}{1 + \beta^2}.\tag{14}$$

For the ngEHT analysis challenges, we select R_{high} values of 160 and 40 and source inclinations of 163° and 50° for the M87* and Sgr A* simulations, respectively. We assume a thermal relativistic Maxwell–Jüttner distribution for describing the electron energy distribution in the Sgr A* model, and a hybrid thermal+non-thermal $\kappa-$distribution (e.g., [47,48]) for the M87* model. The model images are shown in Roelofs et al. [4].

2.3. Two-Temperature Physics

Two-temperature GRMHD (or 2t-GRMHD) simulations (e.g., [49,50]) evolve the ion and electron entropy equation separately and, hence, provide the ion and electron temperature in a self-consistent manner. The main advantage of this method is that we remove the electron temperature as a free parameter when constructing images. However, we do have to make a choice about the sub-grid prescription that determines the fraction of local dissipative energy that heats the electrons. There are two heating mechanisms that are thought to be applicable to global accretion flows: turbulent heating [51,52], and magnetic reconnection [53,54]. For the ngEHT analysis challenges, we focus only on one simulation with reconnection heating, from Mizuno et al. [55], as heating via magnetic reconnection in equatorial current sheets formed in magnetically arrested flows, captured by the Rowan et al. [54] heating model, is arguably more important than heating via small-scale turbulent eddies that are more prevalent in weakly magnetized disks.

We assume that the number densities and velocities of ions and electrons are equal, i.e., $n_i = n_e = n$ and $u_i^\mu = u_e^\mu = u^\mu$, thus maintaining charge neutrality. The electron entropy equation is provided as

$$\partial_\mu\left(\sqrt{-g}\rho u^\mu s_e\right) = \frac{\sqrt{-g}(\Gamma_e - 1)}{\rho^{\Gamma_e - 1}} f_e Q,\tag{15}$$

where the electron entropy is $s_e = p_e/\rho^{\Gamma_e}$, with p_e and Γ_e as the electron pressure and adiabatic index. The total heating rate Q is calculated by comparing the total internal energy of the gas and the internal energy obtained from the electron entropy conservation Equation (see [49] for more details). The fraction of dissipative heating that contributes to electron heating is provided by f_e. For this particular simulation, f_e is designed to capture electron/ion heating via magnetic reconnection from Rowan et al. [54]:

$$f_e = \frac{1}{2}\exp\left[-\frac{1 - \beta/\beta_{max}}{0.8 + \sigma_h^{0.5}}\right],\tag{16}$$

where $\beta_{max} = 1/4\sigma_h$, defined using the hot gas magnetization $\sigma_h = b^2/\rho E_{th}$ and the specific gas enthalpy $E_{th} = 1 + \Gamma_{gas} p_g/[\rho(\Gamma_{gas} - 1)]$.

The 2t-GRMHD simulation from Mizuno et al. [55] assumes modified Kerr–Schild coordinates and a black hole spin of 0.9375. The grid resolution is $384 \times 192 \times 192$. The accretion mode is of a magnetically arrested flow and the simulation is raytraced (using BHOSS) once the near-horizon flow has reached steady state. The target source is M87*, assuming a black hole mass of $M_{BH} = 6.5 \times 10^9 M_\odot$ and distance of 16.9 Mpc [1]. The accretion rate is normalized such that the 230 GHz compact flux density is 0.8 Jy. We assume a thermal electron distribution everywhere except in the jet sheath where we adopt a $\kappa-$distribution. More details about the image are provided in Roelofs et al. [4].

2.4. Radiative GRMHD

Two temperature GRMHD simulations do not include radiative cooling and, hence, are thought to be appropriate for low luminosity supermassive black holes such as M87* and Sgr A*. To verify this assumption, we consider a two-temperature radiative GRMHD (2t-GRRMHD hereafter) simulations from Chael et al. [25]. This simulation accounts for self-consistent radiation physics, incorporating both particle heating via magnetic reconnection (as in Section 2.3) and radiative cooling via bremsstrahlung, synchrotron, Compton, and Coulomb losses. The simulation is run using the 2t-GRRMHD code KORAL [39,56,57], which evolves a two-temperature magnetized fluid and treats the radiation field as a second fluid [58]. The conservation equations solved in 2t-GRRMHD are different from that of GRMHD:

$$(T^\mu_\nu + R^\mu_\nu)_{;\,\mu} = 0, \tag{17}$$

where R^μ_ν is the frequency-integrated radiation field, defined as,

$$R^\mu_\nu = \frac{4}{3}\overline{E}u^\mu_{\mathrm{R}}u_{\nu\mathrm{R}} + \frac{1}{3}\overline{E}\delta^\mu_\nu. \tag{18}$$

Here, the radiation field is described by its rest-frame energy density $\overline{E}$ and four-velocity u^μ_{R} following the M1 closure scheme. The ion and electron entropy equations are

$$T_{\mathrm{e}}(n s_{\mathrm{e}} u^\mu)_{;\mu} = f_{\mathrm{e}}q^{\mathrm{v}} + q^{\mathrm{C}} - G, \tag{19}$$

$$T_{\mathrm{i}}(n s_{\mathrm{i}} u^\mu)_{;\mu} = (1 - f_{\mathrm{e}})q^{\mathrm{v}} - q^{\mathrm{C}}, \tag{20}$$

where q^{v} is the dissipative heating rate and q^{C} is the Coulomb coupling rate that captures the exchange of energy between ions and electrons. The heating fraction f_{e} is from Rowan et al. [54] (see Chael et al. [57] for more details), same as the 2t-GRMHD simulation. Finally, G is the radiative cooling rate [39]. For further details about the equations, see Sądowski et al. [39], Chael et al. [57], Sądowski et al. [58].

The simulation assumes a black hole spin of $a = 0.9375$ and mass $M_{\mathrm{BH}} = 6.2 \times 10^9 M_\odot$, targeting M87*. The gas density is scaled to physical CGS units such that the compact emission at 230 GHz is roughly 0.98 Jy [59,60]. The simulation uses modified Kerr–Schild coordinates with a grid resolution of $N_r \times N_\theta \times N_\varphi = 288 \times 224 \times 128$. See Chael et al. [25] for more details about the simulation, while not utilized for the ngEHT analysis challenges, we included this simulation in this work since this model captures the coupling between gas and radiation, necessary for black holes accreting close the Eddington limit. Further, this model has been used in previous ngEHT reference array papers [3,5].

3. Results

We perform a series of comparisons focused on the time-evolution of horizon-scale quantities and radial dependence of disk and jet properties. The diagnostics are chosen such that any trends we find can inform EHT/ngEHT science applications, such as horizon-scale morphology and variability of the accretion flow. Further, the quantities are similar to those reported in the GRMHD code comparison project [26] and, thus, can be directly compared. There is a total of five models: three (2t-radiative) GRMHD simulations targeting M87 *, and one RIAF solution and one GRMHD simulation for Sgr A *. We further note that all three numerical simulations of M87 * have the same BH spin, favoring direct comparisons of the horizon-scale gas properties.

3.1. Temporal Behavior of Horizon Fluxes

We calculate the mass, magnetic, angular momentum and energy fluxes in the radial direction as follows:

$$\text{Mass}: \quad \dot{M} = \int_0^{2\pi} \int_0^{\pi} (-\rho\, u^r)\, \sqrt{-g}\, d\theta\, d\varphi, \tag{21}$$

$$\text{Magnetic}: \quad \Phi = \frac{\sqrt{4\pi}}{2} \int_0^{2\pi} \int_0^{\pi} |B^r|\, \sqrt{-g}\, d\theta\, d\varphi, \tag{22}$$

$$\text{Ang.Mom.}: \quad \dot{J} = \int_0^{2\pi} \int_0^{\pi} T^r_{\varphi}\, \sqrt{-g}\, d\theta\, d\varphi, \tag{23}$$

$$\text{Energy}: \quad \dot{E} = \int_0^{2\pi} \int_0^{\pi} (T^r_t)\, \sqrt{-g}\, d\theta\, d\varphi, \tag{24}$$

where all quantities are calculated at the event horizon radius $r_{\mathrm{hor}} = r_g(1 + \sqrt{1 - a^2})$. We note that there could be a substantial contribution of density floors when calculating the mass accretion rate for MAD systems. However, this radius was chosen for simplicity when comparing to previous simulations in the literature. Figure 1 shows the mass accretion rate $\dot{M}$ in units of solar masses per year ($M_\odot/\mathrm{yr}$), the dimensionless magnetic flux $\phi = \Phi/\sqrt{\dot{M} r_g^2 c}$, the outflow power $P_{\mathrm{out}} = \dot{M} c^2 - \dot{E}$, and the specific angular momentum flux $\dot{J}/\dot{M}$ for simulations targeting M87 * and Sgr A *. The RIAF solution being a steady-state solution is excluded from this section (though the hotspot evolves with time). The quantities from the 2t-GRRMHD simulation are only shown for $(11 - 16) \times 10^3\ r_g/c$, i.e., the time period over which the simulation was raytraced in Chael et al. [25]. Remarkably, despite the difference in electron the physics complexity, the simulations behave very similarly. The factor of 2 difference in $\dot{M}$ between the M87* non-radiative simulations and the 2t-GRRMHD simulation can be explained by the lower electron temperatures in the near-horizon accretion flow due to radiative cooling (see Section 3.2) as well as the higher 230 GHz flux normalization used for the radiative model.

The accretion rate in all simulations show large variation with quasi-periodic sharp drops. These drops in $\dot{M}$ occur due to the emergence of magnetic flux eruptions, a characteristic feature of the magnetically arrested disks [61–64]. These eruptions also lower the value of ϕ since magnetic flux bundles escape from the vicinity of the BH, carrying away the magnetic flux accumulated in the BH magnetosphere. We see that ϕ often crosses the magnetic flux saturation value of 50 [13], overwhelming the BH magnetosphere with strong magnetic fields that eventually reconnect and trigger flux eruptions (see [63] for the detailed mechanism). Figure 2 shows a series of equatorial snapshots from the M87* 2t-GRMHD simulation where we follow a particular magnetic flux eruption by the rotating gas that is hot ($T_e \gtrsim 10^{12}$ K), has low density ($\rho < -2$ in code units), and primarily consists of vertical field lines. As these field line bundles move out and interact with the disk, they (1) hinder accretion, lowering $\dot{M}$, (2) remove magnetic flux from near the BH, lowering the jet power, and (3) push gas outwards, reducing the inward angular momentum flux. Curiously, we see larger drops in the specific angular momentum flux for the Sgr A* GRMHD model. This is possibly due to the smaller BH spin ($a = 0.5$ as opposed to 0.9375 for the M87* models) as the weakly powered jet does not carry away angular momentum as efficiently as the higher BH spin models and flux eruptions play a bigger role in regulating disk angular momentum transport. Additionally, the reconnection events that trigger these eruptions accelerate electrons to higher energies, and are, thus, crucial for understanding flare activity in BH sources.

To quantify the time-variability of the horizon-fluxes, we calculate the modulation index MI, which is defined as the ratio of the standard deviation and the mean of the quantity over time [20]. We show the MI for the different fluxes in Table 1. The MI($\dot{M}$) is usually a good proxy for the variability of the sub-millimeter (sub-mm) emission in these slowly accreting optically thin black hole sources (e.g., [65]). The MI($\dot{M}$) values we see from the simulations are $\sim$0.23–0.29 and are larger than expected from Sgr A* 230 GHz lightcurves (where MI $\sim$ 0.1; [66]). This suggests that careful analysis of the electron distribution function is needed to understand if we are substantially over-predicting the 230 GHz lightcurve variability. Further, in general, weakly magnetized accretion flows exhibit lower MI($\dot{M}$) values due to the absence of flux eruptions, which suggests that further study of the accretion mode in Sgr A* is also necessary. It is encouraging to note that our MI values for $\dot{M}$ and ϕ are consistent with the MI values from longer time-evolved GRMHD simulations of $a = 0.9$ BHs in Narayan et al. [27], indicating that our simulations are sufficiently converged with respect to horizon-scale quantities.

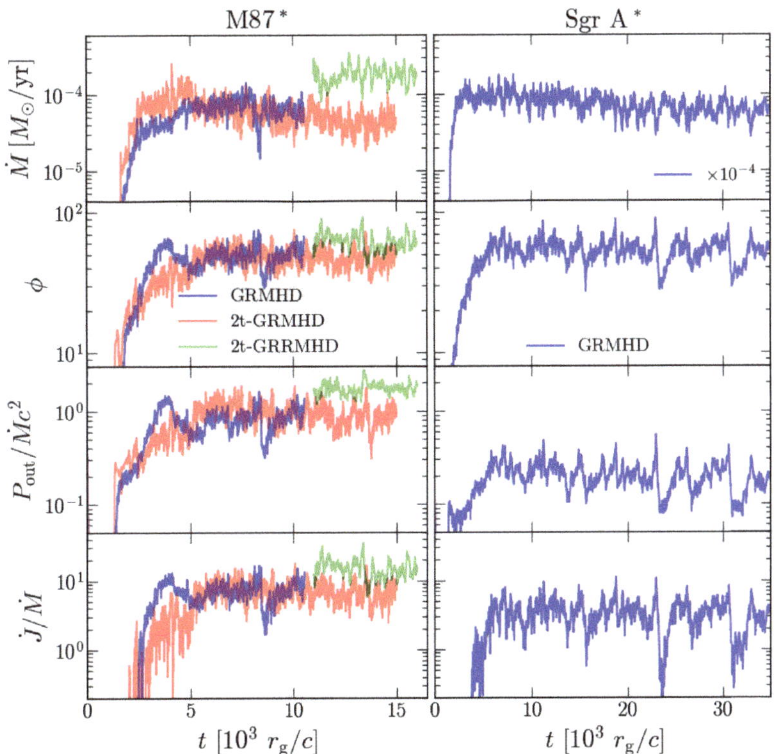

Figure 1. We show the mass accretion rate $\dot{M}$, dimensionless magnetic flux $\phi \equiv \Phi/\sqrt{\dot{M}}$, the outflow efficiency $P_\mathrm{out}/\dot{M}c^2 = 1 - \dot{E}/\dot{M}c^2$, and specific radial flux of the angular momentum $\dot{J}/\dot{M}$ over time. Values are calculated at the event horizon.

Table 1. GRMHD simulations considered in this work. Simulation grid resolution, simulation time period over which raytracing was performed, modulation index (MI) of the mass accretion rate $\dot{M}$, dimensionless magnetic flux ϕ, outflow efficiency $P_\mathrm{out}/\dot{M}c^2$, and the specific angular momentum flux $\dot{J}/\dot{M}$ for each GRMHD model. The MI is calculated over the final 5000 r_g/c in runtime and at the event horizon (see Figure 1).

Model Name	Grid Resolution ($N_r \times N_\theta \times N_\varphi$)	Sim. Time ($\times 10^3\ r_\mathrm{g}/c$)	MI($\dot{M}$)	MI(ϕ)	MI ($P_\mathrm{out}/\dot{M}c^2$)	MI ($\dot{J}/\dot{M}$)
M87* GRMHD	$580 \times 288 \times 512$	5.6–10.6	0.27	0.15	0.26	0.33
M87* 2t-GRMHD	$348 \times 192 \times 192$	10–15	0.29	0.14	0.25	0.31
M87* 2t-GRRMHD	$288 \times 224 \times 128$	11–16	0.28	0.14	0.14	0.31
Sgr A* GRMHD	$348 \times 192 \times 192$	30–35	0.23	0.21	0.39	0.57

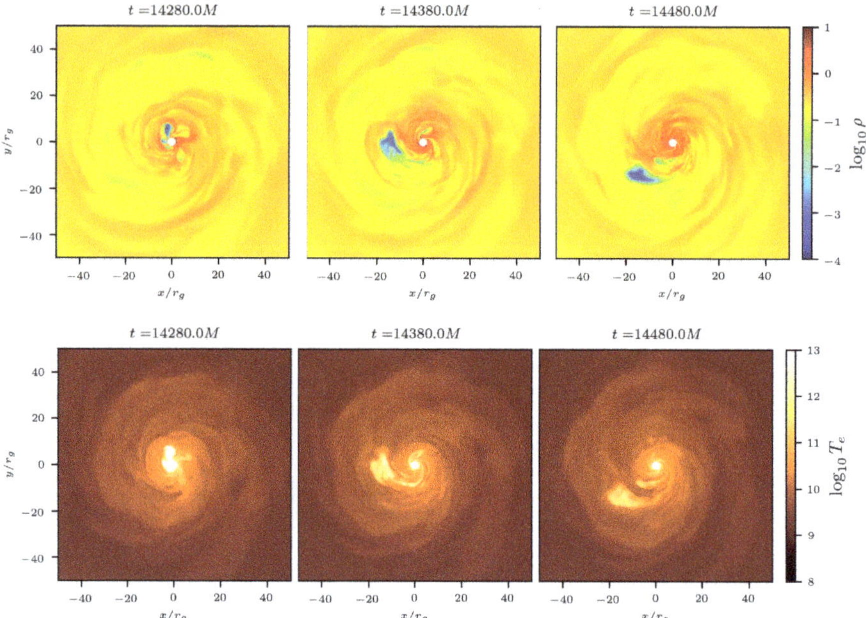

Figure 2. We show equatorial cross-sections of gas density ρ (in arbitrary code units) and electron temperature T_e (in Kelvin) during a magnetic flux eruption in the 2t-GRMHD simulation of M87*. Magnetic flux eruptions, a characteristic feature of magnetically arrested disks, eject bundles of vertical magnetic fields filled with relativistically hot, low-density plasma. These flux bundles spiral around the black hole and may be responsible for high-energy flares [63]. The time unit M is equivalent to r_g/c.

3.2. Disk-Averaged Quantities

Here, we calculate the disk-averaged properties of each model, namely gas density ρ, thermal pressure p_{gas}, magnetic pressure p_{mag}, radial velocity $|v_r|$, azimuthal velocity $|v_\varphi|$, angular momentum u_φ, disk scale height h/r, ion temperature T_i, and the electron temperature T_e. We define disk-averaging of a quantity q as

$$\langle q \rangle (r,t) = \frac{\int_0^{2\pi} \int_0^{\pi} \rho\, q\, \sqrt{-g}\, d\theta\, d\varphi}{\int_0^{2\pi} \int_0^{\pi} \rho\, \sqrt{-g}\, d\theta\, d\varphi}, \tag{25}$$

where $q \in \{\rho, p_{\mathrm{gas}}, p_{\mathrm{mag}}, |v_r|, |v_\varphi|, u_\varphi, h/r, T_i[\mathrm{Kelvin}], T_e[\mathrm{Kelvin}]\}$. Further definitions follow:

$$p_{\mathrm{gas}} = (\Gamma_{\mathrm{ad}} - 1)u, \tag{26}$$
$$p_{\mathrm{mag}} = b^\mu b_\mu / 2, \tag{27}$$
$$|v_i| = \sqrt{v^i v^i g_{ii}}, \tag{28}$$
$$\text{where,}\ v^i = u^i / u^t,$$
$$h/r = |\theta - \pi/2|, \tag{29}$$

where Γ_{ad} and u are the adiabatic index and the internal energy of the gas.

Figures 3 and 4 show the respective disk-averaged radial profiles for each model, including the Sgr A* RIAF solution. The density profiles in the inner few tens of r_g converge roughly to a $n_e \propto r^{-1}$ profile, matching the RIAF density profile as well as the longer time-evolved MAD simulations [64]. The M87* 2t-GRRMHD density is larger by a factor of ≈ 2 from the GRMHD/2t-GRMHD, as is expected from the difference in the mass accretion rate (Figure 1). The 2t-GRRMHD simulation exhibits a slightly more magnetized inflow within the inner 2 $r_{\rm g}$, but, overall, the GRMHD simulations have a similar plasma-$\beta \equiv p_{\rm gas}/p_{\rm mag}$ disk profile. The stronger magnetic field seen in the 2t-GRRMHD model could explain the higher values of the horizon magnetic flux seen in Figure 1. The RIAF model assumes a constant disk plasma-$\beta = 10$, (see Section 2.1), which is substantially higher when compared to the MAD GRMHD models. This value of plasma-β is chosen in order to match the observed 230 GHz flux density of Sgr A*. As we see from the disk scale height in Figure 4, the RIAF model has a much thicker disk than the GRMHD models, and, therefore, produces a lot more sub-mm emission even with a low electron temperature and weak magnetic field strength.

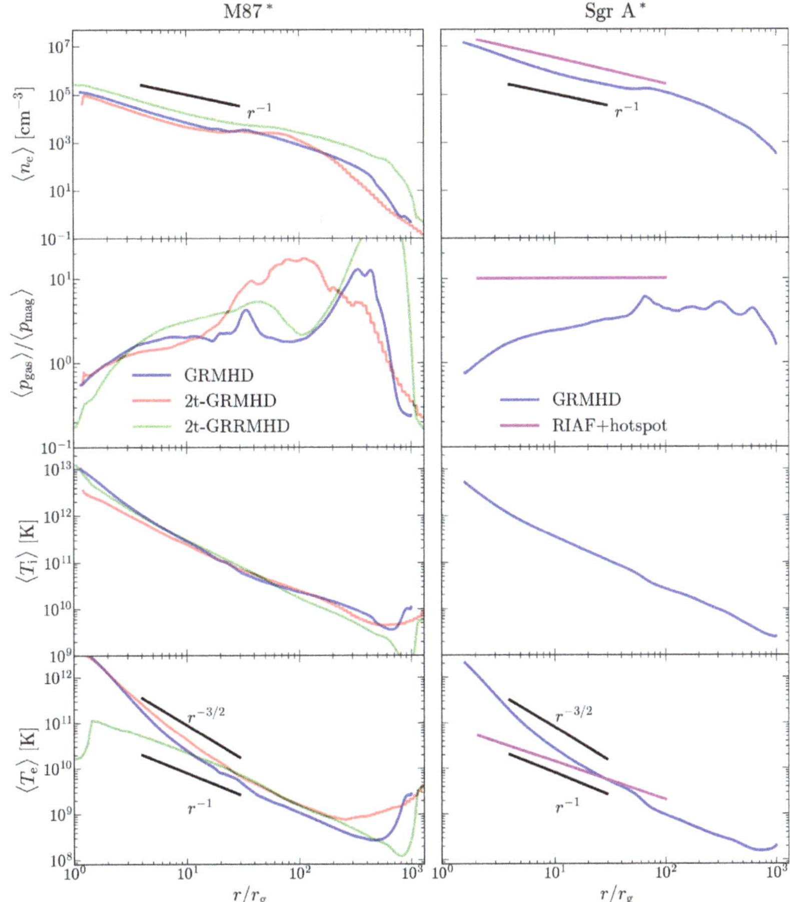

Figure 3. We show the radial profiles of gas density ρ, plasma$-\beta$, proton temperature $T_{\rm p}$, and electron temperature $T_{\rm e}$. Quantities are disk-averaged and time-averaged over the raytracing period.

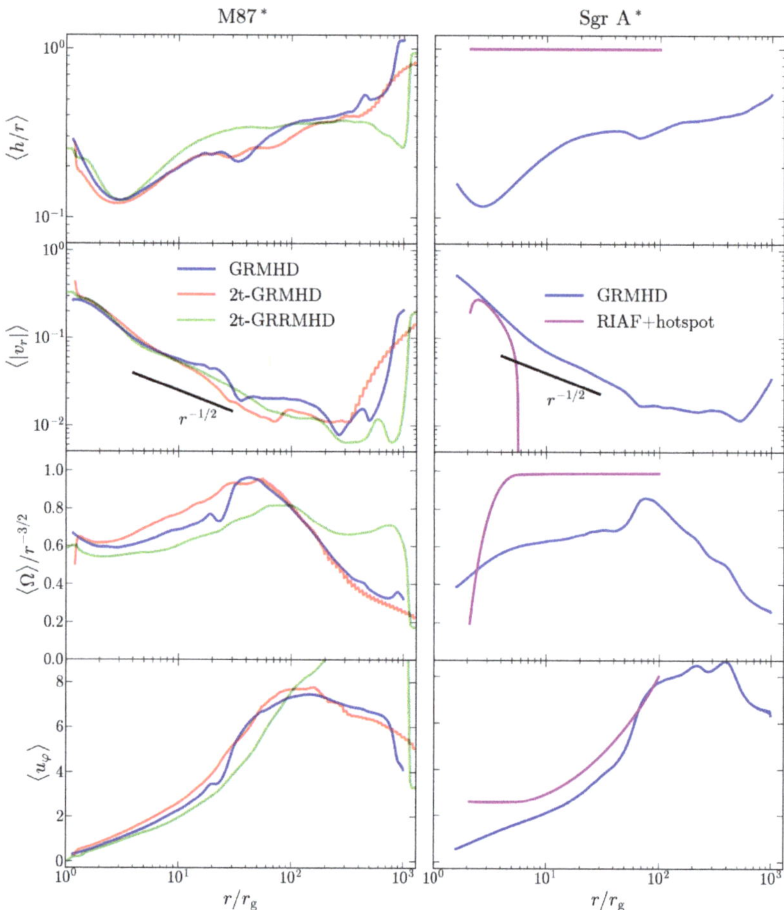

Figure 4. We show the radial profiles of disk scale height h/r, radial velocity $|v_r|$, angular velocity Ω, and specific angular momentum u_φ. Quantities are disk-averaged and time-averaged over the raytracing period.

Next, we see that the disk-averaged electron temperature T_e in the 2t-GRRMHD M87* model is more than an order of magnitude lower than the other GRMHD models within the inner 10 r_g, but actually matches the Sgr A* RIAF T_e profile and has a shallower profile $T_e \propto r^{-1}$ instead of $r^{-3/2}$. It is also interesting to note that the disk ion temperatures T_i are very similar in all the GRMHD simulations shown here. Therefore, despite the same reconnection-driven heating mechanism captured in both the 2t-GRMHD and the 2t-GRRMHD models, the radiative cooling of hot electrons plays a crucial role in determining the eventual T_e. Due to the low T_e, the required accretion rate normalization is higher in the 2t-GRRMHD, as we noted in the previous subsection.

In Figure 4, we show the average disk scale height h/r, the radial and angular velocities (v_r and Ω and the specific angular momentum u_φ). The MAD simulations all show very similar disk properties. The $\langle h/r \rangle \approx 0.1$–0.3 with a sharp increase within 3 r_g where the inflow becomes vertically supported by strong poloidal magnetic fields. The radial velocity has a profile of $r^{-1/2}$, similar to the scaling relation found in ADAF solutions assuming a constant viscosity parameter α [67,68]. The α parameter profile depends on how magnetized the accretion flow is, with $\alpha \propto r^{-1}$ for the weakly magnetized flows and close to constant for the MAD-like flows (e.g., [64,69]). We also see highly sub-Keplerian angular velocity profiles in the GRMHD models, typical for magnetically supported disks. For the RIAF model, the RIAF disk is not infalling and has a constant Keplerian angular velocity. Instead, the hotspot, added to the RIAF solution, undergoes shearing and disappears into the BH with a radial velocity similar to the values found in the GRMHD MAD disks. This occurs because the hotspot is designed to travel along plunging geodesics (see Section 2.1), similar to rapid gas infall close to the BH in the GRMHD models. The angular momentum in the GRMHD models looks sub-Keplerian as expected for MADs.

3.3. Jet Properties

Here, we calculate the radial profiles of the jet half width R_{jet} and Lorentz factor γ:

$$R_{\mathrm{jet}}(r,t) = \sqrt{\frac{\int_0^{2\pi} \int_0^{\pi} \sqrt{-g/g_{rr}}\,(\mu > 2)\,d\theta\,d\varphi}{2\pi}},$$
(30)

$$\gamma(r,t) = \frac{\int_0^{2\pi} \int_0^{\pi} (\mu > 2)\,\alpha u^t \sqrt{-g}\,d\theta\,d\varphi}{\int_0^{2\pi} \int_0^{\pi} (\mu > 2)\,\sqrt{-g}\,d\theta\,d\varphi},$$
(31)

where $\mu = -T^r_t/(\rho u^r)$ is the specific radial energy density and $\alpha = 1/\sqrt{-g^{tt}}$. We calculate the average jet radius R_{jet} by integrating over the total surface area covered by the two jets over a shell at a particular radius, and assume that the total surface area is approximately $2\pi R_{\mathrm{jet}}^2$. We define the jet boundary as $\mu > 2$, i.e., the region over which the jet still remains highly energized. This definition of the jet boundary is quite similar to the condition $\sigma = 1$, which is widely used in the literature (e.g., [27]). Since $\mu = \gamma(\sigma + E_{\mathrm{th}})$, our condition $\mu > 2$ also incorporates regions where the jet might not be magnetically dominated but is relativistically hot or fast. Since we restrict our jet profiles to within $r \lesssim 10^3\,r_g$, the jet radius is primarily determined by the jet magnetization.[3]

Figure 5 shows the jet radius R_{jet} and Lorentz factor γ as a function of the radial distance from the BH for the M87* GRMHD and 2t-GRRMHD models as well as the Sgr A* GRMHD model. The M87* jet radius from our models matches the observed jet width from M87 (gray circles) quite well, with the radial profile roughly proportional to $r^{0.625}$, which is the fitted powerlaw for the M87 jet [70], though the index value has also been reported to be slightly smaller in some works (0.57; [71,72]). The power law index of 0.625 is larger than that found using the $\sigma = 1$ condition from Narayan et al. [27], where the authors found a power law index of 0.428 for their MAD spin $a = 0.9$ GRMHD model. It is possible that we find larger jet radii as we incorporate a part of the hot jet sheath region within our definition of R_{jet} (as suggested by Figure 7 in [73]). For the Sgr A* model, we also find a similar R_{jet} profile. While there are no detections of an extended jet in Sgr A* (e.g., [74]), semi-analytical and GRMHD models of Sgr A* largely favor a jet component from a spinning BH (e.g., [20,75]).

We also show the Lorentz factor γ in Figure 5. Mostly, the jets accelerate to $\gamma \approx 3$–4 by $10^3\,r_g$ in all of our GRMHD models. It is difficult to compare our γ profiles with values inferred from observations of the M87 jet (e.g., [76]) since our γ values are biased toward the jet spine while the observations generally capture the velocities of localized features in the sub-relativistic jet sheath/disk wind, especially at small distances from the BH. Indeed, both simulations and observations show that the jet Lorentz factor varies greatly as a function of jet radius (e.g., see [73]). We speculate that a better approach might be to calculate emissivity-weighted Lorentz factors in order to compare to the measured γ from M87. Since our focus is on the comparison between GRMHD simulations, we leave direct comparisons to observational data to future work.

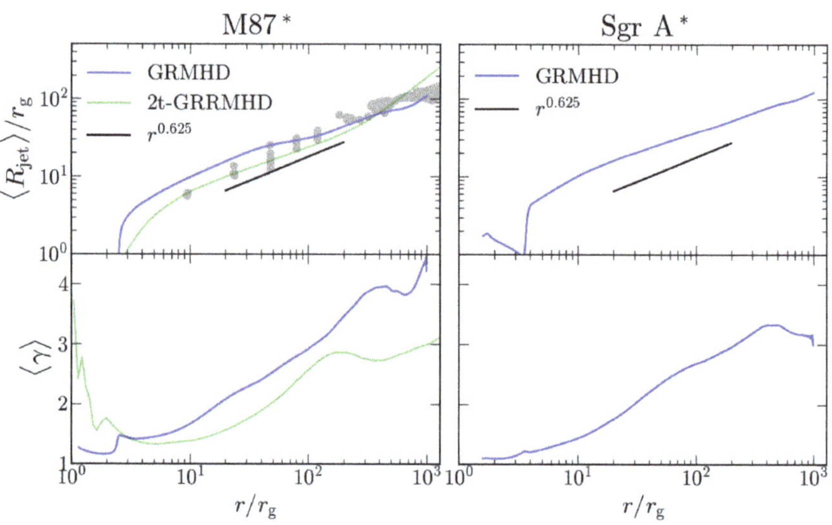

Figure 5. We show the jet radius R_{jet} and the jet Lorentz factor γ from the M87* GRMHD and 2t-GRRMHD models, and the Sgr A* GRMHD model. The gray circles indicates the deprojected jet radius of the M87 jet assuming a BH mass of $6.2 \times 10^9 M_\odot$ and a source inclination of $14°$ [70]. The data points are a compilation of various papers [59,60,71,77–79].

3.4. Axisymmetrized Profiles

In the previous sections, we found that the largest differences between the GRMHD models occur in electron temperature distribution. Figure 6 shows the time and azimuthally averaged 2D vertical plots of gas density n_e and electron temperature T_e. We show the normalized n_e so as to capture the relative change in the disk/wind density distribution, which would provide us information about the disk structure. The large difference in disk scale height is immediately apparent between the RIAF and the MAD GRMHD models (also see Figure 4). The presence of a prominent wide jet component in MADs squeezes the inner disk and pushes against the accretion flow, a feature which is not captured in the constant h/r RIAF model. However, the RIAF model roughly reproduces the density profile of the disk midplane region, suggesting that the RIAF model could represent non/weakly jetted, quasi-spherical accretion flows quite well. For sources such as M87*, where we see a prominent jet component, the density gradient in the vertical direction is expected to be steeper as strong magnetic stresses power winds carry gas away from the disk (e.g., [64]).

Overall, the disk/wind density distribution among the GRMHD models look similar with small differences in the lateral extension of the wind region and the steepness of the vertical gradient in density. For example, if we compare the 2t-GRRMHD model with the other two simulations, the density in the wind region is larger in the radiative model. The reason for the shallow vertical density profile in the 2t-GRRMHD model is unclear since the weakly magnetized thick disk simulations tell us that radiative cooling would lead to the loss of gas pressure in the disk and would result in the disk collapsing to a relatively dense structure in the midplane (e.g., [80,81]). However, in the presence of strong poloidal magnetic fields, i.e., in the MAD state, the plasma-β decreases to $\beta \approx 0.2$–1 in the disk midplane (see Figure 3, third row, left panel), and can reach even lower values in the upper layers of the accretion flow. The high magnetic pressure could help support the disk against collapse while sufficiently strong magnetic stresses could power disk winds. Such behavior is also seen in recent GRMHD simulations of near-Eddington, geometrically thin, strongly magnetized disks, where the inner disk (or corona) has a larger h/r than the outer disk due to magnetic pressure support [82–84]. To verify how radiative cooling affects the inner disk/wind structure in highly sub-Eddington accretion flows such as M87* and Sgr A*, we require longer 2t-GRRMHD simulations such that the disk is in an inflow–outflow equilibrium out to a radius of at least $50\,r_{\mathrm{g}}$.

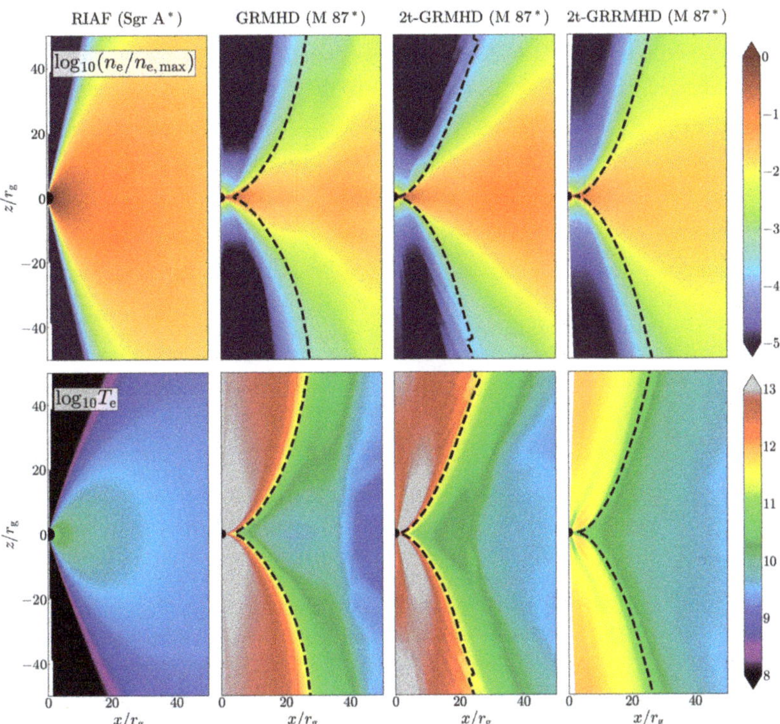

Figure 6. We show t- and φ-averaged data: electron number density n_{e} (top row) and temperature T_{e} (bottom row). We also denote the jet boundary with $\sigma = 1$ (black lines). The time-averaging is performed over the $5000\,r_{\mathrm{g}}/c$ for each model. RIAF plots are for Sgr A* while the rest are for M87. The Sgr A* GRMHD model produces similar plots of n_{e} and T_{e} as the M87* model, and, hence, we do not show it here.

The 2D temperature plot of the RIAF model also looks vastly different in the inner disk ($r \lesssim 20\,r_{\mathrm{g}}$) when compared to the GRMHD and 2t-GRMHD simulations, but is similar to the temperature distribution in the 2t-GRRMHD disk midplane (also seen in the T_{e} plot of Figure 3). The RIAF model does not capture gas heating in the jet sheath region (the region just outside of the jet boundary indicated by the $\sigma = 1$ dashed line) and, therefore, T_{e} drops as we move away from the midplane toward the poles. In the GRMHD models, the jet sheath is as hot, if not hotter, than the inner accretion flow as temperatures reach $T_{\mathrm{e}} > 10^{11}$ K. For the GRMHD simulation, the electron temperature is provided as a fraction of the fluid temperature, where the fraction depends on how magnetized the gas is in the region, as per the $R - \beta$ prescription from Equation (14). For the M87* model, we chose a R_{high} value of 160 to have a jet-dominated sub-mm image. This choice of R_{high} suppresses the electron temperature in the disk, focusing higher temperatures in the jet sheath. Comparing the GRMHD model with the 2t-GRMHD model, the jet sheath region exhibits very similar T_{e} values, but the disk midplane is hotter by a factor of a few in the 2t-GRMHD model. We note that this difference in T_{e} in the midplane is more noticeable in the 2D plot rather than in the disk-averaged T_{e} profile shown in Figure 3 as the upper layers of the disk become substantially hotter in the GRMHD model.

For the radiative 2t-GRRMHD model, the inner regions of the disk are cooler as electrons heated by magnetic reconnection quickly cool via synchrotron and Compton losses. From Figure 3, the drop in T_e for the 2t-GRRMHD model is shown to be as large as an order of magnitude when compared to the (2t-) GRMHD models. Another interesting feature is that the hot region ($T_e > 10^{11}$ K) in the jet sheath is much narrower in the 2t-GRRMHD model, which could have a significant bearing on the ray-traced image, possibly producing a thinner jet sheath. Finally, the difference in T_e in the jet body between the GRMHD models is due to the different density/internal energy floor setups used by the corresponding codes. Since the gas in the jet sheath and the jet body undergo mixing due to boundary instabilities (e.g., [73,85]), it is possible that the choice of floors could affect the overall electron temperature in the jet sheath. Such a study is outside the scope of our paper and is left to future work.

3.5. Orbiting Hotspot in a RIAF Model

High-energy flares are commonly observed in AGNs, with GeV and MeV flares seen in M87* (e.g., [86,87]) and quasi-daily nIR and X-ray flares in Sgr A* (e.g., [44,88–94]). A number of attempts have been performed to explain the origin of flaring, such as magnetic reconnection in turbulent gas flows in the disk and the jet [65,73,95–97] and magnetic flux eruptions [61,63,98,99]. For Sgr A*, the semi-analytical models found that we require high-energy electrons, assumed to be accelerated via an ad hoc process such as a large magnetic reconnection event or shocks, to describe the large flaring events [100–102]. The near-infrared observations from the GRAVITY telescope provided further evidence for orbiting hotspot features in the accretion flow [103] that may be linked to acceleration events. It has also been recently shown that orbiting hotspots can be used to model double-peaked X-ray flares [94,104] and prominent Stokes Q-U loops in the sub-mm emission of Sgr A* [105]. These results provide us with considerable motivation to test the capability of the ngEHT to detect hotspot formation in accretion flows around black holes.

Instead of isolating a particular magnetic flux eruption event in our simulations, we added a shearing hotspot to the RIAF solution as detailed in Section 2.1. Figure 7 shows the temporal evolution of the azimuthally averaged electron number density of the hotspot. We begin with a Gaussian distribution of gas that undergoes shearing as the gas falls in closer to the BH. The overall density normalization is much lower than in the RIAF disk since the optically thin hotspot gas produces a large enough non-thermal synchrotron emissivity. The hotspot is evolved over $800\, r_g/c$, but the gas distribution comes to a near-steady-state profile within the first $200\, r_g/c$, which is roughly one hour for Sgr A*. The shearing of the hotspot gas has a significant impact on the evolution of the 230 GHz image [4,31]. From Figure 4 (right column), we see that the radial velocity matches the disk-averaged gas velocity from the GRMHD model, showing nearly free-fall speeds, while the azimuthal velocity becomes highly sub-Keplerian. The velocity profiles show that our hotspot model should be able to reproduce the expected hotspot motion from the GRMHD models, and is ideal for investigating multiwavelength flare lightcurves. Emami et al. [106], a companion paper, goes into further details about how current dynamical reconstruction techniques can be used to trace out the motion and morphology of the shearing hotspot in the context of ngEHT observations. These hotspot models and reconstruction methods would be integral in deciphering the more complex gas dynamics of magnetic flux eruption events in MADs, which have been shown to produce a significant variation in the image structure of M87* at 230 GHz (e.g., [107]).

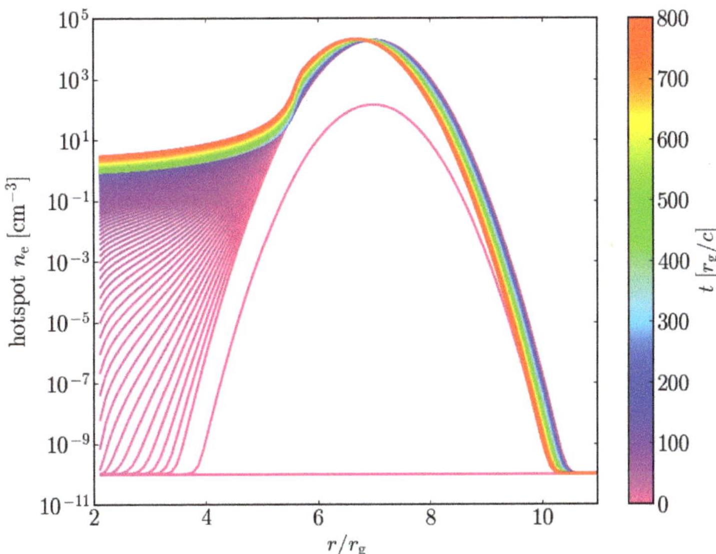

Figure 7. We show the φ−averaged hotspot electron number density as a function of radius and time. The hotspot falls into the BH and becomes sheared over time.

4. Conclusions

In this work, for the first time, we compare a series of numerical solutions with increasing complexity that were specifically constructed to understand the accretion flow around the supermassive black holes M87* and Sgr A*. We include a time-independent radiatively inefficient accretion flow (RIAF) model as well as fully 3D GRMHD simulations of accreting black holes, incorporating the effects of electron heating and cooling losses via two-temperature and radiation physics. In addition, each of our simulations are run with different GRMHD codes, which is similar to the approach of another community-wide code comparison effort [26]. We found that the simulations exhibit remarkably similar properties given that the simulations incorporate varying levels of complexity in electron physics. The notable exception is the electron temperature, where radiative cooling decreases the temperature by a factor of $\lesssim 10$ within the inner 10 gravitational radii, the region that produces the bulk of the 230 GHz emission in M87*, one of the two primary targets of the EHT and the ngEHT (the other being Sgr A*). The main goal of this work is to understand the variation in the underlying accretion flow and jet properties in our models since synthetic ray-traced images constructed from these models are used as "truth" images of M87* and Sgr A* for the ngEHT Analysis Challenges [4]. The ngEHT Analysis Challenges are an effort to determine how much information about the accretion flow and jet dynamics we can glean from the proposed ngEHT reference array, and what modifications in the image reconstruction tools are necessarily required to decode future ngEHT observational data.

Our paper deals with numerical models designed to investigate hotspot evolution, turbulent inspiralling gas flows, and extended powerful jets, targeting M87* and Sgr A*. We restricted our model set to the community-standard setup: a rotating, geometrically thick, optically thin torus of magnetized gas around a spinning black hole, which is the fiducial model choice of the EHT [18–20]. This model choice leaves out the exploration of multiple new setups of black hole accretion, such as quasi-spherical wind-fed inflows (e.g., [108,109]), strongly wind-fed accretion (e.g., [110,111]), geometrically thin accretion disks (e.g., [83,84,112]), puffy radiation-dominated super-Eddington disks (e.g., [113,114]), and misaligned accretion disks (e.g., [24,40,115,116]). Apart from varying the accretion mode, the high resolution of the images from EHT and ngEHT could potentially help distinguish between different space–time metrics [117]. To date, only a limited number of non-Kerr GRMHD simulations have only been performed (e.g., [118–120]). The future of numerical studies is bright, given their rising popularity in the astrophysics community and the increase in computational resources. The breadth of the current investigations in accretion physics would result in a plethora of variable structures that should be thoroughly studied keeping the observational capabilities of the ngEHT in mind.

Author Contributions: Conceptualization, K.C.; Methodology, K.C., P.T., Y.M. and F.R.; Software, K.C., A.C., P.T. and Y.M.; Formal analysis, K.C. and P.T.; Investigation, K.C., Y.M., R.E., C.F., A.R. (Angelo Ricarte), L.B., F.R., P.A., A.F., J.K., N.K., G.L., H.M., N.P., A.R. (Alexander Raymond), E.T. and J.V.; Writing—original draft, K.C. and P.T.; Writing—review & editing, K.C., A.C. and A.R. (Angelo Ricarte); Visualization, K.C.; Supervision, M.D.J. and S.S.D. All authors have read and agreed to the published version of the manuscript.

Funding: We thank the National Science Foundation (AST-1716536, AST-1935980, and AST-2034306) and the Gordon and Betty Moore Foundation (GBMF-10423) for financially supporting this work. This work was supported in part by the Black Hole Initiative, which is funded by grants from the John Templeton Foundation (JTF-61497) and the Gordon and Betty Moore Foundation (GBMF-8273) to Harvard University. K.C. is also supported in part by the Black Hole PIRE program (NSF grant OISE-1743747). R.E. acknowledges the support by the Institute for Theory and Computation at the Center for Astrophysics as well as grant numbers 21-atp21-0077, NSF AST-1816420, and HST-GO-16173.001-A for very generous supports. H.M. received financial support for this research from the International Max Planck Research School (IMPRS) for Astronomy and Astrophysics at the Universities of Bonn and Cologne. This research is supported by the DFG research grant "Jet physics on horizon scales and beyond" (Grant No. FR 4069/2-1), the ERC synergy grant "BlackHoleCam: Imaging the Event Horizon of Black Holes" (Grant No. 610058), and ERC advanced grant "JETSET: Launching, propagation and emission of relativistic jets from binary mergers and across mass scales" (Grant No. 884631). J.K. acknowledges funding by the Deutsche Forschungsgemeinschaft (DFG, German Research Foundation) under Germany´s Excellence Strategy—EXC 2094—390783311. Y.M. is supported by the National Natural Science Foundation of China (No. 12273022) and the Shanghai pilot program of international scientists for basic research (No. 22JC1410600).

Data Availability Statement: The data shown in this study has been generated using the following codes: BHAC [38], BHOSS [41], H-AMR [37], IPOLE [42], KORAL [39], with access provided upon request to the corresponding authors.

Acknowledgments: We thank the anonymous referees for their suggestions that greatly improved the text. This research was enabled by support provided by grant no. NSF PHY-1125915 along with INCITE and ASCR Leadership Computing Challenge (ALCC) programs under the award PHY129, using resources from the Oak Ridge Leadership Computing Facility, Summit, which is a US Department of Energy office of Science User Facility supported under contract DE-AC05- 00OR22725, as well as Calcul Quebec and Compute Canada under award xsp-772-ab (PI: D. Haggard).

Conflicts of Interest: The authors declare no conflict of interest. The funders had no role in the design of the studyl in the collection, analyses, or interpretation of data; in the writing of the manuscript, or in the decision to publish the results.

Notes

1. https://www.ngeht.org/ (accessed on 22 November 2022).
2. https://challenge.ngeht.org/ (accessed on 22 November 2022).
3. Specific enthalpy includes the rest-mass energy contribution in our definition from Section 2.3.

References

1. Akiyama, K. et al. [Event Horizon Telescope Collaboration]. First M87 Event Horizon Telescope Results. I. The Shadow of the Supermassive Black Hole. *Astrophys. J. Lett.* **2019**, *875*, L1.
2. Akiyama, K. et al. [Event Horizon Telescope Collaboration]. First Sagittarius A* Event Horizon Telescope Results. I. The Shadow of the Supermassive Black Hole in the Center of the Milky Way. *Astrophys. J. Lett.* **2022**, *930*, L12. [CrossRef]

3. Doeleman, S.; Blackburn, L.; Dexter, J.; Gomez, J.L.; Johnson, M.D.; Palumbo, D.C.; Weintroub, J.; Farah, J.R.; Fish, V.; Loinard, L.; et al. Studying Black Holes on Horizon Scales with VLBI Ground Arrays. *Bulletin of the AAS*, 2019, Volume 51. Available online: https://baas.aas.org/pub/2020n7i256 (accessed on 22 November 2022).

4. Roelofs, F.; Blackburn, L.; Lindahl, G.; Doeleman, S.S.; Johnson, M.D.; Arras, P.; Chatterjee, K.; Emami, R.; Fromm, C.; Fuentes, A.; et al. The ngEHT Analysis Challenges. *Galaxies*, **2022**, *11*, 12. [CrossRef]

5. Raymond, A.W.; Palumbo, D.; Paine, S.N.; Blackburn, L.; Córdova Rosado, R.; Doeleman, S.S.; Farah, J.R.; Johnson, M.D.; Roelofs, F.; Tilanus, R.P.J.; et al. Evaluation of New Submillimeter VLBI Sites for the Event Horizon Telescope. *Astrophys. J. Suppl. Ser.* **2021**, *253*, 5.

6. Shakura, N.I.; Sunyaev, R.A. Reprint of 1973A&A....24..337S. Black holes in binary systems. Observational appearance. *Astron. Astrophys.* **1973**, *500*, 33–51.

7. Rees, M.J.; Phinney, E.S.; Begelman, M.C.; Blandford, R.D. Ion-supported tori and the origin of radio jets. *Nature* **1982**, *295*, 17–21. [CrossRef]

8. Narayan, R.; Yi, I.; Mahadevan, R. Explaining the spectrum of Sagittarius A* with a model of an accreting black hole. *Nature* **1995**, *374*, 623–625. [CrossRef]

9. Quataert, E.; Gruzinov, A. Convection-dominated Accretion Flows. *Astrophys. J.* **2000**, *539*, 809–814. [CrossRef]

10. Narayan, R.; Igumenshchev, I.V.; Abramowicz, M.A. Magnetically Arrested Disk: An Energetically Efficient Accretion Flow. *PASJ* **2003**, *55*, L69–L72.

11. McKinney, J.C. General relativistic magnetohydrodynamic simulations of the jet formation and large-scale propagation from black hole accretion systems. *Mon. Not. R. Astron. Soc.* **2006**, *368*, 1561–1582.

12. Komissarov, S.S.; Barkov, M.V.; Vlahakis, N.; Königl, A. Magnetic acceleration of relativistic active galactic nucleus jets. *Mon. Not. R. Astron. Soc.* **2007**, *380*, 51–70. [CrossRef]

13. Tchekhovskoy, A.; Narayan, R.; McKinney, J.C. Efficient generation of jets from magnetically arrested accretion on a rapidly spinning black hole. *Mon. Not. R. Astron. Soc.* **2011**, *418*, L79–L83.

14. Narayan, R.; Sądowski, A.; Penna, R.F.; Kulkarni, A.K. GRMHD simulations of magnetized advection-dominated accretion on a non-spinning black hole: Role of outflows. *Mon. Not. R. Astron. Soc.* **2012**, *426*, 3241–3259. [CrossRef]

15. Tchekhovskoy, A.; Bromberg, O. Three-dimensional relativistic MHD simulations of active galactic nuclei jets: Magnetic kink instability and Fanaroff–Riley dichotomy. *Mon. Not. R. Astron. Soc.* **2016**, *461*, L46–L50. [CrossRef]

16. Yuan, F.; Quataert, E.; Narayan, R. Nonthermal Electrons in Radiatively Inefficient Accretion Flow Models of Sagittarius A*. *Astrophys. J.* **2003**, *598*, 301–312.

17. Lucchini, M.; Krauß, F.; Markoff, S. The unique case of the active galactic nucleus core of M87: A misaligned low-power blazar? *Mon. Not. R. Astron. Soc.* **2019**, *489*, 1633–1643.

18. Akiyama, K. et al. [Event Horizon Telescope Collaboration]. First M87 Event Horizon Telescope Results. V. Physical Origin of the Asymmetric Ring. *Astrophys. J. Lett.* **2019**, *875*, L5.

19. Akiyama, K. et al. [Event Horizon Telescope Collaboration]. First M87 Event Horizon Telescope Results. VIII. Magnetic Field Structure near The Event Horizon. *Astrophys. J. Lett.* **2021**, *910*, L13.

20. Akiyama, K. et al. [Event Horizon Telescope Collaboration]. First Sagittarius A* Event Horizon Telescope Results. V. Testing Astrophysical Models of the Galactic Center Black Hole. *Astrophys. J. Lett.* **2022**, *930*, L16. [CrossRef]

21. Igumenshchev, I.V.; Narayan, R.; Abramowicz, M.A. Three-dimensional Magnetohydrodynamic Simulations of Radiatively Inefficient Accretion Flows. *Astrophys. J.* **2003**, *592*, 1042–1059.

22. Blandford, R.D.; Znajek, R.L. Electromagnetic extraction of energy from Kerr black holes. *Mon. Not. R. Astron. Soc.* **1977**, *179*, 433–456. [CrossRef]

23. Gammie, C.F.; McKinney, J.C.; Tóth, G. HARM: A Numerical Scheme for General Relativistic Magnetohydrodynamics. *Astrophys. J.* **2003**, *589*, 444–457.

24. Fragile, P.C.; Blaes, O.M.; Anninos, P.; Salmonson, J.D. Global General Relativistic Magnetohydrodynamic Simulation of a Tilted Black Hole Accretion Disk. *Astrophys. J.* **2007**, *668*, 417–429.

25. Chael, A.; Narayan, R.; Johnson, M.D. Two-temperature, Magnetically Arrested Disc simulations of the jet from the supermassive black hole in M87. *Mon. Not. R. Astron. Soc.* **2019**, *486*, 2873–2895.

26. Porth, O.; Chatterjee, K.; Narayan, R.; Gammie, C.F.; Mizuno, Y.; Anninos, P.; Baker, J.G.; Bugli, M.; Chan, C.k.; Davelaar, J.; et al. The Event Horizon General Relativistic Magnetohydrodynamic Code Comparison Project. *Astrophys. J. Suppl. Ser.* **2019**, *243*, 26.

27. Narayan, R.; Chael, A.; Chatterjee, K.; Ricarte, A.; Curd, B. Jets in magnetically arrested hot accretion flows: Geometry, power, and black hole spin-down. *Mon. Not. R. Astron. Soc.* **2022**, *511*, 3795–3813.

28. Janssen, M.; Falcke, H.; Kadler, M.; Ros, E.; Wielgus, M.; Akiyama, K.; Baloković, M.; Blackburn, L.; Bouman, K.L.; Chael, A.; et al. Event Horizon Telescope observations of the jet launching and collimation in Centaurus A. *Nat. Astron.* **2021**, *5*, 1017–1028.

29. Broderick, A.E.; Loeb, A. Imaging optically thin hotspots near the black hole horizon of Sgr A* at radio and near-infrared wavelengths. *Mon. Not. R. Astron. Soc.* **2006**, *367*, 905–916.

30. Broderick, A.E.; McKinney, J.C. Parsec-scale Faraday Rotation Measures from General Relativistic Magnetohydrodynamic Simulations of Active Galactic Nucleus Jets. *Astrophys. J.* **2010**, *725*, 750–773.

31. Tiede, P.; Pu, H.Y.; Broderick, A.E.; Gold, R.; Karami, M.; Preciado-López, J.A. Spacetime Tomography Using the Event Horizon Telescope. *Astrophys. J.* **2020**, *892*, 132.

32. Broderick, A.E.; Fish, V.L.; Doeleman, S.S.; Loeb, A. Evidence for Low Black Hole Spin and Physically Motivated Accretion Models from Millimeter-VLBI Observations of Sagittarius A*. *Astrophys. J.* **2011**, *735*, 110.
33. Broderick, A.E.; Fish, V.L.; Johnson, M.D.; Rosenfeld, K.; Wang, C.; Doeleman, S.S.; Akiyama, K.; Johannsen, T.; Roy, A.L. Modeling Seven Years of Event Horizon Telescope Observations with Radiatively Inefficient Accretion Flow Models. *Astrophys. J.* **2016**, *820*, 137.
34. Pu, H.Y.; Akiyama, K.; Asada, K. The Effects of Accretion Flow Dynamics on the Black Hole Shadow of Sagittarius A*. *Astrophys. J.* **2016**, *831*, 4.
35. Broderick, A.; Blandford, R. Covariant magnetoionic theory—II. Radiative transfer. *Mon. Not. R. Astron. Soc.* **2004**, *349*, 994–1008.
36. Jones, T.W.; O'Dell, S.L. Transfer of polarized radiation in self-absorbed synchrotron sources. I. Results for a homogeneous source. *Astrophys. J.* **1977**, *214*, 522–539. [CrossRef]
37. Liska, M.T.P.; Chatterjee, K.; Issa, D.; Yoon, D.; Kaaz, N.; Tchekhovskoy, A.; van Eijnatten, D.; Musoke, G.; Hesp, C.; Rohoza, V.; et al. H-AMR: A New GPU-accelerated GRMHD Code for Exascale Computing with 3D Adaptive Mesh Refinement and Local Adaptive Time Stepping. *Astrophys. J. Suppl. Ser.* **2022**, *263*, 26. [CrossRef]
38. Porth, O.; Olivares, H.; Mizuno, Y.; Younsi, Z.; Rezzolla, L.; Moscibrodzka, M.; Falcke, H.; Kramer, M. The black hole accretion code. *Comput. Astrophys. Cosmol.* **2017**, *4*, 1.
39. Sądowski, A.; Narayan, R.; Tchekhovskoy, A.; Zhu, Y. Semi-implicit scheme for treating radiation under M1 closure in general relativistic conservative fluid dynamics codes. *Mon. Not. R. Astron. Soc.* **2013**, *429*, 3533–3550.
40. Liska, M.; Hesp, C.; Tchekhovskoy, A.; Ingram, A.; van der Klis, M.; Markoff, S. Formation of precessing jets by tilted black hole discs in 3D general relativistic MHD simulations. *Mon. Not. R. Astron. Soc.* **2018**, *474*, L81–L85. [CrossRef]
41. Younsi, Z.; Porth, O.; Mizuno, Y.; Fromm, C.M.; Olivares, H. Modelling the polarised emission from black holes on event horizon-scales. *Proc. IAU* **2020**, *14*, 9–12. [CrossRef]
42. Mościbrodzka, M.; Gammie, C.F. IPOLE - semi-analytic scheme for relativistic polarized radiative transport. *Mon. Not. R. Astron. Soc.* **2018**, *475*, 43–54.
43. Gebhardt, K.; Adams, J.; Richstone, D.; Lauer, T.R.; Faber, S.M.; Gültekin, K.; Murphy, J.; Tremaine, S. The Black Hole Mass in M87 from Gemini/NIFS Adaptive Optics Observations. *Astrophys. J.* **2011**, *729*, 119.
44. Do, T.; Witzel, G.; Gautam, A.K.; Chen, Z.; Ghez, A.M.; Morris, M.R.; Becklin, E.E.; Ciurlo, A.; Hosek, Matthew, J.; Martinez, G.D.; et al. Unprecedented Near-infrared Brightness and Variability of Sgr A *. *Astrophys. J. Lett.* **2019**, *882*, L27. [CrossRef]
45. Abuter, R. et al. [GRAVITY Collaboration]. A geometric distance measurement to the Galactic center black hole with 0.3% uncertainty. *Astron. Astrophys.* **2019**, *625*, L10.
46. Mościbrodzka, M.; Falcke, H.; Shiokawa, H. General relativistic magnetohydrodynamical simulations of the jet in M87. *A&A* **2016**, *586*, A38. [CrossRef]
47. Xiao, F. Modelling energetic particles by a relativistic kappa-loss-cone distribution function in plasmas. *Plasma Phys. Control. Fusion* **2006**, *48*, 203–213. [CrossRef]
48. Davelaar, J.; Mościbrodzka, M.; Bronzwaer, T.; Falcke, H. General relativistic magnetohydrodynamical κ-jet models for Sagittarius A*. *Astron. Astrophys.* **2018**, *612*, A34.
49. Ressler, S.M.; Tchekhovskoy, A.; Quataert, E.; Chandra, M.; Gammie, C.F. Electron thermodynamics in GRMHD simulations of low-luminosity black hole accretion. *Mon. Not. R. Astron. Soc.* **2015**, *454*, 1848–1870.
50. Dexter, J.; Jiménez-Rosales, A.; Ressler, S.M.; Tchekhovskoy, A.; Bauböck, M.; de Zeeuw, P.T.; Eisenhauer, F.; von Fellenberg, S.; Gao, F.; Genzel, R.; et al. A parameter survey of Sgr A * radiative models from GRMHD simulations with self-consistent electron heating. *Mon. Not. R. Astron. Soc.* **2020**, *494*, 4168–4186.
51. Howes, G.G. A prescription for the turbulent heating of astrophysical plasmas. *Mon. Not. R. Astron. Soc.* **2010**, *409*, L104–L108.
52. Kawazura, Y.; Barnes, M.; Schekochihin, A.A. Thermal disequilibration of ions and electrons by collisionless plasma turbulence. *Proc. Natl. Acad. Sci. USA* **2019**, *116*, 771–776.
53. Werner, G.R.; Uzdensky, D.A.; Begelman, M.C.; Cerutti, B.; Nalewajko, K. Non-thermal particle acceleration in collisionless relativistic electron-proton reconnection. *Mon. Not. R. Astron. Soc.* **2018**, *473*, 4840–4861.
54. Rowan, M.E.; Sironi, L.; Narayan, R. Electron and Proton Heating in Transrelativistic Magnetic Reconnection. *Astrophys. J.* **2017**, *850*, 29.
55. Mizuno, Y.; Fromm, C.M.; Younsi, Z.; Porth, O.; Olivares, H.; Rezzolla, L. Comparison of the ion-to-electron temperature ratio prescription: GRMHD simulations with electron thermodynamics. *Mon. Not. R. Astron. Soc.* **2021**, *506*, 741–758.
56. Sądowski, A.; Narayan, R. Photon-conserving Comptonization in simulations of accretion discs around black holes. *Mon. Not. R. Astron. Soc.* **2015**, *454*, 2372–2380.
57. Chael, A.; Rowan, M.; Narayan, R.; Johnson, M.; Sironi, L. The role of electron heating physics in images and variability of the Galactic Centre black hole Sagittarius A*. *Mon. Not. R. Astron. Soc.* **2018**, *478*, 5209–5229.
58. Sądowski, A.; Wielgus, M.; Narayan, R.; Abarca, D.; McKinney, J.C.; Chael, A. Radiative, two-temperature simulations of low-luminosity black hole accretion flows in general relativity. *Mon. Not. R. Astron. Soc.* **2017**, *466*, 705–725. [CrossRef]
59. Doeleman, S.S.; Fish, V.L.; Schenck, D.E.; Beaudoin, C.; Blundell, R.; Bower, G.C.; Broderick, A.E.; Chamberlin, R.; Freund, R.; Friberg, P.; et al. Jet-Launching Structure Resolved Near the Supermassive Black Hole in M87. *Science* **2012**, *338*, 355.

60. Akiyama, K.; Lu, Ru.; Fish, V.L.; Doeleman, S.S.; Broderick, A.E.; Dexter, J.; Hada, K.; Kino, M.; Nagai, H.; Honma, M.; et al. 230 GHz VLBI Observations of M87: Event-horizon-scale Structure during an Enhanced Very-high-energy γ-Ray State in 2012. *Astrophys. J.* **2015**, *807*, 150.

61. Porth, O.; Mizuno, Y.; Younsi, Z.; Fromm, C.M. Flares in the Galactic Centre - I. Orbiting flux tubes in magnetically arrested black hole accretion discs. *Mon. Not. R. Astron. Soc.* **2021**, *502*, 2023–2032.

62. Begelman, M.C.; Scepi, N.; Dexter, J. What really makes an accretion disc MAD. *Mon. Not. R. Astron. Soc.* **2022**, *511*, 2040–2051.

63. Ripperda, B.; Liska, M.; Chatterjee, K.; Musoke, G.; Philippov, A.A.; Markoff, S.B.; Tchekhovskoy, A.; Younsi, Z. Black Hole Flares: Ejection of Accreted Magnetic Flux through 3D Plasmoid-mediated Reconnection. *Astrophys. J. Lett.* **2022**, *924*, L32.

64. Chatterjee, K.; Narayan, R. Flux Eruption Events Drive Angular Momentum Transport in Magnetically Arrested Accretion Flows. *Astrophys. J.* **2022**, *941*, 30. [CrossRef]

65. Chatterjee, K.; Markoff, S.; Neilsen, J.; Younsi, Z.; Witzel, G.; Tchekhovskoy, A.; Yoon, D.; Ingram, A.; van der Klis, M.; Boyce, H.; et al. General relativistic MHD simulations of non-thermal flaring in Sagittarius A*. *Mon. Not. R. Astron. Soc.* **2021**, *507*, 5281–5302.

66. Wielgus, M.; Marchili, N.; Martí-Vidal, I.; Keating, G.K.; Ramakrishnan, V.; Tiede, P.; Fomalont, E.; Issaoun, S.; Neilsen, J.; Nowak, M.A.; et al. Millimeter Light Curves of Sagittarius A* Observed during the 2017 Event Horizon Telescope Campaign. *Astrophys. J. Lett.* **2022**, *930*, L19.

67. Narayan, R.; Yi, I. Advection-dominated accretion: A self-similar solution. *Astrophys. J. Lett.* **1994**, *428*, L13–L16.

68. Narayan, R.; Mahadevan, R.; Quataert, E. Advection-dominated accretion around black holes. In *Theory of Black Hole Accretion Disks*; Abramowicz, M.A., Bjornsson, G.; Pringle, J.E., Eds.; Cambridge University Press: Cambridge, UK, 1998; pp. 138–182. [CrossRef]

69. Liska, M.; Tchekhovskoy, A.; Quataert, E. Large-scale poloidal magnetic field dynamo leads to powerful jets in GRMHD simulations of black hole accretion with toroidal field. *Mon. Not. R. Astron. Soc.* **2020**, *494*, 3656–3662. [CrossRef]

70. Nakamura, M.; Asada, K.; Hada, K.; Pu, H.Y.; Noble, S.; Tseng, C.; Toma, K.; Kino, M.; Nagai, H.; Takahashi, K.; et al. Parabolic Jets from the Spinning Black Hole in M87. *Astrophys. J.* **2018**, *868*, 146. [CrossRef]

71. Asada, K.; Nakamura, M. The Structure of the M87 Jet: A Transition from Parabolic to Conical Streamlines. *ApJL* **2012**, *745*, 5. [CrossRef]

72. Nokhrina, E.E.; Gurvits, L.I.; Beskin, V.S.; Nakamura, M.; Asada, K.; Hada, K. M87 black hole mass and spin estimate through the position of the jet boundary shape break. *Mon. Not. R. Astron. Soc.* **2019**, *489*, 1197–1205.

73. Chatterjee, K.; Liska, M.; Tchekhovskoy, A.; Markoff, S.B. Accelerating AGN jets to parsec scales using general relativistic MHD simulations. *Mon. Not. R. Astron. Soc.* **2019**, *490*, 2200–2218.

74. Issaoun, S.; Johnson, M.D.; Blackburn, L.; Brinkerink, C.D.; Mościbrodzka, M.; Chael, A.; Goddi1, C.; Martí-Vidal, I.; Wagner, J.; Doeleman, S.S. The Size, Shape, and Scattering of Sagittarius A* at 86 GHz: First VLBI with ALMA. *Astrophys. J.* **2019**, *871*, 30.

75. Markoff, S.; Bower, G.C.; Falcke, H. How to hide large-scale outflows: Size constraints on the jets of Sgr A *. *Mon. Not. R. Astron. Soc.* **2007**, *379*, 1519–1532.

76. Mertens, F.; Lobanov, A.P.; Walker, R.C.; Hardee, P.E. Kinematics of the jet in M87 on scales of 100–1000 Schwarzschild radii. *Astron. Astrophys.* **2016**, *595*, A54. [CrossRef]

77. Hada, K.; Kino, M.; Doi, A.; Nagai, H.; Honma, M.; Hagiwara, Y.; Giroletti, M.; Giovannini, G.; Kawaguchi, N. The Innermost Collimation Structure of the M87 Jet Down to 10 Schwarzschild Radii. *Astrophys. J.* **2013**, *775*, 70. [CrossRef]

78. Nakamura, M.; Asada, K. The Parabolic Jet Structure in M87 as a Magnetohydrodynamic Nozzle. *Astrophys. J.* **2013**, *775*, 118. [CrossRef]

79. Hada, K.; Kino, M.; Doi, A.; Nagai, H.; Honma, M.; Akiyama, K.; Tazaki, F.; Lico, R.; Giroletti, M.; Giovannini, G.; et al. High-sensitivity 86 GHz (3.5 mm) VLBI Observations of M87: Deep Imaging of the Jet Base at a Resolution of 10 Schwarzschild Radii. *Astrophys. J.* **2016**, *817*, 131. [CrossRef]

80. Fragile, P.C.; Meier, D.L. General Relativistic Magnetohydrodynamic Simulations of the Hard State as a Magnetically Dominated Accretion Flow. *Astrophys. J.* **2009**, *693*, 771–783.

81. Yoon, D.; Chatterjee, K.; Markoff, S.B.; van Eijnatten, D.; Younsi, Z.; Liska, M.; Tchekhovskoy, A. Spectral and imaging properties of Sgr A * from high-resolution 3D GRMHD simulations with radiative cooling. *Mon. Not. R. Astron. Soc.* **2020**, *499*, 3178–3192.

82. Sadowski, A. Magnetic flux stabilizing thin accretion discs. *Mon. Not. R. Astron. Soc.* **2016**, *462*, 960–965. . mnras/stw1852. [CrossRef]

83. Liska, M.T.P.; Musoke, G.; Tchekhovskoy, A.; Porth, O.; Beloborodov, A.M. Formation of Magnetically Truncated Accretion Disks in 3D Radiation-Transport Two-Temperature GRMHD Simulations. *arXiv* **2022**, arXiv:2201.03526.

84. Mishra, B.; Fragile, P.C.; Anderson, J.; Blankenship, A.; Li, H.; Nalewajko, K. The Role of Strong Magnetic Fields in Stabilizing Highly Luminous Thin Disks. *Astrophys. J.* **2022**, *939*, 31.

85. Wong, G.N.; Du, Y.; Prather, B.S.; Gammie, C.F. The Jet-disk Boundary Layer in Black Hole Accretion. *Astrophys. J.* **2021**, *914*, 55.

86. Aharonian, F.; Akhperjanian, A.G.; Bazer-Bachi, A.R.; Beilicke, M.; Benbow, W.; Berge, D.; Bernlöhr, K.; Boisson, C.; Bolz, O.; Borrel, V.; et al. Fast Variability of Tera-Electron Volt γ Rays from the Radio Galaxy M87. *Science* **2006**, *314*, 1424–1427.

87. Acciari, V.A.; Aliu, E.; Arlen, T.; Aune, T.; Beilicke, M.; Benbow, W.; Boltuch, D.; Bradbury, S.M.; Buckley, J.H.; Bugaev, V.; et al. Veritas 2008-2009 Monitoring of the Variable Gamma-ray Source M 87. *Astrophys. J.* **2010**, *716*, 819–824.

88. Baganoff, F.K.; Bautz, M.W.; Brandt, W.N.; Chartas, G.; Feigelson, E.D.; Garmire, G.P.; Maeda, Y.; Morris, M.; Ricker, G.R.; Townsley, L.K.; et al. Rapid X-ray flaring from the direction of the supermassive black hole at the Galactic Centre. *Nature* **2001**, *413*, 45–48.

89. Eckart, A.; Baganoff, F.K.; Schödel, R.; Morris, M.; Genzel, R.; Bower, G.C.; Marrone, D.; Moran, J.M.; Viehmann, T.; Bautz, M.W.; et al. The flare activity of Sagittarius A*. New coordinated mm to X-ray observations. *Astron. Astrophys.* **2006**, *450*, 535–555. [CrossRef]

90. Hornstein, S.D.; Matthews, K.; Ghez, A.M.; Lu, J.R.; Morris, M.; Becklin, E.E.; Rafelski, M.; Baganoff, F.K. A Constant Spectral Index for Sagittarius A* during Infrared/X-ray Intensity Variations. *Astrophys. J.* **2007**, *667*, 900–910.

91. Nowak, M.A.; Neilsen, J.; Markoff, S.B.; Baganoff, F.K.; Porquet, D.; Grosso, N.; Levin, Y.; Houck, J.; Eckart, A.; Falcke, H.; et al. Chandra/HETGS Observations of the Brightest Flare Seen from Sgr A *. *Astrophys. J.* **2012**, *759*, 95.

92. Neilsen, J.; Nowak, M.A.; Gammie, C.; Dexter, J.; Markoff, S.; Haggard, D.; Nayakshin, S.; Wang, Q.D.; Grosso, N.; Porquet, D.; et al. A Chandra/HETGS Census of X-ray Variability from Sgr A * during 2012. *Astrophys. J.* **2013**, *774*, 42.

93. Witzel, G.; Martinez, G.; Hora, J.; Willner, S.P.; Morris, M.R.; Gammie, C.; Becklin, E.E.; Ashby, M.L.N.; Baganoff, F.; Carey, S.; et al. Variability Timescale and Spectral Index of Sgr A * in the Near Infrared: Approximate Bayesian Computation Analysis of the Variability of the Closest Supermassive Black Hole. *Astrophys. J.* **2018**, *863*, 15.

94. Haggard, D.; Nynka, M.; Mon, B.; de la Cruz Hernandez, N.; Nowak, M.; Heinke, C.; Neilsen, J.; Dexter, J.; Fragile, P.C.; Baganoff, F.; et al. Chandra Spectral and Timing Analysis of Sgr A *'s Brightest X-ray Flares. *Astrophys. J.* **2019**, *886*, 96.

95. Dodds-Eden, K.; Sharma, P.; Quataert, E.; Genzel, R.; Gillessen, S.; Eisenhauer, F.; Porquet, D. Time-Dependent Models of Flares from Sagittarius A*. *Astrophys. J.* **2010**, *725*, 450–465.

96. Dibi, S.; Markoff, S.; Belmont, R.; Malzac, J.; Barrière, N.M.; Tomsick, J.A. Exploring plasma evolution during Sagittarius A* flares. *Mon. Not. R. Astron. Soc.* **2014**, *441*, 1005–1016.

97. Nathanail, A.; Fromm, C.M.; Porth, O.; Olivares, H.; Younsi, Z.; Mizuno, Y.; Rezzolla, L. Plasmoid formation in global GRMHD simulations and AGN flares. *Mon. Not. R. Astron. Soc.* **2020**, *495*, 1549–1565.

98. Dexter, J.; Tchekhovskoy, A.; Jiménez-Rosales, A.; Ressler, S.M.; Bauböck, M.; Dallilar, Y.; de Zeeuw, P.T.; Eisenhauer, F.; von Fellenberg, S.; Gao, F.; et al. Sgr A * near-infrared flares from reconnection events in a magnetically arrested disc. *Mon. Not. R. Astron. Soc.* **2020**, *497*, 4999–5007.

99. Scepi, N.; Dexter, J.; Begelman, M.C. Sgr A * X-ray flares from non-thermal particle acceleration in a magnetically arrested disc. *Mon. Not. R. Astron. Soc.* **2022**, *511*, 3536–3547.

100. Markoff, S.; Falcke, H.; Yuan, F.; Biermann, P.L. The Nature of the 10 kilosecond X-ray flare in Sgr A *. *Astron. Astrophys.* **2001**, *379*, L13–L16. [CrossRef]

101. Dibi, S.; Markoff, S.; Belmont, R.; Malzac, J.; Neilsen, J.; Witzel, G. Using infrared/X-ray flare statistics to probe the emission regions near the event horizon of Sgr A *. *Mon. Not. R. Astron. Soc.* **2016**, *461*, 552–559.

102. Gutiérrez, E.M.; Nemmen, R.; Cafardo, F. A Nonthermal Bomb Explains the Near-infrared Superflare of Sgr A *. *Astrophys. J. Lett.* **2020**, *891*, L36.

103. Bauböck, M. et al. [Gravity Collaboration]. Modeling the orbital motion of Sgr A *'s near-infrared flares. *Astron. Astrophys.* **2020**, *635*, A143. [CrossRef]

104. Ball, D.; Özel, F.; Christian, P.; Chan, C.K.; Psaltis, D. A Plasmoid model for the Sgr A * Flares Observed With Gravity and CHANDRA. *Astrophys. J.* **2021**, *917*, 8.

105. Wielgus, M.; Moscibrodzka, M.; Vos, J.; Gelles, Z.; Martí-Vidal, I.; Farah, J.; Marchili, N.; Goddi, C.; Messias, H. Orbital motion near Sagittarius A*. Constraints from polarimetric ALMA observations. *Astron. Astrophys.* **2022**, *665*, L6.

106. Emami, R.; Tiede, P.; Doeleman, S.S.; Roelofs, F.; Wielgus, M.; Blackburn, L.; Liska, M.; Chatterjee, K.; Ripperda, B.; Fuentes, A.; et al. Tracing Hot Spot Motion in Sagittarius A* Using the Next-Generation Event Horizon Telescope (ngEHT). *Galaxies* **2023**, *11*, 23. [CrossRef]

107. Gelles, Z.; Chatterjee, K.; Johnson, M.; Ripperda, B.; Liska, M. Relativistic Signatures of Flux Eruption Events near Black Holes. *Galaxies* **2022**, *10*, 107. [CrossRef]

108. Ressler, S.M.; White, C.J.; Quataert, E.; Stone, J.M. Ab Initio Horizon-scale Simulations of Magnetically Arrested Accretion in Sagittarius A* Fed by Stellar Winds. *Astrophys. J. Lett.* **2020**, *896*, L6.

109. Lalakos, A.; Gottlieb, O.; Kaaz, N.; Chatterjee, K.; Liska, M.; Christie, I.M.; Tchekhovskoy, A.; Zhuravleva, I.; Nokhrina, E. Bridging the Bondi and Event Horizon Scales: 3D GRMHD Simulations Reveal X-shaped Radio Galaxy Morphology. *Astrophys. J. Lett.* **2022**, *936*, L5. [CrossRef]

110. Cruz-Osorio, A.; Sánchez-Salcedo, F.J.; Lora-Clavijo, F.D. Relativistic Bondi-Hoyle-Lyttleton accretion in the presence of small rigid bodies around a black hole. *Mon. Not. R. Astron. Soc.* **2017**, *471*, 3127–3134.

111. Kaaz, N.; Murguia-Berthier, A.; Chatterjee, K.; Liska, M.; Tchekhovskoy, A. Jet Formation in 3D GRMHD Simulations of Bondi-Hoyle-Lyttleton Accretion. *arXiv* **2022**, arXiv:2201.11753.

112. Avara, M.J.; McKinney, J.C.; Reynolds, C.S. Efficiency of thin magnetically arrested discs around black holes. *Mon. Not. R. Astron. Soc.* **2016**, *462*, 636–648.

113. Sadowski, A.; Narayan, R. Three-dimensional simulations of supercritical black hole accretion discs - luminosities, photon trapping and variability. *Mon. Not. R. Astron. Soc.* **2016**, *456*, 3929–3947.

114. Curd, B.; Narayan, R. GRRMHD simulations of tidal disruption event accretion discs around supermassive black holes: Jet formation, spectra, and detectability. *Mon. Not. R. Astron. Soc.* **2019**, *483*, 565–592.
115. White, C.J.; Quataert, E.; Blaes, O. Tilted Disks around Black Holes: A Numerical Parameter Survey for Spin and Inclination Angle. *Astrophys. J.* **2019**, *878*, 51. [CrossRef]
116. Chatterjee, K.; Younsi, Z.; Liska, M.; Tchekhovskoy, A.; Markoff, S.B.; Yoon, D.; van Eijnatten, D.; Hesp, C.; Ingram, A.; van der Klis, M.B.M. Observational signatures of disc and jet misalignment in images of accreting black holes. *Mon. Not. R. Astron. Soc.* **2020**, *499*, 362–378.
117. Akiyama, K. et al. [Event Horizon Telescope Collaboration]. First Sagittarius A* Event Horizon Telescope Results. VI. Testing the Black Hole Metric. *Astrophys. J. Lett.* **2022**, *930*, L17. [CrossRef]
118. Mizuno, Y.; Younsi, Z.; Fromm, C.M.; Porth, O.; De Laurentis, M.; Olivares, H.; Falcke, H.; Kramer, M.; Rezzolla, L. The current ability to test theories of gravity with black hole shadows. *Nat. Astron.* **2018**, *2*, 585–590.
119. Olivares, H.; Younsi, Z.; Fromm, C.M.; De Laurentis, M.; Porth, O.; Mizuno, Y.; Falcke, H.; Kramer, M.; Rezzolla, L. How to tell an accreting boson star from a black hole. *Mon. Not. R. Astron. Soc.* **2020**, *497*, 521–535.
120. Nampalliwar, S.; Yfantis, A.I.; Kokkotas, K.D. Extending GRMHD for thin disks to non-Kerr spacetimes. *Phys. Rev. D* **2022**, *106*, 063009. [CrossRef]

Article

Tracing Hot Spot Motion in Sagittarius A* Using the Next-Generation Event Horizon Telescope (ngEHT)

Razieh Emami [1,*], Paul Tiede [1,2], Sheperd S. Doeleman [1,2], Freek Roelofs [1,2], Maciek Wielgus [3], Lindy Blackburn [1,2], Matthew Liska [1], Koushik Chatterjee [1,2], Bart Ripperda [4,5], Antonio Fuentes [6], Avery E. Broderick [7,8], Lars Hernquist [1], Charles Alcock [1], Ramesh Narayan [1,2], Randall Smith [1], Grant Tremblay [1], Angelo Ricarte [1,2], He Sun [9], Richard Anantua [10], Yuri Y. Kovalev [3,11,12], Priyamvada Natarajan [2,13,14] and Mark Vogelsberger [15]

1   Center for Astrophysics | Harvard & Smithsonian, 60 Garden Street, Cambridge, MA 02138, USA
2   Black Hole Initiative, Harvard University, 20 Garden Street, Cambridge, MA 02138, USA
3   Max-Planck-Institut für Radioastronomie, Auf dem Hügel 69, D-53121 Bonn, Germany
4   School of Natural Sciences, Institute for Advanced Study, 1 Einstein Drive, Princeton, NJ 08540, USA
5   NASA Hubble Fellowship Program, Einstein Fellow
6   Instituto de Astrofísica de Andalucía-CSIC, Glorieta de la Astronomía s/n, E-18008 Granada, Spain
7   Perimeter Institute for Theoretical Physics, 31 Caroline Street North, Waterloo, ON N2L 2Y5, Canada
8   Department of Physics and Astronomy, University of Waterloo, 200 University Avenue West, Waterloo, ON N2L 3G1, Canada
9   National Biomedical Imaging Center, College of Future Technology, Peking University, Beijing 100871, China
10   Department of Physics & Astronomy, The University of Texas at San Antonio, One UTSA Circle, San Antonio, TX 78249, USA
11   Lebedev Physical Institute of the Russian Academy of Sciences, Leninsky Prospekt 53, 119991 Moscow, Russia
12   Moscow Institute of Physics and Technology, Institutsky per. 9, 141700 Dolgoprudny, Russia
13   Department of Astronomy, Yale University, New Haven, CT 06511, USA
14   Department of Physics, Yale University, New Haven, CT 06520, USA
15   Department of Physics, Kavli Institute for Astrophysics and Space Research, Massachusetts Institute of Technology, Cambridge, MA 02139, USA
*   Correspondence: razieh.emami_meibody@cfa.harvard.edu

Citation: Emami, R.; Tiede, P.; Doeleman, S.S.; Roelofs, F.; Wielgus, M.; Blackburn, L.; Liska, M.; Chatterjee, K.; Ripperda, B.; Fuentes, A.; et al. Tracing Hot Spot Motion in Sagittarius A* Using the Next-Generation Event Horizon Telescope (ngEHT). *Galaxies* **2023**, *11*, 23. https://doi.org/10.3390/galaxies11010023

Academic Editor: Bidzina Kapanadze

Received: 13 November 2022
Revised: 11 January 2023
Accepted: 17 January 2023
Published: 29 January 2023

Abstract: We propose the tracing of the motion of a shearing hot spot near the Sgr A* source through a dynamical image reconstruction algorithm, StarWarps. Such a hot spot may form as the exhaust of magnetic reconnection in a current sheet near the black hole horizon. A hot spot that is ejected from the current sheet into an orbit in the accretion disk may shear and diffuse due to instabilities at its boundary during its orbit, resulting in a distinct signature. We subdivide the motion into two different phases: the first phase refers to the appearance of the hot spot modeled as a bright blob, followed by a subsequent shearing phase. We employ different observational array configurations, including EHT (2017, 2022) and the next-generation Event Horizon Telescope (ngEHTp1, ngEHT) arrays, with several new sites added, and make dynamical image reconstructions for each of them. Subsequently, we infer the hot spot angular image location in the first phase, followed by the axes ratio and the ellipse area in the second phase. We focus on the direct observability of the orbiting hot spot in the sub-mm wavelength. Our analysis demonstrates that for this particular simulation, the newly added dishes are better able to trace the first phase as well as part of the second phase before the flux is reduced substantially, compared to the EHT arrays. The algorithm used in this work can be easily extended to other types of dynamics, as well as different shearing timescales. More simulations are required to prove whether the current set of newly proposed sites are sufficient to resolve any motions near variable sources, such as Sgr A*.

Keywords: Sgr A*; hot spot; dynamical image reconstruction; StarWarps; time-variability; EHT; ngEHT

1. Modellng Flares in Sgr A* with Hot Spots

The recent resolved images of Sagittarius A* (Sgr A*) taken by the Event Horizon Telescope (EHT) [1–10] revealed rapid structural variability in the resolved supermassive black hole (SMBH) source at the galactic center [11,12]. These findings complement the reported variability in this compact source across the electromagnetic spectrum [13], in the mm/sub-mm scale [8,14–21], in near-infrared (NIR) [22–24], and in X-rays [25–30]. Later work found evidence for an NIR and sub-mm correlation and a possible adiabatic expansion of the emitting region during a flare [13,31–35]. During flare events, the flux density observed in NIR and X-rays increases by 1-2 orders of magnitude, which roughly aligns with theoretical expectations, e.g., [36]. The flares seem to originate from a compact region near the innermost stable circular orbit (ISCO) [37,38]. In particular, ref. [38] recently reported an orbiting hot spot detection in the unresolved light curve data at the EHT observing frequency following an X-ray flare.

On the theoretical front, there have been various explorations trying to model these flares (hot spots) in a general relativistic magnetohydrodynamical (GRMHD) framework or through semi-analytic models. In the former case, magnetic reconnection and the flux eruption [36,39] are good candidates to produce such flares in a form of a hot spot region orbiting around the SMBH, arising from the local energy injection accelerating the electrons within the accretion disk. In the latter case [40–44], the hot spot may be embedded within a geometrically thick, hot, and optically thin, radiatively inefficient accretion flow (RIAF; [45–48]), expected to be characteristic of low-luminosity SMBHs such as Sgr A*.

2. Dynamical Formation of a Hot Spot in the Simulations

The formation of hot spots has been reported in general relativistic magnetohydrody-namics (GRMHD) simulations. In these simulations, as the gas near the black hole becomes more magnetized, reaching the magnetically arrested disk (MAD; [49]) state, horizontal fields squeeze the accretion flow, thereby forming a thin equatorial current sheet [50]. This current sheet is potentially unstable to tearing instabilities and the formation of plasmoids via reconnection. Plasmoids are blobs of plasma confined by a helical field, consisting of particles energized by the reconnection of the surrounding magnetized plasma.

In a scenario proposed by [50], an equatorial reconnection layer transforms the horizontal field at the jet base into a vertical field that is injected into the accretion disk. The flux tube of the vertical field is filled with non-thermal leptons originating from the jet's magnetized plasma and accelerated by the reconnection event. The resulting low-density hot spot, is contained by the vertical magnetic field and subsequently orbits around the black hole and is conjectured to power the NIR emission trailing a large X-ray flare. Figure 1 presents the dynamical formation of the hot spot filled with low-density plasma contained by vertical field from a HAMR simulation (using a GPU-accelerated and advanced custom-built adaptive mesh refinement (AMR) code) [50,51]. In Figure 1, ρ and T refer to the mass density and the temperature, respectively.

Large plasmoids, formed due to mergers of smaller plasmoids in reconnection layers, have also been conjectured as a model for orbiting hot spots. The growth and propagation of plasmoids is still an ongoing area of research, especially in full 3D GRMHD. Because of the potential of these plasmoids to carry non-thermal electrons (as magnetic reconnection can drive particle acceleration), several works have tried to model plasmoid evolution as spherical or shearing hot spots around black holes [42,43,52–57].

The main difference between the vertical flux tube scenario and an individual large plasmoid as a hot spot model is twofold: a plasmoid consists of a dominantly helical field and is shown to mainly orbit along the jet sheath [58,59], whereas a large flux tube formed as reconnection exhaust consists of a vertical field and orbits in the accretion disk. Recent observations of orbiting hot spots suggest a dominant vertical field component [37,38] associated with the motion, which implies that a vertical field flux tube may be more realistic as the source of emission instead of a large individual plasmoid. On the other hand, in a different scenario, an apparent hot spot observed at mm wavelengths could correspond simply to a local density maximum, possibly originating in the turbulent accretion flow or related to an infalling clump of matter [60].

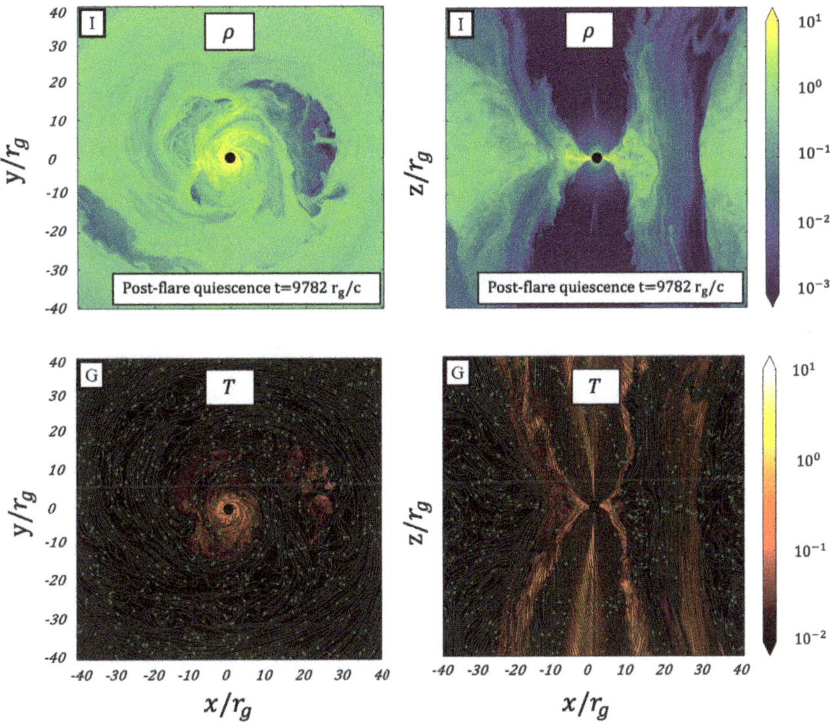

Figure 1. The formation of plasmoids, in HAMR simulation [50], due to a tearing instability in a very thin equatorial current sheet formed as a consequence of squeezing the accretion under the conversion of horizontal field lines to vertical ones. In this scenario, the field conversion is owing to the reconnection and results in an exhausted low-density hot spot confined by the vertical field.

3. Semi-Analytic Simulation of a Shearing Hot Spot

There have been a variety of different hot spot models. The original studies [40–42] only focused on the coherent motion of a spherical Gaussian hot spot. Others [34] extended this model by adding the adiabatic expansion and [61] considered a 2D shearing hot spot, ignoring the radiative transfer effects. More recently, ref. [43] extended this model further and included both the shearing and the expansion of a 3D hot spot, additionally incorporating polarized radiative transfer effects such as absorption and Faraday rotation, while ref. [44] focused on employing a full polarized radiative transfer to the coherent Gaussian hot spot model. To describe shearing and expansion, ref. [43] assumes that the hot spot electron number density follows the continuity equation and travels along a prescribed velocity field u^μ. To make the solution to the continuity equation tractable, the velocity field is restricted to have no vertical motion, be stationary, and be axisymmetric with respect to the spin axis. Given these restrictions, the continuity equation describing the electron number density n_e can be solved semi-analytically using the equation [43]:

$$n_e(\tau, x^\mu) = n_{e0}(y^\mu) \frac{u^r(y^\mu)}{u^r(x^\mu)} \times \sqrt{\frac{g(y^\mu)}{g(x^\mu)}} \, , \tag{1}$$

where $n_{e0}(y^\mu)$ refers to the initial proper density of a hot spot, y^μ describes its initial location, x^μ refers to its subsequent position, and $g(x^\mu)$ denotes the metric tensor determinant evaluated at the location x^μ. Note that $y^\mu = \phi_{-\tau}(x^\mu)$, where ϕ_τ is the velocity field flow found by integrating $\dot{x}^\mu = u^\mu$ for τ units of proper time. For this work, we used the hot spot simulation from [62], representing an orbiting feature around a black hole with a spin of 0.1 and an inclination of 160°. The hot spot was initialized in the equatorial plane at a radius of $7M$ with a Gaussian profile with a full-width-half-max (FWHM) of $1.18M$. We used the semi-analytical RIAF model from [63], with a Keplerian velocity field to represent the quiescent background accretion flow. We note that the hot spot parameters used are consistent with the constraints provided by [37,38].

Figure 2 presents the appearance of the shearing hot spot at a few different times in the image space (top row) and in the visibility space using EHT2017, EHT2022, phase I of ngEHT (ngEHTp1), and the full ngEHT array (ngEHT), respectively.

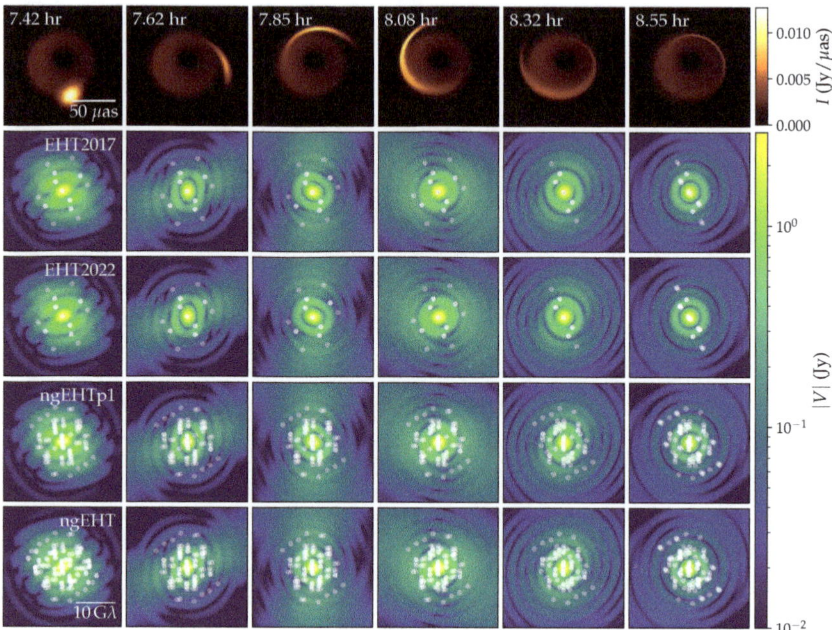

Figure 2. Simulation used for tracing hot spot motion. The top row shows 6 frames from the simulation equally spaced from 7.42 UT to 8.55 UT. The following 4 rows show the visibility amplitudes of the movie frames for different array configurations (array baselines are shown with white dots).

In this paper, we will analyze how well different EHT/ngEHT arrays are able to dynamically reconstruct hot spot motion from Figure 2 using the StarWarps algorithm [64] (see below for more details). To accomplish this analysis, we perform synthetic observations using four different observational array configurations. These configurations include both of the current EHT coverage (EHT2017, EHT2022), as well as the next-generation Event Horizon Telescope (ngEHT) arrays (ngEHTp1, ngEHT), see Table 1.

Table 1. Array configurations used for EHT and ngEHT coverage and simulated data sets. The Greenland Telescope (GLT) is part of the EHT array (from 2018 onwards), but has not been used in our simulations since it is too far north to observe Sgr A*. ngEHT configurations assume the participation of existing EHT (2022) sites, as well as the addition of existing/repurposed dishes at HAY (37 m), OVRO (10 m), and GAM (15 m), or hypothetical 6m dishes at new site locations. On-sky bandwidth is assumed to be 4 GHz for EHT(2017), 8 GHz for EHT(2022), and 16 GHz at both 230 and 345 GHz for ngEHT. The ngEHT Phase 1 and 2 station acronyms relate to nearby observatories or geographical locations: Owens Valley Radio Observatory in California, USA (OVRO); Haystack Observatory in Westford, MA, USA; La Palma, part of the Canary Islands, Spain (CNI); National Astronomical Observatory in Baja California, Mexico (BAJA); Las Campanas Observatory in Chile (LAS); the German Antarctic Receiving Station O'Higgins in Antarctica (GARS); the Gamsberg in Namibia (GAM); Cerro Catedral in R'io Negro in Argentina (CAT); the La Paz region in Bolivia (BOL); and the Espírito Santo region in Brazil (BRZ). See [65] for more site details and locations on a map.

Array	Sites Used for Simulated Data							
EHT(2017)	ALMA	APEX	SMA	JCMT	SMT	LMT	PV	SPT
EHT(2022)	EHT(2017)+	KP	NOEMA					
ngEHTp1	EHT(2022)+	OVRO	HAY	CNI	BAJA	LAS		
ngEHT	ngEHTp1+	GARS	GAM	CAT	BOL	BRZ		

4. Creating Synthetic Data for EHT/ngEHT

To make the synthetic data for the dynamical image reconstruction, we used the `eht-imaging` package [66–68].

We consider four different array configurations, including the EHT(2017), EHT(2022), ngEHTp1, and ngEHT. Representative April weather is used to simulate station performance, along with random (uncalibrated) absolute atmospheric phase and $\sim$10% amplitude gain systematic error. Table 1 contains a list of stations used for each array configuration.

Before generating the synthetic data, we scatter the movie frames using the interstellar scattering model for Sgr A* from [69], as implemented in `eht-imaging`. Figure 3 presents the uv-coverage of the above arrays. The top row presents the EHT2017 (left panel) and EHT2022 (right panel) uv-coverage, while the bottom row shows the uv-coverage for ngEHTp1 (left panel) and ngEHT (right panel), respectively.

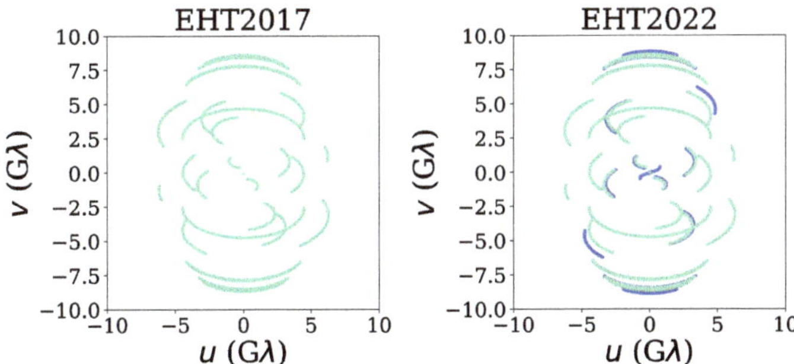

Figure 3. *Cont.*

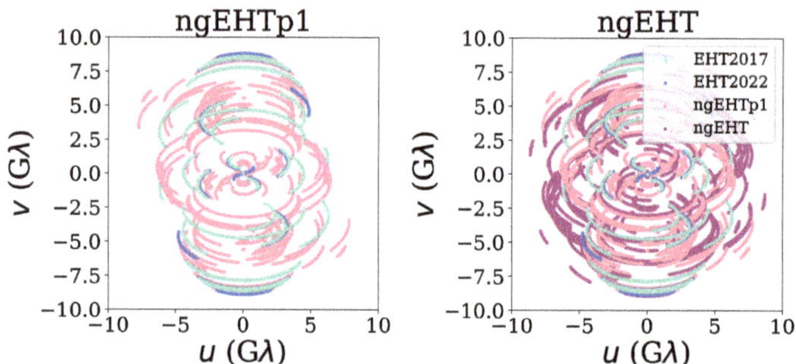

Figure 3. The uv-coverage for different observational arrays. The top row presents the uv-coverage for EHT(2017) (**left panel**) and EHT(2022) (**right panel**), while the bottom row shows the coverage for ngEHTp1 (**left**) and the full ngEHT (**right**) arrays, respectively. It is seen that adding the new sites/dishes significantly improves the uv-coverage in the observational array. As shown below, the improved coverage is significantly useful in tracing the orbital motion of the hot spot.

5. Dynamical Reconstruction Using the `StarWarps` Code

Since the gravitational time scale for Sgr A* is $t_g = GM/c^3 \simeq 20$ sec, the image structure varies greatly throughout an observation. Consequently, the static image assumption [4] breaks down, making the standard imaging approaches inadequate. Constructing a time variable image is further complicated by the fact that as the image changes, so does the telescope, due to the rotation of the Earth. Therefore, any dynamical imaging algorithm needs to disentangle the coverage evolution from the dynamical evolution of the source. To solve this problem, we use `StarWarps` [64]. `StarWarps` is a novel dynamical imaging algorithm that models the Very Long Baseline Interferometry (VLBI) observations using a Gaussian Markov Model to model the image variability. By using a specific model for the variability `StarWarps` is able to simultaneously reconstruct both the image and its motion, allowing the algorithm to disentangle the variations due to the telescope from the image itself. In more detail, `StarWarps` reconstructs an N-dimensional (hereafter referred to as N-D) image vector $X = \{x_1, x_2, \ldots, x_N\}$, where N denotes the total number of unique timestamps of the observation and x_t is a flattened vector of the image pixel intensities. The data vector $Y = \{y_1, y_2, \ldots, y_N\}$ is given by the sparse interferometric measurements for the telescope. Note that each snapshot of data y_t may itself be a vector since each timestamp typically has many different telescopes observing the source. The `StarWarps` dynamical imaging model is given in terms of the following *potentials*, φ:

$$\varphi_{y_t|x_t} = \mathcal{N}_{y_t}(f_t(x_t), R_t), \tag{2}$$

$$\varphi_{x_1} = \mathcal{N}_{x_1}(\mu_t, \Lambda_t), \tag{3}$$

$$\varphi_{x_t|x_{t-1}} = \mathcal{N}_{x_t}(Ax_{t-1}, Q), \tag{4}$$

where $\mathcal{N}(x, \Sigma)$ denotes a multivariate Gaussian with mean μ and covariance Σ. The potential $\varphi_{y_t|x_t}$ is given by the likelihood of the problem, where f_t denotes the function that transforms from the image pixel intensities x_t to the observed data products y_t. The covariance R_t is given by the thermal noise from the telescope, plus an additional systematic factor, which we discuss below.

The potential φ_{x_t} denotes the image prior, where we set $\Lambda_t = \text{diag}[\mu_t]^T \Lambda' \text{diag}[\mu_t]$ to ensure that the image intensities are positive. The mean image μ_t is typically chosen to be a simple image (e.g., a ring) whose structure depends on the problem. We specify the μ and Λ we use below.

Finally, $\varphi_{x_t|x_{t-1}}$ specifies the dynamical aspects of the model. To describe the dynamic, `StarWarps` breaks it down into two separate components: a fixed global variability plus a correlated stochastic piece. The fixed global variability is given by the matrix A and specifies a constant linear evolution of the problem. If A is the identity matrix, then there is no global evolution; instead, all dynamics are given by the second stochastic part. To describe the stochastic evolution, `StarWarps` assumes that the $x_t = x_{t-1} + \varepsilon$, where ε is drawn from a multivariate Gaussian with zero mean and covariance Q. This model thus describes a linear evolution of the image I with correlated fluctuations that can additionally modify the image. Note that `StarWarps` reduces to static imaging when A is the identity matrix and $Q = 0$. For simplicity, we set $A = \mathbb{1}$, the identity matrix, and leave exploring the impact of non-trivial A to future work.

By multiplying these three potentials together, we can form the Joint distribution of `StarWarps` as:

$$p(X, Y; A = \mathbb{1}) \propto \prod_{t=1}^{N} \varphi_{y_t|x_t} \prod_{t=1}^{N} \varphi_{x_t} \prod_{t=2}^{N} \varphi_{x_t|x_{t-1}}. \tag{5}$$

`StarWarps` solves for N-D image array $X = \{x_t\}_t$ by using the N-D observed data points $Y = \{y_t\}_t$. Recall that f_t, R_t are known from the data, while μ_t, Λ_t, and Q are hyperparameters of the model.

For our reconstructions, we used the bispectrum (bs), log closure amplitude (logcam), and closure phase (cphase) as the data products. In addition to the thermal noise of the telescope for a baseline i, σ_i, we include an additional error budget equal to 2% of the visibility amplitudes added in quadrature:

$$\sigma_i^2 \to \sigma_i^2 + (0.02)^2 |V_i|^2 \tag{6}$$

where V_i is the measured visibility for baseline i. This additional error is added to capture the expected magnitude of the unknown systematic errors (e.g., related to averaging over a wide frequency band or resulting from imperfect polarimetric leakage calibration), see [2]. In addition, we include a set of weights $w^{(i)}$ for each data product that effectively modifies the covariance $R_t^{(i)} \to (w^{(i)})^{-1} R_t^{(i)}$ for each data product i. Table 2 gives the weights for each data product.

We set our prior image, $\mu = \mu_t$, to be an infinitely thin ring with the typical diameter of Sgr A*, which is then blurred with a 25 μ as FWHM. To specify Λ', we follow the procedure in [64] and define it using a power spectrum P given by:

$$P(u, v) = \frac{c_0 \epsilon^a}{(u^2 + v^2 + \epsilon)^a}, \tag{7}$$

where $a = 0.3$, $c_0 = 0.3$, and $\epsilon = 10^{-3}$. Λ' is then equal to the Fourier transform of Equation (7). Finally, we set $Q = \beta_Q \mathbb{1}$, where $\beta_Q = 5 \times 10^{-6}$.

Table 2. Down rows present the data terms as well as different χ^2s for the quality of the reconstructed images using `StarWarps` for different arrays. From top to bottom, we present EHT(2017), EHT(2022), ngEHTp1, and the full ngEHT. Bs, logcam, and cphase refer to the data weights associated with the bs, log closure amplitude, and the closure phase, respectively.

Obs	Bs	logcam	cphase	χ^2_{cphase}	χ^2_{logcamp}	χ^2_{camp}
EHT(2017)	1.0	1.0	1.0	0.67	1.16	1.37
EHT(2022)	1.0	1.4	1.7	0.59	0.63	0.77
ngEHTp1	1.2	1.5	1.5	1.14	1.5	1.84
ngEHT	1.0	1.0	1.0	1.17	1.51	1.90

6. Reconstructing the Motion of the Hot Spot in Different Arrays

Here, we use the `StarWarps` code to make a dynamical image reconstruction of the orbiting hot spot using different observational arrays. Table 2 presents the χ^2 of different arrays. In this paper, we focus on the improvements in movie reconstructions using different EHT/ngEHT array configurations, and future work will focus on improving various StarWarps hyperparameters, as well as considering different simulations.

Figure 4 compares the original hot spot with the reconstructed images from different arrays at few time snapshots. Down rows correspond to different times, while different columns (from left to right) show the original hot spot, EHT(2017), EHT(2022), ngEHTp1, and ngEHT, respectively. It is generally seen that the ngEHT arrays are doing a better job at reconstructing the hot spot's motion. However, the quality of the reconstructed images reduces around T = 7.62 UT when the hot spot moves behind the source, leading to a flux reduction.

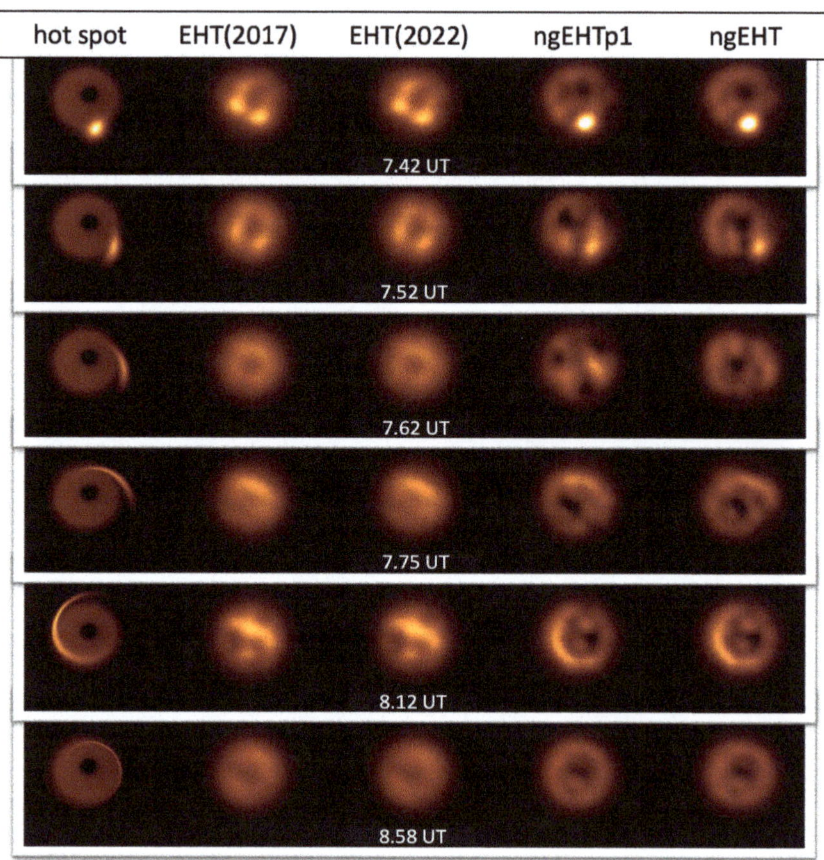

Figure 4. Dynamical reconstruction of the orbital motion of the shearing hot spot using few different observational arrays. Each row corresponds to a different time shown in the middle column. The leftmost column refers to the original hot spot model, the second and third columns describe the reconstructed image using the EHT(2017) and EHT(2022) arrays, and the last two columns refer to the reconstruction made using ngEHT phase I (ngEHTp1), as well as the full ngEHT array (ngEHT), respectively. From the figure, we can see that both ngEHT arrays can trace the motion of the hot spot for most snapshots. However, near T = 7.62 UT, the reconstruction degrades and does not recover the correct hot spot emission. This degradation occurs as the hot spot moves behind the black hole and occurs during a flux density minimum in the light curve. Additionally, past $8.5UT$, the hot spot becomes very dim as it has been substantially sheared out, making reconstructions very difficult.

Nxcorr vs. Nrmse of the Reconstructed and Ground Truth Image

To make a more quantitative comparison between the reconstructed and the ground truth images, here we infer the normalized cross-correlation (hereafter Nxcorr), as well as the normalized root-mean-squared error (hereafter Nrmse), between the reconstructed image and its ground truth image:

- Nxcorr: We make use of [67,70], defining the Nxcorr as:

$$\text{Nxcorr}(X, Y) = \frac{1}{N} \sum_i \frac{(X_i - \langle X \rangle)(Y_i - \langle Y \rangle)}{\sigma_X \sigma_Y}, \tag{8}$$

where X refers to the restructured image, while Y describes the ground truth image of the hot spot. Furthermore, N stands for the number of the pixels in the image, and $\langle \rangle$ refers to the mean pixel value of the image. Finally, σ_i describes the standard deviation of pixel values in image i. Nxcorr determines the similarities between two images. A perfect correlation between the images leads to 1, while a complete anti-correlation between them gives rise to a value of -1 for Nxcorr.

- Nrmse is defined as [67]:

$$\text{Nrmse} = \frac{\sum_i |X_i - Y_i|^2}{\sum_i |X_i|^2}. \tag{9}$$

where, unlike the case of Nxcorr, two completely similar (different) images X and Y have 0 (1) value Nrmse.

Figure 5 presents the Nxcorr and Nrmse for reconstructed images computed using different arrays. From the plot, it is inferred that:

○ Since the background RIAF is dominated in some snapshots, it is seen that we have a globally good correlation between the images.

○ This, however, becomes worse when the hot spot appears and becomes sheared down, in which it is seen that we have a some levels of suppression (enhancement) of Nxcorr (Nrmse) for some cases.

○ The aforementioned suppression (enhancement) is, however, minimal for the ngEHT array compared with the EHT(2017) and EHT(2022).

○ Consequently, we conclude that the ngEHT array helps a lot in improving the quality of the reconstructed image.

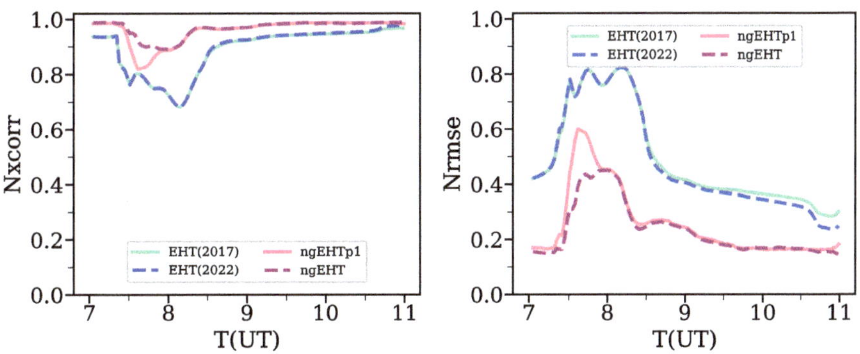

Figure 5. Nxcorr (**left panel**) and the Nrmse (**right panel**) for reconstructed shearing hot spot using different observational arrays. During the shearing phase of the hot spot, there is some suppression (enhancement) of Nxcorr (Nrmse) from the pure background RIAF. The deviation is, however, minimal for ngEHT array compared with both of EHT (2017) and EHT (2022) arrays.

7. Tracking the Angular Location of the Hot Spot

The motion of a shearing hot spot can be subdivided to two distinct phases. The first phase corresponds to a bright (compact) blob that initially appears and starts orbiting around the black hole. The first phase occurs initially up to 7.65 UT. From 7.65 UT and onward, the hot spot is in the second phase, where the hot spot shears out along the differentially rotating velocity field. We use two distinct metrics to trace these distinct evolution phases.

• *First phase* (<7.65 UT): In phase 1, due to the hot spot's compact nature, we track the peak intensity I divided by the initial intensity I_0 of either the original hot spot or the reconstructed intensity for the reconstructed movies. Furthermore, we also follow the peak intensity's angular location Φ.

Figure 6 compares the time evolution of the normalized peak intensity (top row), as well as the angular location of the intensity maximum, referred (bottom row) between the original hot spot (black solid line) and the reconstructed values from different observational arrays during the first phase of evolution. This includes EHT2017 (cyan solid line) and EHT2022 (dashed blue line) arrays (left panel), as well as the ngEHTp1 (pink solid line) and ngEHT (dashed magenta line) arrays (right panel), respectively. Note that to make the figure, we smoothed each curve with a Gaussian filter with a 4 mins standard deviation. From the plot, it is inferred that the reconstructed shape of the intensity and the phase are closer to the original hot spot for ngEHT arrays compared with the EHT ones. Furthermore, in the phase plot on the bottom row, it is seen that in original time the phase is very close to the original hot spot, where the level of agreement is higher in ngEHT arrays than the EHT ones.

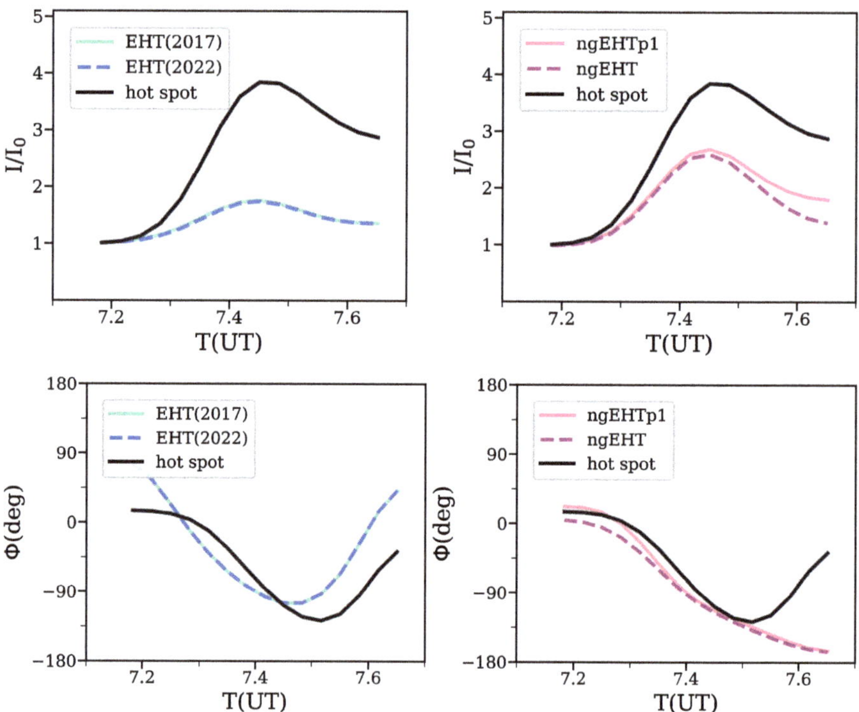

Figure 6. The time evolution of the intensity-ratio (top) and phase of the bright-spot (bottom) of the hot spot from the first phase. In each row, the left (right) panel presents EHT (ngEHT) arrays. Overlaid on the plot, we also present the corresponded parameters in the original hot spot. It is seen that ngEHT arrays are perform relatively better in reconstructing the orbital parameters of the hot spot in the first phase.

• *Second phase* (**>7.65 UT**): To extract the hot spot motion in the second phase, we model it as a stretched ellipse and infer its axes ratio as well as the ellipse area over time. Since the background is dominated by the RIAF model, to extract the ellipsoidal motion, in each snapshot, we first find out the points with an intensity above 80% of the intensity max on that snapshot. We then compute the ellipticity as (b/a), where a and b are the associated semi-major and semi-minor ellipse axes, respectively. The ellipse area is then estimated as πab.

Figure 7 presents the time evolution of the ellipticity (top row) and the ellipse area (bottom row) using different observational arrays. Overlaid on the plot, we also show the corresponding values for the hot spot model (black lines). To make the plot more readable, we first smooth all the curves using a Gaussian filter with a standard deviation of 4 mins. For each data set, we show two curves. The higher-transparency curve is just the value from the smoothed reconstruction. The lower-transparency curve is similar, but there are some snapshots with a relatively poor reconstructed image. The origin of these poor reconstructions are multifaceted and are likely due to both coverage limitations and limitations of the imaging algorithms and the specific hyperparameters chosen for StarWarps. However, we note that importantly both the recovered hot spot ellipticity and area are more accurately recovered with the ngEHT phase 1 and 2 arrays compared to the EHT arrays.

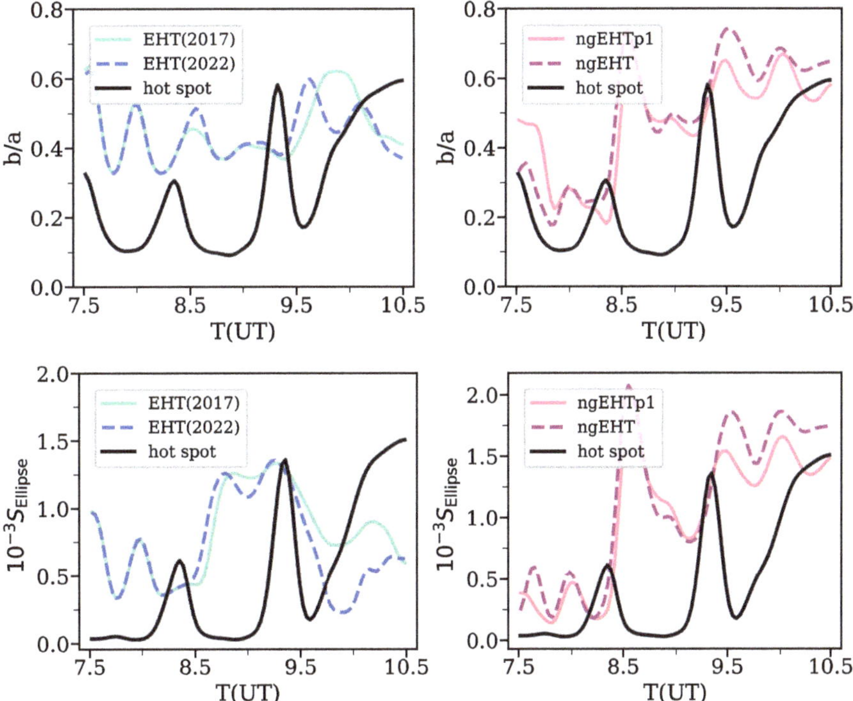

Figure 7. The time evolution of the ellipticity (**top row**) and the ellipse area (**bottom row**) of the shearing hot spot from the second phase. Overlaid on the figure, we also present the original hot spot, different EHT (**left**) and ngEHT (**right**) arrays. It is seen that up until the time that the hot spot's motion decays significantly, ngEHT does a relatively good job in recovering the actual motion.

Figure 8 presents the extracted elliptical motion for the original hot spot (red color map), as well as the ngEHT (blue color map) at a few different snapshots. To make the plot more readable, we skip showing the trajectory for the EHT arrays and the ngEHTp1 array. During phase 1, when the hot spot is more compact, the reconstruction appears to perform better when tracking the position of the hot spot. After ~8.5 UT, the reconstructed hot spot appears to be mostly static. This region corresponds to phase 2 of the hot spot motion and is when the hot spot is almost sheared out, making reconstructions difficult.

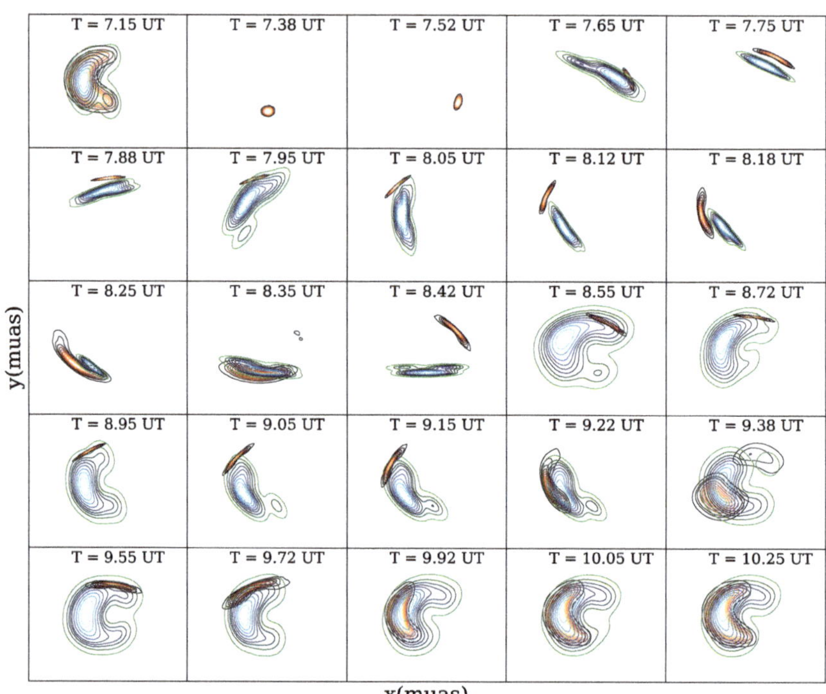

Figure 8. Comparison between the extracted trajectory of the original hot spot (red color map) and from the ngEHT reconstructed image (blue color map) at 25 different snapshots. Overall ngEHT can track the motion of the hot spot quite well. Its performance is, however, much better in some snapshots than the others. The plot shows the KDE of points with an intensity above 80% of the intensity max.

8. Conclusions

We made an in-depth study of tracing the motion of a shearing hot spot simulation, proposed in [43], using the `StarWarps` package [64], a dynamical image reconstruction algorithm employing different observational arrays, including both the EHT and ngEHT arrays. We subdivided the dynamical orbital motion of the hot spot to two distinct phases, which are also observed in GRMHD simulations (see Figure 1), and traced the motion in each of these phases, respectively. The first phase focuses on the appearance of the hot spot and its initial motion when it is ejected from the reconnection layer, while the second phase explores the shearing of the hot spot (potentially driven by Rayleigh–Taylor instabilities at the hot spot boundary during its orbit), being modeled with a re-shaping ellipse. Leptons originating from the jet, accelerated through an equatorial reconnection layer, may end up in the orbiting hot spot confined by a vertical magnetic field. They can then go through a secondary acceleration phase due to the shearing motion. It is conjectured in [50] that such accelerated leptons in the hot spot can power NIR flares and potentially concurrent sub-mm emission. We proposed a novel algorithm to trace the orbital phase in the first phase and the axes ratio as well as the ellipse area in the second phase. Furthermore, we inferred the Nxcorr and Nrmse for different observational arrays. Our analysis showed that while EHT arrays might have some difficulties in locating the hot spot in the first phase appropriately, the motion of which becomes even harder to trace in the shearing phase, adding more sites to the array, as is planned in the ngEHT, substantially helps to improve the quality of the reconstructed image in both phases. Consequently, we propose to use the ngEHT to trace the hot spot motion. With the demonstrated image fidelity improvement, a detailed study of the hot spot properties, such as its orbit and size, becomes possible, offering a new insight into the physical mechanisms responsible for the flaring activity of Sgr A*. While the analysis conducted in this work is limited to a single hot spot simulation, the dynamical reconstruction and feature extraction algorithms used in this study can be extended to other types of dynamics. This is left to a future work. Furthermore, in this study, we did not aim to show which of the new sites in ngEHT are more important in improving the quality of the reconstructed images. Work is in progress where we add sites in order and check out the importance of individual ones as well.

In this work, we only addressed the issues related to the total intensity modeling of hot spots that could be observed with the ngEHT. However, hot spots emerging in the accretion flow may indicate significant fractional linear polarization. Since mm wavelength radiation in Sgr A* originates through the synchrotron process, this allows us to probe the magnetic field geometry with hot spots through imaging of the linear polarization, e.g., [44]. Indeed, the linear polarization observations of unresolved Sgr A* provided a strong argument for an orbiting hot spot model of flares [37,38]. While more comprehensive studies are necessary to address this subject, it is clear that resolving the polarized structure of the source with the ngEHT will vastly improve our understanding of the magnetic field geometry and the time-evolution.

Author Contributions: Conceptualization, R.E., P.T., F.R., S.S.D., A.E.B., M.V., P.N., L.B., A.F., A.R., R.A., H.S., A.E.B. Methodology, R.E. Software, A.F., P.T., F.R., L.B., B.R., M.L., K.C., A.E.B. Validation, R.E. Formal analysis, R.E., P.T., F.R., A.F. Investigation, R.E., P.T., L.B., M.W., Resources, P.T., A.F., L.B. Data curation, R.E., P.T., F.R. Writing, R.E., R.A., A.R., M.W., L.B., Y.Y.K. Original draft preparation, R.E. Review and editing, R.E., P.T., A.R., M.W., B.R. Supervision, S.S.D., R.N., G.T., C.A., L.H., R.S., M.L. Funding, R.N., G.T., R.S., M.L. All authors have read and agreed to the published version of the manuscript.

Funding: Razieh Emami acknowledges the support by the Institute for Theory and Computation at the Center for Astrophysics, as well as grant numbers 21-atp21-0077, NSF AST-1816420, and HST-GO-16173.001-A for the very generous supports. We thank the supercomputer facility at Harvard where most of the simulation work was carried out. P.T. and F.R. was supported by NSF grants AST-1935980 and AST-203430. A.R. and K.C. acknowledge support by the National Science Foundation under Grant No. OISE 1743747. P.T., A.R., and K.C. acknowledge the support of grants from the Gordon and Betty Moore Foundation and the John Templeton Foundation. P.N. gratefully acknowledges support at the Black Hole Initiative (BHI) at Harvard as an external PI with grants from the Gordon and Betty Moore Foundation and the John Templeton Foundation. Support for this work was provided by NASA through the NASA Hubble Fellowship grant HST-HF2-51518.001-A awarded by the Space Telescope Science Institute, which is operated by the Association of Universities for Research in Astronomy, Incorporated, under NASA contract NAS5-26555.

Data Availability Statement: The data presented in this study are available on request from the corresponding author.

Acknowledgments: It is a great pleasure to acknowledge Katherine L. Bouman and Michael Johnson for very fruitful conversations. We thank the referees for their very constructive comments.

Conflicts of Interest: The authors declare no conflict of interest.

References

1. Akiyama, K.; Alberdi, A.; Alef, W.; Algaba, J.C.; Anantua, R.; Asada, K.; Azulay, R.; Bach, U.; Baczko, A.K.; Ball, D.; et al. First Sagittarius A* Event Horizon Telescope Results. I. The Shadow of the Supermassive Black Hole in the Center of the Milky Way. *Astrophys. J. Lett.* **2022**, *930*, L12. [CrossRef]
2. Akiyama, K.; Alberdi, A.; Alef, W.; Algaba, J.C.; Anantua, R.; Asada, K.; Azulay, R.; Bach, U.; Baczko, A.K.; Ball, D.; et al. First Sagittarius A* Event Horizon Telescope Results. II. EHT and Multiwavelength Observations, Data Processing, and Calibration. *Astrophys. J. Lett.* **2022**, *930*, L13. [CrossRef]
3. Akiyama, K.; Alberdi, A.; Alef, W.; Algaba, J.C.; Anantua, R.; Asada, K.; Azulay, R.; Bach, U.; Baczko, A.K.; Ball, D.; et al. First Sagittarius A* Event Horizon Telescope Results. III. Imaging of the Galactic Center Supermassive Black Hole. *Astrophys. J. Lett.* **2022**, *930*, L14. [CrossRef]
4. Akiyama, K.; Alberdi, A.; Alef, W.; Algaba, J.C.; Anantua, R.; Asada, K.; Azulay, R.; Bach, U.; Baczko, A.K.; Ball, D.; et al. First Sagittarius A* Event Horizon Telescope Results. IV. Variability, Morphology, and Black Hole Mass. *Astrophys. J. Lett.* **2022**, *930*, L15. [CrossRef]
5. Akiyama, K.; Alberdi, A.; Alef, W.; Algaba, J.C.; Anantua, R.; Asada, K.; Azulay, R.; Bach, U.; Baczko, A.K.; Ball, D.; et al. First Sagittarius A* Event Horizon Telescope Results. V. Testing Astrophysical Models of the Galactic Center Black Hole. *Astrophys. J. Lett.* **2022**, *930*, L16. [CrossRef]
6. Akiyama, K.; Alberdi, A.; Alef, W.; Algaba, J.C.; Anantua, R.; Asada, K.; Azulay, R.; Bach, U.; Baczko, A.K.; Ball, D.; et al. First Sagittarius A* Event Horizon Telescope Results. VI. Testing the Black Hole Metric. *Astrophys. J. Lett.* **2022**, *930*, L17. [CrossRef]
7. Farah, J.; Galison, P.; Akiyama, K.; Bouman, K.L.; Bower, G.C.; Chael, A.; Fuentes, A.; Gómez, J.L.; Honma, M.; Johnson, M.D.; et al. Selective Dynamical Imaging of Interferometric Data. *Astrophys. J. Lett.* **2022**, *930*, L18. [CrossRef]
8. Wielgus, M.; Marchili, N.; Martí-Vidal, I.; Keating, G.K.; Ramakrishnan, V.; Tiede, P.; Fomalont, E.; Issaoun, S.; Neilsen, J.; Nowak, M.A.; et al. Millimeter Light Curves of Sagittarius A* Observed during the 2017 Event Horizon Telescope Campaign. *Astrophys. J. Lett.* **2022**, *930*, L19. [CrossRef]
9. Georgiev, B.; Pesce, D.W.; Broderick, A.E.; Wong, G.N.; Dhruv, V.; Wielgus, M.; Gammie, C.F.; Chan, C.k.; Chatterjee, K.; Emami, R.; et al. A Universal Power-law Prescription for Variability from Synthetic Images of Black Hole Accretion Flows. *Astrophys. J. Lett.* **2022**, *930*, L20. [CrossRef]
10. Broderick, A.E.; Gold, R.; Georgiev, B.; Pesce, D.W.; Tiede, P.; Ni, C.; Moriyama, K.; Akiyama, K.; Alberdi, A.; Alef, W.; et al. Characterizing and Mitigating Intraday Variability: Reconstructing Source Structure in Accreting Black Holes with mm-VLBI. *Astrophys. J. Lett.* **2022**, *930*, L21. [CrossRef]
11. Doeleman, S.; Blackburn, L.; Dexter, J.; Gomez, J.L.; Johnson, M.D.; Palumbo, D.C.; Weintroub, J.; Farah, J.R.; Fish, V.; Loinard, L.; et al. Studying Black Holes on Horizon Scales with VLBI Ground Arrays. *arXiv* **2019**, arXiv:1909.01411.
12. Johnson, M.; Haworth, K.; Pesce, D.W.; Palumbo, D.C.M.; Blackburn, L.; Akiyama, K.; Boroson, D.; Bouman, K.L.; Farah, J.R.; Fish, V.L.; et al. Studying black holes on horizon scales with space-VLBI. *arXiv* **2019**, arXiv:1909.01405.
13. Witzel, G.; Martinez, G.; Willner, S.P.; Becklin, E.E.; Boyce, H.; Do, T.; Eckart, A.; Fazio, G.G.; Ghez, A.; Gurwell, M.A.; et al. Rapid Variability of Sgr A* across the Electromagnetic Spectrum. *Astrophys. J.* **2021**, *917*, 73,
14. Doeleman, S.S.; Weintroub, J.; Rogers, A.E.E.; Plambeck, R.; Freund, R.; Tilanus, R.P.J.; Friberg, P.; Ziurys, L.M.; Moran, J.M.; Corey, B.; et al. Event-horizon-scale structure in the supermassive black hole candidate at the Galactic Centre. *Nature* **2008**, *455*, 78–80.
15. Doeleman, S. Approaching the event horizon: 1.3mmλ VLBI of SgrA*. *J. Phys. Conf. Ser.* **2008**, *131*, 012055.
16. Fish, V.L.; Doeleman, S.S.; Broderick, A.E.; Loeb, A.; Rogers, A.E.E. Detecting Flaring Structures in Sagittarius A* with (Sub)Millimeter VLBI. *arXiv* **2008**, arXiv:0807.2427.
17. Marrone, D.P.; Baganoff, F.K.; Morris, M.R.; Moran, J.M.; Ghez, A.M.; Hornstein, S.D.; Dowell, C.D.; Muñoz, D.J.; Bautz, M.W.; Ricker, G.R.; et al. An X-ray, Infrared, and Submillimeter Flare of Sagittarius A*. *Astrophys. J.* **2008**, *682*, 373–383.
18. Doeleman, S.S.; Fish, V.L.; Broderick, A.E.; Loeb, A.; Rogers, A.E.E. Detecting Flaring Structures in Sagittarius A* with High-Frequency VLBI. *Astrophys. J.* **2009**, *695*, 59–74.
19. Akiyama, K.; Kino, M.; Sohn, B.; Lee, S.; Trippe, S.; Honma, M. Long-term monitoring of Sgr A* at 7 mm with VERA and KaVA. *Proc. Int. Astron. Union* **2013**, *9*, 288–292. [CrossRef]
20. Johnson, M.D.; Fish, V.L.; Doeleman, S.S.; Broderick, A.E.; Wardle, J.F.C.; Marrone, D.P. Relative Astrometry of Compact Flaring Structures in Sgr A* with Polarimetric Very Long Baseline Interferometry. *Astrophys. J.* **2014**, *794*, 150.
21. Fish, V.L.; Johnson, M.D.; Lu, R.S.; Doeleman, S.S.; Bouman, K.L.; Zoran, D.; Freeman, W.T.; Psaltis, D.; Narayan, R.; Pankratius, V.; et al. Imaging an Event Horizon: Mitigation of Scattering toward Sagittarius A*. *Astrophys. J.* **2014**, *795*, 134.
22. Genzel, R.; Schödel, R.; Ott, T.; Eckart, A.; Alexander, T.; Lacombe, F.; Rouan, D.; Aschenbach, B. Near-infrared flares from accreting gas around the supermassive black hole at the Galactic Centre. *Nature* **2003**, *425*, 934–937.
23. Eckart, A.; Schödel, R.; Meyer, L.; Trippe, S.; Ott, T.; Genzel, R. Polarimetry of near-infrared flares from Sagittarius A*. *Astron. Astrophys.* **2006**, *455*, 1–10. [CrossRef]

24. Do, T.; Witzel, G.; Gautam, A.K.; Chen, Z.; Ghez, A.M.; Morris, M.R.; Becklin, E.E.; Ciurlo, A.; Hosek, M., Jr.; Martinez, G.D.; et al. Unprecedented Near-infrared Brightness and Variability of Sgr A*. *Astrophys. J. Lett.* **2019**, *882*, L27.

25. Baganoff, F.K.; Bautz, M.W.; Brandt, W.N.; Chartas, G.; Feigelson, E.D.; Garmire, G.P.; Maeda, Y.; Morris, M.; Ricker, G.R.; Townsley, L.K.; et al. Rapid X-ray flaring from the direction of the supermassive black hole at the Galactic Centre. *Nature* **2001**, *413*, 45–48.

26. Porquet, D.; Predehl, P.; Aschenbach, B.; Grosso, N.; Goldwurm, A.; Goldoni, P.; Warwick, R.S.; Decourchelle, A. XMM-Newton observation of the brightest X-ray flare detected so far from Sgr A*. *Astron. Astrophys.* **2003**, *407*, L17–L20. [CrossRef]

27. Andrés, A.; van den Eijnden, J.; Degenaar, N.; Evans, P.A.; Chatterjee, K.; Reynolds, M.; Miller, J.M.; Kennea, J.; Wijnands, R.; Markoff, S.; et al. A Swift study of long-term changes in the X-ray flaring properties of Sagittarius A. *Mon. Not. R. Astron. Soc.* **2022**, *510*, 2851–2863.

28. Haggard, D.; Nynka, M.; Mon, B.; de la Cruz Hernandez, N.; Nowak, M.; Heinke, C.; Neilsen, J.; Dexter, J.; Fragile, P.C.; Baganoff, F.; et al. Chandra Spectral and Timing Analysis of Sgr A*'s Brightest X-ray Flares. *Astrophys. J.* **2019**, *886*, 96.

29. Kusunose, M.; Takahara, F. Synchrotron Blob Model of Infrared and X-ray Flares from Sagittarius A*. *Astrophys. J.* **2011**, *726*, 54.

30. Karssen, G.D.; Bursa, M.; Eckart, A.; Valencia-S, M.; Dovčiak, M.; Karas, V.; Horák, J. Bright X-ray flares from Sgr A*. *Mon. Not. R. Astron. Soc.* **2017**, *472*, 4422–4433.

31. Yusef-Zadeh, F.; Bushouse, H.; Dowell, C.D.; Wardle, M.; Roberts, D.; Heinke, C.; Bower, G.C.; Vila-Vilaró, B.; Shapiro, S.; Goldwurm, A.; et al. A Multiwavelength Study of Sgr A*: The Role of Near-IR Flares in Production of X-ray, Soft γ-ray, and Submillimeter Emission. *Astrophys. J.* **2006**, *644*, 198–213.

32. Yusef-Zadeh, F.; Roberts, D.; Wardle, M.; Heinke, C.O.; Bower, G.C. Flaring Activity of Sagittarius A* at 43 and 22 GHz: Evidence for Expanding Hot Plasma. *Astrophys. J.* **2006**, *650*, 189–194.

33. Eckart, A.; Schödel, R.; García-Marín, M.; Witzel, G.; Weiss, A.; Baganoff, F.K.; Morris, M.R.; Bertram, T.; Dovčiak, M.; Duschl, W.J.; et al. Simultaneous NIR/sub-mm observation of flare emission from Sagittarius A*. *Astron. Astrophys.* **2008**, *492*, 337–344. [CrossRef]

34. Eckart, A.; Baganoff, F.K.; Morris, M.R.; Kunneriath, D.; Zamaninasab, M.; Witzel, G.; Schödel, R.; García-Marín, M.; Meyer, L.; Bower, G.C.; et al. Modeling mm- to X-ray flare emission from Sagittarius A*. *Astron. Astrophys.* **2009**, *500*, 935–946.

35. Boyce, H.; Haggard, D.; Witzel, G.; Fellenberg, S.v.; Willner, S.P.; Becklin, E.E.; Do, T.; Eckart, A.; Fazio, G.G.; Gurwell, M.A.; et al. Multiwavelength Variability of Sagittarius A* in 2019 July. *Astrophys. J.* **2022**, *931*, 7.

36. Dexter, J.; Tchekhovskoy, A.; Jiménez-Rosales, A.; Ressler, S.M.; Bauböck, M.; Dallilar, Y.; de Zeeuw, P.T.; Eisenhauer, F.; von Fellenberg, S.; Gao, F.; et al. Sgr A* near-infrared flares from reconnection events in a magnetically arrested disc. *Mon. Not. R. Astron. Soc.* **2020**, *497*, 4999–5007.

37. Gravity Collaboration; Abuter, R.; Amorim, A.; Bauböck, M.; Berger, J.P.; Bonnet, H.; Brandner, W.; Clénet, Y.; Coudé Du Foresto, V.; de Zeeuw, P.T.; et al. Detection of orbital motions near the last stable circular orbit of the massive black hole SgrA*. *Astron. Astrophys.* **2018**, *618*, L10.

38. Wielgus, M.; Moscibrodzka, M.; Vos, J.; Gelles, Z.; Martí-Vidal, I.; Farah, J.; Marchili, N.; Goddi, C.; Messias, H. Orbital motion near Sagittarius A*. Constraints from polarimetric ALMA observations. *Astron. Astrophys.* **2022**, *665*, L6.

39. Yuan, F.; Narayan, R. Hot Accretion Flows Around Black Holes. *Annu. Rev. Astron. Astrophys.* **2014**, *52*, 529–588.

40. Dovčiak, M.; Karas, V.; Yaqoob, T. An Extended Scheme for Fitting X-ray Data with Accretion Disk Spectra in the Strong Gravity Regime. *Astrophys. J. Suppl. Ser.* **2004**, *153*, 205–221.

41. Broderick, A.E.; Loeb, A. Imaging bright-spots in the accretion flow near the black hole horizon of Sgr A*. *Mon. Not. R. Astron. Soc.* **2005**, *363*, 353–362.

42. Broderick, A.E.; Loeb, A. Imaging optically-thin hotspots near the black hole horizon of Sgr A* at radio and near-infrared wavelengths. *Mon. Not. R. Astron. Soc.* **2006**, *367*, 905–916.

43. Tiede, P.; Pu, H.Y.; Broderick, A.E.; Gold, R.; Karami, M.; Preciado-López, J.A. Spacetime Tomography Using the Event Horizon Telescope. *Astrophys. J.* **2020**, *892*, 132.

44. Vos, J.; Mościbrodzka, M.A.; Wielgus, M. Polarimetric signatures of hot spots in black hole accretion flows. *Astron. Astrophys.* **2022**, *668*, A185.

45. Rees, M.J.; Begelman, M.C.; Blandford, R.D.; Phinney, E.S. Ion-supported tori and the origin of radio jets. *Nature* **1982**, *295*, 17–21. [CrossRef]

46. Narayan, R.; Yi, I. Advection-dominated Accretion: A Self-similar Solution. *Astrophys. J. Lett.* **1994**, *428*, L13.

47. Narayan, R.; Yi, I. Advection-dominated Accretion: Self-Similarity and Bipolar Outflows. *Astrophys. J.* **1995**, *444*, 231.

48. Porth, O.; Chatterjee, K.; Narayan, R.; Gammie, C.F.; Mizuno, Y.; Anninos, P.; Baker, J.G.; Bugli, M.; Chan, C.k.; Davelaar, J.; et al. The Event Horizon General Relativistic Magnetohydrodynamic Code Comparison Project. *Astrophys. J. Suppl. Ser.* **2019**, *243*, 26.

49. Narayan, R.; Igumenshchev, I.V.; Abramowicz, M.A. Magnetically Arrested Disk: An Energetically Efficient Accretion Flow. *Publ. Astron. Soc. Jpn.* **2003**, *55*, L69–L72.

50. Ripperda, B.; Liska, M.; Chatterjee, K.; Musoke, G.; Philippov, A.A.; Markoff, S.B.; Tchekhovskoy, A.; Younsi, Z. Black Hole Flares: Ejection of Accreted Magnetic Flux through 3D Plasmoid-mediated Reconnection. *Astrophys. J. Lett.* **2022**, *924*, L32.

51. Liska, M.T.P.; Chatterjee, K.; Issa, D.; Yoon, D.; Kaaz, N.; Tchekhovskoy, A.; van Eijnatten, D.; Musoke, G.; Hesp, C.; Rohoza, V.; et al. H-AMR: A New GPU-accelerated GRMHD Code for Exascale Computing with 3D Adaptive Mesh Refinement and Local Adaptive Time Stepping. *Astrophys. J. Suppl. Ser.* **2022**, *263*, 26.

52. Broderick, A.E.; Loeb, A. Testing General Relativity with High-Resolution Imaging of Sgr A*. *J. Phys. Conf. Ser.* **2006**, *54*, 448–455.
53. Meyer, L.; Eckart, A.; Schodel, R.; Duschl, W.J., 3rd; Dovciak, M.; Karas, V. A two component hot spot/ring model for the NIR flares of Sagittarius A*. *J. Phys. Conf. Ser.* **2006**, *54*, 443–447.
54. Meyer, L.; Eckart, A.; Schödel, R.; Dovčiak, M.; Karas, V.; Duschl, W.J. The orbiting spot model gives constraints on the parameters of the supermassive black hole in the Galactic Center. *Proc. Int. Astron. Union* **2006**, *2*, 407–408.
55. Zamaninasab, M.; Eckart, A.; Meyer, L.; Schödel, R.; Dovciak, M.; Karas, V.; Kunneriath, D.; Witzel, G.; Gießübel, R.; König, S.; et al. An evolving hot spot orbiting around Sgr A*. *J. Phys. Conf. Ser.* **2008**, *131*, 012008.
56. Broderick, A.E.; Fish, V.L.; Doeleman, S.S.; Loeb, A. Evidence for Low Black Hole Spin and Physically Motivated Accretion Models from Millimeter-VLBI Observations of Sagittarius A*. *Astrophys. J.* **2011**, *735*, 110,
57. Younsi, Z.; Wu, K. Variations in emission from episodic plasmoid ejecta around black holes. *Mon. Not. R. Astron. Soc.* **2015**, *454*, 3283–3298.
58. Nathanail, A.; Fromm, C.M.; Porth, O.; Olivares, H.; Younsi, Z.; Mizuno, Y.; Rezzolla, L. Plasmoid formation in global GRMHD simulations and AGN flares. *Mon. Not. R. Astron. Soc.* **2020**, *495*, 1549–1565.
59. Ripperda, B.; Bacchini, F.; Philippov, A.A. Magnetic Reconnection and Hot Spot Formation in Black Hole Accretion Disks. *Astrophys. J.* **2020**, *900*, 100. [CrossRef]
60. Moriyama, K.; Mineshige, S.; Honma, M.; Akiyama, K. Black Hole Spin Measurement Based on Time-domain VLBI Observations of Infalling Gas Clouds. *Astrophys. J.* **2019**, *887*, 227.
61. Zamaninasab, M.; Eckart, A.; Witzel, G.; Dovciak, M.; Karas, V.; Schödel, R.; Gießübel, R.; Bremer, M.; García-Marín, M.; Kunneriath, D.; et al. Near infrared flares of Sagittarius A*. Importance of near infrared polarimetry. *Astron. Astrophys.* **2010**, *510*, A3.
62. Roelofs, F.; Blackburn, L.; Lindahl, G.; Doeleman, S.S.; Johnson, M.D.; Arras, P.; Chatterjee, K.; Emami, R.; Fromm, C.; Fuentes, A.; et al. The ngEHT Analysis Challenges. *Galaxies* **2022**, *11*, 12.
63. Broderick, A.E.; Fish, V.L.; Johnson, M.D.; Rosenfeld, K.; Wang, C.; Doeleman, S.S.; Akiyama, K.; Johannsen, T.; Roy, A.L. Modeling Seven Years of Event Horizon Telescope Observations with Radiatively Inefficient Accretion Flow Models. *Astrophys. J.* **2016**, *820*, 137.
64. Bouman, K.L.; Johnson, M.D.; Dalca, A.V.; Chael, A.A.; Roelofs, F.; Doeleman, S.S.; Freeman, W.T. Reconstructing Video from Interferometric Measurements of Time-Varying Sources. *arXiv* **2017**, arXiv:1711.01357.
65. Raymond, A.W.; Palumbo, D.; Paine, S.N.; Blackburn, L.; Córdova Rosado, R.; Doeleman, S.S.; Farah, J.R.; Johnson, M.D.; Roelofs, F.; Tilanus, R.P.J.; et al. Evaluation of New Submillimeter VLBI Sites for the Event Horizon Telescope. *Astrophys. J. Suppl. Ser.* **2021**, *253*, 5.
66. Palumbo, D.; Johnson, M.; Doeleman, S.; Chael, A.; Bouman, K. Next-Generation Event Horizon Telescope Developments: New Stations for Enhanced Imaging. American Astronomical Society Meeting Abstracts 2018; Volume 231, p. 347.21. Available online: https://ui.adsabs.harvard.edu/abs/2018AAS...23134721P/abstract (accessed on 1 January 2018).
67. Chael, A.; Bouman, K.; Johnson, M.; Blackburn, L.; Shiokawa, H. Eht-Imaging: Tools For Imaging and Simulating Vlbi Data. *Zenodo* **2018**. [CrossRef]
68. Chael, A.; Chan, C.K.; Klbouman; Wielgus, M.; Farah, J.R.; Palumbo, D.; Blackburn, L.; Aviad; Dpesce; Quarles, G.; et al. achael/eht-imaging: V1.2.4. *Zenodo*. 2022. Available online: https://doi.org/10.5281/zenodo.6519440 (accessed on 12 November 2022).
69. Johnson, M.D.; Narayan, R.; Psaltis, D.; Blackburn, L.; Kovalev, Y.Y.; Gwinn, C.R.; Zhao, G.Y.; Bower, G.C.; Moran, J.M.; Kino, M.; et al. The Scattering and Intrinsic Structure of Sagittarius A* at Radio Wavelengths. *Astrophys. J.* **2018**, *865*, 104.
70. Event Horizon Telescope Collaboration; Akiyama, K.; Alberdi, A.; Alef, W.; Asada, K.; Azulay, R.; Baczko, A.K.; Ball, D.; Baloković, M.; Barrett, J.; et al. First M87 Event Horizon Telescope Results. IV. Imaging the Central Supermassive Black Hole. *Astrophys. J. Lett.* **2019**, *875*, L4.

Article

Probing Plasma Composition with the Next Generation Event Horizon Telescope (ngEHT)

Razieh Emami [1,*], Richard Anantua [2], Angelo Ricarte [1,3], Sheperd S. Doeleman [1,3], Avery Broderick [4,5], George Wong [6,7], Lindy Blackburn [1,3], Maciek Wielgus [8], Ramesh Narayan [1,3], Grant Tremblay [1], Charles Alcock [1], Lars Hernquist [1], Randall Smith [1], Matthew Liska [1], Priyamvada Natarajan [3,9], Mark Vogelsberger [10], Brandon Curd [1,3] and Joana A. Kramer [8]

1   Center for Astrophysics | Harvard & Smithsonian, 60 Garden Street, Cambridge, MA 02138, USA
2   Department of Physics & Astronomy, The University of Texas at San Antonio, One UTSA Circle, San Antonio, TX 78249, USA
3   Black Hole Initiative, Harvard University, 20 Garden Street, Cambridge, MA 02138, USA
4   Perimeter Institute for Theoretical Physics, 31 Caroline Street North, Waterloo, ON N2L 2Y5, Canada
5   Department of Physics and Astronomy, University of Waterloo, 200 University Avenue West, Waterloo, ON N2L 3G1, Canada
6   School of Natural Sciences, Institute for Advanced Study, 1 Einstein Drive, Princeton, NJ 08540, USA
7   Princeton Gravity Initiative, Princeton University, Princeton, NJ 08544, USA
8   Max Planck Institute for Radio Astronomy, Auf dem Huegel 69, D-53121 Bonn, Germany
9   Departments of Astronomy & Physics, Yale University, New Haven, CT 06511, USA
10 Department of Physics, Kavli Institute for Astrophysics and Space Research, Massachusetts Institute of Technology, Cambridge, MA 02139, USA
*   Correspondence: razieh.emami_meibody@cfa.harvard.edu

Citation: Emami, R.; Anantua, R.; Ricarte, A.; Doeleman, S.S.; Broderick, A.; Wong, G.; Blackburn, L.; Wielgus, M.; Narayan, R.; Tremblay, G.; et al. Probing Plasma Composition with the Next Generation Event Horizon Telescope (ngEHT). *Galaxies* **2023**, *11*, 11. https://doi.org/10.3390/galaxies11010011

Academic Editor: Bidzina Kapanadze

Received: 14 November 2022
Revised: 27 December 2022
Accepted: 4 January 2023
Published: 10 January 2023

Abstract: We explore the plasma matter content in the innermost accretion disk/jet in M87* as relevant for an enthusiastic search for the signatures of anti-matter in the next generation of the Event Horizon Telescope (ngEHT). We model the impact of non-zero positron-to-electron ratio using different emission models, including a constant electron to magnetic pressure (constant β_e model) with a population of non-thermal electrons as well as an R-beta model populated with thermal electrons. In the former case, we pick a semi-analytic fit to the force-free region of a general relativistic magnetohydrodynamic (GRMHD) simulation, while in the latter case, we analyze the GRMHD simulations directly. In both cases, positrons are being added at the post-processing level. We generate polarized images and spectra for some of these models and find out that at the radio frequencies, both of the linear and the circular polarizations are enhanced with every pair added. On the contrary, we show that, at higher frequencies, a substantial positron fraction washes out the circular polarization. We report strong degeneracies between different emission models and the positron fraction, though our non-thermal models show more sensitivities to the pair fraction than the thermal models. We conclude that a large theoretical image library is indeed required to fully understand the trends probed in this study, and to place them in the context of a large set of parameters which also affect polarimetric images, such as magnetic field strength, black hole spin, and detailed aspects of the electron temperature and the distribution function.

Keywords: M87*; plasma composition; EHT; ngEHT

1. Plasma Composition: Observational Studies

Large scale jets are launched and are collimated by their surrounding accretion flow through extracting the black hole rotational energy by purely electromagnetic mechanisms [1–5]. At radio frequencies, jets are visible from the sub-pc to the kpc scales through the synchrotron emission from the ultra-relativistic electrons gyrating inside the magnetic field. Consequently, jets link the accretion physics to the particle acceleration [6,7]. On the theoretical front, significant progress has been made most recently in modeling the jet formation and its morphology where the current GRMHD simulations are able to recover the observed Lorentz factor. There are many factors that contribute to the jet formation, collimation, acceleration and propagation. Of key importance is the matter content of the jet; either in terms of the normal plasma (composed of electron-protons) or the pair plasma (composed of electron-positron pairs), they lead to very distinct observational signatures. Recently, there have been some theoretical studies [8,9] on probing the observational signatures of different plasma components.

Despite the fundamental importance of plasma composition, up until now, there have not been any conclusive observations that clearly favor either normal or pair plasma. Below, we mention a few different efforts.

Radio-loud quasars and the active galactic nuclei (AGN) make roughly 10% of the population of jets and exhibit very powerful jets, propagating hundreds of kpc away before being disrupted [10]. Consequently, they can be good candidates for searching the matter content of the jet.

Giant elliptical galaxy M87 at the center of the Virgo cluster contains a very spectacular extragalactic jet (firstly discovered by [11]), with a relatively low radio luminosity. The synchrotron and the inverse Compton emissions from the jet bases produce radiation at other frequencies including the optical, X-rays and the γ-rays far outside the host galaxy. Consequently, M87 has been subject to wide studies at multi-frequencies from the radio to γ-rays [12], as well as the X-rays [13]. Furthermore, an extended study has also been conducted on its jet collimation using the Very Long Baseline Interferometry (VLBI) at sub-millimeter wavelengths [14–22]. The synchrotron spectrum of M87 jet was first studied in [23]. The combination of all of these studies makes M87 a very valuable source to probe the matter content of the jet.

Reynolds et al. [10] used the historical data from the VLBI observations of M87 at 5 GHz and probed the physical properties of the jet as well as its matter content using the synchrotron self-absorption theory [6] relevant for the radio emission from the compact core of M87. They put constraints on the magnetic field and the particle density of the jet and eventually on the matter content of the jet. Their results strongly favored a pair dominated plasma; although, they did not yet rule out the possibility of a normal plasma. As they argued, a multi-frequency analysis of the jet may unambiguously put constraints on the matter content of the jet.

The quasar 3C279 is another example of a luminous object in the sky, located at redshift $z = 0.538$, already observed with the EHT [24]. It is luminous from radio to γ-ray wavelengths. At the radio frequency, the VLBI observations demonstrate a very bright and unresolved core along with a jet extended to the kpc scales. There are superluminal motions associated with this source in the jet, with velocities raging from 4–15 times the speed of light, indicating a relativistic bulk speed in the jet. Consequently, the emitted radiation from the jet at different wavelengths are boosted by the Doppler effect. Owing to its relatively high flux, 3C279 is an ideal source to probe the physics of the extragalactic source. To probe the jet decomposition in quasar 3C279, [25] used the circular polarization from the observations of 3C279 at 15 GHz. The circular polarization is produced through the Faraday conversion, requiring the energy distribution of particles being extended to lower energies. Combining their final results with other extragalactic sources, such as M87, they concluded that, in general, the extragalactic jets might be primarily composed of pairs. Furthermore, they argued that since the jet densities should be rather low, the pair-dominated jet points us to a picture in which photon cascades or the Pion decay are the main origin of the radiating particles in the jet.

The quasar 3C345 at redshift $z = 0.594$ is a core-dominated radio source with a prominent pc-scale jet, emitting X-rays through the Synchrotron self-Compton (SSC) process. This source has been monitored at 5 GHz every year since 1977. Higher frequency monitoring of this source, at both 10.5 and 22 GHz, are conducted more frequently. [26] combined the constraints on the electron number density from the Synchrotron Self-Absorption with the kinetic luminosity constraints and concluded that C2, C3, C4, C5 and C7 components of quasar 3C345 are predominantly made of pairs rather than normal plasma.

Despite the fact that in all of the above sources probed using the aforementioned methods, pair plasma is favored against the normal plasma; this conclusion does not hold for other sources. For example, the radio galaxy 3C 120, located at $z = 0.033$, presents a strong interaction between the jet and its interstellar cloud in which the matter content of the jet is mixed with the dense surrounding thermal gas. If the jet in this source is mostly composed of pairs, their positrons would enter the cloud and be thermalized through the ionized energy lost. The annihilation rate of such positrons in the cloud would then have to be proportional to the positron density of the jet. The observations of narrow emission lines would then inform us about the matter content of the jet. [27] made an exploration of the matter content of the jet in this source through their annihilation line flux, using the hard X-ray and soft γ-ray spectrum from the SPI spectrometer on INTEGRAL. Their spectral analysis failed to detect any lines and thus could not consistently constrain the positron to electron ration in this source.

More statistically, Ref. [28] used a sample of radio-load quasars and addressed the puzzle of plasma matter content. Out of the possibility of a normal or a pair plasma and based on the annihilation constraints, combined with the assumption that pairs are originated from the inner part of the accretion, they favored an electron-proton plasma.

In summary, there are some controversial results in the observational searches for the plasma matter content. Consequently, the key question of the plasma matter content remains elusive.

Being mindful of the current status of searches for positrons, in what follows, we review different theoretical studies that motivate a pair plasma. We will then focus on two sets of toy models, one based on a semi-analytic type of models, while the other from the GRMHD simulations. In the latter case, we make the polarized images for one snapshot. Work is in progress to extend this analysis to the case with a time-averaged polarized image set for GRMHD simulations. In both cases, we add positrons at a post-processing level using different ray-tracing codes. We apply the most recent polarimetric constraints, on the fractional linear polarization, from the EHTC [29,30] and put constraints on the plasma matter contents.

The paper is organized as follows. Section 2 reviews theoretical approaches in making the pair-plasma. Section 3 describes the radiative transfer approach in making the polarized images of BH. Section 4 models the impact of positrons in a semi-analytic approach. Section 5 introduces a GRMHD method to deal with the impact of positrons. Section 6 provides a conclusion of the paper.

2. Theoretical Approaches in Creating Pairs

There are two distinct classes of models which lead to pair creation through photon annihilation, known as the Briett-Wheeler process [31]. The electron-positron pairs could either be created from a coherent, steady state and large-scale mechanisms (as relevant in the gap systems) [32–37] or, rather, originate from an incoherent, transient and small-scale approach (being appropriate in pair drizzle systems) [38]. However, while in the gap approach, the high energy photons have an energy orders of magnitude above the rest-mass energy of electrons ($\simeq$MeV); in the pair drizzle approach, the host photons have an energy of roughly MeV. Despite the distinct features of the gap and drizzle models, they could be thought of as the continuum distribution in two different ends for the energy spectrum of the created photons.

In the following, we review these scenarios in more depth:

1. Gap models make high-energy photons in coherent regions with $\mathbf{E} \cdot \mathbf{B} \neq 0$, which accelerate the leptons and make the pair cascades. Broderick and Tchekhovskoy [39] showed that the stagnation surface, in these gap models, defined as the boundary between the material falling back to the black hole and outflows (forming the jet), would be a natural site for the pair creation followed by the particle acceleration. They showed that un-

screened electric fields lead to the production of photons from the accretion flow, inside the jet. These photons would then make non-thermal particles through their inverse Compton scattering. As they argued, this method leads to a population of non-thermal particles consistent with the most recent sub mm-VLBI observations of M87. In this method, there are two distinct mechanisms in making/propagating the pairs. In the gap region, particles are highly accelerated, limited by the inverse Compton cooling, followed by the net charge separation. They emit photons, which then make electron-positron pairs through the Compton scattering off the ambient soft photons. In this picture, particles moving inward are accreted by the central BH while those travelling outward are making the jets. In such spark gaps, positron densities may exceed the Goldreich-Julian value required to screen electric fields, limiting the efficiency of pair cascades once enough positrons are formed.

2. Pair drizzle models predict a steady and smooth background population of photons, being created from the high-energy part of electron distribution function in the near horizon plasma. Such $\sim$ MeV photons make e^-/e^+ pairs throughout their interactions. Mościbrodzka et al. [38] estimated the pair creation rate based on the non-relativistic GRMHD simulations. Wong et al. [40] extended this approach to include a radiative-based GRMHD simulation from `ebh-light` code [41–43]. They used an axisymmetric model and explored different mass accretion rates corresponded to an optically thin and geometrically thick accretion flow of SANE simulation. Figure 1 presents the logarithmic ratio of available pair to the Goldreich-Julian density (see Equation (33) of [40] for more details) (left panel) for a GRMHD model with $a = 0.5$ with the accretion rate $\dot{m}/\dot{M}_{\mathrm{Edd}} = 1.1 \times 10^{-5}$. In the left panel, it is evident that the available pair ratio is much above unity in the disk and in the funnel jet, demonstrating that these locations might be a good site for pairs to appear. The right panel also shows the logarithmic rate of the pair production density in the same model. It is seen that the rate of pair production is enhanced in the inner part of the disk and in the equatorial plane, but diminishes significantly at larger radii, e.g. $x > 5\mathrm{GM}/c^2$ as well as the high scale heights, e.g. $z > 3\mathrm{GM}/c^2$.

Consequently, we conclude that while in gap models, pairs are being mostly created in the jet funnel, pair drizzle models make them more in the accretion disk.

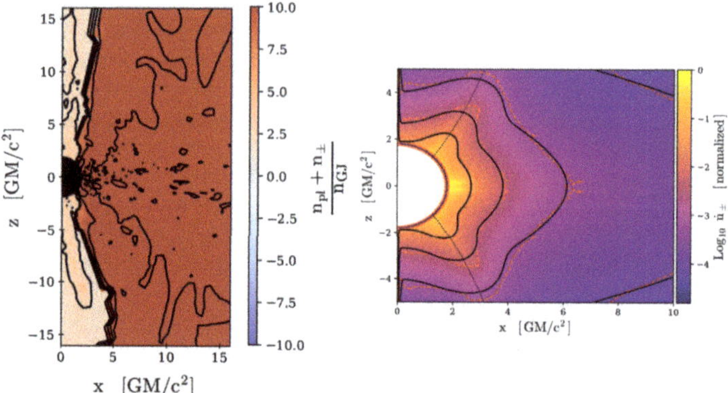

Figure 1. (**Left panel**) The ratio of the available pair density to the Goldreich-Julian density for a GRMHD simulation model with a = 0.5. (**Right panel**) The time-averaged logarithmic rate of the pair production density taken from [40] as a function of the position. In each panel, the horizontal axis presents the radial coordinate while the vertical axis shows the height above the disk mid-plane. From the plot, it is inferred that the pair production rate is enhanced in the inner part of the accretion disk and in the equatorial plane. Dashed red lines and solid black lines compare the numeric and the fits to the contours of the rate of the number density. See [40] for more details.

3. Radiative Transfer of Pair Plasma

Due to their opposite charge, positrons gyrate in a magnetic field in the opposite direction to electrons. This has significant effects on the radiative transfer coefficients, namely the synchrotron emission/absorption and Faraday rotation. Compared to an ionic plasma, a pair plasma produces no Faraday rotation, nor does it produce any circular polarization from synchrotron radiation. On the other hand, linear polarization emission and Faraday conversion persist. Consequently, polarization can be used to place constraints on the plasma composition and has fuelled a decades-long debate about the composition of astrophysical jets [6,25,44].

In general relativistic radiative transfer GRRT, the radiative transfer coefficients (see e.g., [45], for their derivation and definitions) are modified via:

$$
\begin{aligned}
j_{I,Q,U} &\to (1 + f_{\text{pos}}) j_{I,Q,U}, \\
j_V &\to (1 - f_{\text{pos}}) j_V, \\
\alpha_{I,Q,U} &\to (1 + f_{\text{pos}}) \alpha_{I,Q,U}, \\
\alpha_V &\to (1 - f_{\text{pos}}) \alpha_V, \\
\rho_{Q,U} &\to (1 + f_{\text{pos}}) \rho_{Q,U}, \\
\rho_V &\to (1 - f_{\text{pos}}) \rho_V.
\end{aligned}
\tag{1}
$$

where j_i and α_i are the emission and absorption coefficients, ρ_i refer to the rotativities, and f_{pos} describes the positron-to-electron ratio. Qualitatively, increasing the positron fraction tends to increase the linear polarization fraction by reducing the Faraday rotation depth. The circular polarization fraction and the image morphology may also change significantly as the balance between circular polarization generated by emission and Faraday conversion changes. Since the Faraday conversion, Faraday rotation, and circular polarization emission coefficients evolve differently with the frequency, this may lead to large changes in the polarized spectrum. Consequently, a multi-frequency analysis might be very helpful in putting constraints on the plasma composition.

Below, we use the above prescription and analyze the impact of positrons in polarized images from a semi-analytical model in Section 4 as well as two sets of GRMHD simulation in Section 5. As we will describe, in both cases, we have added the positrons at the ray-tracing level by the aforementioned algorithm. Throughout our analysis, we focus on M87*.

4. Positron Effects in the Semi-Analytical Models

To probe the observational signatures of a pair plasma in the polarized emission from the jet/accretion in M87*, we make use of a semi-analytical approach [9], as a self-similar jet model focusing on the force-free regions of a Blandford-Znajek outflow model in [8,46] with several extensions including a general relativistic ray-tracing from the GRTRANS code [45], as well as the usage of non-thermal distribution to describe the number density of pairs. In this model, the number density of pairs is mapped to total electron/positron pressure, with an overall electron to magnetic pressure β_e. Furthermore, we assume a self-similar parabolic jet profile $\zeta = s^2/z$, in the cylindrical coordinate. We use some fitting formulae for the ζ, using the magnetic flux $\Phi_B(\zeta)$, the line angular speed $\Omega_B(\zeta)$ and the velocity component along with the z-coordinate, $v_z(\zeta)$ being extracted from a HARM simulation [47]. In more detail, for the GRMHD, we use a magnetically arrested disk (MAD) simulation with $a/M = 0.92$ and infer the fitting formulas at $z = 50M$. Note that these models are idealized and axis-symmetric, and polarized signatures are sensitive to additional parameters, including spin, magnetic field state, and the electron-to-ion temperature ratio [30]. Nevertheless, this simple model can be used to explore and illustrate the qualitative changes as positrons are added to the accretion flow.

Next, we extrapolate these quantities to larger radii, taking advantage of the self-similarity in ζ; see Equations (18)–(26) of [9] for more details.

As already mentioned above, we use a non-thermal, power-law distribution for the emitting electrons, with a Lorentz factor between $\gamma_{\min}$ and $\gamma_{\max}$:

$$N_{e^-}(\gamma) = \begin{cases} N_0 \gamma^{-p} & \gamma_{\min} \leq \gamma \leq \gamma_{\max} \\ 0 & \text{otherwise} \end{cases}, \tag{2}$$

where $N_0 = n_{e^-}(p-1)/\left(\gamma_{\min}^{1-p} - \gamma_{\max}^{1-p}\right)$ refers to the overall normalization of the electron distribution and with n_{e^-} describing the total number density of electrons as given by Equation (28) of [9]. We implement this model to GRTRANS, modifying its equations for the emissivity, absorption and the Faraday terms to include the contribution of positrons. Below, we describe the results including the polarized images, the spectra and the multi-frequency analysis.

4.1. Polarized Images

In what follows, we compare polarized intensity maps of models with varying ratios n_{e+}/n_0 of added electron-positron pairs to original plasma electrons and mechanisms for accelerating electrons and positrons to relativistic speeds such that they radiate synchrotron radiation at 230 GHz. The color maps show the 230 GHz flux (intensity times solid angle of a pixel) for total intensity I, linearly polarized intensity P and circularly polarized intensity V for synthetic images of our emission models. The polarization serves as a key discriminant for degenerate total intensity images whose electrons and positrons contribute differently to intrinsic circular polarization and Faraday rotativity radiative transfer coefficients.

Figure 2 presents the polarized image of the constant β_e model at 230 GHz for the case with no positrons (top panel) and the one with 50% of positrons (bottom panel). (Note the circularly polarized 230 GHz intensities are shown in terms of the brightness temperature T_b that a black-body would possess to give the same luminosity volumetrically as the maps display at the observed frequency.)From the plot, it is seen that, in this model, adding the positrons significantly increases the linear and the circular polarization. Consequently, we expect that the EHT polarimetric constraints disfavor the case with significant pair fraction in the constant β_e model.

Next, we explore the impact of changing the parameters of the emission model. Figures 3 and 4 show the linear and circular polarization maps for different models' parameters. In each figure, in the top panel, we change $\beta_e = 10^{-6}, 10^{-4}, 10^{-1}$, while in the bottom row, we alter the slope $p = 2.5, 3.0, 3.5$, while fixing $f_{\text{pos}} = 0.1$.

From the plots, it is inferred that increasing the β_e as well as p suppress the linear and circular polarization, since increasing the β_e and p enhances the Faraday rotation, which itself leads to scrambling of the EVPA and, thus, the suppression of the linear and the circular polarization. In [9], we made an in-depth exploration of the impact of changing the β_e and non-thermal power-law slope p on the SED, see Figures 9 and 11 of [9] and tables 2 and 4 for more details. Finally, it is seen that the shape of the circular polarization also changes when we vary the parameters of the emission model. This implies that a direct detection of the circular polarization map might be useful to break the degeneracy between the physical parameters of the model as well as the matter content of the emitting plasma.

We emphasize that the polarization maps here use the standard EHT point spread function based on a 20 μ as beam. However, ngEHT may be able to focus on half this scale, In [48], intensity maps are shown to possess higher maxima when they are more highly resolved owing to an increase in pixels per plane. The effect of partitioning the intensity into refined bins can also slightly shift the location of intensity maxima. These trends hold for the polarized images as well.

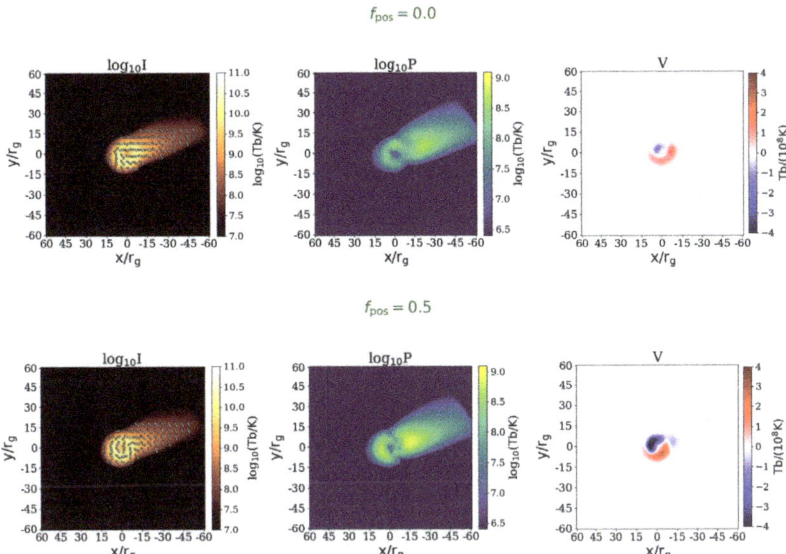

Figure 2. Down rows present the polarized map of M87* inferred from the constant β model at 230 GHz. We have fixed $\beta_e = 0.01$, and the power-law index at $p = 3.5$. From the (**top**) to (**bottom**), we present the images for $f_{\text{pos}} = 0.0, 0.5$, respectively.

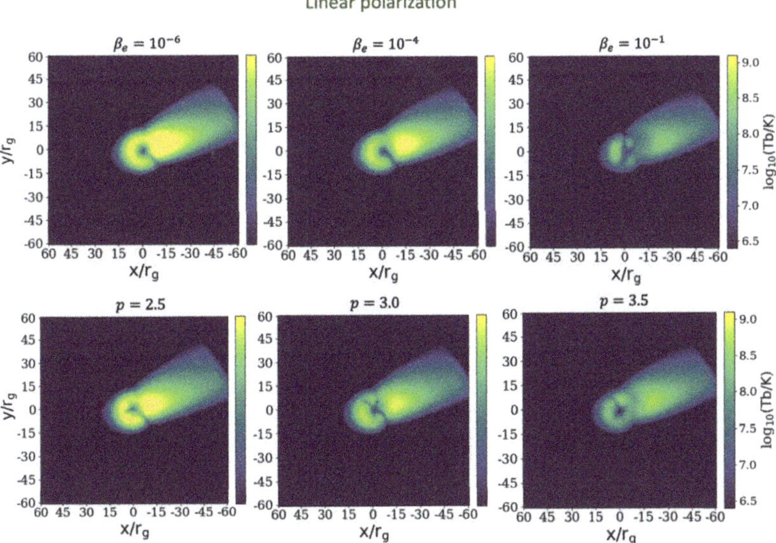

Figure 3. The (log)-linear polarization map for the constant β model with $f_{\text{pos}} = 0.1$. (**Top panel**) shows the case with $p = 3.5$. From the left to right we change $\beta_e = 10^{-6}, 10^{-4}, 10^{-1}$. (**Bottom panel**) presents the case with fixed $\beta_e = 10^{-2}$, also from the left to right, we change $p = (2.5, 3.0, 3.5)$. These images are corresponded to $\nu = 230$ GHz.

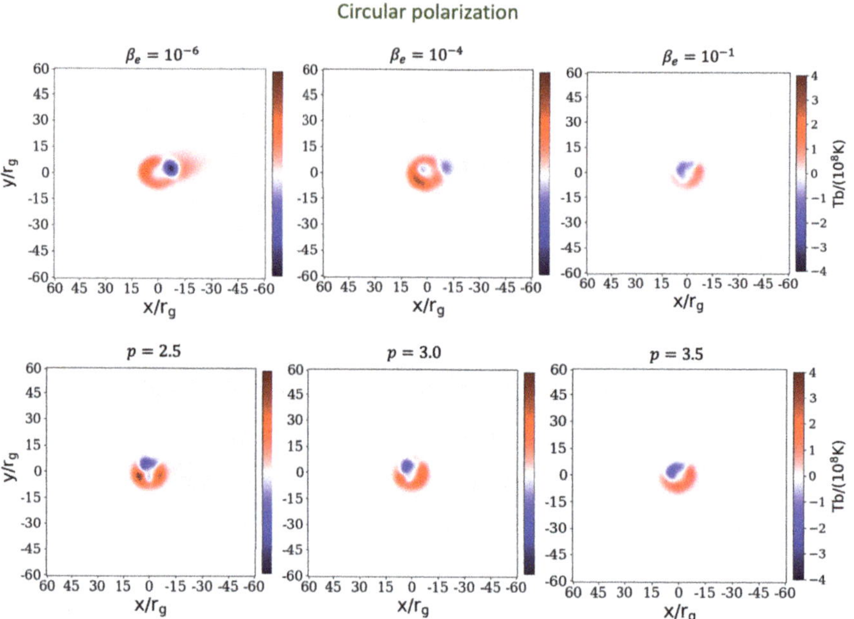

Figure 4. The circular polarization map for the constant β model with $f_{\mathrm{pos}} = 0.1$. (**Top panel**) shows the case with $p = 3.5$. From the left to right we change $\beta_e = 10^{-6}, 10^{-4}, 10^{-1}$. (**Bottom panel**) presents the case with fixed $\beta_e = 10^{-2}$, also from the left to right, we change $p = (2.5, 3.0, 3.5)$. These images are corresponded to $\nu = 230$ GHz.

4.2. Spectral Analysis

In Figure 5, from the left to right panels, we analyze the spectrum of the total intensity, the linear and circular polarization, respectively. Motivated by a wide model survey performed in [9] and since $\beta_e = 10^{-2}$ and $p = 3.5$ perform well in satisfying the EHT polarimetric results, we choose these values and only vary the $f_{\mathrm{pos}} = (0.0, 0.1, 1.0)$. Overlaid in each panel, we present the observational data points from [49–51] as well as the most recent results from the EHT observation [9,30]. Furthermore, we fix the model parameters to match the observed flux at $\nu = 230$ GHz. Consequently, the impact of fluid composition is somewhat renormalized to get the flux right and, thus, does not have any other impact on the total flux at 230 GHz.

As expected, the total flux (left panel) is off at higher frequencies. This implies that, while the current toy model describes the radio observations quite well, a more complicated model would be required to match the flux at higher frequencies.

The fractional linear polarization, from the middle panel, shows more sensitivity to altering the positron fraction. Based on our simple semi-analytic model, it is inferred that models with higher f_{pos} sit above the current EHT constraints. In Emami et al. [9], we made a detailed survey of different models and realized that varying the β_e and p do not change the above conclusion, implying that there are severe constraints for the constant β_e model. However, we emphasize that this conclusion may not be easily generalized to other accretion flow models. For example, as we will see below, certain thermal models provide less severe constraints.

Finally, the circular polarization, in the right panel, establishes more interesting dependencies to the positron fraction. It is explicitly seen that circular polarization is not only too sensitive to the positron fraction at the radio frequencies, but it also shows distinct features at higher frequencies. Consequently, a multi-frequency analysis should break the degeneracies in searching for positrons. While the current polarimetric results only cover

230 GHz, we hope that adding more frequencies to the ngEHT improves these constraints further. In the next section, we briefly discuss this while leaving an in-depth analysis to a future work.

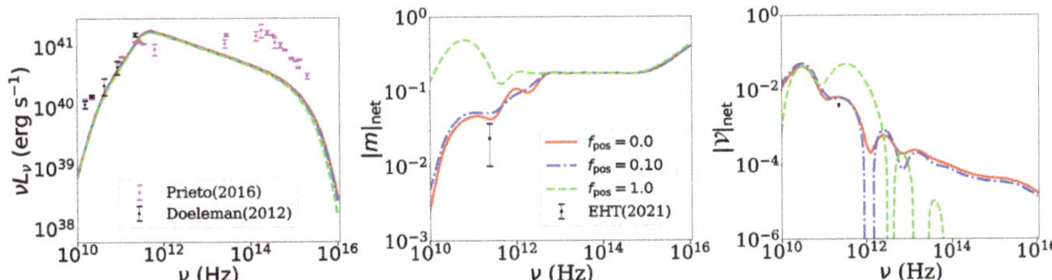

Figure 5. The spectrum of the best bet model for the jet model [9] with $\beta_e = 10^{-2}$ and $p = 3.5$. From the (**left**) to (**right**), we present the total intensity, the linear and the circular polarization, respectively. Overlaid on each panel, we also present the observational data points.

4.3. Multi-Frequency Analysis

In Figure 6 down rows present a multi-frequency image analysis; the top, middle and bottom rows present 86, 345 and 690 GHz, respectively. In each row, from the left to right, we present the intensity, linear and the circular polarizations. From the plot, it is inferred that increasing the frequency washes out the linear and circular polarization substantially. Furthermore, the image is also core shifted at higher frequencies where the larger-scale patterns in the electric vector polarization angles (EVPAs) are boosted to the central part of the image. Finally, the shape of the circular polarization is also altered at higher frequencies. As the morphology of the polarized images strongly depends on the frequency, a simultaneous observation at the above frequencies would be very capable of breaking the model degeneracies. Consequently, we propose to use the ngEHT to break the model degeneracies.

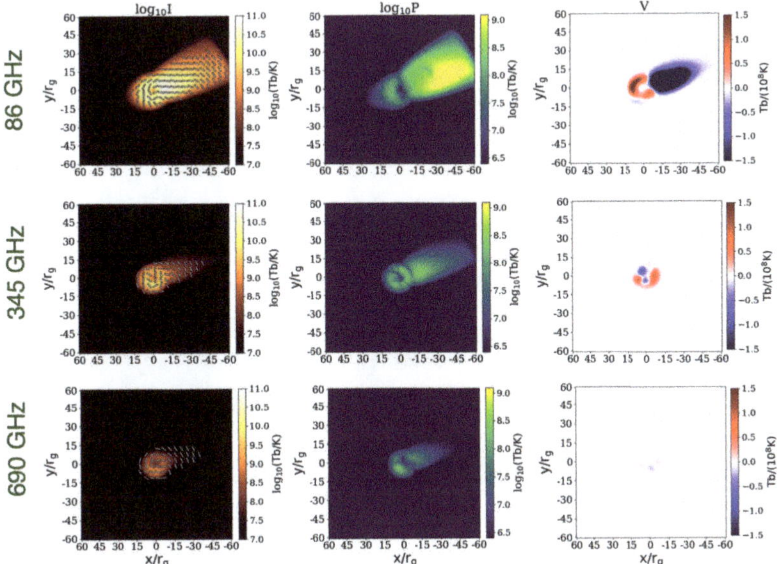

Figure 6. Down rows present the polarized map ((**left**) to (**right**) the intensity, linear and circular polarization) of M87* in constant beta model at 86, 345 and 690 GHz, respectively. The plot is for $f_{pos} = 0.1$, $p = 3.5$ and $\beta = 10^{-2}$.

5. Positrons in GRMHD KHARMA Simulation

Next, we generalize the above semi-analytic approach to a direct GRMHD simulation approach (Anantua et al. in prep.), where positrons are still being added during a post-processing step to KHARMA simulation [52]. These simulations are run assuming an adiabatic index of 5/3, although there is no evidence that changing the adiabatic index to values as low as 4/3 significantly impacts our images. While an in-depth analysis of the impact of positrons is left to a separate work, here we aim to take the first look at the possible importance of changing the accretion type as well as the electron emission profiles from non-thermal models to that of thermal models.

In Figures 7 and 8, we present a MAD model with $a = 0.94$ and a SANE model with $a = +0.5$, respectively. In both of these figures, we use a familiar turbulent heating-based emission model in which the ion-to-electron temperature ratio $R = (1 + \beta^2)^{-1}R_{\text{low}} + \beta^2(1 + \beta^2)^{-1}R_{\text{high}}$ approaches parametrically determined values $1/R_{\text{low}}$ in the low=β polar outflow and $1/R_{\text{high}}$ in the high-β accretion disk. More explicitly, we choose, $R_{\text{low}} = 1$ and $R_{\text{high}} = 20$. It is seen that the addition of positrons (where electron-positron pair number density n_{pair} is varied relative to the original electron number density n_0) has remarkably little effect on the linear polarization pattern. A more significant difference is seen in circular polarization, however, which historically has been promising to use to test plasma composition models [25]. This is because only ionic plasma can produce the circular polarization via direct synchrotron emission, and a pair plasma does not perform Faraday rotation, which can affect linear polarization that goes through Faraday conversion [44]. Since large-scale circularly polarized emission originates from direct synchrotron in this model [53], the large-scale emission disappears when a significant population of pairs is added. Note that polarized EHT observations at present only weakly and indirectly constrain spin, using models without positrons; so, we explore several different values in this text [30]. In the top row, we show the result of a normal ray tracing, without the addition of pairs, using IPOLE [54], while in the bottom one, we present the case with 100 times more electron-positron pairs, added at a post-processing level, to the pre-existing electron number density in KHARMA GRMHD simulation. This approximates a full pair plasma while preserving charge neutrality. From left to right, different columns present the total intensity, linear and the circular polarization, respectively.

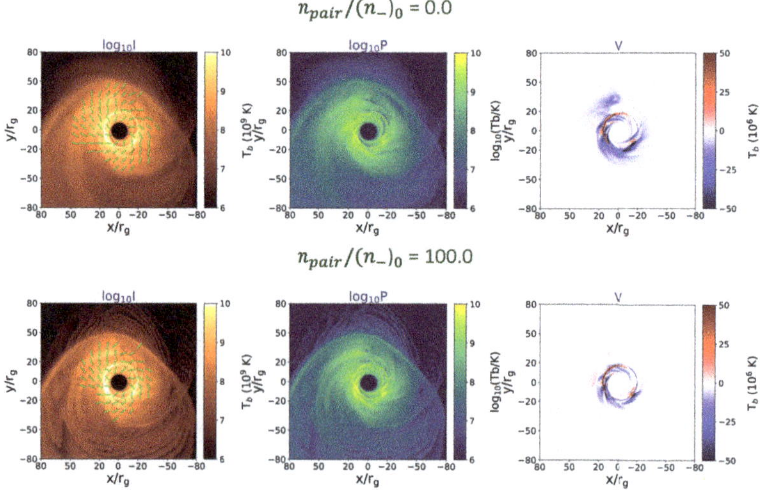

Figure 7. A map from MAD simulations with $a = +0.94$ at $R_{\text{high}} = 20$ in R-beta model from KHARMA simulations. The top row refers to the case with no positrons while the bottom one describes the case with 100% positrons compared with the original number of electrons. In each row, from the (**left**) to (**right**), we present the images of total intensity, the linear and the circular polarization, respectively. We have used $\nu = 230$ GHz to make these images.

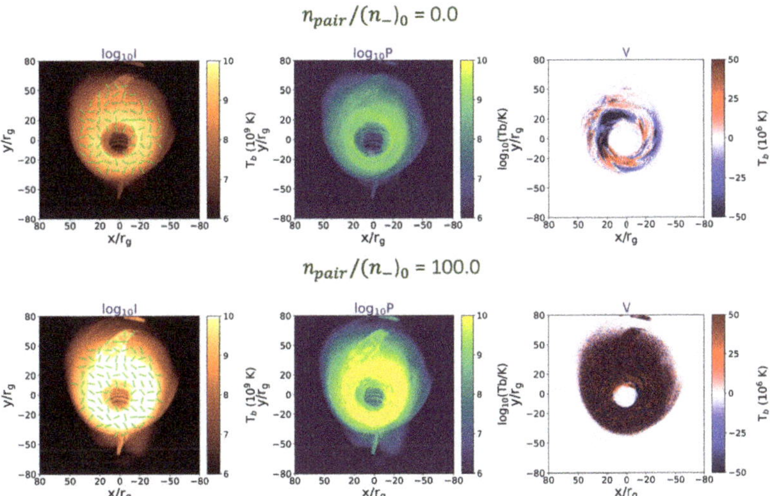

Figure 8. A map from SANE simulations with a = +0.5 at R_{high} = 20 in R-beta model from KHARMA simulations. The top row refers to the case with no positrons while the bottom one describes the case with 100% positrons compared with the original number of electrons. In each row, from the (**left**) to (**right**), we present the images of total intensity, the linear and the circular polarization, respectively. We have used ν = 230 GHz to make these images.

In Figure 8, we see that the linear polarization pattern of the SANE model is much more strongly affected by the addition of pairs than in the MAD model. This is because the model is intrinsically Faraday-thick, resulting in a significant scrambling. The addition of pairs dramatically decreases the Faraday depth, resulting in an ordered linear polarization pattern. Faraday rotation may also have an indirect effect on circular polarization by scrambling the linear polarization that would be transformed into circular polarization. As we see here, the addition of pairs dramatically increases the resolved circular polarization by removing this scrambling effect.

Comparing the current GRMHD simulation results with that of Section 4, it is inferred that changing the emission model significantly affects the morphology of the polarized images.

6. Conclusions

There have been many outstanding observational claims for a substantially pair-dominated plasma in the radio-load quasars and in AGNs. Motivated by this, in this manuscript, we studied the theoretical signatures of having non-zero positrons in the polarized images in a force-free, semi-analytic jet model (Section 4) as well as a snapshot of the KHARMA GRMHD simulations (Section 5), where in both cases, positrons are added at the post-processing level. While the Faraday rotation diminishes by increasing the positron fraction, the Faraday Conversion is boosted linearly. We showed that the role of positrons is also mixed with different emission models in a degenerate picture. A multi-frequency analysis in the radio band, 86–690 GHz as relevant for the ngEHT, or a much wider frequency search in the near infrared and the X-ray, may however break the degeneracy. Increasing the frequency, there is a core shift in images that wash out the power at larger scales and squeeze them more to the central part of the image. Consequently, we argue that lower frequencies may give rise to a better chance to search for positrons. Therefore, 86 or 230 GHz from the ngEHT could be very relevant frequencies for looking for a pair plasma in the heart of M87*. Finally, we emphasize that the positron fraction is just one of many physical parameters that affect the polarized images. In a future work, we aim for a broader investigation including a large image library probing different GRMHD models and emission prescriptions to determine how well these results may be generalized.

Galaxies 2023, 11, 11

Author Contributions: Conceptualization, R.E., R.A., A.R., S.S.D., A.B., M.V., P.N., B.C., J.A.K. Methodology, R.E. Software, R.E., R.A., A.R. Validation, R.E. Formal analysis, R.E., R.A., A.R. Investigation, R.E., G.W., M.W., Resources, R.A., A.R. Data curation, R.E., R.A., G.W. Writing, R.E., R.A., A.R., M.W., L.B. Original draft preparation, R.E. Review and editing, R.E., R.A., A.R., M.W. Supervision, S.S.D., R.N., G.T., C.A., L.H., R.S., M.L. Funding, R.N., G.T., R.S., M.L. All authors have read and agreed to the published version of the manuscript.

Funding: Razieh Emami acknowledges the support of the Institute for Theory and Computation at the Center for Astrophysics as well as grant numbers 21-atp21-0077, NSF AST-1816420 and HST-GO-16173.001-A for their very generous support. This work was supported in part by the Black Hole Initiative, which is supported by grants from the Gordon and Betty Moore Foundation and the John Templeton Foundation. The opinions expressed in this publication are those of the author(s) and do not necessarily reflect the views of the Moore or Templeton Foundations.

Data Availability Statement: The data presented in this study are available on request from the corresponding author.

Acknowledgments: We greatly appreciate positrons for most likely making the AGN jets and we wish to discover them using the ngEHT! Special thanks to the referees as well for very constructive comments.

Conflicts of Interest: The authors declare no conflict of interest.

References

1. Blandford, R.D.; Znajek, R.L. Electromagnetic extraction of energy from Kerr black holes. *Mon. Not. R. Astron. Soc.* **1977**, *179*, 433–456. [CrossRef]
2. Tchekhovskoy, A.; Narayan, R.; McKinney, J.C. Black Hole Spin and The Radio Loud/Quiet Dichotomy of Active Galactic Nuclei. *Astrophys. J.* **2010**, *711*, 50–63.
3. Tchekhovskoy, A.; Narayan, R.; McKinney, J.C. Efficient generation of jets from magnetically arrested accretion on a rapidly spinning black hole. *Mon. Not. R. Astron. Soc.* **2011**, *418*, L79–L83,
4. Tchekhovskoy, A.; McKinney, J.C. Prograde and retrograde black holes: whose jet is more powerful? *Mon. Not. R. Astron. Soc.* **2012**, *423*, L55–L59.
5. Tchekhovskoy, A. Launching of Active Galactic Nuclei Jets. In *The Formation and Disruption of Black Hole Jets*; Astrophysics and Space Science Library; Contopoulos, I., Gabuzda, D., Kylafis, N., Eds.; Springer: Cham, Switzerland, 2015; Volume 414, p. 45. [CrossRef]
6. Blandford, R.D.; Königl, A. Relativistic jets as compact radio sources. *Astrophys. J.* **1979**, *232*, 34–48. [CrossRef]
7. Blandford, R.D.; Payne, D.G. Hydromagnetic flows from accretion disks and the production of radio jets. *Mon. Not. R. Astron. Soc.* **1982**, *199*, 883–903. [CrossRef]
8. Anantua, R.; Emami, R.; Loeb, A.; Chael, A. Determining the Composition of Relativistic Jets from Polarization Maps. *Astrophys. J.* **2020**, *896*, 30.
9. Emami, R.; Anantua, R.; Chael, A.A.; Loeb, A. Positron Effects on Polarized Images and Spectra from Jet and Accretion Flow Models of M87* and Sgr A*. *Astrophys. J.* **2021**, *923*, 272,
10. Reynolds, C.S.; Fabian, A.C.; Celotti, A.; Rees, M.J. The matter content of the jet in M87: evidence for an electron-positron jet. *Mon. Not. R. Astron. Soc.* **1996**, *283*, 873–880.
11. Curtis, H.D. The Planetary Nebulae. *Publ. Lick Obs.* **1918**, *13*, 55–74.
12. Abramowski, A.; Acero, F.; Aharonian, F.; Akhperjanian, A.G.; Anton, G.; Balzer, A.; Barnacka, A.; Barres de Almeida, U.; Becherini, Y.; Becker, J.; et al. The 2010 Very High Energy γ-Ray Flare and 10 Years of Multi-wavelength Observations of M 87. *Astrophys. J.* **2012**, *746*, 151,
13. Biretta, J.A.; Stern, C.P.; Harris, D.E. The Radio to X-ray Spectrum of the M87 Jet and Nucleus. *Astron. J.* **1991**, *101*, 1632. [CrossRef]
14. Palmer, H.P.; Rowson, B.; Anderson, B.; Donaldson, W.; Miley, G.K. Radio Diameter Measurements with Interferometer Baselines of One Million and Two Million Wavelengths. *Nature* **1967**, *213*, 789–790. [CrossRef]
15. Reid, M.J.; Schmitt, J.H.M.M.; Owen, F.N.; Booth, R.S.; Wilkinson, P.N.; Shaffer, D.B.; Johnston, K.J.; Hardee, P.E. VLBI observations of the nucleus and jet of M 87. *Astrophys. J.* **1982**, *263*, 615–623. [CrossRef]
16. Kovalev, Y.Y.; Lister, M.L.; Homan, D.C.; Kellermann, K.I. The Inner Jet of the Radio Galaxy M87. *Astrophys. J.* **2007**, *668*, L27–L30.
17. Kuo, C.Y.; Asada, K.; Rao, R.; Nakamura, M.; Algaba, J.C.; Liu, H.B.; Inoue, M.; Koch, P.M.; Ho, P.T.P.; Matsushita, S.; et al. Measuring Mass Accretion Rate onto the Supermassive Black Hole in M87 Using Faraday Rotation Measure with the Submillimeter Array. *Astrophys. J.* **2014**, *783*, L33.
18. Kino, M.; Takahara, F.; Hada, K.; Akiyama, K.; Nagai, H.; Sohn, B.W. Magnetization Degree at the Jet Base of M87 Derived from the Event Horizon Telescope Data: Testing the Magnetically Driven Jet Paradigm. *Astrophys. J.* **2015**, *803*, 30.
19. Mertens, F.; Lobanov, A.P.; Walker, R.C.; Hardee, P.E. Kinematics of the jet in M 87 on scales of 100-1000 Schwarzschild radii. *A&A* **2016**, *595*, A54.
20. Walker, R.C.; Hardee, P.E.; Davies, F.B.; Ly, C.; Junor, W. The Structure and Dynamics of the Subparsec Jet in M87 Based on 50 VLBA Observations over 17 Years at 43 GHz. *Astrophys. J.* **2018**, *855*, 128.

21. Kim, J.Y.; Krichbaum, T.P.; Lu, R.S.; Ros, E.; Bach, U.; Bremer, M.; de Vicente, P.; Lindqvist, M.; Zensus, J.A. The limb-brightened jet of M87 down to the 7 Schwarzschild radii scale. *A&A* **2018**, *616*, A188.
22. Chael, A.; Narayan, R.; Johnson, M.D. Two-temperature, Magnetically Arrested Disc simulations of the jet from the supermassive black hole in M87. *Mon. Not. R. Astron. Soc.* **2019**, *486*, 2873–2895.
23. Meisenheimer, K.; Roeser, H.J.; Schloetelburg, M. The synchrotron spectrum of the jet in M87. *A&A* **1996**, *307*, 61.
24. Kim, J.Y.; Krichbaum, T.P.; Broderick, A.E.; Wielgus, M.; Blackburn, L.; Gómez, J.L.; Johnson, M.D.; Bouman, K.L.; Chael, A.; Akiyama, K.; et al. Event Horizon Telescope imaging of the archetypal blazar 3C 279 at an extreme 20 microarcsecond resolution. *A&A* **2020**, *640*, A69. [CrossRef]
25. Wardle, J.F.C.; Homan, D.C.; Ojha, R.; Roberts, D.H. Electron-positron jets associated with the quasar 3C279. *Nature* **1998**, *395*, 457–461. [CrossRef]
26. Hirotani, K.; Iguchi, S.; Kimura, M.; Wajima, K. Pair Plasma Dominance in the Parsec-Scale Relativistic Jet of 3C 345. *Astrophys. J.* **2000**, *545*, 100–106.
27. Marscher, A.P.; Jorstad, S.G.; Gómez, J.L.; McHardy, I.M.; Krichbaum, T.P.; Agudo, I. Search for Electron-Positron Annihilation Radiation from the Jet in 3C 120. *Astrophys. J.* **2007**, *665*, 232–236. [CrossRef]
28. Celotti, A.; Fabian, A.C. The kinetic power and luminosity of parsec-scale radio jets - an argument for heavy jets. *Mon. Not. R. Astron. Soc.* **1993**, *264*, 228–236. [CrossRef]
29. Event Horizon Telescope Collaboration; Akiyama, K.; Algaba, J.C.; Alberdi, A.; Alef, W.; Anantua, R.; Asada, K.; Azulay, R.; Baczko, A.K.; Ball, D.; et al. First M87 Event Horizon Telescope Results. VII. Polarization of the Ring. *Astrophys. J.* **2021**, *910*, L12.
30. Event Horizon Telescope Collaboration; Akiyama, K.; Algaba, J.C.; Alberdi, A.; Alef, W.; Anantua, R.; Asada, K.; Azulay, R.; Baczko, A.K.; Ball, D.; et al. First M87 Event Horizon Telescope Results. VIII. Magnetic Field Structure near The Event Horizon. *Astrophys. J.* **2021**, *910*, L13.
31. Breit, G.; Wheeler, J.A. Collision of Two Light Quanta. *Phys. Rev.* **1934**, *46*, 1087–1091. [CrossRef]
32. Beskin, V.S.; Istomin, Y.N.; Parev, V.I. Filling the Magnetosphere of a Supermassive Black-Hole with Plasma. *Soviet Astron.* **1992**, *36*, 642.
33. Hirotani, K.; Okamoto, I. Pair Plasma Production in a Force-free Magnetosphere around a Supermassive Black Hole. *Astrophys. J.* **1998**, *497*, 563–572. [CrossRef]
34. Ford, A.L.; Keenan, B.D.; Medvedev, M.V. Electron-positron cascade in magnetospheres of spinning black holes. *Phys. Rev. D* **2018**, *98*, 063016.
35. Levinson, A.; Cerutti, B. Particle-in-cell simulations of pair discharges in a starved magnetosphere of a Kerr black hole. *A&A* **2018**, *616*, A184.
36. Chen, A.Y.; Yuan, Y.; Yang, H. Physics of Pair Producing Gaps in Black Hole Magnetospheres. *Astrophys. J.* **2018**, *863*, L31.
37. Parfrey, K.; Philippov, A.; Cerutti, B. First-Principles Plasma Simulations of Black-Hole Jet Launching. *Phys. Rev. Lett.* **2019**, *122*, 035101.
38. Mościbrodzka, M.; Gammie, C.F.; Dolence, J.C.; Shiokawa, H. Pair Production in Low-luminosity Galactic Nuclei. *Astrophys. J.* **2011**, *735*, 9.
39. Broderick, A.E.; Tchekhovskoy, A. Horizon-scale Lepton Acceleration in Jets: Explaining the Compact Radio Emission in M87. *Astrophys. J.* **2015**, *809*, 97.
40. Wong, G.N.; Ryan, B.R.; Gammie, C.F. Pair Drizzle around Sub-Eddington Supermassive Black Holes. *Astrophys. J.* **2021**, *907*, 73.
41. Ryan, B.R.; Dolence, J.C.; Gammie, C.F. General Relativistic Radiation Magnetohydrodynamics with Monte Carlo Transport. *Astrophys. J.* **2015**, *807*, 31. [CrossRef]
42. Ryan, B.R.; Ressler, S.M.; Dolence, J.C.; Tchekhovskoy, A.; Gammie, C.; Quataert, E. The Radiative Efficiency and Spectra of Slowly Accreting Black Holes from Two-temperature GRRMHD Simulations. *Astrophys. J.* **2017**, *844*, L24.
43. Ryan, B.R.; Ressler, S.M.; Dolence, J.C.; Gammie, C.; Quataert, E. Two-temperature GRRMHD Simulations of M87. *Astrophys. J.* **2018**, *864*, 126.
44. Wardle, J.F.C.; Homan, D.C. Theoretical Models for Producing Circularly Polarized Radiation in Extragalactic Radio Sources. *Astrophys. Space Sci.* **2003**, *288*, 143–153. [CrossRef]
45. Dexter, J. A public code for general relativistic, polarised radiative transfer around spinning black holes. *Mon. Not. R. Astron. Soc.* **2016**, *462*, 115–136.
46. Blandford, R.; Anantua, R. The Future of Black Hole Astrophysics in the LIGO-VIRGO-LPF Era. *J. Phys. Conf. Ser.* **2017**, *840*, 012023. [CrossRef]
47. McKinney, J.C.; Tchekhovskoy, A.; Blandford, R.D. General relativistic magnetohydrodynamic simulations of magnetically choked accretion flows around black holes. *Mon. Not. R. Astron. Soc.* **2012**, *423*, 3083–3117. [CrossRef]
48. Anantua, R.; Dúran, J.; Ngata, N.; Oramas, L.; Emami, R.; Ricarte, A.; Curd, B.; Röder, J.; Broderick, A.; Wayland, J.; et al. Emission Modeling in the EHT-ngEHT Age. *Galaxies* **2023**, *11*, 4.
49. Doeleman, S.S.; Weintroub, J.; Rogers, A.E.E.; Plambeck, R.; Freund, R.; Tilanus, R.P.J.; Friberg, P.; Ziurys, L.M.; Moran, J.M.; Corey, B.; et al. Event-horizon-scale structure in the supermassive black hole candidate at the Galactic Centre. *Nature* **2008**, *455*, 78–80. [CrossRef]
50. Doeleman, S.S.; Fish, V.L.; Schenck, D.E.; Beaudoin, C.; Blundell, R.; Bower, G.C.; Broderick, A.E.; Chamberlin, R.; Freund, R.; Friberg, P.; et al. Jet-Launching Structure Resolved Near the Supermassive Black Hole in M87. *Science* **2012**, *338*, 355. [CrossRef]

51. Prieto, M.A.; Fernández-Ontiveros, J.A.; Markoff, S.; Espada, D.; González-Martín, O. The central parsecs of M87: jet emission and an elusive accretion disc. *Mon. Not. R. Astron. Soc.* **2016**, *457*, 3801–3816. [CrossRef]
52. Prather, B.; Wong, G.; Dhruv, V.; Ryan, B.; Dolence, J.; Ressler, S.; Gammie, C. iharm3D: Vectorized General Relativistic Magnetohydrodynamics. *J. Open Source Softw.* **2021**, *6*, 3336. [CrossRef]
53. Ricarte, A.; Qiu, R.; Narayan, R. Black hole magnetic fields and their imprint on circular polarization images. *Mon. Not. R. Astron. Soc.* **2021**, *505*, 523–539.
54. Mościbrodzka, M.; Gammie, C.F. IPOLE - semi-analytic scheme for relativistic polarized radiative transport. *Mon. Not. R. Astron. Soc.* **2018**, *475*, 43–54.

Article

How Spatially Resolved Polarimetry Informs Black Hole Accretion Flow Models

Angelo Ricarte [1,2,*], **Michael D. Johnson** [1,2], **Yuri Y. Kovalev** [3,4,5], **Daniel C. M. Palumbo** [1,2] **and Razieh Emami** [1]

1 Center for Astrophysics | Harvard & Smithsonian, 60 Garden Street, Cambridge, MA 02138, USA
2 Black Hole Initiative, 20 Garden Street, Cambridge, MA 02138, USA
3 Max–Planck–Institut für Radioastronomie, Auf dem Hügel 69, 53121 Bonn, Germany
4 Lebedev Physical Institute of the Russian Academy of Sciences, Leninsky Prospect 53, 119991 Moscow, Russia
5 Moscow Institute of Physics and Technology, Institutsky per. 9, 141700 Dolgoprudny, Russia
* Correspondence: angelo.ricarte@cfa.harvard.edu

Abstract: The Event Horizon Telescope (EHT) Collaboration has successfully produced images of two supermassive black holes, enabling novel tests of black holes and their accretion flows on horizon scales. The EHT has so far published total intensity and linear polarization images, while upcoming images may include circular polarization, rotation measure, and spectral index, each of which reveals different aspects of the plasma and space-time. The next-generation EHT (ngEHT) will greatly enhance these studies through wider recorded bandwidths and additional stations, leading to greater signal-to-noise, orders of magnitude improvement in dynamic range, multi-frequency observations, and horizon-scale movies. In this paper, we review how each of these different observables informs us about the underlying properties of the plasma and the spacetime, and we discuss why polarimetric studies are well-suited to measurements with sparse, long-baseline coverage.

Keywords: interferometry; polarimetry; black holes; magnetohydrodynamics; radiative transfer; accretion; Messier 87; Sagittarius A*

Citation: Ricarte, A.; Johnson, M.D.; Kovalev, Y.Y.; Palumbo, D.C.M.; Emami, R. How Spatially Resolved Polarimetry Informs Black Hole Accretion Flow Models. *Galaxies* **2023**, *11*, 5. https://doi.org/10.3390/galaxies11010005

Academic Editor: Phil Edwards

Received: 4 November 2022
Revised: 19 December 2022
Accepted: 22 December 2022
Published: 25 December 2022

1. Simulating Black Hole Accretion Flows

The Event Horizon Telescope (EHT) collaboration has produced the first images of supermassive black holes (SMBHs), ushering in a new era of spatially resolved astrophysics at the event horizon [1–14]. The images have been very constraining for general relativistic magnetohydrodynamics (GRMHD) models, which evolve plasma in a Kerr spacetime under the assumptions of ideal MHD. EHT science has focused mainly on constraining three free parameters: spin, the magnetic field state, and $R_{\rm high}$, which is related to the ion-to-electron temperature ratio [5,8,13]. The SMBH spin, which we will denote as $a_\bullet$, is the dimensionless angular momentum of a SMBH described by a Kerr metric that can vary between $|a_\bullet| \in [0, 1)$. A SMBH's spin reflects its recent assembly history and affects its accretion and feedback processes see [15]. Meanwhile, the accretion flow's magnetic field structure may vary between "MAD" and "SANE" states. In a Magnetically Arrested Disk (MAD), the magnetic flux at the horizon saturates, and the magnetic fields grow dynamically important, resulting in azimuthal asymmetries including flux eruption events [16–18]. This contrasts with "Standard and Normal Evolution" (SANE), where the magnetic fields remain turbulent and dynamically unimportant [19,20]. Finally, the ratio of the ion to electron temperature in different regions is highly uncertain, since the mean free path of the particles is much larger than the size of the system, and ions are heated more efficiently [21–23]. EHT studies have encapsulated this uncertainty with the post-processing parameters $R_{\rm low}$ and $R_{\rm high}$, which describe the asymptotic ion to electron temperature ratio at low and high plasma β, respectively, [24], where β is the ratio of gas to magnetic pressure. Less thoroughly studied parameters include the electron distribution function (eDF) [13,25–28], the detailed particle composition of the plasma [29–31], and the tilt of a potentially misaligned disk [32,33], which are the subject of many recent and ongoing studies.

In this paper, we review how properties of the spacetime and the plasma become imprinted onto multifrequency polarimetric observables accessible to the EHT and ngEHT. In Figure 1, we plot a single GRMHD snapshot, ray-traced at 214 and 228 GHz with IPOLE [34] and scaled to Sagittarius A*. This particular model is a MAD model with $a_\bullet = 0$, $R_{\mathrm{high}} = 40^1$, and a viewing angle of $50°$. To date, the EHT has produced total intensity maps for both M87* and Sgr A* and a linear polarization map of M87*. The remaining maps have yet to be generated for EHT data, and will be explained in detail in this paper. In brief, circular polarization may arise from both Faraday conversion and intrinsically emitted synchrotron, e.g., [35], and is especially sensitive to the details of the underlying magnetic field geometry [36–38] and plasma composition [29,30,39]. Meanwhile, the spectral index is the logarithmic derivative of the flux or intensity with respect to frequency, $\alpha \equiv d \log I / d \log \nu$; $I \propto \nu^\alpha$, which helps break degeneracies between number density, temperature, and magnetic field strength [40,41]. Finally, rotation measure is the derivative of the electric vector position angle (EVPA or χ) with respect to observing wavelength squared RM $\equiv d\chi / d\lambda^2$, which encodes Faraday rotation. Since colder electrons Faraday rotate more efficiently, the RM offers a glimpse into a colder population of electrons, which may exist at large number densities but may be too cold to contribute significantly to the intensity [42–44]. As discussed in the other contributions to this special issue, the ngEHT will enable access to these observables with higher image dynamic range than currently published EHT results (increasing from $\sim$10 to $\gtrsim$10^3), improved spatial resolution (decreasing from about 20 μas to $\sim$10–15 μas), and time-resolved images of the dynamical activity in both M87* and Sgr A* over hundreds-to-thousands of gravitational timescales. This will result in movies of both the accretion disks and relativistic jets near SMBHs. In this article, we discuss what physical information each of these maps carry.

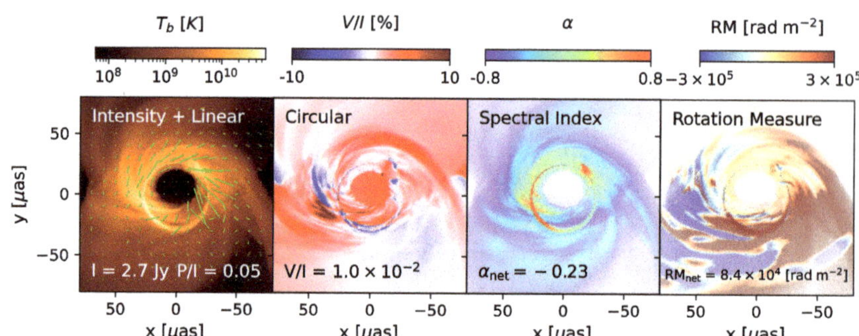

Figure 1. A single GRMHD snapshot ray-traced and scaled to Sgr A* properties, with three decades in dynamic range shown. In the leftmost panel, ticks representing the linear polarization have lengths scaled proportionally to the total linearly polarized intensity. To date, total intensity maps have been produced for both Sgr A* and M87*, a linear polarization map has been produced for M87*, and the remaining observables have yet to be generated for either source. In the era of ngEHT, we will have access to each of these observables with improved dynamic range and time-domain information, which will greatly inform models of the black hole accretion flow. Note that finite spatial resolution and other data corruptions have not been taken into account.

2. Total Intensity and Spectral Index

In the millimeter, we observe Sgr A* and M87* near the peak of emission from synchrotron radiation, where the flow transitions from optically thick to optically thin. Here, the emissivity scales approximately as $j_\nu \propto nB^2\Theta_e^{5/2}$, where n is the electron number density, B is the magnetic field strength, and Θ_e is the electron temperature^2 (in units of the electron rest mass energy). Each of these quantities can vary by orders of magnitude among different models, and thus even the total flux is informative for jointly constraining these parameters. To match the total flux of a given system, the fluid in ideal GRMHD simulations can be rescaled via $n \to \mathcal{M}n$, $B \to \sqrt{\mathcal{M}}B$, and $u \to \mathcal{M}u$, where u is the internal energy and $\mathcal{M}$ is a scalar. After doing so, both MAD and SANE simulations are capable of matching the total flux of EHT sources at a single frequency, as well as broad image characteristics such as the image size [5,13]. However, this rescaling causes SANE simulations to typically have orders of magnitude larger number density than MADs, due to their intrinsically weaker magnetic fields and lower temperatures [8,13]. Consequently, any additional observables sensitive to these variables immediately help break degeneracies and distinguish between models.

For example, the degeneracies between n, B, and Θ_e can be partially resolved with the spectral index, $\alpha \equiv d\log I / d\log \nu$. Spectral index is mainly sensitive to the optical depth τ_ν as well as the temperature and magnetic field strength in the combination $B\Theta_e^2$ (on which the critical synchrotron frequency is dependent), e.g., [45]. Ricarte et al. [41] show that GRMHD models span a wide range of spectral indices, and that SANE models typically exhibit more negative spectral indices than MADs at a fixed optical depth due to their lower temperatures. Although lacking by construction in ideal GRMHD simulations, spectral index can also provide insight into synchrotron cooling processes that affect the temperature.

In the ngEHT era, multi-frequency VLBI will enable spatially resolved spectral index maps. Since the most important parameters (Θ_e, B, and τ_ν) all decline with radius, spectral index maps should generically grow more negative as radius increases. Equivalently, the image becomes smaller as the frequency grows larger, e.g., [46]. One example from Ricarte et al. [41] is shown in Figure 2: a MAD simulation of Sgr A* with $a_\bullet = 0$, $R_{\mathrm{high}} = 40$, and a non-thermal "kappa" electron distribution function[3] with $\kappa = 5$ [47,48], inclined at $50°$. The true spectral index map across 214 to 228 GHz is shown in the top central panel, while a one-zone analytic prediction is shown in the top right, using a $\kappa = 5$ eDF combined with the plasma variables computed in the bottom row. To obtain the analytic prediction, each pixel is treated as a one-zone model using plasma properties computed via an emissivity-weighted average along the geodesic. In this simulation, both τ_ν and B decline with radius, but Θ_e stays in a relatively narrow range. Thus, the decline of α with radius can be attributed to a decline in τ_ν and B. Models also exhibit a generic spike in α in the photon ring, whose geodesics plunge into regions with strong magnetic fields and acquire a larger optical depth due to their longer path lengths in the emitting region.

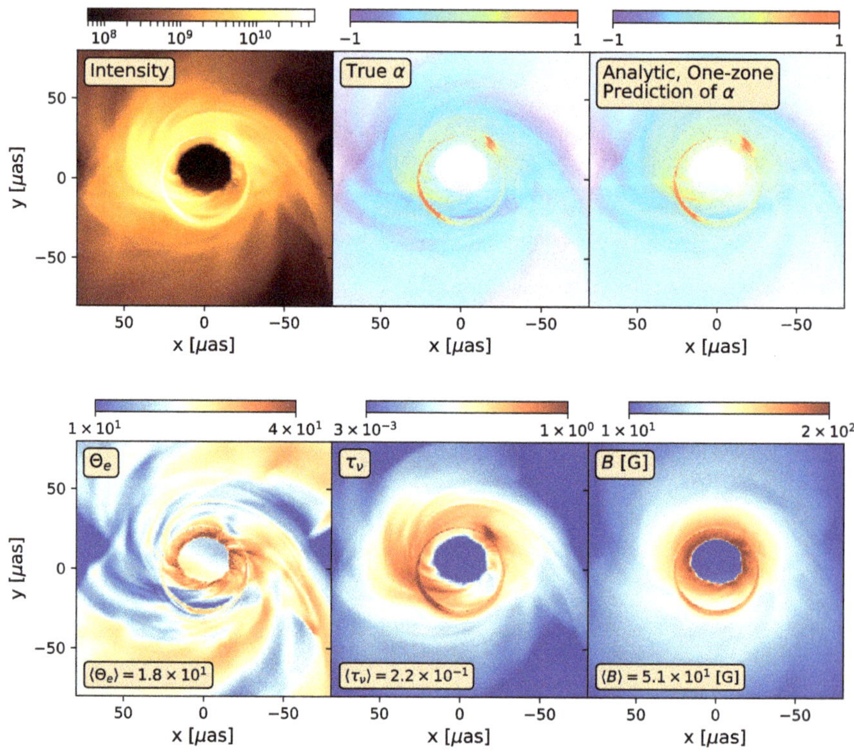

Figure 2. Intensity and spectral index map of a MAD model of Sgr A* adapted from Figure 4 of [41]. The top left panel plots total intensity in log scale averaged between 214 and 228 GHz, the top center panel plots the spectral index across this bandwidth calculated by ray tracing the image at two different frequencies, and the top right panel plots an analytic prediction of the spectral index in each pixel obtained by combining the three quantities in the bottom panel: electron temperature, optical depth, and magnetic field strength, each computed by performing an emissivity-weighted average long each geodesic. The excellent agreement between the true spectral index map and the analytic prediction illustrates the power of spectral index maps to jointly constrain these plasma quantities.

3. Linear Polarization

In flat space, a uniform parcel of synchrotron emitting optically thin plasma intrinsically produces a linear polarization fraction of $\approx 70\%$, with an orientation perpendicular to its magnetic field projected onto the plane perpendicular to the photon wave-vector, e.g., [49]. The observed orientation of this linear polarization gets modified by two effects: achromatic effects of propagation through the curved space-time, and chromatic "Faraday" effects of propagation through a magnetized plasma (recent theoretical investigations in the vicinity of a black hole include [50–53]). Both rotate and potentially scramble the electric vector position angle (EVPA, or χ). Thus, resolved images of linear polarization can inform us about the magnetic field geometry, the magnitude of Faraday effects, and potentially the space-time itself. In Figure 3, we plot the linear polarization pattern of a ring of emission moving clockwise on the sky around a Schwarzschild black hole using the model of Narayan et al. [51]. Here, only the direct ($n = 0$) sub-image is included. The ring is located at a radius of $4.5\,GM_\bullet/c^2$ and has a Gaussian width of $2\,GM_\bullet/c^2$. The angular momentum vector of the ring projected onto the sky has a position angle of $288°$ East of North and is viewed at an inclination of $17°$, consistent with that of M87* [54]. In the toroidal and radial field cases, the resulting linear polarization pattern is simply perpendicular to the magnetic fields projected onto the line-of-sight. In the idealized vertical case, the EVPA pattern becomes more sensitive to the direction of the fluid's motion [8].

Figure 3. Polarization pattern of a ring of emission around a Schwarzschild black hole threaded with magnetic fields of different geometries: toroidal, radial, and vertical adapted from Figure 3 of [8]. The toroidal and radial magnetic field cases clearly illustrate the fact that synchrotron emission is polarized perpendicular to the magnetic field projected onto the sky. The orientation of the ticks in the vertical field case encodes the direction of the fluid's motion [8], chosen here to be clockwise on the sky. These maps were computed using the analytic ring model of Narayan et al. [51]. Here, the color map encodes the total intensity, and unlike in Figure 1, the linear polarization ticks do not scale with the polarized intensity.

EHT studies have identified the linear polarization fraction (on both resolved and unresolved scales) as well as the morphology of polarization ticks as important observables for theoretical interpretation. This "twistiness" can be quantified by the complex number β_2, the rotationally symmetric component of a Fourier decomposition of the polarization pattern [55]. M87* and Sgr A* both exhibit percent level spatially-unresolved linear polarization fractions at 1.3 mm and $\sim$10–20% level resolved polarization fractions, e.g., [7,56], much lower than the ideal value of 70% for a uniform parcel of emitting plasma, e.g., [57–59], as do other AGN observed with cm-VLBI, e.g., [60,61]. This could be due to both Faraday depolarization and disordered magnetic field structures. For the EHT observations of M87*, Event Horizon Telescope Collaboration et al. [8] found that the low polarization fraction was attributable mainly to Faraday rotation in GRMHD models. Combining resolved linear polarization information with an upper limit on circular polarization, Event Horizon Telescope Collaboration et al. [8] found that MAD models were favored over their SANE counterparts for M87*, which could not be concluded based on total intensity alone. Fundamentally, this can be attributed to linear polarization's sensitivity to the geometry of the magnetic field, as well as Faraday rotation's sensitivity to cooler electrons that may otherwise be invisible. MAD models tend to have more ordered fields with stronger poloidal components, which produces twistier polarization patterns. Meanwhile, SANE models tend to require orders of magnitude larger mass density to compensate for their intrinsically weaker magnetic fields and lower temperatures, resulting in much greater Faraday depths. In retrograde systems, images can exhibit flips in the handedness of the polarization spiral with radius, quantified as a sign flip in the imaginary component of β_2. This results from a flip in the angular velocity of inflowing streams due to frame dragging [62].

Sgr A* also exhibits interesting time variability in linear polarization, especially during flares, which are accompanied by large polarization fractions, swings in polarization angle, and "Q-U loops" on the timescale of hours [63–67]. These can be interpreted as the motion of hotspots or other structures as they light up different parts of the magnetic field structure during their orbit [52,68–72]. The hotspots themselves may originate from "flux eruption events" and magnetic reconnections that occur naturally in MAD accretion flows [73–75].

Thus, time variability of linear polarization offers unique insights into the magnetic field structure and direction of orbital motion that could potentially be linked to the inclination and spin of Sgr A*. The GRAVITY Collaboration has detected centroid motion coincident with a flare [66,76]. Spatially resolved movies created by the ngEHT would help test the hotspot interpretation, motivating high-cadence monitoring of this source.

4. Rotation Measure

In an ionic plasma, circularly polarized waves of opposite handedness propagate at different speeds, resulting in a circular birefringence effect known as Faraday rotation. The EVPA of propagating emission rotates an amount sensitive to the density, temperature, and line-of-sight magnetic field. As examined in several studies, internal Faraday rotation is important for depolarizing and scrambling images of GRMHD models of black hole accretion flows [8,42–44]. The magnitude of Faraday rotation has a wavelength-squared dependence, thus it is useful observationally to define the rotation measure $\mathrm{RM} = d\chi/d\lambda^2$, which offers insights into physical parameters of the Faraday rotating plasma. For a linearly polarized emitter entirely behind a uniform Faraday screen, the RM is related to the properties of the screen via

$$\mathrm{RM} = 8.1 \times 10^5 \ \mathrm{rad \ m^{-2}} \int_{\mathrm{source}}^{\mathrm{observer}} f_{\mathrm{rel}}(\Theta_e) \frac{n_e}{1 \ \mathrm{cm^{-3}}} \frac{B_{||}}{G} \frac{ds}{\mathrm{pc}}, \tag{1}$$

where n_e is the electron number density, $B_{||}$ is the component of the magnetic field parallel to the photon wave-vector, and f_{rel} is a correction term suppressing Faraday rotation at relativistic temperatures [77]. For relativistic plasmas, $f_{\mathrm{rel}}(\Theta_e) \approx \log(\Theta_e)/(2\Theta_e^2)$, while f_{rel} asymptotes to 1 as $\Theta_e \to 0$. Here $\Theta_e \equiv k_B T_e/m_e c^2$, k_B is the Boltzmann constant, T_e is the electron temperature, m_e is the electron rest mass, and c is the speed of light [78]. In GRMHD models, the plasma responsible for synchrotron emission is sometimes completely separate from the plasma responsible for Faraday rotation. For example, some large R_{high} SANE models exhibit a cold Faraday rotating midplane sandwiched between emission from their hot jet sheaths [44]. Moreover, while the EVPA probes the magnetic field as projected onto the sky, Faraday rotation is sensitive to the magnetic field along the line-of-sight. Hence, rotation measure and linear polarization can offer a view into electron populations that may otherwise be undetectable from total intensity alone.

At the time of writing, narrow observing bandwidths inhibit our ability to create spatially resolved rotation measure maps with currently published EHT data, but spatially unresolved RM measurements at millimeter wavelengths exist for Sgr A*, the core of M87*, and several other LLAGN, e.g., [79–84]. Note also that rotation measure from AGN generally increases with increasing frequency, e.g., [85] and can reach values of the order of $10^7 \mathrm{rad \ m^{-2}}$ [86] due to the opacity effect probing regions close to the central engine at the ngEHT frequencies. Without spatial resolution, unresolved rotation measure measurements are difficult to interpret because the assumptions underlying Equation (1) are not believed to generally hold. In GRMHD models, Faraday rotation occurs co-spatially with the plasma, can vary by orders of magnitude in different locations, and can also flip sign across the image due to turbulence [44]. As a result, unresolved EVPA measurements may exhibit significant temporal variation and not strictly follow a λ^2 law. Figure 4 plots a rotation measure map of a MAD GRMHD model of M87*, with the spatially unresolved RM written at the bottom of each panel. This turbulence can explain the intra-week time variability of the RM observed for M87* [59]. On the other hand, Sgr A* has exhibited a constant sign of RM for decades, suggesting the existence of a more stable (but still variable) foreground Faraday screen [84]. Spatially resolved rotation measure maps could help disentangle the Faraday screen and give insights into both the turbulence of the accretion flow and the magnetic field structure of jets. This may be of increased importance, since EHT observations of Sgr A* indicate that GRMHD models are too variable [13].

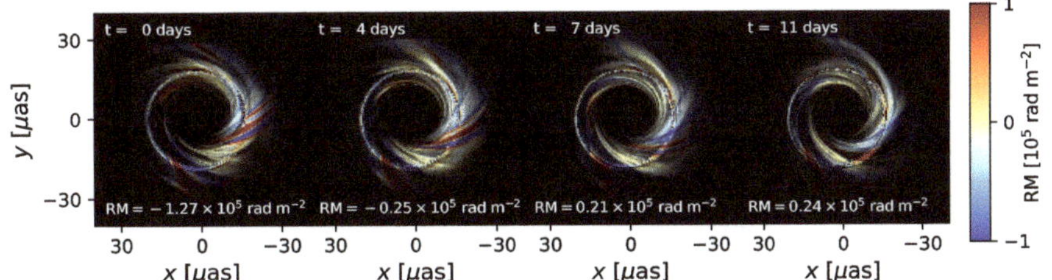

Figure 4. Rotation measure map of a MAD simulation of M87* adapted from Figure 13 of Ricarte et al. [44]. Both positive and negative RM regions are simultaneously present, reflecting flips in the line-of-sight magnetic field direction due to turbulence in the accretion flow. The motion of these structures produces a time variable spatially unresolved RM, written at the bottom of each panel.

5. Circular Polarization

Circular polarization, Stokes V, can be generated both intrinsically through synchrotron emission or through Faraday conversion, which exchanges linear and circular polarization states, e.g., [35,78]. Circular polarization fractions are much lower than their linear counterparts for both Sgr A* ($V/I \approx -1\%$) and M87* ($|V/I| \lesssim 0.8\%$), making it more challenging to study than linear polarization. In addition, the circular feed basis used for EHT sites makes it more challenging to construct circularly polarized images. However, Stokes V has the potential to encode not only the magnetic field direction and geometry, but also the plasma composition.

Unlike the near unity linear polarization fractions produced by a uniform parcel of plasma, intrinsically emitted Stokes V is typically produced at the $\sim 1\%$ level for plasma parameters appropriate for M87* or Sgr A*. The sign of intrinsically emitted circular polarization generated encodes the sign of the magnetic field along the line of sight, following the right hand rule. Stokes V can also be generated via Faraday conversion, which exchanges linear and circular polarization states. The sign of circular polarization generated by conversion depends on the relative orientation of the EVPA with respect to the local magnetic field. Specifically, Stokes V generated by conversion inherits the sign of Stokes U, defined in the local plasma frame. Thus, Stokes V from Faraday conversion is sensitive to the line of sight "twist" in the magnetic field as well as any Faraday rotation affecting the EVPA of the linear polarization that gets converted. Interestingly, Faraday conversion has no effect in a unidirectional magnetic field lacking Faraday rotation.

Both intrinsic synchrotron emission and Faraday conversion are believed to be important for generating circular polarization on event horizon scales [36–38]. For Faraday conversion, both the line of sight twist in the magnetic field and Faraday rotation are important for determining the orientation of the linear polarization that is converted into circular. Small amounts of Faraday rotation can affect the relative alignment between linear polarization and the converting magnetic field. Large amounts of Faraday rotation can lead to depolarization by randomizing the sign of Stokes U that is converted into Stokes V.

Figure 5 depicts a cartoon of a typical helical field geometry as well as the circular polarization produced by a model that exhibits this structure reproduced from [37]. In the left panel, the background colors depict the line of sight direction of the magnetic field, viewed edge-on. In the time averaged image of V/I viewed at $90°$, this structure produces a "four quadrants" pattern. Outside the immediate vicinity of the event horizon, Stokes V originates from intrinsic emission in this model, which reflects the line-of-sight direction of the magnetic field see also [38]. Another interesting feature that arises due to a generic helical field geometry is the successive sign flipping of sub-images in the photon ring also discussed in [36], which can be explained by Faraday conversion and parallel transport in a generic helical field geometry viewed face-on [37]. In this particular model, the spatially unresolved Stokes V is surprisingly dominated by this sign-flipped photon ring.

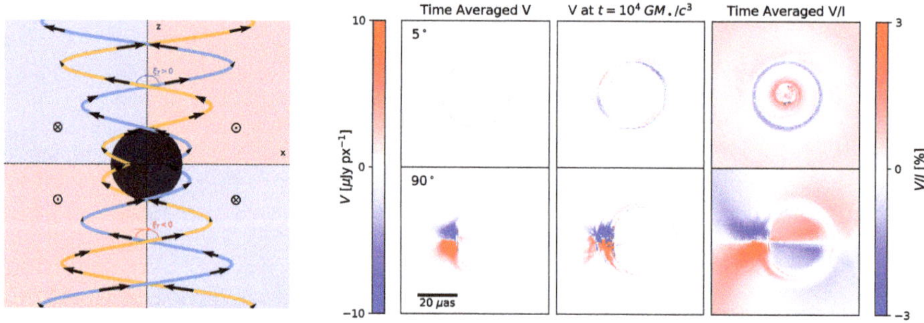

Figure 5. Maps of circular polarization encode properties of the geometry of the magnetic field, both its line-of-sight direction and twist. A cartoon of a generic helical field geometry is depicted on the left. On the right, we plot the circular polarization of a MAD model of M87* at two inclinations. Both are reproduced from Ricarte et al. [37]. The top row depicts a 5° viewing angle, and the bottom row depicts a 90° viewing angle. The first column shows the time averaged circularly polarized image, the second column shows the same at a single snapshot, and the third column shows fractional circular polarization. For face-on viewing angles, the photon ring exhibits an interesting sign flip due to Faraday conversion and the sourcing of photons from the opposite side of the disk. For edge-on viewing angles, circular polarization exhibits a "four quadrants" pattern that reflects the line-of-sight magnetic field direction.

Finally, circular polarization is strongly affected by plasma composition, and can potentially be used to distinguish pair plasmas from ionic plasmas, e.g., [39]. Most models used to study EHT images have contained only ionic plasma. Electron-positron pairs can also be naturally produced on event horizon scales, but their abundance is highly theoretically uncertain. In a pair plasma, intrinsic circularly polarized emission and Faraday rotation both vanish, but Faraday conversion persists. Intuitively, a pair plasma with equal parts positively and negatively charged particles should not gyrate in a preferred direction. This can cause dramatic differences in images of circular polarization, and potentially also those of linear polarization [29,30].

6. Scattering

A major challenge for studies of Sgr A* with the EHT and ngEHT is interstellar scattering by dilute plasma in the ionized interstellar medium. This scattering arises from variations in the index of refraction of the plasma because of fluctuations in the electron number density see, e.g., [87]. In particular, the line of sight to Sgr A* is heavily scattered by plasma in the spiral arms of the Milky Way [88], resulting in angular broadening that is approximately three orders of magnitude larger than median values for lines of sight at higher galactic latitudes. The effects of scattering are two-fold: (1) small-scale modes in the scattering material result in diffractive "blurring," described by a convolution with an anisotropic kernel, and (2) large-scale modes in the scattering material result in refractive "substructure," described by additive image noise with a slowly falling power spectrum. For detailed discussion of the scattering of Sgr A*, see Psaltis et al. [89], Johnson et al. [90,91].

While the scattering severely affects images of Sgr A*, many polarimetric properties of the images are comparatively immune because the scattering is not significantly birefringent. For example, in the case of purely diffractive scattering, the image-integrated fractional polarization is independent of scattering. More generally, the interferometric fractional polarization $\breve{m}(\mathbf{u}) \equiv \tilde{P}(\mathbf{u})/\tilde{I}(\mathbf{u})$ is independent of diffractive scattering because convolution is multiplicative in the visibility domain and is identical for all Stokes parameters, thereby canceling in the quotient. Here, $\mathbf{u}$ is the dimensionless baseline vector, projected orthogonal to the line of sight and measured in units of the observing wavelength, which corresponds to the sampled image wavevector in Fourier space; $\tilde{F}$ denotes the Fourier transform of some quantity F.

Figure 6 shows an example GRMHD snapshot before and after scattering, while the images look substantially different, key polarimetric observables such as the β_2 mode, which is highly constraining for GRMHD models [8] and carries information about black hole spin [55], are almost unaffected by scattering. Likewise, certain interferometric observables, such as the interferometric fractional polarization, are only mildly affected by scattering (see Figure 7). For additional discussion of how the deterministic frequency dependence of scattering can be used for scattering mitigation on images, see [56]; for discussion of how the lack of birefringence can be used to study the relative power spectra in different polarization modes, see Ni et al. (in prep).

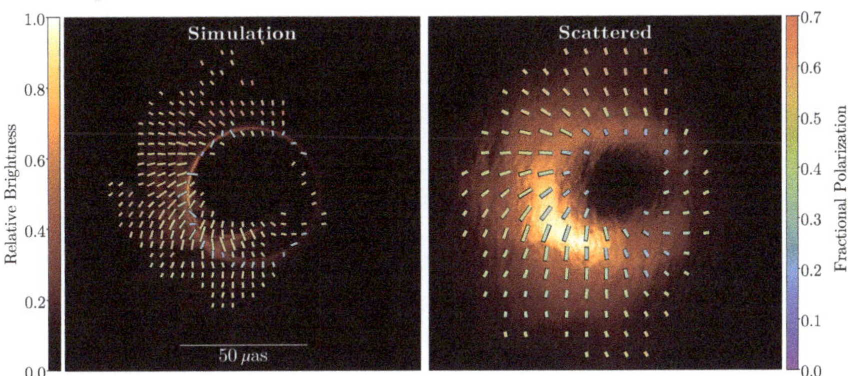

Figure 6. GRMHD model of Sgr A* at 230 GHz before (**left**) and after (**right**) including the effects of interstellar scattering. This simulation is a MAD with $a_\bullet = 0.7$, $R_{\text{high}} = 20$, and $i = 30°$. The background image shows total intensity with respect to the image peak, while the ticks show the polarization magnitude and direction, colored by fractional polarization, while scattering severely affects the image, key polarimetric measures are nearly immune to scattering. For example, the unresolved fractional polarization is 10.5% before scattering and is 10.6% after scattering. Likewise, the β_2 mode in polarization [55] has $|\beta_2| = 0.40$ and $\arg(\beta_2) = 52.1°$ before scattering, and $|\beta_2| = 0.37$ and $\arg(\beta_2) = 51.0°$ after scattering.

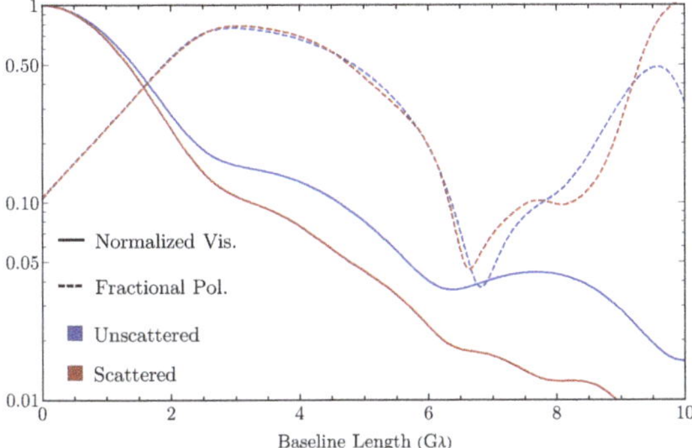

Figure 7. Interferometric properties of the GRMHD model shown in Figure 6. Solid lines show the normalized intensity $|\tilde{I}(\mathbf{u})/\tilde{I}(\mathbf{0})|$ before (blue) and after (red) scattering, and the dashed lines show the interferometric fractional polarization magnitude $|\tilde{m}(\mathbf{u})|$. For these curves, baselines are oriented along the East–West direction: $\mathbf{u} = (u, 0)$. Over the full range of baseline lengths accessible from the ground, the fractional polarization is largely immune to scattering, while diffractive scattering causes a substantial reduction in the flux on long baselines.

7. Studying Polarimetry with Interferometry

Thus, far, the discussion has focused on the polarimetric properties of simulated black hole images or images reconstructed from interferometric visibilities. In practice, since the measured visibilities are actually samples of the Fourier transform, image reconstruction can introduce significant systematic uncertainties. Reconstruction methods must find images consistent with incomplete and noisy information in Fourier space, to which there can be multiple families of solutions. Images cannot be constructed at all without sufficient uv coverage (or strict image priors). Thus, it can be useful to study signals in their native visibility space.

The visibility-domain response of polarimetric observables has some key differences from that of total intensity. For instance, the visibility amplitude for total intensity is guaranteed to be maximal for the zero baseline because the image is positive. However, because the Stokes parameters Q, U, and V are not constrained to be positive, their visibility functions may not be maximal on the zero baseline. This simple property can be used to make powerful inferences from sparse measurements (e.g., from the EHT or the space VLBI project *RadioAstron*). For instance, a single measurement of $|\tilde{V}(\mathbf{u})| > |\tilde{V}(\mathbf{0})|$ would demonstrate that the image does not have uniform *sign* of circular polarization. Likewise, if $|\tilde{P}(\mathbf{u})| \neq |\tilde{P}(-\mathbf{u})|$, where $\tilde{P}(\mathbf{u}) \equiv \tilde{Q}(\mathbf{u}) + i\tilde{U}(\mathbf{u})$, then the linear polarization field *must* have variations in direction. This test can be performed with as few as two stations, but each must have dual polarization receivers. In addition, because polarization breaks the trivial baseline reflection degeneracy that is present in total intensity, the polarimetric structure can be compared on baselines $\mathbf{u}$ and $-\mathbf{u}$, effectively doubling the angular resolution relative to total intensity alone!

Another significant difference between the total intensity and polarization is that the linear polarization images can have changes in both amplitude and *direction*, allowing it to have significant image substructure relative to the intensity image. In the visibility domain, this substructure translates to a relative increase in the power on long baselines in polarization versus total intensity. In the limit of a heavily resolved source, the polarimetric signal may exceed that of the total intensity, even for a source with a low fractional polarization!

This can be quantified using the interferometric fractional polarization, $\breve{m}(\mathbf{u}) \equiv [\tilde{Q}(\mathbf{u}) + i\tilde{U}(\mathbf{u})]/\tilde{I}(\mathbf{u})$. On a zero baseline, this complex quantity corresponds to the unresolved fractional polarization, $\breve{m}(\mathbf{0}) = (Q_\text{tot} + iU_\text{tot})/I_\text{tot}$, where the "tot" subscript denotes an image-integrated quantity. However, unlike the image fractional polarization, $|\breve{m}(\mathbf{u})|$ can exceed unity on long baselines. This was found out from ground-based observations of for Sgr A* [92] as well as within the *RadioAstron* Space VLBI survey of AGN [93]. These observations had one common feature: a very high angular resolution corresponding to several Gλ spatial scales. In general, for a heavily resolved source with polarized substructure, we expect $|m(\mathbf{u})|$ to generically grow with increasing baseline length.

From a calibration perspective, the interferometric fractional polarization has the benefit of properties analogous to VLBI closure quantities, since the source of rapidly varying gains at VLBI sites (e.g., changing atmospheric delay and reference frequency errors) are equivalent for both polarization feeds. In addition, the interferometric fractional polarization is resilient to the effects of interstellar scattering, which is likewise not significantly birefringent (see Figure 7). Finally, $|m(\mathbf{u})|$ is a useful observable to measure the relative coherence of the polarization field. For a perfectly uniform polarization field, $|m(\mathbf{u})|$ will be independent of baseline length. However, for a polarization field that varies on spatial scales much smaller than the the beam size of the interferometer, $|m(\mathbf{u})|$ will grow roughly as $1/|\tilde{I}(\mathbf{u})|$, as the observations resolve the structure in total intensity without resolving the structure in polarization.

In Figure 8, we plot total intensity and linear polarization maps of two models of M87* with very different polarization characteristics. The top model is a MAD simulation with $a_\bullet = 0.9$ and $R_\text{high} = 1$, which exhibits an ordered polarization pattern due to ordered magnetic fields and little Faraday rotation. Meanwhile, the bottom model is a

SANE simulation with $a_{\bullet} = -0.3$ and $R_{\text{high}} = 40$. As mentioned in Section 4, SANE simulations tend to have much larger Faraday depths than MADs, causing this model's linear polarization to be much more disordered. The pixel-to-pixel intensity weighted Faraday depth $\langle \tau_F \rangle$ is written at the bottom of each panel.

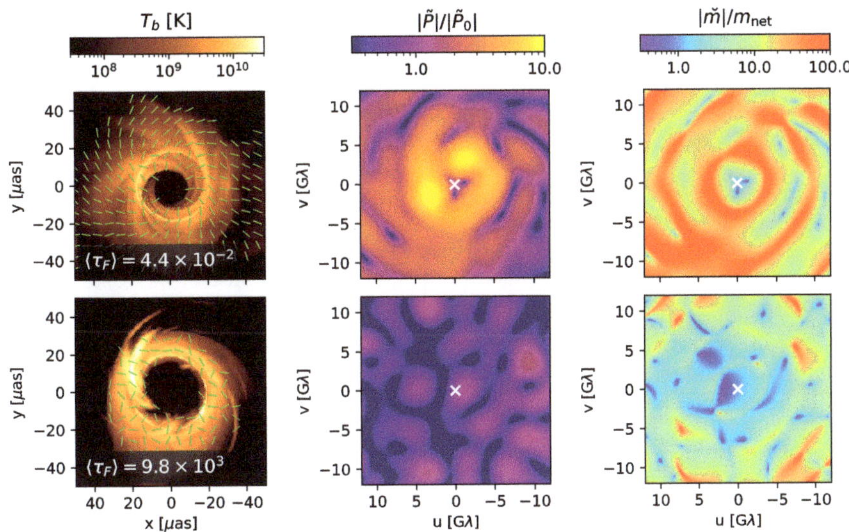

Figure 8. Two GRMHD models imaged at 228 GHz and corresponding maps of linear polarization in visibility space. The top row corresponds to a MAD model of M87* with $a_{\bullet} = 0.9$ and $R_{\text{high}} = 1$, while the bottom row corresponds to a SANE model with $a_{\bullet} = -0.3$ and $R_{\text{high}} = 40$. Due to a much larger Faraday depth, written at the bottom of the images, the SANE model exhibits a much more disordered linear polarization pattern. In the ordered model, measures of the linear polarization rise dramatically with radius in the Fourier domain, while the disordered model is characterized by blobs with a coherence length corresponding to the size of the image.

These characteristics are reflected in their Fourier space maps of $|\tilde{P}|$ and $|\tilde{m}|$, shown in the second and third columns, which can be directly sampled using an interferometer. Here, $\tilde{P} \equiv \tilde{Q} + i\tilde{U}$ and $\tilde{m} \equiv \tilde{P}/\tilde{I}$, where $\sim$ denotes a Fourier transform. For the MAD model, $\tilde{P}$ and $\tilde{m}$ both rise dramatically with radius in Fourier space in the central region. $\tilde{P}$ rises because the linear polarization is higher on resolved scales than a spatially unresolved measurement would suggest. The rotational symmetry of the polarization pattern causes substantial cancellation of polarization without spatial resolution. $\tilde{m}$ also rises for this reason, and also because $\tilde{I}$ exhibits nulls in Fourier space that do not necessarily coincide with the nulls in $\tilde{P}$. Meanwhile, the disordered SANE simulation exhibits a mottled pattern in $\tilde{P}$ with a characteristic length scale corresponding to the size of the image. These phenomena should not change much qualitatively as a function of wavelength in the sub-millimeter.

The previous discussion has focused on the relationship between expected image features and their appearance in the (Fourier-conjugate) interferometric visibility domain. However, a crucial consideration is how to study frequency-dependent effects, such as spectral index and rotation measure, using interferometry. Because the fringe spacing $u \propto \nu$, interferometric measurements across multiple frequencies necessarily mix the effects of a changing dimensionless baseline with those of a changing image.

Specifically, the polarized visibility $\tilde{P}$ on a physical vector baseline $\mathbf{b}$ at an observing frequency ν is

$$\tilde{P}(\mathbf{b}, \nu) = \int d^2\boldsymbol{\theta}\, P(\boldsymbol{\theta}, \nu) e^{-2\pi i \frac{\nu}{c}\boldsymbol{\theta}\cdot\mathbf{b}} \tag{2}$$

$$\Rightarrow \nu\partial_\nu \tilde{P}(\mathbf{b}, \nu) = \int d^2\boldsymbol{\theta}\, [\nu\partial_\nu P(\boldsymbol{\theta}, \nu) - 2\pi i\boldsymbol{\theta}\cdot\mathbf{u}P(\boldsymbol{\theta}, \nu)] e^{-2\pi i\boldsymbol{\theta}\cdot\mathbf{u}}, \tag{3}$$

where ∂_ν denotes a partial derivative with respect to ν. The first term in the square brackets of Equation (3) accounts for the frequency dependence of the image, while the second accounts for the changing dimensionless baseline with frequency. For the first term, an image with spectral index α has $P \propto \nu^\alpha$, so $|\nu\partial_\nu P| \sim |\alpha P|$; likewise, an image with finite rotation measure has $P \propto e^{2i\lambda^2 \mathrm{RM}}$, so $|\nu\partial_\nu P| \sim 4|\mathrm{RM} \times \lambda^2 P|$. Roughly speaking, we expect that the relative dominance or subdominance of spectral index versus RM are independent of baseline length, so the relative effects on long baselines are likely similar to those for unresolved measurements of a source. For instance, the effects of rotation measure for observations of Sgr A* at millimeter wavelengths are likely to vastly dominate the effects of spectral index. Sgr A* has $\alpha = 0.0 \pm 0.1$ [94] but $4 \times \mathrm{RM} \times \lambda^2 \approx -2.7$ [59]. The second term gives a relative contribution that increases as the image is increasingly resolved. It becomes significant when the spanned frequencies change the baseline length by the inverse field-of-view, F.

8. Discussion and Conclusions

The EHT and upcoming ngEHT enable us to probe accreting supermassive black holes via a variety of multi-frequency polarimetric observables. In this contribution, we have discussed the many ways in which the physical properties of underlying accretion flow are mapped onto these observables. Total intensity and spectral index encode the density, temperature, and magnetic field strength of emitting plasma in different regions. Linear polarization encodes the geometry of the magnetic field, and its depolarization via Faraday rotation offers an observational probe into what may be otherwise invisible cool electrons, and a different component of the magnetic field vectors. Rotation measure maps probe this cooler Faraday rotating electron population directly, and can probe the magnetic field direction, which can reveal turbulent structures. Finally, circular polarization encodes both overall geometry and direction of the magnetic field via emission, Faraday rotation, and Faraday conversion. We have discussed that even if imaging proves prohibitively challenging for some datasets, constraining information exists already in visibility space. For some models, low polarization fractions in spatially unresolved measurements hide large polarization fractions in spatially resolved measurements. In the ngEHT era, we will have access to not only single snapshots, but also movies, with a much larger dynamic range in intensity than is presently possible with the EHT. Multiple snapshots will also enable cleaner theoretical connections via time averaging, e.g., [95,96]. This will enable an unprecedented deluge of data about the nearest SMBHs that will help us understand their accretion and feedback processes.

Author Contributions: Conceptualization, A.R., M.D.J., and Y.Y.K.; methodology, A.R., M.D.J., Y.Y.K., and D.C.M.P.; software, A.R., M.D.J., and D.C.M.P.; investigation, A.R., M.D.J., and D.C.M.P.; writing—original draft preparation, A.R., M.D.J., and Y.Y.K.; writing—review and editing, A.R., M.D.J., Y.Y.K., D.C.M.P., and R.E.; visualization, A.R., M.D.J., and D.C.M.P. All authors have read and agreed to the published version of the manuscript.

Funding: We thank the National Science Foundation (AST-1716536, AST-1935980, AST-2034306, AST-1816420, and OISE-1743747) for financial support of this work. YYK was supported by the Russian Science Foundation grant 21-12-00241. RE acknowledges the support from NASA via grant HST-GO-16173.001-A. This work was supported in part by the Black Hole Initiative, which is funded by grants from the John Templeton Foundation and the Gordon and Betty Moore Foundation to Harvard University. The opinions expressed in this publication are those of the author(s) and do not necessarily reflect the views of the Moore or Templeton Foundations.

Data Availability Statement: The data underlying the figures of this article can be provided upon reasonable request to the corresponding author.

Conflicts of Interest: The authors declare no conflicts of interest.

Notes

[1] That is, the ion-to-electron temperature ratio asymptotes to a value of 40 in high-β regions, which typically occur in the mid-plane. This is a typical value in EHT GRMHD studies, e.g., [5,8,13].

[2] Θ_e is the temperature normalized by the electron rest mass energy, $\Theta_e = k_B T / m_e c^2$, where k_B is the Boltzmann constant, T is the temperature in Kelvin, m_e is the electron rest mass, and c is the speed of light.

[3] These distributions are characterized by a thermal core with the addition of a high energy power-law tail, with slope $p = \kappa - 1$.

References

1. Event Horizon Telescope Collaboration; Akiyama, K.; Alberdi, A.; Alef, W.; Asada, K.; Azulay, R.; Baczko, A.K.; Ball, D.; Baloković, M.; Barrett, J.; et al. First M87 Event Horizon Telescope Results. I. The Shadow of the Supermassive Black Hole. *ApJ* **2019**, *875*, L1. [CrossRef]
2. Event Horizon Telescope Collaboration; Akiyama, K.; Alberdi, A.; Alef, W.; Asada, K.; Azulay, R.; Baczko, A.K.; Ball, D.; Baloković, M.; Barrett, J.; et al. First M87 Event Horizon Telescope Results. II. Array and Instrumentation. *ApJ* **2019**, *875*, L2. [CrossRef]
3. Event Horizon Telescope Collaboration; Akiyama, K.; Alberdi, A.; Alef, W.; Asada, K.; Azulay, R.; Baczko, A.K.; Ball, D.; Baloković, M.; Barrett, J.; et al. First M87 Event Horizon Telescope Results. III. Data Processing and Calibration. *ApJ* **2019**, *875*, L3. [CrossRef]
4. Event Horizon Telescope Collaboration; Akiyama, K.; Alberdi, A.; Alef, W.; Asada, K.; Azulay, R.; Baczko, A.K.; Ball, D.; Baloković, M.; Barrett, J.; et al. First M87 Event Horizon Telescope Results. IV. Imaging the Central Supermassive Black Hole. *ApJ* **2019**, *875*, L4. [CrossRef]
5. Event Horizon Telescope Collaboration; Akiyama, K.; Alberdi, A.; Alef, W.; Asada, K.; Azulay, R.; Baczko, A.K.; Ball, D.; Baloković, M.; Barrett, J.; et al. First M87 Event Horizon Telescope Results. V. Physical Origin of the Asymmetric Ring. *ApJ* **2019**, *875*, L5. [CrossRef]
6. Event Horizon Telescope Collaboration; Akiyama, K.; Alberdi, A.; Alef, W.; Asada, K.; Azulay, R.; Baczko, A.K.; Ball, D.; Baloković, M.; Barrett, J.; et al. First M87 Event Horizon Telescope Results. VI. The Shadow and Mass of the Central Black Hole. *ApJ* **2019**, *875*, L6. [CrossRef]
7. Event Horizon Telescope Collaboration; Akiyama, K.; Algaba, J.C.; Alberdi, A.; Alef, W.; Anantua, R.; Asada, K.; Azulay, R.; Baczko, A.K.; Ball, D.; et al. First M87 Event Horizon Telescope Results. VII. Polarization of the Ring. *ApJ* **2021**, *910*, L12. [CrossRef]
8. Event Horizon Telescope Collaboration; Akiyama, K.; Algaba, J.C.; Alberdi, A.; Alef, W.; Anantua, R.; Asada, K.; Azulay, R.; Baczko, A.K.; Ball, D.; et al. First M87 Event Horizon Telescope Results. VIII. Magnetic Field Structure near The Event Horizon. *ApJ* **2021**, *910*, L13. [CrossRef]
9. Akiyama, K.; Alberdi, A.; Alef, W.; Algaba, J.C.; Anantua, R.; Asada, K.; Azulay, R.; Bach, U.; Baczko, A.K.; Ball, D.; et al. First Sagittarius A* Event Horizon Telescope Results. I. The Shadow of the Supermassive Black Hole in the Center of the Milky Way. *ApJ* **2022**, *930*, L12. [CrossRef]
10. Akiyama, K.; Alberdi, A.; Alef, W.; Algaba, J.C.; Anantua, R.; Asada, K.; Azulay, R.; Bach, U.; Baczko, A.K.; Ball, D.; et al. First Sagittarius A* Event Horizon Telescope Results. II. EHT and Multiwavelength Observations, Data Processing, and Calibration. *ApJ* **2022**, *930*, L13. [CrossRef]
11. Akiyama, K.; Alberdi, A.; Alef, W.; Algaba, J.C.; Anantua, R.; Asada, K.; Azulay, R.; Bach, U.; Baczko, A.K.; Ball, D.; et al. First Sagittarius A* Event Horizon Telescope Results. III. Imaging of the Galactic Center Supermassive Black Hole. *ApJ* **2022**, *930*, L14. [CrossRef]
12. Akiyama, K.; Alberdi, A.; Alef, W.; Algaba, J.C.; Anantua, R.; Asada, K.; Azulay, R.; Bach, U.; Baczko, A.K.; Ball, D.; et al. First Sagittarius A* Event Horizon Telescope Results. IV. Variability, Morphology, and Black Hole Mass. *ApJ* **2022**, *930*, L15. [CrossRef]
13. Akiyama, K.; Alberdi, A.; Alef, W.; Algaba, J.C.; Anantua, R.; Asada, K.; Azulay, R.; Bach, U.; Baczko, A.K.; Ball, D.; et al. First Sagittarius A* Event Horizon Telescope Results. V. Testing Astrophysical Models of the Galactic Center Black Hole. *ApJ* **2022**, *930*, L16. [CrossRef]
14. Akiyama, K.; Alberdi, A.; Alef, W.; Algaba, J.C.; Anantua, R.; Asada, K.; Azulay, R.; Bach, U.; Baczko, A.K.; Ball, D.; et al. First Sagittarius A* Event Horizon Telescope Results. VI. Testing the Black Hole Metric. *ApJ* **2022**, *930*, L17. [CrossRef]
15. Ricarte, A.; Tiede, P.; Emami, R.; Tamar, A.; Natarajan, P. The ngEHT's Role in Measuring Supermassive Black Hole Spins. *arXiv* **2022**, arXiv:2211.03910.
16. Bisnovatyi-Kogan, G.S.; Ruzmaikin, A.A. The Accretion of Matter by a Collapsing Star in the Presence of a Magnetic Field. *Ap&SS* **1974**, *28*, 45–59. [CrossRef]
17. Igumenshchev, I.V.; Narayan, R.; Abramowicz, M.A. Three-dimensional Magnetohydrodynamic Simulations of Radiatively Inefficient Accretion Flows. *ApJ* **2003**, *592*, 1042–1059. [CrossRef]
18. Narayan, R.; Igumenshchev, I.V.; Abramowicz, M.A. Magnetically Arrested Disk: An Energetically Efficient Accretion Flow. *PASJ* **2003**, *55*, L69–L72. [CrossRef]
19. Narayan, R.; Sądowski, A.; Penna, R.F.; Kulkarni, A.K. GRMHD simulations of magnetized advection-dominated accretion on a non-spinning black hole: Role of outflows. *MNRAS* **2012**, *426*, 3241–3259. [CrossRef]

20. Sądowski, A.; Narayan, R.; Penna, R.; Zhu, Y. Energy, momentum and mass outflows and feedback from thick accretion discs around rotating black holes. *MNRAS* **2013**, *436*, 3856–3874. [CrossRef]

21. Shapiro, S.L.; Lightman, A.P.; Eardley, D.M. A two-temperature accretion disk model for Cygnus X-1: Structure and spectrum. *ApJ* **1976**, *204*, 187–199. [CrossRef]

22. Rees, M.J.; Begelman, M.C.; Blandford, R.D.; Phinney, E.S. Ion-supported tori and the origin of radio jets. *Nature* **1982**, *295*, 17–21. [CrossRef]

23. Narayan, R.; Yi, I.; Mahadevan, R. Explaining the spectrum of Sagittarius A* with a model of an accreting black hole. *Nature* **1995**, *374*, 623–625. [CrossRef]

24. Mościbrodzka, M.; Falcke, H.; Shiokawa, H. General relativistic magnetohydrodynamical simulations of the jet in M 87. *A&A* **2016**, *586*, A38. [CrossRef]

25. Özel, F.; Psaltis, D.; Narayan, R. Hybrid Thermal-Nonthermal Synchrotron Emission from Hot Accretion Flows. *ApJ* **2000**, *541*, 234–249. [CrossRef]

26. Mao, S.A.; Dexter, J.; Quataert, E. The impact of non-thermal electrons on event horizon scale images and spectra of Sgr A*. *MNRAS* **2017**, *466*, 4307–4319. [CrossRef]

27. Cruz-Osorio, A.; Fromm, C.M.; Mizuno, Y.; Nathanail, A.; Younsi, Z.; Porth, O.; Davelaar, J.; Falcke, H.; Kramer, M.; Rezzolla, L. State-of-the-art energetic and morphological modelling of the launching site of the M87 jet. *Nat. Astron.* **2022**, *6*, 103–108. [CrossRef]

28. Fromm, C.M.; Cruz-Osorio, A.; Mizuno, Y.; Nathanail, A.; Younsi, Z.; Porth, O.; Olivares, H.; Davelaar, J.; Falcke, H.; Kramer, M.; et al. Impact of non-thermal particles on the spectral and structural properties of M87. *A&A* **2022**, *660*, A107. [CrossRef]

29. Anantua, R.; Emami, R.; Loeb, A.; Chael, A. Determining the Composition of Relativistic Jets from Polarization Maps. *ApJ* **2020**, *896*, 30. [CrossRef]

30. Emami, R.; Anantua, R.; Chael, A.A.; Loeb, A. Positron Effects on Polarized Images and Spectra from Jet and Accretion Flow Models of M87* and Sgr A*. *ApJ* **2021**, *923*, 272. [CrossRef]

31. Wong, G.N.; Gammie, C.F. Effects of Hydrogen vs. Helium on Electromagnetic Black Hole Observables. *arXiv* **2022**, arXiv:2207.13705.

32. Fragile, P.C.; Blaes, O.M.; Anninos, P.; Salmonson, J.D. Global General Relativistic Magnetohydrodynamic Simulation of a Tilted Black Hole Accretion Disk. *ApJ* **2007**, *668*, 417–429. [CrossRef]

33. Liska, M.; Hesp, C.; Tchekhovskoy, A.; Ingram, A.; van der Klis, M.; Markoff, S.B.; Van Moer, M. Disc tearing and Bardeen-Petterson alignment in GRMHD simulations of highly tilted thin accretion discs. *MNRAS* **2021**, *507*, 983–990. [CrossRef]

34. Mościbrodzka, M.; Gammie, C.F. IPOLE—semi-analytic scheme for relativistic polarized radiative transport. *MNRAS* **2018**, *475*, 43–54. [CrossRef]

35. Wardle, J.F.C.; Homan, D.C. Theoretical Models for Producing Circularly Polarized Radiation in Extragalactic Radio Sources. *Ap&SS* **2003**, *288*, 143–153. [CrossRef]

36. Mościbrodzka, M.; Janiuk, A.; De Laurentis, M. Unraveling circular polarimetric images of magnetically arrested accretion flows near event horizon of a black hole. *MNRAS* **2021**, *508*, 4282–4296. [CrossRef]

37. Ricarte, A.; Qiu, R.; Narayan, R. Black hole magnetic fields and their imprint on circular polarization images. *MNRAS* **2021**, *505*, 523–539. [CrossRef]

38. Tsunetoe, Y.; Mineshige, S.; Ohsuga, K.; Kawashima, T.; Akiyama, K. Polarization images of accretion flow around supermassive black holes: Imprints of toroidal field structure. *PASJ* **2021**, *73*, 912–928. [CrossRef]

39. Wardle, J.F.C.; Homan, D.C.; Ojha, R.; Roberts, D.H. Electron-positron jets associated with the quasar 3C279. *Nature* **1998**, *395*, 457–461. [CrossRef]

40. Bower, G.C.; Dexter, J.; Asada, K.; Brinkerink, C.D.; Falcke, H.; Ho, P.; Inoue, M.; Markoff, S.; Marrone, D.P.; Matsushita, S.; et al. ALMA Observations of the Terahertz Spectrum of Sagittarius A*. *ApJ* **2019**, *881*, L2. [CrossRef]

41. Ricarte, A.; Gammie, C.; Narayan, R.; Prather, B.S. Probing Plasma Physics with Spectral Index Maps of Accreting Black Holes on Event Horizon Scales. *arXiv* **2022**, arXiv:2202.02408.

42. Mościbrodzka, M.; Dexter, J.; Davelaar, J.; Falcke, H. Faraday rotation in GRMHD simulations of the jet launching zone of M87. *MNRAS* **2017**, *468*, 2214–2221. [CrossRef]

43. Jiménez-Rosales, A.; Dexter, J. The impact of Faraday effects on polarized black hole images of Sagittarius A*. *MNRAS* **2018**, *478*, 1875–1883. [CrossRef]

44. Ricarte, A.; Prather, B.S.; Wong, G.N.; Narayan, R.; Gammie, C.; Johnson, M.D. Decomposing the internal faraday rotation of black hole accretion flows. *MNRAS* **2020**, *498*, 5468–5488. [CrossRef]

45. Pandya, A.; Zhang, Z.; Chandra, M.; Gammie, C.F. Polarized Synchrotron Emissivities and Absorptivities for Relativistic Thermal, Power-law, and Kappa Distribution Functions. *ApJ* **2016**, *822*, 34. [CrossRef]

46. Blandford, R.D.; Königl, A. Relativistic jets as compact radio sources. *ApJ* **1979**, *232*, 34–48. [CrossRef]

47. Vasyliunas, V.M. A survey of low-energy electrons in the evening sector of the magnetosphere with OGO 1 and OGO 3. *J. Geophys. Res.* **1968**, *73*, 2839–2884. [CrossRef]

48. Xiao, F. Modelling energetic particles by a relativistic kappa-loss-cone distribution function in plasmas. *Plasma Phys. Control. Fusion* **2006**, *48*, 203–213. [CrossRef]

49. Rybicki, G.B.; Lightman, A.P. *Radiative Processes in Astrophysics*; John Wiley & Sons, Inc.: New York, NY, USA, 1986.

50. Himwich, E.; Johnson, M.D.; Lupsasca, A.; Strominger, A. Universal polarimetric signatures of the black hole photon ring. *Phys. Rev. D.* **2020**, *101*, 084020. [CrossRef]

51. Narayan, R.; Palumbo, D.C.M.; Johnson, M.D.; Gelles, Z.; Himwich, E.; Chang, D.O.; Ricarte, A.; Dexter, J.; Gammie, C.F.; Chael, A.A.; et al. The Polarized Image of a Synchrotron-emitting Ring of Gas Orbiting a Black Hole. *ApJ* **2021**, *912*, 35. [CrossRef]

52. Gelles, Z.; Himwich, E.; Johnson, M.D.; Palumbo, D.C.M. Polarized image of equatorial emission in the Kerr geometry. *Phys. Rev. D.* **2021**, *104*, 044060. [CrossRef]

53. Palumbo, D.C.M.; Wong, G.N. Photon Ring Symmetries in Simulated Linear Polarization Images of Messier 87*. *ApJ* **2022**, *929*, 49. [CrossRef]

54. Walker, R.C.; Hardee, P.E.; Davies, F.B.; Ly, C.; Junor, W. The Structure and Dynamics of the Subparsec Jet in M87 Based on 50 VLBA Observations over 17 Years at 43 GHz. *ApJ* **2018**, *855*, 128. [CrossRef]

55. Palumbo, D.C.M.; Wong, G.N.; Prather, B.S. Discriminating Accretion States via Rotational Symmetry in Simulated Polarimetric Images of M87. *ApJ* **2020**, *894*, 156. [CrossRef]

56. Johnson, M.D. Stochastic Optics: A Scattering Mitigation Framework for Radio Interferometric Imaging. *ApJ* **2016**, *833*, 74. [CrossRef]

57. Pacholczyk, A.G. *Radio Astrophysics. Nonthermal Processes in Galactic and Extragalactic Sources*; W. H. Freeman and Company: New York, NY, USA, 1970.

58. Jones, T.W.; Hardee, P.E. Maxwellian synchrotron sources. *ApJ* **1979**, *228*, 268–278. [CrossRef]

59. Goddi, C.; Martí-Vidal, I.; Messias, H.; Bower, G.C.; Broderick, A.E.; Dexter, J.; Marrone, D.P.; Moscibrodzka, M.; Nagai, H.; Algaba, J.C.; et al. Polarimetric Properties of Event Horizon Telescope Targets from ALMA. *ApJ* **2021**, *910*, L14. [CrossRef]

60. Jorstad, S.G.; Marscher, A.P.; Lister, M.L.; Stirling, A.M.; Cawthorne, T.V.; Gear, W.K.; Gómez, J.L.; Stevens, J.A.; Smith, P.S.; Forster, J.R.; et al. Polarimetric Observations of 15 Active Galactic Nuclei at High Frequencies: Jet Kinematics from Bimonthly Monitoring with the Very Long Baseline Array. *AJ* **2005**, *130*, 1418–1465. [CrossRef]

61. Lister, M.L.; Aller, M.F.; Aller, H.D.; Hodge, M.A.; Homan, D.C.; Kovalev, Y.Y.; Pushkarev, A.B.; Savolainen, T. MOJAVE. XV. VLBA 15 GHz Total Intensity and Polarization Maps of 437 Parsec-scale AGN Jets from 1996 to 2017. *ApJS* **2018**, *234*, 12. [CrossRef]

62. Ricarte, A.; Palumbo, D.C.M.; Narayan, R.; Roelofs, F.; Emami, R. Observational Signatures of Frame Dragging in Strong Gravity. *arXiv* **2022**, arXiv:2211.01810.

63. Marrone, D.P.; Moran, J.M.; Zhao, J.H.; Rao, R. The Submillimeter Polarization of Sgr A*. *J. Phys. Conf. Ser.* **2006**, *54*, 354–362. [CrossRef]

64. Trippe, S.; Paumard, T.; Ott, T.; Gillessen, S.; Eisenhauer, F.; Martins, F.; Genzel, R. A polarized infrared flare from Sagittarius A* and the signatures of orbiting plasma hotspots. *MNRAS* **2007**, *375*, 764–772. [CrossRef]

65. Zamaninasab, M.; Eckart, A.; Dovčiak, M.; Karas, V.; Schödel, R.; Witzel, G.; Sabha, N.; García-Marín, M.; Kunneriath, D.; Mužić, K.; et al. Near-infrared polarimetry as a tool for testing properties of accreting supermassive black holes. *MNRAS* **2011**, *413*, 322–332. [CrossRef]

66. GRAVITY Collaboration; Abuter, R.; Amorim, A.; Bauböck, M.; Berger, J.P.; Bonnet, H.; Brandner, W.; Clénet, Y.; Coudé Du Foresto, V.; de Zeeuw, P.T.; et al. Detection of orbital motions near the last stable circular orbit of the massive black hole SgrA*. *A&A* **2018**, *618*, L10. [CrossRef]

67. Wielgus, M.; Moscibrodzka, M.; Vos, J.; Gelles, Z.; Martí-Vidal, I.; Farah, J.; Marchili, N.; Goddi, C.; Messias, H. Orbital motion near Sagittarius A*. Constraints from polarimetric ALMA observations. *A&A* **2022**, *665*, L6. [CrossRef]

68. Broderick, A.E.; Loeb, A. Imaging bright-spots in the accretion flow near the black hole horizon of Sgr A*. *MNRAS* **2005**, *363*, 353–362. [CrossRef]

69. GRAVITY Collaboration.; Jiménez-Rosales, A.; Dexter, J.; Widmann, F.; Bauböck, M.; Abuter, R.; Amorim, A.; Berger, J.P.; Bonnet, H.; Brandner, W.; et al. Dynamically important magnetic fields near the event horizon of Sgr A*. *A&A* **2020**, *643*, A56. [CrossRef]

70. Vos, J.; Moscibrodzka, M.; Wielgus, M. Polarimetric signatures of hot spots in black hole accretion flows. *arXiv* **2022**, arXiv:2209.09931.

71. Hu, Z.; Hou, Y.; Yan, H.; Guo, M.; Chen, B. Electromagnetic radiations and polarized images of synchrotron radiations in curved spacetime. *arXiv* **2022**, arXiv:2203.02908.

72. Lee, T.; Hu, Z.; Guo, M.; Chen, B. Circular orbits and polarized images of charged particles orbiting Kerr black hole with a weak magnetic field. *arXiv* **2022**, arXiv:2211.04143.

73. Ripperda, B.; Bacchini, F.; Philippov, A.A. Magnetic Reconnection and Hot Spot Formation in Black Hole Accretion Disks. *ApJ* **2020**, *900*, 100. [CrossRef]

74. Dexter, J.; Tchekhovskoy, A.; Jiménez-Rosales, A.; Ressler, S.M.; Bauböck, M.; Dallilar, Y.; de Zeeuw, P.T.; Eisenhauer, F.; von Fellenberg, S.; Gao, F.; et al. Sgr A* near-infrared flares from reconnection events in a magnetically arrested disc. *MNRAS* **2020**, *497*, 4999–5007. [CrossRef]

75. Chatterjee, K.; Narayan, R. Flux eruption events drive angular momentum transport in magnetically arrested accretion flows. *arXiv* **2022**, arXiv:2210.08045.

76. GRAVITY Collaboration.; Bauböck, M.; Dexter, J.; Abuter, R.; Amorim, A.; Berger, J.P.; Bonnet, H.; Brandner, W.; Clénet, Y.; Coudé Du Foresto, V.; et al. Modeling the orbital motion of Sgr A*'s near-infrared flares. *A&A* **2020**, *635*, A143. [CrossRef]

77. Gardner, F.F.; Whiteoak, J.B. The Polarization of Cosmic Radio Waves. *ARA&A* **1966**, *4*, 245. [CrossRef]

78. Jones, T.W.; Odell, S.L. Transfer of polarized radiation in self-absorbed synchrotron sources. I. Results for a homogeneous source. *ApJ* **1977**, *214*, 522–539. [CrossRef]
79. Marrone, D.P.; Moran, J.M.; Zhao, J.H.; Rao, R. An Unambiguous Detection of Faraday Rotation in Sagittarius A*. *ApJ* **2007**, *654*, L57–L60. [CrossRef]
80. Plambeck, R.L.; Bower, G.C.; Rao, R.; Marrone, D.P.; Jorstad, S.G.; Marscher, A.P.; Doeleman, S.S.; Fish, V.L.; Johnson, M.D. Probing the Parsec-scale Accretion Flow of 3C 84 with Millimeter Wavelength Polarimetry. *ApJ* **2014**, *797*, 66. [CrossRef]
81. Kuo, C.Y.; Asada, K.; Rao, R.; Nakamura, M.; Algaba, J.C.; Liu, H.B.; Inoue, M.; Koch, P.M.; Ho, P.T.P.; Matsushita, S.; et al. Measuring Mass Accretion Rate onto the Supermassive Black Hole in M87 Using Faraday Rotation Measure with the Submillimeter Array. *ApJ* **2014**, *783*, L33. [CrossRef]
82. Park, J.; Hada, K.; Kino, M.; Nakamura, M.; Ro, H.; Trippe, S. Faraday Rotation in the Jet of M87 inside the Bondi Radius: Indication of Winds from Hot Accretion Flows Confining the Relativistic Jet. *ApJ* **2019**, *871*, 257. [CrossRef]
83. Marti-Vidal, I.; Muller, S. Submillimeter polarization and variability of quasar PKS 1830-211. *A&A* **2019**, *621*, A18. [CrossRef]
84. Bower, G.C.; Broderick, A.; Dexter, J.; Doeleman, S.; Falcke, H.; Fish, V.; Johnson, M.D.; Marrone, D.P.; Moran, J.M.; Moscibrodzka, M.; et al. ALMA Polarimetry of Sgr A*: Probing the Accretion Flow from the Event Horizon to the Bondi Radius. *ApJ* **2018**, *868*, 101. [CrossRef]
85. Kravchenko, E.V.; Kovalev, Y.Y.; Sokolovsky, K.V. Parsec-scale Faraday rotation and polarization of 20 active galactic nuclei jets. *MNRAS* **2017**, *467*, 83–101. [CrossRef]
86. Martí-Vidal, I.; Muller, S.; Vlemmings, W.; Horellou, C.; Aalto, S. A strong magnetic field in the jet base of a supermassive black hole. *Science* **2015**, *348*, 311–314. [CrossRef] [PubMed]
87. Rickett, B.J. Radio propagation through the turbulent interstellar plasma. *ARA&A* **1990**, *28*, 561–605. [CrossRef]
88. Bower, G.C.; Deller, A.; Demorest, P.; Brunthaler, A.; Eatough, R.; Falcke, H.; Kramer, M.; Lee, K.J.; Spitler, L. The Angular Broadening of the Galactic Center Pulsar SGR J1745-29: A New Constraint on the Scattering Medium. *ApJ* **2014**, *780*, L2. [CrossRef]
89. Psaltis, D.; Johnson, M.; Narayan, R.; Medeiros, L.; Blackburn, L.; Bower, G. A Model for Anisotropic Interstellar Scattering and its Application to Sgr A*. *arXiv* **2018**, arXiv:1805.01242.
90. Johnson, M.D.; Narayan, R.; Psaltis, D.; Blackburn, L.; Kovalev, Y.Y.; Gwinn, C.R.; Zhao, G.Y.; Bower, G.C.; Moran, J.M.; Kino, M.; et al. The Scattering and Intrinsic Structure of Sagittarius A* at Radio Wavelengths. *ApJ* **2018**, *865*, 104. [CrossRef]
91. Johnson, M.D.; Kovalev, Y.Y.; Lisakov, M.M.; Voitsik, P.A.; Gwinn, C.R.; Bruni, G. First Space-VLBI Observations of Sagittarius A*. *ApJ* **2021**, *922*, L28. [CrossRef]
92. Johnson, M.D.; Fish, V.L.; Doeleman, S.S.; Marrone, D.P.; Plambeck, R.L.; Wardle, J.F.C.; Akiyama, K.; Asada, K.; Beaudoin, C.; Blackburn, L.; et al. Resolved magnetic-field structure and variability near the event horizon of Sagittarius A*. *Science* **2015**, *350*, 1242–1245. [CrossRef]
93. Kovalev, Y.Y.; Kardashev, N.S.; Sokolovsky, K.V.; Voitsik, P.A.; An, T.; Anderson, J.M.; Andrianov, A.S.; Avdeev, V.Y.; Bartel, N.; Bignall, H.E.; et al. Detection statistics of the RadioAstron AGN survey. *Adv. Space Res.* **2020**, *65*, 705–711. [CrossRef]
94. Wielgus, M.; Marchili, N.; Martí-Vidal, I.; Keating, G.K.; Ramakrishnan, V.; Tiede, P.; Fomalont, E.; Issaoun, S.; Neilsen, J.; Nowak, M.A.; et al. Millimeter Light Curves of Sagittarius A* Observed during the 2017 Event Horizon Telescope Campaign. *ApJ* **2022**, *930*, L19. [CrossRef]
95. Gold, R.; McKinney, J.C.; Johnson, M.D.; Doeleman, S.S. Probing the Magnetic Field Structure in Sgr A* on Black Hole Horizon Scales with Polarized Radiative Transfer Simulations. *ApJ* **2017**, *837*, 180. [CrossRef]
96. Pushkarev, A.B.; Aller, M.F.; Aller, H.D.; Homan, D.C.; Kovalev, Y.Y.; Lister, M.L.; Pashchenko, I.N.; Savolainen, T.; Zobnina, D. MOJAVE. XXI. Persistent Linear Polarization Structure in Parsec-scale AGN Jets. *arXiv* **2022**, arXiv:2209.04842.

Article

Modeling Reconstructed Images of Jets Launched by SANE Super-Eddington Accretion Flows around SMBHs with the ngEHT

Brandon Curd [1,2,*], **Razieh Emami** [2], **Freek Roelofs** [1,2] and **Richard Anantua** [3]

1. Black Hole Initiative, Harvard University, 20 Garden Street, Cambridge, MA 02138, USA
2. Center for Astrophysics, Harvard & Smithsonian, 60 Garden Street, Cambridge, MA 02138, USA
3. Department of Physics & Astronomy, The University of Texas at San Antonio, One UTSA Circle, San Antonio, TX 78249, USA
* Correspondence: brandon.curd@cfa.harvard.edu

Abstract: Tidal disruption events (TDEs) around supermassive black holes (SMBHs) are a potential laboratory to study super-Eddington accretion disks and sometimes result in powerful jets or outflows which may shine in the radio and sub-millimeter bands. In this work, we modeled the thermal synchrotron emission of jets by general relativistic radiation magneto-hydrodynamics (GRRMHD) simulations of a BH accretion disk/jet system which assumed the TDE resulted in a magnetized accretion disk around a BH accreting at $\sim$12–25 times the Eddington accretion rate. Through synthetic observations with the Next Generation Event Horizon Telescope (ngEHT) and an image reconstruction analysis, we demonstrate that TDE jets may provide compelling targets within the context of the models explored in this work. In particular, we found that jets launched by a SANE super-Eddington disk around a spin $a_* = 0.9$ reach the ngEHT detection threshold at large distances (up to 100 Mpc in this work). A two-temperature plasma in the jet or weaker jets, such as a spin $a_* = 0$ model, requires a much closer distance, as we demonstrate detection at 10 Mpc for limiting cases of $a_* = 0$, $\mathcal{R} = 1$ or $a_* = 0.9$, $\mathcal{R} = 20$. We also demonstrate that TDE jets may appear as superluminal sources if the BH is rapidly rotating and the jet is viewed nearly face on.

Keywords: accretion disk; relativistic Jet; GRMHD

Citation: Curd, B.; Emami, R.; Roelofs, F.; Anantua, R. Modeling Reconstructed Images of Jets Launched by SANE Super-Eddington Accretion Flows around SMBHs with the ngEHT. *Galaxies* **2022**, *10*, 117. https://doi.org/10.3390/galaxies10060117

Academic Editor: Bidzina Kapanadze

Received: 14 November 2022
Accepted: 7 December 2022
Published: 13 December 2022

Publisher's Note: MDPI stays neutral with regard to jurisdictional claims in published maps and institutional affiliations.

1. Introduction

Tidal disruptions of stars by supermassive black holes (SMBHs), or tidal disruption events (TDEs), have recently become a regularly observed transient phenomenon. Stars which enter the tidal radius

$$R_t \simeq 7 \times 10^{12} \left(\frac{M_{\mathrm{BH}}}{10^6 M_\odot} \right)^{1/3} \left(\frac{M_*}{M_\odot} \right)^{-1/3} \left(\frac{R_*}{R_\odot} \right) [\mathrm{cm}] \tag{1}$$

of the central SMBH in their host galaxy will be disrupted [1,2], either partially or fully, depending on the orbit and equation of state of the star [3,4]. The bound stream of gas returns towards the BH delivering mass at the fallback rate ($\dot{M}_{\mathrm{fb}}$). Apsidal precession of the returning stream leads to self-intersection with material that has yet to pass through pericenter and leads to dissipation and disk formation. Dissipation due to self-intersection may also be a source of early emission in a TDE [5]. After the initial rise to peak, the fallback rate follows a power law behavior which can be approximated as

$$\dot{M}_{\mathrm{fb}} = \dot{M}_{\mathrm{fb,peak}} \left(\frac{t}{t_{\mathrm{fb}}} \right)^{-5/3} . \tag{2}$$

Here, $\dot{M}_{\text{fb,peak}}$ is the peak mass fallback rate:

$$\frac{\dot{M}_{\text{fb,peak}}}{\dot{M}_{\text{Edd}}} \approx 133 \left(\frac{M_{\text{BH}}}{10^6 M_\odot} \right)^{-3/2} \left(\frac{M_*}{M_\odot} \right)^2 \left(\frac{R_*}{R_\odot} \right)^{-3/2} \tag{3}$$

making the "frozen in" approximation as in Stone et al. [6], and

$$t_{\text{fb}} = 3.5 \times 10^6 \, \text{s} \left(\frac{M_{\text{BH}}}{10^6 M_\odot} \right)^{1/2} \left(\frac{M_*}{M_\odot} \right)^{-1} \left(\frac{R_*}{R_\odot} \right)^{3/2}, \tag{4}$$

is the fallback time, which is the orbital time of the most bound part of the stream. Of note is the fact that the mass fallback rate can greatly exceed the Eddington mass accretion rate $\dot{M}_{\text{Edd}}$. Since the mass accretion rate is expected to be similar in magnitude to the mass fallback rate, it is possible that TDEs result in a super-Eddington accretion disk, which are geometrically thick, radiatively inefficient accretion disks [7,8]. The exact power law behavior varies with the properties and orbit of the star [9].

TDEs are typically seen as optical/X-ray transients [10,11], but several TDEs have resulted in outflows or jets which shine in the radio bands [12]. In the most common case in which no relativistic jet is launched (commonly referred to as "non-jetted" TDEs), the X-ray and optical/UV luminosity follows a roughly $t^{-5/3}$ decline, similar to the fallback rate. If the TDE leads to prompt disk formation, the X-rays are thought to arise from an accretion disk, and the optical/UV emission arises from a large scale reprocessing layer [13,14]. TDEs have also been observed to launch relativistic X-ray jets in a few cases. These jetted TDEs have been argued to arise due to a magnetically arrested disk (MAD) [15] forming around the BH during the TDE [13,16,17], which leads to jet production via the Blandford–Znajek (BZ) mechanism [18] extracting spin energy from the BH. Alternatively, these powerful relativistic jets may be produced thanks to radiative acceleration of gas through a narrow funnel region [19,20].

It is important to note that the BZ mechanism is the strong field limit of the magnetic Penrose process (MPP), but the MPP can operate with weaker magnetic field strengths [21,22]. Note as well that the magnetic Penrose process, as its name suggests, is itself the magnetic flavor of the Penrose process (PP) [23]. While the PP relies on the existence of negative energy orbits in the vicinity of the BH, the MPP relies on negative energy orbits and quadrupole electric fields produced by twisting of the magnetic field lines threading the BH horizon. A key distinction, and why the BZ mechanism is often cited while interpreting astrophysical jets, is that the energy extraction from the MPP can exceed 100 percent, whereas the maximum energy extraction from a PP is $\sim$20% [21,22].

A handful of non-jetted TDEs have been observed to produce radiowaves peaking at tens of GHz with $L_{\text{radio}} \sim 10^{37-39} \text{erg s}^{-1}$. This emission is thought to arise from an outflow launched by the TDE with velocity $v \sim 0.1c$ shocking on the gas surrounding the BH. Meanwhile, jetted TDEs produce bright radio emission peaking at $L_{\text{radio}} \sim 10^{40-42} \text{erg s}^{-1}$. The appearance of radio emission is often delayed by several weeks from the initial appearance of the optical/UV/X-ray emission in non-jetted TDEs, which hints at some connection to the disk formation process to the occurrence of outflows.

When $\dot{M}_{\text{fb}}$ rises to peak, it has previously been assumed that by this stage a circularized accretion disk has formed [13,17]. The first direct demonstration of circularization near the peak fallback rate was recently demonstrated in a numerical simulation by Steinberg and Stone [5]. Multiple authors have argued in favor of a picture in which an inner accretion flow is surrounded by a quasi-spherical reprocessing layer, since this naturally explains the sometimes delayed appearance of X-ray emission in optical/UV-discovered TDEs [13,14,24]. This picture naturally arises if the accretion flow is actually super-Eddington [13], which has motivated multiple studies of GRRMHD simulations magnetized of super-Eddington disks. Motivated by this fact and the demonstration that a "standard and normal evolution" (SANE) [15] super-Eddington accretion disk also leads to viewing angle effects which may explain the behavior in non-jetted TDEs [17], we conducted a study of the outflows launched by SANE models in Curd et al. [14] and studied their radio-submm emission. We focus on these SANE models in this work as well. The BZ mechanism also acts in the SANE model [14,25]; however, the disk and BH have a much lower magnetic flux and thus a substantially lower jet power. Although BZ driven outflows in SANE disks can be relativistic, they are weakly comptonized and do not appear as powerful, beamed X-ray sources like the presently known jetted TDEs [14].

The Next Generation Event Horizon Telescope (ngEHT) will provide more baseline coverage and faster response times than the previous mission. An estimate for ngEHT is to double the antenna sites [26] of its 20 µas predecessor EHT, and thus the number of possible baseline pairs and triads of sites available for imaging jet/accretion flow/black hole systems will scale combinatorially (the number of baselines grows with the number of antennae as $(N(N-1)/2)$. This could allow for interesting sources, such as jets from nearby tidal disruption events, to be imaged directly. In our previous work [14], we provided the first demonstration that SANE super-Eddington accretion flows can produce radio emission which is bright enough at 230 GHz to be detected and resolved. Here we take things a step further and produce reconstructed images assuming such jets happen in the nearby universe.

The detection rate of TDEs in the optical/UV/X-ray is expected to grow rapidly once the Large Synoptic Survey Telescope comes online [27,28]. Assuming rapid follow-up of TDEs in radio-submm bands finds detectable emission, this could provide a large number of targets for the ngEHT. As we demonstrated in Curd et al. [14], some models produced detectable emission, even at ~180 Mpc. At this distance, a conservative estimate of the volume integrated TDE rate suggests more than 200 TDEs per year assuming volumetric TDE rates based on Stone and Metzger [29]. Even at < 40 Mpc, we estimate that several TDEs should occur per year (see Figure 2 in Curd et al. [14]), which suggests some nearby TDEs may become targets of opportunity during the ngEHT mission.

We stress that jets such as those in our first work on the subject of jets from SANE models of TDE accretion disks [14] do not resemble any previously detected radio TDEs. This may suggest most, or even all, TDEs do not form accretion disks which resemble SANE models to begin with. However, the number of TDEs that have appeared in the radio-submm in the first place is extremely small as of this writing; fewer than twenty radio TDEs have been reported. Furthermore, magnetic fields were certainly present in the forming disk, albeit dynamically subdominant to hydrodynamic effects early in the disk formation [30,31]. Nevertheless, it is possible that after the disk circularizes, which Steinberg and Stone [5] suggests may take tens of days, the magnetic field builds up in a dynamo effect similar to Sadowski et al. [30]. In this case, one would almost certainly expect the magnetic field to become dynamically important, in which case a magnetized outflow may be launched as in Curd et al. [14]. In addition, if the currently known jetted TDEs are indeed MAD disks around rapidly rotating BHs, it stands to reason that a sub-class of less strongly magnetized disks which could be described by a SANE model should exist. TDEs continue to surprise observers in terms of the range of behavior, so such a jet formation channel may yet be discovered.

2. Numerical Methods

2.1. GRRHMD Simulations

Throughout this work, we often use gravitational units to define length and time. In particular, we use the gravitational radius

$$r_g = \frac{GM_{\mathrm{BH}}}{c^2} \tag{5}$$

and the gravitational time

$$t_g = \frac{GM_{\mathrm{BH}}}{c^3}, \tag{6}$$

where M_{BH} is the mass of the black hole (BH). We also adopt the following definition for the Eddington mass accretion rate:

$$\dot{M}_{\mathrm{Edd}} = \frac{L_{\mathrm{Edd}}}{\eta_{\mathrm{NT}} c^2}, \tag{7}$$

where $L_{\mathrm{Edd}} = 1.25 \times 10^{38} \, (M_{\mathrm{BH}}/M_\odot) \, \mathrm{erg \, s}^{-1}$ is the Eddington luminosity and η_{NT} is the radiative efficiency of a Novikov–Thorne thin disk around a BH with spin parameter a_* [32].

We conducted an imaging analysis of GRRMHD simulations presented in Curd et al. [14]. In particular, we analyzed the most massive BH models m7a0.0-HR and m7a0.9-HR, which are $M_{\mathrm{BH}} = 10^7 \, M_\odot$ BHs of spin $a_* = 0$ and 0.9 BHs. We specify the simulation diagnostics relevant for this work in Table 1. The simulations were conducted in 2D (r, ϑ) coordinates on a $N_r \times N_\vartheta = 640 \times 256$ grid with added resolution near the poles to adequately resolve both the disk and jet. The radial grid cells were logarithmically spaced with a maximum domain radius of $R_{\mathrm{max}} = 10^5 \, r_g$ to capture the large-scale features of the jet.

Table 1. We tabulate the mass accretion rate $\dot{M}$, jet efficiency $\eta_{\mathrm{jet}} \equiv L_{\mathrm{jet}}/\dot{M}c^2$ (where L_{jet} is the jet power as defined in [14]), and total simulation duration t_{sim} for each KORAL simulation. Note that $\dot{M}$ and η_{jet} are time averaged over the final 50,000 t_g of each simulation.

Model	$\dot{M}$ ($\dot{M}_{\mathrm{Edd}}$)	η_{jet}	t_{sim} (t_g)
m7a0.0-HR	12	0.24%	83,000
m7a0.9-HR	25	1.15%	81,200

On horizon scales, gas is flowing across the BH horizon in an accretion disk due to angular momentum transport driven by the magneto-rotational instability. The disk is optically thick and turbulent, and gas inside of the disk is advected with the gas across the BH horizon. However, an optically thin funnel above and below the disks exists. Here, radiation can escape freely and pushes on gas, accelerating a significant outflow. In addition, the funnel is magnetized and sometimes exhibits magnetization parameter $\sigma = b^2/\rho c^2 > 1$, where b is the magnetic field strength and ρ is the mass density of the plasma. In the jet, magnetic energy is partially converted into kinetic energy, as it contributes to accelerating gas into an outflow.

In both simulations, radiative and Poynting acceleration drove fast outflows. The jet reached relativistic speeds with Lorentz factor $\gamma > 5$ for model m7a0.9-HR, which was likely due to the BZ effect extracting spin energy from the BH, which may produce roughly $\sim$1 percent of the jet efficiency, even though the magnetic flux threading the black hole was well below the MAD limit [14,25]. The primary sites of dissipation were the jet head and internal shocks inside of the jet. Internal shocks were due to fast- and slow-moving gas interacting downstream of the jet head, in addition to recollimation shocks. As we show in Figure 1, this resulted in a hot, magnetized jet which reached large scales ($r > 30,000 \, r_g$) by the end of the simulation. The $a_* = 0.9$ model had a significantly more magnetized jet and also produced a more powerful jet by a significant fraction (see Table 1).

The simulations were conducted in 2D to overcome the substantial numerical requirements of resolving the accretion disk and jet in a global GRRMHD simulation. While SANE super-Eddington disks can be described by an axisymmetric flow quite well [33], the jet may undergo 3D instabilities which cannot be captured in our numerical setup. In particular, the jet may become kink-unstable as it propagates [34]. This will certainly change the morphological features, and additional dissipation of magnetic energy along the jet may occur. If the dissipation is extreme, especially at large scales such as those we consider here, the synchrotron emission which we describe in the next section may differ from our analysis of 2D jets.

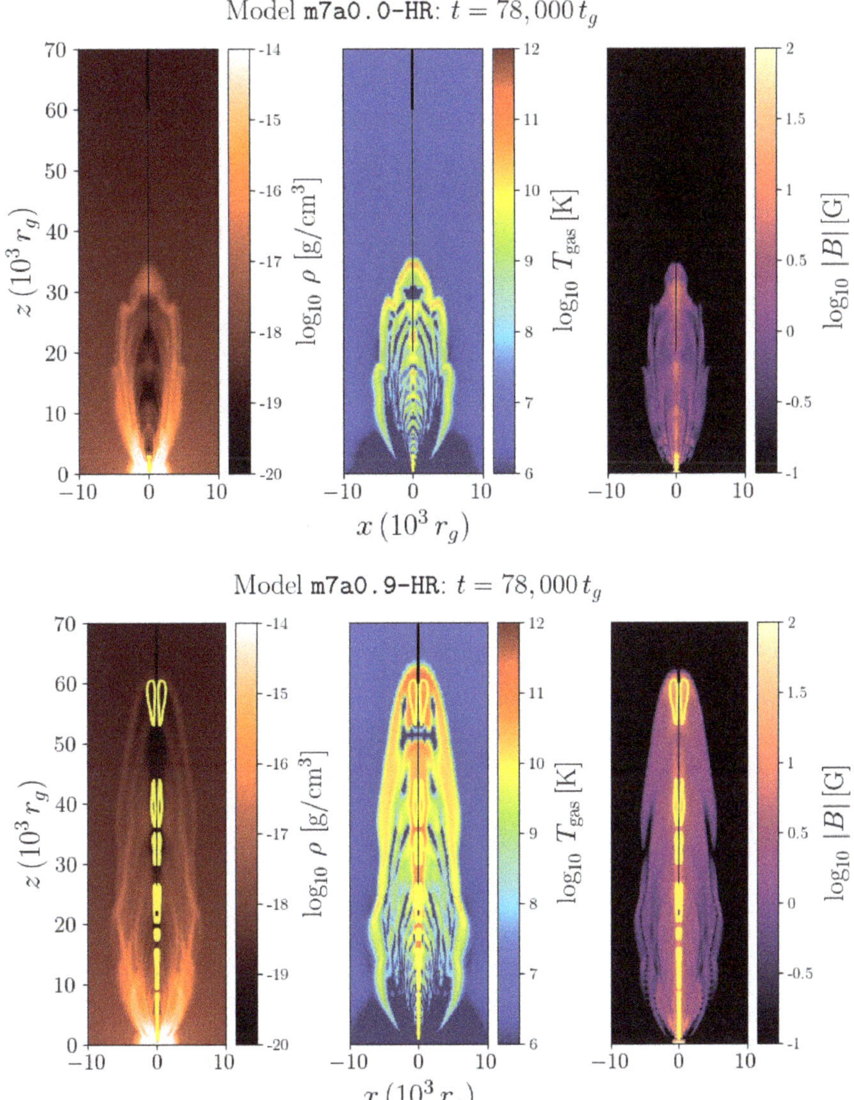

Figure 1. Here we show snapshots of the GRRMHD KORAL simulations that we post-processed with ipole. All data are shown for $t = 78,000\,t_g$ in both m7a0.0-HR (**top**) and m7a0.9-HR (**bottom**). The colors indicate the gas density ρ (**left**), gas temperature T_{gas} (**middle**), and magnetic field strength $|B|$ (**right**); the yellow contours indicate the $\sigma = 1$ boundary, in which we set $\rho = 0$ in the ray tracing step to prevent emission.

2.2. 230 GHz Emission

We post-processed the KORAL simulation data with the general relativistic ray tracing (GRRT) code ipole [35–37], which includes synchrotron and Bremsstrahlung emission and absorption. Since the KORAL simulation was conducted in 2D, we first copied the 2D data onto a full 3D (r, ϑ, ϕ) grid with 32 cells in azimuth which spanned $\phi = 0 - 2\pi$ before ray tracing. We did this by simply copying the original 2D data at each ϕ cell such that the new 3D data maintained the assumed axisymmetry in ϕ. The electron distribution function was assumed to be thermal. Ohmura et al. [38,39] demonstrated that large-scale active galactic nuclei (AGN) jets can produce a two-temperature plasma. Motivated by their findings and the possibility that a two-temperature plasma will be produced due to shocks at the jet head and within the jet itself, we tested a simple two-temperature jet model by scaling the electron temperature relative to the ion temperature via the plasma temperature ratio:

$$\mathcal{R} = \frac{T_i}{T_e},\tag{8}$$

where T_i is the temperature of the ions and T_e is the temperature of the electrons. Note that T_i was obtained directly from the KORAL simulation by setting $T_i = T_{\text{gas}}$.

The peak of the radio-submm spectra in m7a0.0-HR is lower than that of m7a0.9-HR, so increasing $\mathcal{R}$ has a much more significant impact on the 230 GHz emission and can make the jet undetectable even at 10 Mpc for values of $\mathcal{R} > 2$ [14]. It is possible that a non-thermal electron distribution will have greater high-energy emission even as $\mathcal{R}$ increases, but we saved an exploration of non-thermal electron models for a future analysis.

Each model was imaged at 230 GHz. For both models, we imaged the simulation at times $t = 38{,}000\,t_g$ and $t = 78{,}000\,t_g$ for a difference in observing times of ~23 days. We chose a distance $D = 10$ Mpc, $\mathcal{R} = 1$, observing angles relative to the jet axis (z in Figure 1) of $\theta = 10, 45$, and $90°$, respectively. Note that we use θ for the observer angle, and ϑ is the polar angle in the KORAL grid coordinates. For model m7a0.9-HR, we also tested limiting cases $D = 100$ Mpc, $\mathcal{R} = 1$, and $D = 10$ Mpc, $\mathcal{R} = 20$, imaged at $\theta = 90°$. The total 230 GHz flux of each ray-traced model is tabulated in Table 2.

We show a full library of each of the ipole images convolved with a Gaussian beam with a 20 μas full width at half maximum (FWHM) in Figures A1 and A3.

Table 2. Here we tabulate the 230 GHz flux density for each model given a specific time, viewing angle θ, distance D, and temperature ratio $\mathcal{R}$.

Model	Time (t_g)	Distance (Mpc)	$\mathcal{R}$	$F_{230\,\text{GHz}}$ (Jy) $\theta = 10°$	$\theta = 45°$	$\theta = 90°$
m7a0.0-HR	38,000	10	1	0.219	0.214	0.074
	78,000	10	1	0.014	0.013	0.006
m7a0.9-HR	38,000	10	1	2.001	4.452	6.036
	78,000	10	1	11.968	26.780	35.092
	38,000	10	20	-	-	0.190
	78,000	10	20	-	-	0.485
	38,000	100	1	-	-	0.060
	78,000	100	1	-	-	0.351

2.3. Synthetic ngEHT Observations and Image Reconstruction

In order to test to what extent the jet features in our models can be observed, we simulated observations with a potential ngEHT array, consisting of the 2022 EHT stations and 11 additional stations, selected from Raymond et al. [40] and similar to the ngEHT reference array used in the ngEHT Analysis Challenges [41]. The new dishes were assumed to have a diameter of 10 m and a receiver temperature of 50 K, with the array operating at a bandwidth of 8 GHz. For each image, we simulated a 24-hour observation with a 50% duty cycle with this array, using the ngehtsim [1] library, which makes use of eht-imaging [42,43]; see also [44]. The atmospheric opacity was set to reflect a good day in April, using the top 1σ quantile from the MERRA-2 data interpolated and integrated for each site on a 3 h cadence for a 10-year period [45,46]. Thermal noise was added to the complex visibilities, visibility phases were randomized, and no systematic visibility amplitude errors were added to the data.

We subsequently used the regularized maximum likelihood framework in eht-imaging to produce image reconstructions, with maximum entropy and (squared) total variation regularizers, fitting to visibility amplitudes and closure phases; see [42,43,47]. After establishing a set of well-performing imaging parameters on the m7a0.9-HR, $\mathcal{R} = 1$ model at $t_g = 78{,}000$ and a distance of 10 Mpc (Figure 2), we applied the same script to all other simulated datasets.

3. Results

In this section, we comment on the detectability of our models and then compare the ipole images with the reconstructed images. We comment on features which may be of interest in terms of the broader study of astrophysical jets.

3.1. Reconstructed Images

In the ray-traced image (i.e., see the left panel in Figure 2), the jet head produces bright emission as it shocks on the circumnuclear medium (CNM). In addition, various shocks occur within the jet due to both slow- and fast-moving components colliding radially and due to recollimation shocks. This leads to dissipation within the jet and bright "bubbles" of emission at 230 GHz. As we show in the right panel of Figure 2, the jet head and the structures in the jet are faithfully reproduced in the reconstruction for favorable viewing angles ($\theta = 45°$ and $90°$) as long as the source is nearby ($D = 10$ Mpc). Jets viewed near $\theta = 10°$ are dominated by emission from the jet head, and distinguishing internal jet features would be unlikely. This can be seen by comparing the base images with the full library of reconstructed images for models m7a0.0-HR and m7a0.9-HR (Figures A1–A4).

For distant sources ($D = 100$ Mpc), distinguishing internal features is impossible, and only the jet head can be fully distinguished in the reconstruction (right panel in Figure 3). This would still allow for the jet motion to be tracked, but detailed information is lost.

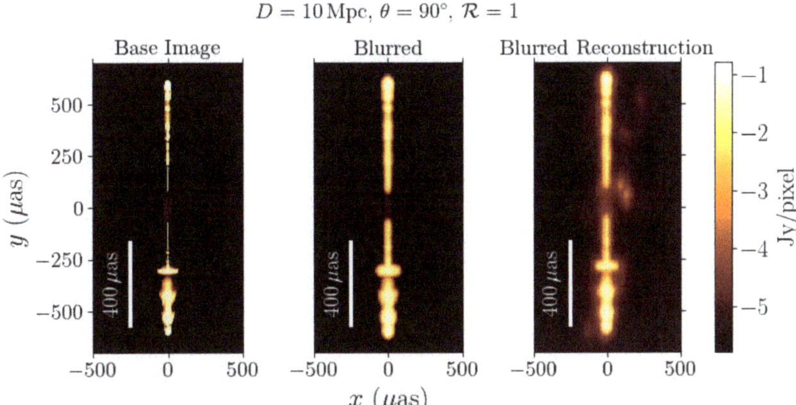

Figure 2. Model m7a0.9-HR at $t = 78{,}000\,t_g$ imaged at $D = 10$ Mpc with $\theta = 90°$ and $\mathcal{R} = 1$. We show the base ipole image with no blurring (**left**), the base ipole image blurred via convolution with a 20 μas FWHM Gaussian beam (**middle**), and the reconstructed image blurred using the same Gaussian beam (**right**).

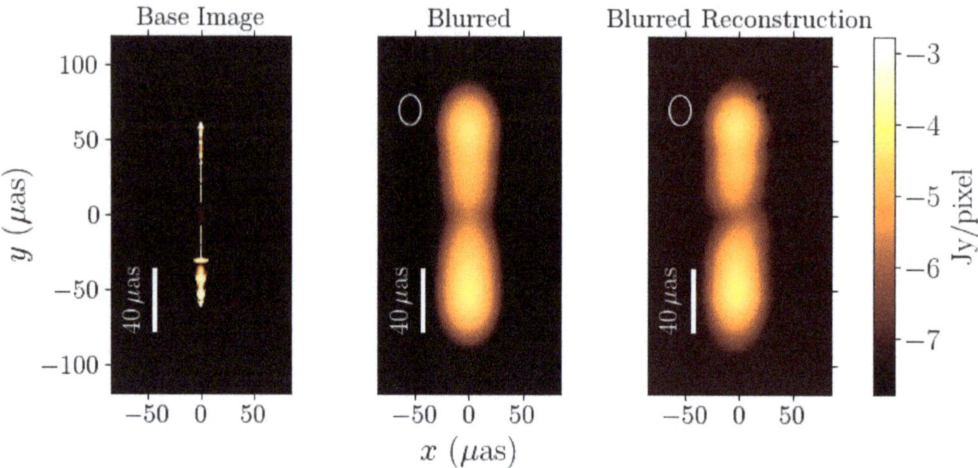

Model `m7a0.9-HR`: 230 GHz Emission

$$D = 100\,\mathrm{Mpc},\ \theta = 90°,\ \mathcal{R} = 1$$

Figure 3. The same as Figure 2 but for model m7a0.9-HR at $t = 78{,}000\,t_g$ imaged at $D = 100$ Mpc with $\theta = 90°$ and $\mathcal{R} = 1$.

3.2. Tracking Jet Motion

In this subsection, we demonstrate that the original ray-traced images and the reconstructed images allow for the jet motion to be tracked and yield similar results for the time evolution of the jet. The jet features are approximately Lorentzian, so we fit Lorentzian profiles to the image to find the position of the top and bottom jet in both the base images and the reconstructed images. We detail the peak finding algorithm in Appendix B. Since we cannot properly center the jet in the reconstructed images (there is no bright, central radiation from the near BH), we only measured the distance between the two jet peaks y_1 and y_2, respectively. We define the apparent jet length as:

$$l_{\mathrm{jet}} = |y_2 - y_1|. \tag{9}$$

Note that we have not differentiated the "top" or "bottom" jet here, as we are only concerned with the total distance between the jet heads. We obtained errors on the jet length from the error estimates of the jet head locations using standard error propagation analysis:

$$\delta l_{\mathrm{jet}} = \sqrt{\delta y_1^2 + \delta y_2^2}. \tag{10}$$

We computed the relative difference between the jet lengths for the ray-traced (l_{jet}) and reconstructed ($l_{\mathrm{jet,rec}}$) images in order to quantify the extent to which measurements of the jet length and velocity agree. We found that, generally, the ray-traced and reconstructed images yield similar jet lengths within 3 standard deviations (Figure 4). In general, there are much larger errors on the fit for the Lorentzian profile's center at steep angles (i.e., see the relative difference for $\theta = 10°$). The agreement is also effected by how bright the source is, as illustrated by the shift to the right for model m7a0.9-HR when $D = 100$ Mpc. We compiled estimates of the jet lengths for each image in Table 3.

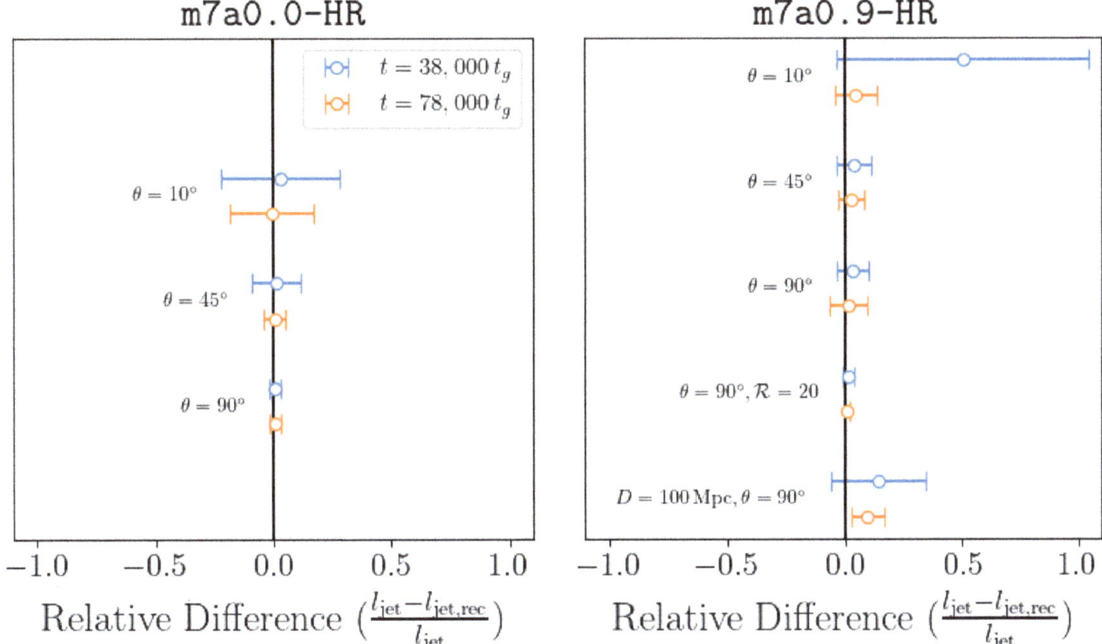

Relative Difference $\left(\dfrac{l_{\mathrm{jet}} - l_{\mathrm{jet,rec}}}{l_{\mathrm{jet}}}\right)$ Relative Difference $\left(\dfrac{l_{\mathrm{jet}} - l_{\mathrm{jet,rec}}}{l_{\mathrm{jet}}}\right)$

Figure 4. Here we show the relative difference (open circles) and the error within 3 standard deviations (horizontal bars) between the jet lengths obtained from the ray-traced images (l_{jet}) and the reconstructed images ($l_{\mathrm{jet,rec}}$). Models m7a0.0-HR (**left panel**) and m7a0.9-HR (**right panel**) are shown. For each choice of viewing angle θ, distance to the source D, and $\mathcal{R}$, we show the data at $t = 38,000\,t_g$ (blue) and $t = 78,000\,r_g$. We indicate the viewing angle next to each pairing of error bars. For each model, except where explicitly indicated to be otherwise, we show $D = 10$ Mpc and $\mathcal{R} = 1$. In general, there is a very small relative difference when the jet is viewed edge on. As the viewing angle approaches face on or the jet is placed at a larger distance, the relative difference increases along with the error. In all but one image, even when the relative difference shifts away from zero, the zero relative difference line is within 3 standard deviations, which suggests excellent agreement between jet lengths derived from the raw ipole and reconstructed images.

Since we lack centering information in the reconstructed images, we chose to estimate the jet velocity perpendicular to the line of sight by assuming both the top and bottom jet have the same speed. Then, the velocity of the jet in the source's frame is:

$$v = \frac{1}{2} \frac{|l_{\mathrm{jet}}(t_2) - l_{\mathrm{jet}}(t_1)|}{t_2 - t_1}. \tag{11}$$

Note that we used the same expression to derive the velocity in the reconstructed images (v_{rec}) but replaced l_{jet} with $l_{\mathrm{jet,rec}}$ in Equation (11). Very long baseline interferometry (VLBI) observations can provide the apparent motion of the jet, which may be superluminal due to relativistic effects. To account for this, we assumed the jet has a velocity of v_{rec} and then estimated the apparent velocity ($v_{\mathrm{app,rec}}$) via the time in the observer's frame (t'_1, t'_2):

$$v_{\mathrm{app,rec}} = \frac{1}{2} \frac{|l_{\mathrm{jet,rec}}(t'_2) - l_{\mathrm{jet,rec}}(t'_1)|}{t'_2 - t'_1} = v_{\mathrm{rec}}\left(1 - \frac{v_{\mathrm{rec}}}{c}\frac{\cos\theta}{\sin\theta}\right)^{-1}. \tag{12}$$

We used the relationship

$$t_2' - t_1' = (t_2 - t_1)\left(1 - \frac{v_{\text{rec}}}{c}\frac{\cos\theta}{\sin\theta}\right) \tag{13}$$

in the last expression. Note that the division by $\sin\theta$ was to account for the fact the v_{rec} measures the velocity parallel to the line of sight with no time delay effects while we require an estimate of the velocity along the jet axis (which we can obtain since the geometry is fully known). We only present the apparent velocity for the reconstructed images ($v_{\text{app,rec}}$), since this represents an estimate of what VLBI observations would truly be.

Table 3. Here we tabulate the estimated jet length for each model at each time for various choices of the distance D, viewing angle θ, and plasma temperature ratio $\mathcal{R}$. We compare the jet length as computed from the base ipole image (l_{jet}) and the reconstructed image ($l_{\text{jet,rec}}$).

Model	Time (t_g)	Distance (Mpc)	θ	$\mathcal{R}$	l_{jet} (r_g)	$l_{\text{jet,rec}}$ (r_g)
m7a0.0-HR	38,000	10	10°	1	5091^{+272}_{-272}	4927^{+317}_{-317}
	78,000	10	10°	1	9042^{+356}_{-356}	9104^{+407}_{-407}
	38,000	10	45°	1	$22{,}006^{+123}_{-123}$	$21{,}774^{+731}_{-731}$
	78,000	10	45°	1	$39{,}832^{+318}_{-318}$	$39{,}678^{+501}_{-501}$
	38,000	10	90°	1	$28{,}658^{+148}$	$28{,}463^{+173}_{-173}$
	78,000	10	90°	1	$57{,}068^{+401}_{-401}$	$56{,}844^{+237}_{-237}$
m7a0.9-HR	38,000	10	10°	1	8457^{+446}_{-446}	4191^{+726}_{-726}
	78,000	10	10°	1	$17{,}585^{+233}_{-233}$	$16{,}837^{+726}_{-726}$
	38,000	10	45°	1	$33{,}128^{+336}_{-336}$	$31{,}941^{+683}_{-683}$
	78,000	10	45°	1	$73{,}535^{+890}_{-890}$	$71{,}855^{+1058}_{-1058}$
	38,000	10	90°	1	$48{,}468^{+306}_{-306}$	$46{,}972^{+1028}_{-1028}$
	78,000	10	90°	1	$111{,}503^{+1085}_{-1085}$	$110{,}327^{+2657}_{-2657}$
	38,000	10	90°	20	$49{,}457^{+150}_{-150}$	$48{,}930^{+358}_{-358}$
	78,000	10	90°	20	$119{,}897^{+173}_{-173}$	$119{,}256^{+415}_{-415}$
	38,000	100	90°	1	$46{,}715^{+400}_{-400}$	$40{,}244^{+2714}_{-2714}$
	78,000	100	90°	1	$109{,}091^{+959}_{-959}$	$98{,}822^{+2075}_{-2075}$

We used the data at $t_1 = 38{,}000\,t_g$ and $t_2 = 78{,}000\,t_g$ for each model to estimate the jet velocity. We tabulated the estimated velocity and apparent velocity for each model in Table 4. We found excellent agreement between the velocities derived from the ray-traced and reconstructed images. It is interesting to note the apparently faster jet for $\mathcal{R} = 20$. We suspect the increased speed is due to the emitting material being dominated by material near the jet axis rather than some of the slower moving material around the jet head, which does not produce much emission at 230 GHz as the electron temperature is reduced. The more powerful jet model m7a0.9-HR demonstrates that such jets may appear as superluminal sources, as we found a maximum $v_{\text{app,rec}} \approx 1.526c$ at $\theta = 10°$.

Table 4. Here we tabulate the estimated jet velocity for each model for different choices of the distance D, viewing angle θ, and plasma temperature ratio $\mathcal{R}$. The velocities shown were calculated using the base ipole images (v), the reconstructed images (v_{rec}), and the reonstructed images, while accounting for the possibility of superluminal motion ($v_{\mathrm{app,rec}}$).

Model	Distance (Mpc)	θ	$\mathcal{R}$	v (c)	v_{rec} (c)	$v_{\mathrm{app,rec}}$ (c)
m7a0.0-HR	10	10°	1	$0.049^{+0.006}_{-0.006}$	$0.052^{+0.006}_{-0.006}$	$0.074^{+0.009}_{-0.009}$
	10	45°	1	$0.223^{+0.005}_{-0.005}$	$0.224^{+0.011}_{-0.011}$	$0.288^{+0.014}_{-0.014}$
	10	90°	1	$0.355^{+0.005}_{-0.005}$	$0.355^{+0.004}_{-0.004}$	$0.355^{+0.004}_{-0.004}$
m7a0.9-HR	10	10°	1	$0.114^{+0.006}_{-0.006}$	$0.158^{+0.011}_{-0.011}$	$1.526^{+0.102}_{-0.102}$
	10	45°	1	$0.505^{+0.012}_{-0.012}$	$0.499^{+0.016}_{-0.016}$	$0.996^{+0.031}_{-0.031}$
	10	90°	1	$0.788^{+0.014}_{-0.014}$	$0.792^{+0.036}_{-0.036}$	$0.792^{+0.036}_{-0.036}$
	10	90°	20	$0.88^{+0.002}_{-0.002}$	$0.879^{+0.007}_{-0.007}$	$0.879^{+0.007}_{-0.007}$
	100	90°	1	$0.780^{0.013}_{-0.013}$	$0.732^{+0.043}_{-0.043}$	$0.732^{+0.043}_{-0.043}$

4. Discussion

4.1. Extracting Jet Physics from VLBI Images

A key feature of the jets in our models is the bright "bubbles" (or knots) of 230 GHz emission, which appear to correlate with recollimation shocks. Such structures have been seen in VLBI images of various AGN jets [48–50]. Previous simulations of jets in various astrophysical contexts have demonstrated that recollimation occurs when there is a pressure mismatch between the jet and the surrounding medium, which could be a static atmosphere or a slower-moving jet sheath [51–54]. The number of recollimation shocks along the jet axis is dependent on the properties of the jet and medium. It is therefore possible that direct VLBI of TDE jets will allow in-depth modeling of jet launching and could also aid in constraining the properties of the surrounding medium. For instance, one work successfully applied simulations of MHD jets to constrain properties of BL Lacartae, which is a blazar jet with recollimation features [55].

We suggest that a similar approach may be applied in TDE jets. With a suitable exploration of the parameter space, it is conceivable that an analysis similar to that of [55] could be applied to TDE jets in cases where VLBI is possible. A broader exploration of TDE jets through various simulation methodologies is strongly suggested. We plan to explore the effects of the ambient medium, magnetic field strength, and disk accretion rate on the jet properties in the case of a SANE, super-Eddington disk in a future work.

4.2. Proposed Observational Methodology

Our synthetic ngEHT observations demonstrate that a 24-hour observation may be sufficient to study both the structure and/or motion of newly born TDE jets. This is much shorter than the fallback time, which is on the order of month(s), and the duration of radio emissions in several TDEs, which can sometimes be visible for years [12]. Our suggested observational strategy is conducting rapid followup of newly discovered optical/X-ray TDEs when they are near the peak of their emission in order to study both the early- and late-time properties of their jets (if present). Observations with a single telescope at 230 GHz could be conducted to search for TDEs emitting in the radio. If emission at 230 GHz is detected, we suggest that the ngEHT conduct VLBI follow-up of targets within no more than a month. Our imaging simulations were done assuming a full ngEHT array consisting of the 2022 EHT stations plus 11 additional sites, but depending on the target, not all sites may need to be available in order to obtain a high-fidelity image reconstruction.

Unlike many other EHT/ngEHT targets, TDEs will appear randomly across the sky, and the ngEHT will need to be capable of follow-up observations in the order of a week to weeks. Our current modeling of jets and outflows from super-Eddington disks is too sparse to let us make predictions regarding how long the jets will be visible at 230 GHz. However, if radio TDE observations are any indicator, emissions may persist for many months [12].

TDEs provide an excellent laboratory for studying jet/accretion/black hole systems across a wide range of accretion states over a relatively short period of time ($\sim$1-a few years). In several cases, TDEs have shown state transitions after several hundred days in the X-ray range, which are likely associated with the evolution of the disk as the mass accretion rate declines. Stone and Metzger [29] argue, for instance, that the transition from a thick, super-Eddington disk to a thin disk can explain the jet shut-off in jetted TDEs, such as *Swift* J1644+57 and *Swift* J2058+05 [56,57], but recent simulations [58,59] demonstrated that MAD is possible even for thinner accretion disks. As such, long-term VLBI monitoring is strongly suggested. as this would allow for (1) the radio-submm emission of the outflows to be characterized and compared to the behavior of the accretion flow, and (2) the direct study of how the jet evolves morphologically as the disk state changes.

Another attractive potential target which we have yet to attribute a self-contained study to is jetted TDEs. These TDEs are extremely rare, and current observations suggest only about 1% of all TDEs will produce powerful relativistic jets. These jets will produce extremely bright radiation in the X-ray and the radio-submm regions. However, most have been distant due to the lower probability of their occurrence. Should a jetted TDE occur nearby enough for VLBI to resolve the jet, we strongly suggest such jets be treated as targets for the ngEHT.

Lastly, a recent TDE AT2018hyz showed a late outflow ($\sim$3 years after the initial outburst) and brightened in the radio over several hundred days [60]. Unlike many other radio TDEs, a relatively bright 240 GHz emission was detected. If AT2018hyz is in fact a jet instead of a spherical outflow, Cendes et al. [60] estimates that the velocity could reach $\lesssim$0.6c. AT2018hyz is a relatively nearby TDE at $\sim$204 Mpc, but the flux density at the time of detection ($\sim$0.2 mJy at 240 GHz) makes it too dim for ngEHT follow-up. However, placing AT2018hyz at $\sim$50 Mpc would shift the flux density to $\sim$5 mJy, which is the minimum estimated flux density required for an ngEHT VLBI detection. Future TDEs will likely be monitored across the radio-submm, so nearby targets of opportunity such as late radio TDEs like AT2018hyz should be considered, should they show significant radio emission.

While the EHT operated using dedicated time, the ngEHT will include more arrays and some telescopes may be sequestered for targeting of opportune VLBI observations (Doeleman et al., in prep). Development and proposals for the ngEHT are still underway, but we expect that fast (within 1-several weeks) response coverage of sources of interest in order to conduct VLBI of TDEs (and other transient sources) should be possible (Shepard Doeleman, private communication). The Atacama Large Millimeter Array (or ALMA), the most sensitive antenna, will maintain sustained coverage [40]. As such, we anticipate that at least unresolved radio data can be captured, which would prove equally invaluable to the study of radio TDEs.

5. Conclusions

In this work, we have demonstrated through a synthetic imaging analysis that TDE jets resembling the GRRMHD models presented in Curd et al. [14] are compelling ngEHT targets. We also confirmed that the detection limits considered in Curd et al. [14] are roughly applicable, as m7a0.0-HR did not produce detectable emission at a distance of 100 Mpc in our imaging analysis.

Various shock features in the jet are visible for the 10 Mpc images we consider, and studying the jet morphology in these cases could aid in characterizing the environment of the BH. Most TDEs that occur during the ngEHT mission will be farther away, but the apparent motion, which may be superluminal, can be extracted in such cases.

We suggest that the ngEHT be utilized for radio follow-up of TDEs. Our models study the birth of a TDE jet in the first ~ 48 days after the disk forms under the assumption that the disk is SANE, super-Eddington, and threaded by a dynamically important magnetic field. However, TDEs which occur nearby and have jet properties similar to those of jetted TDEs, such as *Swift* J1644+57 or AT2018hyz, may provide interesting targets.

Author Contributions: Conceptualization, R.E. and R.A.; Formal analysis, B.C. and F.R.; Investigation, B.C.; Writing—original draft, B.C., F.R. and R.A.; Writing—review & editing, B.C. All authors have read and agreed to the published version of the manuscript.

Funding: Brandon Curd was supported by NSF grant AST-1816420, and made use of computational support from NSF via XSEDE/ACCESS resources (grant TG-AST080026N). Razieh Emami acknowledges the support by the Institute for Theory and Computation at the Center for Astrophysics and grants 21-atp21-0077, NSF AST-1816420, and HST-GO-16173.001-A for very generous support. Freek Roelofs was supported by NSF grants AST-1935980 and AST-2034306. This work was supported by the Black Hole Initiative at Harvard University, made possible through the support of grants from the Gordon and Betty Moore Foundation and the John Templeton Foundation. The opinions expressed in this publication are those of the authors and do not necessarily reflect the views of the Moore or Templeton Foundations.

Data Availability Statement: The data presented in this study are available on request from the corresponding author.

Acknowledgments: We graciously acknowledge Koushik Chatterjee, Lani Oramas, Joaquin Duran, and Hayley West for fruitful discussions that were helpful in the preparation of this work.

Conflicts of Interest: The authors declare no conflict of interest.

Appendix A. Full Image Library

In Figures A1–A4, we show the full library of images analyzed in this work. All of the base images were ray traced at $\nu = 230$ GHz and then convolved with a Gaussian beam with a FWHM of 20 μas. Similarly, we blurred the reconstructed image using the same beam for comparison. Note that for Figures A2 and A4, we also shifted the reconstructed image to be approximately centered for comparison with the base image. In general, bright features are represented quite well in the reconstruction; however, some noise is introduced in dimmer sources (i.e., see the reconstruction of m7a0.9-HR at $D = 100$ Mpc).

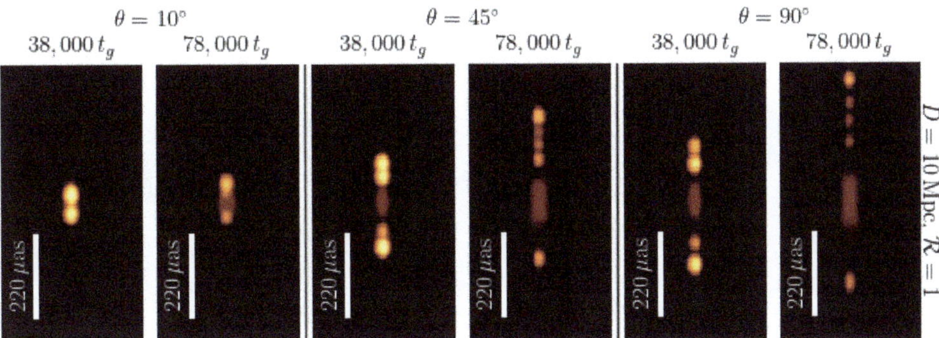

Figure A1. Here we show the full library of base ipole images for model m7a0.0-HR. The time of each column is indicated at the top; the distance D and plasma temperature ratio $\mathcal{R}$ for each row are indicated on the right. The angle of the observer relative to the jet axis is indicated above each set of two rows. Each image spans 550×770 μas^2 and is blurred by convolving the base image with a Gaussian beam with a FWHM of 20 μas. The color scale is logarithmic, spanning three orders of magnitude, and each image uses the same maximum for the intensity scale.

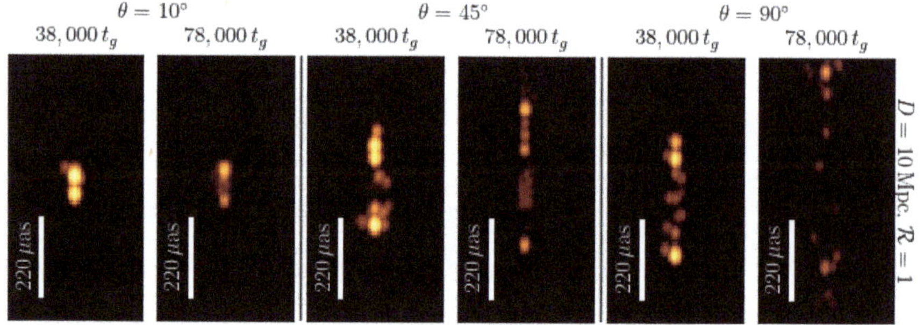

Model `m7a0.0-HR`: 230 GHz Blurred Reconstruction

Figure A2. The same as Figure A1 but showing the reconstructed images. Note that the intensity scale for each panel is the same as the corresponding panel in Figure A1 for comparison.

Model `m7a0.9-HR`: 230 GHz Blurred Images

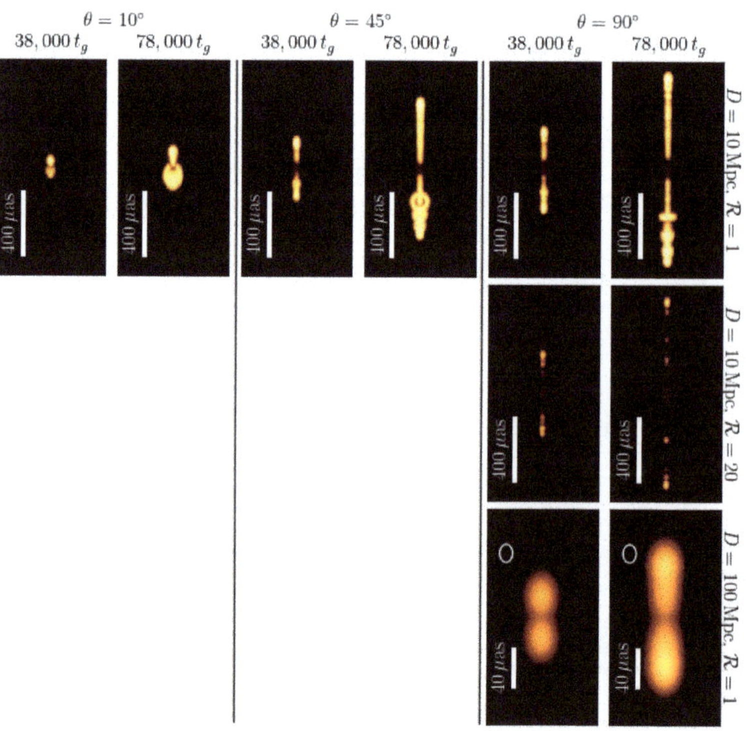

Figure A3. Here we show the full library of base ipole images for model m7a0.9-HR. Each image in the top and middle rows spans $1000 \times 1400\,\mu\mathrm{as}^2$, and the bottom row spans $170 \times 238\,\mu\mathrm{as}^2$. Each image was blurred by convolving the base image with a Gaussian beam with a FWHM of $20\,\mu\mathrm{as}$ (indicated by the white circle in the bottom right panel). The time of each column is indicated at the top; the distance D and plasma temperature ratio $\mathcal{R}$ for each row are indicated on the right. The angle of the observer relative to the jet axis is indicated above each set of two rows. The color scale is logarithmic, spanning three orders of magnitude in each image. We used the same color scale for the $D = 10$ Mpc images, but reduced the maximum by an order of magnitude in the $D = 100$ Mpc images to better show the image features.

Model `m7a0.9-HR`: 230 GHz Blurred Reconstruction

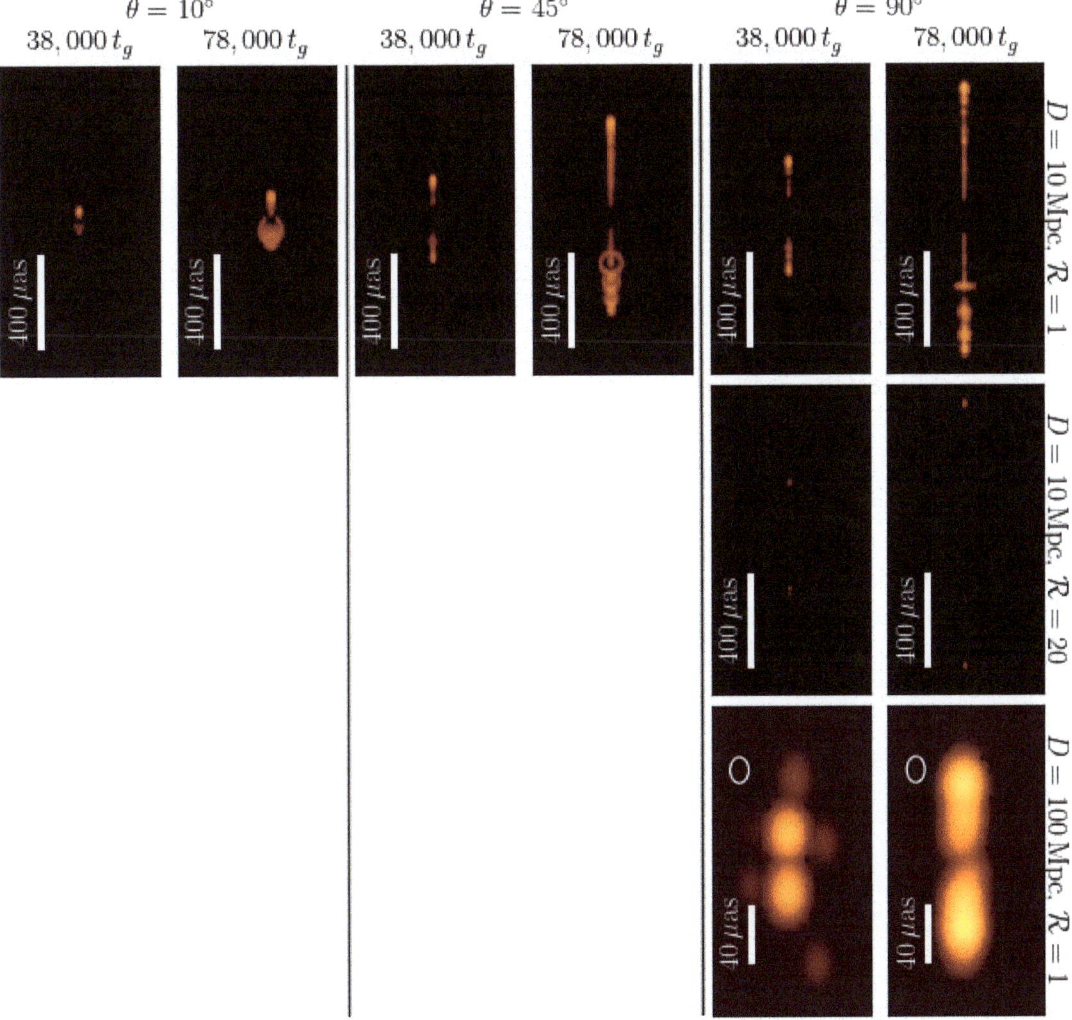

Figure A4. The same as Figure A3 but showing the reconstructed images. Note that the intensity scale for each panel is the same as the corresponding panel in Figure A3 for comparison.

Appendix B. Fitting Procedure for Jet Head Position

Here we describe the algorithm implemented to estimate the positions of the top and bottom jet heads in each image. In order to track the jet's motion, we first smoothed the data (either the base image or the reconstructed image) with a Gaussian beam which assumed an angular resolution for the VLBI observations of $\Delta\theta = 20\,\mu$as. We then binned the data along the symmetry axis of the jet (y) by summing the flux along each row (x).

Since the images appear to be roughly Lorentzian, we first attempted to fit a double Lorentzian function of the form:

$$f_d(y) = a_1 \frac{w_1^2}{(y - y_{\text{cen},2})^2 + w_1^2} + a_2 \frac{w_2^2}{(y - y_{\text{cen},2})^2 + w_2^2}, \tag{A1}$$

where a is the amplitude, w is the width, and y_{cen} is the center defining the curve. We took y_{cen} as a measure of the jet head location. If this fitting procedure did not produce a good fit, we found the peaks by performing a two-step fitting procedure in which we fit a single Lorentzian:

$$f_1(y) = a_1 \frac{w_1^2}{(y - y_{\text{cen},1})^2 + w_1^2}, \tag{A2}$$

and then subtracted the fit $f_1(y)$ from the data and then fit the second peak with:

$$f_2(y) = a_2 \frac{w_2^2}{(y - y_{\text{cen},2})^2 + w_2^2}. \tag{A3}$$

We implemented the Python package SciPy [61] to optimize the curve(s) and estimate the jet head positions and errors. We show an example of the data and the Lorentzian fit for both the base image and the reconstruction in Figure A5. The reconstruction tends to be a bit broader, but the fitting procedure works equally well for all images and reconstructed images.

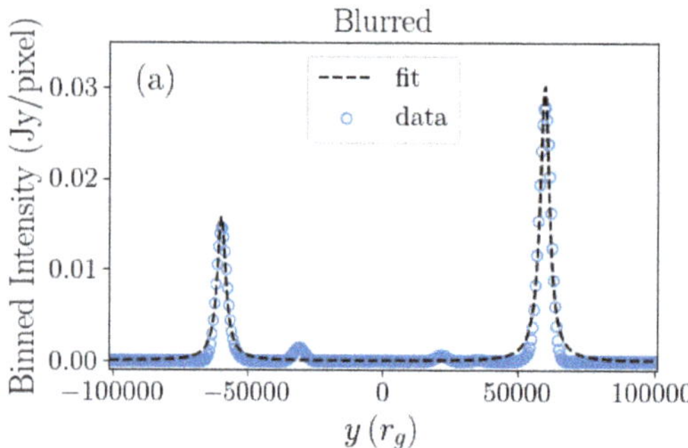
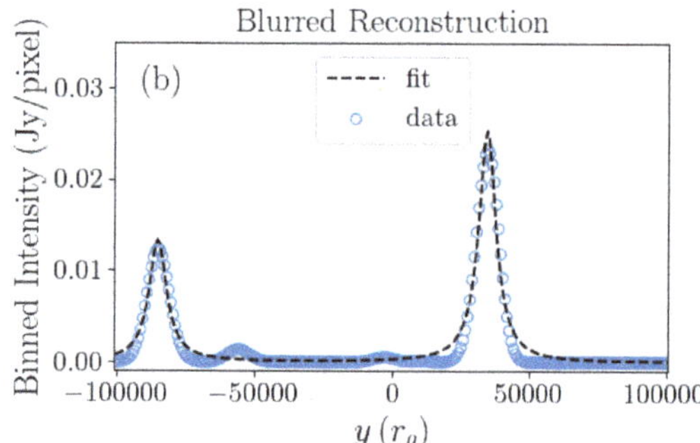

Figure A5. We demonstrate the fit performance using a ray-traced image of model m7a0.9-HR at $t = 78{,}000\,t_g$, $\theta = 90°$, $D = 10\,\text{Mpc}$, and $\mathcal{R} = 20$. We show the x-binned data and a double Lorentzian (Equation (A1)) fit for (**a**) the blurred base image and (**b**) the blurred reconstruction.

Notes

[1]   https://github.com/Smithsonian/ngehtsim (accessed on 1 June 2022).

References

1.   Hills, J.G. Possible power source of Seyfert galaxies and QSOs. *Nature* **1975**, *254*, 295–298. [CrossRef]
2.   Rees, M.J. Tidal disruption of stars by black holes of 10^6–10^8 solar masses in nearby galaxies. *Nature* **1988**, *333*, 523–528. [CrossRef]
3.   Guillochon, J.; Ramirez-Ruiz, E. Hydrodynamical Simulations to Determine the Feeding Rate of Black Holes by the Tidal Disruption of Stars: The Importance of the Impact Parameter and Stellar Structure. *Astrophys. J.* **2013**, *767*, 25. [CrossRef]
4.   Mainetti, D.; Lupi, A.; Campana, S.; Colpi, M.; Coughlin, E.R.; Guillochon, J.; Ramirez-Ruiz, E. The fine line between total and partial tidal disruption events. *Astron. Astrophys.* **2017**, *600*, A124. [CrossRef]
5.   Steinberg, E.; Stone, N.C. The Origins of Peak Light in Tidal Disruption Events. *arXiv* **2022**, arXiv:2206.10641. [CrossRef]
6.   Stone, N.; Sari, R.; Loeb, A. Consequences of strong compression in tidal disruption events. *MNRAS* **2013**, *435*, 1809–1824. [CrossRef]
7.   Abramowicz, M.A.; Calvani, M.; Nobili, L. Thick accretion disks with super-Eddington luminosities. *Astrophys. J.* **1980**, *242*, 772–788. [CrossRef]

8.   Abramowicz, M.A.; Czerny, B.; Lasota, J.P.; Szuszkiewicz, E. Slim Accretion Disks. *Astrophys. J.* **1988**, *332*, 646. [CrossRef]

9.   Golightly, E.C.A.; Nixon, C.J.; Coughlin, E.R. On the Diversity of Fallback Rates from Tidal Disruption Events with Accurate Stellar Structure. *Astrophys. J.* **2019**, *882*, L26. [CrossRef]

10. Komossa, S. Tidal disruption of stars by supermassive black holes: Status of observations. *J. High Energy Astrophys.* **2015**, *7*, 148–157. [CrossRef]

11. Gezari, S. Tidal Disruption Events. *ARA&A* **2021**, *59*, 21–58 . [CrossRef]

12. Alexander, K.D.; van Velzen, S.; Horesh, A.; Zauderer, B.A. Radio Properties of Tidal Disruption Events. *Space Sci. Rev.* **2020**, *216*, 81. [CrossRef]

13. Dai, L.; McKinney, J.C.; Roth, N.; Ramirez-Ruiz, E.; Miller, M.C. A Unified Model for Tidal Disruption Events. *Astrophys. J.* **2018**, *859*, L20. [CrossRef]

14. Curd, B.; Emami, R.; Anantua, R.; Palumbo, D.; Doeleman, S.; Narayan, R. Jets from SANE Super-Eddington Accretion Disks: Morphology, Spectra, and Their Potential as Targets for ngEHT. *arXiv* **2022**, arXiv:2206.06358.

15. Gammie, C.F.; McKinney, J.C.; Tóth, G. HARM: A Numerical Scheme for General Relativistic Magnetohydrodynamics. *Astrophys. J.* **2003**, *589*, 444–457. [CrossRef]

16. Tchekhovskoy, A.; Metzger, B.D.; Giannios, D.; Kelley, L.Z. Swift J1644+57 gone MAD: The case for dynamically important magnetic flux threading the black hole in a jetted tidal disruption event. *MNRAS* **2014**, *437*, 2744–2760. [CrossRef]

17. Curd, B.; Narayan, R. GRRMHD simulations of tidal disruption event accretion discs around supermassive black holes: Jet formation, spectra, and detectability. *MNRAS* **2019**, *483*, 565–592. [CrossRef]

18. Blandford, R.D.; Znajek, R.L. Electromagnetic extraction of energy from Kerr black holes. *MNRAS* **1977**, *179*, 433–456. [CrossRef]

19. Abramowicz, M.A.; Ellis, G.F.R.; Lanza, A. Relativistic Effects in Superluminal Jets and Neutron Star Winds. *Astrophys. J.* **1990**, *361*, 470. [CrossRef]

20. Coughlin, E.R.; Begelman, M.C. Structured, relativistic jets driven by radiation. *MNRAS* **2020**, *499*, 3158–3177. . 10.1093/mnras/staa3026. [CrossRef]

21. Dadhich, N.; Tursunov, A.; Ahmedov, B.; Stuchlík, Z. The distinguishing signature of magnetic Penrose process. *MNRAS* **2018**, *478*, L89–L94,

22. Stuchlík, Z.; Kološ, M.; Kovář, J.; Slaný, P.; Tursunov, A. Influence of Cosmic Repulsion and Magnetic Fields on Accretion Disks Rotating around Kerr Black Holes. *Universe* **2020**, *6*, 26. [CrossRef]

23. Penrose, R. Gravitational Collapse: The Role of General Relativity. *Nuovo Cim. Riv. Ser.* **1969**, *1*, 252.

24. Thomsen, L.L.; Kwan, T.M.; Dai, L.; Wu, S.C.; Roth, N.; Ramirez-Ruiz, E. Dynamical Unification of Tidal Disruption Events. *Astrophys. J.* **2022**, *937*, L28. [CrossRef]

25. Penna, R.F.; Narayan, R.; Sądowski, A. General relativistic magnetohydrodynamic simulations of Blandford-Znajek jets and the membrane paradigm. *MNRAS* **2013**, *436*, 3741–3758.

26. Doeleman, S.; Blackburn, L.; Dexter, J.; Gomez, J.L.; Johnson, M.D.; Palumbo, D.C.; Weintroub, J.; Farah, J.R.; Fish, V.; Loinard, L.; et al. Studying Black Holes on Horizon Scales with VLBI Ground Arrays. *Bull. Am. Astron. Soc.* **2019**, *51*, 256.

27. Ivezić, Ž.; Kahn, S.M.; Tyson, J.A.; Abel, B.; Acosta, E.; Allsman, R.; Alonso, D.; AlSayyad, Y.; Anderson, S.F.; Andrew, J.; et al. LSST: From Science Drivers to Reference Design and Anticipated Data Products. *Astrophys. J.* **2019**, *873*, 111. [CrossRef]

28. Bricman, K.; Gomboc, A. The Prospects of Observing Tidal Disruption Events with the Large Synoptic Survey Telescope. *Astrophys. J.* **2020**, *890*, 73. [CrossRef]

29. Stone, N.C.; Metzger, B.D. Rates of stellar tidal disruption as probes of the supermassive black hole mass function. *MNRAS* **2016**, *455*, 859–883. [CrossRef]

30. Sadowski, A.; Tejeda, E.; Gafton, E.; Rosswog, S.; Abarca, D. Magnetohydrodynamical simulations of a deep tidal disruption in general relativity. *Mon. Not. Roy. Astron. Soc.* **2016**, *458*, 4250–4268. [CrossRef]

31. Curd, B. Global simulations of tidal disruption event disc formation via stream injection in GRRMHD. *Mon. Not. Roy. Astron. Soc.* **2021**, *507*, 3207–3227. [CrossRef]

32. Novikov, I.D.; Thorne, K.S. Astrophysics of black holes. In *Black Holes (Les Astres Occlus)*; Gordon & Breach: New York, NY, USA, 1973; pp. 343–450.

33. Sądowski, A.; Narayan, R.; Tchekhovskoy, A.; Abarca, D.; Zhu, Y.; McKinney, J.C. Global simulations of axisymmetric radiative black hole accretion discs in general relativity with a mean-field magnetic dynamo. *MNRAS* **2015**, *447*, 49–71.

34. Barniol Duran, R.; Tchekhovskoy, A.; Giannios, D. Simulations of AGN jets: Magnetic kink instability versus conical shocks. *MNRAS* **2017**, *469*, 4957–4978.

35. Mościbrodzka, M.; Gammie, C.F. IPOLE - semi-analytic scheme for relativistic polarized radiative transport. *MNRAS* **2018**, *475*, 43–54. [CrossRef]

36. Yarza, R.; Wong, G.N.; Ryan, B.R.; Gammie, C.F. Bremsstrahlung in GRMHD Models of Accreting Black Holes. *Astrophys. J.* **2020**, *898*, 50. [CrossRef]

37. Wong, G.N.; Prather, B.S.; Dhruv, V.; Ryan, B.R.; Mościbrodzka, M.; Chan, C.k.; Joshi, A.V.; Yarza, R.; Ricarte, A.; Shiokawa, H.; et al. PATOKA: Simulating Electromagnetic Observables of Black Hole Accretion. *Astrophys. J.* **2022**, *259*, 64. [CrossRef]

38. Ohmura, T.; Machida, M.; Nakamura, K.; Kudoh, Y.; Asahina, Y.; Matsumoto, R. Two-Temperature Magnetohydrodynamics Simulations of Propagation of Semi-Relativistic Jets. *Galaxies* **2019**, *7*, 14. [CrossRef]

39. Ohmura, T.; Machida, M.; Nakamura, K.; Kudoh, Y.; Matsumoto, R. Two-temperature magnetohydrodynamic simulations for sub-relativistic active galactic nucleus jets: Dependence on the fraction of the electron heating. *MNRAS* **2020**, *493*, 5761–5772. [CrossRef]

40. Raymond, A.W.; Palumbo, D.; Paine, S.N.; Blackburn, L.; Córdova Rosado, R.; Doeleman, S.S.; Farah, J.R.; Johnson, M.D.; Roelofs, F.; Tilanus, R.P.J.; et al. Evaluation of New Submillimeter VLBI Sites for the Event Horizon Telescope. *Astrophys. J.* **2021**, *253*, 5. [CrossRef]

41. Roelofs, F. et al. [Black Hole Initiative at Harvard University]. 2022, in preparation.

42. Chael, A.A.; Johnson, M.D.; Narayan, R.; Doeleman, S.S.; Wardle, J.F.C.; Bouman, K.L. High-resolution Linear Polarimetric Imaging for the Event Horizon Telescope. *Astrophys. J.* **2016**, *829*, 11. [CrossRef]

43. Chael, A.A.; Johnson, M.D.; Bouman, K.L.; Blackburn, L.L.; Akiyama, K.; Narayan, R. Interferometric Imaging Directly with Closure Phases and Closure Amplitudes. *Astrophys. J.* **2018**, *857*, 23. [CrossRef]

44. Doeleman, S. et al. [Harvard-Smithsonian Center for Astrophysics]. 2022, in preparation.

45. Gelaro, R.; McCarty, W.; Suárez, M.J.; Todling, R.; Molod, A.; Takacs, L.; Randles, C.A.; Darmenov, A.; Bosilovich, M.G.; Reichle, R.; et al. The Modern-Era Retrospective Analysis for Research and Applications, Version 2 (MERRA-2). *J. Clim.* **2017**, *30*, 5419–5454. [CrossRef]

46. Paine, S. The Am Atmospheric Model. 2019. Available online: https://zenodo.cern.ch/record/3406496#.Y5fRyH1ByUk (accessed on 1 June 2022).

47. Event Horizon Telescope Collaboration.; Akiyama, K.; Alberdi, A.; Alef, W.; Asada, K.; Azulay, R.; Baczko, A.K.; Ball, D.; Baloković, M.; Barrett, J.; et al. First M87 Event Horizon Telescope Results. IV. Imaging the Central Supermassive Black Hole. *Astrophys. J.* **2019**, *875*, L4. [CrossRef]

48. Jorstad, S.G.; Marscher, A.P.; Lister, M.L.; Stirling, A.M.; Cawthorne, T.V.; Gear, W.K.; Gómez, J.L.; Stevens, J.A.; Smith, P.S.; Forster, J.R.; et al. Polarimetric Observations of 15 Active Galactic Nuclei at High Frequencies: Jet Kinematics from Bimonthly Monitoring with the Very Long Baseline Array. *Astrophys. J.* **2005**, *130*, 1418–1465. [CrossRef]

49. Lister, M.L.; Aller, M.F.; Aller, H.D.; Homan, D.C.; Kellermann, K.I.; Kovalev, Y.Y.; Pushkarev, A.B.; Richards, J.L.; Ros, E.; Savolainen, T. MOJAVE. X. Parsec-scale Jet Orientation Variations and Superluminal Motion in Active Galactic Nuclei. *Astrophys. J.* **2013**, *146*, 120. [CrossRef]

50. Cohen, M.H.; Meier, D.L.; Arshakian, T.G.; Homan, D.C.; Hovatta, T.; Kovalev, Y.Y.; Lister, M.L.; Pushkarev, A.B.; Richards, J.L.; Savolainen, T. Studies of the Jet in Bl Lacertae. I. Recollimation Shock and Moving Emission Features. *Astrophys. J.* **2014**, *787*, 151. [CrossRef]

51. Kohler, S.; Begelman, M.C.; Beckwith, K. Recollimation boundary layers in relativistic jets. *MNRAS* **2012**, *422*, 2282–2290. [CrossRef]

52. Lazzati, D.; Morsony, B.J.; Blackwell, C.H.; Begelman, M.C. Unifying the Zoo of Jet-driven Stellar Explosions. *Astrophys. J.* **2012**, *750*, 68. [CrossRef]

53. Mizuno, Y.; Gómez, J.L.; Nishikawa, K.I.; Meli, A.; Hardee, P.E.; Rezzolla, L. Recollimation Shocks in Magnetized Relativistic Jets. *Astrophys. J.* **2015**, *809*, 38. [CrossRef]

54. Hervet, O.; Meliani, Z.; Zech, A.; Boisson, C.; Cayatte, V.; Sauty, C.; Sol, H. Shocks in relativistic transverse stratified jets. A new paradigm for radio-loud AGN. *Astron. Astrophys.* **2017**, *606*, A103. [CrossRef]

55. Gómez, J.L.; Lobanov, A.P.; Bruni, G.; Kovalev, Y.Y.; Marscher, A.P.; Jorstad, S.G.; Mizuno, Y.; Bach, U.; Sokolovsky, K.V.; Anderson, J.M.; et al. Probing the Innermost Regions of AGN Jets and Their Magnetic Fields with RadioAstron. I. Imaging BL Lacertae at 21 Microarcsecond Resolution. *Astrophys. J.* **2016**, *817*, 96. [CrossRef]

56. Zauderer, B.A.; Berger, E.; Margutti, R.; Pooley, G.G.; Sari, R.; Soderberg, A.M.; Brunthaler, A.; Bietenholz, M.F. Radio Monitoring of the Tidal Disruption Event Swift J164449.3+573451. II. The Relativistic Jet Shuts Off and a Transition to Forward Shock X-Ray/Radio Emission. *Astrophys. J.* **2013**, *767*, 152. [CrossRef]

57. Pasham, D.R.; Cenko, S.B.; Levan, A.J.; Bower, G.C.; Horesh, A.; Brown, G.C.; Dolan, S.; Wiersema, K.; Filippenko, A.V.; Fruchter, A.S.; et al. A Multiwavelength Study of the Relativistic Tidal Disruption Candidate Swift J2058.4+0516 at Late Times. *Astrophys. J.* **2015**, *805*, 68. [CrossRef]

58. Curd, B.; Narayan, R. GRRMHD Simulations of MAD Accretion Disks Declining from Super-Eddington to Sub-Eddington Accretion Rates. *arXiv* **2022**, arXiv:2209.12081.

59. Liska, M.T.P.; Musoke, G.; Tchekhovskoy, A.; Porth, O.; Beloborodov, A.M. Formation of Magnetically Truncated Accretion Disks in 3D Radiation-transport Two-temperature GRMHD Simulations. *Astrophys. J.* **2022**, *935*, L1. [CrossRef]

60. Cendes, Y.; Berger, E.; Alexander, K.D.; Gomez, S.; Hajela, A.; Chornock, R.; Laskar, T.; Margutti, R.; Metzger, B.; Bietenholz, M.F.; et al. A Mildly Relativistic Outflow Launched Two Years after Disruption in Tidal Disruption Event AT2018hyz. *Astrophys. J.* **2022**, *938*, 28. [CrossRef]

61. Virtanen, P.; Gommers, R.; Oliphant, T.E.; Haberland, M.; Reddy, T.; Cournapeau, D.; Burovski, E.; Peterson, P.; Weckesser, W.; Bright, J.; et al. SciPy 1.0: Fundamental Algorithms for Scientific Computing in Python. *Nat. Methods* **2020**, *17*, 261–272. [CrossRef]

Article

Spectral Line VLBI Studies Using the ngEHT

Dong-Jin Kim [1,2,*,†] and Vincent Fish [1,†]

1   Massachusetts Institute of Technology Haystack Observatory, Westford, MA 01886, USA
2   Max-Planck-Institut für Radioastronomie, Auf dem Hügel 69, 53121 Bonn, Germany
*   Correspondence: dongjink@mit.edu
†   These authors contributed equally to this work.

Abstract: Spectroscopy in the mm/sub-mm wavelength range is a powerful tool to study the gaseous medium in various astrophysical environments. The next generation Event Horizon Telescope (ngEHT) equipped with a wide-bandwidth backend system has great potential for science using high angular resolution spectroscopy. Spectral line VLBI studies using the ngEHT will enable us to scrutinize compact astrophysical objects obscured by an optically thick medium on unprecedented angular scales. However, the capabilities of ngEHT for spectroscopy and specific scientific applications have not been properly envisioned. In this white paper, we briefly address science cases newly achievable via spectral line VLBI observations in the mm/sub-mm wavelength ranges, and suggest technical requirements to facilitate spectral line VLBI studies in the ngEHT era.

Keywords: very long baseline interferometry (1769); radio astronomy (1338); millimeter astronomy (1061); submillimeter astronomy (1647); radio telescopes (1360); high angular resolution (2167)

1. Introduction

The next generation Event Horizon Telescope (ngEHT) is primarily designed to image the continuum emission from the vicinity of the photon ring of supermassive black holes (SMBHs) with a better imaging performance than that of the current EHT array in terms of sensitivity and sampling in the spatial-frequency domain. A major upgrade of the ngEHT would be a new front and back end system, supporting a bandwidth of 8 GHz per sideband, dual polarization, and simultaneous dual band 230/345 GHz capability with a recording rate of 256 Gbps [1]. This new capability provokes interest for spectral line studies using the ngEHT for two reasons: (1) VLBI observations with such a wide bandwidth enable spectral line studies without additional frequency tunings at each station, and (2) the ngEHT offers several orders of magnitude higher angular resolutions than those available with single-dish telescopes or connected arrays, such as SMA, NOEMA, and ALMA. Microarcsecond scale angular resolution in spectral line studies would open new doorways to explore intriguing physical phenomena in compact astrophysical systems, such as the core of star-forming regions (SFRs), circumstellar envelope (CSE) of evolved stars, and circumnuclear disk (CND) of active galactic nuclei (AGNs).

Spectral line studies with a mm/sub-mm VLBI array and their possible scientific impacts have attracted the attention of researchers in recent years, especially due to the significant increase in sensitivity provided by the Atacama Large Millimeter/submillimeter Array (ALMA) Phasing System [2–5]. ALMA has already offered a prototype spectral line VLBI capability in Band 3 (3.5 mm) in conjunction with the Global mm-VLBI Array (GMVA) and anticipates offering more flexible spectral line VLBI modes in additional bands in the near future. However, much work is still needed to utilize its unique capability for VLBI studies. One of the most pressing tasks is to categorize possible candidates and types of spectral lines suited for mm/sub-mm VLBI experiments. In this context, here we highlight the types of spectral line sources available for VLBI studies and their potential use in various scientific applications.

Citation: Kim, D.-J.; Fish, V. Spectral Line VLBI Studies Using the ngEHT. *Galaxies* 2023, 11, 10. https://doi.org/10.3390/galaxies11010010

Academic Editor: Luigina Feretti

Received: 18 November 2022
Revised: 30 December 2022
Accepted: 1 January 2023
Published: 6 January 2023

2. Molecular Maser Lines

Microwave amplification by stimulated emission of radiation (maser) is a non-thermal process in the interstellar medium (ISM) that can result in remarkably bright spectral line emission. Molecular maser lines were discovered in both galactic and extragalactic sources [6–8]. The physical sizes of masing clouds measured by interferometric observations range from sub-AU to a few AU for galactic masers and sub-pc to pc scales for extragalactic masers [9,10]. Considering the distance and physical size of maser sources, their angular scale is typically a milli-arcsecond (mas). For instance, a galactic maser source with a physical scale of 1 AU corresponds to an angular scale of 1 mas at a source distance of $\sim$1000 pc. In this case, a proper motion of $10\,\mathrm{km\,s^{-1}}$ ($\sim$2.1 AU yr^{-1}) corresponds to 2.1 mas yr^{-1} with an additional annual parallax of up to 1 mas. Monitoring VLBI observations at cm/mm wavelengths over a few months allow us to measure the proper motion or annual parallax of nearby galactic maser sources. VLBI observations at shorter wavelengths (mm/sub-mm) achieve higher angular resolutions, which can enlarge the number of spectral line source for astrometry, such as more distant galactic maser sources (>a few kpc) or sporadic maser features that last only a few weeks or less. In addition, mm/sub-mm maser transitions are theorized to have distinct excitation conditions, and hence they can unveil different part of the gaseous medium that cannot be traced by conventional maser transitions available at cm wavelengths [11,12].

2.1. Circumstellar Envelope of AGB Stars

In the late stages of stellar evolution, the outer layers of intermediate mass stars (1–$8\,\mathrm{M_\odot}$) expand to a few AU in radius, pulsating with a period of up to several years [13]. They are called asymptotic giant branch (AGB) stars, and their strong stellar winds (10–$30\,\mathrm{km\,s^{-1}}$) form a thick circumstellar envelope (CSE) [14]. AGB stars show morphological transitions of CSEs on relatively short time scales (10^5–10^6 yr), evolving into planetary nebulae (PNs) [14]. Despite such drastic changes, CSEs are known as the cradle of complex molecular species owing to their dense and cool environments, which are well-suited for molecular synthesis [15]. The characteristics of CSEs vary depending on their chemical and physical environments. For instance, the chemical composition of AGB stars determines the strength of stellar winds driven by heated dust [16]. Meanwhile, the presence of a companion star or giant planet likely plays an important role in the morphological evolution of CSEs, such as spiral patterns, collimated jets, and multi-polar structures [17,18]. To investigate the evolution of CSEs, multiple different types of AGB stars need to be studied separately, tracing stellar winds from the inner to the outer part of the CSEs.

In oxygen-rich (O-rich) AGB stars, SiO masers show a symmetric ring-like structure within a few stellar radii (R_*), while H_2O masers are located farther out up to a few tens of stellar radii with either symmetric or asymmetric spatial distributions [19–21]. To understand the development of asymmetric structure, it is necessary to probe the region between symmetric SiO masers and asymmetric H_2O masers clouds. According to radiative transfer models, the sub-mm H_2O maser emission at 321 GHz traces denser and warmer regions (2–$5\,R_*$) compared to physical conditions required for masing at 22 GHz (2–$10\,R_*$) [11,22]. VLBI observations of 321 GHz H_2O masers using the 345 GHz ngEHT receivers can test this theoretical prediction, constraining the size of masing clouds down to sub-AU scales (e.g., 15 µas $\sim$ 0.015 AU at 1 kpc distance) and also fill in the gap in our understanding of the development of stellar winds in CSEs. In contrast, carbon-rich (C-rich) AGB stars mainly form carbon-based molecules in their CSEs, and thus maser lines from oxygen-bearing molecules (e.g., H_2O, SiO, and OH) are barely detected in their CSEs. Spectral line surveys have detected HCN and SiS maser lines in the sub-mm wavelength range, and these lines seem widespread in C-rich AGB stars [23,24]. The CSEs of C-rich AGB stars are supposed to generate stronger stellar winds than those in O-rich AGB stars, but much is unknown about their kinematics due to the lack of high-resolution observations [16]. VLBI observations of HCN masers in C-rich AGB stars would shed light on this matter by spatially resolving maser spots and measuring their proper motions.

2.2. Star-Forming Regions

Gravitational collapse of clouds in the ISM triggered by internal or external physical conditions forms dense and compact clumps suited for star formation. Since star-forming regions (SFRs) are optically thick, much is still unknown about the process of star formation and differences in regions of high- and low-mass star formation [25] (and references therein). A variety of maser lines observed in massive SFRs (e.g., OH, H_2O, class I and II CH_3OH masers) are unique tools to classify different stages or classes of star formation. For instance, two distinct types of methanol masers excited collisionally (class I) or radiatively (class II) trace outer and inner regions of SFRs, respectively [26,27]. VLBI observations of class II methanol masers have shown complex inner structure in SFRs, including a ring-like structure [28–30]. On the contrary, class I methanol masers have revealed outflow features at a large distance from the central object [31]. Proper motion measurements via multi-epoch VLBI campaigns show diverse kinematics within scales of a few hundred AU, such as infall, outflow, and rotation. On the other hand, maser lines are rarely detected in low- and intermediate-mass SFRs compared to those in high-mass SFRs [32,33]. H_2O and methanol masers detected in some low-mass SFRs likely trace the inner parts of outflows [34]. Observations in the sub-mm wavelength range have detected new types of H_2O and CH_3OH maser lines in SFRs, but they have not been spatially resolved due to insufficient angular resolutions. [35–38]. Their excitation conditions are different from those of cm maser lines, suggesting the possibility to probe different parts of SFRs via VLBI observations [27]. Kinematics of clumpy molecular clouds revealed by VLBI observations will advance our understanding on the process of star-formations.

2.3. Circumnuclear Gas of AGNs

Extraordinarily bright maser lines (e.g., H_2O and OH) have been detected in a number of active galactic nuclei (AGNs) at cm wavelengths. VLBI observations show that they originate in either the accretion disk or outflow in the circumnuclear region of AGNs [10]. Measurements of the angular scale and rotational curve of the disk masers in AGNs provide an accurate measurement of the mass of the SMBHs [39,40]. The host galaxies of H_2O megamaser sources are commonly Seyfert galaxies, having relatively small SMBH masses (10^6–10^8 $M_\odot$) [41]. H_2O masers are very rare in elliptical galaxies. NGC 1052 and NGC 4261 are two such cases where the line profiles and VLBI observations indicate that the 22 GHz H_2O masers are likely associated with the ambient ISM interacting with radio jets [42,43]. Sub-mm H_2O masers have been detected in a number of AGNs, but their fine structure has not been spatially resolved [44–46]. VLBI observations of those sub-mm masers have potential to explore nuclear region of AGNs, such as molecular outflows, inner structure of molecular accretion disks, or jet-ISM interactions. We note that VLBI observations of 22 GHz H_2O masers towards the nearby AGN NGC 3079 show clumpy clouds with size in the range of 0.002 to 0.02 pc [9]. The ngEHT observations at 345 GHz offer an angular resolution of $\sim$15 μas (15 μas = $\sim$0.0015 pc at 20 Mpc distance), which can resolve a compact circumnuclear gas cloud in a nearby AGN.

3. Atomic Maser Lines

Ionized H II regions are known as the site of radio recombination lines (RRLs). In ionized regions, hydrogen RRLs sometimes show maser features in the sub-mm wavelength range [47]. Thanks to their high brightness temperatures, atomic maser lines are promising candidates for spectral line VLBI observations. The ionized region is commonly identified by optical emission lines (e.g., $H\alpha$), but a thick ambient medium often absorbs or reflects the optical emission from the ionized H II regions. Atomic maser lines in the radio wavelength range are optically thin, and thus help scrutinize gas kinematics in the obscured ionized region.

3.1. Post-AGB Stars and Pre-Planetary Nebulae

At the end of the AGB phase, the central star starts to ionize its surrounding materials. The ionized core region is the key to understand the development of asymmetric multipolar jets or winds seen in PNs, but a dense dusty envelope heavily obscures the central region, impeding investigations of the core region of CSEs during the post-AGB and pre-planetary nebulae (pPNe) phases. η Carinae is the first RRL maser source detected in AGB stars [48]. ALMA observations of H30α emission show that a slightly blueshifted narrow line feature in the H30α spectrum is extended, but broad line features are compact [49,50]. More RRL maser sources have been detected in post-AGB stars, and they are characterized by different line profiles [51]. Follow-up high-resolution observations towards those RRL masers are necessary to clarify their origins and possible connections with bipolar/multipolar outflows in the early phase of post-AGB stars.

3.2. Ultra-Compact H II Regions of Massive Stars

Strong radiation from newborn massive stars forms ionized regions. In the process of massive star formation, the ambient medium is thick and dense enough to cover the ionized region. This type of object is known as an ultra-compact (UC) H II region. Unveiling the kinematics of UC H II region is one of the most important topics in the study of star-formation. Theoretical studies point out that feedback from young stars prevents further matter accretion in high-mass star formations. However, somehow high-mass stars do exist in various star-forming regions. Studying hyper-compact (HC) H II regions might hint how matter keeps accreting during the early stage of high-mass star formation [52,53].

RRLs appear in ionized gas, such as UC H II regions. Since ionized gas also radiates continuum emission, line/continuum ratio is an important parameter to detect RRL. At sub-mm wavelengths, the line/continuum ratio increases remarkably and maser emission can arise in certain local conditions. MWC 349 is the high-mass star-forming region, showing RLL maser at sub-mm wavelengths [54]. The H30α maser emission in MWC 349 traces the innermost regions of the cocoon. Submillimeter Array (SMA) observations of RRL maser reveal an ionized accretion disk in MWC 349 [55]. More RRL masers sources have been detected in UC H II region. The double-peaked RRL maser spectrum with very broad line widths (up to a few hundreds km s^{-1}) suggests that RRL masers are associated with a bi-conical or collimated radio jet [56,57]. High-angular resolution VLBI observations towards RRL maser lines are needed to unveil the nuclear region of very young massive stars.

4. Molecular Absorption Lines

Molecular clouds become visible in absorption against the featureless continuum emission of the AGN jet. Absorption studies allow us to detect even low abundance molecular species regardless of their cosmological distance if the continuum jet is bright enough, effectively using the jet as a strong flashlight to illuminate the material. The absorption technique can be used to access any intervening molecular clouds lying on the line of sight. Spectral-line VLBI observations towards molecular absorption lines spatially resolve the structure and kinematics of the obscuring molecular clouds, and therefore can clarify the origin of obscuring molecular clouds. Gravitationally lensed blazars (GLB) and circumnuclear gas in radio AGNs are well-known such systems, showing molecular absorption lines from the intervening molecular clouds. Previous absorption studies focused on cm wavelengths hint interaction between radio jets and ambient medium [58,59], but observable molecular species and angular resolutions are limited. In the sub-mm range, various molecular species in different chemical environments are detectable in absorption with improved angular resolutions. Therefore, mm/sub-mm absorption line studies are desirable to advance our understanding on the chemistry of extragalactic ISM and kinematics of circumnuclear molecular gas in radio AGNs.

4.1. Gravitationally Lensed Blazars

PKS 1830-211 is one of the most well-known gravitationally lensed blazar systems, consisting of a distant blazar at a redshift of 2.5 and a face-on spiral galaxy at a redshift of 0.89 [60]. The blazar serves as a background continuum source for the intervening galaxy, and thus various molecular lines have been detected in absorption. With bright continuum, absorption line observations allow us to detect rare molecular species compared to observations using emission lines. Indeed, over 50 molecular species have been detected in the intervening spiral galaxy via absorption lines [60–64]. However, observations using single-dish telescopes or connected arrays tend to underestimate absorption depths due to low filling factors, since the beam size is larger than the size of absorbed area [65]. Absorption line observations using VLBI can spatially resolve individual continuum components related to absorption features, hence providing a better physical constraint on the obscuring molecular gas. Furthermore, VLBI monitoring of molecular absorption lines can provide a unique tool to study a structural variation of the background blazar through absorption line variability at multiple frequency bands [66].

4.2. Circumnuclear Gas in AGNs

AGNs are among the most powerful energy sources in the Universe. The radio jet outflows triggered by AGN activity are believed to play an important role in galaxy formation and evolution [67–69]. However, the nature of different types of radio jets and their influence on their host galaxies (i.e., radio jet feedback) are still poorly understood. As a matter reservoir in the vicinity of the central engine, the circumnuclear gas is the key to study different accretion environments and radio jet feedback in AGNs harboring radio jets (i.e., radio AGNs). Although radio AGNs tend to be gas-poor and distant, molecular absorption line observations using bright continuum radio jets enable us to study the circumnuclear gas. Sparse molecular species can be detected in absorption, providing a unique tool to study the small-scale ($<$pc) structure of the circumnuclear gas via spectral-line VLBI observations. Such high resolution observations are particularly important to reveal the onset of radio jet feedback or the fate of a fueling flow in the vicinity of the central engine. Despite the advantages of absorption line study, only a handful of radio AGNs have been identified as molecular absorption systems. In the sub-mm wavelength range, plenty of molecular lines are available for absorption line observations. Indeed, recent ALMA observations shows molecular absorption lines in radio AGNs [70–72]. VLBI observations towards sub-mm molecular absorption lines are the next step to clarify the origin of obscuring gas in radio AGNs and their roles in AGN fueling and feedback.

5. Technical Overview for Spectroscopy

5.1. Spectral Line Sensitivity

New ngEHT stations may have an aperture size of 10 m or less, resulting in a lower spectral line sensitivity on new baselines than for the current EHT array. Assuming a conventional quantization loss in the data recording process ($\eta_s = 0.8$), the expected spectral line sensitivities of ngEHT$_{6m}$-ALMA and ngEHT$_{6m}$-ngEHT$_{6m}$ baselines are 88 mJy and 1040 mJy, respectively with an integration time of 15 min and a spectral resolution of 1 MHz at 345 GHz (see Table 1). An order of magnitude higher spectral line sensitivity can be achievable by adding ALMA as an anchor station. Therefore, the optimal configuration of the ngEHT for spectral line VLBI observations would be including NOEMA and ALMA as anchor stations. This will provide exquisite spectral line sensitivity for sources in both the northern and southern hemispheres. Despite its relatively small aperture size, additional ngEHT stations will eventually improve the total spectral line sensitivity of ngEHT array by increasing the number of baselines ($N_{baseline} = N_{ant}(N_{ant}-1)/2$). Considering the expected spectral line sensitivity, a minimum peak flux density of a narrow maser emission line (FWHM < 10 km s^{-1}) would be >10 Jy and >20 Jy at 230 GHz and 345 GHz, respectively for imaging the spectral line with the standalone ngEHT array. Table 2 shows a representative sub-mm spectral line sources mentioned in Sections 2–4 with line parameters.

Table 1. Single-dish spectral line sensitivity of ngEHT stations with an integration time of 15 min at 230 GHz and 345 GHz, respectively.

Antenna	Aperture Size	1σ at 230 GHz 1 MHz (1.3 km s^{-1})	1σ at 345 GHz 1 MHz (0.8 km s^{-1})
ngEHT-Small	6 m	474 mJy	832 mJy
ngEHT-Large	10 m	282 mJy	496 mJy
NOEMA	15 m $\times$ 12 $\sim$ 26 m	9.6 mJy	15 mJy
ALMA	12 m $\times$ 50 $\sim$ 42 m	2.8 mJy	6 mJy

Note: Aperture efficiencies are 0.8, 0.67, 0.8 at 230 GHz and 0.8, 0.62, 0.7 at 345 GHz for ngEHT, NOEMA, and ALMA stations, respectively. The same amount of precipitable water vapor (PWV) was assumed at all stations (3 mm at 230 GHz and 1 mm at 345 GHz). Reference: https://www.iram.fr/GENERAL/NOEMA-Phase-A.pdf. Accessed on 30 December 2022.

Table 2. Different types of spectral line sources at sub-mm wavelengths.

Type	Name	Source	Transition	Peak Flux (Jy beam^{-1})	Line Width (km s^{-1})	Beam Size (arcsec $\times$ arcsec)
Molecular maser	VY Cma [a]	AGB	$H_2O\ 10_{29}$–9_{36}	500	< 10	0.75 $\times$ 0.75
	G358.93-0.03 [b]	SFR	$CH_3OH\ 13_{-1}$–14_{-2}	270	< 5	0.46 $\times$ 0.42
	Circinus [c]	AGN	$H_2O\ 10_{29}$–9_{36}	1.5	< 10	0.33 $\times$ 0.21
Atomic maser	η Carinae (core) [d]	post-AGB	H30α	5.5	28	0.09 $\times$ 0.09
	MWC 349 [e]	SFR	H30α	24–40	4–5	1.20 $\times$ 0.90

Type	Name	Source	Transition	Cont/Line (mJy beam^{-1})	Line width (km s^{-1})	Beam size (arcsec $\times$ arcsec)
Molecular absorption	PKS 1830-211 [f]	GLB	$H_2O\ 1_{10}$–1_{01}	700/210 (NE) 570/541 (SW)	< 53 53	0.50 $\times$ 0.50
	NGC 1052 [g]	AGN	CO J=3–2	442/37	169	0.21 $\times$ 0.21

Reference: [a] [73–75]. [b] [76] [c] [46]. [d] [49,50]. [e] [54,55]. [f] [77] [g] [71].

Adding an anchor station, such as ALMA and NOEMA will lower those limits by a factor of 12. Observations of a broad (FWHM $>$ 10 km s^{-1}) absorption or emission line with a moderate spectral resolution will enlarge the number of detectable spectral line sources with the ngEHT. In previous H_2O maser surveys in the cm/mm wavelength ranges, about 2300 H_2O maser sources have been detected (SFR: $\sim$1400, AGB: $\sim$700, and extragalactic: $\sim$180) [38,78]. Assuming 5% detection rate of bright ($>$20 Jy) sub-mm H_2O maser among known maser sources, about 115 sources will be feasible targets for the ngEHT. Observations of other sub-mm maser species, such as CH_3OH and SiO, will increase the number of available targets to a few hundreds. Meanwhile, an increased coherence time by frequency phase transfer (FPT) technique will further improve the spectral line sensitivity of ngEHT [79,80], but a detailed investigation is needed to evaluate the capability of FPT for spectral line VLBI observations.

5.2. Frequency Coverage

A simultaneous triple-band (3 mm, 1 mm, and 0.8 mm) receiver system has been considered for the ngEHT to apply the FPT technique. Simultaneous multi-band VLBI observations can yield several atomic and molecular lines in a single run. This unique capability is beneficial for spectral line VLBI studies in various aspects. For instance, a combined study using multiple VLBI images of different molecular maser lines would provide a comprehensive view of a masing region [21,81]. In case of molecular absorption line studies, observing multiple rotational lines provides solid constraints on the physical quantities of the obscuring gas, such as temperature and column density. We note that the observing frequency of 1 mm and 0.8 mm receivers should have integer frequency ratios of the 3 mm receiver for FPT. This technical requirement would help simultaneous observations of several rotational lines at different bands, such as HCN, HCO+, SiO, and CO. However, simultaneous multi-band observations will result in narrower bandwidth at each band than single-band observations due to a limited recording rate of the backend system. Therefore, we need to consider several spectral line observation modes for individual science projects. For instance, non-integer and wide bandwidth observations are preferred for RRL maser line observations. On the contrary, integer and multi-band observations are preferred for molecular maser or absorption line observations. In practice, offering a flexible frequency setup would not be feasible in the early phase of array operation. Therefore, we suggest two optimal frequency tunings for 8 GHz and 16 GHz bandwidth observations, respectively considering continuum observations with FPT. Specific frequency tunings and distribution of spectral lines are present in Table 3 and Figure 1.

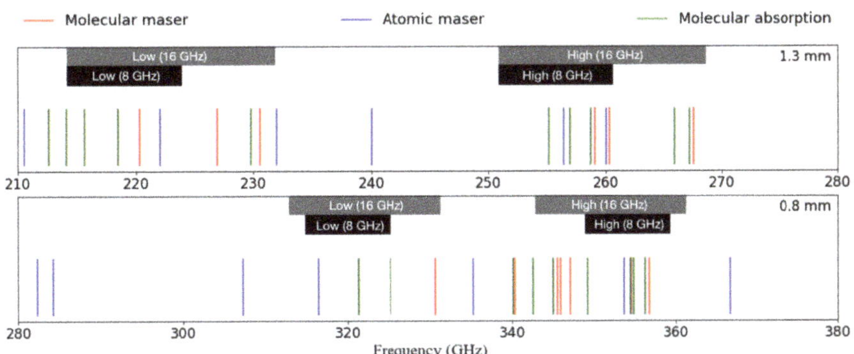

Figure 1. Distribution of spectral lines over the frequency range of 1.3 mm and 0.8 mm. Each color indicates the type of spectral lines. Frequency coverage of 8 GHz and 16 GHz observations is marked with two different frequency tunings (low and high).

Table 3. Suggested frequency tunings with different bandwidths (8 GHz and 16 GHz) at each band for spectral line VLBI observations.

Frequency (GHz)	Bandwidth (GHz)	Frequency (GHz)	Primary Target
230_{low}	8	215–223	SiO, CH_3OH, Hβ
230_{high}	8	252–260	SiO, Hα
345_{low}	8	316–324	H_2O, Hα
345_{high}	8	350–358	HCN, HCO+, Hα, H_2O
230_{low}	16	215–231	CO, SiO, CH_3OH, CN Hβ
230_{high}	16	252–268	SiO, Hα, HCN, HCO+
345_{low}	16	314–330	H_2O Hα
345_{high}	16	344–360	CO, HCN, HCO+, Hα, H_2O, SiO

6. Conclusions

The ngEHT will be equipped with multi-band receivers supporting simultaneous wideband observations. It will not only improve sensitivity for continuum imaging, but also open new opportunities for high angular resolution spectroscopy. In this white paper, we have outlined the types of spectral lines suited for VLBI studies and briefly described several scientific applications newly achievable with the ngEHT. Spectral lines observable with the 230 and 345 GHz receivers are listed in Table 4. Masers appear in various astrophysical environments, and they trace certain chemical and physical environments depending on atomic/molecular species and transitions. Atomic and molecular maser lines observable in the sub-mm wavelength range possibly unveil regions unprobed by previous cm-maser observations. The ngEHT can be used to image sub-mm masers, investigating their peculiar physical environments and kinematics.

Table 4. Selected spectral lines near 230 GHz and 345 GHz and their energy level above ground state. Three different types of spectral lines available for VLBI observations are listed with physical origins.

Type	Frequency (GHz)	Transition	E_{low} (K)	Source
Molecular masers	214.088	^{28}SiO v = 2, J=5–4	3541	O-rich AGB
	215.596	^{28}SiO v = 1, J=5–4	1790	O-rich AGB
	256.898	^{28}SiO v = 2, J=6–5	3551	O-rich AGB
	258.707	^{28}SiO v = 1, J=6-5	1800	O-rich AGB
	299.704	^{28}SiO v = 2, J=7–6	3564	O-rich AGB
	301.814	^{28}SiO v = 1, J=7–6	1813	O-rich AGB
	342.504	^{28}SiO v = 2, J=8–7	3579	O-rich AGB
	344.916	^{28}SiO v = 1, J=8–7	1827	O-rich AGB
	265.853	HCN (0, 1^{1e}, 0), J=3–2	1050	C-rich AGB
	267.199	HCN (0, 1^{1f}, 0), J=3–2	1050	C-rich AGB
	354.460	HCN (0, 1^{1e}, 0), J=4–3	1067	C-rich AGB
	356.256	HCN (0, 1^{1f}, 0), J=4–3	1067	C-rich AGB
	218.440	$CH_3OH\ 4_2–3_1$	24	SFRs (class I)
	229.759	$CH_3OH\ 8_{-1}–7_0$	54	SFRs (class I)
	343.599	$CH_3OH\ 13_{-1}–14_{-2}$	607	SFRs (class II)
	349.107	$CH_3OH\ 14_1–14_0$	243	SFRs (class II)
	321.226	$H_2O\ 10_{29}–9_{36}$	1846	SFRs/AGB/AGNs
	325.153	$H_2O\ 5_{14}–4_{22}$	454	SFRs/AGB/AGNs
	354.809	$H_2O\ 17_{4,13}–4_{7,10}$	5764	SFRs/AGB/AGNs
Atomic masers	210.502	H31α		SFRs
	231.900	H30α		SFRs
	256.302	H29α		SFRs
	284.251	H28α		SFRs
	316.415	H27α		SFRs
	353.623	H26α		SFRs
	222.012	H38β		SFRs
	240.021	H37β		SFRs
	260.033	H36β		SFRs
	282.333	H35β		SFRs
	307.258	H34β		SFRs
	335.207	H33β		SFRs
	366.653	H32β		SFRs
Molecular absorption	220.299	^{13}CO J=2–1	5	AGNs/GLB
	230.538	CO J=2–1	6	AGNs/GLB
	330.588	^{13}CO J= 3–2	16	AGNs/GLB
	345.796	CO J=3–2	17	AGNs/GLB
	265.886	HCN J=3–2	13	AGNs/GLB
	354.505	HCN J=4–3	26	AGNs/GLB
	267.558	HCO+ J=3–2	13	AGNs/GLB
	356.734	HCO+ J=4–3	18	AGNs/GLB
	259.012	H^{13}CN J=3–2	9	AGNs/GLB
	345.340	H^{13}CN J=4–3	17	AGNs/GLB
	260.255	H^{13}CO+ J=3–2	9	AGNs/GLB
	346.998	H^{13}CO+ J=4–3	18	AGNs/GLB
	226.874	CN N=2–1, J=5/2–3/2	5	AGNs/GLB
	340.247	CN N=3–2, J=7/2–5/2	16	AGNs/GLB

Note: Rest frequencies were obtained from the Splatalogue database available at www.splatalogue.net. Accessed on 30 December 2022.

However, VLBI experiments with maser lines could be challenging due to the limited spectral line sensitivity of the ngEHT and the rarity of sub-mm masers compared to cm/mm masers. Absorption line observations using a bright background continuum source (e.g., radio jet) are an alternative approach to enable spectral line VLBI studies towards common molecular clouds regardless of their distance. Molecular absorption from circumnuclear gas in AGNs or the intervening galaxy of a lensed blazar are promising candidates for spectral line VLBI studies. Lastly, we have highlighted the capabilities of ngEHT for spectroscopy. Considering the rest frequencies of atomic and molecular transitions, we suggest frequency tunings to facilitate the operation of ngEHT in spectral line mode.

Author Contributions: Conceptualization, D.-J.K.; methodology, D.-J.K.; writing—original draft, D.-J.K.; writing—review and editing, V.F.; supervision, V.F. All authors have read and agreed to the published version of the manuscript.

Funding: This research was supported by the National Science Foundation through grant numbers AST-1935980 and AST-2034306.

Data Availability Statement: Not applicable.

Conflicts of Interest: The authors declare no conflict of interest.

References

1. Doeleman, S.; Blackburn, L.; Dexter, J.; Gomez, J.L.; Johnson, M.D.; Palumbo, D.C.; Weintroub, J.; Farah, J.R.; Fish, V.; Loinard, L.; et al. Studying Black Holes on Horizon Scales with VLBI Ground Arrays. *Proc. Bull. Am. Astron. Soc.* **2019**, *51*, 256.
2. Fish, V.; Alef, W.; Anderson, J.; Asada, K.; Baudry, A.; Broderick, A.; Carilli, C.; Colomer, F.; Conway, J.; Dexter, J.; et al. High-Angular-Resolution and High-Sensitivity Science Enabled by Beamformed ALMA. *arXiv* **2013**, arXiv:1309.3519.
3. Tilanus, R.P.J.; Krichbaum, T.P.; Zensus, J.A.; Baudry, A.; Bremer, M.; Falcke, H.; Giovannini, G.; Laing, R.; van Langevelde, H.J.; Vlemmings, W.; et al. Future mmVLBI Research with ALMA: A European vision. *arXiv* **2014**, arXiv:1406.4650.
4. Asada, K.; Kino, M.; Honma, M.; Hirota, T.; Lu, R.S.; Inoue, M.; Sohn, B.W.; Shen, Z.Q.; Ho, P.T.P.; Akiyama, K.; et al. White Paper on East Asian Vision for mm/submm VLBI: Toward Black Hole Astrophysics down to Angular Resolution of 1~R $_S$. *arXiv* **2017**, arXiv:1705.04776.
5. Matthews, L.D.; Crew, G.B.; Doeleman, S.S.; Lacasse, R.; Saez, A.F.; Alef, W.; Akiyama, K.; Amestica, R.; Anderson, J.M.; Barkats, D.A.; et al. The ALMA Phasing System: A Beamforming Capability for Ultra-high-resolution Science at (Sub)Millimeter Wavelengths. *Publ. Astron. Soc. Pac.* **2018**, *130*, 015002. [CrossRef]
6. Litvak, M.M. Coherent molecular radiation. *Annu. Rev. Astron. Astrophys.* **1974**, *12*, 97–112. [CrossRef]
7. Reid, M.J.; Moran, J.M. Masers. *Annu. Rev. Astron. Astrophys.* **1981**, *19*, 231–276. [CrossRef]
8. Elitzur, M. Astronomical masers. *Annu. Rev. Astron. Astrophys.* **1992**, *30*, 75–112. [CrossRef]
9. Trotter, A.S.; Greenhill, L.J.; Moran, J.M.; Reid, M.J.; Irwin, J.A.; Lo, K.Y. Water Maser Emission and the Parsec-Scale Jet in NGC 3079. *Astrophys. J.* **1998**, *495*, 740–748. [CrossRef]
10. Lo, K.Y. Mega-Masers and Galaxies. *Annu. Rev. Astron. Astrophys.* **2005**, *43*, 625–676. [CrossRef]
11. Gray, M.D.; Baudry, A.; Richards, A.M.S.; Humphreys, E.M.L.; Sobolev, A.M.; Yates, J.A. The physics of water masers observable with ALMA and SOFIA: Model predictions for evolved stars. *Mon. Not. R. Astron. Soc.* **2016**, *456*, 374–404. [CrossRef]
12. Bergman, P.; Humphreys, E.M.L. Submillimetre water masers at 437, 439, 471, and 474 GHz towards evolved stars. APEX observations and radiative transfer modelling. *Astron. Astrophys.* **2020**, *638*, A19. [CrossRef]
13. Herwig, F. Evolution of Asymptotic Giant Branch Stars. *Annu. Rev. Astron. Astrophys.* **2005**, *43*, 435–479. [CrossRef]
14. Höfner, S.; Olofsson, H. Mass loss of stars on the asymptotic giant branch. Mechanisms, models and measurements. *Astron. Astrophys. Rev.* **2018**, *26*, 1. [CrossRef]
15. Gail, H.P.; Sedlmayr, E. *Physics and Chemistry of Circumstellar Dust Shells*; Cambridge University Press: Cambridge, UK, 2013; ISBN 978-0-5119-8560-7.
16. Woitke, P. Too little radiation pressure on dust in the winds of oxygen-rich AGB stars. *Astron. Astrophys.* **2006**, *460*, L9–L12. [CrossRef]
17. Huang, P.S.; Lee, C.F.; Sahai, R. Evolution from Spherical AGB Wind to Multipolar Outflow in Pre-planetary Nebula IRAS 17150-3224. *Astrophys. J.* **2020**, *889*, 85. [CrossRef]
18. Randall, S.K.; Trejo, A.; Humphreys, E.M.L.; Kim, H.; Wittkowski, M.; Boboltz, D.; Ramstedt, S. Discovery of a complex spiral-shell structure around the oxygen-rich AGB star GX Monocerotis. *Astron. Astrophys.* **2020**, *636*, A123. [CrossRef]
19. Diamond, P.J.; Kemball, A.J.; Junor, W.; Zensus, A.; Benson, J.; Dhawan, V. Observation of a Ring Structure in SiO Maser Emission from Late-Type Stars. *Astrophys. J. Lett.* **1994**, *430*, L61. [CrossRef]
20. Imai, H.; Shibata, K.M.; Marvel, K.B.; Diamond, P.J.; Sasao, T.; Miyoshi, M.; Inoue, M.; Migenes, V.; Murata, Y. The Three-dimensional Kinematics of Water Masers around the Semiregular Variable RT Virginis. *Astrophys. J.* **2003**, *590*, 460–472. [CrossRef]
21. Kim, D.J.; Cho, S.H.; Yun, Y.; Choi, Y.K.; Yoon, D.H.; Kim, J.; Dodson, R.; Rioja, M.J.; Yang, H.; Yoon, S.J. Simultaneous VLBI Astrometry of H_2O and SiO Masers toward the Semiregular Variable R Crateris. *Astrophys. J. Lett.* **2018**, *866*, L19. [CrossRef]
22. Humphreys, E.M.L.; Yates, J.A.; Gray, M.D.; Field, D.; Bowen, G.H. Qualitative reproduction of stellar H_2O maser morphology. I. Results at a single stellar phase. *Astron. Astrophys.* **2001**, *379*, 501–514. [CrossRef]
23. Gong, Y.; Henkel, C.; Ott, J.; Menten, K.M.; Morris, M.R.; Keller, D.; Claussen, M.J.; Grasshoff, M.; Mao, R.Q. SiS in the Circumstellar Envelope of IRC +10216: Maser and Quasi-thermal Emission. *Astrophys. J.* **2017**, *843*, 54. [CrossRef]

24. Menten, K.M.; Wyrowski, F.; Keller, D.; Kamiński, T. Widespread HCN maser emission in carbon-rich evolved stars. *Astron. Astrophys.* **2018**, *613*, A49. [CrossRef]

25. Evans, N.J., II. Physical Conditions in Regions of Star Formation. *Annu. Rev. Astron. Astrophys.* **1999**, *37*, 311–362. [CrossRef]

26. Cragg, D.M.; Johns, K.P.; Godfrey, P.D.; Brown, R.D. Pumping the interstellar methanol masers. *Mon. Not. R. Astron. Soc.* **1992**, *259*, 203–208. [CrossRef]

27. Cragg, D.M.; Sobolev, A.M.; Godfrey, P.D. Models of class II methanol masers based on improved molecular data. *Mon. Not. R. Astron. Soc.* **2005**, *360*, 533–545. [CrossRef]

28. Bartkiewicz, A.; Szymczak, M.; van Langevelde, H.J. Ring shaped 6.7 GHz methanol maser emission around a young high-mass star. *Astron. Astrophys.* **2005**, *442*, L61–L64. [CrossRef]

29. Bartkiewicz, A.; Szymczak, M.; van Langevelde, H.J.; Richards, A.M.S.; Pihlström, Y.M. The diversity of methanol maser morphologies from VLBI observations. *Astron. Astrophys.* **2009**, *502*, 155–173. [CrossRef]

30. Bartkiewicz, A.; Sanna, A.; Szymczak, M.; Moscadelli, L.; van Langevelde, H.J.; Wolak, P. The nature of the methanol maser ring G23.657-00.127. II. Expansion of the maser structure. *Astron. Astrophys.* **2020**, *637*, A15. [CrossRef]

31. Bartkiewicz, A.; van Langevelde, H.J. Masers in star forming regions. In Proceedings of the Cosmic Masers—From OH to H0, Stellenbosch, South Africa, 29 January–3 February 2012; Booth, R.S., Vlemmings, W.H.T., Humphreys, E.M.L., Eds.; Volume 287, pp. 117–126. [CrossRef]

32. Kalenskii, S.V.; Johansson, L.E.B.; Bergman, P.; Kurtz, S.; Hofner, P.; Walmsley, C.M.; Slysh, V.I. Search for Class I methanol masers in low-mass star formation regions. *Mon. Not. R. Astron. Soc.* **2010**, *405*, 613–620. [CrossRef]

33. Bae, J.H.; Kim, K.T.; Youn, S.Y.; Kim, W.J.; Byun, D.Y.; Kang, H.; Oh, C.S. A Multi-epoch, Simultaneous Water and Methanol Maser Survey toward Intermediate-mass Young Stellar Objects. *Astrophys. J. Suppl.* **2011**, *196*, 21. [CrossRef]

34. Moscadelli, L.; Testi, L.; Furuya, R.S.; Goddi, C.; Claussen, M.; Kitamura, Y.; Wootten, A. First results from a VLBA proper motion survey of H2O masers in low-mass YSOs: The Serpens core and RNO 15-FIR. *Astron. Astrophys.* **2006**, *446*, 985–999. [CrossRef]

35. Slysh, V.I.; Kalenskiĭ, S.V.; Val'tts, I.E. Methanol Radio Emission at Millimeter Wavelengths: New Masers at 1.3 and 2.8 Millimeters. *Astron. Rep.* **2002**, *46*, 49–56. [CrossRef]

36. van Kempen, T.A.; Wilner, D.; Gurwell, M. 183 GHz H_2O Maser Emission Around the Low-Mass Protostar Serpens SMM1. *Astrophys. J. Lett.* **2009**, *706*, L22–L26. [CrossRef]

37. Ladeyschikov, D.A.; Bayandina, O.S.; Sobolev, A.M. Online Database of Class I Methanol Masers. *Astron. J.* **2019**, *158*, 233. [CrossRef]

38. Ladeyschikov, D.A.; Sobolev, A.M.; Bayandina, O.S.; Shakhvorostova, N.N. Online Database of Multiwavelength Water Masers in Galactic Star-forming Regions. *Astron. J.* **2022**, *163*, 124. [CrossRef]

39. Kuo, C.Y.; Braatz, J.A.; Condon, J.J.; Impellizzeri, C.M.V.; Lo, K.Y.; Zaw, I.; Schenker, M.; Henkel, C.; Reid, M.J.; Greene, J.E. The Megamaser Cosmology Project. III. Accurate Masses of Seven Supermassive Black Holes in Active Galaxies with Circumnuclear Megamaser Disks. *Astrophys. J.* **2011**, *727*, 20. [CrossRef]

40. Pesce, D.W.; Braatz, J.A.; Condon, J.J.; Greene, J.E. Measuring Supermassive Black Hole Peculiar Motion Using H_2O Megamasers. *Astrophys. J.* **2018**, *863*, 149. [CrossRef]

41. Greene, J.E.; Seth, A.; Kim, M.; Läsker, R.; Goulding, A.; Gao, F.; Braatz, J.A.; Henkel, C.; Condon, J.; Lo, K.Y.; et al. Megamaser Disks Reveal a Broad Distribution of Black Hole Mass in Spiral Galaxies. *Astrophys. J. Lett.* **2016**, *826*, L32. [CrossRef]

42. Claussen, M.J.; Diamond, P.J.; Braatz, J.A.; Wilson, A.S.; Henkel, C. The Water Masers in the Elliptical Galaxy NGC 1052. *Astrophys. J. Lett.* **1998**, *500*, L129–L132. [CrossRef]

43. Wagner, J. 22 GHz water maser search in 37 nearby galaxies. Four new water megamasers in Seyfert 2 and OH maser/absorber galaxies. *Astron. Astrophys.* **2013**, *560*, A12. [CrossRef]

44. Humphreys, E.M.L.; Greenhill, L.J.; Reid, M.J.; Beuther, H.; Moran, J.M.; Gurwell, M.; Wilner, D.J.; Kondratko, P.T. First Detection of Millimeter/Submillimeter Extragalactic H_2O Maser Emission. *Astrophys. J. Lett.* **2005**, *634*, L133–L136. [CrossRef]

45. Humphreys, E.M.L. Submillimeter and millimeter masers. In Proceedings of the Astrophysical Masers and Their Environments, Alice Springs, Australia, 12–16 March 2007; Chapman, J.M., Baan, W.A., Eds.; Volume 242, pp. 471–480. [CrossRef]

46. Hagiwara, Y.; Horiuchi, S.; Imanishi, M.; Edwards, P.G. Second-epoch ALMA Observations of 321 GHz Water Maser Emission in NGC 4945 and the Circinus Galaxy. *Astrophys. J.* **2021**, *923*, 251. [CrossRef]

47. Gordon, M.A.; Sorochenko, R.L. *Radio Recombination Lines. Their Physics and Astronomical Applications*; Springer: Berlin/Heidelberg, Germany, 2002; Volume 282. [CrossRef]

48. Cox, P.; Martin-Pintado, J.; Bachiller, R.; Bronfman, L.; Cernicharo, J.; Nyman, L.A.; Roelfsema, P.R. Millimeter recombination lines toward η Carinae. *Astron. Astrophys.* **1995**, *295*, L39–L42.

49. Abraham, Z.; Falceta-Gonçalves, D.; Beaklini, P.P.B. η Carinae Baby Homunculus Uncovered by ALMA. *Astrophys. J.* **2014**, *791*, 95. [CrossRef]

50. Abraham, Z.; Beaklini, P.P.B.; Cox, P.; Falceta-Gonçalves, D.; Nyman, L.Å. η Carinae: High angular resolution continuum, H30α and He30α ALMA images. *Mon. Not. R. Astron. Soc.* **2020**, *499*, 2493–2512. [CrossRef]

51. Sánchez Contreras, C.; Báez-Rubio, A.; Alcolea, J.; Bujarrabal, V.; Martín-Pintado, J. A pilot search for mm-wavelength recombination lines from emerging ionized winds in pre-planetary nebulae candidates. *Astron. Astrophys.* **2017**, *603*, A67. [CrossRef]

52. Wolfire, M.G.; Cassinelli, J.P. Conditions for the Formation of Massive Stars. *Astrophys. J.* **1987**, *319*, 850. [CrossRef]

53. Cesaroni, R.; Galli, D.; Lodato, G.; Walmsley, M.; Zhang, Q. The critical role of disks in the formation of high-mass stars. *Nature* **2006**, *444*, 703–706. [CrossRef]

54. Martin-Pintado, J.; Thum, C.; Bachiller, R. Time-variable recombination line emission in MWC 349. *Astron. Astrophys.* **1989**, *222*, L9–L11.

55. Weintroub, J.; Moran, J.M.; Wilner, D.J.; Young, K.; Rao, R.; Shinnaga, H. Submillimeter Array Imaging of the Maser Emission from the H30α Radio Recombination Line in MWC 349A. *Astrophys. J.* **2008**, *677*, 1140–1150. [CrossRef]

56. Jiménez-Serra, I.; Martín-Pintado, J.; Báez-Rubio, A.; Patel, N.; Thum, C. Extremely Broad Radio Recombination Maser Lines Toward the High-velocity Ionized Jet in Cepheus A HW2. *Astrophys. J. Lett.* **2011**, *732*, L27. [CrossRef]

57. Jiménez-Serra, I.; Báez-Rubio, A.; Rivilla, V.M.; Martín-Pintado, J.; Zhang, Q.; Dierickx, M.; Patel, N. A New Radio Recombination Line Maser Object toward the MonR2 H II Region. *Astrophys. J. Lett.* **2013**, *764*, L4. [CrossRef]

58. Morganti, R.; Oosterloo, T.A.; Emonts, B.H.C.; van der Hulst, J.M.; Tadhunter, C.N. Fast Outflow of Neutral Hydrogen in the Radio Galaxy 3C 293. *Astrophys. J. Lett.* **2003**, *593*, L69–L72. [CrossRef]

59. Morganti, R.; Tadhunter, C.N.; Oosterloo, T.A. Fast neutral outflows in powerful radio galaxies: A major source of feedback in massive galaxies. *Astron. Astrophys.* **2005**, *444*, L9–L13. [CrossRef]

60. Wiklind, T.; Combes, F. The redshift of the gravitational lens of PKS1830-211 determined from molecular absorption lines. *Nature* **1996**, *379*, 139–141. [CrossRef]

61. Muller, S.; Guélin, M.; Dumke, M.; Lucas, R.; Combes, F. Probing isotopic ratios at z = 0.89: Molecular line absorption in front of the quasar PKS 1830-211. *Astron. Astrophys.* **2006**, *458*, 417–426. [CrossRef]

62. Muller, S.; Guélin, M. Drastic changes in the molecular absorption at redshift z = 0.89 toward the quasar PKS 1830-211. *Astron. Astrophys.* **2008**, *491*, 739–746. [CrossRef]

63. Henkel, C.; Braatz, J.A.; Menten, K.M.; Ott, J. The kinetic temperature of a molecular cloud at redshift 0.9: Ammonia in the gravitational lens PKS0-211. *Astron. Astrophys.* **2008**, *485*, 451–456. [CrossRef]

64. Menten, K.M.; Güsten, R.; Leurini, S.; Thorwirth, S.; Henkel, C.; Klein, B.; Carilli, C.L.; Reid, M.J. Submillimeter water and ammonia absorption by the peculiar z $\approx$ 0.89 interstellar medium in the gravitational lens of the PKS 1830-211 system. *Astron. Astrophys.* **2008**, *492*, 725–730. [CrossRef]

65. Muller, S.; Beelen, A.; Guélin, M.; Aalto, S.; Black, J.H.; Combes, F.; Curran, S.J.; Theule, P.; Longmore, S.N. Molecules at z = 0.89. A 4-mm-rest-frame absorption-line survey toward PKS 1830-211. *Astron. Astrophys.* **2011**, *535*, A103. [CrossRef]

66. Martí-Vidal, I.; Muller, S.; Combes, F.; Aalto, S.; Beelen, A.; Darling, J.; Guélin, M.; Henkel, C.; Horellou, C.; Marcaide, J.M.; et al. Probing the jet base of the blazar PKS 1830-211 from the chromatic variability of its lensed images. Serendipitous ALMA observations of a strong gamma-ray flare. *Astron. Astrophys.* **2013**, *558*, A123. [CrossRef]

67. Harrison, C.M.; Alexander, D.M.; Mullaney, J.R.; Swinbank, A.M. Kiloparsec-scale outflows are prevalent among luminous AGN: Outflows and feedback in the context of the overall AGN population. *Mon. Not. R. Astron. Soc.* **2014**, *441*, 3306–3347. [CrossRef]

68. Crenshaw, D.M.; Kraemer, S.B.; Schmitt, H.R.; Jaffé, Y.L.; Deo, R.P.; Collins, N.R.; Fischer, T.C. The Geometry of Mass Outflows and Fueling Flows in the Seyfert 2 Galaxy MRK 3. *Astron. J.* **2010**, *139*, 871–877. [CrossRef]

69. Woo, J.H.; Bae, H.J.; Son, D.; Karouzos, M. The Prevalence of Gas Outflows in Type 2 AGNs. *Astrophys. J.* **2016**, *817*, 108. [CrossRef]

70. Nagai, H.; Onishi, K.; Kawakatu, N.; Fujita, Y.; Kino, M.; Fukazawa, Y.; Lim, J.; Forman, W.; Vrtilek, J.; Nakanishi, K.; et al. The ALMA Discovery of the Rotating Disk and Fast Outflow of Cold Molecular Gas in NGC 1275. *Astrophys. J.* **2019**, *883*, 193. [CrossRef]

71. Kameno, S.; Sawada-Satoh, S.; Impellizzeri, C.M.V.; Espada, D.; Nakai, N.; Sugai, H.; Terashima, Y.; Kohno, K.; Lee, M.; Martín, S. A Massive Molecular Torus inside a Gas-poor Circumnuclear Disk in the Radio Galaxy NGC 1052 Discovered with ALMA. *Astrophys. J.* **2020**, *895*, 73. [CrossRef]

72. Rose, T.; Edge, A.C.; Combes, F.; Hamer, S.; McNamara, B.R.; Russell, H.; Gaspari, M.; Salomé, P.; Sarazin, C.; Tremblay, G.R.; et al. A molecular absorption line survey towards the AGN of Hydra-A. *Mon. Not. R. Astron. Soc.* **2020**, *496*, 364–380. [CrossRef]

73. Yates, J.A.; Cohen, R.J.; Hills, R.E. Submillimetre water masers in circumstellar envelopes. *Mon. Not. R. Astron. Soc.* **1995**, *273*, 529–548. [CrossRef]

74. Menten, K.M.; Melnick, G.J. 321 GHz Submillimeter Water Masers around Evolved Stars. *Astrophys. J.* **1991**, *377*, 647. [CrossRef]

75. Richards, A.M.S.; Impellizzeri, C.M.V.; Humphreys, E.M.; Vlahakis, C.; Vlemmings, W.; Baudry, A.; De Beck, E.; Decin, L.; Etoka, S.; Gray, M.D.; et al. ALMA sub-mm maser and dust distribution of VY Canis Majoris. *Astron. Astrophys.* **2014**, *572*, L9. [CrossRef]

76. Brogan, C.L.; Hunter, T.R.; Towner, A.P.M.; McGuire, B.A.; MacLeod, G.C.; Gurwell, M.A.; Cyganowski, C.J.; Brand, J.; Burns, R.A.; Caratti o Garatti, A.; et al. Sub-arcsecond (Sub)millimeter Imaging of the Massive Protocluster G358.93-0.03: Discovery of 14 New Methanol Maser Lines Associated with a Hot Core. *Astrophys. J. Lett.* **2019**, *881*, L39. [CrossRef]

77. Muller, S.; Combes, F.; Guélin, M.; Gérin, M.; Aalto, S.; Beelen, A.; Black, J.H.; Curran, S.J.; Darling, J.; V-Trung, D.; et al. An ALMA Early Science survey of molecular absorption lines toward PKS 1830-211. Analysis of the absorption profiles. *Astron. Astrophys.* **2014**, *566*, A112. [CrossRef]

78. Kuo, C.Y.; Hsiang, J.Y.; Chung, H.H.; Constantin, A.; Chang, Y.Y.; Cunha, E.d.; Pesce, D.; Chien, W.T.; Chen, B.Y.; Braatz, J.A.; et al. A More Efficient Search for H_2O Megamaser Galaxies: The Power of X-Ray and Mid-infrared Photometry. *Astrophys. J.* **2020**, *892*, 18. [CrossRef]

79.   Rioja, M.J.; Dodson, R. Precise radio astrometry and new developments for the next-generation of instruments. *Astron. Astrophys. Rev.* **2020**, *28*, 6. [CrossRef]
80.   Rioja, M.; Dodson, R.; Asaki, Y. Transformative Multi-Frequency Phase Transfer Solutions for ngEHT. *Galaxies* **2023**, *11*, 16.
81.   Dodson, R.; Rioja, M.J.; Jung, T.H.; Sohn, B.W.; Byun, D.Y.; Cho, S.H.; Lee, S.S.; Kim, J.; Kim, K.T.; Oh, C.S.; et al. Astrometrically Registered Simultaneous Observations of the 22 GHz H_2O and 43 GHz SiO Masers toward R Leonis Minoris Using KVN and Source/Frequency Phase Referencing. *Astron. J.* **2014**, *148*, 97. [CrossRef]

MDPI

St. Alban-Anlage 66

4052 Basel

Switzerland

www.mdpi.com

Galaxies Editorial Office

E-mail: galaxies@mdpi.com

www.mdpi.com/journal/galaxies